Introduction to Food Process Engineering

FOOD PRESERVATION TECHNOLOGY SERIES

Series Editor

Gustavo V. Barbosa-Cánovas

Introduction to Food Process Engineering
Editors: Albert Ibarz and Gustavo V. Barbosa-Cánovas

Shelf Life Assessment of Food
Editor: Maria Cristina Nicoli

Cereal Grains: Laboratory Reference and Procedures Manual
Sergio O. Serna-Saldivar

Advances in Fresh-Cut Fruits and Vegetables Processing
Editors: Olga Martín-Belloso and Robert Soliva-Fortuny

Cereal Grains: Properties, Processing, and Nutritional Attributes
Sergio O. Serna-Saldivar

Water Properties of Food, Pharmaceutical, and Biological Materials
Editors: Maria del Pilar Buera, Jorge Welti-Chanes, Peter J. Lillford, and Horacio R. Corti

Food Science and Food Biotechnology
Editors: Gustavo F. Gutiérrez-López and Gustavo V. Barbosa-Cánovas

Transport Phenomena in Food Processing
Editors: Jorge Welti-Chanes, Jorge F. Vélez-Ruiz, and Gustavo V. Barbosa-Cánovas

Unit Operations in Food Engineering
Albert Ibarz and Gustavo V. Barbosa-Cánovas

Engineering and Food for the 21st Century
Editors: Jorge Welti-Chanes, Gustavo V. Barbosa-Cánovas, and José Miguel Aguilera

Osmotic Dehydration and Vacuum Impregnation: Applications in Food Industries
Editors: Pedro Fito, Amparo Chiralt, Jose M. Barat, Walter E. L. Spiess, and Diana Behsnilian

Pulsed Electric Fields in Food Processing: Fundamental Aspects and Applications
Editors: Gustavo V. Barbosa-Cánovas and Q. Howard Zhang

Trends in Food Engineering
Editors: Jorge E. Lozano, Cristina Añón, Efrén Parada-Arias, and Gustavo V. Barbosa-Cánovas

Innovations in Food Processing
Editors: Gustavo V. Barbosa-Cánovas and Grahame W. Gould

Introduction to Food Process Engineering

Albert Ibarz
Gustavo V. Barbosa-Cánovas

CRC Press
Taylor & Francis Group
Boca Raton London New York

CRC Press is an imprint of the
Taylor & Francis Group, an informa business

CRC Press
Taylor & Francis Group
6000 Broken Sound Parkway NW, Suite 300
Boca Raton, FL 33487-2742

International Standard Book Number-13: 978-1-4398-0918-1 (Hardback)

Library of Congress Cataloging-in-Publication Data

Ibarz, Albert.
Introduction to food process engineering / authors, Albert Ibarz, Gustavo V. Barbosa-Canovas.
pages cm. -- (Food preservation technology series)
Includes bibliographical references and index.
ISBN 978-1-4398-0918-1 (hardback)
1. Food industry and trade. I. Barbosa-Canovas, Gustavo V. II. Title.

TP370.I229 2013
338.4'7664--dc23 2013045326

Visit the Taylor & Francis Web site at
http://www.taylorandfrancis.com

and the CRC Press Web site at
http://www.crcpress.com

Albert dedicates this book to his family and

Gustavo to Kezban Candoğan

Contents

Preface

Proper design and analysis of food processing operations require the utilization of solid engineering principles. It is therefore essential to develop an adequate engineering platform to meet global and specific requirements in order to achieve safe, efficient, and reliable food processes at a reasonable cost. Recently, other important requirements have emerged in the food industry, such as making processes part of a sustainable environment, minimizing carbon footprints, and incorporating, whenever possible, "green" technologies. In other words, what may have been considered an optimal process in the past must now be discarded because it does not meet the new expectations and standards. Due to globalization, raw materials and intermediate food products are coming from all around the world, and food manufacturers are constantly changing suppliers due to costs, regulations, political issues, and ethical concerns. As a consequence, the versatility of some unit operations, as well as the processing lines associated with them, should be designed to be much more flexible than in the past. On top of that, the availability of a significant number of new technologies is, essentially, forcing the development of modern food processing facilities forward in a different fashion than had been used in the past. These new processing scenarios have led to innovative approaches to integrate, in a very logical and harmonious manner, some of the concepts mentioned. As an illustration of the new waves of innovation impacting the processing of food, it is worth mentioning a European project, "Food Factory of the Future," which is integrating traditional practices to process foods with new trends and expectations, such as offering foods that promote health and well-being. Another effort in that direction is a recently published book, *Food Engineering Interfaces* (J.M. Aguilera, G.V. Barbosa-Cánovas, R. Simpson, and J. Welti-Chanes), which presents analysis of disciplines disconnected from food processing in the past but that are now becoming pivotal for the advancement of this very important discipline.

Introduction to Food Process Engineering will be very helpful in building a strong background, in an organized manner, toward the better understanding of most of the unit operations encountered in today's food industry. A large segment of people involved with these issues will benefit from this book, including food engineers from industry, as well as instructors and students at the undergraduate and graduate level.

In 2002, CRC Press published *Unit Operations in Food Engineering*, which has been quite successful. After considering different options for updating and expanding the content of this book, the authors and publisher decided to develop two books based on the previous one: *Introduction to Food Process Engineering* and *Unit Operations in Food Engineering*, Second Edition. The idea was to capture all of the basic theoretical and fundamental aspects needed to provide the basic background for those interested in food process engineering in the first one. The second book covers, in great detail, most of the unit operations of interest to today's food industry. These two books were not written as a sequence; rather, they are two distinctive stand-alone books while, at the same time, there is a significant similarity between the two in terms of terminology, definitions, units, symbols, and nomenclature.

This book includes 25 comprehensive chapters compiled by the authors after many years of experience dealing with all the addressed subjects. Chapters 1 and 2 set the scope of the book and describe the basic tools that will be used throughout it in terms of units and how the dimensional analysis will be developed. Chapters 3 through 7 address transport phenomena as well as the physicochemistry of food systems. Chapters 8 and 9 cover air–water mixtures and water activity; both topics are essential to the better understanding of unit operations such as dehydration. Chapters 10 through 15 deal with food properties, including optical, thermal, mechanical, electrical, as well as physical and chemical properties of food powders. Chapters 15 through 18 focus on the three heat transfer mechanisms: conduction, convection, and radiation. Chapter 19 presents an in-depth analysis of chilling and freezing, together with different refrigeration configurations. Chapter 20 covers, in great detail, the thermal processing of food, including the basic concepts, and provides an introduction to canning and aseptic processing. Chapter 21 introduces some of the most promising novel technologies for treating foods, such as high pressure, microwaves, ohmic heating, and pulsed electric fields. Chapters 22 and 23 cover the fundamentals of the most important water removal operations, that is, concentration and dehydration. Finally, Chapter 24 deals with the hygienic design of food facilities and equipment, while Chapter 25 covers many aspects of food packaging.

To facilitate the understanding of the topics covered in this book, each chapter concludes with a set of solved problems. We hope this book will be useful as a reference for food engineers and as a text for advanced undergraduate and graduate students in food engineering. We also hope this book will be a meaningful addition to the literature dealing with food processing operations.

Albert Ibarz
Gustavo V. Barbosa-Cánovas

Acknowledgments

The authors wish to express their gratitude to the following institutions and individuals who contributed to making this book possible:

- The University of Lleida and Washington State University (WSU) for providing resources and a superb working environment for the development of this book.
- Jeannie Bagby for her excellent in-house copyediting of the entire manuscript prior to submission for publication. Her professionalism and dedication are truly commendable.
- Víctor Falguera and Ilce Medina-Meza, our young colleagues, who spent a significant amount of time revising the entire manuscript and contributing very valuable suggestions.

Authors

Albert Ibarz serves as professor of food technology at the University of Lleida, Lleida, Spain. He earned his PhD in chemical engineering at the University of Barcelona, Barcelona, Spain and his technical engineering degree in agricultural and food industries at Polytechnic University of Catalonia, Barcelona, Spain. He has been a visiting professor at Universidad Nacional del Sur (Bahía Blanca, Argentina, 1989) and Universidad de las Américas (Puebla, Mexico, 1999–2001), adjunct faculty professor at Washington State University (Pullman, Washington, 1995), and honorary professor at Universidad Nacional del Santa (Chimbote, Perú, 2004). He was also named Doctor Honoris Causa by Universidad Nacional de Trujillo (Trujillo, Perú, 2013).

Dr. Ibarz has held important administration positions at Polytechnic University, Catalonia (1987–1991), Superior Technical School of Agrarian Engineering, Lleida (1984–1985), and University of Lleida (1999–2003). He has published more than 200 scientific articles on rheology, chemistry, biochemistry, photochemistry kinetics, and UV treatments and has presented at more than 150 national and international congresses.

Dr. Ibarz is the coauthor of *Unit Operations in Food Engineering* (1999; Spanish edition, 2005), *Experimental Methods in Food Engineering* (2000), and *Unit Operations in Food Engineering* (2003). He has also authored/coauthored eight book chapters, most notably "Newtonian and Non-Newtonian Flow" in *Food Engineering—Encyclopedia of Life Support Systems* (UNESCO). He has translated 11 books on diverse areas of food engineering and has coordinated and edited "Reports de Recerca of Catalunya" (1996–2002) and "Enginyeria Agronòmica, Forest i Alimentària" (Institut d'Estudis Catalans). Dr. Ibarz has served on many scientific committees of national and international congresses and has presided over the organizing committee of the Second Spanish Congress of Food Engineering (Lleida, 2002).

Gustavo V. Barbosa-Cánovas received his BS in mechanical engineering from the University of Uruguay, Montevideo, Uruguay and his MS and PhD in food engineering from the University of Massachusetts-Amherst, Amherst, Massachusetts where he was a Fulbright Scholar. He was recently awarded a Honoris Causa Doctorate at Polytechnic University of Cartagena, Spain. He was assistant professor at the University of Puerto Rico (1985–1990), where he was granted two National Science Foundation awards for research productivity. At present, he serves as professor of food engineering and director of the Center for Nonthermal Processing of Food at Washington State University. Dr. Barbosa-Cánovas has been editor of the journal of *Food Science and Technology International* (*FSTI*), as well as *Innovative Food Science and Emerging Technologies* (*IFSET*), and the Food Engineering Theme in the *Encyclopedia of Life Support Systems* (*EOLSS*) published by UNESCO. He is editor in chief of the Food Engineering Book Series (Springer), as well as the Food Preservation Technology Series (CRC Press). He has edited several books on food engineering topics and has authored books on the following topics: food powders, food plant design, dehydration of foods, preservation of foods by pulsed electric fields, unit operations in food engineering, and nonthermal preservation of foods. He also serves as international consultant for the United Nation's Food and Agriculture Organization (FAO) and several major international food companies. Dr. Barbosa-Cánovas is immediate past president of the International Society of Food Engineering (ISFE) and chair of the Scientific Council of the International Union of Food Science and Technology (IUFoST). He has received several prestigious awards, including the IFT Nicholas Appert Award (highest award in food science and technology) and the IFT International Award, and is an IFT, IFST, and IUFoST Fellow, as well as a member of the Uruguayan Academy of Engineering. He recently received the Sahlin Award for Research, Scholarship and Arts at WSU (highest research award at this university), as well as a Fulbright Fellowship.

1 Introduction to Unit Operations
Fundamental Concepts

1.1 PROCESS

The word *process* refers to the set of activities or industrial operations that modify the properties of raw materials, with the purpose of obtaining products to satisfy the needs of society. Such modifications of natural raw materials are designed to obtain products with greater acceptance in the market or with better possibilities for storage and transport.

The primary needs of human beings, whether for the individual or society as a whole, did not change much through history; the three basics of food, clothing, and housing were needed by prehistoric humans as well as by modern ones for survival.

The fulfillment of these necessities is carried out by employing, transforming, and consuming the resources available in our natural surroundings. In the early stages of humankind's social development, natural products were used directly or with only small physical modifications.

This simple productive scheme changed as society developed, in a way such that at the present time raw materials are often not directly used to satisfy necessities, but rather they are subjected to physical and chemical transformations that convert them into products with different properties.

In this way, the raw materials not only directly fulfill the necessities of consumers but also constitute the basis for the products derived from the manipulation of such raw materials.

1.2 FOOD PROCESS ENGINEERING

By analogy with other engineering branches, different definitions of food process engineering can be given. Thus, according to one definition, "Food Process Engineering includes the part of human activity in which the knowledge of physical, natural and economic sciences is applied to agricultural products as related to their composition, energetic content or physical state" (CDOChE, 1961).

Food process engineering can also be defined as "the science of conceiving, calculating, designing, building and running the facilities where the transformation processes of agricultural products, at the industrial level (as economically as possible), are carried out" (Cathala, 1951).

So, an engineer in the food industry should know the basic principles of process engineering and be able to develop new production techniques for agricultural products. In the same way, he or she should be capable of designing the equipment that should be used in a given process.

The main objective of food process engineering is to study the principles and laws governing the physical, chemical, or biochemical stages of different processes and the apparatus or equipment by which such stages are industrially carried out. The studies should be focused on the transformation processes of agricultural raw materials into final products or on the conservation of materials and products.

1.3 TRANSFORMATION AND COMMERCIALIZATION OF AGRICULTURAL PRODUCTS

For the efficient commercialization of agricultural products, it should be easy for them to be handled and placed in the market.

As a general rule, products obtained directly from the harvest cannot be commercialized as they are, but rather they have to undergo certain transformations. Products that can be directly used should be adequately packaged according to the requirements of the market. These products are generally used as food, and they should be conveniently prepared for use.

One problem during handling of agricultural products is their transport from the fields to the consumer. Since many agricultural products have a short shelf life, treatment and preservation methods that allow their later use should be developed. For this reason, many of these products cannot be directly used as food, but they serve as raw material for obtaining other products.

Developed countries tend to elaborate such products in the harvest zone, avoiding perishable products that deteriorate during transport from the production zone to the processing plant.

1.4 FLOWCHARTS AND DESCRIPTION OF SOME FOOD PROCESSES

Food processes are usually schematized by means of flowcharts. These flowcharts are diagrams of all the processes that indicate different manufacturing steps, as well as the flow of materials and energy in the process.

There are different types of flowcharts, with *blocks* or *rectangles* being most used. In these charts, each stage of the process is represented by a block or rectangle connected by arrows to indicate the way in which the materials flow. The stage represented is noted within the rectangle.

Other types of flowcharts are *equipment* and *instrumentation.* Figures 1.1 through 1.3 show some flowcharts of food processes.

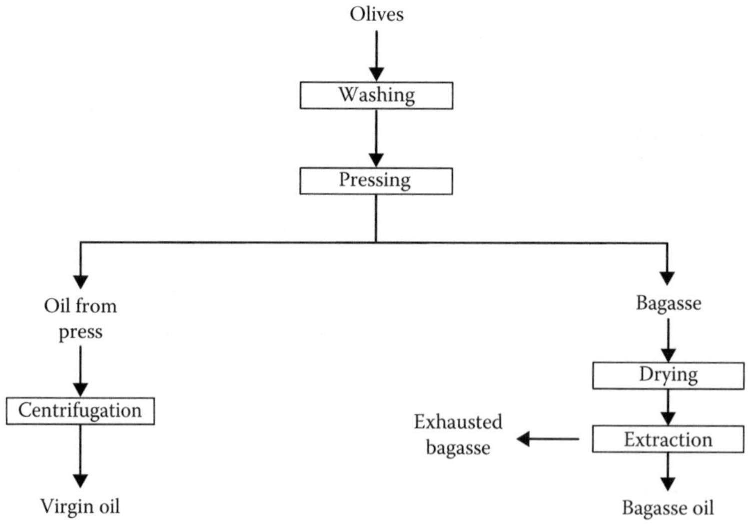

FIGURE 1.1 Extraction of olive oil.

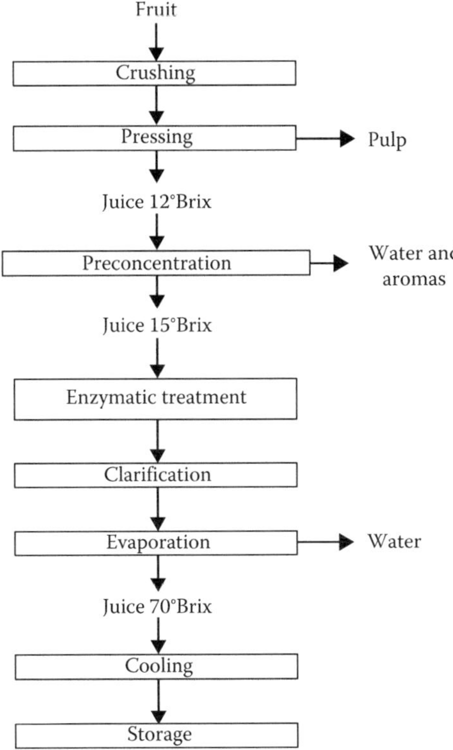

FIGURE 1.2 Production of fruit concentrated juices.

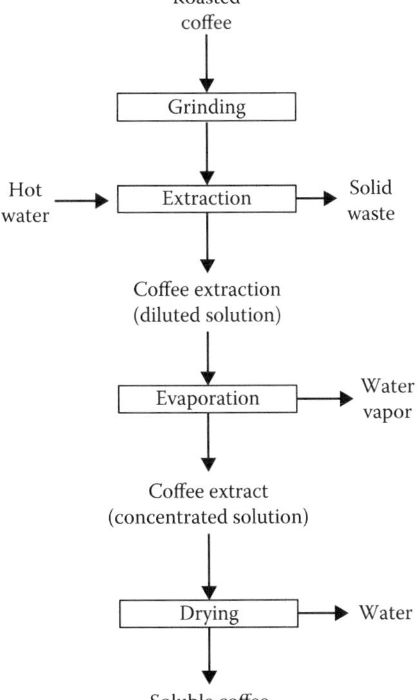

FIGURE 1.3 Elaboration of soluble coffee.

1.5 STEADY AND UNSTEADY STATES

A system is said to be under steady state when all of the physical variables remain constant and invariable along time, at any point of the system; however, they may be different from one point to another.

On the other hand, when not only the characteristic intensive variables of the operation vary through the system at a given moment, but also the variables corresponding to each system's point vary along time, the state is called unsteady.

The physical variables to consider may be mechanical or thermodynamic. Among the former variables are volume, velocity, etc., while the thermodynamic variables are viscosity, concentration, temperature, pressure, etc.

1.6 DISCONTINUOUS, CONTINUOUS, AND SEMICONTINUOUS OPERATIONS

The operations carried out in industrial processes may be performed in three different ways. In a *discontinuous* operation, the raw material is loaded in the equipment; after performing the required transformation, the obtained products are unloaded. These operations, also called *batch* or *intermittent*, are carried out in the following steps:

1. Loading of equipment with raw materials
2. Preparation of conditions for transformation
3. Required transformation
4. Unloading of products
5. Cleaning of equipment

The batch operation takes place under an unsteady state, since its intensive properties vary along time. An example of this batch process is the crushing of oily seeds to obtain oil.

In *continuous* operations, the loading, transformation, and unloading stages are performed simultaneously. Equipment cleaning is carried out every given time, depending on the nature of the process and the materials used. To carry out the cleaning, production must be stopped. Continuous operations take place under steady state, in such a way that the characteristic intensive variables of the operation may vary at each point of the system but do not vary along time. It is difficult to reach an absolute steady state, since there may be some unavoidable fluctuations. One clear example of a continuous operation might be the rectification of an alcohol–water mixture.

In some cases, it is difficult to maintain a continuous operation. This type of operation is called *semicontinuous.* A semicontinuous operation may occur by loading some materials in the equipment that will remain there for a given time in a discontinuous way, while other materials enter or exit continuously. Sometimes it is necessary to unload those accumulated materials. For example, in the extraction of oil by solvents, flour might be loaded while the solvent is fed in a continuous way; after some time, the flour runs out of oil and must be replaced.

The different ways of operation have advantages and disadvantages.

Advantages of continuous operation:

1. Loading and unloading stages are eliminated.
2. Automation of the operation is allowed, thus reducing the work force.
3. Composition of products is more uniform.
4. Better use of thermal energy is achieved.

Disadvantages of continuous operation:

1. Raw materials should have a uniform composition to avoid operation fluctuations.
2. It is usually expensive to start the operation, so stops should be avoided.

3. Fluctuations in product demand require availability of considerable quantities of raw materials and products in stock.
4. Due to automation of operation, equipment is more expensive and delicate.

Continuous operation is performed under an unsteady state during starts and stops, but once it is adequately running, it may be considered to be working under steady state. However, this is not completely true, since there could be fluctuations due to variations in the composition of the raw materials and due to modifications of external agents.

When selecting a form of operation, the advantages and disadvantages of each type should be considered. However, when low production is required, it is recommended to work under discontinuous conditions, and when high production is required, it is more profitable to operate under a continuous condition.

1.7 UNIT OPERATIONS: CLASSIFICATION

When analyzing the flow charts of different processes described in the sections that follow, it can be observed that some of the stages are found in all of them. These stages are referred to as basic or unit operations, in common with many industrial processes. The individual operations have common techniques and are based on the same scientific principles, making the study of these operations and the treatment of these processes easier.

There are different types of unit operations, depending on the nature of the transformation performed; thus, physical, chemical, and biochemical stages can be distinguished:

1. Physical stages: grinding, sieving, mixture, fluidization, sedimentation, flotation, filtration, rectification, absorption, extraction, adsorption, heat exchange, evaporation, and drying
2. Chemical stages: refining and chemical peeling
3. Biochemical stages: fermentation, sterilization, pasteurization, and enzymatic peeling

Hence, the group of physical, chemical, and biochemical stages that take place in the transformation processes of agricultural products constitute the so-called unit operations of the food industry. The purpose of the unit operations is the separation of two or more substances present in a mixture or the exchange of a property due to a gradient. Separation is achieved by means of a separating agent that is different, depending on the transferred property.

Unit operations can be classified into different groups depending on the transferred property, since the possible changes that a body may undergo are defined by variations in its mass, energy, or velocity.

Thus, unit operations are classified under

- Mass transfer unit operations
- Heat transfer unit operations
- Momentum transfer unit operations

Besides the unit operations considered in each mentioned group, there exist those of simultaneous heat and mass transfer and other operations that cannot be classified in any of these groups, which are called complementary unit operations.

All of the unit operations grouped in these sections are found in physical processes; however, certain operations that include chemical reactions can be included.

1.7.1 MOMENTUM TRANSFER UNIT OPERATIONS

These operations study the processes in which two phases at different velocities are in contact. The operations included in this section are generally divided into three groups: internal circulation of fluids, external circulation of fluids, and solids movement within fluids.

Internal circulation of fluids: This refers to the study of the movement of fluids through the interior of the tubing; it also includes the study of equipment used to impel the fluids (pumps, compressors, blowers, and fans) and the mechanisms used to measure the properties of fluids (diaphragms, venturi meters, rotameters, etc.).

External circulation of fluids: The fluid circulates through the external part of a solid. This circulation includes the flow of fluids through porous fixed beds, fluidized beds (fluidization), and pneumatic transport.

Solids movement within fluids: This is the base for separation of solids within a fluid. This type of separation includes sedimentation, filtration, and ultrafiltration, among others.

1.7.2 Mass Transfer Unit Operations

These operations are controlled by the diffusion of a component within a mixture. Some of the operations included in this group are

Distillation: Separation of one or more components by taking advantage of their vapor pressure differences.

Absorption: A component of a gas mixture is absorbed by a liquid, according to the solubility of the gas in the liquid. Absorption may occur with or without chemical reaction. The opposite process is *desorption.*

Extraction: Based on the dissolution of a mixture (liquid or solid) in a selective solvent, which can be *liquid–liquid* or *solid–liquid.* The latter is also called washing, lixiviation, etc.

Adsorption: Also called sorption, adsorption involves the elimination of one or more components of a fluid (liquid or gas) by retention on the surface of a solid.

Ionic exchange: Substitution of one or more ions of a solution with another exchange agent.

1.7.3 Heat Transfer Unit Operations

These operations are controlled by temperature gradients. They depend on the mechanism by which heat is transferred, conduction, convection, or radiation:

Conduction: In continuous material media, heat flows in the direction of temperature decrease, and there is no macroscopic movement of mass.

Convection: The enthalpy flow associated to a moving fluid is called convective flow of heat. Convection can be natural or forced.

Radiation: Energy transmission by electromagnetic waves. No material media is needed for transmission.

Thermal treatments (sterilization and pasteurization), evaporation, heat exchangers, ovens, solar plates, etc. are studied based on these heat transfer mechanisms.

1.7.4 Simultaneous Mass–Heat Transfer Unit Operations

In these operations, a concentration and a temperature gradient exist at the same time:

Humidification and dehumidification: Include the objectives of humidification and dehumidification of a gas and cooling of a liquid.

Crystallization: Formation of solid glassy particles within a homogeneous liquid phase.

Dehydration: Elimination of a liquid contained within a solid. The application of heat changes the liquid, contained in a solid, into a vapor phase. In *freeze-drying*, the liquid in the solid phase is removed by sublimation, that is, by changing it into a vapor phase.

Cryoconcentration: Concentration of food solutions by removing water by freezing.

1.7.5 Complementary Unit Operations

A series of operations exist that are not included in this classification, because they are not based on any of the transport phenomena cited previously. They are auxillary operations, whose purpose is to condition the raw materials and solid products, in order to have the right size for subsequent treatments. These operations include *grinding*, *milling*, *sieving*, and *mixing* of solids and pastes.

1.8 MATHEMATICAL SETUP OF THE PROBLEMS

The problems that are set up in the study of unit operations are very diverse, although in all of them the conservation laws (mass, energy, momentum, and stoichiometric) of chemical reactions apply.

Applying the conservation laws to a given problem is to perform a balance of the "property" studied in such a problem. In a general way, the expression of the mass, energy, and momentum balances related to the unit time can be expressed as

$$\begin{pmatrix}\text{Property entering}\\ \text{the system}\end{pmatrix} = \begin{pmatrix}\text{Property exiting}\\ \text{the system}\end{pmatrix} + \begin{pmatrix}\text{Property that}\\ \text{accumulates}\end{pmatrix}$$

What enters into the system of a given component is equal to what exits, in addition to what is accumulated. In a schematic way,

$$\mathbf{E} = \mathbf{S} + \mathbf{A}$$

In the cases where a chemical reaction exists, when carrying out a balance for a component, an additional generation term may appear. In these cases, the balance expression will be

$$\mathbf{E} + \mathbf{G} = \mathbf{S} + \mathbf{A}$$

When solving a given problem, a certain number of unknown quantities or variables (**V**) are present, and a set of relationships or equations (**R**) are obtained from the balances. According to values of **V** and **R**, the following cases can arise:

- If $\mathbf{V} < \mathbf{R}$, the problem is established incorrectly, or one equation is repeated.
- If $\mathbf{V} = \mathbf{R}$, the problem has only one solution.
- If $\mathbf{V} > \mathbf{R}$, different solutions can be obtained, finding the best solution by optimizing the process.

There are $\mathbf{F} = \mathbf{V} - \mathbf{R}$ design variables.

The different types of problems presented depend on the type of equation obtained when performing the corresponding balances. Thus, there are the following equations:

- **Algebraic equations** that have an easy mathematical solution, which is obtained by analytical methods.
- **Differential equations** that are usually obtained for unsteady continuous processes. The solution of the mathematical model established with the balances can be carried out through analytical or approximate methods. In some cases, differential equations may have an analytical solution. However, in those cases when it is not possible to analytically solve the mathematical model, it is necessary to appeal to approximate methods of numerical integration (digital calculus) or graphic (analogic calculus).
- **Equations in finite differences** that are solved by means of analogic computers that give the result in a graphic form. In some cases the exact solution can be obtained by numerical methods.

2 Unit Systems, Dimensional Analysis, and Similarities

2.1 MAGNITUDE AND UNIT SYSTEMS

Physical magnitudes are those properties that can be measured and expressed with a number and a unit.

The physical properties of a system are related by a series of physical and mechanical laws. Some magnitudes may be considered as fundamental and others as derived. The fundamental magnitudes vary from one system to another.

Generally, time and length are taken as fundamental. The unit systems need a third fundamental magnitude, which may be mass or force. Those unit systems that have mass as the third fundamental magnitude receive the name of *absolute unit systems*, while those that have force as a fundamental unit are called *technical unit systems*. There are also *engineering unit systems* that consider as fundamental magnitudes length, time, mass, and force.

2.1.1 Absolute Unit Systems

There are three absolute unit systems: the cgs (CGS), the Giorgi (MKS), and the English (FPS). In all of these, the fundamental magnitudes are length, mass, and time. The different units for the three mentioned systems are shown in Table 2.1. In these three systems, force is a derived unit that is defined beginning with the three fundamental units. The force and energy units are detailed in Table 2.2.

When the properties to measure involve temperature changes, it is necessary to define a temperature unit. For the CGS and MKS systems, the unit of temperature is Celsius (°C), while for the English system it is Fahrenheit (°F). Temperature changes are also associated with heat, which is another physical magnitude with its own units, well defined for each unit system.

It must be noted that heat units are defined independently of work units. Later, it will be shown that relating work and heat requires a factor called the mechanical equivalent of heat.

2.1.2 Technical Unit Systems

Among the most used technical systems are the metric system and the English system. In both, the fundamental magnitudes are length, force, and time. In regard to temperature, the unit of the metric system is the Celsius degree, and that of the English system is the Fahrenheit degree. Table 2.3 shows the fundamental units of the metric and English systems.

In engineering systems, mass is a derived magnitude, which in each system is

Metric system: 1 TMU (technical mass unit)
English system: 1 slug

TABLE 2.1
Absolute Unit Systems

	System		
Magnitude	**cgs (CGS)**	**Giorgi (MKS)**	**English (FPS)**
Length (L)	1 centimeter (cm)	1 meter (m)	1 foot (ft)
Mass (M)	1 gram (g)	1 kilogram (kg)	1 pound-mass (lb)
Time (T)	1 second (s)	1 second (s)	1 second (s)

TABLE 2.2
Units Derived from Absolute Systems

	System		
Magnitude	**cgs (CGS)**	**Giorgi (MKS)**	**English (FPS)**
Force	1 dyne	1 newton (N)	1 poundal
Energy	1 erg	1 joule (J)	1 (pound) (foot)

TABLE 2.3
Technical Unit Systems

	System	
Magnitude	**Metric**	**English**
Length (L)	1 meter (m)	1 foot (ft)
Force (F)	1 kilogram force (kp or kgf)	1 pound force
Time (T)	1 second (s)	1 second (s)
Temperature (θ)	1 Celsius degree (°C)	1 Fahrenheit degree (°F)

2.1.3 Engineering Unit Systems

Until now only unit systems that consider three magnitudes as fundamental have been described. However, in the engineering systems, four magnitudes are considered as basic: length, time, mass, and force. Table 2.4 presents the different units for the metric and English engineering systems.

When defining mass and force as fundamental, an incongruity may arise, since these magnitudes are related by the dynamics basic principle. To avoid this incompatibility, a correction or proportionality factor (g_c) should be inserted. The equation of this principle would be

$$g_c \times \text{Force} = \text{Mass} \times \text{Acceleration}$$

TABLE 2.4
Engineering Unit Systems

	System	
Magnitude	**Metric**	**English**
Length (L)	1 meter (m)	1 foot (ft)
Mass (M)	1 kilogram (kg)	1 pound-mass (lb)
Force (F)	1 kilogram force (kp or kgf)	1 pound force (lbf)
Time (T)	1 second (s)	1 second (s)
Temperature (θ)	1 Celsius degree (°C)	1 Fahrenheit degree (°F)

Observe that g_c has mass units (acceleration/force). The value of this correction factor in the engineering systems is as follows:

$$\text{Metric system: } g_c = 9.81\frac{(\text{kg mass})(\text{meter})}{(\text{kg force})(\text{second})^2} = 9.81 \text{ kg m/kg s}^2$$

$$\text{English system: } g_c = 32.17\frac{(\text{lb mass})(\text{foot})}{(\text{lb force})(\text{second})^2} = 32.17 \text{ lb ft/lbf s}^2$$

2.1.4 International Unit System

It was convenient to unify the use of the unit systems when the Anglo-Saxon countries incorporated the metric decimal system. With that purpose, the MKS was adopted as the International System and was denoted as SI. Although the obligatory nature of the system is recognized, other systems are still used; however, at present many engineering journals and books are edited only in SI, making it more and more acceptable over other unit systems. Table 2.5 presents the fundamental units of this system, along with some supplementary and derived units.

Sometimes the magnitude of a selected unit is too large or too small, making it necessary to adopt prefixes to indicate multiples and submultiples of the fundamental units. Generally, it is advisable to use these multiples and submultiples as powers of 10^3. Following is a list of the multiples and submultiples most often used, as well as the name and symbol of each.

Prefix	Multiplication Factor	SI Symbol
yotta	10^{24}	Y
zetta	10^{21}	Z
exa	10^{18}	E
peta	10^{15}	P
tera	10^{12}	T
giga	10^{9}	G
mega	10^{6}	M
kilo	10^{3}	k
hecto	10^{2}	h
deca	10^{1}	da
deci	10^{-1}	d
centi	10^{-2}	c
milli	10^{-3}	m
micro	10^{-6}	μ
nano	10^{-9}	n
pico	10^{-12}	p
femto	10^{-15}	f
atto	10^{-18}	a
zepto	10^{-21}	z
yocto	10^{-24}	y

It is interesting to note that in many problems concentration is expressed by using molar units. The molar unit most frequently used is the *mole*, defined as the quantity of substance whose mass in grams is numerically equal to its molecular weight.

2.1.5 Thermal Units

As we know, heat is a form of energy; in this way the dimension of both is ML^2T^{-2}. However, in some systems temperature is taken as dimension. In such cases, the heat energy can be expressed as

TABLE 2.5
International Unit System

Magnitude	Unit	Abbreviation	Dimension
Length	Meter	m	L
Mass	Kilogram	kg	M
Time	Second	s	T
Force	Newton	N	MLT^{2}
Energy	Joule	J	$ML^{2}T^{-2}$
Power	Watt	W	$ML^{2}T^{-3}$
Pressure	Pascal	Pa	$ML^{-1}T^{-2}$
Frequency	Hertz	Hz	T^{-1}

proportional to the product (mass × time × temperature). The proportionality constant is the specific heat, which depends on the material and varies from one to another. The amount of heat is defined as a function of the material. With water taken as a reference and the specific heat being the unit

$$\text{Heat} = \text{Mass} \times \text{Specific heat} \times \text{Temperature}$$

The heat unit depends on the unit system adopted. Thus,

- Metric system
  Calorie: Heat needed to raise the temperature of a gram of water from 14.5°C to 15.5°C.
- English systems
  Btu (British thermal unit): Quantity of heat needed to raise the temperature of a pound of water 1°F (from 60°F to 61°F).
  Chu (Celsius heat unit or pound calorie): Quantity of heat needed to raise the temperature of 1 lb of water 1°C.
- SI
  Joule: Since heat is a form of energy, its unit is the joule.

The calorie can be defined as a function of the joule:

$$1 \text{ cal} = 4.185 \text{ J}$$

Since heat and work are two forms of energy, it is necessary to define a factor that relates them. For this reason, the denominated mechanical equivalent of heat (Q) is defined in such way that

$$Q \times \text{Heat energy} = \text{Mechanical energy}$$

so

$$Q = \frac{\text{Mechanical energy}}{\text{Heat energy}} = \frac{MLT^{-2}L}{M\theta} = L^2T^{-2}\theta^{-1}$$

2.1.6 Unit Conversion

The conversion of units from one system to another is easily carried out if the quantities are expressed as a function of the fundamental units: mass, length, time, and temperature. The so-called *conversion factors* are used to convert the different units. The conversion factor is the number of units of a certain system contained in one unit of the correspondent magnitude of another system. The most common conversion factors for the different magnitudes are given in Table 2.6.

TABLE 2.6
Conversion Factors

Mass	
1 lb	0.4536 kg
	(1/32.2) slug
Length	
1 in.	2.54 cm
1 ft	0.3048 m
1 mile	1,609 m
Surface	
1 in.^{2}	645.2 mm^2
1 ft^2	0.09290 m^2
Volume and capacity	
1 ft^3	0.02832 m^3
1 gal (imperial)	4.546 L
1 gal (the United States of America)	3.786 L
1 barrel	159.241 L
Time	
1 min	60 s
1 h	3,600 s
1 day	86,400 s
Temperature difference	
1°C = 1 K	1.8°F
Force	
1 poundal (pdl)	0.138 N
1 lbf	4.44 N
	4.44×10^5 dyne
	32.2 pdl
1 dyne	10^{-5} N
Pressure	
1 technical atmosphere (at)	1 kgf/cm^2
	14.22 psi
1 bar	100 kPa
1 mmHg (torr)	133 Pa
	13.59 kgf/cm^2
1 psi (lb/in.^{2})	703 kgf/m^2
Energy, heat, and power	
1 kilocalorie (kcal)	4,185 J
	426.7 kgm
1 erg	10^{-7} J
1 Btu	1,055 J
Energy, heat, and power	
1 Chu	0.454 kcal
	1.8 Btu
1 horse vapor (CV)	0.736 kW
	75 kgm/s
1 horsepower (hp)	0.746 kW
	33,000 ft · lb/min
	76.04 kgm/s
1 kilowatt (kW)	1,000 J/s
	1.359 CV

(continued)

TABLE 2.6 (continued)
Conversion Factors

1 kilowatt hour (kW h)	3.6×10^6 J
	860 kcal
1 atm. liter	0.0242 kcal
	10.333 kg m
Viscosity	
1 poise (P)	0.1 Pa s
1 pound/(ft h)	0.414 mPa s
1 stoke (St)	10^{-4} m^2/s
Mass flow	
1 lb/h	0.126 g/s
1 ton/h	0.282 kg/s
Mass flux	
1 lb/(ft^2 h)	1.356 g/s m^2
Thermal magnitudes	
1 Btu/(h ft^2)	3.155 W/m^2
1 Btu/(h ft^2 °F)	5.678 W/(m^2 K)
1 Btu/lb	2.326 kJ/kg
1 Btu/(lb °F)	4.187 kJ/(kg K)
1 Btu/(h ft °F)	1.731 W/(m K)

When converting units, it is necessary to distinguish the cases in which only numerical values are converted from those in which a formula should be converted.

1. Conversion of numerical values
   When it is necessary to convert numerical values from one unit to another, the equivalencies between them, given by the conversion factors, are used directly.
2. Conversion of units of a formula
   In this case, the constants that appear in the formula usually have dimensions. To apply the formula in units different from those given, only the constant of the formula should be converted. In cases in which the constant is dimensionless, the formula can be directly applied using any unit system.

2.2 DIMENSIONAL ANALYSIS

The application of equations deducted from physical laws is one method that can be used to solve a determined problem. However, it may be difficult to obtain equations of this type; therefore, in some cases it will be necessary to use equations derived in an empirical form.

In the first case, the equations are homogeneous from a dimensional point of view. That is, their terms have the same dimensions, and the possible constants that may appear will be dimensionless. This type of equation can be applied in any unit system when using coherent units for the same magnitudes. On the contrary, equations experimentally obtained may not be homogeneous regarding the dimensions, since it is normal to use different units for the same magnitude.

The objective of dimensional analysis is to relate the different variables involved in the physical processes. For this reason, the variables are grouped in dimensionless groups or rates, allowing discovery of a relationship among the different variables. Table 2.7 shows the dimensionless modules usually found in engineering problems. Dimensional analysis is an analytical method in which, once the variables that intervene in a physical phenomenon are known, an equation to bind them can be established. That is, dimensional analysis provides a general relationship

TABLE 2.7
Dimensionless Modules

Modules	Expression	Equivalence
Biot (Bi)	$\frac{hd}{k}$	
Bodenstein (Bo)	$\frac{vd}{D}$	(Re)(Sc)
Euler (Eu)	$\frac{\Delta P}{\rho v^2}$	
Froude (Fr)	$\frac{d_P N}{g}$	
Graetz (Gz)	$\frac{\rho v d^2 \hat{C}_P}{kL}$	(Re)(Pr)(d/L)
Grashof (Gr)	$\frac{g\beta d^3 \Delta T \rho^2}{\eta^2}$	
Hedstrom (He)	$\frac{d\sigma_0\rho}{\eta'}$	
Nusselt (Nu)	$\frac{hd}{k}$	
Peclet (Pe)	$\frac{\rho v d\hat{C}_P}{k}$	(Re)(Pr)
Power (Po)	$\frac{P}{d_P N^5 \rho}$	
Prandtl (Pr)	$\frac{\hat{C}_P\eta}{k}$	
Reynolds (Re)	$\frac{\rho v d}{\eta}$	
Schmidt (Sc)	$\frac{\eta}{\rho D}$	
Sherwood (Sh)	$\frac{k_g d}{D}$	
Stanton (St)	$\frac{h}{\hat{C}_P \rho v}$	(Nu)[(Re)(Pr)]$^{-1}$
Weber (We)	$\frac{\rho l v^2}{\sigma}$	

among the variables that should be completed with the assistance of experimentation to obtain the final equation that binds all the variables.

2.2.1 Buckingham's π Theorem

Every term that has no dimensions is defined as *factor* π. According to Bridgman, there are three fundamental principles of dimensional analysis:

1. All of the physical magnitudes may be expressed as power functions of a reduced number of fundamental magnitudes.
2. The equations that relate physical magnitudes are dimensionally homogeneous; this means that the dimensions of all their terms must be equal.

3. If an equation is dimensionally homogeneous, it may be reduced to a relation among a complete series of dimensionless rates or groups. These rates or groups induce all the physical variables that influence the phenomenon, the dimensional constants that may correspond to the selected unit system, and the universal constants related to the phenomenon being treated.

This principle is denoted as *Buckingham's π theorem*. A series of dimensionless groups is complete if all of the groups are independent among them, and any other dimensionless group that can be formed will be a combination of two or more groups from the complete series.

Because of Buckingham's π theorem, if the series $q_1, q_2, \ldots, q_n$ is the set of n independent variables that define a problem or a physical phenomenon, then there will always exist an explicit function of the type:

$$f(q_1, q_2, \ldots, q_n) = 0 \tag{2.1}$$

In this way, a number of dimensionless factor p can be defined with all the variables; hence

$$\pi_1 = q_1^{a_1}, q_2^{a_2}, \ldots, q_n^{a_n}$$

$$\pi_2 = q_1^{b_1}, q_2^{b_2}, \ldots, q_n^{b_n}$$

$$\cdots\cdots\cdots\cdots\cdots\cdots\cdots\cdots$$

$$\cdots\cdots\cdots\cdots\cdots\cdots\cdots\cdots$$

$$\pi_i = q_1^{p_1}, q_2^{p_2}, \ldots, q_n^{p_n}$$

Thus, i factors π are obtained, each of them being a function of the variables raised to a power that may be positive, negative, or null. The number of dimensionless factors π will be i, where $i = n - k$, with n being the number of independent variables and k is the characteristic of the matrix formed by the exponents of the dimensional equations of the different variables and constants in relation to a defined unit system.

These i dimensionless factors $\pi_1, \pi_2, \ldots, \pi_i$ will be related by means of a function

$$f(\pi_1, \pi_2, \ldots, \pi_i) = 0 \tag{2.2}$$

that can be applied in any unit system. Sometimes it is difficult to find this type of relationship, so a graphical representation that relates the different parameters is used as an alternative.

2.2.2 Dimensional Analysis Methods

The three main methods for dimensional analysis are *Buckingham's method, Rayleigh's method*, and *the method of differential equations*. The first two methods will be studied in detail, and the third method will be briefly described.

2.2.2.1 Buckingham's Method

The variables that may influence the phenomenon studied are listed first. The dimensional equations of the different variables are established, as well as the dimensional constants. In the case of variables having the same dimensions, only one of them is chosen. The rest of the variables are divided by the chosen variable, obtaining dimensionless groups that will be added to the total obtained. These dimensionless rates are the so-called form factors. The next step is to build the matrix with the exponents of the magnitudes corresponding to the different variables and dimensionless constants. Thus, for the case of n variables $q_1, q_2, \ldots, q_n$ and the constant g_c,

$$\begin{array}{c} \\ L \\ M \\ F \\ T \\ \theta \end{array} \begin{array}{c} q_1, q_2 \ldots, q_n \quad g_c \\ \left(\begin{array}{l} \text{Matrix of the exponents} \\ \text{of the magnitudes for the} \\ \text{dimensional variables} \\ \text{and constants} \end{array} \right) \end{array}$$

The determinant k is obtained from this matrix. This value represents the minimum number of variables and constants that do not form a dimensionless group.

Next, the i dimensionless groups or factors (π) are formed. Each group is formed by the product of $k + 1$ factors q, each factor being a dimensionless variable or constant, raised to powers that should be determined. There will be k dimensionless factors that will be the variables that make the matrix to be of determinant k, in addition to each of the $n - k$ remaining variables with a unit exponent. In this way, the factors π will be

$$\pi_1 = q_1^{a_1} \cdot q_2^{a_2} \ldots q_k^{a_k} \cdot q_{k+1}$$

$$\pi_2 = q_1^{b_1} \cdot q_2^{b_2} \ldots q_n^{b_n} \cdot q_{k+2}$$

....

....

$$\pi_i = q_1^{p_1} \cdot q_2^{p_2} \ldots q_n^{p_n} \cdot q_{k+i-n}$$

In this set of equations, the magnitudes ($q_1, q_2, \ldots, q_k$), variables, and constants contain as a whole the total of fundamental magnitudes of the unit system selected. The set of exponents $a_1, b_1, \ldots, p_1$; $a_2, b_2, \ldots p_2$, etc., should be such that the groups lack dimensions.

Since the factors π lack dimensions, the magnitudes of each variable are set up in the different dimensionless collections, grouping each magnitude in such a way that it is raised to a power that is a combination of the exponents of the variables in which this magnitude appears. This combination of exponents should be equal to zero, obtaining for each factor π a system of k equations and k unknowns. The systems are solved to determine the exponents of the variables in each dimensionless group.

With these i factors and those possible form factors, the following function may be obtained:

$$f(\pi_1, \pi_2, \ldots, \pi_i, \ldots) = 0 \tag{2.3}$$

In some cases there is no need to form the matrix of exponents or to determine its determinant, since by simple observation the minimum number of variables and constants that do not form a dimensionless group can be found.

2.2.2.2 Rayleigh's Method

As in the previous method, initially, the physical variables that participate in the process should be identified. Next, one of the variables, generally the variable of greatest interest, is expressed in an analytical way as an exponent function of the remaining variables:

$$q_1 = K \cdot q_2^{a_2} q_3^{a_3} \ldots q_n^{a_n} \tag{2.4}$$

where K is a dimensionless constant.

The variables and possible dimensionless constants are substituted by the magnitudes of a unit system, applying the homogeneity conditions for each magnitude. In this way, as many equations as there are fundamental magnitudes in the selected unit system are obtained. If p is the number of such magnitudes, the number of unknowns in this equation system will be $n - 1$:

Equations p
Unknowns $n - 1$ exponents

Since the number of unknowns is greater than the number of equations, $(n - 1) - p$ exponents are chosen, while the rest are set up as a function of the exponents. In this way, all the exponents may be established as a function of those $(n - 1 - p)$ selected and are substituted in the last equation. The variables and dimensional constants are grouped in such a way that groups are raised to the same power. Thus, $(n - 1 - p)$ groups that have risen to the selected power are obtained, in addition to one group whose power is the unit. In this way, the relationship among dimensionless groups is obtained:

$$\pi_1 = f(\pi_2, \pi_3, \ldots, \pi_i) \tag{2.5}$$

2.2.2.3 Method of Differential Equations

This method is based on the differential equations of momentum, mass, and energy conservation, which can be applied to a given problem, as well as those that can be obtained from the boundary conditions.

Since this case begins with equations that are dimensionally homogeneous, if all the terms in each equation are divided by any of them, as many dimensionless groups as terms had the equation minus one can be obtained.

The advantage of this method compared to the other two methods is that it is less probable that variables that may have influence on a determined problem will be omitted. This can only occur in the case in which an incorrect equation is employed. Also, this method supplies a more intuitive approach to the physical meaning of the resultant dimensionless groups (Dickey and Fenic, 1976).

2.3 SIMILARITY THEORY

For design and building of industrial equipment, there are two methods based on the construction of models. These models may be mathematical or empirical. The industrial equipment is called **prototype**:

- **Mathematical models**: Beginning with theoretical aspects it is possible, sometimes, to design and build a prototype applicable to industrial scale. In practice, this happens on scarce occasions.
- **Empirical models**: In this case, experimentation in reduced models is necessary, following the directions given by the dimensional analysis.

The values of the prototype are calculated from the values found in the model. A series of similarity criteria must be met to pass from the model to the prototype.

The main difference between the two models relies on the fact that the mathematical model is applicable at any scale, while for the application of the empirical model a series of similarity criteria between the model and the prototype should be in agreement.

In a general way, the similarity criteria can be expressed according to the following linear equation:

$$m' = km \tag{2.6}$$

where m and m' are measures of the same magnitude in the model and in the prototype, respectively. The proportionality constant k receives the name of *scale factor*.

This type of similarity is applicable to the different magnitudes that encircle the system, such as geometry, force profile, velocity, temperature, and concentration. Therefore, the different similarity criteria will be as follows:

- **Geometric similarity** refers to the proportionality between the dimensions of the model and the prototype.
- **Mechanical similarity**, whether static, kinematic, or dynamic, refers to the proportionality between deformations, velocities, and forces, respectively.
- **Thermal similarity** refers to whether it exists proportionally between the temperatures.
- **Concentration similarity** refers to chemical processes, which requires proportionality between concentrations and compositions.

Geometric similarity is a previous requisite for the other criteria. In general, each similarity results in a requisite for the following similarities.

Geometric and mechanical similarities will be studied next, and thermal and concentration similarities will be briefly discussed. The thermal, concentration, and chemical similarities may be achieved by working at equal temperature and concentration.

2.3.1 Geometric Similarity

There is geometric similarity between two systems when each point of one of them has a correspondent point in the other one. Correspondent points of two systems are those points for which there is a constant ratio for correspondent coordinates.

Figure 2.1 presents two pieces of cylindrical tubing, with radii r_1 and r_2 and lengths L_1 and L_2; the points P_1 and P_2 will correspond if their radial and axial coordinates have a constant ratio:

$$\frac{x_1}{x_2} = \frac{y_1}{y_2} = \frac{r_1}{r_2} = \frac{L_1}{L_2} = k \tag{2.7}$$

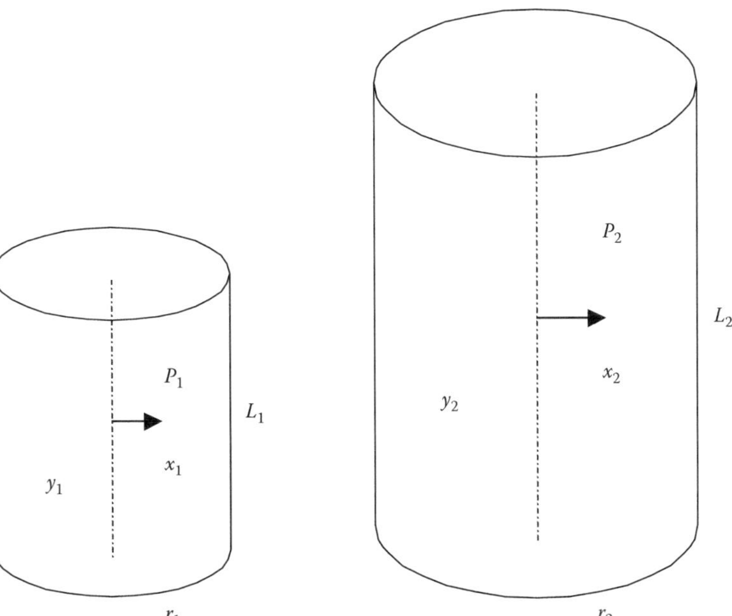

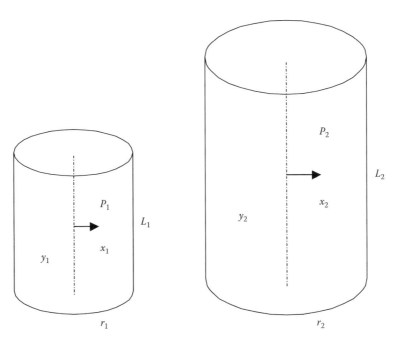

FIGURE 2.1 Geometric similarity. Similar figures.

Another form used to define geometric similarity is to employ ratios between dimensions that belong to the same system, and that receive the name of shape factors. Thus,

$$\frac{r_1}{L_1} = \frac{r_2}{L_2} = \omega \tag{2.8}$$

2.3.2 Mechanical Similarity

2.3.2.1 Static Similarity

Static similarity links the proportionality of the deformations. However, this type of similarity can be neglected if materials with enough resistance are employed. When constant tension is applied to solid bodies and geometric similarity is maintained, static similarity exists.

2.3.2.2 Kinematic Similarity

Once the model and the prototype are similar, the proportionality ratios between velocities and times should be sought. In this way, kinematic similarity complies when

$$\frac{v_1}{v_2} = C \tag{2.9}$$

where v_1 and v_2 are the velocities for correspondent points of the model and the prototype.

2.3.2.3 Dynamic Similarity

Dynamic similarity implies equality of all the rates and dimensionless numbers among the significant forces that intervene in analyzed systems. Different equalities that depend on the forces that act on the systems should be complied with in order to have dynamic similarity. Thus, if inertia and friction forces act, then the equality of the Reynolds number should conform. If gravity forces also act, then the equality of Froude's module should be complied with. It will be necessary to obey the equality of Weber's module when free liquid–gas surfaces are present, since superficial tensions appear. When pressure forces exist, the equality of Euler's number should be complied with.

A practical study of dynamic similarity is presented in the following texts. Consider a mass differential in each of the systems (model and prototype) with different density and viscosity. Inertia forces that move the mass act in both systems, and, additionally, both are subjected to friction forces. As shown in Figure 2.2, the considered mass is enclosed in a cubic volume that moves within a fluid describing a path L.

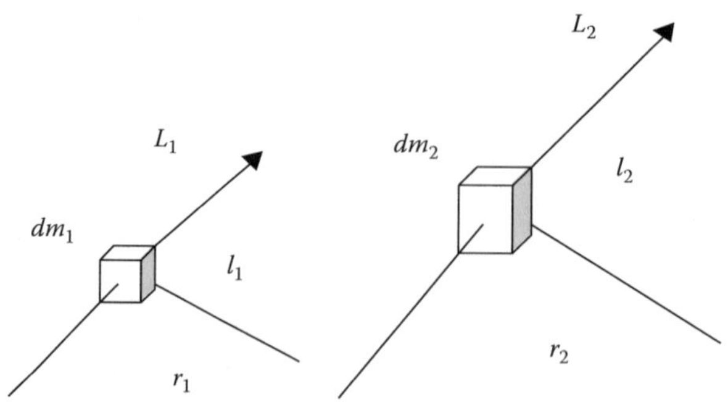

FIGURE 2.2 Displacement of a mass within a fluid: dynamic similarity.

Inertia forces are normal forces (F_n); hence,

$$F_n = \frac{v^2 dm}{r} = \frac{v^2 \rho dV}{r} = \frac{v^2 \rho (dl)^3}{r} \tag{2.10}$$

The friction forces will be tangential, and according to Newton's law,

$$F_t = \eta \, dA \frac{dv}{dr} = \eta (dl)^2 \frac{dv}{dr} \tag{2.11}$$

where
- l is the length
- A is the area
- η is the viscosity

For dynamic similarity to exist, then

$$\frac{F_{n1}}{F_{t1}} = \frac{F_{n2}}{F_{t2}}$$

or

$$\left(\frac{F_n}{F_t}\right)_1 = \left(\frac{F_n}{F_t}\right)_2 \tag{2.12}$$

Since

$$\frac{F_n}{F_t} = \frac{v^2 \rho (dl)^3}{r} \frac{1}{\eta (dl)^2 (dv/dr)} = \frac{v^2 \rho \, dl}{r} \frac{1}{\eta (dv/dr)} \tag{2.13}$$

then

$$\frac{v_1^2 \rho_1 \, dl_1}{r_1} \frac{1}{\eta_1 \left(dv_1/dr_1\right)} = \frac{v_2^2 \rho_2 \, dl_2}{r_2} \frac{1}{\eta_2 \left(dv_2/dr_2\right)}$$

Taking geometric and kinematic similarity ratios into account,

$r_i = \omega l_i$		
$l_i = kl_j$	$dl_i = kdl_j$	
$r_i = kr_j$	$dr_i = kdr_j$	
$v_i = Cv_j$	$dv_i = Cdv_j$	$v_i^2 = Cv_j^2$

Substituting these relationships into the left-hand side of the last equality, the following can be obtained:

$$\frac{\rho_1 v_1 k}{\eta_1} = \frac{\rho_2 v_2}{\eta_2}$$

but $k = l_1/l_2$; hence,

$$\frac{\rho_1 v_1 l_1}{\eta_1} = \frac{\rho_2 v_2 l_2}{\eta_2} \tag{2.14}$$

that is, $(\mathrm{Re})_1 = (\mathrm{Re})_2$

which points out that the equality of the Reynolds module conforms to the model and prototype only when inertia and friction forces exist.

If, in addition, gravity forces act on the systems, then the obtained relations are taken ($(\mathrm{Re})_1 = (\mathrm{Re})_2$), and also the following equality is confirmed:

$$\left(\frac{F_n}{F_G}\right)_1 = \left(\frac{F_n}{F_G}\right)_2 \tag{2.15}$$

Gravity force is defined as

$$F_G = g\,dm = g\rho(dl)^3 \tag{2.16}$$

Considering Equations 2.10 and 2.16, the relationship between the inertia and gravity forces will be defined by

$$\frac{F_n}{F_G} = \frac{v^2}{rg} \tag{2.17}$$

and since $r_i = \omega l_i$, then the following relationship applies:

$$\frac{v_1^2}{l_1 g} = \frac{v_2^2}{l_2 g} \tag{2.18}$$

this indicates equality in Froude's module. It must be noted that g cannot be simplified, since a dimensional expression would be obtained. When there are gravity forces, Froude's modules of the model and the prototype are equal.

In the case in which surface tension forces (F_s) intervene, the new relationship to comply with is

$$\left(\frac{F_n}{F_s}\right)_1 = \left(\frac{F_n}{F_s}\right)_2 \tag{2.19}$$

The surface tension forces are given by

$$F_s = \sigma l \tag{2.20}$$

where σ is the surface tension. In this way,

$$\frac{F_n}{F_s} = \frac{\rho v^2 (dl)^3}{r} \frac{1}{\sigma l} \tag{2.21}$$

from which the following equation is obtained:

$$\frac{\rho_1 v_1^2 (dl_1)^3}{r_1} \frac{1}{\sigma_1 l_1} = \frac{\rho_2 v_2^2 (dl_2)^3}{r_2} \frac{1}{\sigma_2 l_2} \tag{2.22}$$

Taking into account the geometric similarity relationships,

$$r_i = \omega l_i$$
$$l_1 = kl_2 \qquad dl_1 = kdl_2$$

and substituting them into the last equality and simplifying it, the following can be obtained:

$$\frac{\rho_1 v_1^2 l_1}{\sigma_1} = \frac{\rho_2 v_2^2 l_2}{\sigma_2} \tag{2.23}$$

this indicates that Weber's modules are equal. That is, to have dynamic similarity when surface forces act, Weber's modules must coincide in the model and prototype.

Finally, the case in which there are pressure forces due to pressure differences is examined later. The new relation to be conformed with is

$$\left(\frac{F_P}{F_N}\right)_1 = \left(\frac{F_P}{F_N}\right)_2 \tag{2.24}$$

The pressure forces are defined by

$$F_P = \Delta p l^2 \tag{2.25}$$

consequently,

$$\frac{F_P}{F_N} = \Delta p l^2 \frac{r}{\rho v^2 (dl)^3} \tag{2.26}$$

The combination of Equations 2.24 and 2.26 gives

$$\Delta p_1 l_1^2 \frac{r_1}{\rho_1 v_1^2 (dl_1)^3} = \Delta p_2 l_2^2 \frac{r_2}{\rho_2 v_2^2 (dl_2)^3} \tag{2.27}$$

and according to geometric similarity,

$$r_i = wl_i$$
$$l_1 = kl_2 \qquad dl_1 = kdl_2$$

When substituting them in the last equality, and then simplifying it, the following can be obtained:

$$\frac{\Delta p_1}{\rho_1 v_1^2} = \frac{\Delta p_2}{\rho_2 v_2^2} \tag{2.28}$$

this indicates that there is equality in Euler's module. That is, when there is dynamic similarity, Euler's modules of the model and prototype coincide.

Not all the forces described here are always present, so there will only be dynamic similarity in those equalities between the dimensionless modules of the model and prototype referring to the force or forces acting on the system.

PROBLEMS

2.1 A fluid food has a viscosity of 6 P. Obtain the value for viscosity in the SI and in the absolute English system.

By definition, 1 P is one gram per centimeter and second: 1 P = g/(cm s)

Conversion to the SI:

$$6\frac{\text{g}}{\text{cm s}}\frac{1\,\text{kg}}{10^3\,\text{g}}\frac{100\,\text{cm}}{1\,\text{m}} = 0.6\frac{\text{kg}}{\text{m s}} = 0.6\,\text{Pa s}$$

In the SI, the viscosity unit is Pa s that is equivalent to kg/(m s).

Conversion to the absolute English system:

$$6\frac{\text{g}}{\text{cm s}}\frac{1\,\text{lb}}{453.5\,\text{g}}\frac{30.48\,\text{cm}}{1\,\text{ft}} = 0.403\,\text{lb/ft s}$$

2.2 Empirical equations are used, in many cases, to calculate individual heat transfer coefficients. Thus, the following expression may be used for the circulation of water in a cylindrical pipe:

$$h = 160(1+0.01t)\frac{(v_m)^{0.8}}{(d_i)^{0.2}}$$

where

h is the film coefficient in Btu/(h ft^2 °F)
v_m is the mean velocity of water in ft/s
d_i is the interior diameter of the pipe in inches (in.)
t is the water temperature in °F

Perform the adequate unit conversion so this equation can be used in the SI.

When the temperature does not appear as an increment, but as temperature in absolute terms, it is recommended to convert it as follows:

$$t°\text{F} = 1.8t°\text{C} + 32°\text{C}$$

hence,

$$h = 160(1.32+0.018t)\frac{(v_m)^{0.8}}{(d_i)^{0.2}}$$

Simplifying

$$h = 211.2\frac{(v_m)^{0.8}}{(d_i)^{0.2}} + 2.88\frac{t(v_m)^{0.8}}{(d_i)^{0.2}}$$

in which the temperature t is expressed in °C.

The units of the two coefficients that appear in this new equation are

$$211.2 \frac{\text{Btu}}{\text{h ft}^2\,°\text{F}} \frac{(\text{in.})^{0.2}}{(\text{ft/s})^{0.8}}$$

$$2.88 \frac{\text{Btu}}{\text{h ft}^2\,°\text{F}} \frac{(\text{in.})^{0.2}}{°\text{C}(\text{ft/s})^{0.8}}$$

The next step is the conversion of these coefficients to the SI:

$$211.2 \frac{\text{Btu}}{\text{h ft}^2\,°\text{F}} \frac{(\text{in.})^{0.2}}{(\text{ft/s})^{0.8}} \frac{5.678\,\text{W/(m}^2\,°\text{C)}}{1\,\text{Btu/(h ft}^2\,°\text{F)}} \left(\frac{0.0254\,\text{m}}{1\,\text{in.}}\right)^{0.2} \left(\frac{1\,\text{ft}}{0.3048\,\text{m}}\right)^{0.8}$$

$$= 1488 \frac{\text{W}}{\text{m}^2\,°\text{C}} \frac{\text{m}^{0.2}}{(\text{m/s})^{0.8}}$$

$$2.88 \frac{\text{Btu}}{\text{h ft}^2\,°\text{F}} \frac{(\text{in.})^{0.2}}{°\text{C}(\text{ft/s})^{0.8}} \frac{5.678\,\text{W/(m}^2\,°\text{C)}}{1\,\text{Btu/(h ft}^2\,°\text{F)}} \left(\frac{0.0254\,\text{m}}{1\,\text{in.}}\right)^{0.2} \left(\frac{1\,\text{ft}}{0.3048\,\text{m}}\right)^{0..8}$$

$$= 20.3(\text{W/m}^2\,°\text{C}) \frac{\text{m}^{0.2}}{°\text{C}(\text{m/s})^{0.8}}$$

So the resulting equation is

$$h = 1488 \frac{(v_m)^{0.8}}{(d_i)^{0.2}} + 20.3 \frac{t(v_m)^{0.8}}{(d_i)^{0.2}}$$

which when rearranged gives

$$h = 1488(1 + 0.01364t) \frac{(v_m)^{0.8}}{(d_i)^{0.2}}$$

with units h in W/(m² °C), v_m in m/s, d_i in m, and t in °C.

2.3 Use a dimensional analysis to obtain an expression that allows the calculation of the power of a stirrer as a function of the variables that could affect it. It is known, from experimental studies, that the stirring power depends on the diameter of the stirrer (D), on its rotation velocity (N), on the viscosity (η) and density (ρ) of the fluid being stirred, and on the gravity acceleration (g).

The power of the stirrer, called P, can be expressed as a function of the other variables: $P = f(D, \eta, g, \rho, N)$

Applying Rayleigh's method, $P = K \cdot D^a \cdot \eta^b \cdot g^c \cdot \rho^d \cdot N^e$.

The number of fundamental magnitudes is three: length (L), mass (M), and time (T). The number of variables is 6. Three equations with five unknowns can be obtained. The number of factors π will be 6 − 3 = 3:

	P	D	η	g	ρ	N
M	1	0	1	0	1	0
L	2	1	−1	1	−3	0
T	−3	0	−1	−2	0	−1
	1	a	b	c	d	e

$$\frac{ML^2}{T^3} = K^0 L^a \left(\frac{M}{LT}\right)^b \left(\frac{L}{T^2}\right)^c \left(\frac{M}{L^3}\right)^d \left(\frac{L}{T}\right)^e$$

Mass (M): $1 = b + d$
Length (L): $2 = a - b + c - 3d$
Time (T): $-3 = -b - 2c - e$

Since there are three equations with five unknowns, two of them can be fixed. If b and c are fixed, the other unknowns can be set up as a function of b and c:

$$d = 1 - b$$

$$e = 3 - b - 2c$$

$$a = 2 + b - c + 3 - 3b = 5 - 2b - c$$

If a, d, and e are substituted in the equation in which power is a function of the different variables, we obtain $P = K \cdot D^{5-2b-c} \cdot \eta^b \cdot g^c \cdot \rho^{1-b} \cdot N^{3-b-2c}$.

The variables are grouped in such way that they have the same exponent:

$$P = KD^5 \rho N^3 \left(\frac{\rho N D^2}{\eta}\right)^{-b} \left(\frac{DN^2}{g}\right)^{-c}$$

$$\frac{P}{\rho N^3 D^5} = K \left(\frac{\rho N D^2}{\eta}\right)^{-b} \left(\frac{DN^2}{g}\right)^{-c}$$

It can be observed that three dimensionless constants were obtained:

Power module or number: $(\text{Po}) = \dfrac{P}{\rho N^3 D^5}$

Reynolds number: $(\text{Re}) = \dfrac{\rho N D^2}{\eta}$

Froude's number: $(\text{Fr}) = \dfrac{DN^2}{g}$

The last number expresses the ratio of dynamic action/gravity action.

In general, the power module can be expressed as a function of Reynolds and Froude's numbers, according to an expression of the type:

$$\text{Po} = K(\text{Re})^m(\text{Fr})^n$$

2.4 When a fluid circulates through a pipe, mechanical energy losses occur due to friction with the pipe walls' mechanical energy losses by mass unit of the fluid ($\hat{E}_f$) depending on the characteristics of the pipe (internal diameter, roughness, and length), on the properties of the circulation fluid (density and viscosity), as well as on the circulation velocity (v). Use Buckingham's method to deduce an expression that allows calculation of $\hat{E}_f$ as a function of the mentioned variables.

The number of variables is 7, and the number of fundamental magnitudes is 3. Therefore, the number of dimensionless factors π is $\pi = 7 - 3 = 4$.

Energy losses due to friction can be expressed as a function of the remaining variables: $\hat{E}_f = K\rho^a(d_i)^b v^c l^d \varepsilon^e \eta^f$.

Writing all this information in matrix form gives

	ρ	d_i	v	L	ε	η	$\hat{E}_f$
M	1	0	0	0	0	1	1
L	−3	1	1	1	1	−1	2
T	0	0	−1	0	0	−1	−2
	a	*b*	*c*	1	1	1	1

We will work with three variables that are fundamental, so we establish them in such a way as to obtain a matrix determinant different from zero. We obtain that the rank of the matrix is equal to three.

Factors π

1. $\pi_1 = \rho^a(d_i)^b v^c \hat{E}_f$

   The fundamental magnitudes are mass, length, and time:

   Mass (M): $0 = a$
   Length (L): $0 = -3a + b + c + 2$
   Time (T): $0 = -c - 2$

   When solving the system, the values obtained for *a*, *b*, and *c* are $a = 0$; $b = 0$; $c = -2$. Hence, the factor π_1 is $\pi_1 = \hat{E}_f/v^2$.

2. $\pi_2 = \rho^a(d_i)^b v^c l$

   Mass (M): $0 = a$
   Length (L): $0 = -3a + b + c + 1$
   Time (T): $0 = -c$

   When solving the system, the values obtained for *a*, *b*, and *c* are $a = 0$; $b = -1$; $c = 0$. Hence, the factor π_2 is $\pi_2 = l/d_i$.

3. $\pi_3 = \rho^a(d_i)^b v^c \varepsilon$

   Mass (M): $0 = a$
   Length (L): $0 = -3a + b + c + 1$
   Time (T): $0 = -c$

   When solving the system, the values obtained for *a*, *b*, and *c* are $a = 0$; $b = -1$; $c = 0$. Hence, the factor π_3 is $\pi_3 = \varepsilon/d_i$.

4. $\pi_4 = \rho^a(d_i)^b v^c \eta$

   Mass (M): $0 = a + 1$
   Length (L): $0 = -3a + b + c - 1$
   Time (T): $0 = -c - 1$

   When solving the system, the values obtained for *a*, *b*, and *c* are $a = -1$; $b = -1$; $c = -1$. Hence, the factor π_4 is $\pi_4 = \eta/(\rho v d_i) = (\mathrm{Re})^{-1}$. Since it is a dimensionless factor, it can be considered that its value is the Reynolds number $\pi_4 = \mathrm{Re}$.

   According to Buckingham's π theorem, one of these dimensionless factors can be expressed as a function of the other three factors. Therefore, it can be written that

$$\pi_1 = f(\pi_2, \pi_3, \pi_4)$$

that is,

$$\frac{\hat{E}_f}{v^2} = \phi\left(\frac{l}{d_i}, \frac{\varepsilon}{d_i}, \mathrm{Re}\right)$$

rearranging gives

$$\hat{E}_f = v^2\phi\left(\frac{l}{d_i},\frac{\varepsilon}{d_i},\mathrm{Re}\right)$$

It is known that mechanical energy losses per mass unit are proportional to the length, so $\hat{E}_f = v^2\frac{l}{d_i}\phi\left(\frac{\varepsilon}{d_i},\mathrm{Re}\right)$

It is obtained that $\hat{E}_f$ is directly proportional to the squared velocity and to the length and inversely proportional to the pipe diameter. Likewise, it depends on a function ϕ' that depends on the Reynolds number and the so-called relative roughness, ε/d_i.

Experimentation should be performed to complete this expression. However, the function ϕ' can be substituted by a factor f, called the *friction factor*, in such a way that the energy losses by friction could be obtained from

$$\hat{E}_f = fv^2\frac{l}{d_i}$$

where the factor f is a function of the Reynolds number and refers to the relative roughness.

2.5 One of the most frequently used devices for batch fermentations is the stirred tank. The power that should be applied to the stirrer (P) is a function of its rotation velocity (N) and diameter (D), density (ρ) and viscosity of the substrate, gravity acceleration (g), and time (t) since the beginning of operation. Demonstrate, using Rayleigh's and Buckingham's methods, that the power module (Po) is a function of the Reynolds (Re) number, Froude's (Fr) number, and time module (Nt), that is,

$$(\mathrm{Po}) = \Phi[(\mathrm{Re}), (\mathrm{Fr}), (\mathrm{Nt})]$$

The modules (Po), (Re), and (Fr) are defined by the following expressions:

$$(\mathrm{Po}) = \frac{P}{\rho N^3 D^5} \quad (\mathrm{Re}) = \frac{D^2 N\rho}{\eta} \quad (\mathrm{Fr}) = \frac{DN^2}{g}$$

The number of variables is 7, while the number of fundamental magnitudes is 3; therefore, the number of dimensionless factors π is 4.

Rayleigh's Method

The power of the stirrer can be expressed in relation to the function of the other variables:

$$P = KN^a\rho^b\eta^c D^d g^e t^f \tag{P.2.1}$$

	P	***N***	***ρ***	***η***	***D***	***g***	***t***
M	1	0	1	1	0	0	0
L	2	0	−3	−1	1	1	0
T	−3	−1	0	−1	0	−2	1
	1	*a*	*b*	*c*	*d*	*e*	*f*

$$\frac{ML^2}{T^3} = K^0\left(\frac{1}{T}\right)^a\left(\frac{M}{L^3}\right)^b\left(\frac{M}{LT}\right)^c(L)^d\left(\frac{L}{T^2}\right)^e(T)^f$$

Mass (M): $1 = b + c$
Length (L): $2 = -3b - c + d + e$
Time (T): $-3 = -a - c - 2e + f$

Since three equations in six unknowns resulted, three of them should be fixed. If c, e, and f are fixed, the other variables are set up as a function of c, e, and f:

$$b = 1 - c$$

$$d = 5 - e - 2c$$

$$a = 3 - c - 2e + f$$

Substituting a, b, and d in Equation P.2.1 leads to

$$P = KN^{3-c-2e+f}\rho^{1-c}\eta^{c}D^{5-2c-e}g^{e}t^{f}$$

The variables are grouped in such way that they have the same exponent:

$$P = KD^5\rho N^3\left(\frac{\eta}{\rho ND^2}\right)^c\left(\frac{g}{DN^2}\right)^e(\mathrm{Nt})^f$$

therefore,

$$\frac{P}{\rho N^3 D^5} = K\left(\frac{\rho ND^2}{\eta}\right)^{-c}\left(\frac{DN^2}{g}\right)^{-e}(\mathrm{Nt})^f$$

Taking into account the definitions of power, Reynolds, and Froude's modules, this expression is converted into $(\mathrm{Po}) = K(\mathrm{Re})^{-c}(\mathrm{Fr})^{-e}(\mathrm{Nt})^f$
or: $(\mathrm{Po}) = \Phi[(\mathrm{Re}), (\mathrm{Fr}), (\mathrm{Nt})]$, which is the focus of our research.

Buckingham's Method
The matrix of the exponents is formed:

	P	N	D	g	ρ	η	t
M	1	0	0	0	1	1	0
L	2	0	1	1	3	−1	0
T	−3	−1	0	−2	0	−1	1

We look for the rank of the matrix, which can be a maximum of 3. There is a determinant different from zero, formed by the columns N, D, and ρ:

$$\begin{array}{c} \quad N \quad D \quad \rho \\ \text{Det}\begin{vmatrix} 0 & 0 & 1 \\ 0 & 1 & -3 \\ -1 & 0 & 0 \end{vmatrix} = 1 \end{array}$$

Therefore, the rank of the matrix is 3. *N*, *D*, and ρ are chosen as fundamental variables:

	N	*D*	ρ	*P*	η	*G*	*t*
M	0	0	1	1	1	0	0
L	0	1	−3	2	−1	1	0
T	−1	0	0	−3	−1	−2	1
	a	*d*	*b*	1	1	1	1

The dimensionless factors π are determined one by one:

$$\pi_1 = N^a \rho^b D^d P$$

Mass (M): $0 = b + 1$
Length (L): $0 = d - 3b + 2$
Time (T): $0 = -a - 3$

When solving this system, we obtain $b = -1$; $d = -5$; $a = -3$
so the dimensionless factor π_1 is $\pi_1 = P/N^3D^5\rho$

$$\pi_2 = N^a \rho^b D^d \eta$$

Mass (M): $0 = b + 1$
Length (L): $0 = d - 3b - 1$
Time (T): $0 = -a - 1$

When solving this system, we obtain $b = -1$; $d = -2$; $a = -1$
so the dimensionless factor π_2 is $\pi_2 = \eta/ND^2\rho$

$$\pi_3 = N^a \rho^b D^d g$$

Mass (M): $0 = b$
Length (L): $0 = d - 3b + 1$
Time (T): $0 = -a - 2$

When solving this system, we obtain $b = 0$; $d = -1$; $a = -2$
so the dimensionless factor π_3 is $\pi_3 = g/N^2D$

$$\pi_4 = N^a \rho^b D^d t$$

Mass (M): $0 = b$
Length (L): $0 = d - 3b$
Time (T): $0 = -a + 1$

When solving this system, we obtain $b = 0$; $d = 0$; $a = 1$
so the dimensionless factor π_4 is π_4 = Nt

Applying Buckingham's theorem, $\pi_1 = F(\pi_2, \pi_3, \pi_4)$

Hence, (Po) = Φ[(Re), (Fr), (Nt)], which is the focus of our research.

2.6 A cylindrical 5 m diameter tank that stores concentrated juice (54°Brix) has a 20 cm diameter drainage, placed 50 cm away from the lateral wall of the deposit, and a nozzle 10 cm above the bottom of the tank. When draining the tank, a vortex is formed in such a way that when the liquid level is not high enough, the vortex can reach the outlet pipe, and air can be drawn off together with the juice. The tank should operate at a draining flow of 20 m^3/s. With the purpose of predicting what should be the minimum level of juice in the tank to avoid the vortex reaching the drainage, a study will be performed with a reduced scale model that will operate with water. Determine the dimensions the model should have, as well as the operation conditions.

Data: It may be supposed that the shape of the vortex depends only on the draining rate and the quantity of liquid in the tank.

Properties of the juice: Density 1250 kg/m^3. Viscosity 50 mPa s

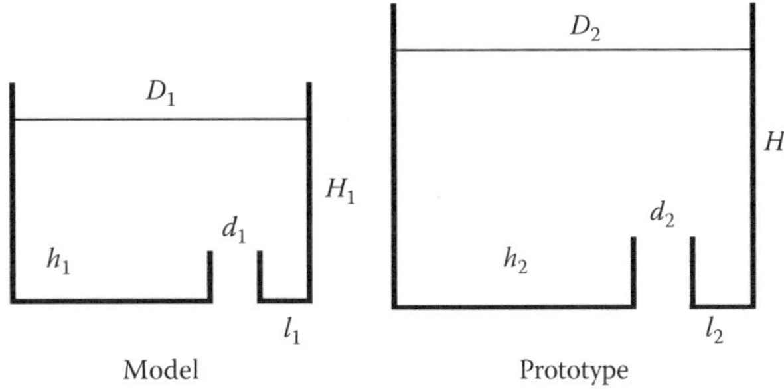

H_2 is the level of the juice in the tank required for the vortex not to reach the drainage. Subscript 1 indicates model and subscript 2 the prototype.

Geometric Similarity

$$\frac{d_1}{d_2} = \frac{D_1}{D_2} = \frac{l_1}{l_2} = \frac{h_1}{h_2} = \frac{H_1}{H_2} = k$$

Dynamic Similarity

Since there are inertia, friction, and gravity forces, it should be understood that Reynolds and Froude's numbers for the model and for the prototype should be equal:

$$(\text{Re})_1 = (\text{Re})_2 \quad \text{and} \quad (\text{Fr})_1 = (\text{Fr})_2$$

From the equality of Reynolds numbers,

$$\frac{\rho_1 v_1 d_1}{\eta_1} = \frac{\rho_2 v_2 d_2}{\eta_2}$$

hence,

$$\frac{d_1 v_1}{d_2 v_2} = \frac{\eta_1 \rho_2}{\eta_2 \rho_1}$$

From the equality of Froude's numbers,

$$\frac{(v_1)^2}{g d_1} = \frac{(v_2)^2}{g d_2}$$

therefore,

$$\frac{v_1}{v_2} = \left(\frac{d_1}{d_2}\right)^{1/2}$$

When combining the obtained expressions

$$\frac{d_1\left(d_1\right)^{1/2}}{d_2\left(d_2\right)^{1/2}}=\frac{\eta_1\rho_2}{\eta_2\rho_1}$$

Rearranging gives

$$\frac{d_1}{d_2}=\frac{\left(\eta_1\rho_2\right)^{2/3}}{\left(\eta_2\rho_1\right)^{2/3}}$$

From the data given in the problem and from the properties of water,

$\eta_1 = 1$ mPa s $\quad \rho_1 = 1000$ kg/m^3
$\eta_2 = 50$ mPa s $\quad \rho_2 = 1250$ kg/m^3

it is obtained that the *geometric similarity ratio* is $k = d_1/d_2 = 0.0855$.

This factor allows us to obtain the dimensions of the model:

$$d_1 = kd_2 = 0.0855\cdot 0.2\text{ m} = 0.017\text{ m}$$

$$h_1 = kh_2 = 0.0855\cdot 0.1\text{ m} = 0.0085\text{ m}$$

$$D_1 = kd_2 = 0.085\cdot 5\text{ m} = 0.428\text{ m}$$

$$l_1 = kl_2 = 0.0855\cdot 0.5\text{ m} = 0.0428\text{ m}$$

Volumetric flow (q) is the product of the linear velocity (v) times the cross-sectional area (s). Since it is expressed as a function of the diameter of the pipe, we have

$$q = vs = v\frac{\pi}{4}d^2$$

so the flow rate between the model and the prototype will be

$$\frac{q_1}{q_2}=\frac{v_1\left(d_1\right)^2}{v_2\left(d_2\right)^2}=\left(\frac{d_1}{d_2}\right)^{5/2}=k^{5/2}$$

Therefore, the volumetric flow to drain the model is

$$q_1 = k^{5/2}q_2 = (0.0855)^{5/2}(20\text{ m}^3/\text{s}) = 0.0428\text{ m}^3/\text{s}$$

The minimum level of the liquid in the tank can be expressed as a function of the level in the model, from the value of geometric similarity rate:

$$H_2=\frac{H_1}{k}=\frac{H_1}{0.0855}=11.7H_1$$

The value of H_1 can be experimentally obtained in the laboratory, working with the model. Then with H_1 value, it is possible to obtain the value of H_2, which is the minimum height the juice should reach in the prototype so that the vortex does not reach the drainage pipe.

3 Introduction to Transport Phenomena

3.1 HISTORICAL INTRODUCTION

It should be recalled that the term *unit operation* was established in 1916 by Professor Arthur D. Little, Massachusetts Institute of Technology (MIT). Because of its historical and conceptual value, it is interesting to recall its definition:

> ..... [E]very chemical process conducted at any scale can be decomposed in an ordered series of what can be called Unit Operations, as pulverization, drying, crystallization, filtration, evaporation, distillation, etc. The amount of these Unit Operations is not very large and, generally, only few of them take place in a determined process. (Costa et al., 1984)

This simplification reduced the complexity of the study of processes, since from the almost infinite set of processes that can be imagined, it will only be necessary to study the set of existing unit operations. A given process will be formed by a combination of unit operations.

A new development and growth stage began with the systematic study of these unit operations, adding new operations and generalizing, for many of them, their didactic presentation, through dimensional analysis and experimental study. It was a phase of reasoned empiricism in which theory and practice were skillfully combined, and during which the theoretical fundamentals of the different operations were gradually established. This traditional concept of unit operations has been one of the main factors of the extraordinary success of this branch of engineering in the past.

Continuing with the systematization effort, a new generalization period began, grouping unit operations according to the general principles on which they are based. In this way, they are classified within the following sections:

- Fluid treatment
- Mass transfer, by multiple contacts
- Energy and mass transfer, by continuous contact

Later, and because of a better knowledge of the fundamentals of the unit operations, it was noticed that all of them are based on the three following phenomena:

- Momentum transfer
- Energy transfer
- Mass transfer

In all three cases, the flow of the transferred property is directly proportional to an impelling force (velocity, temperature, or concentration gradient) and inversely proportional to a resistance that depends on the properties of the system and operation conditions. Therefore, it is possible to develop an abstract doctrine body from which the three transport phenomena mentioned can be taken as particular cases.

The fact that the physical, chemical, or biochemical stages that constitute the processes of agricultural industries are the same as those found in the processes of chemical industries makes the knowledge and advances of the unit operations of chemical engineering applicable to the agricultural and food industries whenever they are adapted to the characteristics of raw material (particularly natural products, which are generally perishable) and to the particular conditions (hygiene, cleanliness, etc.) normally required by agricultural–industrial processes. The importance of the concepts of process engineering is that they unify the techniques of what are normally considered separate industries. In this way, the basic principles that are common to all food industries are unified in a logical way, in spite of their apparent diversity.

3.2 TRANSPORT PHENOMENA

All the physical stages encountered in industrial processes are based on the three following phenomena: momentum, energy, and mass transfer. In nonequilibrium condition of a given process, it is expected equilibrium will be reached by at least two of the previously mentioned phenomena.

Some examples to explain this definition are as follows:

1. If in a fluid stream there are two points at which their velocities are different, the system will evolve to counteract this velocity difference by means of a momentum transfer.
2. If there are zones of different temperature in a solid, there will be heat transfer from the hottest zone to the coldest one, so that the system tends to thermal equilibrium.
3. If inside the same phase there is a concentration difference between two points, a mass transfer will tend to equilibrate this concentration difference.

As mentioned, in all processes, at least two out of the three phenomena are simultaneously present, but there are some operations in which normally one phenomenon predominates. Thus, momentum transport predominates in operations of fluid transport, sedimentation, filtration, etc.; heat transfer dominates in the design of heat exchangers, condensers, etc.; and mass transfer dominates in operations such as absorption, solvent extraction, and distillation. There are operations, such as air–water interaction, drying, and crystallization, in which mass and heat transfer phenomena are equally important.

3.3 CIRCULATION REGIMES: REYNOLD'S EXPERIMENT

Before studying the mechanisms of transport phenomena, it is interesting to experimentally prove the existence of such mechanisms. When studying circulation regimes, the experiment carried out by Reynolds in 1833 is extremely important. This experiment consisted of circulating water through transparent tubing with constant cross section, varying the circulation velocity of the liquid by means of a valve placed at the entrance of the tubing. In the middle of the tubing, and in the entrance section, a colored solution was introduced. The variation of the colored vein was observed along the tubing at different circulation velocities of the liquid.

In this type of experiment, if the velocity of the liquid is low, the colored vein does not lose its identity while circulating through the center of the tubing, and although a low and progressive increase of the vein's width can be observed, mixture in a transversal sense is not seen, indicating that the flow takes place as parallel streams with no interference between them. Mass exchange occurs only at the molecular level. As the velocity of the liquid increases, oscillations in the colored filament appear until, at a certain velocity, it breaks into whirlpools and transversally inundates the conduction. Different images of the Reynolds experiment can be observed in Figure 3.1.

Based on these observations, it can be deduced that there are two types of well-differentiated circulation regimes in which distinctive types of mass transfer mechanisms are different.

For low velocities, it can be inferred that the liquid moves in a horizontal way with concentric parallel layers and without transversal movement. This regime is called *laminar* and is characterized by the absence of a global movement perpendicular to the main direction of the stream.

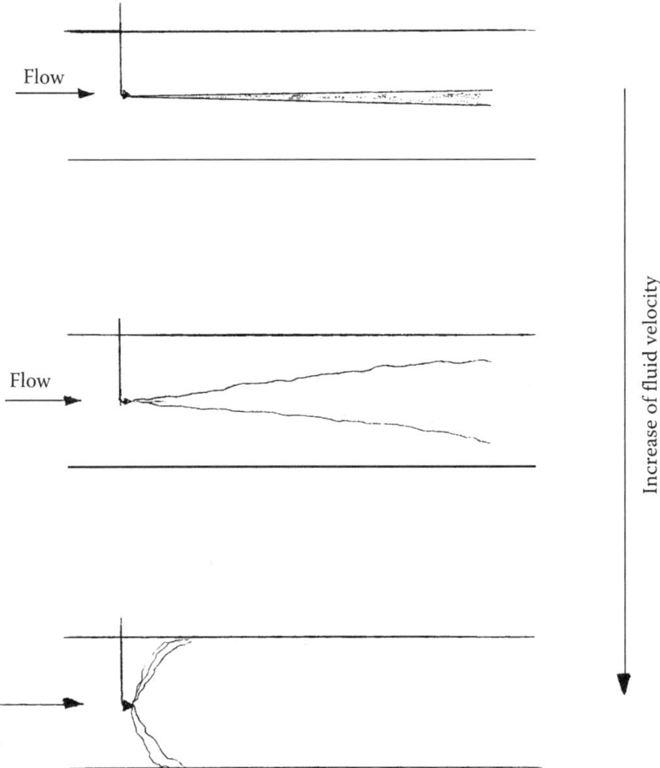

FIGURE 3.1 Reynolds experiment.

For high velocities, there is movement of the liquid in a transversal way of macroscopic proportions. This type of circulation regime is called *turbulent* and is characterized by the rapid movement of the liquid in the form of whirlpools with no preestablished direction in the transversal section of the tubing.

The only variable modified in these experiments was the velocity, but there may be other variables that could alter the regime, such as diameter of the tubing and the nature of the liquid. Hence, for a better study of circulation regimes, a dimensionless module that relates the magnitudes that characterize the circulation phenomenon is defined, setting up the boundaries of the different regimes. This kind of module is called the *Reynolds number*, which represents the quotient between the inertia and viscosity forces of the moving fluid. In the case of a cylindrical conduit and a Newtonian fluid, it is

$$\mathrm{Re} = \frac{\rho v d}{\eta} \tag{3.1}$$

where

Re is the Reynolds number (dimensionless)
ρ is the density of the fluid (kg/m^3)
v is the mean velocity of the fluid (m/s)
d is the diameter of the conduit (m)
η is the viscosity of the fluid (Pa s)

The numerical value of the Reynolds number (Re) is used to define the type of circulation regime of a fluid stream. It has been observed that for values lower than a given Re, called critical,

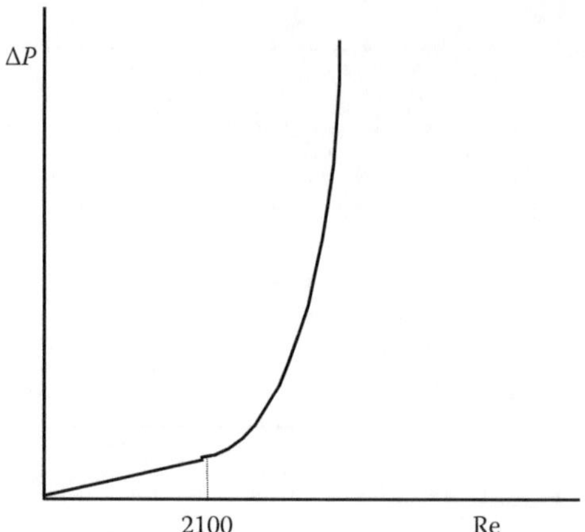

FIGURE 3.2 Variation of the pressure drop in relation to the Reynolds number.

the oscillations of the fluid are unstable, and any perturbation disappears quickly. For values higher than this critical Re, the oscillations become stable and of greater amplitude, giving place to a high radial mixture. For Newtonian fluids the critical Re value is 2100. Values lower than 2100 indicate a laminar regime, while for greater values there is a range of Re number values called transition, in which metastable phenomena could appear. The regime is completely turbulent for values higher than 10,000. This study can also be applied to momentum transport and heat transfer.

If the pressure drop (ΔP) in a fluid between the inlet and outlet sections of a conduit is measured in the Reynolds experiment, an increase with the volumetric flow of the fluid will be observed. Figure 3.2 shows the variation of the pressure drop as a function of the Reynolds number. The loss of charge of the fluid is a result of the energy consumed by the fluid, because a momentum transport took place. It can be seen in this figure that, starting from Re = 2100, the loss of charge increases rapidly, which favors the momentum transport process, since radial components appear in the velocity of the fluid particles.

In order to study heat transfer, we can consider a tubing in which water with mass flow rate (w) at a temperature t_1 enters, and the same volume of water at temperature t_2, due to the heat flow rate gained by the fluid through the tubing wall, exits. The total heat gained will be

$$q = wC_p\,(t_1 - t_2) \tag{3.2}$$

When the mass flow rate (w) varies, the exit temperature (t_2) will also vary. A graphical representation of the variation of gained heat, the Reynolds number will yield a graphic similar to that in Figure 3.2. It can be observed that for a laminar fluid, the exchanged heat increases in direct proportion to the mass of the fluid; however, starting at the critical value of the Reynolds number, a sharp increase is observed. In a laminar regime, heat transfer occurs in radial form, that is, molecule by molecule, but in a turbulent regime there are streams or whirlpools that favor radial heat transport.

3.4 TRANSPORT PHENOMENA MECHANISMS

The mechanism of energy transfer by means of electromagnetic waves is called *radiation*, and it can be performed through vacuum, which does not require a material media for transmission. However, other forms of energy transfer and of momentum transport are associated in one way or another to material movement, although there is not a net transfer. Thus, in heat transfer by conduction in a continuous material media, there is no material movement at the macroscopic level, although there is movement at the molecular level due to the motion of free electrons (in metals) or to the vibration of molecules or ions in solids.

Because these different transport phenomena are associated, it is interesting to study them as a whole. Transport of the three previously mentioned properties may take place by means of two well-differentiated mechanisms:

- Molecular transport
- Turbulent transport

In molecular transport, the transfer of the property is carried out molecule by molecule, by movement of the individual molecules, or by interactions among them. Turbulent transport is produced when large groups of molecules move as aggregates or whirlpools, transporting with them momentum, mass, or energy. These aggregates serve as transportation media, and they transfer the property to other groups of molecules that interact with them. Molecular transport can happen alone, while turbulent transport never occurs in isolation but is always accompanied by molecular transport.

3.4.1 Mass Transfer

To study the mechanism of mass transfer, suppose a given component of the considered material is transferred from one point to another in the studied system. This mass transfer may take place according to two mechanisms, by molecular flow or by *forced convective* flow. When there is a concentration gradient of the considered component between two points of the system, mass transfer is produced by molecular flow. However, when the entire mass moves from one point to another, the transfer is produced by *forced convective* flow.

According to the physical nature of the media, different situations can occur and the mass transfer is carried out by one or by both of the transport mechanisms considered:

1. When there is no concentration gradient of the considered component, and if the medium is a fluid, there can only be *convective* transport. Usually, this type of problem is studied as momentum and not as mass transport.
2. When there is a concentration gradient of the component, and the medium is a fluid in repose, the mass transfer is carried out by molecular flow, due only to molecular diffusion. Thus, if we consider a beaker filled with water with a crystal of colorant placed on the bottom, the crystal dissolves, gradually diffusing throughout the beaker, since the concentration around the crystal is greater than in other zones. This diffusion will take place until equilibrium is reached.
3. When there is a concentration gradient and the medium is a fluid moving in a laminar regime, mass transfer is carried out by the two mechanisms. Recalling the Reynolds experiment, the colorant is injected into an entrance point P of the tubing (Figure 3.3). Upon exit at point Q, the colorant is transferred from the entrance to the exit by *forced convection* flow and from the center of the conduit to point Q by molecular flow.

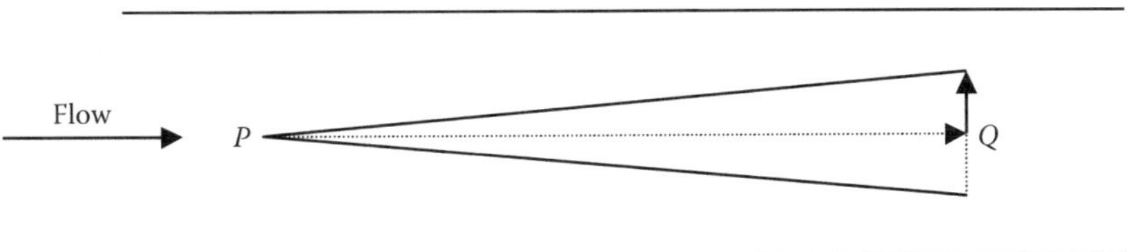

FIGURE 3.3 Simultaneous molecular and forced convective flows.

4. When the medium is a fluid in which there are turbulence and concentration gradients, the mechanisms of molecular and *forced convection* mass transport occur simultaneously. Although the phenomenon is complex, it is similar to case (3), using an analogous model. In this case, we consider an effective diffusion that brings together the molecular diffusion due to the gradient and the denominated turbulent diffusion, passing from P to Q by means of turbulent transport by whirlpools.

3.4.2 Energy Transfer

As mentioned at the beginning of this section, energy transfer by radiation occurs via a different mechanism than do those of conduction and convection. It is interesting to mention some aspects of energy transfer by these two mechanisms.

Conduction supposes a molecule-to-molecule flow of energy, due to temperature gradients, by mechanisms that depend on the physical nature of the medium. By analogy with mass diffusion, the principle of these mechanisms can be explained at an atomic–molecular level; however, these mechanisms differ, because in conduction there is no net mass flux.

If the considered medium is a fluid, and if there is a temperature gradient, in many cases a density difference will be noticeable. Therefore, there will be a mass flow due to flotation forces associated with an energy flow of the natural convection type. Forced convection exists as well, and, as in natural convection, is due to the associated energy of moving fluids; however, in this case the energy applied to move the fluid comes from mechanical devices. Besides convection, energy transfer by conduction will also occur but is less important. In a general way, in fluid media energy transfer is studied as a convection phenomenon, bringing together convection and conduction.

3.4.3 Momentum Transfer

To study the mechanisms of momentum transfer, a study analogous to the mass transfer study can be carried out. Molecular and forced convection momentum flows can also be considered.

3.4.4 Velocity Laws

In the mechanisms of molecular transport, property transfer occurs due to a potential gradient, which can be a concentration, temperature, or velocity gradient, depending on whether the transferred property is mass, energy, or momentum, respectively.

In molecular transport, the flux of the property is proportional to the potential gradient. The proportionality constant is an intensive property of the media. Depending on the nature of the property, the proportionality constant receives different names, as well as the laws of each of the transport phenomena:

1. Fick's law

$$(\text{Flux of mass}) = (\text{Diffusivity})\,(\text{Concentration gradient})$$

2. Fourier's law

$$(\text{Flux of thermal energy}) = (\text{Thermal conductivity})\,(\text{Temperature gradient})$$

3. Newton's law

$$(\text{Flux of momentum}) = (\text{Viscosity})\,(\text{Velocity gradient})$$

When the transport regime is turbulent, a laminar regime subsists in the inner part of the whirlpools, involving molecular transport in this region; normally, the parameters include both phenomena.

TABLE 3.1
Coupled Phenomena

Potential	Flow Density: Mass	Energy	Momentum
Concentration gradient	Diffusion (*Fick's law*)	Thermodiffusion (*Dufour's effect*)	
Temperature gradient	Thermodiffusion (*Soret's effect*)	Thermal conductivity (*Fourier's law*)	
Velocity gradient			Molecular transport of momentum (*Newton's law*)

3.4.5 Coupled Phenomena

The velocity laws for mass, thermal energy, and momentum transfer are expressed in a similar manner, in which the flux of the considered property is proportional to the gradient of the impelling force:

$$\vec{J} = k\vec{X} \tag{3.3}$$

where
$\vec{J}$ is the flux (quantity of property/m^2 s)
k is the proportionality constant
$\vec{X}$ is the potential gradient

This equation is applicable to all systems that are not in equilibrium. The flow of an equal property may be caused by various simultaneous potential gradients, or a given gradient can generate diverse flows. Coupled phenomena or processes are those that occur in systems in which different flows and gradients take place simultaneously. Thus, for example, a temperature gradient, in addition to causing an energy flow, can cause a mass flow (Soret's effect of thermodiffusion) (Table 3.1).

Onsegar generalized the latter expression to a system with R flows and S gradients as

$$\vec{J}_i = \sum_{j=1}^{S} k_{ij}\vec{X}_j \quad \text{for } i = 1, 2, \ldots R \tag{3.4}$$

This equation indicates that each flow J_j depends not only on its combined gradient but can also depend on other gradients acting on the system ($X_{j \neq i}$).

4 Momentum, Energy, and Mass Transfer

4.1 INTRODUCTION

The latest applications of new and improved traditional food preservation processes have generated the need for increased knowledge of the phenomenological and engineering principles that are the basis of the correct application of factors that produce stability and maintain the quality of transformed and processed products. This need for knowledge has given the field of food engineering a new identity at both the research and industrial levels. Understanding the transport phenomena that govern the engineering analysis and design of food preservation processes is a key element in improving processing conditions and the employment of energy resources and to increasing the quality of the product.

Momentum, energy, and mass transfer, where all of them are irreversible, have a similar mathematical form. This is very helpful because by analyzing one of them it facilitates the analyses of the others. At the same time, because of these similarities, it facilitates the understanding of an overall process that includes more than one of these transfer mechanisms.

4.2 MOMENTUM TRANSFER: NEWTON'S LAW OF VISCOSITY

Consider a static fluid between two parallel slabs, of area A separated by a distance y. If at a given time ($t = 0$) the lower slab begins to move at a velocity v, a time will come when the velocity profile is stable, as shown in Figure 4.1. Once the steady state is reached, a force F must continue to be applied to maintain the motion of the lower slab. Assuming that the circulation regime is laminar, the force per unit of area that should be applied is proportional to the velocity/distance ratio, according to the following equation:

$$\frac{F}{A} = \eta \frac{v}{y} \tag{4.1}$$

The proportionality constant η is called the *viscosity* of the fluid.

For certain applications it is convenient to express the latter equation in a more explicit way. The shearing stress exerted in the direction x on the fluid's surface, located at a distance y, by the fluid that is in the region where y is smaller, is designated by τ_{YX}. If the velocity component in the direction x has a value v_x, the last equation may be expressed as

$$\tau_{YX} = -\eta \frac{dv_X}{dy} \tag{4.2}$$

That is, the shear stress or force per unit area is proportional to the local velocity gradient. This equation is the expression of *Newton's law* of viscosity. All fluids that follow this law are denominated *Newtonian fluids*.

The shearing stress or shear stress τ_{YX} in a Newtonian fluid, for a distance y from the limit surface, is a measure of the velocity of momentum transport per unit area in a direction perpendicular

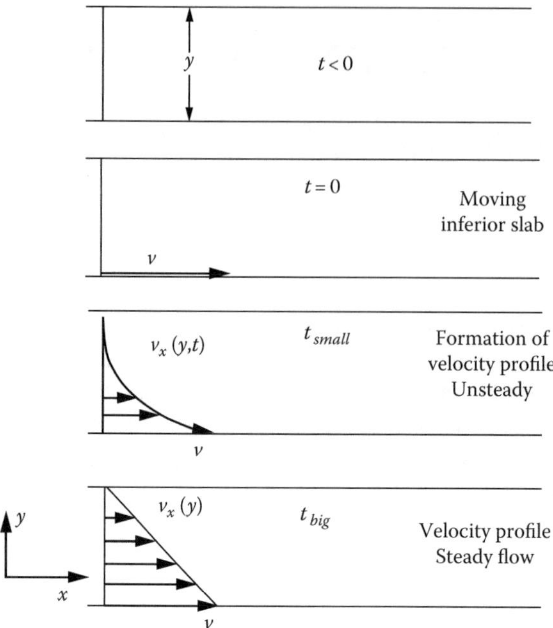

FIGURE 4.1 Velocity profile.

to the surface. The momentum flux is the amount of momentum per unit area and unit time, which corresponds to a force per unit area. For this reason, τ_{YX} can be interpreted as flux of momentum x in the direction y.

According to Equation 4.2, the momentum flux goes in the direction of the negative velocity gradient. That is, the momentum is transferred from the fastest moving fluid to the slowest one. Also, shear stress acts in such a direction that it opposes the motion of the fluid.

Another way to express Newton's law of viscosity is

$$\tau_{YX} = -\nu \frac{d(\rho v_X)}{dy} \tag{4.3}$$

where ν is the kinematic viscosity, which is the viscosity divided by the density:

$$\nu = \frac{\eta}{\rho} \tag{4.4}$$

4.3 ENERGY TRANSFER: FOURIER'S LAW OF HEAT CONDUCTION

To study heat transfer, a solid material with the form of a parallelepiped with surface A and thickness y is considered. Initially, the temperature of the slab is T_0. At a given time ($t = 0$), the lower part of the slab suddenly reaches a temperature T_1, higher than T_0, and remains constant over time. The temperature will vary along the slab until reaching the steady state, in which a temperature profile such as that shown in Figure 4.2 is reached. To maintain this profile, a heat flow rate $\dot{Q}$ should be sent through the slab.

Values of the temperature difference ($\Delta T = T_1 - T_0$) small enough comply with the following relation:

$$\frac{\dot{Q}}{A} = k \frac{\Delta T}{y} \tag{4.5}$$

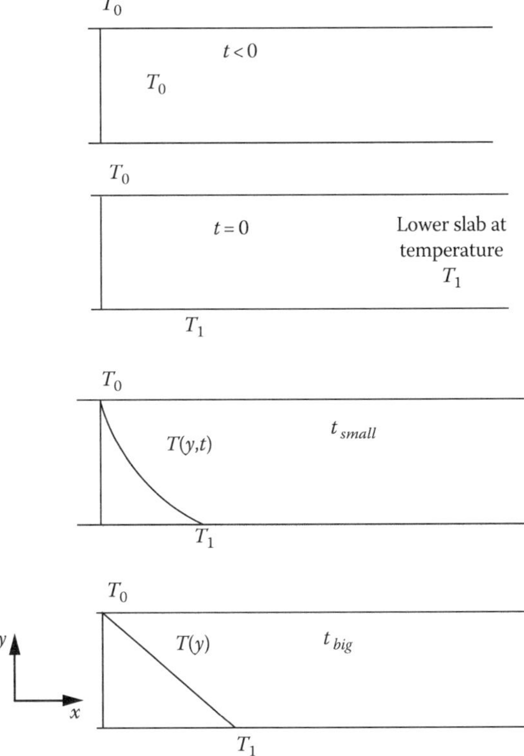

FIGURE 4.2 Development of the temperature profile.

This equation indicates that the heat flow per unit area is proportional to the increase of temperature with distance y. This proportionality constant k is called *thermal conductivity* of the solid slab. This equation also applies for liquids and gases placed between slabs, provided that there are no convection or radiation processes. Therefore, this equation is applicable to all heat conduction processes in solids, liquids, and gases.

For certain applications it is convenient to express this equation in a differential form. That is, if the thickness y of the solid tends to zero, the limit form of Equation 4.5 will be

$$q_Y = -k\frac{dT}{dy} \tag{4.6}$$

where q_Y represents the heat flux in the direction y. This equation expresses that heat flux is proportional to the negative gradient of temperature and is the 1D form of *Fourier's law* of heat conduction.

For other directions, the equations are analogous to that described for direction y. Therefore, Fourier's law is expressed as

$$\vec{q} = -k\vec{\nabla}T \tag{4.7}$$

that is, the heat flux vector is proportional to the temperature gradient and with opposite direction. It is assumed that the medium is isotropic; that is, the thermal conductivity has the same value in all of the directions of the material.

The units in which heat conductivity is normally expressed are W/(m K) or kcal/(h m °C), and its dimensions are $[k] = \mathrm{MLT}^{-3}\theta^{-1}$.

Besides the thermal conductivity, in the latter equations *thermal diffusivity* can be used, defined according to the following equation:

$$\alpha = \frac{k}{\rho \hat{C}_P} \tag{4.8}$$

where $\hat{C}_P$ is the specific heat of the material.

Taking into account the definition of thermal diffusivity, the expression of Fourier's law for one direction is

$$q_Y = -\alpha \frac{d\left(\rho \hat{C}_P T\right)}{dy} \tag{4.9}$$

for an isotropic material in which ρ and $\hat{C}_P$ are constants.

4.4 MASS TRANSFER: FICK'S LAW OF DIFFUSION

Fick's law of diffusion refers to the movement of a substance through a binary mixture due to the existence of a concentration gradient. The movement of a substance within a binary mixture from high concentration points to points with lower concentrations may be easily deduced by recalling the dissolution of a color crystal in water, as described in Chapter 3. The diffusion of a component due to the existence of a concentration gradient receives the name of *ordinary diffusion*. There are also other types of diffusion, according to the property that confers the movement to the components of the mixture; thus, if movement is due to a pressure gradient, it is called *pressure diffusion*; if it is due to a thermal gradient, the diffusion is *thermal*; when there is an inequality in the external forces that causes such movement, it is called *forced diffusion*.

The study of diffusion is more complicated than the cases of momentum and energy transport, since diffusion involves the movement of a species within a mixture. In a diffusive mixture, the velocities of the individual components are different, and they should be averaged to obtain the local velocity of the mixture that is required to define diffusion velocities. To achieve the expression of Fick's law, it is convenient to define the different forms for expressing concentrations, velocities, and flux densities.

The different forms for defining the concentration of a mixture with *n* components include

- **Mass concentration** ρ_i: the mass of species *i* per unit of mixture volume.
- **Molar concentration** C_i: the number of moles of the species *i* per unit of mixture volume. $C_i = \rho_i/M_i$ in which M_i is the molecular weight of species *i*.
- **Mass fraction** w_i: the mass concentration of the species *i* divided by the total molar density of the mixture: $w_i = \rho_i/\rho$.
- **Molar fraction** X_i: the molar concentration of the species *i* divided by the total molar density (global concentration) of the mixture: $X_i = C_i/C$.

In the considered mixture, each component moves at a different velocity. If a component *i* has a velocity v_i with respect to steady coordinate axes, the different types of velocity are defined as follows:

- **Mean mass velocity** v

$$\vec{v} = \frac{\sum_{i=1}^{n} \rho_i \vec{v}_i}{\sum_{i=1}^{n} \rho_i} = \frac{\sum_{i=1}^{n} \rho_i \vec{v}_i}{\rho} = \sum_{i=1}^{n} w_i \vec{v}_i \tag{4.10}$$

- **Mean molar velocity v^***

$$\vec{v}^* = \frac{\sum_{i=1}^{n} C_i \vec{v}_i}{\sum_{i=1}^{n} C_i} = \frac{\sum_{i=1}^{n} C_i \vec{v}_i}{C} = \sum_{i=1}^{n} X_i \vec{v}_i \tag{4.11}$$

When considering flow systems, it is convenient to refer to the velocity of the component i with respect to v or v^*, rather than referring to steady coordinate axes. In this way, we obtain the so-called *diffusion velocities* that represent the movement of the species i with respect to the movement of the fluid stream:

- *Diffusion velocity of component i with respect to $\vec{v}$*: the velocity of component i with respect to an axis system that moves at velocity v; it is given by the difference $(v_i - v)$.
- *Diffusion velocity of the component i with respect to $\vec{v}^*$*: the velocity of component i with respect to an axis system that moves at velocity v^*; it is given by the difference $(v_i - v^*)$.

The flux can be of mass or molar, and it is a vectorial magnitude defined by the mass or moles that cross through a unit area per unit time. The motion can be referred to as steady axis or as axis moving at velocity v or v^*. In this way, the different forms for expressing the flux of a component i in a mixture made of n components will be

1. *With respect to steady axis*
   a. Mass flux

$$\vec{m}_i = \rho_i \vec{v}_i \tag{4.12}$$

   b. Molar flux

$$\vec{N}_i = \vec{C}_i \vec{v}_i \tag{4.13}$$

2. *For moving axis*

   a. Diffusion mass flux in relation to velocity $\vec{v}$

$$\vec{j}_i = \rho_i \left(\vec{v}_i - \vec{v} \right) \tag{4.14}$$

   b. Diffusion mass flux in relation to velocity $\vec{v}^*$

$$\vec{j}_i^* = \rho_i \left(\vec{v}_i - \vec{v}^* \right) \tag{4.15}$$

   c. Diffusion molar flux in relation to velocity $\vec{v}$

$$\vec{J}_i = C_i \left(\vec{v}_i - \vec{v} \right) \tag{4.16}$$

   d. Diffusion molar flux in relation to velocity $\vec{v}^*$

$$\vec{J}_i^* = C_i \left(\vec{v}_i - \vec{v}^* \right) \tag{4.17}$$

Some of these expressions are rarely applied, such as J_i and j_i^*. Most often used in engineering is the molar flux, referred to as steady axis N_i.

Once having been reviewed, the different forms in which concentrations, velocities, and flux can be expressed, mass transfer is studied next. Consider a binary mixture with components A and B, in which the diffusion of one of the components occurs due to a concentration gradient of the considered component. Just as for momentum and energy transfer, viscosity and thermal conductivities are defined as proportionality factors between momentum flux and the velocity gradient for viscosity (Newton's law of viscosity) and between heat flux and temperature gradient for thermal conductivity (Fourier's law of heat conduction); in an analogous way the diffusivity $D_{AB} = D_{BA}$ in a binary mixture is defined as the proportionality factor between the mass flux and the concentration gradient, according to the following equation:

$$\vec{J}_A^* = -CD_{AB}\vec{\nabla}X_A \tag{4.18}$$

that is, *Fick's first law of diffusion for molar flux*. In addition to the concentration gradient, the temperature, pressure, and external forces contribute to the diffusion flux, although their effect is small compared to that of the concentration gradient. This indicates that the diffusion molar flux related to the velocity v^* is proportional to the gradient of the molar fraction. The negative sign expresses that this diffusion takes place from higher to lower concentration zones.

It is easy to deduce that the diffusion sum of a binary mixture is zero, that is,

$$\vec{J}_A^* + \vec{J}_B^* = 0$$

since, if a component diffuses to one side, the other component diffuses in the opposite way.

When the global concentration is constant, or if a chemical reaction does not exist, or when there is a chemical reaction, but the number of moles does not vary, Equation 4.18 can be transformed into

$$\vec{J}_A^* = -D_{AB}\vec{\nabla}C_A \tag{4.19}$$

There are other ways to express Fick's first law according to the flux and the correspondent concentration gradient considered. Thus, for the mass flux, in reference to the steady axis, the gradient considered is that of the mass fraction, so this law is expressed by the following equation:

$$\vec{m}_A = w_A\left(\vec{m}_A + \vec{m}_B\right) - \rho D_{AB}\vec{\nabla}w_A \tag{4.20}$$

Of all the possible expressions for Fick's first law of diffusion, the one with the greatest importance is that related to the molar flux in steady or fixed axis:

$$\vec{N}_A = X_A\left(\vec{N}_A + \vec{N}_B\right) - CD_{AB}\vec{\nabla}X_A \tag{4.21}$$

It can be observed that N_A is the result of two vectorial magnitudes, $X_A(N_A + N_B)$: the molar flux by convective transport, resulting the global motion of the fluid, and $(-CD_{AB}\nabla X_A)$, due to the molecular transport, according to the definition of J_A^*.

The units of diffusivity are unit area per unit time and can be expressed in cm^2/s or m^2/h.

There is a lack of diffusivity data for most mixtures, so it is necessary to use estimated values in many calculations in which diffusivities are required. When possible, experimental data should employed, since, generally, they are safer. The magnitude order of diffusivity is given next for various cases found in practice:

Gas–gas diffusion	0.776–0.096 cm²/s
Liquid–liquid diffusion	2×10^{-5}–0.2×10^{-5} cm²/s
Diffusion of a gas in a solid	0.6×10^{-8}–8.5×10^{-11} cm²/s
Diffusion of a solid in another solid	2.5×10^{-15}–1.3×10^{-30} cm²/s

Just as viscosity and thermal conductivity of a pure fluid are only a function of the temperature and pressure, the diffusivity D_{AB} for a binary mixture is a function of temperature, pressure, and composition.

4.5 GENERAL EQUATION OF VELOCITY

The dimensions of the diffusivity D_{AB} are surface per unit time: $D_{AB} = L^2T^{-1}$. If Newton's law of viscosity is considered, the dimensions do not correspond to those of diffusivity: $[\eta] = ML^{-1}T^{-1}$. However, if the kinematic viscosity υ is taken, that is, the relationship between dynamic viscosity and density, it turns out that its dimensions are surface per time unit: $[\upsilon] = L^2T^{-1}$, that is, the dimensions coincide with those of diffusivity. So, η also receives the name of *momentum diffusivity*.

In an analogous way, in Fourier's law of heat conduction, the dimensions of thermal conductivity are $[k] = MLT^{-3}T^{-1}$. However, if the relation $\alpha = k/(\rho\hat{C}_P)$ is considered, its dimensions coincide with those of diffusivity: $[\alpha] = L^2T^{-1}$, where α is denominated *thermal diffusivity*.

The analogy of these three magnitudes D_{AB}, v, and α is deduced from the velocity equations for the three transport phenomena in unidirectional systems:

Fick's law	$j_{AB} = -D_{AB}\dfrac{d}{dy}(\rho_A)$
Newton's law	$\tau_{YX} = -\nu\dfrac{d}{dy}(\rho \cdot v_X)$
Fourier's law	$q_Y = -\alpha\dfrac{d}{dy}(\rho\hat{C}_P\theta)$

where ρ and $\hat{C}_P$ remain constant.

In these equations the media is assumed to be isotropic, that is, the material's viscosity, thermal conductivity, and diffusivity have the same value in any direction. This is acceptable for fluids and for most homogeneous solids.

Considering only one direction, the previous equations can be grouped into one expression:

$$\Phi_y = -\delta\frac{d\xi}{dy} \tag{4.22}$$

where

Φ_y is the flux y of any of the three properties in the direction y
ξ is the concentration per unit volume of the considered property
δ is the proportionality factor, called *diffusivity*

Analog equations will result for the other directions in such a way that the experimental equations of Fick, Newton, and Fourier can be grouped into three general equations, one for each direction. These equations constitute the expressions of molecular transport of the three studied properties.

The analogies as expressed by the previous equations for momentum, energy, and mass cannot be applied to 2D and 3D problems, since shear stresses τ are grouped in a tensorial magnitude, while j_A and q are vectorial magnitudes of three components.

If vectorial notation is used to generalize these equations, for direction x we obtain

$$\vec{\tau}_X = -\eta\vec{\nabla}(v_X) \tag{4.23a}$$

$$\vec{q} = -k\vec{\nabla}T_X \tag{4.24}$$

$$\vec{j}_{AB} = -D_{AB}\frac{d}{dy}(\rho_A) \tag{4.25}$$

For the y and z directions, the only equation that changes is that related to Newton's law; thus, the equations used for direction y and z will be

$$\tau_Y = -\eta\nabla(v_Y) \tag{4.23b}$$

$$\tau_Z = -\eta\nabla(v_Z) \tag{4.23c}$$

Therefore, generalizing, it is obtained as

$$\vec{\Phi} = -\delta\vec{\nabla}\xi \tag{4.26}$$

which expresses that the flux vector is of opposite direction and it is proportional to the gradient of the driving force or potential (for the experimental laws) and the property concentration (in the case of more strict laws).

5 Macroscopic Balances

5.1 INTRODUCTION

In the study of unit operations, it is necessary to set up mathematical models. The tools utilized to obtain these models are conservation equations, which are usually called *balances*. In general, the equations obtained starting from these balances do not have great mathematical complexity and they are easy to solve; only in nonstationary (or unsteady) processes and in those where chemical reactions could take place some additional complexity could be found.

On the one hand, for the definition of the state of any system, it is necessary to set certain macroscopic variables that do not depend on the size of the system in the study. These variables are defined as *intensive variables*, such as pressure, temperature, concentration, and the speed of a moving fluid. *Extensive variables*, on the other hand, depend on the size of the system under study, such as mass and volume. The relationship between two extensive variables is an intensive variable; therefore, for example, density is an intensive variable, that is, the relationship between two extensive variables, the mass and the volumc.

In any unit operation, the system evolves from one state to another, and the intensive variables are related with four properties: molecular scale, mass of each component, energy, and the system's quantum molecules. These properties are extensive, since they depend on the size of the system in consideration.

When the balance is carried out on a system with a finite volume, it is called *macroscopic*, but if the balance is carried out on a system with a differential volume, the balance is called *microscopic*. This chapter will present the macroscopic balances of mass and energy applied to a system with a finite volume.

If a system like the one shown in Figure 5.1 is considered, where both in and out currents from the two extensive properties of mass and energy can exist, the typical balance of the considered property could be expressed as follows:

$$(In\ flow) + (Generation) = (Out\ flow) + (Accumulation)$$

In many cases this general equation can also be reduced to the following form:

$$(Accumulation) = (In - Out) + (Generation)$$

where

the *Accumulation* term represents the variation of the property in the system by unit of time
the term (*In* – *Out*) represents the net flow entering the system
the term *Generation* is the quantity of property generated in the system by unit of time

It can be observed that each of these terms has units of property per unit of time. It is important to notice that in the entering net flow, entering currents will be positive and exiting currents will be negative. In addition, the generation term could be positive or negative, relying on the property that is generated or dissipated, respectively.

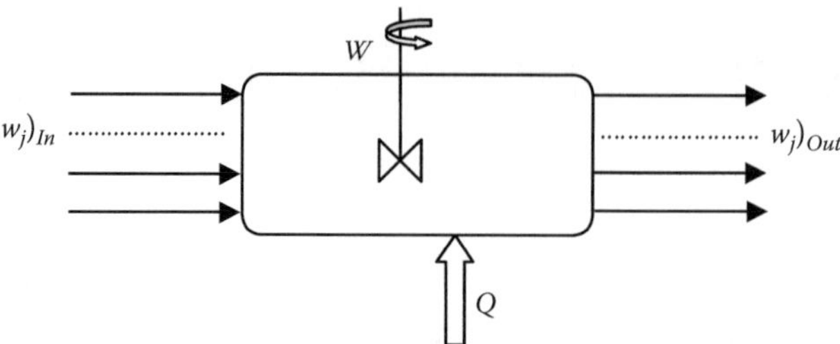

FIGURE 5.1 System where macroscopic balances are in place.

5.2 MACROSCOPIC MASS BALANCE

The mass balance is a generalization of the material conservation principle for open systems. Basically, the material conservation law establishes that the material is not created nor is it destroyed; it can only be transformed. This law applies to several unit operations and processes of food engineering, since it is not suitable for nuclear reactions, in which part of the material is converted to energy according to Einstein's equation: $E = mc^2$.

The analysis of a mass balance allows us to know the flows and compositions of the entering currents and the exiting currents from a system, as well as the total quantity and the average composition of the inner system at a determined instant. The mass balances could be carried out in mass or molar units, depending on the type of problem that is set up.

If the system is formed by a mixture of n components ($i = 1, 2, 3\ldots, n$), and they are considered p currents ($j = 1, 2, 3\ldots, p$), carrying out a mass balance for a component i of the mixture will give

$$\frac{d(M_i)}{dt} = \sum_{j=1}^{p} w_{ij} + R_i \tag{5.1}$$

where

M_i is the mass of component i in the system
R_i is the generation of component i by unit of time
w_{ij} is the flow of i component in the j current

The currents that enter the system are considered positives and those that exit are considered negatives.

If there are n components, it will result in n similar equations for each of the components; the n equations will be mathematically independent. The dimensions of each of the terms of Equation 5.1 are mass per unit of time (kg/s or kmol/s in the International System).

Furthermore, to calculate the generation term, it is necessary to use the intensive reaction rate (r_i), which is defined as the variation of the material per unit of volume in the unit of time owed to the reaction

$$r_i = \frac{dR_i}{dV} \tag{5.2}$$

In many cases this reaction rate is constant for the whole system volume; for example, in the case of perfectly agitated tanks, this will give

$$r_i = \frac{R_i}{V} \tag{5.3}$$

After substituting this last equation in Equation 5.1 for component i, the mass balancc can now be written as

$$\frac{d(M_i)}{dt} = \sum_{j=1}^{p} w_{ij} + r_i V \tag{5.4}$$

Equations 5.1 and 5.4 represent the mass balances for a component i in molar units, respectively. In practice, in order to make their applications simpler, this equation could be expressed in a different manner according to the utilized units. Therefore, in mass units

$$\frac{d(V\rho_i)}{dt} = \sum_{j=1}^{p} q_j \rho_{ij} + r_{iM} V \tag{5.5}$$

where

- V is the total volume of the inner system
- ρ_i and ρ_{ij} are the densities of the component i of the inner system and the component j current, respectively
- q_j is the volumetric flow rate for the j current
- r_{iM} is the mass generation rate of component i by unit of volume and time

In the case that molar units are utilized, it is convenient to express Equation 5.5 in the following form:

$$\frac{d(VC_i)}{dt} = \sum_{j=1}^{p} q_j C_{ij} + r_i V \tag{5.6}$$

where

- C_i is the molar concentration of component i in the system
- C_{ij} is the molar concentration of component i in the j current
- q_j is the volumetric flow rate of j current
- r_i is the molar generation intensive rate of the component i by unit of volume and unit of time

The following definitions could be set as

$$M = \sum_{i=1}^{n} M_i \quad w_j = \sum_{i=1}^{n} w_{ij}$$

If all components of the system are considered, Equation 5.4 can be written as

$$\frac{d(M)}{dt} = \sum_{j=1}^{p} w_j \tag{5.7}$$

representing the total macroscopic mass balance.

However, it can be observed that the reaction term does not appear; this is because the total material generated in the system is null. If the summation would have been applied to Equation 5.6, which is expressed in molar units, this simplification could not be done, since the total number of moles could vary. Only in the case that there is no variation in the total mole number will the reaction term be null.

From this analysis it can be observed that $n + 1$ equations can be set up, that is, n for the components plus the total balance. For these equations only n will be mathematically independent.

In practice there are particular cases in which not all of the terms exposed in the equations will appear. Therefore, in the case where the process develops in a stationary regime, the accumulation term disappears, since variables are time independent. These time-independent variables could vary with position; however, the accumulation term is still zero. For this case, the balance of the i component will remain in the following form:

$$\sum_{j=1}^{p} w_{ij} + r_i V = 0 \tag{5.8}$$

Also, if there is no chemical reaction, the summation of mass flows for in and out currents is equal to zero:

$$\sum_{j=1}^{p} w_{ij} = 0 \tag{5.9}$$

where the mass flow rate for a j current is

$$w_j = \rho_j q_j = \rho_j v_j S_j$$

where

q_j is the volumetric flow rate
ρ_j the density for the j current
v_j is the average velocity
S_j is the in or out cross section for the j current

Such consideration could also bring up the case of a system in where there are no mass currents out, such as

$$\frac{d(M_i)}{dt} = r_i V \tag{5.10}$$

which indicates that the accumulation of component i in the system is due only to generation by reaction of component i.

5.3 MACROSCOPIC ENERGY BALANCE

The flow and composition of different currents can be obtained from the macroscopic mass balances; this calculation will give information about the system's entering and exiting currents. In many situations the information is not enough to know all of the associate variables at different currents. For example, in many food systems, it is necessary to know the temperature of each current, the necessary energy quantity in order to work a pump or a compressor, and the vapor necessities for heating an evaporator. Therefore, it is necessary to carry out the energy balances. In spite of the fact that the balances of mass and energy are considered separately, neither the energy nor the mass is conserved independently, although a combination of both is conserved. In practice it could be considered, if there are no mistakes, that in the food engineering processes, the energy is conserved, and as has been noted previously, this is not true for nuclear reactions.

The forms of energy are various; in a general manner energy can be defined as the capacity to produce work. The several forms of energy can be converted from one form to another; therefore, it is necessary to note the types of energy that should be kept in mind for energy balances. These energy types are internal energy, potential energy, kinetic energy, heat, and work. The first three types are state functions, and it will be necessary to define a state of reference, while heat and work depend on the evolution of the process in study.

Internal energy (U): The material is formed by a group of particles (atoms, molecules, ions, etc.) in continuous movement. Internal energy is the sum of the rotation, translation, and vibration energy of these particles, depending on temperature.

For a system that contains a mass M, the internal energy is a function of temperature (T), quantity of the material (M), and specific volume ($\hat{V}$). The specific volume is assumed to have little influence; therefore, the internal energy could be written as

$$U = \int_M \hat{C}_v T dM$$

where $\hat{C}_v$ is the specific heat at constant volume.

Potential energy (ϕ): This is the type of energy that is stored by the system as a function of position under a force field. This field could be gravitational, electric, and magnetic. Since the last two are not usually significant in the regular processes of food engineering, only the gravitational field will be considered. Thus, for a system of mass M,

$$\phi = \int_M gz dM$$

where

g is the gravity acceleration

z is the position in the gravitational field or the height from the level taken as reference

Kinetic energy (K): Kinetic energy is the energy stored by the system due to speed (v); it can be defined as

$$K = \int_M \frac{v^2}{2} dM$$

Heat ($\dot{Q}$): Heat or thermal energy is a form of energy that depends on the temperature difference between the system and its surroundings and through the area that it is transmitted. When heat is transmitted to the system, it is considered to be positive, while the heat that the system yields to its surroundings is considered to be negative. Heat is an energy that is not interchanged through the in or out currents due to mass flow rate. A simple equation that allows the calculation of heat flow rate is

$$\dot{Q} = UA(T_e - T)$$

where

U it is the global transmission heat coefficient (not to be confused with the internal energy, which is symbolized by the same letter)

A is the exchange area

T_e and T are the external temperature and the system's temperature, respectively

Work (W): In many industrial processes the mechanical work is due to expansion or compression of a fluid. The calculation of the work of a fluid in expansion is obtained from the expression

$$W = \int_x F\,dx = \int_x -pS\,dx = -\int_V p\,dV$$

where
F is the force
p is the pressure
S is the section in which the force is applied
V is the volume

The negative sign designates that the work done by the system is considered negative, while the work transmitted to the system is positive.

Once the several forms of energy are defined, the macroscopic energy balance can be set. In the general expression, the generation term does not appear, since in habitual processes in food engineering the energy is neither created nor destroyed; it only transforms between the different existing types. Therefore, the macroscopic energy balance is set as

$$(Accumulation) = (Inlet - Outlet)$$

where the second term expresses the net entrance of energy experienced by the system.

After being exposed to several energy types, the system can only accumulate internal, potential, and kinetic energies, which are state functions. However, work and the heat are not state functions; their value depends on a consecutive process in order to pass from an initial state to an end state. When a great amount of work is conducted by heat, it is considered energy in traffic, which will increase some of the state function energies (internal, potential, or kinetic).

As soon as the energy enters the system, there are two types of energy to be considered: the energy that accompanies the in and out material currents and the energy that enters from the surroundings of the system.

Therefore, the macroscopic energy balance, applied to a system such as the one shown in Figure 5.1, will be expressed as

$$\frac{d(U+\phi+K)}{dt} = \dot{Q} + \dot{W} + \sum_j w_j\left(\hat{U}_j + \hat{\phi}_j + \hat{K}_j\right) + \sum_j p_j S_j v_j \qquad (5.11)$$

where
w_j is the mass flow rate of j current
$\hat{U}_j$, $\hat{\phi}_j$, and $\hat{K}_j$ are internal, potential, and kinetic energies by unit of mass of j current
p_j is the pressure that the j current exerts on the S_j section
v_j is the average speed of this current

The addition of the second member of Equation 5.11 represents the net energy entrance due to the work flow that is exerted by the different material current. This term should not be confused with the workflow due to pumps, compressors, turbines, etc., which enters or leaves the system without being associated with the j material current.

As the volumetric flow rate of the j current, (q_j) could be express as

$$q_j = S_j v_j \tag{5.12a}$$

$$q_j = \frac{w_j}{\rho_j} = w_j \hat{V}_j \tag{5.12b}$$

where $\hat{V}_j$ is the specific volume of the j current.

If it is kept in mind that the enthalpy by unit of mass is defined as

$$\hat{H}_j = \hat{U}_j + p_j \hat{V}_j$$

the previous equation (Equation. 5.11) can also be expressed as

$$\frac{d(U+\phi+K)}{dt} = \dot{Q} + \dot{W} + \sum_j w_j \left(\hat{H}_j + \hat{\phi}_j + \hat{K}_j\right) \tag{5.13}$$

This result corresponds to the macroscopic energy balance, which indicates that the energy accumulation speed of a system is equal to the summation of heat and work flows that is given to or transmitted from the system, plus the net entering enthalpy, potential energy, and kinetic energy. No matter how the terms are expressed in Equation 5.13, they are an energy flow (power), being their units in the International System J/s or watt (W).

Equation 5.13 uses absolute values of internal, potential, and kinetic energies and enthalpy. All of these are state functions, and it is necessary to define a reference state, which results in a macroscopic energy balance that uses relative magnitudes.

If $\hat{H}^*$, $\hat{\phi}^*$, and $\hat{K}^*$ are defined as reference values for enthalpy, potential energy, and kinetic energy by mass unit and are multiplied by their sum for each one of the terms of the global mass balance in mass units (Equation 5.7), the following is obtained:

$$\frac{d\left[M\left(\hat{H}^* + \hat{\phi}^* + \hat{K}^*\right)\right]}{dt} = \sum_j w\left(\hat{H}^* + \hat{\phi}^* + \hat{K}^*\right)$$

or

$$\frac{d(H^* + \phi^* + K^*)}{dt} = \sum_j w_j \left(\hat{H}^* + \hat{\phi}^* + \hat{K}^*\right)$$

If from Equation 5.13 this last expression is subtracted, it is obtained as

$$\frac{d\left(U - H^* + \phi - \phi^* + K - K^*\right)}{dt} = Q + W + \sum_j w_j \left(\hat{H} - \hat{H}^* + \hat{\phi} - \hat{\phi}^* + \hat{K} - \hat{K}^*\right)_j$$

But keeping in mind that

$$U = H - pV$$

$$\frac{dU}{dt} = \frac{dH}{dt} - \frac{d(pV)}{dt}$$

it is obtained as

$$\frac{d(H-H^*+\phi-\phi^*+K-K^*)}{dt}=\dot{Q}+\dot{W}+\frac{d(pV)}{dt}+\sum_j w_j\left(\hat{H}-\hat{H}^*+\hat{\phi}-\hat{\phi}^*+\hat{K}-\hat{K}^*\right)_j \tag{5.14}$$

In this last equation relative magnitudes are used, but it should be kept in mind that in their application, mass magnitudes should be utilized, although in the absence of reactions, molar units could be used.

In the case of stationary regime, all of the variables are independent of time, and therefore, their variation with time is null. If Equation 5.13 is used,

$$\sum_j w_j\left(\hat{H}_j+\hat{\phi}_j+\hat{K}_j\right)+\dot{Q}+\dot{W}=0 \tag{5.15}$$

With relative magnitudes and mass flow rates (Equation 5.14), the equation that is obtained is

$$\sum_j w_j\left(\hat{H}-\hat{H}^*+\hat{\phi}-\hat{\phi}^*+\hat{K}-\hat{K}^*\right)_j+\dot{Q}+\dot{W}=0 \tag{5.16}$$

In the case of no presence of entering or exiting work, for absolute magnitudes,

$$\sum_j w_j\left(\hat{H}_j+\hat{\phi}_j+\hat{K}_j\right)+\dot{Q}=0 \tag{5.17}$$

and for relative magnitudes,

$$\sum_j w_j\left(\hat{H}-\hat{H}^*+\hat{\phi}-\hat{\phi}^*+\hat{K}-\hat{K}^*\right)_j+\dot{Q}=0 \tag{5.18}$$

In food processes, the corresponding terms for kinetic and potential energy are generally small and opposite to the enthalpy term; therefore, they can be neglected. For example, the temperature increase of 1°C of a determined water mass is equivalent to a rise of 430 m of the same water mass or giving a speed of 330 km/h. In the case that there is a phase change, the differences between the enthalpy term and the potential and kinetic energies are yet more evident. Therefore, the evaporation of determined water mass is equal to raise this mass to a height of 230 km or giving it a speed of 8820 km/h. In this case, considering an adiabatic process and neglecting the terms of potential and kinetics, for absolute magnitudes, the following expression is obtained:

$$\sum_j w_j\hat{H}_j=0 \tag{5.19}$$

and for relative magnitudes, the following is obtained:

$$\sum_j w_j\left(\hat{H}-\hat{H}^*\right)_j=0 \tag{5.20}$$

which belong together to simple enthalpy balances. In this way, in a stationary state, for adiabatic processes and in those in which the system does not give or communicate work, it can be defined that the addition of the enthalpy of entering material current will be equal to the addition of the enthalpies of material current exiting from the system.

As mentioned previously, in many industrial food processes, the kinetic and potential terms can be neglected, and if machine work does not exist in a stationary state, the equation can be written as

$$\sum_j w_j \hat{H}_j + \dot{Q} = 0$$

The enthalpy is a state function; it is not possible to calculate its absolute value, since it is necessary to establish a reference state. This state corresponds to the free elements of all substances for a reference pressure and a reference temperature. Generally, this reference state is defined at 1 atm and 278 K. Reaction enthalpies are included in $\hat{H}_j$, since this includes the formation enthalpies of all of the components in the j current.

Enthalpy depends on the chemical composition, aggregation state, temperature, and pressure, although in practice it could be considered that enthalpy is independent of the pressure and even more so in the case of industrial food processes. For a pure compound, it is assumed that if the variation of the enthalpy with the pressure is negligible, then the expression reduces to

$$\hat{H} = \hat{H}_f^* + \int_{T^*}^{T} \hat{C}_p \, dT = \hat{H}_f^* + \hat{C}_p (T - T^*) \tag{5.21}$$

where

T^* is the reference temperature
$\hat{C}_p$ is the specific heat at constant pressure
$\hat{H}_f^*$ is the formation enthalpy at the reference temperature

If there is a state change, this expression will change. Therefore, in the case of a change from a state a to another state b,

$$\hat{H} = \hat{H}_f^* + \left.\hat{C}_p\right)_a (T_e - T^*) + \lambda_{a \to b} + \left.\hat{C}_p\right)_b (T - T_e) \tag{5.22}$$

where

$\lambda_{a \to b}$ is the latent heat for the change from a state to b state
T_e is the temperature to which the change is produced, indicating the subscripts a and b the corresponding states

If in each of the j currents there exists n components, and the mixture enthalpy is neglected, the enthalpy of each of the currents could be expressed as a function of the enthalpy for each of components ($\hat{H}_i$), according to the expression

$$\hat{H} = \sum_{i=1}^{n} \frac{\hat{V}}{\hat{V}_i} \hat{H}_i \tag{5.23}$$

where $\hat{V}$ and $\hat{V}_i$ are the total specific volume of the mixture and of each component, respectively. If this last equation is expressed as a function of the reference state, it is defined as

$$\hat{H} = \sum_{i=1}^{n} \frac{\hat{V}}{\hat{V}_i} \hat{H}_{fi}^* + \sum_{i=1}^{n} \frac{\hat{V}}{\hat{V}_i} \hat{C}_{pi}^* (T - T^*) \tag{5.24}$$

In this equation it was assumed that there is no phase change, but an expression such as that given in Equation 5.22 should be used. Equation 5.24 could also be used with mass magnitudes such as molars, while keeping in mind that the relationship between the specific volumes is a mass fraction, and if this equation is in molar units, the fraction to use will be a molar faction.

Therefore, for a process that is carried out in a stationary state and given that machine work does not exist, the energy balance equation is obtained:

$$\sum_{j=1}^{p} w_j \left[\sum_{i=1}^{n} \frac{\hat{V}}{\hat{V}_i} \hat{C}_{pi}(T - T^*) + \sum_{i=1}^{n} \frac{\hat{V}}{\hat{V}_i} \hat{H}_{fi}^* \right] + \dot{Q} = 0$$

then

$$\dot{Q} = -\sum_{j=1}^{p} w_j \left[\sum_{i=1}^{n} \frac{\hat{V}}{\hat{V}_i} \hat{C}_{pi}(T - T^*) + \sum_{i=1}^{n} \frac{\hat{V}}{\hat{V}_i} \hat{H}_{fi}^* \right] \tag{5.25}$$

where the negative sign in the second member indicates that it is a net entry. If it exit minus entrance is considered, the sign would be positive.

If the enthalpy of all entering (H_E) and exiting (H_S), currents are defined with reference temperature T^* as

$$H_E = \sum_{j=1}^{p} w_{jE} \sum_{i=1}^{n} \frac{\hat{V}_E}{\hat{V}_{iE}} \hat{C}_{PiE}(T - T^*)$$

$$H_S = \sum_{j=1}^{p} w_{jS} \sum_{i=1}^{n} \frac{\hat{V}_S}{\hat{V}_{iS}} C_{PiS}(T - T^*)$$

where the subscripts E and S are entrance and exit, respectively.

Also, ΔH_R^* is defined as the standard enthalpy for chemical reactions:

$$\Delta H_R^* = \sum_{j=1}^{p} w_{jS} \sum_{i=1}^{n} \frac{V_S}{V_{iS}} H_{fi}^* - \sum_{j=1}^{p} w_{jE} \sum_{i=1}^{n} \frac{V_E}{V_{iE}} H_{fi}^*$$

From Equation 5.25 it is obtained as

$$(H - H_E) + \Delta H_R^* = \dot{Q} \tag{5.26}$$

The chemical reaction enthalpy, for a given temperature and pressure, is obtained as the difference between the summation of product formation enthalpies and the summation of reagent formation enthalpies, although this result could also be obtained from the difference between the reagent combustion heats and the product combustion heats for the considered reaction.

In the case that the enthalpies of entering and exiting currents are equal, the heat flow will be equal to the enthalpy reaction term:

$$\dot{Q} = \Delta H_R^*$$

If the term of reaction enthalpies is negative (exothermic reaction), the heat flow is negative ($Q < 0$), which indicates that it is the heat that should be eliminated from the system. In the case of endothermic reaction, the heat flow result will be positive ($Q > 0$), which assumes a contribution of energy to the system.

PROBLEMS

5.1 A tank of 1 m^3 perfectly agitated with a spillway contains 800 kg of an apple clarified juice of 50°Brix. At a determined instant the tank is fed by 50 L/min of water. Determine the time that delays in overflowing the tank, and in that instant determine the concentration of the steam that leaves from the spillway. Calculate the additional necessary time so that the exiting current has a content of solubility in solids of 12°Brix.

All processes are carried out at 20°C, and at this temperature the variation of density with the concentration can be expressed by means of the following equation:

$$\rho = 999 + 3.8C + 0.018C^2 \text{ kg/m}^3$$

where C is expressed in °Brix.

In Figure P.5.1 the stirred tank in which initially M_0 kg exists and occupies a volume V_0 is shown. It is supposed that the water has a density of 1000 kg/m^3. The density of 50°Brix juice is 1234 kg/m^3.

The initial volume that the juice occupies in the tank is

$$V = \frac{M_0}{\rho_0} = \frac{800\,\text{kg}}{1234\,\text{kg/m}^3} = 0.648\,\text{m}^3$$

Global mass balance:

$$\frac{dM}{dt} = w_1$$

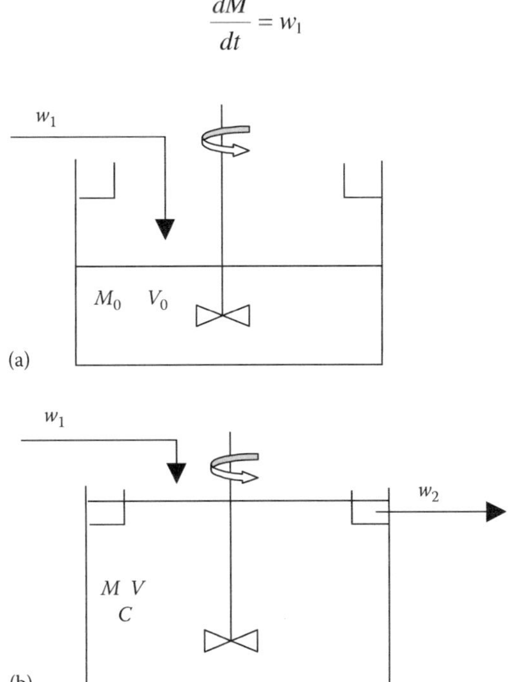

FIGURE P.5.1 Perfectly agitated tank (initial conditions).

This equation could be integrated with the initial condition of time $t = 0$; the mass is the initial $M = M_0$, obtaining

$$M = M_0 + w_1 t$$

Component (soluble solids) balance:

$$\frac{d(MC)}{dt} = w_1 \times 0 \text{ since the current } w_1 \text{ does not contain soluble solids:}$$

$$C\frac{dM}{dt} + M\frac{dC}{dt} = 0$$

Keeping in mind that $M = M_0 + w_1 t$, it is obtained as

$$Cw_1 + (M_0 + w_1 t)\frac{dC}{dt} = 0$$

then rearranged gives an equation in separable variables:

$$\frac{dC}{C} = -\frac{dt}{\left((M_0/w_1) + t\right)}$$

This equation is integrated with the initial condition $t = 0$, $C = C_0$, obtaining

$$t = \frac{M_0}{w_1}\left(\frac{C_0}{C} - 1\right) \quad \text{or} \quad C = \frac{M_0 C_0}{M_0 + w_1 t} = \frac{M_0 C_0}{M}$$

Initially, there is 800 kg of juice in the tank occupying 0.648 m^3. When it overflows the tank is full and the volume is 1 m^3; added water is

(1 − 0.648) = 0.352 m^3; then there is 352 kg of water.

Thus, the total mass when the tank is full is $M = 800 + 352 = 1152$ kg.

The concentration at the moment that overflow begins is

$$C = \frac{M_0 C_0}{M} = \frac{(800\,\text{kg})(0.5)}{(1152\,\text{kg})} = 0.3472\,(34.72°\text{Brix})$$

The necessary time is

$$t = \frac{M_0}{w_1}\left(\frac{C_0}{C} - 1\right) = \frac{(800\,\text{kg})}{(50\,\text{kg/min})}\left(\frac{50}{34.72} - 1\right) \cong 7.04\,\text{min}$$

Once the overflow begins, the concentration of the stream that leaves the tank (w_2) (Figure P.5.2) is the same as that of the inner the tank (perfectly mixed).

Soluble solids balance:
It is important to point out that in spite of the fact that the tank volume is 1 m^3, its total mass could vary due to density variation in time. However, at any moment, the juice density content in the tank coincides with the density of the exiting current.

Global mass balance:
This equation is set out as a function of the density.

The agitated tank volume is constant (V = 1 m^3), the density of the current that exits the tank coincides with the inner tank ($\rho = \rho_2$), and the volumetric flow rates entering and exiting from the tank are equal ($q_1 = q_2 = q$); thus, the previous expression is converted to

$$V\frac{d(\rho)}{dt} = q\rho_1 - q\rho$$

This is an equation in separable variables that can be integrated with the initial condition $t = 0$; at the moment the tank starts overflowing, the density becomes $\rho = \rho_0 = 1152.5$ kg/m^3, obtaining

$$t = \frac{V}{q}\ln\left(\frac{\rho_1 - \rho_0}{\rho_1 - \rho}\right)$$

If it is desired to know the time that it takes for the current overflow to reach 12°Brix, the adequate data in function of the density in the previous equation become

for C = 12°Brix, the density of the supreme is ρ = 1047 kg/m^3.

Substituting data are as follows:

$$t = \frac{(1\,\text{m}^3)}{(50\times 10^{-3}\,\text{m}^3/\text{min})}\ln\left(\frac{1000-1152.5}{1000-1047}\right) = 23.54\,\text{min}$$

Therefore, when the tank starts to overflow, 23.54 min should pass, so that soluble solid exiting the current is 12°Brix.

5.2 Juices for consumption are usually marketed with a 12°Brix in soluble solid content, and generally, they are obtained from concentrated juices by means of dilution with water. A 50 kg/min stream from a juice of 70°Brix is fed to a tank perfectly agitated of 2 m^3, where it is diluted with a water stream. Calculate the water flow that should be introduced into the tank if it is desired to obtain 100 L/min of a 12°Brix juice. The process is carried out

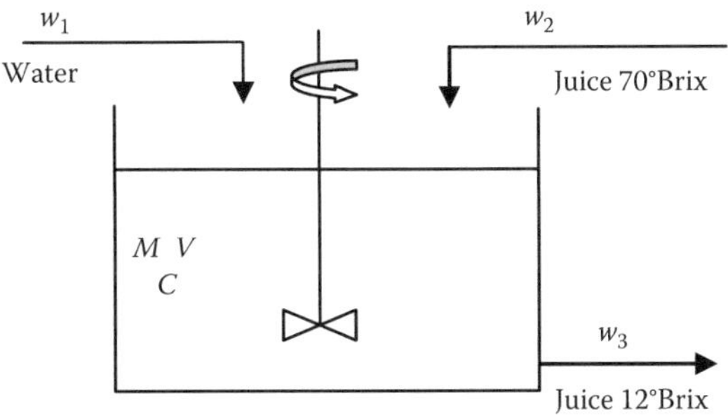

FIGURE P.5.2 Perfectly agitated tank.

isothermally at 20°C, and the temperature of the juice density varies with the soluble solid content according to the expression

$$\rho = 999 + 3.8C + 0.018C^2 \text{ kg/m}^3, \quad \text{where } C \text{ is expressed in °Brix.}$$

The tank operates in a stationary state, there is no accumulation term, and, as there is no generation, the entering and exiting terms are equal. The process illustrated in Figure P.5.3 represents this case.

$$\textit{Global balance}: \quad w_1 + w_2 = w_3$$

$$\textit{Component balance}: \quad q_1\rho_1 + q_2\rho_2 = q_3\rho_3$$

As the densities are a function of the soluble solid content (°Brix), it is obtained as

$$\begin{array}{lll} & C = 12°\text{Brix} & \rho_3 = 1047\,\text{kg/m}^3 \\ \text{For} & C = 70°\text{Brix} & \rho_1 = 1353.20\,\text{kg/m}^3 \\ & \text{Water} & \rho_2 = 1000\,\text{kg/m}^3 \end{array}$$

And as for the 70°Brix juice, the current is

$$q_2 = \frac{w_2}{\rho_2} = \frac{(50\,\text{kg/min})}{(1353.2\,\text{kg/m}^3)} = 36.95 \times 10^{-3}\,\text{m}^3/\text{min}$$

After substituting all the data in the component balance,

$$q_1(1000\,\text{kg/m}^3) + (36.95 \times 10^{-3}\,\text{m}^3/\text{min})(1353.2\,\text{kg/m}^3) = (100 \times 10^{-3}\,\text{m}^3/\text{min})(1047\,\text{kg/m}^3)$$

obtaining

$$q_1 = 54.7 \times 10^{-3}\,\text{m}^3/\text{min} = 54.7\,\text{L/min}$$

5.3 One of the most common deteriorations in the manufacturing and storage of clarified juices is nonenzymatic browning due to the Maillard reaction. The browning grade is measured from absorbance at 420 nm (A_{420}) of 12°Brix juice, and it is considered as not fit for commercialization if it exceeds the value of 0.210. A company that manufactures clarified pear juices stores them in the form of 72°Brix concentrated juice, under refrigeration at 4°C. The color apparition rate can be described by means of a first kinetic order with a kinetic constant $k = 1.2 \times 10^{-3}$ days^{-1}. For a storage tank of 50 m^3 that contains pear juice with an initial $A_{420} = 0.105$ (for the 12°Brix diluted juice), calculate the commercial life of this juice if it is assumed that the inner of the tank does not have concentration gradients.

Inside the tank there are no concentration gradients, it is assumed that have an intensive color apparition rate, which is the same at any point in the tank, that is, the system behaves as a perfect mixing tank.

Since there are no entering or exiting currents, the accumulation color term should be the same as the generation term. The balance for the colored components can be written as

$$\frac{d(VA)}{dt} = rV$$

Since the volume of tank is constant, $d(A)/dt = r$.
The reaction is of first order:

$$\frac{dA}{dt} = kA$$

This is an equation in separable variables that can be integrated with the initial condition, for $t = 0$ the absorbance is $A = A_0$, obtaining

$$\ln\left(\frac{A}{A_0}\right) = kt$$

from which the time can be calculated.

As the juice can only be marketed for a maximal absorbance of $A = 0.210$, the corresponding time to this value is

$$t = \frac{1}{1.2 \times 10^{-3}\,\text{days}^{-1}} \ln\left(\frac{0.210}{0.105}\right) = 577\,\text{days}$$

5.4 A pear juice of 15°Brix, at 50°C, containing 10% insoluble solids is fed to an ultrafiltration module at a flow rate of 20 kg/min, in order to obtain 15 kg/min of pear-clarified juice. This clarified juice feeds an evaporator in which part of the water is eliminated, and a 72°Brix concentrated juice stream is obtained. Calculate (a) the solid content in the retentate stream in the ultrafiltration module, (b) the product flow rate, and (c) evaporation rate.

Figure P.5.4 shows a scheme of the process with the different currents that take part in the process.

The process is continuous, and it will be assumed that it operates in a stationary regime; therefore, the accumulation term is zero, and since there is no generation, the entering current is equal to the exiting current.

Mass balance in an ultrafiltration module:

$$\textit{Global balance}: \quad w_1 + w_2 = w_3$$

$$\textit{Insoluble solids balance}: \quad w_1X_1 = w_2X_2 + w_3X_3$$

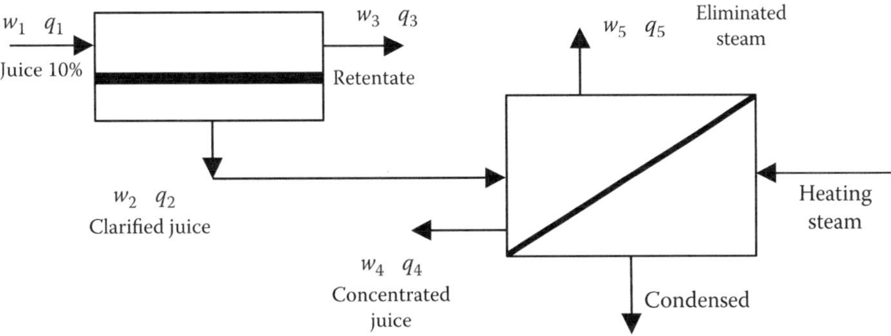

FIGURE P.5.4 Preparation of concentrated orange juice using ultrafiltration and evaporation stages.

where X is the insoluble solids mass fraction. It is important to point out that clarified juice does not possess insoluble solids, then $X_2 = 0$.

Substituting this data shows a system of two equations with two unknown variables:

$$20\,\text{kg/min} = 15\,\text{kg/min} + w_3$$

$$(20\,\text{kg/min})(0.10) = (15\,\text{kg/min})(0) + w_3 X_3$$

after resolving this system, it is obtained as

$$w_3 = 5\text{ kg/min} \quad \text{and} \quad X_3 = 0.40\ (40\%)$$

Mass balance in evaporator:

$$\textit{Global balance}: \qquad w_2 = w_4 + w_5$$

$$\textit{Soluble solids balance}: \qquad w_2 X_2 = w_4 X_4 + w_5 X_5$$

The mass fraction for current w_5 is zero ($X_5 = 0$), since the vapor that leaves from evaporator does not contain soluble solids.

Substituting this data shows a system of two equations with two unknown variables:

$$15\,\text{kg/min} = w_4 + w_5$$

$$(15\,\text{kg/min})(0.15) = w_4(0.7) + w_5(0)$$

after solving this system, it is obtained as

$$\textit{Concentrated juice flow}: \qquad w_4 = 3.125\,\text{kg/min}$$

$$\textit{Water eliminated (in vapor form)}: \qquad w_5 = 11.875\,\text{kg/min}$$

5.5 In the manufacturing of preservatives in syrup, it is necessary to prepare a sweetened solution that accompanies fruit slices. To accomplish this, 400 kg of sugar is utilized and then mixed with water in a jacketed sidelong-agitated vessel, obtaining a 40°Brix solution. Initially the sweetened solution is at room temperature (20°C), and it receives heat through the lateral jacketed wall at a rate of 100 kW/m^2. The vessel is a cylinder of 1 m of diameter and a height of 1.5 m. When the sweetened solution has reached 50°C, the heating operation is stopped, and the sweetened solution is utilized in the filling process of the cans that contain fruit slices. Assuming that under work conditions the specific heat (= 2.85 kJ/(kg °C)) and density (ρ = 1175 kg/m^3) of sweetened solution does not vary, calculate the necessary time so that sweetened solution reaches 50°C.

Figure P.5.5 shows a scheme of the jacketed agitated vessel. To obtain a solution of 40°Brix using 400 kg of sugar, 600 kg of water is required.

The volume occupied by 1000 kg of the sweetened solution is

$$V = \frac{m}{\rho} = \frac{(1000\,\text{kg})}{(1175\,\text{kg/m}^3)} = 0.851\,\text{m}^3$$

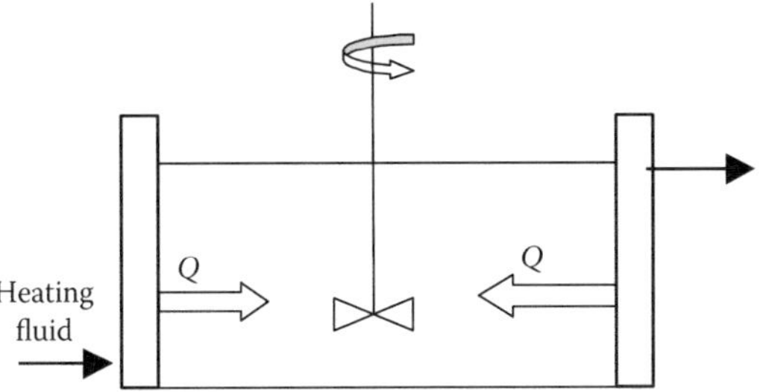

FIGURE P.5.5 Jacketed and perfectly agitated tank.

The height reached by the solution in the agitated tank is

$$h = \frac{V}{(\pi/4)D_T^2} = \frac{(0.851\,\text{m}^3)}{(\pi/4)(\,\text{m})^2} = 1.084\,\text{m}$$

where D_T is the vessel diameter.

In spite of the fact that the vessel height is 1.5 m, the sweetened solution only receives heat sidelong for the corresponding area to the height that reaches the sweetened solution.

Heat transmission surface is as follows:

$$A = \pi D_T h = \pi(1\,\text{m})(1.084\,\text{m}) = 3.405\,\text{m}^2$$

The heat flow that the sweetened solution receives through the lateral area is

$$\dot{Q} = (100\,\text{kW/m}^2)(3.405\,\text{m}^2) = 340.5\,\text{kW}$$

Assuming that there are no convection or radiation heat losses, from the energy balance, it is obtained that the accumulation term is equal to the lateral wall-entering heat:

$$\frac{d\left(m\hat{C}_p T\right)}{dt} = \dot{Q}$$

As the mass contained in the agitated tank does not vary and the specific heat is considered constant, the previous equation is converted to

$$m\hat{C}_p \frac{d(T)}{dt} = \dot{Q}$$

This is an equation in separable variables that can be integrated with the initial condition for $t = 0$ and with the initial temperature of sweetened solution of $T_0 = 20°\text{C}$. For the temperature of the inner vessel, this equation can be written as

$$T = T_0 + \frac{\dot{Q}}{m\hat{C}_p} t$$

From this equation, time can be obtained:

$$t = \frac{m\hat{C}_p}{\dot{Q}}(T - T_0)$$

Substituting the data in this last equation,

$$t = \frac{(1000\,\text{kg})(2.85\,\text{kJ/kg °C})}{(340.5\,\text{kJ/s})}(50°\text{C} - 20°\text{C}) = 251\,\text{s}$$

Therefore, 251 s will be necessary for the sweetened solution to reach 50°C, which is the same as 4.2 min.

5.6 To an evaporator feed a stream of 10,000 kg/h of raw milk at 25°C, whose water content is 87%, it is desired to concentrate its water content to 20%. A heating fluid is utilized along with a reheated steam of 2 atm at 135°C; thus, it exits from the heating chamber at a saturation temperature in the form of liquid water. Calculate the heating steam flow, concentrated milk flow and water steam flow evaporated from the milk. Assume that there is no raising boiling point and that the temperature of the stream that leaves from the evaporation chamber is 100°C.

Data. For the 2 atm steam: Latent heat 2202 kJ/kg. Condensation temperature 120°C. Specific heat 1.95 kJ/(kg °C).

For steam that evaporates at 100°C: Latent evaporation heat 2250 kJ/kg.

The specific heat of the milk can be obtained from the equation

$$C_p = 0.83 + 3.35X_{Water}\ \text{kJ/(kg °C)}$$

where X_{Water} is the water mass fraction in the milk.

In order to solve this problem, it is necessary to carry out mass and energy balances in the system represented in Figure P.5.6.

Mass balances:

$$\textit{Global balance:}\quad w_F + w_W = w_C + w_V + w_W$$

$$\textit{Component balance:}\quad w_F X_F = w_C X_C$$

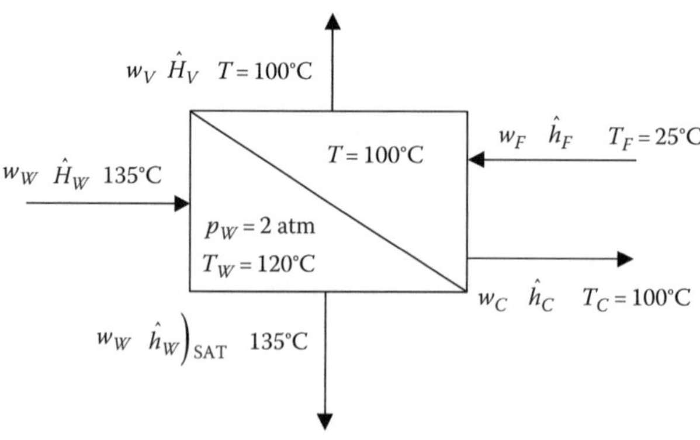

FIGURE P.5.6 Scheme of an evaporator.

The subscripts F, C, V, and W indicate the feed, concentrated, evaporated, and heating steam currents. Upon carrying out solute balance, it is convenient to keep in mind that steam currents do not possess solute, so the solute mass fraction is zero.

Upon considering the milk as a binary mixture, formed by water and solute, the mass fractions of fed and concentrated milk currents are

$$X_F = 1 - 0.87 = 0.13 \quad \text{and} \quad X_C = 1 - 0.20 = 0.80$$

Substituting the data in mass balances,

$$(10{,}000\ \text{kg/h}) = w_C + w_V$$

$$(10{,}000\ \text{kg/h})(0.13) = w_C(0.78)$$

Solving this system of two equations with two unknown variables allows finding their values:

$$\textit{Concentrated milk flow}: \quad w_C = 1625\ \text{kg/h}$$
$$\textit{Eliminated steam flow}: \quad w_V = 8375\ \text{kg/h}$$

Energy balances:
Since this is a process in a stationary state, and there is no generation term, the entering energy is equal to the exiting energy. Thus, by neglecting the kinetic and potential terms,

$$\sum_j w_j \hat{H}_j = 0$$

or similarly

$$w_F \hat{h}_F + w_W \hat{H}_W = w_C \hat{h}_C + w_V \hat{H}_V + w \hat{h}_W$$

$$w_W \left(\hat{H}_W - \hat{h}_W \right) = w_C \hat{h}_C + w_V \hat{H}_V - w_F \hat{h}_F$$

Where
- $\hat{H}_W$ and $\hat{H}_V$ are enthalpies by mass unit of heating steam and eliminated steam currents, respectively
- $\hat{h}_W$ is enthalpy by mass unit of condensed liquid from heating steam (liquid saturated to their boiling point)
- $\hat{h}_F$ and $\hat{h}_C$ are enthalpies by mass unit of fed and concentrated milk currents, respectively

Enthalpies of fed (raw milk) and concentrated milk currents are

$$\textit{Fed}: \quad \hat{h}_F = \hat{C}_p\Big)_F (T_F - T^*)$$

$$\textit{Concentrated}: \quad \hat{h}_C = \hat{C}_p\Big)_C (T_C - T^*)$$

where
- T^* is the reference temperature
- T_F is the raw milk temperature when it enters the evaporator
- T_C is the concentrated milk temperature when it exits from the evaporator and it coincides with the boiling point of solution ($T_C = T_e$)

Enthalpies for the steam currents are

$$\hat{H}_W = \hat{H}_W\big)_{SAT} + \hat{C}_p\big)_{STEAM}\left(T - T_e^{'}\right) = h_W\big)_{SAT} + \lambda_W + \hat{C}_p\big)_{STEAM}\left(T - T_e^{'}\right)$$

$$\hat{H}_V = \hat{H}_V\big)_{SAT} = \hat{C}_p\big)_{Liquid\ water}(T_e - T^*) + \lambda_V + \hat{C}_p\big)_{STEAM}(T_V - T_e)$$

where subscript SAT indicates enthalpies at their saturation point. The temperature of reheated steam is T_R, and the boiling temperature of water at 2 atm is $T_e^{'}$. For the eliminated vapor, T_e is the boiling temperature of the solution in the evaporation chamber and T_V is the eliminated steam temperature, and since there is no raising boiling point, it coincides with T_e ($T_V = T_e$).

Enthalpy for condensed liquid current heating steam is a liquid at its boiling point, which has an enthalpy of a saturated liquid:

$$\hat{h}_W = \hat{h}_W\big)_{SAT}$$

Then the vaporization or condensation latent heat is defined as

$$\left(\hat{H}_W - \hat{h}_W\right)_{SAT} = \lambda_W$$

Therefore, it can be written as

$$\left(\hat{H}_W - \hat{h}_W\right) = \lambda_W + \hat{C}_p\big)_{STEAM}\left(T - T_e^{'}\right)$$

Upon substituting all these expressions, the enthalpy balance equation is obtained:

$$w_W\left(\lambda_W\,\hat{C}_p\big)_{Steam}\left(T_R - T_e^{'}\right)\right) = w_C\,\hat{C}_p\big)_C(T_e - T^*) + w_V\left[\hat{C}_p\big)_{Water}(T_e - T^*) + \lambda_V\right] - w_F\,\hat{C}_p\big)_F(T_F - T^*)$$

As the reference temperature is arbitrary, if it is taking $T^* = T_e$, the previous equation is simplified, and it can be written as

$$w_W\left(\lambda_W\,\hat{C}_p\big)_{Steam}\left(T_R - T_e^{'}\right)\right) = w_V\lambda_V - w_F\,\hat{C}_p\big)_F(T_F - T^*)$$

For example, these equations can be used substituting the following data:

$$\lambda_w = 2202\ \text{kJ/kg} \qquad \lambda_V = 2250\ \text{kJ/kg}$$

$$\hat{C}_p\big)_{Steam} = 1.95\ \text{kJ/(kg °C)}$$

For $X_F = 0.13$ the feed specific heat is $\hat{C}_p)_F = 3.74$ kJ/(kg °C)

$$T_R = 131°\text{C} \quad T_F = 25°\text{C} \quad T_e = 100°\text{C} \quad T_e' = 120°\text{C}$$

$$w_F = 10{,}000 \text{ kg/h} \quad w_V = 8{,}375 \text{ kg/h}$$

Upon substituting all these data in the last equation, it is possible to calculate the heating steam flow, obtaining the following value:

$$w_W = 9753.7 \text{ kg vapor/h.}$$

PROPOSED PROBLEMS

5.1 In the concentration of apple juice, a fresh juice extracted by pressure and sieved contains 10.2% (w/w) of solids. This extraction is fed to an evaporator whose evaporation chamber works under vacuum conditions. In the evaporator, water is eliminated until a juice with 65% solids is obtained. For each 1000 kg/h of fed juice, calculate the quantities for the concentration currents and the water that exits the evaporator as it is eliminated in vapor form.

5.2 In the sugar process it was obtained that 1000 kg/h of a sugary solution containing 15% (w/w) of sugar fed to an evaporator eliminates part of the water in order to obtain a solution with 50% (w/w) of solids. This stream is taken to a crystallizing device where sugar crystals containing 4% moisture are eliminated. The saturated solution that exits the crystallizer contains 37.5% (w/w) of solids, and it is recycled to the evaporator. Calculate the amount of recycled stream and the crystal stream flow.

5.3 A fluid food at 25°C is pumped with a flow of 1.500 kg/h through a heat exchanger, where it warms up to 75°C under pressure. Hot water is used as the heating fluid, and the fluid enters the heat exchanger at 95°C and it exits it at 82°C. The mean specific heat for the fluid food is 4.03 kJ/(kg °C) and for the water it is 4.18 kJ/(kg °C). The fluid food and the hot water are separated by a metallic surface through which the heating process is carried out. Calculate the water flow and the amount of heat added to the fluid food, assuming that there are no heat losses.

5.4 In sugar beet production, a stream of 500 kg/h of a solution that contains 35% (w/w) of sugar is initially obtained. This solution is concentrated in an evaporator to 75% (w/w) of sugar, prior to the crystallization stage. Calculate the concentrated solution flow and water eliminated in the evaporator.

5.5 From a fish sample, it is desirable to obtain fish meal in order to use it as a nutritious protein supplement. In an extraction stage with an organic solvent, the oil contained in 10 fish samples is extracted, obtaining as a by-product a fish cake with 80% water. This moist cake is dried using a continuous dryer, obtaining a final fish flour with 35% (w/w) of water. Before final packaging, this dry product is milled and packed under vacuum. If it is desired to obtain 2000 kg/h of dry flour, calculate the amount of moist cake that should be fed to the dryer.

5.6 For fruit jam production, crushed fruits with 12°Brix in soluble solids content are mixed with sugar and pectin. The amount of sugar and pectin added is 1.25 kg of sugar and 2 g of pectin for each kg of crushed fruit. Once the ingredients have been adequately mixed, they are taken to a perfectly stirred tank evaporator open to the atmosphere, obtaining a jam that contains 65% (w/w) of soluble solids. Taking as a calculation base a feeding of 100 kg of crushed fruit, calculate the amount of the mixture that exits the mixer, the amount of water eliminated in the evaporator, and the final obtained jam.

5.7 Soya seeds are usually processed in order to obtain various food compounds, such as oil and meal for animal feed. Soya seeds contain 16% oil, 10% water, and inert solids. The seeds are crushed and pressed, obtaining an oil stream and a slurry of seed that contains 6% of oil. This last seed stream is fed to an extractor, in which an extraction stage is carried out with organic solvents, producing an underflow stream containing meal with 0.5% oil content, and an overflow stream containing the organic solvent and the extracted oil. In the last stage, the underflow is dried in order to obtain a soya meal containing 5% water. If 20,000 kg of soya seeds are treated in this process, calculate (a) the amount of seeds that exits the press stage, (b) the amount of meal that exits the extractor, and (c) the amount of dried soya meal.

Note: Assume that in the press stage the oil stream does not contain other components. For the extraction stage assume that the underflow does not contain organic solvents. All of the percentages are in weight.

5.8 A solution that contains 7% of certain salts and some impurities is used to obtain 1000 kg/h of crystallized salt. Thus, the diluted solution concentrates until it reaches 40% in an evaporator; later the concentrated solution is cooled in a crystallizer that operates at 10°C, where it is possible to separate the crystallized salt and the mother liquor solution. Ninety percent of the mother liquor solution is recycled to the evaporator, while 10% remains discharged as water waste. Assuming that the solubility of the salt at 10°C is of 10% (p/p), calculate the diluted solution flow that is fed to the system, as well as the evaporated water flow.

5.9 A certain food company produces 100 m^3/h of water waste effluent that contains 1 g/L of a toxic component for the environmental flora and fauna. This toxic compound breaks down spontaneously according to a first-order kinetics ($r = -100$°C kg/m^3 h). Local legislation allows that the water waste can be discharged in the aquifer currents with concentrations lower than 0.01 g/L toxic pollutant. To do so, prior to discharge in the aquifers, the water waste is sent to a perfectly mixed tank so that the pollutant concentration is lowered to the value required by the legislation. Calculate the volume of the tank operating in a stationary state.

5.10 For the elaboration of fruits in syrup, it is necessary to have an 18°Brix sucrose solution. An industry has two 120 L perfectly mixed tanks. In order to obtain the sucrose solution, in each tank 82 L of water is placed, and 18 kg of sugar should be added in each tank. The operator in charge of preparing this solution added 82 L of water in each tank; with some distraction, he has added 36 kg of sugar in one of the tanks, while the other tank only contains water. For the purpose of equaling the concentrations, the tanks are connected by means of two circuits with pumps that transfer the solution of one tank to the other simultaneously. If the flow for both pumps is 50 kg/h, calculate (a) the sugar concentration variation in both tanks and (b) the concentration that is obtained when the product arrives at the stationary state.

Note: Assume that the densities for all the streams are the same, taking their value to the corresponding of water.

5.11 A peach puree stream of 500 kg/h is at 20°C, and it is desired that it be heated in a heat exchanger, where it receives 125,000 kJ through the exchange surface, by condensation of water steam. Calculate the temperature to which the peach puree exits from the heat exchanger. The specific heat of the peach puree is of 3.85 kJ/(kg °C).

5.12 In one of the stages of milk pasteurization, it is desired to cool 5000 kg/h of whole milk from 65°C to 4°C, using a plate heat exchanger. Calculate the amount of heat eliminated by the coolant fluid. Assume that the specific heat of the milk is 3.95 kJ/(kg °C).

5.13 An airstream, at 25°C, is used in a drying process, in which the temperature should exceed 65°C. To heat the air stream, a heat exchanger is used where saturated steam at 150°C is the heating medium. The steam current, once condensed, exits the heater at 135°C. If the air flow rate is 1000 kg/h, calculate the steam flow necessary to carry out the proposed heating.

5.14 Two thousand cans of tomato puree are sterilized in an autoclave that operates at 120°C. Once the sterilization is carried out, the cans cool down in the same autoclave to 35°C. A 22°C water stream is used for this process, leaving the autoclave at 30°C. Calculate the amount of water needed for this cooling process.

Data. It is considered that the heat losses through the autoclave walls by convection and by radiation are 15,000 kJ. Each can contains 300 g of tomato puree, the total weight of a full can being 330 g. The mean specific heat of tomato puree is 3.43 kJ/(kg °C), while for the metal of the can, it can take a value of 0.3 kJ/(kg °C). Inside the autoclave the cans are placed in a metallic rack that weighs 150 kg and whose specific heat is of 0.35 kJ/(kg °C). It can be assumed that at the end of the cooling stage the metallic rack has the same temperature as the water that leaves from the autoclave.

6 Physicochemistry of Food Systems

6.1 INTRODUCTION

Different processes in the food industry can be studied from the laws that govern the different transfers involved in each process. The application of these fundamental equations allows us to obtain the mathematical model on which the process is based. Previous to this study, it has been necessary to define the system in which the process will take place and to determine its type, since systems may be classified as *closed*, *open*, or *isolated*. When a system is defined, the possible mass and energy exchanges with the surroundings must be considered. Those transfers will lead to changes in the system, whose components tend to vary, as much in their composition as in their energy state, even being able to undergo chemical reactions.

When the effects of the energy exchange between a system and the surroundings are studied, changes in their chemical structure may be generated. When structure changes do not exist, the study of a system is based on the principles of a branch of thermodynamics denominated *physical thermodynamics* or *thermophysics*. However, if chemical changes take place, the system is studied by *chemical thermodynamics*. Thus, it will be necessary to define the different laws of thermodynamics, with the purpose of understanding the concepts of temperature, quantity of energy, and energy efficiency involved in a certain process. In addition, different properties and functions that allow the possibility that certain reactions will take place in the considered system should also be defined. In the description of many food processes such as thermal treatments, heating and refrigeration, and in the operation of the devices used in these processes, it is necessary to know the concepts of some properties such as enthalpy or free energy, as well as equilibrium state.

6.2 GENERAL CONCEPTS

6.2.1 Thermodynamic System

In any process, it is necessary to define the system on which a particular study is to be carried out. A thermodynamic system is a macroscopic part of the physical universe separated from the rest of the universe by limits, borders, or walls. The region of the external space of the system is its surroundings.

There are different types of walls. If the wall does not allow matter to pass through it, it is denominated *impermeable*. A wall is considered *rigid* if a system is subjected to its own variations or changes in its surroundings and its volume does not change. An *adiabatic* wall prevents any interaction type between the system and its surroundings; if a wall is not adiabatic, it is denominated *diathermic*.

Systems can be classified depending on the type of the wall. In this way, a system that can exchange matter and energy with its surroundings has walls that are not impermeable, and it is called an *open* system. A system is *closed* when it does not exchange matter, but it can exchange energy, so its walls are impermeable, diathermic, and not rigid. The walls of a system that exchanges neither matter nor energy with its surroundings are impermeable, rigid, and adiabatic, and therefore, the system is denominated *isolated*.

A system is said to be in *thermodynamic equilibrium* when the observable properties of the system do not change with time. In order to have thermodynamic equilibrium, it is necessary that three equilibrium types exist simultaneously: thermal, chemical, and mechanical. *Thermal equilibrium* is reached if the temperature is the same in the entire system. *Chemical equilibrium* exists if the composition of the system does not vary with time. *Mechanical equilibrium* supposes that macroscopic movements do not exist inside the system or between the system and its surroundings.

The *state* of a system represents its condition, being defined by the group of its properties. Relationships usually exist among such properties, and these relationships can generally consist of equations, which correlate some of those properties. In that context, by giving values to a few properties, the remaining properties can be obtained from the defined equations. Thus, the state of a system is determined if the values of the different variables that define it are known, and it can be completely defined by means of four properties or state variables: composition, pressure, volume, and temperature. In the case of homogeneous systems, the composition is fixed, and the system only depends upon the other three properties. If these three properties are specified, all of the other physical properties will be fixed.

If some of the properties that define a system change, its state changes, and it is said that the system has undergone a process. In this way, a *process* can be defined as the transformation from one state to another one, allowing definition of the *stationary* and *transient* states: if a system has the same values of a state in two different times, it is said that it is in stationary state; otherwise, its state is transient. In a thermodynamic process, a variation from a state 1 of the system that is in equilibrium until a final state 2, also in equilibrium, is carried out. Thermodynamic processes can be *reversible* or *irreversible*: a process is considered reversible when it goes through different equilibrium states where the state variables are always defined, although they may vary; otherwise, the process is irreversible.

When a system that experiences different successive processes returns to the conditions of the initial state, it is said that the system has described a *thermodynamic cycle*. In this case, the properties of the system again have the same values as in the initial state, for which it is said that the system has not experienced a state change. The thermodynamic cycles play an important role in many food processes, as in the case of refrigeration operations.

Generally, to schematize the processes, diagrams are usually used, in which some of the properties of the system are correlated. Thus, in Figure 6.1 a *P–V* diagram (pressure–volume) is shown, plotting different system states represented by different points. The change of the system through the different states is carried out by processes, which at the end represent a cycle. The processes that are carried out at constant volume are designated as being *isochoric*, as in the case of processes 2–3 and 4–1. If the process is carried out at constant pressure, it is denominated *isobaric* (process 1–2).

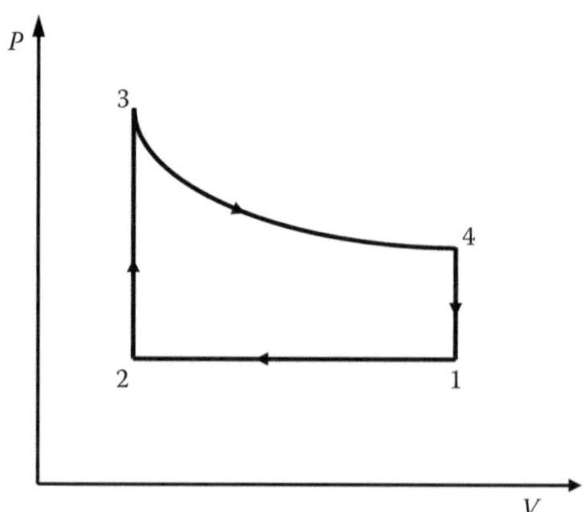

FIGURE 6.1 Thermodynamic cycle.

Moreover, there are *isothermal* processes, in which the temperature does not vary; in those cases in which exchange of heat does not exist, they are denominated *adiabatic* processes.

The physical properties of a system can be classified into two types. The *extensive* properties are those that depend on the quantity of samples of the system, while the *intensive* properties are characteristic of the substance in the system, independent of their quantity. The properties of a system in thermodynamic equilibrium only depend on the state and not on the previous history of the system. Therefore, the variation of any property due to a variation in the thermodynamic state depends only on the initial and final state of the system.

6.2.2 State Equation for Gases

As has been already introduced, in homogeneous systems, it is only necessary to specify the pressure, volume, and temperature to determine the state of the system. However, in practice, experimental results have shown that these three properties are related to each other in a system with defined mass quantity. In these cases, the denominated state equation is defined. This is an equation that ties together the different variables that affect the system. In the case of ideal or perfect gases, the state equation has the following form:

$$PV = nRT \tag{6.1}$$

where

P is the pressure
V is the volume that occupies the system
T is the absolute temperature
n is the number of moles contained in the system
R is the universal constant of gases, whose value is 8.314 J/mol K

As can be observed, in this equation the thermodynamic state of a simple homogeneous system can be completely defined specifying two of these properties.

Although this state equation is the one most used, due to its simplicity, it is not reliable for real gases. In the equation of perfect gases, it is assumed that the gas molecules do not occupy any space; however, in a real system that is not true. This is why other state equations can correlate the different variables of the system, such as the equation of van der Waals:

$$\left(P + \frac{an^2}{V^2}\right)\left(V - nb\right) = nRT \tag{6.2}$$

where

a is the denominated parameter of attraction
b is the repulsion parameter, whose value depends on the considered gas

It is observed that this equation derives from the perfect gases equation, in which it is considered that the molecules occupy a place in the space, leading to the use of $(V - nb)$ instead of the volume V. In addition, it is considered that the molecules can interact with other molecules, being able to display attraction effects, and thus making it necessary to add these effects of attraction, to pressure P quantified as proportional to the inverse of the squared molar volume.

In addition to these equations, other equations can be applied to real gases, such as the virial equation, as well as the equations of Redlich–Kwong, Soave, Peng–Robinson, Elliott, Suresh, Donohue, and BWRS, among others. Nevertheless, due to the simplicity of the perfect gas equation, and due to the good results obtained with its application, it is the more used equation in engineering calculations.

The equations described previously are applied when there is only one component; however, for a mixture of gases, each of the components exercises partial pressure, so the total pressure is the sum of the partial pressures of the components:

$$P = \sum_i p_i \tag{6.3}$$

Partial pressure of each component is obtained from the equation

$$p_i = x_i P \tag{6.4}$$

where x_i is the molar fraction of the component i, which is the relationship between the moles of this component and the total moles:

$$x_i = \frac{n_i}{\sum_i n_i} \tag{6.5}$$

6.2.3 Different Forms of Energy

It is difficult to give a wide definition of energy; however, from a mechanical interpretation, energy can be defined as the ability or aptitude of a material system to develop a mechanical work. However, this definition is not perfect, since it is not adequate to quantitatively determine the value of the defined magnitude. Heat is an energy form, and because not all heat can become work, some authors prefer to define the energy as every property that can become heat.

Different forms of energy can be converted into each other, and it is convenient to remember those energy types that should be considered in the study of a process. These types are internal energy, potential energy, kinetic energy, heat, and work. The first three are state functions, and it will be necessary to define a reference state, while heat and work depend on the evolution of the considered process.

Matter is formed by particles (atoms, molecules, ions, etc.) that are continuously in motion. The *internal energy* (U) is the sum of the rotation, translation, and vibration energy of these particles, and it depends on the temperature. The energy that a system has as a function of their position under a force field (gravitational, electric, and magnetic) receives the name of *potential energy*. For the conventional processes in food engineering, only the gravitational potential field is usually considered. The *kinetic energy* is the energy that the system has due to the speed that it possesses with regard to a coordinate system.

The *heat* or thermal energy is an energy form that depends on the temperature gradient between the system and its surroundings, as well as that of the area through which it is transmitted. The heat that enters the system is considered positive, while the heat that the system gives to the surroundings is negative. If a system with an impermeable, rigid, and diathermic wall, with a different temperature from its surroundings (or another system) is considered, variations are observed in the properties of the system due to the interactions through the wall until thermal equilibrium. In this entire process, an energy exchange has existed in the form of heat; therefore, heat can be defined as the energy form exchanged between the system and its surroundings (or another system) at different temperatures when they are separated by a diathermic wall. The absorbed or transferred heat is proportional to the mass and to the variation of temperature that it experiences; in infinitesimal form, it can be expressed as

$$\delta q = mC_e dT \tag{6.6}$$

where

m is the mass contained in the system

C_e is the specific heat

In many industrial processes, mechanical work appears due to the expansion or the work consumption for compression of a fluid. The calculation of the work of a fluid in expansion is obtained from the expression

$$W = \int_x F\,dx = \int_x -pS\,dx = -\int_V p\,dV \tag{6.7}$$

where
 F is the force
 p is the pressure
 S is the section in which the force is applied
 V is the volume

The negative sign exists because, as in the case of heat, the work that the system carries out is considered negative, while the work that enters the system is positive.

6.3 ZEROTH LAW OF THERMODYNAMICS

When two closed systems are put in contact by means of a rigid and impermeable wall, they form an isolated system. If in addition, the wall is adiabatic, no variations are observed in the state magnitudes of the two systems. However, if the wall is diathermic, variations of these magnitudes are observed, so that they tend to evolve to give a common final state, leading to a state of thermal equilibrium. Thus, the principle known as the zeroth law of thermodynamics can be enunciated as: "When two systems are in thermal equilibrium with a third system, they are in thermal equilibrium with each other."

This statement leads to the introduction of the concept of temperature. The intensive state function that takes the same value for all the states of the system that are in heat equilibrium is denominated temperature. This function is used among all the thermodynamic terms. In addition, two systems that are in thermal equilibrium have the same temperature.

6.4 FIRST PRINCIPLE OF THERMODYNAMICS

This principle relates the energy exchanges that take place between a system and its surroundings or with another system during a thermodynamic process. It is based on the law of mass conservation as a result of the experience, through the historical precedents associated to the impossibility of the perpetual motion and of the perseverance of the mechanic equivalent of heat.

The first principle of thermodynamics establishes that energy can be transformed from one form into another one, but it can neither be created nor destroyed. For a closed system that only exchanges heat and work with its surroundings, this principle can be enunciated in the following way: "The total variation of energy (kinetics, potential, and intern) of a closed system is similar to the sum of the absorbed heat and the released heat."

If the considered system is in rest and in that the preservative forces cannot exercise any type of work, the existence of an extensive state function of the system, denominated internal energy (U) is postulated, which should fulfill the principle of energy conservation. In this way, when a system passes from a state to another one, this first principle can be expressed by means of the equation

$$\Delta U = Q + W \tag{6.8}$$

that indicates that the change experienced by the internal energy of the system is due to the absorbed heat and the work carried out on the system.

This equation can be expressed in differential form

$$dU = \delta Q + \delta W \quad (6.9)$$

It is necessary to mention that the internal energy is a state function, and it is a true differential, while heat and work are not state functions, and that is why its differential form is expressed by δ. Also, keeping in mind that the expansion work is due to the change of volume that experiences the system,

$$\delta W = P\,dV \quad (6.10)$$

There are particular processes in which the previous equation is simplified. Thus, in the case that the process is carried out without change of volume, when applying Equation 6.8 it is obtained that the work is null, with what the change of internal energy that experiences the system is only owed to the involved heat, so

$$Q_V = \Delta U = U_2 - U_1 \quad (6.11)$$

Also, in the case of adiabatic processes (without exchange of heat), the variation of internal energy is due to the work involved in the process:

$$W = \Delta U = U_2 - U_1 \quad (6.12)$$

A process is denominated cyclical, when the final conditions are the same as the initial ones. In those cases, the variation of the internal energy of the system is null, which indicates that the work executed by the system is equal to the heat absorbed, or vice versa, the work exerted on the system is translated in the heat release. It is important to emphasize that in a cyclical process, the involved work and heat depend on the process that was followed, so that both work and heat are not thermodynamic properties of a system.

6.5 ENTHALPY

Generally, most food processes are developed at a constant pressure. In these cases the work carried out by the system can be expressed as

$$\delta W = -P\,dV \quad (6.13)$$

When this equation is substituted in the differential form of the first principle (Equation 6.9), when applying it for a certain process passing from a state 1 to 2, integrating Equation 6.9, it is obtained as

$$\Delta U = U_2 - U_1 = Q_P - P\left(V_2 - V_1\right) \quad (6.14)$$

where Q_P is the heat at constant pressure.

This equation can be rearranged, obtaining

$$Q_P = \left(U_2 + PV_2\right) - \left(U_1 + PV_1\right) \quad (6.15)$$

Defining a new state function, denominated enthalpy (H) is expressed as

$$H = U + PV \quad (6.16a)$$

or in their differential form

$$dH = dU + d(PV) = dU + P\,dV + V\,dP \qquad (6.16b)$$

The enthalpy is a state function that represents the total energy of a system that can become useful work.

When substituting the expression of the enthalpy in Equation 6.15, it is obtained as

$$Q_P = \Delta H = H_2 - H_1 \qquad (6.17a)$$

or in its differential form

$$dQ_P = dH \qquad (6.17b)$$

Also, keeping in mind Equations 6.9 and 6.13 in (6.16b), it is obtained as

$$dH = \delta Q + V\,dP \qquad (6.18)$$

So, for a process at a constant pressure, the heat absorbed or released by the system is similar to the enthalpy change that it undergoes.

Most food processes that involve energy changes are carried out at constant pressure, which is why it is important to evaluate the enthalpic changes that take place, making it possible to determine the heat involved in the process.

6.6 HEAT CAPACITY

The heat absorbed or released by the system, in infinitesimal form, can be expressed according to Equation 6.6. The product of the mass and the specific heat is denominated *heat capacity*:

$$\delta q = C\,dT \qquad (6.19)$$

where C is the heat capacity of the system, being an extensive parameter that is not state function.

There are two very important forms of heat capacity, referred to as constant volume or constant pressure. The heat capacity at constant volume (C_V) is expressed as

$$C_V = \frac{\delta q_V}{dT} = \left(\frac{\partial U}{\partial T}\right)_V \qquad (6.20)$$

which indicates that it can be defined as the variation of internal energy with regard to the temperature when the volume remains constant. As the volume is constant, expansion work and the change of internal energy that experiences the system do not exist; it is equal to the involved heat, being able to be calculated by means of the equation

$$Q_V = \Delta U = \int C_V\,dT \qquad (6.21)$$

The heat capacity at constant pressure (C_P) is expressed as

$$C_P = \frac{\delta q_P}{dT} = \left(\frac{\partial H}{\partial T}\right)_P \qquad (6.22)$$

which indicates that it can be defined as the enthalpy variation with regard to the temperature when the pressure remains constant. In this case, the heat is equal to the enthalpy change that experiences the system, being calculated by means of the expression

$$Q_P = \Delta H = \int C_P \, dT \tag{6.23}$$

In the case of gas ideal systems, the relationship between the heat capacities at constant pressure and volume is given by

$$C_P - C_V = nR \tag{6.24}$$

This expression is denominated Mayer's relationship, indicating that for an ideal gas $C_P > C_V$.

Another important property is the denominated polytropic coefficient (γ), which is the relationship between the heat capacities at constant pressure and volume:

$$\gamma = \frac{C_P}{C_V} \tag{6.25}$$

As has been indicated, heat capacity is an extensive property that depends on the mass contained in the system, so comparing Equations 6.6 and 6.19, a relationship between the heat capacity and the specific heat is obtained:

$$C = mC_e \tag{6.26}$$

The specific heat is a function of temperature, and its variation is much more important in gases than in solids and liquids. Usually, for liquid and solid foods, it is supposed that the specific heat does not much vary with temperature, and it is assumed that it is constant at the usual processing temperatures. Moreover, most processes are carried out at atmospheric pressure; for this reason, it is important to know the specific heats at constant pressure. In food, water content is important, and it can be also observed that the specific heat depends on this water content. The specific heat of water has a value of 4.185 kJ/kg °C, while foods show lower values, although the higher the water content of foods are closest to this value.

The various described state functions can be correlated depending on the process type applied to the system. Thus, in the case of an adiabatic reversible process ($\delta Q = 0$) for an ideal gas, Equation 6.18 can be expressed as

$$C_P \, dT = nR\left(\frac{dP}{P}\right) \tag{6.27}$$

Integrating this equation between an initial state 1 to a final state 2, keeping in mind the Mayer's relationship, it is obtained as

$$\ln\left(\frac{T_2}{T_1}\right) = \left(\frac{\gamma - 1}{\gamma}\right)\ln\left(\frac{P_2}{P_1}\right) \tag{6.28a}$$

or

$$\frac{P_2}{P_1} = \left(\frac{T_2}{T_1}\right)^{(\gamma/(\gamma-1))} \tag{6.28b}$$

Considering the perfect gas equation, the relationship between the pressure and volume can be obtained as

$$P_1V_1^{\gamma} = P_2V_2^{\gamma} \tag{6.29}$$

6.7 SECOND PRINCIPLE OF THERMODYNAMICS: ENTROPY

The first principle of thermodynamics establishes that only the processes in which energy is conserved are possible; however, experience indicates that not all of the processes that satisfy this condition take place. In the case of isolated systems within any process, their energy stays constant, and when they are not in equilibrium, they evolve in some ways until achieving equilibrium, although opposing processes never take place.

The second principle arose from a study of the conversion of heat in work, basically in experiments that were carried out with thermal machines. Through these works the conclusion has been reached that although the whole work could become heat, not all the heat could become work. This result was attributed to the fact that the real processes are irreversible, and a new state function was defined, *entropy* (S), in such a way that this function should increase during an irreversible process in an isolated system until achieving a maximum value when arriving at the final equilibrium state. In addition, with thermal machines it was observed that the entropy variations were directly related with the exchanged heat and the system temperature.

The definition of this state function is difficult, although some authors describe it as a measure of the order or disorder of a system, being a good measure of the non-available energy to give work. There are other interpretations for entropy, possibly correlating it with the spontaneity of the processes, so the spontaneous processes will only take place in cases where enough energy exists so that the entropy increases, which implies that for the spontaneous processes changes are required that range from states of higher energy to lower energy.

There is no exact definition for the second principle of thermodynamics, although a possible definition might be that there is a state function denominated entropy that is an extensive magnitude and that the entropy variation in a system during an infinitesimal process where a certain quantity of heat δq is exchanged with its surroundings (or another system) at a temperature T can be expressed according to Clausius' inequality:

$$dS \geq \frac{\delta q}{T} \tag{6.30}$$

For reversible processes it takes as equality, but in the irreversible ones, the entropy variation will always be higher.

As entropy is a state function, variation that is experienced in a certain process does not depend on the process but only on the initial and final states. For a process developed in a system between an initial (1) and a final (2) state, it can always be a process with reversible exchange of heat, so the entropy change can be obtained from the expression

$$\Delta S = S_2 - S_1 = \int_1^2 \frac{\delta q_{rev}}{T} \tag{6.31}$$

In an isolated system, the energy exchange is null, so that starting from the inequality of Clausius,

$$\Delta S_{isolated\ system} \geq 0 \tag{6.32}$$

which indicates that in an isolated system the entropy can never decrease, and any process always implies an entropy increase until it arrives to a state with the maximum value that belongs together with that of thermodynamic equilibrium.

In the case of closed systems, it can be an exchange of heat and/or work with its surroundings. In this case, the variation of entropy of the universe must be considered instead of the system, so it is accomplished as

$$dS_{universe} = dS_{system} + dS_{surroundings} \tag{6.33}$$

As far as the surroundings are concerned, the entropy change that they experience can be calculated assuming that the heat transfer can be reversible, since the surroundings are very big so that this heat exchange affects significantly its state and its temperature. Thus, the change of entropy of the surrounds can be expressed according to the equation

$$dS_{surroundings} = -\frac{\delta q}{T} \tag{6.34}$$

where the negative sign indicates that the heat is absorbed by the system, being T the temperature of the surroundings. If dS is the variation of the system's entropy, Equation 6.33 is expressed as

$$dS_{universe} = dS - \frac{\delta q}{T} \tag{6.35}$$

For an isothermal process in which the temperature does not change by applying Equation 6.30, it is obtained that the reversible heat exchanged will be

$$\delta q_{reversible} = T\,dS$$

Combining Equations 6.9 and 6.10,

$$\delta W_{rev} = dU - T\,dS \tag{6.36}$$

which indicates that the work is equal to the change of internal energy that experiences the system less the non-available energy.

The calculation of the entropy variation that the systems experience depends on the process that is carried out. These types of changes will be examined later.

As has been indicated previously, if the process is adiabatic, the exchanged heat is null, and if the process is also reversible, the entropy variation is also null, which means that in all reversible and adiabatic process the entropy is constant, and this type of process is isentropic.

For a reversible isothermal process, the entropy variation that experiences the system, from Equation 6.31, is

$$\Delta S = S_2 - S_1 = \int_1^2 \frac{\delta q_{rev}}{T} = \frac{\delta Q_{rev}}{T} \tag{6.37}$$

where δQ_{rev} represents the heat exchanged by the system.

In irreversible isothermal processes, the entropy variation of the system can be calculated assuming that the process is reversible, since the entropy is a state function and the entropy change will be the same in the reversible and irreversible processes.

For isobaric processes at constant pressure, the entropy variation of the system between two temperatures T_1 and T_2 can also be calculated assuming a reversible process, since the entropy is state function. In this case, the entropy variation will be

$$\Delta S = S_2 - S_1 = \int_1^2 \frac{\delta q_{rev}}{T} = \int_1^2 \frac{C_P\, dT}{T} = C_P \ln\left(\frac{T_2}{T_1}\right) \tag{6.38}$$

For an isochoric process, at constant volume, through a similar process it is obtained as

$$\Delta S = S_2 - S_1 = \int_1^2 \frac{\delta q_{rev}}{T} = \int_1^2 \frac{C_V\, dT}{T} = C_V \ln\left(\frac{T_2}{T_1}\right) \tag{6.39}$$

6.8 THERMAL MACHINES: CARNOT'S CYCLE

It has already been mentioned that the work transformation in heat can be made in a total way; however, according to the second principle of thermodynamics, all of the heat cannot become work. In order to obtain work from heat, it is necessary to have a system that absorbs energy in the form of heat from its surroundings during a certain process and give a quantity of work. If this conversion of heat in work is desired to be carried out in an indefinite way, the process that experiences the system must be a cycle. The mechanical device that is used so that the system travels the cycle is denominated thermal machine. During the cycle, the system absorbs heat from its surroundings (Q_A) and gives a part of it in heat form (Q_C), while the rest is given as work (W). The yield of a thermal machine (η) is defined as the work generated by unit of absorbed heat:

$$\eta = \frac{|W|}{Q_A} \tag{6.40}$$

For the work, the absolute value is taken because it is work that the system gives and it has negative sign. As the system completes a cycle, the variation of the state functions is zero, as in the case of the internal energy. When applying the first principle of thermodynamics it is obtained as

$$0 = Q_A - Q_C - W$$

substituting in Equation 6.40,

$$\eta = \frac{|W|}{Q_A} = \frac{|Q_A| - |Q_C|}{Q_A} = 1 - \frac{|Q_C|}{Q_A} \tag{6.41}$$

Figure 6.2 shows a schematic diagram of the energy exchange of a thermal machine where the heat is absorbed from a hot source at a temperature T_1, and it is transferred to a drain that is at a lower temperature T_2. As the system covers a cycle, its entropy variation is null. For both focuses it is assumed that the exchange of heat takes place in a reversible way, since they are systems of very great heat capacity, and its temperature does not change. Therefore, when applying the second principle,

$$\Delta S_{universe} = -\frac{|Q_A|}{T_1} + \frac{|Q_C|}{T_2} \geq 0 \tag{6.42}$$

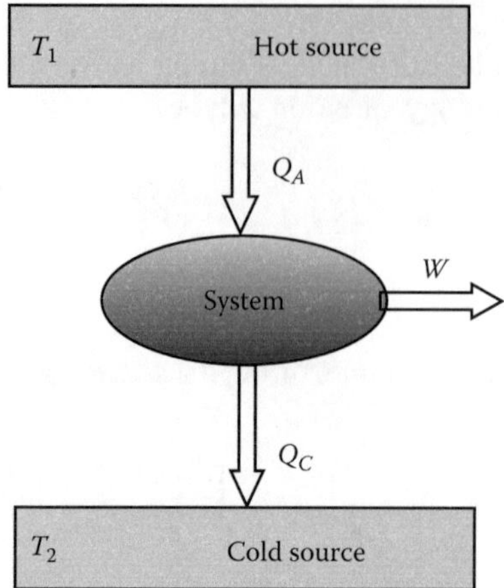

FIGURE 6.2 Energy exchanges for a thermal machine working between two heat sources at a constant temperature.

obtaining

$$|Q_C| \geq |Q_A| \frac{T_2}{T_1} \tag{6.43}$$

The heat given to the cold focus will be minimal, when this last equation is equality, which will indicate that the cycle that the system follows is reversible. In this case the generated work is maximum, as well as the yield of the thermal machine:

$$\eta_{max} == 1 - \frac{|Q_C|_{min}}{Q_A} = 1 - \frac{T_2}{T_1} \tag{6.44}$$

This last equation is the base of Carnot's theorem, which can be enunciated in the following way: No thermal machine that works between two constant focuses of temperature can have a larger yield than a reversible thermal machine working between the same focus temperatures. The reversible thermal machines that work between these two focuses have the same yield, and the yield is also maximum, independent of the nature of the system that the cycle carries out.

The reversible thermal machine that works between two constant focuses of temperature is called a Carnot machine.

6.9 THIRD PRINCIPLE OF THERMODYNAMICS

This principle can be enunciated in different ways, and it is based on the value that the entropy acquires in the absolute zero. The entropy of a system approaches a constant value as the temperature approaches zero. In the case of a perfect crystal at absolute zero, the entropy is exactly equal to zero. According to Nernst–Simon, the third principle of thermodynamics can be enunciated as follows: "It is impossible for any procedure to lead to the isotherm $T = 0$ in a finite number of steps."

Mathematically, this principle can be expressed as

$$\lim_{T \to 0} S = 0 \tag{6.45}$$

6.10 CHEMICAL THERMODYNAMICS

When a food system is being studied, it is important to point out that, in many cases, reactions can happen among the components that comprise the system. To carry out a correct description of food processes, it is necessary to be able to define the equilibrium conditions and, also, to introduce new state functions that allow definition if a certain reaction will be carried out in a spontaneous way. Among equilibriums, it is important to be able to study and define not only the chemical equilibriums but also the equilibrium among phases. In the case of the changes that can go through a food system due to the chemical reactions that it presents, it is necessary to know the irreversibility and spontaneity of these changes and to define new functions. To apply thermodynamic equations to systems in which those chemical reactions take place, it is necessary to introduce composition as a state variable of the system. In addition, it is necessary to introduce the concept of chemical potential of a component as a fundamental thermodynamic variable associated with their composition. Moreover, in closed systems of unsteady composition, it is necessary to obtain the conditions of different equilibrium types.

6.11 HELMHOLTZ FREE ENERGY

For a process carried out in a system, it is important to know the quantity of available work. Also, if in the considered system there are chemical reactions, it is necessary to be able to evaluate how they affect the evolution of the system. Thus, a new state function (A) is defined, denominated Helmholtz energy free, according to the expression

$$A = U - TS \tag{6.46}$$

If a reversible process is considered, from the equation that defines the first principle of thermodynamics, and keeping in mind Equation 6.36, an expression that allows the calculation of maximum work that the system can develop can be obtained:

$$\delta W_{rev} = dU - T\,dS \tag{6.36}$$

If the process is also carried out in an isothermal way, the differential variation of the Helmholtz energy free will be expressed as

$$dA = dU - T\,dS \tag{6.47}$$

obtaining

$$\delta W_{rev} = dA \tag{6.48}$$

This equation represents the condition that should be accomplished in the reversible processes or reactions, and it indicates the condition for a system that is in equilibrium.

The Helmholtz free energy is a state function that gives a measure of the quantity of work that a system can exchange with its surroundings and if the process or reaction will be spontaneous. Thus, for an isothermal process at constant volume, if the variation of A is null ($dA = 0$), the process or reaction is in equilibrium. In the case that $A < 0$, the process or reaction is spontaneous; if $A > 0$, the process and the reaction are not spontaneous.

6.12 GIBBS FREE ENERGY

Many processes and reactions are carried out at constant pressure and temperature, and with the purpose of studying the approach to the equilibrium of these processes or reactions, a new state function that denominated Gibbs free energy is defined, which is mathematically expressed as

$$G = H - TS = U + PV - TS = A + PV \tag{6.49}$$

In processes and reactions at constant pressure and temperature, it is accomplished as

$$dG = dU + P\,dV - T\,dS \tag{6.50}$$

In a reversible process, the network is the difference between the work given by the system and the additional one, so that

$$\delta W_{rev} = -P\,dV + \delta W_{ad} \tag{6.51}$$

If one keeps in mind Equation 6.37, it is obtained as

$$\delta W_{ad} = dU - T\,dS + P\,dV \tag{6.52}$$

Comparing this last equation with (6.50), it is obtained as

$$\delta W_{ad} = dG \tag{6.53}$$

This equation represents an equilibrium condition in a system at constant pressure and temperature. So, if additional work is not carried out, it is obtained that the equilibrium condition is accomplished for $dG = 0$. The processes will be spontaneous if $dG < 0$, while if $dG > 0$, they will not be spontaneous.

Generally, from the Gibbs free energy definition (Equation 6.49), taken in their differential form,

$$dG = dU + P\,dV + V\,dP - T\,dS - S\,dT \tag{6.54}$$

If work is only due to the pressure, when substituting Equation 6.37 in (6.54), it is obtained as

$$dG = V\,dP - S\,dT \tag{6.55}$$

that expressed for the variation of G

$$\Delta G = V\Delta P - S\Delta T \tag{6.56}$$

This equation allows obtaining the Gibbs free energy variation when the processes and reactions are studied in food systems, as well as the influence of the pressure and temperature in relation to the variation of the system with regard to equilibrium.

From Equation 6.55 the processes and reactions can be analyzed depending on how they are carried out. If the process is carried out at constant temperature,

$$dG = V\,dP \tag{6.57}$$

In many food processes, there are cases in which an interaction exists between the food and the atmosphere that surrounds it, and it is important to be able to evaluate these interactions.

These interactions are important when the water vapor surrounding the food is considered, and processes of adsorption and desorption are already presented that affect the drying and humidification characteristics of the food.

If it is considered that the gases behave as ideals, the volume can be given as a function of the pressure and of the number n of moles of the system, so the previous equation can be expressed as

$$dG = \frac{nRT}{P} dP \tag{6.58}$$

The integration of this equation between conditions 1 and 2, with initial P_1 and final P_2 pressures, leads to the equation

$$\Delta G = G_2 - G_1 = nRT \ln\left(\frac{P_2}{P_1}\right) \tag{6.59}$$

Usually, the variation of a state function is referred to as a standard state. If it is considered that this standard state is for the pressure of 1 atm (P^0), the free energy at any other state can be expressed as

$$G = G^0 + nRT \ln\left(\frac{P}{P^0}\right) \tag{6.60}$$

where G^0 is denominated standard Gibbs free energy.

As has been previously indicated, many processes and reactions are carried out at a constant pressure, so that Equation 6.55 is simplified to

$$dG = -S\,dT \tag{6.61}$$

From the definition of Gibbs free energy (Equation 6.49), the entropy can be expressed as

$$S = \frac{H}{T} - \frac{G}{T} \tag{6.62}$$

In this way, for constant pressure,

$$S = -\left(\frac{\partial G}{\partial T}\right)_P = \frac{H}{T} - \frac{G}{T} \tag{6.63}$$

obtaining

$$T\left(\frac{\partial(G/T)}{\partial T}\right)_P = \left(\frac{\partial G}{\partial T}\right)_P - \frac{G}{T} = -\frac{H}{T} \tag{6.64}$$

or (equivalently)

$$\left(\frac{\partial(G/T)}{\partial T}\right)_P = -\frac{H}{T^2} \tag{6.65a}$$

denominated Gibbs–Helmholtz equation, which allows evaluation of the relationship between the variation of Gibbs free energy and the enthalpy variation:

$$\left(\frac{\partial(G/T)}{\partial T}\right)_P = -\frac{\Delta H}{T^2} \tag{6.65b}$$

6.13 CHEMICAL POTENTIAL: PHASE EQUILIBRIUM

In closed systems the only way to exchange energy with the surroundings is through heat or work, but the quantity of matter remains constant. However, in this type of system, whether such systems are homogeneous or heterogeneous, there may be chemical reactions or mass transfers inside one phase or between different phases of the system, which means that the concentration of different chemical species in each phase might not be constant. Thus, it is necessary to consider the composition of each component in each phase as the state variable that allows description of states of equilibrium of the system. To express the composition of the system, the number of moles of each chemical species in each of the phases (n_i^α) is used, where i refers to the chemical species and α to the phase.

The different state functions will be a function of two fundamental variables plus the number of chemical species present in the system. One of these representative state functions of the system is the internal energy, although they could also take the enthalpy, entropy, and Gibbs and Helmholtz free energy. Thus, if it is considered the internal energy,

$$U = \sum_{\alpha=1}^{f} U^\alpha \tag{6.66}$$

where U^α is the internal energy in the phase α, and it will be function of the entropy, volume, and n_i compositions in the considered phase:

$$U^\alpha = U\left(S^\alpha, V^\alpha, n_1^\alpha, n_2^\alpha, \ldots, n_c^\alpha\right) \tag{6.67}$$

From these equations the evolution of the system can be studied; if it is considered a homogeneous phase, there is an only phase, and the exponent of the equations will be neglected, although if different phases exist, for each phase the equations that are obtained next will be applied.

For homogeneous systems of variable composition,

$$dU = \left(\frac{\partial U}{\partial S}\right)_{V,n_i} dS + \left(\frac{\partial U}{\partial V}\right)_{S,n_i} dV + \sum_i \left(\frac{\partial U}{\partial n_i}\right)_{S,V,n_{j\neq i}} dn_i \tag{6.68}$$

In this equation an intensive variable denominated chemical potential for the i species appears, and mathematically it is defined as

$$\mu_i = \left(\frac{\partial U}{\partial n_i}\right)_{S,V,n_{j\neq i}} \tag{6.69}$$

Equation 6.68 can be expressed as

$$dU = T\,dS - P\,dV + \sum_i \mu_i dn_i \tag{6.70}$$

The sum of this last equation represents a new work type that is given the name of *generalized chemistry work*, which is a new form of varying the internal energy of the system.

If instead of using the internal energy, it uses the Gibbs free energy; it can be considered that it is a function of temperature and pressure, as well as of composition, so if there are considered different phases, it is obtained that the total value of this state function is

$$G = \sum_\alpha G^\alpha = \sum_\alpha G\left(T^\alpha, P^\alpha, n_1^\alpha, n_2^\alpha, \ldots, n_c^\alpha\right) \tag{6.71}$$

Carrying out a mathematical treatment similar to the one used for the internal energy, it is obtained as

$$dG = \sum_\alpha \left[\left(\frac{\partial G^\alpha}{\partial T}\right)_{P,n_i} dT + \left(\frac{\partial G^\alpha}{\partial P}\right)_{T,n_i} dP + \sum_i \left(\frac{\partial G^\alpha}{\partial n_i}\right)_{P,T,n_{j\neq i}} dn_i^\alpha\right] \tag{6.72}$$

This equation can be expressed as

$$dG = S\,dT - V\,dP + \sum_\alpha \sum_i \mu_i^\alpha dn_i^\alpha \tag{6.73}$$

For closed systems in thermal and mechanical equilibrium, and because in the equilibrium the variation of the Gibbs free energy is null, it is obtained as

$$\sum_\alpha \sum_i \mu_i^\alpha dn_i^\alpha = 0 \tag{6.74}$$

That is the condition of material equilibrium for closed systems of variable composition in thermal and mechanical equilibrium, and the system should comply when it arrives to the equilibrium state as a consequence of chemical reactions and/or mass transfers between phases of the system. In heterogeneous systems in which only mass transfer between the different phases and expansion work exist, the final state is an equilibrium among phases, which implies that if the system is closed, variation of the mass of the system does not exist, and it is obtained that the equilibrium condition among phases is accomplished if the chemical potential of any one of the components among the different phase is the same in all them:

$$\mu_i^{\alpha_1} = \mu_i^{\alpha_2} = \cdots = \mu_i^{\alpha_f} \tag{6.75}$$

In the case that the chemical potential of one of the components of the system is not the same as in all phases, the system is not in equilibrium. In this case there will be mass transfer between the different phases of the system until the chemical potential becomes the same in all of them.

In systems, whether homogeneous or heterogeneous, in which k chemical reactions and i components exist, the concept of chemical equilibrium should be used, and it should be accomplished that

$$\sum_i \upsilon_{i,k}\mu_i = 0 \tag{6.76}$$

In any thermodynamic process, the variations of the different intensive magnitudes of a system cannot be considered independent from each other. For a system composed of several components, the equation of Gibbs–Duhem should be accomplished:

$$S\,dT - V\,dP + \sum_i n_i d\mu_i = 0 \tag{6.77}$$

In this equation, it can be seen that in a certain process the temperature, the pressure, and the chemical potential cannot be arbitrarily changed at the same time in each and every component of the system.

For a multicomponent system with a certain composition, from Equation 6.77, it is obtained that the chemical potential of any component is expressed as

$$d\mu_i = \overline{V}_i\,dP - \overline{S}_i\,dT \tag{6.78}$$

where $\overline{V}_i$ and $\overline{S}_i$ are the molar volume and molar entropy of the component *i*.

If the temperature is constant, its differential is null, and it is obtained as

$$\left(\frac{\partial \mu_i}{\partial P}\right)_T = \overline{V}_i \tag{6.79}$$

For an ideal gas it is accomplished that $P_i\overline{V}_i = RT$; the chemical potential can be expressed as

$$d\mu_i = \frac{RT}{P_i}dP = RTd\ln P_i \tag{6.80}$$

For a nonideal gas, this equation is not valid; however, a function f can be defined, denominated *fugacity*, so that it is accomplished as

$$d\mu_i = RTd\ln f_i \tag{6.81}$$

Integrating this equation between two states allows us to obtain the variation of chemical potential:

$$\Delta\mu = \mu_2 - \mu_1 = RT\ln\left(\frac{f_2}{f_1}\right) \tag{6.82}$$

With the purpose of expressing the fugacity in any state, it is necessary to define a reference state where the fugacity acquires a specific value. For the standard conditions, this value is represented by f^0, so the variation of potential for the component *i* will be expressed as

$$\Delta\mu_i = \mu_i - \mu_i^0 = RT\ln\left(\frac{f_i}{f^0}\right) \tag{6.83}$$

Generally, the reference fugacity usually takes the value of the vapor pressure of the pure component under vacuum conditions or at 1 atm. The reference value is denominated *standard chemical potential*.

For an ideal gas the fugacity is proportional to the pressure, and this constant of proportionality generally takes the unit value, with which the pressure coincides with the fugacity. For a real gas the constant of proportionality is different from the unit, and the fugacity does not coincide with the pressure; however, when the pressure is very low, the real gas tends to behave as ideal, the relationship between the fugacity and the pressure of the component *i* tends toward the unit value:

$$\lim_{P \to 0} \left(\frac{f_i}{P_i} \right) = 1 \tag{6.84}$$

The relationship between the fugacity of the component *i* and the value at standard conditions is denominated *activity* (a_i) of this component:

$$a_i = \frac{f_i}{f_i^0} \tag{6.85}$$

Thus, when substituting this expression in Equation 6.83, the chemical potential of a component *i* is obtained according to the expression

$$\mu_i = \mu_i^0 + RT \ln a_i \tag{6.86}$$

For ideal gases or real gases at very low pressures, the activity is the relationship between the pressure of the component *i* in the mixture and the vapor pressure of the pure component.

For solids and liquids at a certain temperature, it is likely that they will have a vapor pressure, although their value can be very low; however, pressure will always exist in which a solid or liquid is in equilibrium with its vapor. When being in equilibrium, the chemical potential of both phases is the same, and if the same reference state is taken, the fugacity of the solid (or liquid) will be equal to that of the vapor with the one that is in equilibrium. If the vapor pressure is not very high, the fugacity coincides with the vapor pressure, and the fugacity of the solids or liquids is usually taken as this vapor pressure.

In the case of a liquid solution that is in equilibrium with its vapor phase, the chemical potentials for any *i* component will be the same:

$$\mu_i^L = \mu_i^V$$

where μ_i^L and μ_i^V represent the chemical potentials of the component *i* in liquid and vapor phases, respectively. If the vapor also presents an ideal gas behavior, it will be accomplished that

$$\mu_i^L = \mu_i^G = \mu_i^0 + RT \ln P_i \tag{6.87}$$

where P_i is the partial pressure of the component *i* in the vapor phase. If the behavior is ideal, Raoult's law is also accomplished:

$$P_i = x_i P_i^0 \tag{6.88}$$

where the partial pressure in the vapor phase is proportional to the vapor pressure of the pure component (P_i^0), being the constant of proportionality the molar fraction of the component in liquid phase (x_i).

Combining Equations 6.87 and 6.88, the chemical potential of the component *i* in liquid phase will be

$$\mu_i^L = \mu_i^* + RT \ln x_i \tag{6.89}$$

in the one that is the chemical potential of the pure component in liquid phase, and it is equal to

$$\mu_i^* = \mu_i^0 + RT \ln x_i \tag{6.90}$$

For nonideal solutions, whether gas or liquid, the chemical potential should be calculated from Equation 6.86. Because the behavior is not ideal, an activity coefficient is usually used (γ), so that the activity will be expressed by the equation

$$a_i = \gamma_i x_i \tag{6.91}$$

and substituting in Equation 6.86, it is obtained as

$$\mu_i = \mu_i^0 + RT \ln\left(\gamma_i x_i\right) \tag{6.92}$$

For liquid solutions, the quantity of the solvent of the solution is high in comparison with the solute, and its activity coefficient is practically the unit.

In food systems, one of the most usual components is water, and in the same way that the activity for any one component *I* has been defined, the water activity (a_w) is also defined in a similar way to the definition given in Equation 6.85, as the relationship between fugacity. This is a parameter that is often used in different applications in food engineering. As previously noted, the fugacity is proportional to the pressure, and if the proportionality constant is the unit, the pressure coincides with the fugacity. Thus, the water activity will be expressed as

$$a_w = \frac{P_w}{P_w^0} \tag{6.93}$$

where

P_w is the vapor pressure of the water in equilibrium with the food
P_w^0 is the vapor pressure of pure water to the same temperature of the food

The relative humidity (ϕ) that possesses the air in contact with the food is defined as

$$\phi = \frac{P_w}{P_w^0} \times 100 \tag{6.94}$$

Comparing Equations 6.93 and 6.94, it is observed that the relative humidity is a 100 times the activity of the water. Measuring relative humidity is simple to perform; therefore, the determination of the water activity of foods is also easy.

6.14 PHASE DIAGRAM

A *phase diagram* is the representation of the equilibrium states of a thermodynamic system, classified in accordance with their physical state. In the description of these diagrams, it is necessary to consider the Gibbs phase rule that the sum of the number of present phases in a system plus the freedom degrees equals to the number of components plus two.

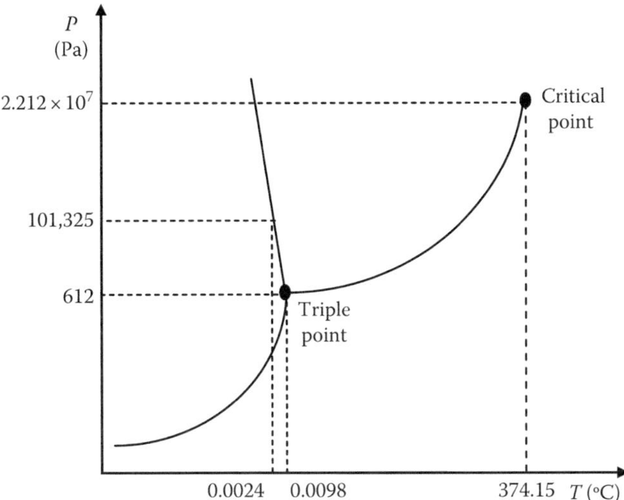

FIGURE 6.3 Phase diagram for water.

For a system with one component, if the equilibrium between two phases is studied, it is obtained that there is only one freedom degree. In this case the phase diagrams are 2D representations, which are built using two intensive state variables, generally temperature and pressure. In this type of diagram, consisting of a single component and two phases, the coexistence of two phases in equilibrium is represented by a line, called the *line of phase change*. In this diagram type the point in which three phases coexist is called the *triple point*, and it is a situation in which freedom degrees do not exist. Another characteristic point of the phase diagram is the *critical point*, where the properties of the two phases in equilibrium tend to coincide when modifying some variables of the system. Figure 6.3 shows a phase diagram for water, where it can be observed that the triple point appears for a temperature of 0.01°C and a pressure of 612 Pa. In addition, a critical point appears in the line of liquid–vapor change of phase for a temperature of 374.15°C and pressure of 2.212×10^7 Pa.

For the transition of two phases in equilibrium, the variation of Gibbs free energy is null, and the phase change is also carried out at constant temperature, and if it is applied in Equation 6.49,

$$\Delta G = \Delta H - T\Delta S = 0$$

So, the entropy variation for a phase change can be expressed as

$$\Delta S = \frac{\Delta H}{T} \tag{6.95}$$

where

ΔH represents the enthalpy of phase change (fusion or vaporization)
T is the temperature to which this phase change takes place

The equilibrium condition between two phases of a pure substance, such as water, at temperature and pressure constants is established according to Equation 6.75. If the equation of Gibbs–Duhem is used (Equation 6.77) for equilibrium in an infinitesimal process on the line of phase change from α_1 to α_2, it is accomplished that

$$V^{\alpha_1} dP - S^{\alpha_1} dT = V^{\alpha_2} dP - S^{\alpha_2} dT \tag{6.96}$$

Reordering this equation appropriately, it is obtained as

$$\frac{dP}{dT} = \frac{S^{\alpha_2} - S^{\alpha_1}}{V^{\alpha_2} - V^{\alpha_1}} = \frac{\Delta S}{\Delta V} \tag{6.97}$$

If Equation 6.95 is recalled, the following equation is obtained:

$$\frac{dP}{dT} = \frac{\Delta H}{T\Delta V} \tag{6.98}$$

The denominated *Clapeyron's equation* is applied to the equilibrium between two phases for fusion and boiling processes.

For two phases, one condensed and one vapor, it can be considered that the volume occupied by the vapor is much greater than that which is occupied by the condensed phase, and the increment of volume practically coincides with the volume of the vapor phase:

$$\Delta V = V^v - V^c \approx V^v$$

where the superscripts refer to the vapor phase (v) and the condensed phase (c).

Furthermore, if it is considered that the vapor in equilibrium behaves as an ideal gas, the volume is a function of the pressure, according to the equation

$$V^v = \frac{RT}{P}$$

Substituting in Clapeyron's equation,

$$\frac{1}{P}\frac{dP}{dT} = \frac{\Delta H_v}{RT^2} \tag{6.99a}$$

or

$$\frac{d\ln P}{dT} = \frac{\Delta H_v}{RT^2} \tag{6.99b}$$

which is known as Clausius–Clapeyron's equation, and it correlates the vapor pressure of a liquid with temperature and vaporization enthalpy of vaporization per mole.

7 Mass Transfer

7.1 INTRODUCTION

In mass transfer operations between different points of a system, transport of the components takes place by means of molecular or convection mechanisms, which implies a displacement of the molecules of the components from the points with higher concentration to those of lower concentration. Globally, the transfer is carried out by concentration gradients; thus, when concentration equilibrium is reached, the mass transfer stops. For this reason, in order to improve transfer performance, it is important to work far from the equilibrium conditions.

An example that allows easy visualization of the mechanisms of mass transfer is illustrated by the case in which a colored drop is introduced inside a receptacle that contains water. In the surroundings of the drop, there is a high concentration of dye that is transferred into the water bulk in all directions. If allowed to remain long enough in the whole container, the color will diffuse, reaching to a situation where the dye concentration is the same in each point of the container, thus reaching a state of equilibrium. In this case the mass transfer has been carried out by means of a molecular mechanism of diffusion. Also, once the water drop is placed in the receptacle and it is agitated by means of an agitator, it is observed that the equilibrium state is rapidly achieved by means of a convection mechanism.

The transfer of a component in the bulk fluid at rest or that in a laminar regime in the perpendicular direction to that of the concentration gradient of the component is carried out randomly due to the motion of the molecules, which constitutes the mechanism of molecular diffusion. In the case that the transfer occurs in a fluid that flows in a turbulent regime, in addition to the molecular diffusion, displacement of the component associated with the global movement of the fluid appears. This motion can be due to differences of density that in turn are due to concentration or temperature changes; this constitutes the mechanism of natural convection. Also, if the motion of the fluid is due to external forces of agitation or pumping, the mechanism is a forced convection.

A characteristic variable in mass transfer is concentration, and it is important to define the different forms of expressing the concentration of a certain component. Concentration can be expressed in mass or molar units. In the case of mass units, the concentration is given by the density (ρ), which is the mass of the component considered by unit of volume, its units being expressed in kg/m^3. Also, it can be expressed as mass fraction (x_w), which is defined as the mass of the component per unit of total mass. In molar units, concentration is expressed as the moles of the component considered by unit of volume, being expressed in mol/m^3. Also, the molar fraction (x) can be defined as the relationship between the number of moles of the component and the number of total moles.

There are occasions in which the diffusion of a component is carried out through a simple phase, although cases are usually presented in which this transfer is given among different phases. For monophase systems in which there is a difference of concentration of a given component, this component will be transferred from the points of greater concentration to those of smaller concentration until the concentration is equal in all points of the considered system. In the case of two-phase systems, the transfer of the component is carried out from one phase to another through an interface. The component should overcome three resistances to the transfer, one in the bulk of the initial phase, the following one through the interface, then to carry out the transfer from the interface to the bulk of the second phase. Generally, the resistance of the interface can be neglected, assuming that the equilibrium is reached instantaneously.

The component transfer from one phase to another concludes when the chemical potentials of the considered component are equal in both phases.

In different processes of mass transfer, interfacial surfaces can be presented with liquids, gases, and solids. However, the interfacial surface solid–solid is neglected, since the mass transfer is extremely slow; the gas–gas interface is also not considered because the phases are usually completely miscible. Therefore, in systems with two phases, the mass transfer usually occurs when gas–liquid interfaces, liquid–liquid interfaces, liquid–solid interfaces, and gas–solid interfaces are presented. There are different separation operations in which the mass is transferred through the mentioned interfaces.

Thus, in the absorption of the components of a gas mixture by means of a liquid, mass transfer is carried out through a gas–liquid interface. Liquid–solid interfaces are presented in operations of adsorption of components contained in liquid mixtures using a solid adsorbent, or in cases of desorption, in which components retained in a solid surface are separated using an appropriate solvent; also, this interface type is found in solid–liquid extractions, where a solute contained in the solid is transferred toward a solvent in contact with the solid, as in the case of extraction of vegetable oils with organic solvent; it is also present in processes of osmotic dehydration, where the water contained in a solid is transferred toward a concentrated solution. Gas–solid interfaces are found in such operations as the adsorption of contained components in a gas mixture using a solid adsorbent, such as the case of purification or drying of gases by means of active coal or molecular resins; it is also in the processes of conventional drying of solid using an airstream.

7.2 MASS TRANSFER BY DIFFUSION

The mass transfer rate by diffusion is first determined by Fick's law, whose expression depends on the reference system and of the used units (Bird et al., 1960). For a component A that is diffusing through a binary mixture A–B, the molar flux relative to an average speed can be expressed as

$$J_A = -cD_{AB}\nabla x_A \tag{7.1a}$$

indicating that the molar flux is proportional to the concentration gradient, where D_{AB} is the proportionality constant denominated diffusivity, whose units in the IS are m^2/s, c being the global molar concentration, and x the molar fraction of the considered component. In the case that the flux was expressed in mass units, the molar concentration should be changed by mass concentration, that is to say, for the density, obtaining the equation

$$J_A = -\rho D_{AB}\nabla x_A^w \tag{7.1b}$$

where

J_A is the mass flux
ρ is the mass density
x_A^w is the mass fraction for the component A

The flux can also be referred to as stationary coordinates: If N_A is designated as the flux referring to stationary coordinates, it is related to the relative flux and an average speed according to the expression

$$J_A = N_A - x_A\sum_{j=1}^{n} N_j \tag{7.2}$$

This indicates that the flux by diffusion (J_A) is the difference between the flux N_A and the global flux of the component A due to the local flux of the mixture.

The molecular diffusion through the solids usually presents difficulties and the transfer rate is low, although it can be facilitated if there is an appropriate concentration gradient at high temperatures. The values of the diffusivities present values of the order of 10^{-12} and 10^{-14} m²/s. The diffusion through liquids is greater, presenting values of diffusivity from 10^{-9} to 10^{-10} m²/s, and they are functions of the composition. For gases, the values of the diffusivity present values from 10^{-5} to 10^{-4} m²/s, and they are usually independent of the concentration.

The diffusivity through liquid solutions is difficult to calculate. However, it could be evaluated by means of the Wilke and Chang correlation:

$$D_S = 7.4\times10^{-8}\frac{T(\psi M)^{1/2}}{\eta\left(\bar{V}_m\right)^{0.6}} \tag{7.3}$$

where

$\bar{V}_m$ is the molar volume of solute at boiling temperature (cm³/mol)
η is the viscosity of solution (mPa s)
ψ is the solvent association parameter (for water 2.6)
T is the absolute temperature

Another form of evaluating the solute diffusivity would be from the equation

$$D_S = 7.7\times10^{-16}\frac{T}{\eta\left(\bar{V}^{1/3} - \bar{V}_0^{1/3}\right)} \tag{7.4}$$

where

the diffusivity D_S is expressed in m²/s
the viscosity η in Pa s
the temperature T in K

For solutes diffusion in water the value of $\bar{V}_0$ is 0.008. The variable $\bar{V}$ is the molecular volume that is defined as the volume in m³ of 1 kmol of substance in liquid form to its boiling temperature. Table 7.1 shows the molecular volume values for different substances.

TABLE 7.1
Molecular Volume Values

Substance	$\bar{V}$(m³/kmol)
Carbon	0.0148
Hydrogen	0.0037
Oxygen	
Double bond	0.0074
In aldehydes and ketones	0.0074
In methyl esters	0.0091
In ethylic esters	0.0099
In ethers and others esters	0.0110
In acids	0.0120
Bounded at S, P, N	0.0083

Source: Perry, R.H. and Chilton, C.H., *Chemical Engineer's Handbook*, McGraw-Hill, New York, 1973.

In many processes the diffusion of a gas component through a porous solid is presented. In this case, the effective diffusivity is much smaller than the one corresponding to diffusion in a free fluid, since the way that the component should travel is through a more tortuous channel, and the distance that the molecules should travel will increase; also, the solid restricts the pass transversal area. If the size of the pores of the solid is within the order of the half path of the diffusing molecules, the denominated Knudsen's diffusion appears, in which the molecules collide preferably with the walls of the pore instead of the other molecules. For a component A, the Knudsen's diffusivity (D_{KA}) can be obtained from the equation

$$D_{KA} = \frac{d_p}{3}\sqrt{\frac{8RT}{\pi M_A}} \tag{7.5}$$

where

d_p is the pore diameter
M_A is the molecular weight for the considered component

7.2.1 1D Diffusion in Binary Mixtures

For a binary mixture formed by two components A and B, two cases can be presented: a diffusion of A through B at the same time that B diffuses through A, which is referred to as counterdiffusion, or A diffuses through a B stagnant film.

7.2.1.1 Diffusion through Gases

7.2.1.1.1 Counterdiffusion

In this case the mass flux with which one of the considered species diffuses can be obtained by integrating Fick's equation, applying the adequate boundary conditions.

If the counterdiffusion is in the bulk gases, it can be considered that they behave as perfect gases, and then the concentration of a component will be expressed as

$$c = \frac{n}{V} = \frac{p}{RT} \tag{7.6a}$$

where

c is the molar concentration
n is the mole
V is the volume
p is the pressure
T is the absolute temperature
R is the gas constant

For mass concentration, the density is expressed according to equation

$$\rho = \frac{m}{V} = \frac{pM}{RT} \tag{7.6b}$$

where

ρ is the mass concentration
m is the mass content
M is the molecular weight

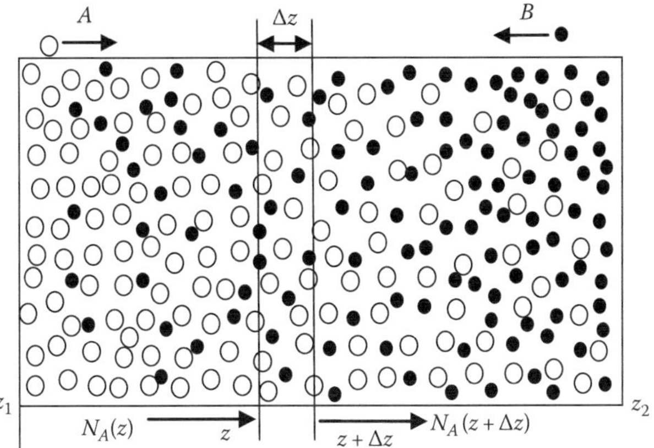

FIGURE 7.1 Transfer of two components in counterdiffusion.

For a binary mixture, the sum of flux relative to the average speed is zero, defined as

$$N_A = -N_B \tag{7.7}$$

From Equation 7.2 it can be obtained as $J_A = N_A$.

In Figure 7.1, there is a position z_1 where the A concentration is greater than the concentration in position z_2, and at the same time, the composition of B is greater in this last position than in z_1, and there will occur a transfer of A from z_1 toward z_2, and a transfer of B from z_2 to z_1. This is the case in which component A is transferred in counterdiffusion with component B. If a balance of A along a thickness film Δz is carried out, the A flow rate that enters is the same as the one that leaves this thickness, and then

$$N_A(z)\cdot S - N_A(z+\Delta z)\cdot S = 0$$

where S is the cross-sectional area. If S is eliminated, and it is divided by Δz, when taking limits for Δz tending to zero, this being the derivative definition, then

$$-\frac{dN_A}{dz} = 0$$

This implies that N_A = constant.

Considering 1D diffusion in the z direction, when substituting in Fick's equation in molar units, it is obtained that

$$N_A = J_A = -D_{AB}\frac{dc_A}{dz} = -D_{AB}\frac{d}{dz}\left(\frac{p_A}{RT}\right) \tag{7.8}$$

Keeping in mind that N_A is constant, this equation can be integrated with the boundary conditions:

For $z = z_1$ $\quad p_A = p_{A1}$

For $z = z_2$ $\quad p_A = p_{A2}$

Obtaining that the transfer flow rate for the component A through the section S can be expressed as

$$N_A \cdot S = \frac{SD_{AB}}{RT}\frac{(p_{A1}-p_{A2})}{(z_2-z_1)} \tag{7.9a}$$

or

$$N_A \cdot S = \frac{(p_{A1}-p_{A2})}{(RT(z_2-z_1)/D_{AB}S)} \tag{7.9b}$$

where the numerator represents the driving force for the mass transfer, while denominator is the resistance to this transfer. For component B the same equations are obtained, but the subscripts are changed from A to B.

7.2.1.1.2 Diffusion through a Stagnant Gas Film

Considering a system (Figure 7.2) in which a component A evaporates from a liquid and it is transferred through a B stagnant film, from a position z_1 to another z_2, for this binary system, Equation 7.2 can be applied, although it should be considered that J_A can be substituted for Equation 7.1:

$$-cD_{AB}\frac{dx_A}{dz} = N_A - x_A(N_A+N_B) \tag{7.10}$$

As the gas film is stagnated, $N_B = 0$. Rearranging the previous equation, it is obtained that

$$N_A = -\frac{cD_{AB}}{1-x_A}\frac{dx_A}{dz} \tag{7.11}$$

If a balance for the A component is carried out in a similar way to the previous case, then

$$-\frac{dN_A}{dz} = 0$$

Substituting the value of N_A given by Equation 7.11,

$$\frac{d}{dz}\left(\frac{cD_{AB}}{1-x_A}\frac{dx_A}{dz}\right) = 0 \tag{7.12}$$

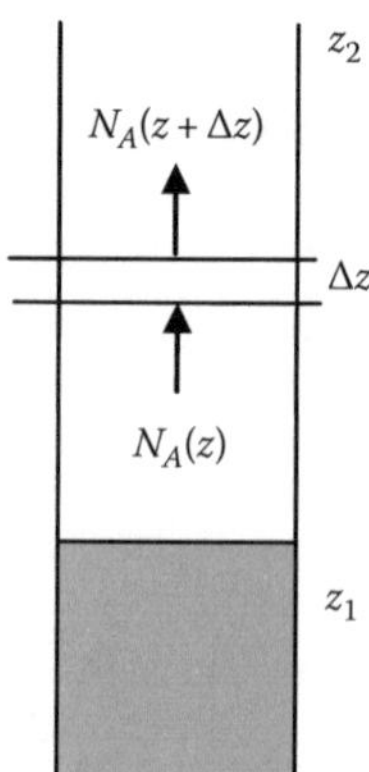

FIGURE 7.2 Transfer of component A through a stagnant film.

Assuming that the mixture consists of ideal gases, and is at constant pressure and temperature, the concentration c has a diffusivity that is considered constant. This equation can be integrated with the boundary conditions:

$$\text{For} \quad z = z_1 \qquad x_A = x_{A1}; \quad x_B = x_{B1}$$
$$\text{For} \quad z = z_2 \qquad x_A = x_{A2}; \quad x_B = x_{B2}$$

The integration leads to an expression that allows us to calculate the flow rate for A transferred:

$$N_A \cdot S = \frac{cD_{AB}S}{RT} \frac{(x_{A1} - x_{A2})}{(z_2 - z_1)(x_B)_{ml}} = \frac{(x_{A1} - x_{A2})}{(RT(z_2 - z_1)(x_B)_{ml}/cD_{AB}S)} \tag{7.13a}$$

where $(x_B)_{ml}$ is the molar fraction mean logarithmic of the component B between positions z_1 and z_2:

$$(x_B)_{ml} = \frac{x_{B2} - x_{B1}}{\ln(x_{B2}/x_{B1})} \tag{7.14}$$

This same equation can be expressed as a function of the total pressure (P) and of the partial pressures of the components:

$$N_A \cdot S = \frac{PD_{AB}S}{RT} \frac{(p_{A1} - p_{A2})}{(z_2 - z_1)(p_B)_{ml}} = \frac{(p_{A1} - p_{A2})}{(RT(z_2 - z_1)(p_B)_{ml}/PD_{AB}S)} \tag{7.13b}$$

where $(p_B)_{ml}$ is the mean logarithmic of the partial pressures of the component B, defined in a similar way to Equation 7.14.

7.2.1.2 Diffusion through Liquids

In this case the mathematical treatment is similar to that for gases, although they can only be used like driving forces in the increments of molar fraction molar for the studied component.

7.2.1.2.1 Counterdiffusion

In this case the flux of the components A and B is the same but of opposite sign. Carrying out a mathematical treatment similar to that of counterdiffusion of gases is obtained in which the molar flow rate for the component A through the cross section S is

$$N_A \cdot S = cD_{AB} \frac{(x_{A2} - x_{A1})}{(z_2 - z_1)} = \frac{(x_{A2} - x_{A1})}{((z_2 - z_1)/cD_{AB}S)} \tag{7.15}$$

where the numerator is the increment of molar fraction that experiences the component A, and it represents the driving force for the mass transfer, while the denominator is the resistance to this transfer; c is the mean global molar concentration. For component B a similar expression is obtained, changing the subscripts B to A.

7.2.1.2.2 Diffusion through a Stagnant Liquid Film

In the case of diffusion of component A through a liquid film containing another component B, the equation that is obtained is similar to that of diffusion through a stagnant gas film, in such a way that the A molar flow rate transferred can be expressed as the product of the flux by the cross section:

$$N_A \cdot S = cD_{AB} \frac{(x_{A1} - x_{A2})}{(z_2 - z_1)(x_B)_{ml}} = \frac{(x_{A1} - x_{A2})}{((z_2 - z_1)(x_B)_{ml}/cD_{AB}S)} \tag{7.16}$$

7.2.2 1D Diffusion of a Component through Solids

The diffusion of a gas component through a solid depends on the geometry of the system. Therefore, three cases of diffusion will be considered: through a plane layer, a cylindrical layer, and a spherical layer.

Figure 7.3 shows a porous plane slab, in which component A diffuses through a cross-sectional area S, with a wall thickness Δz, the volume being $V = S\Delta z$.

Carrying out a component balance for A, assuming that there are no chemical reactions, and that the diffusion is 1D, it is obtained that the accumulation term is equal to the net entrance, and the entrance and exit terms follow Fick's law. When the balance is applied to the volume represented in Figure 7.2, it is obtained as

$$\frac{d}{dt}(C_A V) = N_A(z)\cdot S - N_A(z+\Delta z)\cdot S$$

or similarly as

$$S\Delta z\frac{dC_A}{dt} = N_A(z)\cdot S - N_A(z+\Delta z)\cdot S$$

Dividing this expression by $S\Delta z$, and taking the limit for Δz tending to zero, and keeping in mind that the flux is expressed according to Fick's equation, it is obtained as

$$\frac{dC_A}{dt} = D_A\frac{d^2C_A}{dz^2}$$

This equation can be generalized, considering that the diffusion is carried out in the three directions of Cartesian coordinates. Thus, it is obtained as

$$\frac{dC_A}{dt} = D_A\nabla^2 C_A \qquad (7.17a)$$

where the Laplace operator ∇^2 is expressed as $\nabla^2 = (\partial^2/\partial x^2) + (\partial^2/\partial y^2) + (\partial^2/\partial z^2)$.

This equation can be expressed in cylindrical and spherical coordinates, in such a way that their expressions are

Cylindrical coordinates

$$\frac{dC_A}{dt} = D_A\left[\frac{1}{r}\frac{\partial}{\partial r}\left(r\frac{\partial C_A}{\partial r}\right)+\frac{1}{r^2}\frac{\partial^2 C_A}{\partial \phi^2}+\frac{\partial^2 C_A}{\partial z^2}\right] \qquad (7.17b)$$

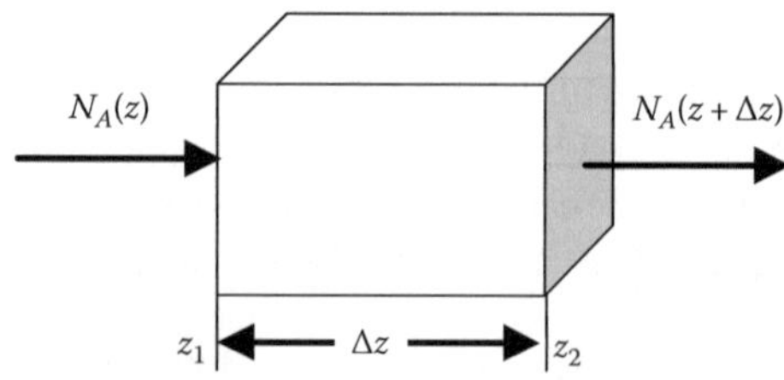

FIGURE 7.3 Diffusion through a plane slab.

Spherical coordinates

$$\frac{\partial C_A}{\partial t} =$$

$$= D_A \left[\frac{1}{r^2} \frac{\partial}{\partial r} \left(r^2 \frac{\partial C_A}{\partial r} \right) + \frac{1}{r^2 sin\theta} \frac{\partial}{\partial \theta} \left(sin\theta \frac{\partial C_A}{\partial \theta} \right) + \frac{1}{r^2 sin^2\theta} \frac{\partial^2 C_A}{\partial \phi^2} \right] \quad (7.17c)$$

7.2.2.1 Plane Slab

Considering a plane layer of thickness e and, if the mass transfer is carried out in a stationary state and in a single direction, from Equation 7.17a it is obtained as

$$\frac{\partial^2 C_A}{\partial x^2} = 0$$

which can be integrated with the boundary conditions:

$$\text{For} \quad z_1 = 0 \quad C_A = C_{A1}$$
$$\text{For} \quad z_2 = e \quad C_A = C_{A2}$$

The integration allows obtaining the concentration profile along the plane layer:

$$C_A = C_{A1} - \frac{(C_{A1} - C_{A2})}{e} z \quad (7.18)$$

A linear variation of the concentration can be observed along the porous solid.

The molar flow rate for the component A through the constant section S will be

$$N_A \cdot S = -SD_A \frac{dC_A}{dz} = SD_A \frac{C_{A1} - C_{A2}}{e} = \frac{C_{A1} - C_{A2}}{(e/SD_A)} \quad (7.19)$$

An important case of gas diffusion through a plane sheet is the transfer through films covering foods. Keeping in mind that for a gas the concentration of a component can be correlated by means of Henry's constant (H):

$$p_A = HC_A \quad (7.20)$$

where Henry's constant, in the International System, has units of Pa m^3/mol. The inverse of this constant represents the *solubility* (s) of a gas component in the solid, so the relationship between the concentration and the solubility is expressed according to the equation

$$C_A = sp_A \quad (7.21)$$

Substituting in Equation 7.19, it is obtained that

$$N_A \cdot S = SD_A s(p_{A1} - p_{A_2})/e \quad (7.22)$$

Many times, instead of using diffusivities, in the gas transfer through solids, the *permeability* of the gas through the solute is defined as the product of the diffusion coefficient by the solubility:

$$P_M = D_A s \tag{7.23}$$

Equation 7.22 can be expressed as

$$N_A \cdot S = SP_M \frac{p_{A1} - p_{A2}}{e} \tag{7.24}$$

This permeability coefficient can be expressed in different ways, with different units. A usual format is to define the permeability as the m^3 of A gas component, measured at 0°C and at 1 atm that diffuses per second through a m^2 of cross section of solid of 1 m of thickness under a difference of pressure of 1 atm. In this way, the units of the permeability coefficient are m^3/s m^2 (atm m).

This process of component transfer has great importance in the preservation of foods, since, for example, to avoid oxidation an appropriate container should be used that avoids oxygen diffusion through the container walls. Also, other gases that should be considered are carbon dioxide, nitrogen, and water vapor whose diffusion through the wall of the container can affect food quality.

7.2.2.2 Cylindrical Layer

Consider a cylindrical layer like the one shown in Figure 7.4, where the component A is diffusing through this layer.

For diffusion in a stationary state, and in the radial direction, from Equation 7.17b, it is obtained as

$$D_A \left[\frac{1}{r} \frac{\partial}{\partial r} \left(r \frac{\partial C_A}{\partial r} \right) \right] = 0$$

and then

$$r \frac{\partial C_A}{\partial r} = \text{constant}$$

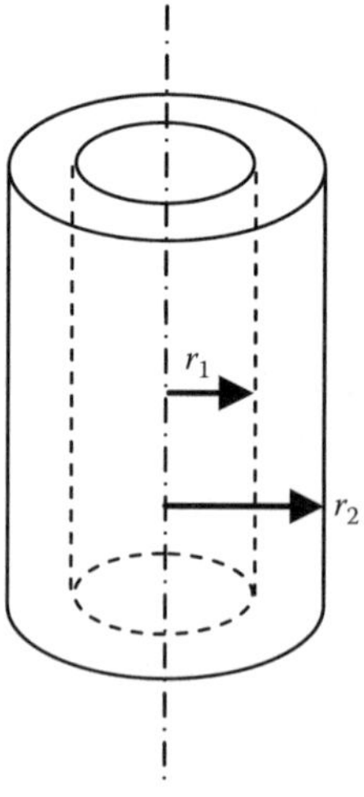

FIGURE 7.4 Diffusion through a cylindrical layer.

This equation can be integrated with the boundary conditions:

$$\text{For} \quad r = r_1 \qquad C_A = C_{A1}$$
$$\text{For} \quad r = r_2 \qquad C_A = C_{A2}$$

Obtaining the concentration profile of the component A as a function of the radius of the cylindrical layer, the equation can be expressed as

$$C_A = C_{A1} - (C_{A1} - C_{A2}) \frac{\ln(r/r_1)}{\ln(r_2/r_1)} \tag{7.25}$$

The molar flow rate of the component A through the constant cross section S, for a cylinder of height L, will be

$$N_A \cdot S = -2\pi r L D_A \frac{dC_A}{dr} = D_A S_{ml} \frac{C_{A1} - C_{A2}}{e} \tag{7.26}$$

where
- e is the thickness of the cylindrical layer
- S_{ml} is the mean logarithmic section between the external and internal sections, defined according to the equations

$$e = r_2 - r_1$$

$$S_{ml} = \frac{S_2 - S_1}{\ln(S_2/S_1)} = \frac{2\pi L(r_2 - r_1)}{\ln(r_2/r_1)} \tag{7.27}$$

7.2.2.3 Spherical Layer

Carrying out a similar reasoning for the plane and cylindrical layers, for diffusion in a stationary state and in a radial direction, from Equation 7.17c, it is obtained as

$$\frac{1}{r^2} \frac{\partial}{\partial r}\left(r^2 \frac{\partial C_A}{\partial r} \right) = 0$$

integrating with the boundary conditions:

$$\text{For} \quad r = r_1 \qquad C_A = C_{A1}$$
$$\text{For} \quad r = r_2 \qquad C_A = C_{A2}$$

A concentration profile along the spherical layer is obtained, according to the expression

$$C_A = C_{A1} - (C_{A1} - C_{A2}) \frac{r_2}{r_2 - r_1} \left(1 - \frac{r_1}{r} \right) \tag{7.28}$$

In this case, the component A flow rate that crosses the spherical layer of thickness $e = (r_2 - r_1)$ will be

$$N_A \cdot S = -4\pi r^2 D_A \frac{dC_A}{dr} = D_A S_{mg} \frac{C_{A1} - C_{A2}}{e} \tag{7.29}$$

where S_{mg} represents the mean geometric section between the external and internal sections, expressed as

$$S_{mg} = \sqrt{S_1 S_2} = 4\pi r_1^2 r_2^2 \tag{7.30}$$

7.3 MASS TRANSFER BY CONVECTION

Mass transfer is not only carried out by molecular diffusion. Many times the mass is transferred by its own global motion associated with a difference of densities or by the effect of an external device by agitation or pumping. When a fluid circulates on the outside of a solid, a boundary layer is formed on the solid that offers the resistance to the mass transfer. As the thickness increases, the boundary layer resistance to the transfer will also increase. This layer thickness is a function of the fluid speed, so it decreases as the speed increases, so that the resistance is bigger in the laminar regime, and the transfer is favored under conditions of the turbulent regime.

In these cases, it is convenient to express the transfer flux for a component according to the expression

$$N_A = k(c_{A1} - c_{A2}) \tag{7.31}$$

where the flux is due to a concentration difference as a driving force, with k being a proportionality constant that is the constant of mass transfer.

The calculation of the mass transfer coefficients can be carried out through dimensional analysis, correlating the different variables that appear in the transfer process. The transfer coefficient depends on the system type, on the circulation speeds, and on the considered geometries. In this way, different dimensionless modules are obtained that lead to a global expression such as

$$\text{Sh} = f(\text{Re}, \text{Sc}) = a(\text{Re})^b(\text{Sc})^c \tag{7.32}$$

This equation indicates the existent relationship among three dimensionless numbers, defined by

$$\text{Sherwood number:} \quad \text{Sh} = \frac{kL}{D_{AB}} \tag{7.33}$$

$$\text{Reynolds number:} \quad \text{Re} = \frac{\rho v L}{\eta} \tag{7.34}$$

$$\text{Schmidt number:} \quad \text{Sc} = \frac{\eta}{\rho D_{AB}} \tag{7.35}$$

In these equations k is the mass transfer coefficient, L is the characteristic length, ρ is the density, η is the viscosity, D_{AB} is the diffusivity, and v the linear speed.

The Sherwood number represents the relationship between the transfer by means of convection and diffusion, while the module of Reynolds indicates the turbulence grade.

Sometimes, a dimensionless mass transfer number (j_D) is used, defined according to the expression

$$j_D = \frac{k}{v}(\text{Sc})^{2/3} = (\text{Sh})(\text{Re})^{-1}(\text{Sc})^{-1/3} \tag{7.36}$$

In practice, different cases and situations may be found, and it is possible to obtain the specific equations that correlate these three modules. These correlations are different depending on whether the circulation regime for the fluid is laminar or turbulent and also on the considered geometry.

Once the different dimensionless numbers that intervene in the process of mass transfer have been correlated (Equation 7.32), it is necessary to carry out experimental adjustments to obtain the parameters of this equation and the exponents of the dimensionless numbers.

7.3.1 Mass Transfer Coefficients

If the mass transfer is carried out by molecular diffusion, in general it is not necessary to have the mass transfer coefficients since the equations outlined in the mass balances can be solved. However, there are some cases in which the resolution of the outlined mathematical model is difficult, and it is necessary to approach expressions that allow the calculation of the mass transfer coefficients in order to appropriately solve the problem of diffusion in laminar flow. Next, some equations will be shown that make the calculation of the mass transfer coefficients possible for different types of geometries.

7.3.1.1 Mass Transfer for Flow Inside Pipes

For fluid flow, for both gases and liquids that circulate inside pipes with values of the Reynolds numbers greater than 2100, the following equation can be used:

$$\mathrm{Sh} = 0.023(\mathrm{Re})^{0.83}(\mathrm{Sc})^{1/3} \tag{7.37}$$

In case the fluid circulates in a laminar flow, the following equation can be used:

$$\mathrm{Sh} = 1.86\left[(\mathrm{Re})(\mathrm{Sc})\left(\frac{d}{L}\right)\right]^{1/3} \tag{7.38}$$

7.3.1.2 Mass Transfer for Flow over Plane Sheets

The mass transfer and vaporization of liquids from a plane surface toward a current that flows parallel to this surface are presented in numerous processes, for example, in food drying and in water evaporation from the surface of the food.

In laminar or transition flow, the values of the Reynolds number are lower than 15,000 for a gas circulating in a parallel way to the plane plates of length L, so the different dimensionless numbers can be correlated according to the expression

$$\mathrm{Sh} = 0.664(\mathrm{Re})^{0.5}(\mathrm{Sc})^{1/3} \tag{7.39}$$

This equation can be applied for Schmidt number values greater than 0.6.

For gases that circulate with Reynolds number values between 15,000 and 300,000, the equation to use is

$$\mathrm{Sh} = 0.036(\mathrm{Re})^{0.8}(\mathrm{Sc})^{1/3} \tag{7.40}$$

For flow of liquids through plane plates, when Reynolds number values vary between 600 and 50,000, the data can be correlated according to the equation

$$\mathrm{Sh} = 0.99(\mathrm{Re})^{0.5}(\mathrm{Sc})^{1/3} \tag{7.41}$$

7.3.1.3 Mass Transfer for Flow over Spheres

Sometimes fluids circulate on the exterior of spherical solids, and in these cases for the transfer from or toward the spherical solid, the following equation can be used:

$$\mathrm{Sh} = 2.0 + \left[0.4(\mathrm{Re})^{0.5} + 0.06(\mathrm{Re})^{2/3}\right](\mathrm{Sc})^{0.4} \tag{7.42}$$

For gases with Reynolds number values lower than 48,000 and Schmidt module of values between 0.6 and 2.77, the following expression can be used:

$$\text{Sh} = 2.0 + 0.552(\text{Re})^{0.53}(\text{Sc})^{1/3} \tag{7.43}$$

For liquids with Reynolds module values lower than 2000,

$$\text{Sh} = 2.0 + 0.95(\text{Re})^{0.5}(\text{Sc})^{1/3} \tag{7.44}$$

For liquids with Reynolds module values between 2,000 and 17,000,

$$\text{Sh} = 0.347(\text{Re})^{0.62}(\text{Sc})^{1/3} \tag{7.45}$$

7.3.2 Global Coefficients for Mass Transfer

For mass transfer among different phases, the component that is transferred should overcome different resistances in such a way that the global transfer is a function of a global driving force that is the difference between the concentrations of the component in both phases, with the proportionality constant being the denominated global coefficient of mass transfer.

In the mathematical treatment of global transfer, the denominated double film theory is usually applied, in which it is assumed that at both sides of the interface, there is a stagnant film that offers true resistance to the transfer of the considered component.

Consider that component A is transferred from phase 1 toward phase 2 (Figure 7.5), from the bulk phase 1 until the interface, and then transferred to phase 2 through an interface toward the bulk of phase 2. Also, it is necessary to know the equilibrium rate between phases or the relationship between the concentrations of component A in both phases. As mentioned before, in the vicinity of the interface, a double film is formed in such a way that the transfers can be expressed as follows:

$$\text{Transfer in phase 1:} \quad N_A = k_{A1}\left(C_{A1} - C_{A1}^i\right) \tag{7.46}$$

$$\text{Transfer in phase 2:} \quad N_A = k_{A2}\left(C_{A2}^i - C_{A2}\right) \tag{7.47}$$

where

k_{Aj} are the mass transfer coefficients in phase j

C_{A1} and C_{A2} are the concentrations of component A in the bulk of phase 1 and 2, respectively

superscript i means the concentration in the interface

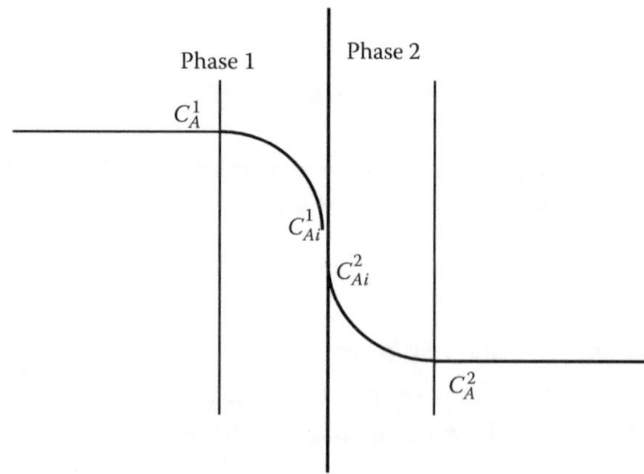

FIGURE 7.5 Diffusion of component A between two phases.

In the previous equations, it is difficult to determine the concentrations of component A in the interface; thus, in many cases the equilibrium concentration is usually used, so the flux for A transferred is expressed as

$$N_A = K_{A1}\left(C_{A1} - C_{A1}^e\right) = K_{A2}\left(C_{A2}^e - C_{A2}\right) \tag{7.48}$$

The driving force in each phase is the difference of concentrations for component A in the bulk of the phase and the corresponding equilibrium concentration, with K_{Aj} being the proportionality coefficient that is denominated as the global coefficient of matter transfer. This global coefficient can be expressed as a function of the individual coefficients.

The flux is the same in both phases:

$$N_A = k_{A1}\left(C_{A1} - C_{A1}^i\right) = k_{A2}\left(C_{A2}^i - C_{A2}\right) = K_{A1}\left(C_{A1} - C_{A1}^e\right) = K_{A2}\left(C_{A2}^e - C_{A2}\right) \tag{7.49a}$$

$$N_A = \frac{\left(C_{A1} - C_{A1}^i\right)}{\left(1/k_{A1}\right)} = \frac{\left(C_{A2}^i - C_{A2}\right)}{\left(1/k_{A2}\right)} = \frac{\left(C_{A1} - C_{A1}^e\right)}{\left(1/K_{A1}\right)} = \frac{\left(C_{A2}^e - C_{A2}\right)}{\left(1/K_{A2}\right)} \tag{7.49b}$$

The equilibrium between phases is usually expressed in the form of an equilibrium curve, as shown in Figure 7.6. Point F represents the conditions of a system in which the concentration in the bulk of phase 1 is C_{A1}, while in the bulk of phase 2 it is C_{A2}. The concentrations in the interface are defined by point i, while points A and B represent the conditions in which C_{A1} is in equilibrium with C_{A2}^e, and C_{A2} is in equilibrium with C_{A1}^e, respectively.

Keeping in mind the equilibrium curve between phases (Figure 7.6), this last equation is transformed to

$$N_A = \frac{\left(C_{A1} - C_{A1}^i\right)}{\left(1/k_{A1}\right)} = \frac{\left(C_{A1}^i - C_{A1}^e\right)}{\left(m/k_{A2}\right)} = \frac{\left(C_{A1} - C_{A1}^e\right)}{\left(1/K_{A1}\right)} = \frac{\left(C_{A1} - C_{A1}^e\right)}{\left(n/K_{A2}\right)} \tag{7.49c}$$

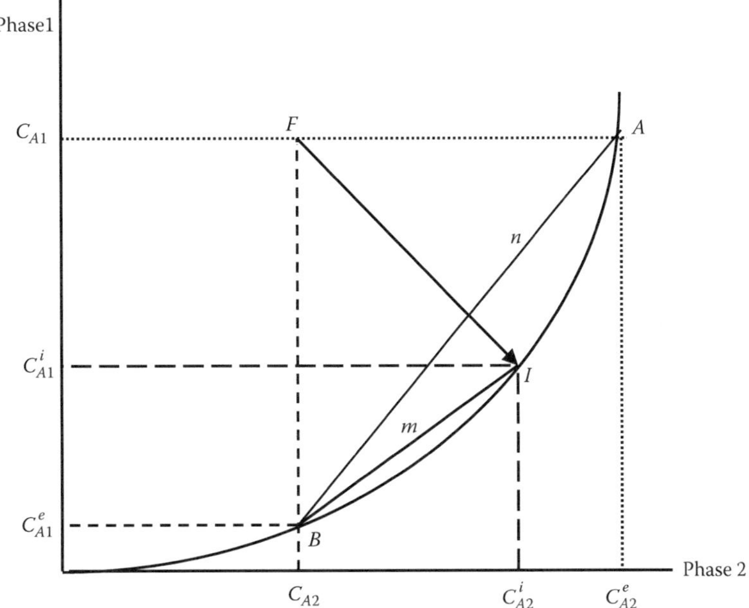

FIGURE 7.6 Equilibrium curve between two phases.

Thus, the correlation between the overall and individual coefficients can be expressed as

$$\frac{1}{k_{A1}} + \frac{m}{k_{A1}} = \frac{1}{K_{A1}} = \frac{n}{K_{A2}} \tag{7.50}$$

Many cases may be found in which the concentration relationship between different phases is not a curve but rather a linear relationship, so K is defined as the equilibrium rate for a component between the two phases in equilibrium:

$$K = \frac{C_{A1}}{C_{A2}} \tag{7.51}$$

In these cases $m = n = K$. Thus, the relationship among the different mass transfer coefficients will be expressed as

$$\frac{1}{k_{A1}} + \frac{K}{k_{A1}} = \frac{1}{K_{A1}} = \frac{K}{K_{A2}} \tag{7.52}$$

A particular case is the absorption of gases by means of liquids in those for which the equilibrium relationship can be expressed by means of the denominated Henry's constant (H). In this case the equilibrium rate between the gas and liquid phases coincides with Henry's constant: $K = H$.

7.4 UNSTEADY MASS TRANSFER

When a contained component in a flowing phase is transferred without chemical reaction through a solid by means of a mass balance, Equation 7.17 is obtained:

$$\frac{dC_A}{dt} = D_A \nabla^2 C_A \tag{7.17a}$$

The Laplace operator (∇^2) has different expressions, depending on the adopted coordinate system. If it is considered that the transfer is carried out in a single direction, in Cartesian coordinates, the previous equation can be expressed as

$$\frac{dC_A}{dt} = D_A \frac{d^2 C_A}{dz^2} \tag{7.53a}$$

For systems in cylindrical and spherical coordinates, the equation would be similar, but the transfer direction takes the radial one. Equation 7.53 can be expressed as a function of concentration, time, and position dimensionless variables:

$$\frac{dY}{d\tau} = D_A \frac{d^2 Y}{dn^2} \tag{7.53b}$$

where the different dimensionless variables are defined as

$$Y = \frac{C_S - C}{C_S - C_0} \tag{7.54}$$

$$\tau = \text{Fo} = \frac{D_A t}{e^2} \tag{7.55}$$

$$n = \frac{z}{e} \tag{7.56}$$

where
- C_S is the concentration of component A in the surface of the solid
- C is the composition at a certain position inside the solid at a certain time
- C_0 is the initial composition
- t is the time
- e is the semi-thickness of the sheet

The variables Y, τ, and n are the dimensionless variables for concentration, time, and position, respectively. It is necessary to point out that the dimensionless time variable τ coincides with Fourier's number. In the cylindrical and spherical geometries, the radius, r_0, substitutes the semi-thickness e.

This equation presents a difficult analytical solution; however, in the case of semi-infinite solids and simple geometries, analytic solutions can be found.

7.4.1 Solids with Simple Geometries

For mass transfer the Biot number is defined, which is a dimensionless module whose expression is

$$\text{Bi} = \frac{K k_F L}{D_A} \tag{7.57}$$

where
- L is the characteristic length
- k_F is the mass transfer coefficient from the bulk flowing phase to the interface
- K is the equilibrium rate between the composition of the component in fluid phase in the interface (C_F^i) and the composition in the solid surface (C_S)

The Biot number is the relationship between the mass transfer for external convection and the diffusion inside the solid. Depending on the value of this module, the mass transfer can be controlled by internal diffusion or the external convection. For values of the Biot number greater than 1, it can be considered that the concentration in the interface coincides with the concentration in the bulk fluid ($C_F = C_F^i$).

Equation 7.53 can be integrated with the previous boundary conditions and with the convection condition:

$$N_A = k_F\left(C_F - C_F^i\right) = -D_A \frac{dC}{dz} \tag{7.58}$$

indicating that the mass transfer flux by convection from the bulk fluid until the interface is equal to the flux inside the solid, at a concentration C_F, is governed by Fick's equation.

The resulting equations of the integration depend on the geometry of the solid and on whether the transfer process is controlled by external convection or internal diffusion; that is to say, it depends on the Biot number value.

7.4.1.1 Internal Resistance Is Negligible

If the Biot number value is greater than 40 (Bi > 40), it is considered that the transfer process is controlled by the external transfer for convection in the fluid. In this case it is possible to obtain the integrated equations for the three types of coordinates:

- Rectangular coordinates
  The local concentration in a certain point of the solid one can be obtained from the following equation:

$$Y = \frac{C_S - C}{C_S - C_0} = \frac{4}{\pi}\sum_{i=0}^{\infty}\frac{(-1)^i}{2i+1}\cos\left[\frac{(2i+1)\pi n}{2}\right]\exp\left[-\left[\frac{(2i+1)^2\pi^2}{4}\text{Fo}\right]\right] \tag{7.59}$$

  To calculate a mean concentration (C_m), referred to any point of the solid, the following equation should be used:

$$Y = \frac{C_S - C_m}{C_S - C_0} = \frac{8}{\pi^2}\sum_{i=0}^{\infty}\frac{1}{(2i+1)^2}\exp\left(-\frac{(2i+1)^2\pi^2}{4}\text{Fo}\right) \tag{7.60}$$

- Cylindrical coordinates
  The local composition of a certain point of the solid can be calculated by means of the equation

$$Y = \frac{C_S - C}{C_S - C_0} = \frac{2}{r_0}\sum_{i=1}^{\infty}\frac{J_0(\beta_i n)^i}{\beta_i J_1(\beta_i r_0)}\exp\left(-\beta_i^2 r_0^2\text{Fo}\right) \tag{7.61}$$

  where J_0 and J_1 are the Bessel functions of the first species of zero and first order, respectively. The Bessel functions are tabulated, and they can be found in Appendix. Also, β_i is the root of the Bessel function. The different roots that make this the Bessel function zero are $\beta_1 n = 2.4048$; $\beta_2 n = 5.5201$; $\beta_3 n = 8.6537$; $\beta_4 n = 11.7915$; and $\beta_5 n = 14.9309$.

  To calculate a mean concentration (C_m) referred to any point of the solid, the following equation should be used:

$$Y = \frac{C_S - C_m}{C_S - C_0} = \frac{4}{\pi^2}\sum_{i=1}^{\infty}\frac{\pi^2}{\beta_i^2 r_0^2}\exp\left(-\beta_i^2 r_0^2\text{Fo}\right) \tag{7.62}$$

- Spherical coordinates
  The concentration in a certain point of the solid is calculated with the equation

$$Y = \frac{C_S - C}{C_S - C_0} = -\frac{2}{\pi}\sum_{i=1}^{\infty}\frac{(-1)^i}{i}\frac{\sin(i\pi n)}{n}\exp\left(-i^2\pi^2\text{Fo}\right) \tag{7.63}$$

  The mean composition is obtained with the equation

$$Y = \frac{C_S - C_m}{C_S - C_0} = \frac{6}{\pi^2}\sum_{i=1}^{\infty}\frac{1}{i^2}\exp\left(-i^2\pi^2\text{Fo}\right) \tag{7.64}$$

7.4.1.2 Internal and External Resistances Not Negligible

In the case that the Biot module value is between 0.1 and 40, both resistances, the internal one and the external one, are significant. As in the previous case, the integrated equations depend on the coordinate type.

- Rectangular coordinates
  For local concentration, the following equation may be applied:

$$Y = \frac{C_S - C}{C_S - C_0} = \sum_{i=1}^{\infty} \frac{2\mathrm{Bi}}{\beta_i^2 + \mathrm{Bi}^2 + \mathrm{Bi}} \frac{\cos(\beta_i n)}{\cos(\beta_i)} \exp\left(-\beta_i^2 \mathrm{Fo}\right) \tag{7.65}$$

  where β_i are the roots of the equation $\beta_i \tan \beta_i = \mathrm{Bi}$.
  The mean concentration can be obtained from the equation

$$Y = \frac{C_S - C_m}{C_S - C_0} = \sum_{i=1}^{\infty} \frac{2(\mathrm{Bi})^2}{\left(\beta_i^2 + \mathrm{Bi}^2 + \mathrm{Bi}\right)\beta_i^2} \exp\left(-\beta_i^2 \mathrm{Fo}\right) \tag{7.66}$$

- Cylindrical coordinates
  The local composition can be calculated by means of the following equation:

$$Y = \frac{C_S - C}{C_S - C_0} = \sum_{i=1}^{\infty} \frac{2\mathrm{Bi}}{\beta_i^2 + \mathrm{Bi}^2} \frac{J_0(\beta_i n)}{J_0(\beta_i)} \exp\left(-\beta_i^2 \mathrm{Fo}\right) \tag{7.67}$$

  where β_i are the roots of the equation $\beta_i J_1(\beta_i)/J_0(\beta_i) = \mathrm{Bi}$.
  The mean concentration can be obtained from equation

$$Y = \frac{C_S - C_m}{C_S - C_0} = \sum_{i=1}^{\infty} \frac{4(\mathrm{Bi})^2}{\left(\beta_i^2 + \mathrm{Bi}^2\right)\beta_i^2} \exp\left(-\beta_i^2 \mathrm{Fo}\right) \tag{7.68}$$

- Spherical coordinates
  The local concentration can be obtained from the following equation:

$$Y = \frac{C_S - C}{C_S - C_0} = \sum_{i=1}^{\infty} \frac{2\mathrm{Bi}}{\beta_i^2 + \mathrm{Bi}^2 + \mathrm{Bi}} \frac{\sin(\beta_i n)}{\sin(\beta_i)} \exp\left(-\beta_i^2 \mathrm{Fo}\right) \tag{7.69}$$

  where β_i are the roots of the equation $\beta_i / \tan \beta_i = 1 - \mathrm{Bi}$.
  The mean concentration can be obtained from the equation

$$Y = \frac{C_S - C_m}{C_S - C_0} = \sum_{i=1}^{\infty} \frac{6(\mathrm{Bi})^2}{\left(\beta_i^2 + \mathrm{Bi}^2 + \mathrm{Bi}\right)\beta_i^2} \exp\left(-\beta_i^2 \mathrm{Fo}\right) \tag{7.70}$$

7.4.2 Semi-Infinite Solids

7.4.2.1 Constant Concentration at the Surface

If the solid through which the component A is transferred initially has a constant composition, C_0, and it is considered that the wall is at a constant concentration, C_S, with the following boundary conditions:

For any value of z and $\tau = 0$, $C = C_0$.
For $z = e$ and any time, $C = C_s$.
For $z = 0$ and any time, $dC/dz = 0$ (optimum condition).

Equation 7.53 can be integrated, obtaining

$$Y = \frac{C_S - C}{C_S - C_0} = fer\left(\frac{z}{2\sqrt{D_A t}}\right) \tag{7.71}$$

where *fer* is the Gauss error function.

7.4.2.2 Convection Condition

In the case that the solid is immersed in a fluid, there is an external convection transfer from the bulk fluid to the interface, while there is a transfer to the wall of the solid phase that is considered to possess a concentration C_S. In this case the integrated equation leads to the expression

$$Y = \frac{C_S - C}{C_S - C_0} = fer\left(\frac{z}{2\sqrt{D_A t}}\right) + \left[\exp\left(\frac{hz}{k} + \frac{h^2 D_{SA} t}{k^2}\right)\right]\left[fer\left(\frac{z}{2\sqrt{D_A t}} + \frac{h\sqrt{D_A t}}{k}\right)\right] \tag{7.72}$$

PROBLEMS

7.1 A recipient contains water at 80°C; on the surface of the water, there is a stagnant air film at 25°C and 1 atm. If at a distance of 1.5 cm of the water surface the water vapor pressure is 2800 Pa, calculate the water vapor flux transferred through the air film. The diffusion coefficient for water vapor is 3×10^{-5} m²/s.

The water, in the form of vapor, is transferred from the liquid surface toward the bulk of the air that surrounds it, and it is assumed that it is stagnated. Therefore, the water vapor flux from the liquid surface toward the air can be calculated by applying Equation 7.13:

$$N_A = \frac{P D_{AB}}{RT} \frac{(p_{A1} - p_{A2})}{(z_2 - z_1)(p_B)_{ml}} \tag{7.13b}$$

where subscripts A and B refer to water vapor and air, respectively.

From the water steam tables it can be obtained that the water vapor partial pressure in contact with the liquid surface at 80°C is

$$p_{A1} = 0.4829\,\text{kgf/cm}^2 \frac{98{,}000\,\text{N/m}^2}{1\,\text{kgf/cm}^2} = 47{,}324.2\,\text{Pa}$$

According to the problem data, $p_{A1} = 2{,}800$ Pa.

The total pressure is 1 atm = 101,325 Pa, and the partial pressures for air will bc

$$p_{B1} = (101{,}325 - 47{,}324.2)\,\text{Pa} = 54{,}000.8\,\text{Pa}$$

$$p_{B2} = (101{,}325 - 2{,}800)\,\text{Pa} = 98{,}525\,\text{Pa}$$

Therefore,

$$(p_B)_{ml} = \frac{p_{B2} - p_{B1}}{\ln\left(p_{B2}/p_{B1}\right)} = \frac{98{,}525 - 54{,}000.8}{\ln(98{,}525/54{,}000.8)} = 74{,}045.15\,\text{Pa}$$

Also, $z_2 - z_1 = 0.015$ m.

The water vapor flux can be obtained by substituting data in Equation 7.13:

$$N_A = \frac{(101{,}325\,\text{Pa})(3\times10^{-5}\,\text{m}^2/\text{s})}{(8{,}314\,\text{Pa}\,\text{m}^3/(\text{kmol K}))(298\,\text{K})}\frac{(47{,}324.2 - 2{,}800\,\text{Pa})}{(0.015\,\text{m})(74{,}045.15\,\text{Pa})} = 4.92\times10^{-5}\,\text{kmol/s}\,\text{m}^2$$

7.2 A food is on a tray of 25 cm × 10 cm that is covered by a polyethylene film of 0.15 mm of thickness. The permeability of the polyethylene for oxygen is 2.2×10^{-8} cm^3/(s cm^2 atm/cm), and the permeability for water vapor is 1.3×10^{-7} cm^3/(s cm^2 atm/cm). The partial pressure of oxygen inside the container is 1 kPa, while in the exterior there is a pressure of 20 kPa. The tray is stored in a room at 20°C and 80% relative humidity, with the water activity for the product inside the container being 0.3. Calculate the oxygen and water vapor flow in g/s that is transferred through the polyethylene film.

It will be assumed that the oxygen and water vapor behave as ideal gases and the perfect gas equation can be applied.

It is assumed that the area of the polyethylene film is the same as that of the tray; thus,

$$S = (0.25\,\text{m})(0.10\,\text{m}) = 0.025\,\text{m}^2$$

The component flow rate transferred through the film is the product of the flux and the section. Thus, Equation 7.24 can be used to calculate this mass flow rate:

$$w_A = N_A \cdot S = S P_M \frac{p_{A1} - p_{A2}}{e} \tag{7.24}$$

From water steam tables, at 20°C, the water vapor pressure is

$$p_w = 0.02383\,\text{kgf/cm}^2\,\frac{98{,}000\,\text{N/m}^2}{1\,\text{kgf/cm}^2} = 2{,}335.3\,\text{Pa}$$

Water vapor partial pressures inside and exterior of the container will be

$$p_{wi} = (0.3)(2335.3\,\text{Pa}) = 700.6\,\text{Pa}$$

$$p_{wo} = (0.8)(2335.3\,\text{Pa}) = 1868.3\,\text{Pa}$$

Permeability will be expressed in adequate units:

Oxygen

$$P_{MO_2} = 2.2\times10^{-8}\ \text{cm}^3/(\text{s}\,\text{cm}^2(\text{atm/cm}))\left(\frac{1\,\text{atm}}{101{,}325\,\text{Pa}}\right)\left(\frac{1\,\text{m}^2}{10^4\,\text{cm}^2}\right) = 2.171\times10^{-17}\ \text{m}^3/(\text{s}\,\text{m}^2(\text{Pa/m}))$$

Water vapor

$$P_{MWV} = 1.3\times10^{-7}\ \text{cm}^3/(\text{s}\,\text{cm}^2(\text{atm/cm}))\left(\frac{1\,\text{atm}}{101{,}325\,\text{Pa}}\right)\left(\frac{1\,\text{m}^2}{10^4\,\text{cm}^2}\right) = 1.283\times10^{-16}\ \text{m}^3/(\text{s}\,\text{m}^2(\text{Pa/m}))$$

Oxygen and water vapor flow rate transferred through the polyethylene film are

Oxygen

$$w_{O_2} = N_{O_2}\cdot S = (2.171\times10^{-17}\ \text{m}^3/(\text{s}\,\text{m}^2(\text{Pa/m})))\frac{(20{,}000-1{,}000)\,\text{Pa}}{0.15\times10^{-3}\,\text{m}}(0.025\,\text{m}^2)$$

$$= 6.875\times10^{-11}\,(\text{m}^3\,\text{O}_2)/\text{s}$$

For oxygen, ideal gas behavior is assumed, and then the oxygen flow rate transferred will be

$$w_{O_2} = 6.875\times10^{-11}\,(\text{m}^3\,\text{O}_2)/\text{s}\left(\frac{10^3\,\text{L}}{1\,\text{m}^3}\right)\left(\frac{1\,\text{mol}}{22.4\,\text{L}}\right)\left(\frac{32\,\text{g}}{1\,\text{mol}}\right) = 9.82\times10^{-8}\,(\text{g}\,\text{O}_2)/\text{s}$$

Water vapor

$$w_{H_2O} = N_{H_2O}\cdot S = (1.283\times10^{-16}\ \text{m}^3/(\text{s}\,\text{m}^2(\text{Pa/m})))\frac{(1868.3-700.6)\,\text{Pa}}{0.15\times10^{-3}\,\text{m}}(0.025\,\text{m}^2)$$

$$= 2.5\times10^{-11}\,(\text{m}^3\,\text{H}_2\text{O})/\text{s}$$

For water vapor, ideal gas behavior is also assumed, and then the transferred water vapor flow rate will be

$$w_{H_2O} = 2.5\times10^{-11}\,(\text{m}^3\,\text{H}_2\text{O})/\,\text{s}\left(\frac{10^3\,\text{L}}{1\,\text{m}^3}\right)\left(\frac{1\,\text{mol}}{22.4\,\text{L}}\right)\left(\frac{18\,\text{g}}{1\,\text{mol}}\right) = 2.0\times10^{-8}\,(\text{g}\,\text{H}_2\text{O})/\text{s}$$

7.3 In a process to obtain powdered milk, an atomization stage is used. Drops of milk that are of 0.8 mm diameter fall in the atomizer chamber, where air circulates at 150°C. The drops and the air circulate in countercurrent with a relative speed of 30 m/s. Assuming that the surface temperature of the drop is 50°C, calculate the mass transfer coefficient in the surface of the milk drops. The diffusion coefficient for the water vapor in the air at 25°C is 2.5×10^{-5} m²/s, and it is proportional to $T^{1.2}$, where T is the absolute temperature. Under the working conditions, an air mean viscosity of 0.022 mPa s can be assumed.

To solve the problem it will be assumed that the air humidity is negligible and that the milk drops do not vary in diameter in the drying process.

The air density will be calculated at the mean temperature between the drop surface and the air (87.5°C):

$$\rho_{Air} = \frac{PM_{Air}}{RT} = \frac{(1\,\text{atm})(28.9\,\text{g/mol})}{(0.08206\,(\text{atm L/mol K}))(273+87.5)\,\text{K}} = 0.977\,\text{g/L} = 0.977\,\text{kg/m}^3$$

The water vapor diffusivity in air at 87.5°C is

$$D_{ij} = (2.5\times10^{-5}\,\text{m}^2/\text{s})\left(\frac{273+87.5}{273+25}\right)^{3/2} = 3.326\times10^{-5}\,\text{m}^2/\text{s}$$

Reynolds number:

$$\text{Re} = \frac{(0.977\,\text{kg/m}^3)(30\,\text{m/s})(0.8\times10^{-3}\,\text{m})}{(0.022\times10^{-3}\,\text{Pa s})} = 1067$$

Schmidt number:

$$\text{Sc} = \frac{(0.022\times10^{-3}\,\text{Pa s})}{(0.977\,\text{kg/m}^3)(3.326\times10^{-5}\,\text{m}^2/\text{s})} = 0.68$$

It is assumed that the liquid milk retains its form, and as the Reynolds number possesses a value lower than 200, in the calculation of the mass transfer coefficient, Equation 7.44 is used; therefore, by substituting the values of the Reynolds and Schmidt numbers in this equation, the Sherwood number can be obtained:

$$\text{Sh} = \frac{kd_p}{D_{ij}} = 2.0 + 0.95(1067)^{0.5}(0.68)^{1/3} = 29.3$$

From this equation the mass transfer coefficient is obtained:

$$k = \text{Sh}\frac{D_{ij}}{d_p} = (29.3)\frac{(3.326\times10^{-5}\,\text{m}^2/\text{s})}{(0.8\times10^{-3}\,\text{m})} = 1.218\,\text{m/s}$$

7.4 It is necessary to dry a piece of meat of 25 × 25 × 2 cm in a drying tunnel where hot air is flowing in a turbulent regime. Initially the meat contains 220 kg water/m^3. The operation conditions allow that the meat surface quickly reaches a humidity content of 60 kg water/m^3. If the water diffusivity in the meat is 1.25×10^{-9} m^2/s, calculate (a) the necessary time so that in the slab center the water content is half of the initial value, and the mean water content of the slab under these conditions and (b) the humidity content in the slab center after 6 h.

a. Because the thickness of the meat slab is much smaller than the other dimensions, it will be assumed that it is an infinite slab of finite thickness. Also, as the surface rapidly reaches a fixed concentration, the external mass transfer coefficient is high, and the Biot number value is high; therefore, the process is controlled by the diffusivity inside the slab.

   For the calculation of the concentration in the center of the slab, Equation 7.59 will be used. Initially, the dimensionless concentration Y is calculated:

$$Y = \frac{C_S - C}{C_S - C_0} = \frac{60 - 0.5\times220}{60-220} = 0.3125$$

The water content in the center of the slab has a dimensionless position number, which will be

$$n = \frac{0}{(e/2)} = \frac{0}{1\,\text{cm}} = 0$$

If it is assumed that Fourier's number is greater than 0.20, from Equation 7.59 only the first adding term of the second member is important; substituting data, it is obtained as

$$Y = 0.3125 = \frac{4}{\pi}\cos\left[\frac{\pi \times 0}{2}\right]\exp\left(-\frac{\pi^2}{4}\text{Fo}\right)$$

From this equation it is possible to obtain Fourier's number value of Fo = 0.5693. Since this value is greater than 0.2, it is correct to neglect the first addition term:

$$\text{Fo} = 0.5693 = \frac{D_A t}{(e/2)^2}$$

It is possible to obtain the time:

$$t = \frac{(\text{Fo})(e/2)^2}{D_A} = \frac{(0.5693)(10^{-2}\,\text{m})^2}{(1.25 \times 10^{-9}\,\text{m/s})} = 45{,}544\,\text{s} = 12\,\text{h}\,39\,\text{min}$$

For the calculation of the mean concentration, Equation 7.60 is used. Since the lapsed time is 45,544 s, Fourier's number is the same as that obtained previously (Fo = 0.5693). Substituting data,

$$Y = \frac{C_S - C_m}{C_S - C_0} = \frac{8}{\pi^2}\sum_{i=0}^{\infty}\frac{1}{(2i+1)^2}\exp\left(-\frac{(2i+1)^2\pi^2}{4}\text{Fo}\right)$$

$$= \frac{8}{\pi^2}\left[\exp\left(-\frac{\pi^2}{4}0.5693\right) + \frac{1}{9}\exp\left(-\frac{9\pi^2}{4}0.5693\right) + \frac{1}{25}\exp\left(-\frac{25\pi^2}{4}0.5693\right) + \cdots\right]$$

$$= \frac{8}{\pi^2}(0.24544 + 3.59 \times 10^{-7} + 2.24 \times 10^{-17} + \cdots) = 0.1990$$

It is observed that only the second member has importance to the first addition, since Fourier's number is greater than 0.2. The mean concentration is obtained from the value of the dimensionless variable Y:

$$Y = 0.1990 = \frac{60 - C_m}{60 - 220}$$

The mean concentration of water in the slab is

$$C_m = 91.84\,\text{kg/m}^3$$

b. After 6 h of drying, Fourier's number has a value of

$$\text{Fo} = \frac{D_A t}{(e/2)^2} = \frac{(1.25\times10^{-9}\ \text{m}^2/\text{s})\big(6\,\text{h}\ (3600\,\text{s}/1\,\text{h})\big)}{(10^{-2}\ \text{m})^2} = 0.27$$

The water content in the middle of the slab is obtained from Equation 7.59, substituting Fourier's module value (Fo = 0.27). Because these are the points located in the center of the slab, the dimensionless position coordinate is null ($n = 0$), and the value of cosine terms is the unit. Substituting data,

$$Y = \frac{C_S - C}{C_S - C_0} = \frac{4}{\pi}\sum_{i=0}^{\infty}\frac{(-1)^i}{2i+1}\cos\left[\frac{(2i+1)\pi n}{2}\right]\exp\left[-\left[\frac{(2i+1)^2\pi^2}{4}\text{Fo}\right]\right]$$

$$= \frac{4}{\pi}\left[\exp\left(-\frac{\pi^2}{4}0.27\right) - \frac{1}{3}\exp\left(-\frac{9\pi^2}{4}0.27\right) + \frac{1}{5}\exp\left(-\frac{25\pi^2}{4}0.27\right) - \cdots\right]$$

$$= \frac{4}{\pi}(0.51366 - 8.30\times10^{-4} + 1.17\times10^{-8} - \cdots) = 0.51283$$

The concentration in the middle of the slab is obtained from the dimensionless position coordinate Y:

$$Y = 0.51283 = \frac{60 - C}{60 - 220}$$

Then, the water content in the middle of the slab is $C = 142.05$ kg/m³.

7.5 A cheese maker industry elaborates pieces in spherical form of 15 cm diameter. In the salted stage, the spherical pieces are introduced in a brine bath that possesses a salt concentration of 20 °Baumé (1150 g/L). The equilibrium rate between the fluid and solid phase possesses a value of 8.75. In order to favor the penetration of salt in the cheese, the brine solution is continuously agitated. The diffusion coefficient for NaCl in the cheese is 6.5 × 10^{-9} m²/s, while the mass transfer constant in the liquid phase has a value of 9.2 × 10^{-5} m²/s. Calculate the salt concentration in the center of the cheese piece, as well as the mean concentration after 2 days.

The Biot number for mass transfer is calculated from Equation 7.57, taking as the characteristic length the radius of the sphere r_0:

$$\text{Bi} = \frac{K k_F r_0}{D_A} = \frac{(8.75)(9.2\times10^{-5}\ \text{m/s})(0.075\,\text{m})}{(6.5\times10^{-9}\ \text{m}^2/\text{s})} = 9288$$

From this high value of the Biot number, it can be assumed that the external mass transfer is very high, and the salt concentration in the interface coincides with bulk in the fluid phase ($C_F - C_F^i$); also, the process is controlled by the diffusivity inside the sphere. From the definition of equilibrium between the fluid and the solid phase, the salt concentration at the cheese surface can be calculated:

$$C_S = \frac{C_F^i}{K} = \frac{C_F}{K} = \frac{1150\,\text{kg/m}^3}{8.75} = 131.43\,\text{kg/m}^3$$

Fourier's number is

$$\text{Fo} = \frac{D_A t}{(r_0)^2} = \frac{(6.5\times10^{-9}\,\text{m}^2/\text{s})(2\,\text{days}(86{,}400\,\text{s}/1\,\text{day}))}{(0.075\,\text{m})^2} = 0.19968$$

The Biot number presents a very high value, so to calculate the local salt composition, Equation 7.63 will be used. As the salt concentration at the center of the sphere must be calculated, the dimensionless position coordinate has null value ($n = 0$). For a value of $n \to 0$, in the previous equation, it is accomplished that

$$\lim \frac{\sin(i\pi n)}{n} = 1$$

The dimensionless composition (Y) is obtained substituting data in Equation 7.63:

$$Y = \frac{C_S - C}{C_S - C_0} = -\frac{2}{\pi}\sum_{i=1}^{\infty}\frac{(-1)^i}{i}\frac{\sin(i\pi n)}{n}\exp(-i^2\pi^2\text{Fo})$$

$$= -\frac{2}{\pi}\left[-\exp(-\pi^2 0.19968) + \frac{1}{2}\exp(-4\pi^2 0.19968) - \frac{1}{3}\exp(-9\pi^2 0.19968) + \cdots\right]$$

$$= -\frac{2}{\pi}(-0.13935 + 1.885\times10^{-4} - 6.605\times10^{-9} + \cdots) = 0.08859$$

Assuming that at the beginning the cheese does not contain salt, the concentration in the sphere center is obtained, finding the dimensionless variable Y:

$$Y = 0.08859 = \frac{131.43 - C}{131.43 - 0}$$

Obtaining the water concentration at the center of the sphere, $C = 119.79$ kg/m^3.

The mean composition can be obtained from Equation 7.64:

$$Y = \frac{C_S - C_m}{C_S - C_0} = \frac{6}{\pi^2}\sum_{i=1}^{\infty}\frac{1}{i^2}\exp(-i^2\pi^2\text{Fo})$$

$$= \frac{6}{\pi^2}\left[\exp(-\pi^2 0.19968) + \frac{1}{4}\exp(-4\pi^2 0.19968) + \frac{1}{9}\exp(-9\pi^2 0.19968) + \cdots\right]$$

$$= \frac{6}{\pi^2}(0.13935 + 9.43\times10^{-5} + 2.20\times10^{-9} + \cdots) = 0.08477$$

obtaining the dimensionless concentration value Y

$$Y = 0.08477 = \frac{131.43 - C_m}{131.43 - 0}$$

The mean concentration will be $C_m = 120.29$ kg/m^3

7.6 The soil of a potato field contains a humidity of 190 kg/m^3. The mean depth at which the potatoes are buried is 10 cm. Due to a change in the atmospheric conditions, a strong dry wind causes the soil surface to immediately acquire a continuous humidity of 110 kg/m^3. The diffusion coefficient of the water in the earth is of 1.25×10^{-8} m^2/s. Potatoes suffer water stress if the humidity of the earth in relation to its surroundings is 150 kg/m^3. Determine whether there is a danger of water stress if the wind blows for 3 days. How long should the air blow so that conditions of water stress take place?

According to the idea that mass transfer in a semi-infinite solid with a boundary condition, the soil surface acquires a constant humidity of 110 kg/m^3. The equation that allows obtaining the humidity content that protects the potatoes is Equation 7.71. First, the dimensionless parameter is calculated:

$$\frac{z}{2\sqrt{D_A t}} = \frac{(0.1\,\text{m})}{2\sqrt{(1.25\times 10^{-8}\,\text{m}^2/\text{s})\left(3\,\text{days}\dfrac{86,400\,\text{s}}{1\,\text{day}}\right)}} = 0.87841$$

From table of the Gauss error function, it is possible to calculate the value

$$fer(0.87841) = 0.786604$$

Substituting this value in Equation 7.71, it is possible to obtain the value that will reach the point located at 10 cm of depth after 2 days:

$$Y = \frac{110 - C}{110 - 190} = fer(0.87841) = 0.786604$$

Obtaining the water content, C = 173 kg/m^3.
For water stress,

$$Y = \frac{110 - 150}{110 - 190} = 0.5$$

Therefore, the variable value that makes the error function value 0.5 should be found in the table of the Gauss error function:

$$Y = 0.5 = fer\left(\frac{z}{2\sqrt{D_A t}}\right)$$

Obtaining

$$\frac{z}{2\sqrt{D_A t}} = 0.47699$$

Substituting data in this equation, it is possible to obtain the time for water stress conditions:

$$\frac{z}{2\sqrt{D_A t}} = \frac{(0.1\,\text{m})}{2\sqrt{(1.25\times 10^{-8}\,\text{m}^2/\text{s})\left(t\,\text{days}\,\dfrac{86,400\,\text{s}}{1\,\text{day}}\right)}} = 0.47699$$

obtaining a time of 10.17 days; that is, 10 days 4 h 11 min.

8 Air–Water Interactions

8.1 INTRODUCTION

The operations involving air–water interactions are based on the mass transfer between two phases. It is assumed that the liquid phase is pure water, while the gas phase consists of an inert gas that contains water vapor.

It is considered that the mass transfer takes place exclusively in a gas phase, so the water vapor migrates from the interface to the gas or from the interface to the liquid. The transfer mechanisms involved in these processes are a combination of turbulent transport and diffusion. When there is a phase change, in addition to mass transfer, there is also heat transfer.

The air–water interactions are present in different processes, where air conditioning and water cooling by evaporation are the most important. Air humidification and dehumidification operations are based on conditioning the air used in a given food preservation process. Also, food drying operations are based on the interaction between air and water, in which the water contained in the food is transferred to the air in the form of vapor.

8.2 PROPERTIES OF HUMID AIR

In this section, a description of the different properties of air–water vapor mixtures that are useful in the calculation of unit operations involving water transfer between liquid and gas phases will be presented. To facilitate calculations, all properties are usually referred to as mass of water per mass of dry air or moles of water per moles of dry air.

Molar Humidity X_m

The moisture content of air is expressed as moles of water per mole of dry air. According to this definition,

$$X_m = \frac{P_V}{P - P_V} \text{ kmol water vapor/kmol dry air} \tag{8.1}$$

where

P_V is the partial pressure of water vapor

P is the total pressure

Absolute Humidity X

The moisture content of air is expressed in kg of water per kg of dry air. Absolute moisture can be expressed as a function of the partial pressure of the water vapor in the air according to the equation

$$X = \frac{P_V M_{WATER}}{(P - P_V) M_{AIR}} \text{ kg water vapor/kg dry air} \tag{8.2}$$

where

$M_{WATER} = 18$ kg/kmol

$M_{AIR} = 28.9$ kg/kmol

Thus, a direct relation between absolute and molar moisture can be found:

$$X = 0.622\frac{P_V}{P - P_V} = 0.622X_m \tag{8.3}$$

When the water vapor partial pressure in the air coincides with the water vapor tension at the temperature of air, it is said that the air is saturated. If P_S is the tension of pure water vapor, then the following is compiled:

$$X_{SATURATION} = 0.622\frac{P_S}{P - P_S} \tag{8.4}$$

Relative Humidity ϕ

Relative humidity is the relation between the amount of water vapor contained in a determined air volume and the amount of water vapor that this air volume could contain if it was saturated. From this definition it is easy to see that

$$\phi = \frac{P_V}{P_S} \tag{8.5}$$

Since for saturated air the water vapor pressure in the air is equal to saturation pressure, the relative humidity is one ($\phi = 1$).

The relation between absolute and relative humidities is obtained when combining Equations 8.3 and 8.5:

$$X = 0.622\frac{P_S\phi}{P - P_S\phi} \tag{8.6}$$

Percentage Humidity Y

The relationship between the absolute and saturation humidity:

$$Y = \frac{X}{X_{SAT}} = \frac{P_V\left(P - P_S\right)}{P_S\left(P - P_V\right)} = \phi\frac{\left(P - P_S\right)}{\left(P - P_V\right)} \tag{8.7}$$

Humid Volume $\hat{V}$

The volume occupied by 1 kg of dry air plus the water vapor that it contains is

$$\hat{V} = \hat{V}_{DRY\ AIR} + \hat{V}_{WATER\ VAPOR}$$

Since all of the variables refer to 1 kg of dry air, the humid volume will contain 1 kg of dry air plus X kg of water vapor. From the equation of the ideal gas, the following is obtained:

$$\hat{V} = \frac{1}{M_{AIR}}\frac{RT}{P} + \frac{X}{M_{WATER}}\frac{RT}{P} = \left(\frac{1}{28.9} + \frac{X}{18}\right)\frac{RT}{P} \tag{8.8}$$

If the air is saturated with humidity, the volume is saturated, and it is obtained by replacing the absolute humidity with the humidity of saturation.

Density of Humid Air ρ

The relationship between the mass of humid air and the volume it occupies is

$$\rho = \frac{m}{V} = \frac{PM_m}{RT} \tag{8.9}$$

where M_m is the mean molecular mass of the dry air and water vapor mixture

$$M_m = X_{WATER} \cdot M_{WATER} + (1 - X_{WATER}) \cdot M_{AIR}$$

X_{WATER} being the molar fraction of the water vapor contained in the air, that is, $X_{WATER} = P_V/P$. So

$$M_m = \frac{P_V}{P}\left(M_{WATER} - M_{AIR}\right) + M_{AIR}$$

If this expression is substituted in Equation 8.9, and considering that $P_V = \phi P_S$, the following expression is obtained:

$$\rho = \frac{P}{RT} M_{AIR} - \frac{\phi P_S}{RT}\left(M_{AIR} - M_{WATER}\right) \tag{8.10}$$

Since the molecular masses of air and of water are 28.9 and 18 kg/kmol, respectively, and the gas constant has a value of 0.082 atm m^3/(kmol K) or of 8.314 Pa m^3/(mol K), Equation 8.10 is transformed into

$$\rho = 352.44\frac{P}{T} - 132.93\frac{\phi P_S}{T} \tag{8.11a}$$

in which the pressures should be expressed in atmospheres, although the following equation can be used:

$$\rho = 0.0035\frac{P}{T} - 0.0013\frac{\phi P_S}{T} \tag{8.11b}$$

if the pressure is expressed in Pa. In both expressions, the density is obtained in kg/m^3.

Humid-Specific Heat $\hat{s}$

The amount of heat needed to increase the temperature by 1°C of 1 kg of dry air plus the quantity of water it contains is

$$\hat{s} = \hat{C}_{AIR} + X\hat{C}_V \tag{8.12}$$

where

$\hat{C}_{AIR}$ is the specific heat of dry air
$\hat{C}_V$ is the specific heat of water vapor

In the most common processes in which air–water interactions are involved, it can be considered that the values of the specific heats vary somewhat, and so mean values are taken:

$$\hat{C}_{AIR} = 1\,\text{kJ/kg}\,°\text{C} \quad \text{and} \quad \hat{C}_V = 1.92\,\text{kJ/kg}\,°\text{C}$$

which allows us to obtain the humid-specific heat by

$$\hat{s} = 1 + 1.92X \text{ kJ/kg dry air °C} \quad (8.13)$$

Humid Enthalpy $\hat{i}_G$

The enthalpy of dry air plus the enthalpy of the water vapor it contains is

$$\hat{i}_G = \hat{i}_{AIR} + X\hat{i}_V \quad (8.14)$$

where

$\hat{i}_{AIR}$ is the enthalpy of 1 kg of dry air
$\hat{i}_V$ is the enthalpy of 1 kg of water vapor

Enthalpies are state functions; therefore, it is necessary to define some reference states. Generally, the enthalpy of liquid water at 0°C and 1 atm is taken as the reference value for water, while for air the reference value corresponds to the enthalpy of dry air at 0°C and 1 atm. For water it would be more adequate to take as the reference state the correspondence to the triple point. However, with this reference state, the water at 0°C and 1 atm has an enthalpy of 0.22 kcal/kg, and since this is a low value, in practice it is usually considered that both reference states coincide.

The enthalpy of 1 kg of dry air with an absolute humidity X and at a temperature T will be calculated by the equation

$$\hat{i}_G = \hat{C}_{AIR}\left(T - T^*\right) + \hat{C}_V X\left(T - T^*\right) + \lambda_0 X$$

where

$T^* = 0°C$ (reference temperature)
$\lambda_0 = 2490$ kJ/kg (latent heat of vaporization for water at 0°C)

When replacing these data, an equation that allows the calculation of the humid enthalpy as a function of temperature and moisture content is obtained:

$$\hat{i}_G = \left(\hat{C}_{AIR} + \hat{C}_V X\right)T + \lambda_0 X \quad (8.15)$$

$$\hat{i}_G = \hat{s}T + \lambda_0 X \quad (8.16)$$

Dew Temperature T_R

The temperature at which the saturation of air takes place, given water content, that is, the temperature at which condensation of the water vapor begins when the humid air undergoes a cooling process.

Adiabatic Saturation Temperature T_S

The temperature at which air is saturated when it is adiabatically cooled. In this adiabatic process, the air undergoes a continuous variation of temperature and moisture content.

Humid Temperature T_H

Also called wet bulb temperature, T_H is the temperature reached by a small water mass under adiabatic and steady conditions, exposed to an airstream of constant temperature, moisture, and velocity (Figure 8.4).

Humid temperature is a function of the temperature T of the air and of the humidity X. Also, from this humid temperature, it is possible to determine the absolute humidity of air. In the air–water system, for air velocities between 4 and 10 m/s, the humid temperature coincides with the saturation temperature.

8.3 MOLLIER'S PSYCHROMETRIC DIAGRAM FOR HUMID AIR

The so-called Mollier diagram for humid air is usually employed for the graphic solution of problems in which air–water interaction processes take place, such as air drying, humidification, and dehumidification processes, among others. There are different types of psychrometric diagrams depending on which variables are correlated; one of the most used diagrams is the X–T chart, in which the absolute humidity against temperature is represented, although a diagram showing the product of humid-specific heat times temperature against absolute humidity ($\hat{s}T$–X chart) is used occasionally.

8.3.1 Psychrometric Chart $\hat{s}T$–X

As indicated, in this chart, which can be found in Appendix A, the values of the product $\hat{s}T$ are represented in the ordinates, while the values of the absolute humidity X are represented in the abscissas (Figure 8.1). In the following, the different straight lines and curves that appear in this chart will be obtained.

8.3.1.1 Isothermic Straight Lines

Equation 8.13 was obtained from the definition of humid-specific heat. If this equation is multiplied by the temperature, the following one is obtained:

$$\hat{s}T = T + 1.92TX \tag{8.17}$$

It can be observed that for each temperature a straight line with slope $1.92T$ and intercept to the origin T can be obtained. It is important to note that for each temperature value, a different straight line is obtained.

8.3.1.2 Isenthalpic Straight Lines

Equation 8.16 is the expression of the humid enthalpy, which when rearranged can be expressed as

$$\hat{s}T = \hat{i}_G - \lambda_0 X \tag{8.18}$$

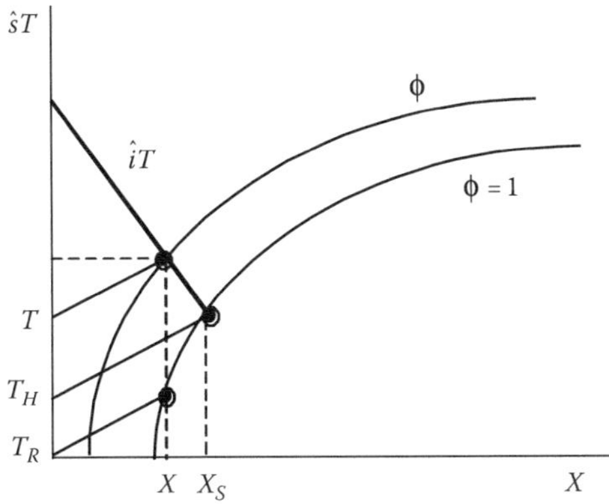

FIGURE 8.1 Psychrometric chart $\hat{s}T$ versus X.

In the $\hat{s}T$–X chart, this equation corresponds to the equation of a straight line with slope λ_0 and intercept to the origin $\hat{i}_G$. It is evident that the isenthalpic lines are parallel straight lines, since their slope is $\lambda_0 = 2490$, but they differ in their humid enthalpy value.

8.3.1.3 Constant Relative Humidity Curves

To obtain this set of curves, it is necessary to combine Equations 8.6 and 8.17:

$$X = 0.622\frac{P_S\phi}{P - P_S\phi} \tag{8.6}$$

$$\hat{s}T = T + 1.92TX \tag{8.17}$$

The procedure to adhere to is as follows:

a. Fix the total pressure for which the chart is built.
b. Fix the relative humidity value.
c. Give values to the temperatures, and by means of the water vapor tables, look for the corresponding values of the vapor pressures.
d. The values of the absolute humidities are obtained, using Equation 8.6, from the different vapor pressure values and the value of the relative humidity fixed in part (b).
e. The values of the product $\hat{s}T$ are calculated, using Equation 8.17, from the temperature values and their corresponding absolute humidities.
f. When plotting the values $\hat{s}T$ against their corresponding absolute humidities X, a curve in which the relative humidity value is constant is obtained.
g. Go back to step (b), changing the relative humidity value and the process with the new value of ϕ. A new curve is obtained from the new value of relative humidity and so on.

Out of the set of curves obtained, the lower curve corresponds to a value $\phi = 1$, that is, the saturation curve. It is important to highlight that this curve is not crossed by the ordinates origin, since for a total pressure of 1 atm and a temperature of 0°C, the vapor pressure of pure water is 610 Pa, corresponding to an absolute humidity of $X = 0.00377$ kg water/kg dry air.

8.3.1.4 Adiabatic Humidification Straight Lines

Suppose a process such as the one shown in Figure 8.2, in which there is a chamber with water at a constant temperature T, with water recirculation entering the system along with the water that is being added, reaching a temperature almost equal to the temperature of the air leaving the chamber. An airflow w' kg dry air/h with a humidity X_1, temperature T_1, and enthalpy $\hat{i}_{G1}$ is introduced into the chamber. Likewise, an airstream with humidity X_2, temperature T_2, and enthalpy $\hat{i}_{G2}$ exists in the chamber. To compensate for the evaporated water, a water stream at temperature T and a flow w' $(X_2 - X_1)$ kg water/h is introduced into the system. Since it is an adiabatic process, there will be no heat exchange with the surroundings.

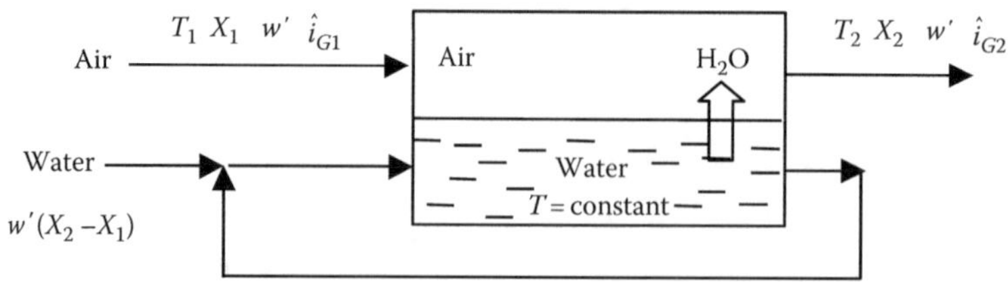

FIGURE 8.2 Adiabatic humidification of air.

When carrying out a global energy balance of the system, we obtain

$$w'\hat{i}_{G1} + w'\hat{C}_L(X_2 - X_1)(T - T^*) = w'\hat{i}_{G2} \tag{8.19}$$

in which $\hat{C}_L$ is the specific heat of water. If $T = 0°C$ is taken as reference temperature, and taking into account that the enthalpy is expressed according to

$$\hat{i}_G = \hat{s}T + \lambda_0 X \tag{8.16}$$

when substituting in Equation 8.19 we obtain

$$\hat{s}_1 T_1 + \lambda_0 X_1 + \hat{C}_L T(X_2 - X_1) = \hat{s}_2 T_2 + \lambda_0 X_2$$

which can be rearranged as

$$\hat{s}_2 T_2 - \hat{s}_1 T_1 = (\hat{C}_L T - \lambda_0)(X_2 - X_1) \tag{8.20}$$

In the $\hat{s}T - X$ chart this is the equation of a straight line with a slope of $(\hat{C}_L T - \lambda_0)$. Generally, the value of the product $\hat{C}_L T$ is much smaller than the latent heat of vaporization, so it can be neglected, and the following is obtained: $\hat{s}_2 T_2 - \hat{s}_1 T_1 = -\lambda_0 (X_2 - X_1)$.

If Equation 8.18 is observed, it can be seen that it corresponds to an isenthalpic line. This indicates that adiabatic humidification straight lines coincide with isenthalpic lines.

8.3.1.5 Nonadiabatic Humidification Straight Lines

A process similar to the previous one is considered, but there is no water recirculation, so to maintain a constant temperature T in this water, a $\dot{Q}_E$ heat flow rate is provided; this should compensate for heat losses by evaporation and for losses from the chamber toward the surrounding $\dot{Q}_S$.

When carrying out a global energy balance, we obtain that

$$w'\hat{i}_{G1} + \hat{C}_L w'(X_2 - X_1)(T - T^*) + \dot{Q}_E = w'\hat{i}_{G2} + \dot{Q}_S \tag{8.21}$$

If the reference temperature is $T^* = 0°C$, and taking into account that

$$\hat{Q}_E = \frac{\dot{Q}_E}{w'(X_2 - X_1)} \quad \text{and} \quad \hat{Q}_S = \frac{\dot{Q}_S}{w'(X_2 - X_1)}$$

are, respectively, provided and lost heat by the system per unit mass, then Equation 8.21 can be rearranged as

$$\hat{Q}_E + (\hat{C}_L T - \hat{Q}_S) = \frac{(\hat{i}_{G2} - \hat{i}_{G1})}{(X_2 - X_1)}$$

defining $\hat{Q}_0 = \hat{C}_L T - \hat{Q}_S$ and taking into account Equation 8.16

$$\hat{Q}_E + \hat{Q}_0 = \lambda_0 + \frac{\hat{s}_2 T_2 - \hat{s}_1 T_1}{X_2 - X_1}$$

or

$$\hat{s}_2 T_2 - \hat{s}_1 T_1 = \left(\hat{Q}_E + \hat{Q}_0 - \lambda_0\right)\left(X_2 - X_1\right) \tag{8.22}$$

which in the $\hat{s}T$–X chart corresponds to the equation of a straight line with slope $(\hat{Q}_E + \hat{Q}_0 - \lambda_0)$ that passes through the points with coordinates $(\hat{s}_1 T_1, X_1)$ and $(\hat{s}_2 T_2, X_2)$.

The quantity $(\hat{Q}_E + \hat{Q}_0)$ represents the amount of heat that should be provided per kg of evaporated water. This is the heat required to carry out the described process, and it is obtained from the slope of the straight line that connects the points that represent the state of the air at the entrance and exit of the system.

8.3.2 Psychrometric Chart *X–T*

This diagram is the representation of the equations obtained in Section 8.2. In this chart the absolute humidity is represented in the ordinates versus the temperature in the abscissas (Figure 8.3).

The curves of constant relative humidity are plotted in analog form as the $\hat{s}T$–X chart. For a total pressure P, a relative humidity value is fixed. The values of the water vapor pressures at different temperatures are determined, and by means of Equation 8.6 the corresponding absolute humidities are calculated at these temperatures. Different curves are obtained with different relative humidity values. The upper curve corresponds to the saturation curve ($\phi = 1$). This curve divides the diagram into two zones; the points placed to the left of the curve represent mixtures of air and liquid water and are unstable fogs of air and water vapor. The points placed to the right of the saturation curve are air–water vapor reheated mixtures.

In the X–T chart, the isenthalpic lines and, consequently, the adiabatic cooling ones, as observed in Equation 8.16, are straight lines of negative slope $-\hat{s}/\lambda_0$.

8.4 WET BULB TEMPERATURE

When a nonsaturated airstream circulates in a water surface, there is a heat transfer from air to water, which causes water evaporation. The water vapor is incorporated into the air. Evaporation causes a decreasing water temperature, whereas air gets closer to saturation conditions. At the end, a steady

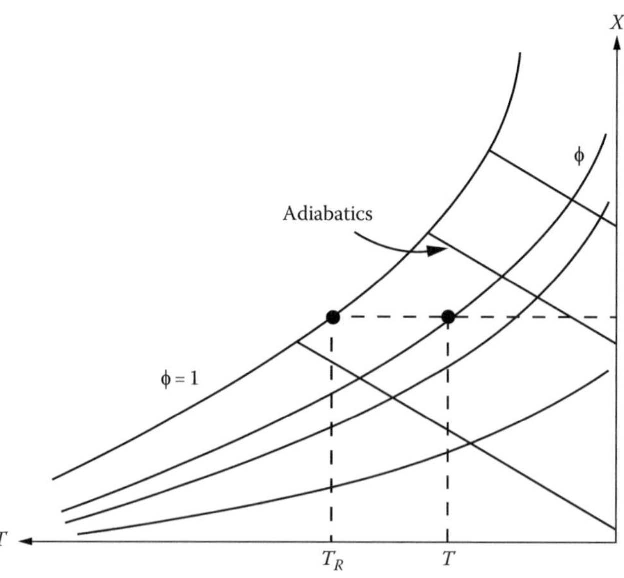

FIGURE 8.3 Psychrometric chart X versus T.

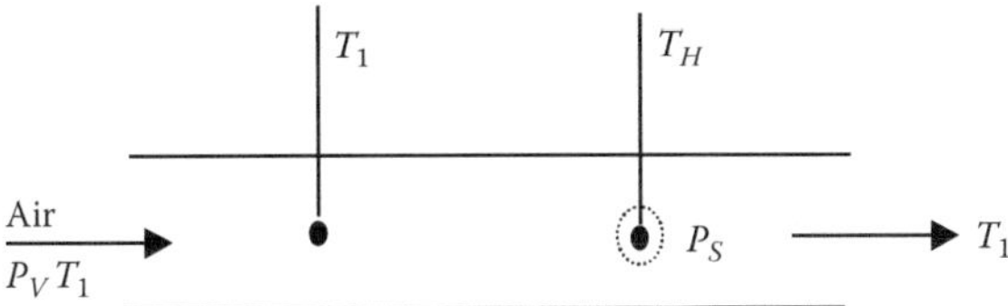

FIGURE 8.4 Wet bulb temperature.

state is reached, in which the heat transfer from the air to water equilibrates with the heat required to evaporate the water. Under these steady conditions, the water is at the *wet bulb temperature*.

This process is similar to the one described next. Consider a nonsaturated airstream humidity such that its vapor pressure (P_V) is smaller than its saturation pressure. A thermometer measures the temperature T_1 of this air, and there is also another thermometer whose bulb is surrounded by a water droplet, as shown in Figure 8.4. The air around the second thermometer is considered to be saturated, that is, the vapor pressure is that of saturation (P_S). Since the vapor pressure of the air is smaller than the saturation pressure ($P_V < P_S$), water evaporates to surroundings of the water droplet, causing a decrease in temperature. Therefore, the second thermometer will show a reading T_H lower than T_1 ($T_H < T_1$), T_H is called the wet bulb temperature, while T_1 is called dry bulb temperature.

The flux of evaporated water is

$$w_{WATER} = K_G \cdot A_G \cdot (P_S - P_V) \tag{8.23}$$

where

K_G is the mass transfer coefficient kg/(m^2 s Pa)
A_G is the area of the water droplet

The heat needed to evaporate the water is the latent heat:

$$\dot{Q}_L = \lambda_S w_{WATER} = \lambda_S K_G A_G \left(P_S - P_V\right) \tag{8.24}$$

in which λ_S is the latent heat of vaporization at the temperature of water.

Heat is transferred from the air to the water droplet, by radiation and convection mechanisms, in such a way that they can be calculated according to the expression

$$\dot{Q}_S = \left(h_C + h_R\right) A_G \left(T_1 - T_H\right) \tag{8.25}$$

in which h_C and h_R are the heat transfer coefficients by convection and radiation, respectively.

When steady conditions are reached, an equilibrium exists in which the heat transferred from the air to the droplet is equal to the heat required for evaporation:

$$\left(\dot{Q}_L = \dot{Q}_S\right): \quad \lambda_S K_G A_G \left(P_S - P_V\right) = \left(h_C + h_R\right) A_G \left(T_1 - T_H\right)$$

from which the following is deduced:

$$P_S - P_V - \frac{\left(h_C + h_R\right)}{\lambda_S K_G}\left(T_H - T_1\right) \tag{8.26}$$

This equation is called *psychrometric equation*.

In this equation, the heat transfer coefficient by convection h_C, as well as the mass transfer coefficient K_G, depends on the thickness of the gas film. Therefore, an increase of the air velocity causes a decrease of this film and an increase in the values of h_C and K_G. For air velocity values from 5 to 8 m/s, and usual work temperatures, the heat transmitted by convection is much higher than by radiation, so it can be considered that the relation $(h_C + h_R)\ K_G$ is practically constant. Under these conditions it is considered that $(h_C + h_R)\ /\lambda_S K_G = 66$ Pa/°C. The wet bulb temperature T_H depends only on the temperature and humidity of the air.

If in Equation 8.6, which gives the absolute humidity in function of the vapor pressure, we consider that this pressure P_V is very small in comparison to the total pressure, and that it is worked at $P = 1$ atm, it can be obtained:

$X = 0.622P_V$, giving P_V in atm

or $X = 0.00614P_V$, if pressure is given in kPa.

When substituting this equation in the psychrometric equation, we obtain

$$\frac{X_S - X_1}{T_H - T_1} = \frac{0.00614(h_C + h_R)}{\lambda_S K_G} \tag{8.27}$$

This equation relates dry and wet bulb temperatures to air humidity.

8.5 ADIABATIC SATURATION OF AIR

In air–water interaction adiabatic processes, it is important to determine the conditions of air after its contact with water. Consider a process like the one shown in Figure 8.5, in which an airstream of w' kg of dry air and with temperature, humidity, and enthalpy T_1, X_1, and $\hat{i}_{G1}$, respectively, is introduced into a chamber with water at a saturation temperature T_S. If the contact time is sufficiently long, and if the air is sprinkled with water, then the airstream leaving the chamber exits at saturation conditions T_S, X_S, $\hat{i}_{GS}$. To compensate for water losses incorporated into the air by evaporation, a water flow equal to $w_L = w'(X_S - X_1)$ at temperature T_S is added continuously.

An energy balance of the system yields the equation

$$w_L \hat{C}_L \left(T_S - T^*\right) + w' \hat{i}_{G1} = w' \hat{i}_{GS}$$

in which the reference temperature is $T^* = 0$°C.

The enthalpy is a function of temperature and moisture content, according to Equation 8.16, so

$$w' \hat{C}_L T_S \left(X_S - X_1\right) + w'(\hat{s}_1 T_1 + \lambda_0 X_1) = w'\left(\hat{s}_S T_S + \lambda_0 X_S\right)$$

The liquid water that evaporates and is incorporated into the air does so at saturation temperature T_S. The humid-specific heat at temperature T_S is given by

$$\hat{s}_S = \hat{s}_1 + \hat{C}_L \left(X_S - X_1\right) \tag{8.28}$$

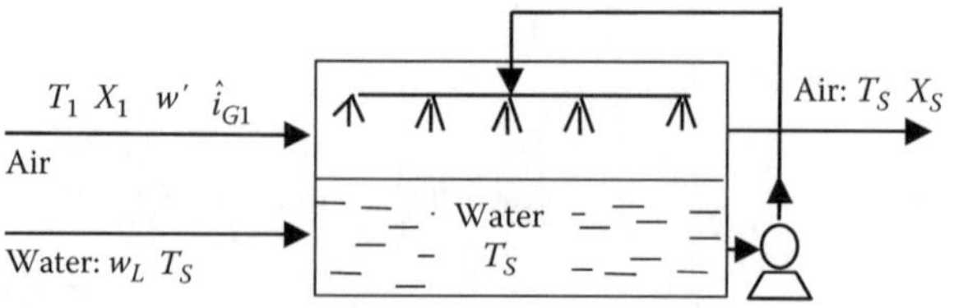

FIGURE 8.5 Installation scheme for adiabatic saturation of air.

hence

$$\hat{C}_L T_S\left(X_S - X_1\right) + \hat{s}_1 T_1 + \lambda_0 X_1 = \hat{s}_1 T_S + \hat{C}_V\left(X_S - X_1\right) T_S + \lambda_0 X_S$$

which when rearranged is $(X_S - X_1)[\hat{C}_L T_S - \hat{C}_V T_S - \lambda_0] = \hat{s}_1(T_S - T_1)$.

Since the latent heat of evaporation at the saturation temperature T_S is given by

$$\lambda_S = \lambda_0 + \hat{C}_V T_S - \hat{C}_L T_S \tag{8.29}$$

one can obtain

$$X_S - X_1 = -\left(\frac{\hat{s}_1}{\lambda_S}\right)\left(T_S - T_1\right) \tag{8.30}$$

which is an equation that allows the calculations in adiabatic saturation processes, in function of the absolute humidity and temperature.

If Equations 8.27 and 8.30 are compared, it can be observed that the adiabatic saturation temperature coincides with that of the wet bulb temperature ($T_S = T_H$) for a total pressure of 101.3 kPa (1 atm) and specific heat given by

$$\hat{s}_1 = 0.00614 \frac{\left(h_C + h_R\right)}{K_G} \tag{8.31}$$

which is an expression known as Lewis relation.

From the last comparison, the following psychrometric relation is obtained:

$$b = 0.00614 \frac{\left(h_C + h_R\right)}{K_G \hat{s}_1} \tag{8.32}$$

that for the air–water system has a value of 1 kJ/(kg °C). For other gas–liquid systems, this psychrometric relation is greater than the unit ($b > 1$), and also the wet bulb temperature is higher than the saturation temperature.

In the $\hat{s}T$–X chart the conditions of air at the entrance of the chamber are given by point 1 (Figure 8.6). The conditions at the exit point are obtained when drawing a straight line with slope $-\hat{s}/\lambda_S$, whose intersection with the saturation curve (point 2) precisely gives the conditions of the air that leaves the chamber.

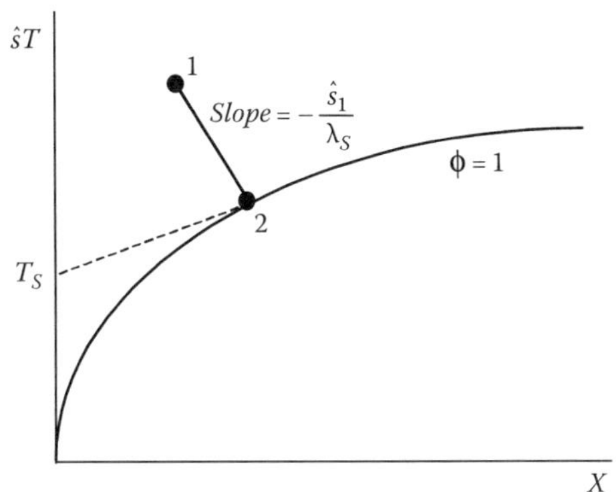

FIGURE 8.6 Adiabatic saturation of air.

Without a psychrometric chart, the adiabatic saturation temperature can be obtained by means of an iterative process, consisting of the following calculation steps:

a. A saturation temperature T_S is supposed.
b. The latent heat λ_S and saturation vapor pressure P_S are calculated according to the temperature earlier.
c. The corresponding saturation humidity X_S is calculated using Equation 8.30.
d. The saturation humidity X'_S is calculated with Equation 8.4.
e. The saturation humidities X_S and X'_S calculated in steps (c) and (d) are compared. If the values coincide, the calculation is assumed to be complete. On the contrary, if the values do not coincide, a new value of the saturation temperature is taken, beginning again with step (a).

PROBLEMS

8.1 A humid air sample has a dry bulb temperature of 75°C and a wet bulb temperature of 45°C. Calculate its absolute and relative humidity, as well as its density.

From the psychrometric equation (8.26), as indicated before, under usual work conditions, it is considered that this equation can be expressed as

$$P_S - P_V = -66(T_H - T_1)$$

if pressure is given in Pa and temperature in °C.

For the dry bulb temperature of 75°C, the saturation pressure is obtained from the saturated water vapor tables (P_V = 38.5 kPa), from which the absolute humidity can be obtained using Equation 8.3:

$$X = 0.622\frac{P_V}{P - P_V} = 0.622\frac{38.5}{101.23 - 38.5} = 0.382\left(\text{kg water/kg dry air}\right)$$

A saturation pressure of 9.8 kPa corresponds to the temperature of 45°C, from which a saturation humidity X_S = 0.067 kg water/kg dry air is obtained.

Applying the psychrometric equation: $P_V - 9800 = 66(75 - 45)$, so $P_V = 7820$ Pa. The absolute humidity can be determined from this pressure and Equation 8.3:

$$X = 0.622\frac{P_V}{P - P_V} = 0.622\frac{7.82}{101.23 - 7.82} = 0.052\left(\text{kg water/kg dry air}\right)$$

The relative humidity is obtained from Equation 8.5:

$$\phi = \frac{P_V}{P_S} = \frac{7{,}820\ \text{Pa}}{38{,}500\ \text{Pa}} = 0.20\ (20\%)$$

The humid-specific volume is calculated by Equation 8.8, in which the absolute humidity is 0.052 kg water/kg dry air, the temperature is 348 K, and a total pressure of 101.23 kPa, while the constant R is 8.314×10^3 Pa m^3/(kmol K). With these values, a humid volume value of 1.072 m^3 dry air/kg is obtained. Hence, the density will be the inverse of this value: 0.933 kg/m^3.

The density can be calculated from Equation 8.11b:

$$\rho = 0.0035\frac{101{,}230}{348} - 0.0013\frac{(0.2)(38{,}500)}{348} = 0.989\ \text{kg/m}^3$$

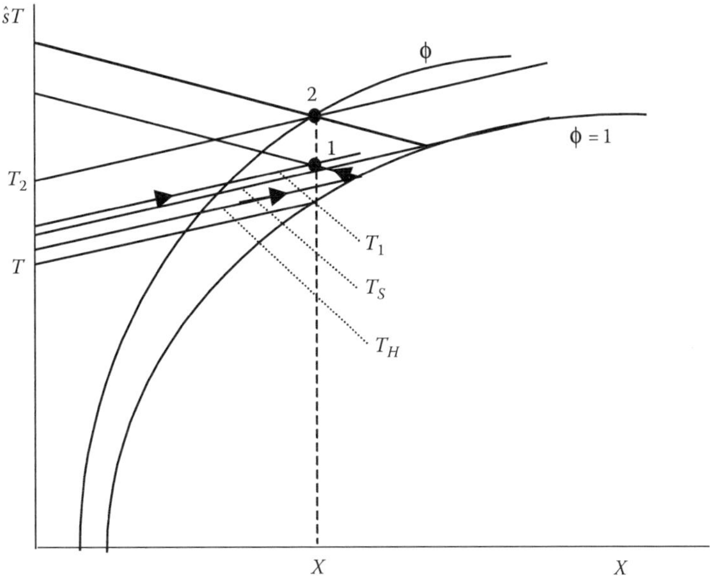

FIGURE P.8.2 Characteristic points in the psycrometric chart for the Problem 8.2.

8.2 A room is maintained at a temperature of 21°C, by means of a radiator, while outside the temperature is 15°C, with a corresponding wet bulb temperature at 10°C. Determine the absolute humidity, water vapor pressure, relative humidity, dew temperature, and adiabatic saturation temperature of the air contained in the room.

With the temperatures T_1 = 15°C and T_H = 10°C, it is possible to locate point 1 in the psychrometric chart (Figure P.8.2). When the room is warmed up to 21°C, new air conditions are obtained, as represented by point 2 in the same chart.

Point 2 is obtained from point 1, by moving up at constant absolute humidity until reaching the 21°C isotherm. The absolute humidity will not vary, since the kg water/kg dry air is constant.

The values asked for in the problem can be obtained from the psychrometric chart:

Absolute humidity: $X = 0.0056$ kg water/kg dry air
Relative humidity: $\phi = 0.37$ (37%)
Dew temperature: $T_R = 5.2$°C
Adiabatic saturation temperature: $T_S = 12.8$°C

The vapor pressure is calculated from Equation 8.3, in which the values of the different variables ($X = 0.0056$ kg water/kg dry air, $P = 101.23$ kPa) are substituted to obtain the vapor pressure $P_V = 903$ Pa.

8.3 The temperature in a warehouse is 30°C, the dew temperature being 12°C of the air contained in that warehouse. Calculate (a) the relative humidity that the air will have if it is cooled down to 16°C and (b) the amount of water that will be eliminated from 570 m^3 of air under the indicated conditions if it is cooled down to 2°C.

In the psychrometric chart, point 1 (Figure P.8.3) represents the conditions of the air in the warehouse, with T_1 = 30°C and T_R = 12°C, and an absolute humidity content of $X_1 = 0.0087$ kg water/kg dry air. When cooling this air down to 16°C, the absolute humidity will not vary ($X_2 = X_1$). Therefore, point 2 is obtained when the vertical straight line of constant absolute humidity X_2 intersects the 16°C isotherm.

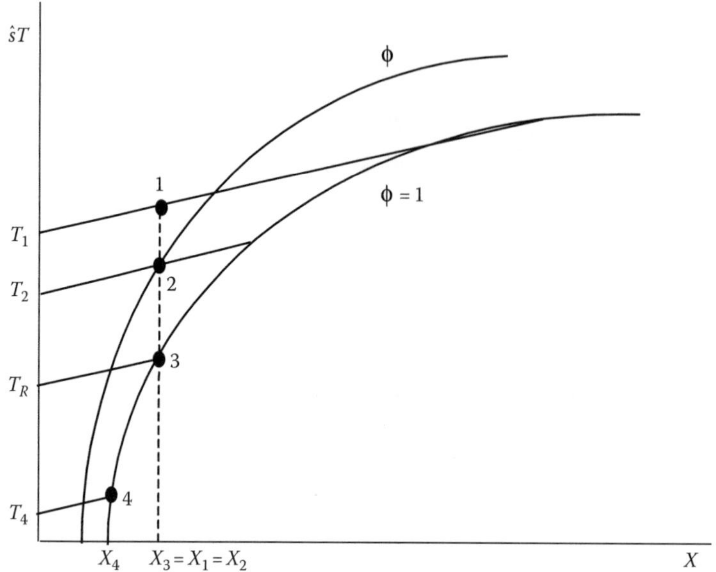

FIGURE P.8.3 Characteristic points in the psycrometric chart for the Problem 8.3.

This intersection allows us to see point 2, which corresponds to a relative humidity curve $\phi_2 = 0.77$ (77%).

The value of this relative humidity can be analytically obtained. Initially, the absolute humidity is calculated from Equation 8.3, using the vapor pressure obtained from water vapor tables for $T_R = 12°C$, being $P_V = 1390$ Pa. These values are substituted in Equation 8.3:

$$X_1 = 0.622\frac{1.39\times10^3}{\left(101.23-1.39\right)\times10^3} = 0.00866\left(\text{kg water/kg dry air}\right)$$

From water vapor tables, the saturation pressure for 16°C is $P_S = 1832$ Pa. Since absolute humidity X_2 is the same as X_1, it is possible to determine the relative humidity from Equation 8.6:

$$0.0087 = 0.622\frac{1.83\times10^3\phi}{(101.23-1.83\phi)\times10^3}$$

Hence, the relative humidity for the conditions of point 2 is $\phi_2 = 0.76$ (76%).

To calculate the kg of water condensed when the temperature decreases to 2°C, absolute humidities at 16°C (X_3) and at 2°C (X_4) are previously determined. The difference ($X_3 - X_4$) gives the amount of water eliminated per each kg of dry air. Therefore, we first calculate the kg of dry air contained in the 570 m^3 of humid air.

The humid volume occupied by 1 kg of dry air at 30°C is calculated from Equation 8.8:

$$\hat{V} = \left(\frac{1}{28.9}+\frac{0.0087}{18}\right)\frac{\left(8.314\times10^3\right)\left(303\right)}{101.23\times10^3} = 0.873\left(\text{kg humid air/kg dry air}\right)$$

So, the kg of dry air contained in the 570 m^3 of air will be

$$570\,\text{kg dry air}\frac{1\,\text{kg dry air}}{0.873\,\text{kg humid air}} = 653\,\text{kg dry air}$$

The absolute humidity of point 3 coincides with that of points 1 and 2, while absolute humidity of point 4 is obtained from the psychometric chart:

$X_3 = X_1 = X_2 = 0.0087$ kg water/kg dry air
$X_4 = 0.0042$ kg water/kg dry air

This allows us to calculate the water eliminated when the air is cooled down to 2°C as

$$653\,\text{kg dry air}\left(0.0087-0.0042\right)(\text{kg water/kg dry air}\ ) = 2.94\,\text{kg of water}$$

8.4 The atmospheric pressure of a July day was 1 atm, the temperature of the air was 32°C, and its relative humidity was 30%. Determine (a) the amount of water contained in the air in a 162 m^3 room, (b) the weight of the air–water vapor mixture found in the room, (c) why eggs "sweat" when taken out of the refrigerator (the temperature inside the refrigerator was 8°C), and (d) the absolute humidity of the air in the room, once saturated, if it was isothermally humidified.

a. For the temperature T_1 = 32°C and a relative humidity $\phi = 0.3$, point 1 can be fixed in the psychrometric chart (Figure P.8.4), corresponding to an absolute humidity of $X_1 = 0.0088$ kg water/kg dry air. The wet volume can be calculated from Equation 8.8:

$$\hat{V} = \left(\frac{1}{28.9}+\frac{0.0088}{18}\right)\frac{\left(8.314\times10^{3}\right)\left(305\right)}{101.23\times10^{3}} = 0.879\,\text{m}^3\ \text{humid air/kg dry air}$$

Since the volume of the room is 162 m^3, and it is humid air, the amount of dry air in the room will be

$$162\,\text{m}^3\ \text{wet air}\frac{1\,\text{kg dry air}}{0.879\,\text{m}^3\ \text{humid air}} = 184.3\ \text{kg dry air}$$

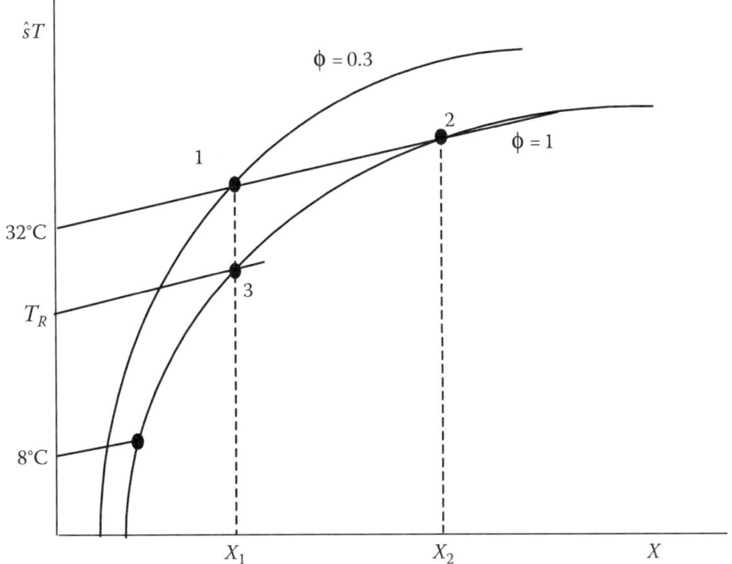

FIGURE P.8.4 Characteristic points in the psycrometric chart for the Problem 8.4.

The amount of water in the room can be calculated multiplying this value by the absolute humidity of the air:

$$(184.3\,\text{kg dry air})0.0088(\text{ kg water/kg dry air}) = 1.622\text{ kg water}$$

b. The weight of the air–water mixture is the sum of the values obtained in the last section:

$$184.3\text{ kg dry air} + 1.62\text{ kg water} = 185.92\text{ kg humid air}$$

c. From the psychrometric chart, for T_1 = 32°C and ϕ = 0.3, the absolute humidity is X_1 = 0.0088 kg water/kg dry air. For this absolute humidity (X_1) and for a relative humidity ϕ = 1, point 3 can be obtained, where the 12.3°C isotherm crosses; that is, the corresponding dew temperature is T_R = 12.3°C. Therefore, when cooling the egg to 8°C $T_{EGG} < T_R$, water will condense on the surface of the egg.
d. When air is isothermally humidified until saturation, in the psychometric chart, the 32°C isotherm is followed to go from point 1 to point 2. Hence, the conditions of point 2 are T_2 = 32°C and ϕ_2 = 1. The correspondent abscissa to point 2 is the value of the absolute humidity of the air under these new conditions: X_2 = 0.030 kg water/kg dry air.

8.5 In a dryer, 100 kg/h of water are eliminated from a wet material, using an airstream at 24°C with an absolute humidity of 0.01 kg water/kg dry air. This air is heated up to 69°C before entering the dryer. At the exit of the dryer, there is a thermometer indicating a dry bulb temperature of 54°C and a wet bulb temperature of 38°C. (a) Determine the consumption of air. (b) At the entrance of the dryer, the water in the material is at 24°C, while the vapor that leaves the dryer is at 54°C. Calculate the heat flux that should be supplied to the dryer. Determine, as well, the heat that should be supplied to the preheater.

All of the transformations undergone by the air throughout the process can be represented in a psychrometric chart (Figure P.8.5). Initially, the air has a temperature T_1 = 24°C and absolute humidity X_1 = 0.01 kg water/kg dry air, which allows us to determine point 1. In the heater, the air increases its temperature up to T_2 = 69°C, although its absolute humidity remains the same ($X_2 = X_1$), determining in this way point 2. The conditions of the air that leaves the dryer are T_H = 38°C and T_3 = 54°C, so point 3 is obtained by drawing the isotherm that passes at 38°C,

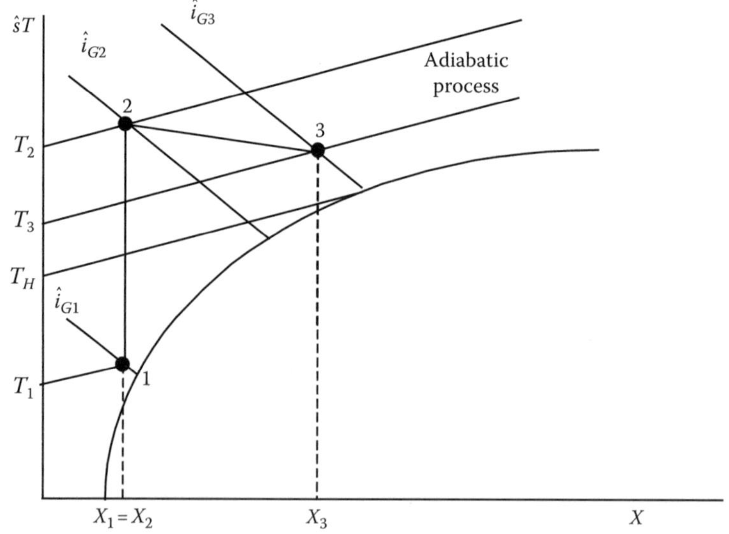

FIGURE P.8.5 Characteristic points in the psycrometric chart for the Problem 8.5.

when this isotherm intersects the curve $\phi = 1$, then the isenthalpic is plotted, and the intersection of the isenthalpic line with the 54°C isotherm allows us to determine point 3, with a corresponding absolute humidity of $X_3 = 0.0375$ kg water/kg dry air.

a. If the flow of dry air that circulates through all the process is w', from a water balance in the dryer we obtain

$$w'(X_3 - X_2) = 100 \text{ kg evaporated water/h}$$

$$w' = \frac{100 \text{ kg evaporated water/h}}{(0.0375 - 0.01) \text{ kg water/kg dry air}} = 3636.36 \text{ kg dry air/h}$$

The humid airflow rate entering the heater is the same that leaves it, that is,

$$w_1 = w_2 = w'(1 + X_2) = 3636.36\ (1 + 0.01) = 3672.73 \text{ kg humid air/h}$$

b. It is assumed that the dry material leaves the dryer at a temperature of 21°C, so the enthalpy associated with this material coincides at the entrance and at the exit of the dryer. The reference temperature in this case is $T^* = 0$°C. If Q_S is the heat flow to the dryer, the following is obtained when an energy balance is performed:

$$\dot{Q}_S + w'\,\hat{i}_{G2} w_{WATER} \hat{C}_P T_{WATER} = w'\hat{i}_{G3}$$

The enthalpies $\hat{i}_{G2}$ and $\hat{i}_{G3}$ are obtained from the psychrometric chart, and their values are

$$\hat{i}_{G2} = 95.42 \text{ kJ/kg}$$

$$\hat{i}_{G3} = 151.5 \text{ kJ/kg}$$

When substituting the data in the last equation, the following is obtained:

$$\dot{Q}_S + (3636.36)(95.42) + (100)(4.18)(21) = (3636.36)(151.5)$$

$$\dot{Q}_S = 1.95 \times 10^5 \text{ kJ/h} = 54.2 \text{ kW}$$

If $\dot{Q}_C$ is the heat flow rate supplied to the heater, from an energy balance, we obtain

$$\dot{Q}_C = w\left(\hat{i}_{G2} - \hat{i}_{G1}\right)$$

The value of $\hat{i}_{G1}$ is obtained from the psychometric chart: $\hat{i}_{G1} = 49.38$ kJ/kg. This value could also be obtained from Equations 8.13 and 8.16:

$$\hat{s} = 1 + 1.92X_1 = 1 + 1.92(0.01) = 1.0192 \text{ kJ/(kg °C)}$$

$$\hat{i}_G = \hat{s}T + \lambda_0 X_1 = (1.0192)(24) + (2490)(0.01) = 49.36 \text{ kJ/kg}$$

Hence, the heat supplied to the air in the heater will be

$$\dot{Q}_C = (3{,}636.36)(95.42 - 49.36) = 167{,}490\,\text{kJ/h}$$

$$\dot{Q}_C = 167{,}490\,\text{kJ/h} = 46.5\,\text{kW}$$

PROPOSED PROBLEMS

8.1 A humid air possesses a dry temperature of 65°C and a bulb temperature of 40°C. Calculate its absolute and relative humidity, as well as its density.

8.2 A radiator heater maintains a camera at 22°C, the temperature of the exterior being 12°C. Determine the absolute and relative humidity, the dew temperatures and adiabatic saturation temperature, and the water vapor pressure for the contained air inside the chamber.

8.3 Certain air possesses a relative humidity of 35%. It is desirable to cool 1000 kg of this air from 60°C up to 20°C. Calculate (a) the absolute humidity of the air after being cooled, as well as its dew temperature; (b) the amount of water that is condensed in this cooling process; and (c) the total eliminated heat.

8.4 In the drying process of a food material, 50 kg/h of water of this food is eliminated. An air current at 20°C is used, with an absolute humidity of 0.01 kg water/kg dry air. Before entering the dryer, the air warms up to 65°C. The air leaving the dryer possesses a dry temperature of 50°C and a bulb temperature of 35°C. Calculate the air consumption necessary for the drying process, as well as the heat flow rate that should be given in the heater. If the food material to the dryer entrance is of 20°C and the leaving vapor temperature is 50°C, calculate the heat that should be given to this dryer. Assume that the solid fraction of the food does not experience change in its temperature.

8.5 A fruit reception warehouse has a thermometer with a dry temperature of 25°C, while the humid temperature is 10°C. With the purpose of conditioning this warehouse appropriately, the air cools down up to 15°C; for these new conditions, calculate the new relative humidity that the air possesses. In the case of 600 m^3 of air with the previous conditions cooling down up to 5°C, determine the water vapor amount that will be eliminated in form of liquid water.

9 Water Activity

9.1 INTRODUCTION

Water is one of the main components of food, and it plays a very important role in the rate of deteriorative processes in foods. Water has a great influence on the development of different types of microorganisms as well as on many of the chemical and enzymatic reactions that can be present in food.

Historically, it is well known that foods with high water content deteriorate rapidly; therefore, strategies and processes to decrease water content in order to preserve food for longer periods of time have been studied. However, water content is not the only indicative factor of deterioration, but if two different food products with the same water content undergo different deterioration processes, their preservation period is also different. That is why the water content for a food can serve as an indicative factor of its perishability, but other factors should also be taken into account such as the interaction of water with the food matrix. It is at this point where it is necessary to introduce a new concept of water availability so that the different types of deterioration (microbial, chemical, and enzymatic) can be included; this new concept is denominated water activity. This new factor provides a measure of the water available in a food so that it can participate in the processes of microbial and chemical deterioration.

A fundamental aspect in food preservation is to not only know the quantity of water that it contains, but also how it is bound to the food matrix. The term water activity was introduced in the 1950s to describe the state of water in foods.

9.2 DEFINITION OF WATER ACTIVITY

Water activity is defined as the relationship between the vapor pressure of water in a system and the vapor pressure of pure water at the same temperature or relative humidity of the air surrounding the system. In Chapter 6 the thermodynamics of water activity (a_w) was presented, and mathematically it is expressed as

$$a_w = \frac{P_w}{P_w^0} \tag{9.1}$$

where

P_w is the vapor pressure of water in equilibrium with the food
P_w^0 is the vapor pressure of pure water at the same temperature

Water activity is a key factor in microbial growth and the development of many reactions in food, as well as the heat resistance of many microorganisms. Generally, for values of water activity lower than 0.90, bacterial growth is inhibited, while for yeasts the value of the activity should be lower than 0.87, and in the case of mushrooms the limit is 0.80. For nonenzymatic browning reactions, it is observed that water activity has a fundamental influence; so for water activity values around 0.70–0.80, the browning rate presents a maximum. Also, water activity can be used in the prediction of different thermodynamic properties, such as the freezing and boiling points, as well as the sorption heat (Rahman and Labuza, 2002).

Water activity in foods can be affected by different factors, such as the components, the structural characteristics of the food, as well as the pressure and temperature. The composition and type of components of food affects water activity, since, for example, at low values of water activity proteins and starch adsorb more water than fatty matters or crystalline substances. The presence of electrolytes also has an important influence on water adsorption processes by food, as they will affect their own water activity.

One factor that affects water the most is temperature; it already affects the mobility of water molecules and the dynamic equilibrium between the vapor phase and adsorbed phases. Clausius–Clapeyron's equation (Equation 6.95) is useful to determine the effect of temperature in which water interacts with food.

Water activity influences certain reactions with foods. At low values of a_w, water is closely bound to the adsorption polar points, forming a monolayer on the food surface, thereby preventing the reactants from solvating. For high values, the water is in multilayers or in a condensed phase in the food capillaries, and then its mobility increases. Anything that makes the availability of water and the reactants greater will guarantee enough mobility to allow interactions between them, in this way creating certain reactions. If the activity possesses very high values, there could be a case where the reactants are much diluted and the reaction possibility among them decreases, since it diminishes the probability of interactions between reactants. However, each specific reaction has a specific interval of water activity that is more susceptible to the reaction that is being carried out. Figure 9.1 shows the influence of water activity on different types of reactions.

In this figure it can be observed that for most chemical and microbial growth reactions, the reaction rate diminishes as water activity decreases. However, for lipid oxidation, a different behavior to that of other types of deteriorations is observed, where it can be seen that for very low water activity values, the oxidation rate shows an appreciable increase; this could be due to the possible catalytic activity of metallic ions when the water of the hydrate sphere is eliminated around the ions. The biggest reaction rate observed in all of the considered reactions can be attributed to high values of water activity due to the greater mobility of the reactants, which causes them to interact better among themselves. Understanding water activity is an important factor in guaranteeing food stability as it will be shown later.

In solid systems with low humidity content, the debate as to whether the property of water activity or the state of the system, which is dictated by the glass transition temperature, most affects the chemical reaction rate in food has been settled. Below the glass transition temperature, it is difficult for the reactants that intervene in a certain reaction to diffuse and interaction among them is not achieved, so that the reaction is not complete; thus, foods in a glassy state are considered to be very stable. Water activity is a property of the solvent, while the glassy state is related to the structure of the food matrix, and both are necessary to understand the reactions that take place in a food under different conditions.

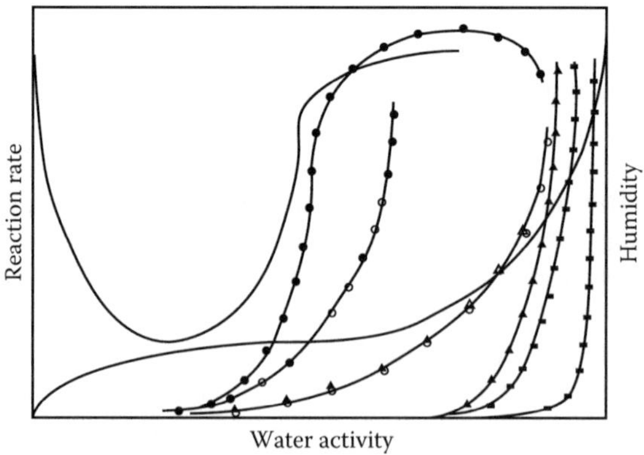

FIGURE 9.1 Stability of a food product. Lipid oxidation (—). Nonenzymatic browning (•). Hydrolytic reactions (○). Enzymatic activity (□). Fungi growth (Δ). Yeast growth (■). Bacterial growth (□). Sorption isotherm (→). (Adapted from Welti, J. and Vergara, F., Actividad de Agua. Concepto y aplicación en alimentos con alto contenido de humedad, in *Temas en Tecnología de Alimentos*, J.M. Aguilera, ed., CYTED (Programa Iberoamericano de Ciencia y Tecnología para el Desarrollo), Puebla, México, 1997.)

9.3 METHODS TO MEASURE WATER ACTIVITY

Diverse reviews of the different methods used in the measurement of water activity in foods have been published (Troller and Christian, 1978; Rockland and Nishi, 1980; Gal, 1983; Troller, 1983; Schurer, 1985; van der Berg, 1985; Leung, 1986). There are different methods for measuring the water activity in foods, although there have been considerable variations in their precision according to the method used. In general, the methods for measuring water activity can be classified as methods of the colligative properties, methods of the sorption rate in equilibrium, and methods that are based on the hygroscopicity of salts or hydrometric instrumental methods (Rahman, 1995). The methods of colligative properties are based on measuring the vapor pressure, the boiling point, and the freezing point.

Equilibrium sorption rate methods are based on the equilibrium of a food sample at constant temperature in atmospheres controlled by different relative humidity. When placing the food in an atmosphere at a certain relative humidity, a flow of water appears between the food and its surroundings until arriving at equilibrium, so that in the end the food possesses a water activity similar to the relative humidity of the atmosphere. With this method the sorption isotherms of the food can be obtained. This is a commonly used method, although it presents the inconvenience that equilibrium can create high relative humidity for long periods of time, and microbial growth in the food can take place.

The hygroscopicity method of salts is based on the fact that water vapor will condense in a salt crystal in an atmosphere that has a greater relative humidity than the critical humidity of each salt. The relative humidity can be considered by observing the change of color of certain salts in the headspace on a sample.

The hydrometric instrumental method is based on the determination of the humidity of the atmosphere that surrounds a food when they are in equilibrium, using a hygrometer. The hygrometers used in this measurement are based on the frost point, the dew point, dry and wet temperature, electric resistance or capacitance of a material, and the expansion capacity of the material.

9.3.1 Measure of Vapor Pressure

This method is based on the measurement of food vapor pressure, and with the value of the vapor pressure for pure water at the same temperature, Equation 9.1 is used to calculate the water activity value. To do so, the food is placed under conditions that allow the food's equilibrium with its surroundings, and once in equilibrium the pressure is measured with a pressure gauge or transducer. This measurement can be affected by factors such as size of the sample, equilibrium time, temperature, and volume of the atmosphere that surrounds the sample, among others. This method is not advisable for foods that contain an appreciable quantity of active respiration, since the atmosphere that surrounds the sample will be affected.

9.3.2 Cryoscopic and Boiling Increments

Cryoscopic decrease and boiling increase, as well as the change that can experience other colligative properties, are correlated with water activity. Several researchers have used these methods for the determination of water activity (Strong et al., 1970; Ferro-Fontán and Chirife, 1981). This method has been successfully applied for liquid foods at high water activities. The equations that are used for water activity calculation are the following:

Cryoscopic decrease

$$\log a_w = -0.004207\Delta T_f - 2.1\times 10^{-6}\Delta T_f^2 \tag{9.2}$$

Cooling increase

$$\log a_w = -0.01526\Delta T_b + 4.862\times 10^{-5}\Delta T_b^2 \tag{9.3}$$

where ΔT_f and ΔT_b are cryoscopic decrease and boiling increase, respectively.

9.3.3 Osmotic Pressure

Another colligative property that allows calculation of the water activity of aqueous solutions is osmotic pressure (π), using the equation

$$\pi = \frac{RT}{V_w}\ln(a_w) \tag{9.4}$$

where

V_w is the water molar volume in solution
T is the absolute temperature
R is the gas constant

For ideal solutions, osmotic pressure can be expressed as a function of the water molar fraction in solution, then

$$\pi = \frac{RT}{V_w}\ln(X_w) \tag{9.5}$$

For real solutions, osmotic pressure can be obtained from the equation

$$\pi = \phi\upsilon RT\frac{C_S}{C_w V_w} \tag{9.6}$$

where

ϕ is the osmotic coefficient
υ is the number of ions in solution
C_S and C_w are the solute and water molar concentration, respectively

When comparing Equations 6.4 and 6.6, the following is obtained:

$$\ln(a_w) = \phi\upsilon\frac{C_S}{C_w} \tag{9.7}$$

9.3.4 Dew Point Hygrometer

The measure of water activity is based on the water condensation, in its dew point, for an air–water vapor mixture, on a flat and cold surface, similar to a mirror. The formation of dew on the mirror is detected photoelectrically. This method allows the determination of water activity in the interval from 0.72 to 1.

9.3.5 Thermocouple Psychrometer

This measuring method of water activity is based on the decrease of wet temperature. The sample is placed in a chamber until equilibrium is reached with the atmosphere that surrounds the sample. The chamber has a thermocouple inserted on which water condenses. Later, the water on the thermocouple evaporates, and in this way a decrease of temperature will take place, relating the evaporation rate for water with the relative humidity of the atmosphere in equilibrium with the sample.

9.3.6 Electrical and Filament Hygrometers

This apparatus contains a filament recovered by a hygroscopic salt, so this coating adsorbs the water vapor liberated by the sample, which causes a change in the conductance or capacitance of the filament that is correlated with the water activity. In this type of hygrometer, there is a possibility that some volatile compounds of food can affect the filament and thus the reading of the water activity given might not be correct.

Filament hygrometers basically consist of hairs or filaments of synthetic fiber that shrink or stretch depending on the relative humidity to which they are exposed. The variation of filament length is correlated with the relative humidity of the atmosphere and with water activity of the food. The biggest disadvantage in this instrument type is the length of time needed to reach equilibrium, although its main advantage is the low cost of the device.

9.3.7 Isopiestic Methods

For water activity calculation of any sample, the measurement takes place inside a desiccator together with a reference material. The sample is set to a certain temperature and a certain time so that equilibrium is reached. Once equilibrium conditions are achieved, the humidity of the reference material is determined, and by means of the sorption curve of this material at the work temperature, its water activity is determined. For example, if the reference material is in equilibrium, the activity of both is the same. Microcrystalline cellulose, potato starch, filter paper, and certain proteins are usually used as reference materials.

Another isopiestic method is graphic interpolation. In this method solutions of sulfuric acid of different concentrations are used to induce atmospheres with different relative humidity. Solutions of salts can also be used, which give rise to different relative humidity. Table 9.1 shows the values of relative humidity generated by saturated solutions of different salts. The sample is left to reach equilibrium in these atmospheres with different relative humidity. The gain or loss of humidity is

TABLE 9.1
Relative Humidity Generated by Saturated Salt Solutions

	Relative Humidity (%)				
Salt	**10°C**	**15°C**	**20°C**	**25°C**	**30°C**
LiBr	7.1	6.9	6.6	6.4	6.2
NaOH	—	9.6	8.9	8.2	7.6
LiCl	11.3	11.3	11.3	11.3	11.3
$KC_2H_3O_2$	23.5	23.5	23.0	22.5	22.0
$MgCl_2$	33.5	33.0	33.0	33.0	32.5
K_2CO_3	44.0	43.5	43.0	43.0	43.0
NaBr	60.0	59.0	58.0	57.5	56.5
$CuCl_2$	68.0	68.0	68.0	67.5	67.0
KI	72.0	71.0	70.0	69.0	68.0
NaCl	76.0	75.5	75.5	75.5	75.0
$(NH_4)_2SO_4$	81.0	80.5	80.5	80.0	80.0
KCl	87.0	86.0	85.0	84.5	84.0
$KC_7H_5O_2$	88.0	88.0	88.0	88.0	88.0
KNO_3	95.5	95.0	94.0	93.0	92.0
K_2SO_4	98.0	98.0	97.5	97.0	97.0

Source: Rockland, L.B. and Nishi, S.K., *Food Technol.*, 34(4), 42, 1980.

represented graphically versus the water activity for each sample in equilibrium. The water activity is estimated when there is no change in the weight of the sample.

9.4 PREDICTION OF WATER ACTIVITY IN BINARY SOLUTIONS

In the previous section, different experimental methods to obtain water activity of foods have been described. However, and without the necessity of appealing to laboratory experiments, there are theoretical and semiempirical models that allow calculation of the water activity of solutions from equations, although it is necessary to know the solute type possessed by the food solution. The water activity for a sample of pure water is 1; however, the solute presence causes the water activity to decrease its value. Assuming that a solution behaves ideally, to calculate the water activity, Raoult's law can be applied. However, for concentrated solutions and for electrolyte solutes, it is not possible to apply this law, but there are different equations that allow the calculation of water activity.

9.4.1 Ideal Behavior

The diluted binary solutions generally show an ideal behavior, so it is possible to apply Raoult's law for a component i of the mixture:

$$P_i = x_i P_i^0 \tag{9.8}$$

where
P_i is the partial pressure of the component i
P_i^0 is the vapor pressure of the pure component i
x_i is the molar fraction for component i in the mixture

If the considered component is water, keeping in mind the definition of water activity (Equation 9.1), we can write

$$a_w = x_w = \frac{P^w}{P_0^w} = \frac{n_w}{n_w + n_S} \tag{9.9}$$

where n_w and n_S are the water and solute mol number in solution, respectively. This equation can be applied to diluted solutions and for electrolytes with a concentration lower than 1 M. It can also be applied to solutions of glucose up to 4 M, although it is not applicable for most food solutions.

9.4.2 Nonideal Behavior

Most food solutions behave in a nonideal way, and thus it is not possible to use Equation 9.9 to calculate water activity. In this case, water activity is not equal to the molar fraction, but rather an activity coefficient should be introduced (γ_i), so we can use the following equation:

$$a_w = \gamma_i x_w \tag{9.10}$$

It is observed that the activity coefficient is of great importance in the calculation of water activity. That is why different equations have been developed for the calculation of this coefficient, thus giving place to diverse expressions that allow the calculation of water activity.

Also, the solute type of solutions has great importance, since the behavior of the solution is different depending upon whether the solute is an electrolyte. In the following sections, different equations are shown, depending on the solute type.

9.4.2.1 Nonelectrolytic Solutions

A nonelectrolytic solute in water is not dissociated. These solutes, denominated moisturizers, have the capacity to bind water, which gives rise to many food products that can maintain a good texture. Different equations that allow the calculation of water activity are shown next.

9.4.2.1.1 Norrish's Equation

It is assumed that the coefficient of water activity, γ_w, is a function of the solute molar fraction (x_S), according to the equation (Hildebrant and Scott, 1962)

$$\log(\gamma_w) = Kx_S^2 \tag{9.11}$$

where K is a proportionality constant that depends on the solute type. In Table 9.2 values of K are shown for different solute types.

TABLE 9.2
***K* Values for Norrish's Equation**

Substance	K
Sugars	
Sucrose	6.47
Maltose	4.54
Glucose	2.25
Xylose	1.54
Lactose	10.2
Lactulose	8.0
Polyols	
Sorbitol	1.65
Erythritol	1.34
Glycerol	1.16
Mannitol	0.91
Arabitol	1.41
Ribitol	1.49
Xylitol	1.66
Propylene glycol	4.04
PEG-200	6.10
PEG-400	26.6
PEG-600	56.0
Amino acids and amides	
α-Amino-*n*-butyric acids	2.59
β-Alanine	2.52
Lactamide	−0.705
Glycolamide	−0.743
Urea	−2.02
Glycine	−0.868
Organic acids	
Citric acid	6.2
Malic acid	1.82
Tartaric acid	4.68

Sources: Chirife, J. et al., *J. Food Technol.*, 15, 59, 1980; Chirife, J. and Favetto, G.J., Fundamental aspects of food preservation by combined methods, International Union of Food Science and Technology—CYTED D—Univ. de las Américas, Puebla, México, 1992.

Combining Equations 9.10 and 9.11, an expression is obtained that allows calculation of the water activity:

$$a_w = x_w \exp\left(-Kx_S^2\right) \tag{9.12}$$

such that it is the denominated Norrish's equation. It is observed that this equation is applicable for much diluted solutions, the solute fraction molar is practically zero, and the equation coincides with Raoult's law.

9.4.2.1.2 Money and Born Equation

For sweetened solutions used in the elaboration of marmalades and sweet shop products, in order to calculate water activity, the empiric equation can be used (Money and Born, 1951):

$$a_w = \frac{1}{1+0.27n_S} \tag{9.13}$$

where n_S is the number of solute moles (sugar) for each 100 g of water.

9.4.2.1.3 Grover's Equation

For candies there is an empiric equation (Barbosa-Cánovas and Vega-Mercado, 1996) to determine water activity as a function of composition. The equation is

$$a_w = 1.04 - 0.1\sum_i s_i c_i + 0.0045\sum_i \left(s_i c_i\right)^2 \tag{9.14}$$

where

c_i is the concentration of component i
s_i is a factor of equivalent conversion of sucrose of each of the candy ingredients

In this way, for the inverted sugar, the factor is 1.3, as well as for lactose and gelatin, being 0.8 for corn syrup 45DE.

9.4.2.2 Electrolytic Solutions

Electrolytes are compounds that in solution are dissociated into ions. The ions that are present in solution affect water activity because the solution osmotic pressure is increased when there are more components in the solution. Osmotic pressure and water activity are correlated by Equation 9.4. In the case that there are electrolytes, an osmotic coefficient appears (Φ) that is defined for nonideal solutions according to the following equation:

$$\Phi = -\frac{1000}{M_w \sum_i \nu m_i} \ln\left(a_w\right) \tag{9.15}$$

where

M_w is the molecular weight of the solvent
ν is the number of ions present in the solution
m_i is the molal concentration for the species i (moles of the species i for kg of solvent)

For diluted solutions, where an infinite dilution of all the solutes can be considered, the osmotic coefficient tends to zero.

For aqueous solutions, the molecular weight value for the solvent is 18.02 kg/kmol, and if the value of osmotic coefficient is known, the water activity of a solution of electrolytes can be obtained from Equation 9.15:

$$a_w = \exp\left(-0.01802\Phi \sum_i \nu m_i\right) \tag{9.16}$$

In order to calculate the osmotic coefficient, there are a number of different equations that will be discussed as follows:

9.4.2.2.1 Bromley's Equation

The osmotic coefficient can be evaluated starting from the following equation:

$$\Phi = 1 + 2.303\left[F_1 + (0.06 + 0.6B)F_2 + 0.5BI\right] \tag{9.17}$$

where

I is the ionic strength

F_1, F_2, and B are parameters that can be calculated starting from different equations

The ionic strength is calculated by means of the following equations:

$$I = \frac{1}{2}\sum_i m_i z_i^2 \tag{9.18}$$

$$F_1 = |z_+ z_-| A_\gamma \left[\frac{1 + 2\left(1 + I^{1/2}\right)\ln\left(1 + I^{1/2}\right) - \left(1 + I^{1/2}\right)}{I\left(1 + I^{1/2}\right)}\right] \tag{9.19}$$

$$F_2 = |z_+ z_-| \left[\frac{(1 + 2aI)}{a(1 + aI)^2} - \frac{\ln(1 + aI)}{a^2 I}\right] \tag{9.20}$$

where

$$a = \frac{1.5}{|z_+ z_-|} \tag{9.21}$$

$|z_+ z_-|$ = absolute value for the product of ionic charges

$A_\gamma = 0.511$ (at 25°C)

B parameter is characteristic for each electrolyte, and Table 9.3 shows values of this parameter for different salts.

TABLE 9.3
Parameters of Bromley's and Pitzer's Equations

Electrolytes	B	$\beta(0)$	$\beta(1)$	$C_{\pm}$
NaCl	0.0574	0.0765	0.2664	0.00127
LiCl	0.1283	0.1494	0.3074	0.00359
KCl	0.0240	0.0483	0.2122	−0.0008
HCl	0.1433	0.1775	0.2945	0.0008
KOH	0.1131	0.1298	0.3200	0.0041
NaOH	0.0747	0.0864	0.2530	0.0044
KH_2PO_4		−0.0678	−0.1042	−0.1124
NaH_2PO_4	−0.0460	−0.0533	0.0396	0.00795

Sources: Bromley, L.A., *AIChE J.*, 19(2), 313, 1973; Pitzer, K.S. and Mayorga, G., *J. Phys. Chem.*, 77(19), 2300, 1973.

9.4.2.2.2 Pitzer's Equation

The osmotic coefficient can be calculated by means of Pitzer's equation:

$$\Phi = 1 - \left|z_+ z_-\right| F + 2m\left(\frac{\nu_+\nu_-}{\nu_+ + \nu_-}\right) B_{\pm} + 2m^2\left[\frac{\left(\nu_+\nu_-\right)^{3/2}}{\nu_+ + \nu_-}\right] C_{\pm} \tag{9.22}$$

where

$$F = -A\frac{I^{1/2}}{\left(1 + bI^{1/2}\right)} \tag{9.23}$$

$$B_{\pm} = \beta(0) + \beta(1)\exp\left(-\alpha I^{1/2}\right) \tag{9.24}$$

A = Debye–Hückel's coefficient at 25°C taking the value of 0.392.

In these equations $\beta(0)$, $\beta(1)$, and $C_{\pm}$ are characteristic, all solutes take the value of 1.2, while I is the ionic strength that is calculated using Equation 9.18. Table 9.3 shows the different parameters of Pitzer's equation for different solutes.

9.5 PREDICTION OF WATER ACTIVITY IN MULTICOMPONENT SOLUTIONS

To maintain the quality of many foods, it is necessary to add compounds that decrease the water activity of a food. Thus, it is common to find solutions that are formed with more than two components, making it possible to obtain different levels of water activity depending on the types of used solutes. Many researchers have discussed the equations that allow prediction of water activity of multicomponent solutions (Norrish, 1966; Pitzer and Kim, 1974; Ferro-Fontán et al., 1980, 1981; Ferro-Fontán and Chirife, 1981; Teng and Seow, 1981).

9.5.1 Norrish's Generalized Equation

Starting from the Norrish's equation for binary mixtures of nonelectrolytes, a generalized equation for multicomponent mixtures can be obtained:

$$\ln a_w = \ln x_w - \left[(K_1)^{1/2} x_1 + (K_2)^{1/2} x_2 + \cdots + (K_n)^{1/2} x_n \right]^2 \tag{9.25}$$

where K_1 and x_1 are the binary coefficients and molar fraction of the solute, respectively, for each component i of the mixture. In Table 9.2 the values of K are shown for different compounds.

9.5.2 Ross's Equation

In not very concentrated multicomponent solutions, the interactions among different solutes can be considered to be negligible or that the interactions among the different components are neglected when they are averaged (Ross, 1975). Therefore, the equation obtained for water activity of these systems is

$$a_w = \prod_i (a_w)_i = (a_w)_1 (a_w)_2 (a_w)_3 \cdots \tag{9.26}$$

This equation shows that the water activity for solutions is the product of the water activities of each of the components in a binary mixture at the same concentration and temperature as that of the multicomponent system. For water activity values greater than 0.8, the error calculating a_w is smaller than 2% (Bone et al., 1975; Ross, 1975).

9.5.3 Ferro-Fontán, Benmergui, and Chirife Equation

For strong electrolytes, Ferro-Fontán et al. (1980) developed a model starting from Ross's equation:

$$a_w = \prod_i \left[a_{wi}(I) \right]^{I_i/I} \tag{9.27}$$

$$I_i = 0.5 \nu_i m_i \left| z_+ z_- \right| \tag{9.28}$$

where

I_i is the ionic strength of the component i in the mixture
I is the total ionic strength of the multicomponent solution
$a_{wi}(I)$ is the water activity of a binary solution of ionic strength I at the same total ionic strength of the multicomponent solution
ν_i are the total solute ions i in the solution
m_i is the solute molality
z_+ and z_- are ion charges

This equation gives good results when it is applied to mixtures of strong electrolytes in the interval of water activity for foods of intermediate humidity.

For multicomponent mixtures of nonelectrolyte solutes, Ferro-Fontán and Chirife (1981) proposed the following model:

$$a_w = \prod_i \left[a_{wi}(m)\right]^{m_i/m} \tag{9.29}$$

where
m_i is the molality of the component i
m is the total molality of the mixture

9.5.4 Ferro-Fontán, Chirife, and Boquet Equation

For nonelectrolyte mixtures, Ferro-Fontán et al. (1981) proposed the following equation:

$$\left(a_w\right)_m = x_w \exp\left(-K_M x_M^2\right) \tag{9.30}$$

where
x_M is the molar fraction of all the solutes
K_M is a constant that is calculated by means of the equations

$$K_M = \sum_s K_S C_S \frac{M}{M_S} \tag{9.31}$$

$$M = \left[\sum_s \frac{C_S}{M_S}\right]^{-1} \tag{9.32}$$

where
K_S is Norrish's constant for each solute (Table 9.2)
C_S is the weight ratio of each solute with regard to the total mixture
M_S is the molecular weight of each component

9.5.5 Pitzer and Kim Equation

For multicomponent mixtures, the osmotic coefficient can be obtained from the following equation:

$$\Phi = 1 + \frac{1}{\sum_i m_i}\left\{\left(IF' - F\right) + \sum_i \sum_j m_i m_j \left(\lambda_i + I\lambda'_{ij}\right) + 2\sum_i \sum_j \sum_k m_i m_j m_k \mu_{ijk}\right\} \tag{9.33}$$

where
$F' = dF/dI$
$\lambda'_{ij} = d\lambda_{ij}/dI$
F is given in Equation 9.23
m is the molality of a certain ion (i, j, or k)
λ_{ij} and μ_{ijk} are the second and third virial coefficients, respectively

In the case that there are two solutes in the multicomponent system, the previous equation transforms

$$\Phi = 1 + \frac{1}{\sum_i m_i}\left\{2IF + 2\sum_i \sum_j m_i m_j \left[B_{ij} + C_{ij}\frac{\sum m(z_i + z_j)}{(z_i z_j)^{1/2}}\right]\right\} \tag{9.34}$$

where
B_{ij} can be calculated through Equation 9.24
F can be calculated through Equation 9.23
the ionic strength I can be calculated through Equation 9.18

9.5.6 Lang and Steinberg Equation

Non-soluble components such as starch, casein, and soy flour affect water activity of the multicomponent mixture. To describe the data of the sorption isotherms, Smith's equation can be used:

$$M = a_i + b_i \ln(1 - a_w) \tag{9.35}$$

where a_i and b_i are constants and they can be determined by intercepting the slope of the straight line with that representing the humidity content M of the sample versus $\ln(1 - a_w)$.

Lang and Steinberg (1981) based their calculations on an equation to obtain an expression that allows calculation of the global water activity of the mixture:

$$\log(1 - a_w) = \frac{MW - \sum_i a_i w_i}{\sum_i b_i w_i} \tag{9.36}$$

where
a_i and b_i are the coefficients of Smith's equation for each component i of the mixture
M is the water content of the mixture
W is the total dry solids of the mixture
w_i is the weight of dry material of each component i

This model can predict water activities of multicomponent mixtures ranging from 0.3 to 0.95.

9.5.7 Salwin and Slawson Equation

From the sorption isotherms of the components, the water activity of dry mixtures can be predicted through the following equation (Salwin and Slawson, 1959):

$$a_w = \frac{\sum_i a_i s_i w_i}{\sum_i s_i w_i} \tag{9.37}$$

where
a_i is the initial water activity of the component i in the mixture
w_i is the dry weight of the component i
s_i is the slope of the tangent straight line of the sorption isotherm of the component i at the temperature of the mixture

This model assumes that there are no significant interactions that could alter the isotherms of the ingredients.

9.6 SORPTION ISOTHERMS

The water contained in a food interacts with the solid matrix in different ways. Thus, there are water molecules that are strongly retained and their elimination is difficult in the drying process. Generally, water in foods is considered as "bound water" or "free water." Free water is water that can be eliminated easily in the drying processes and it behaves like pure water. Bound water is defined as water that possesses less vapor pressure and less mobility, and it reduces the freezing point below pure water (Leung, 1986; Okos et al., 1992). There are different approaches to define bound water, although the most used definition is non-freezable at very low temperatures.

At a certain temperature, water activity is related to the water content in food through a sorption isotherm. In equilibrium conditions water activity coincides with the relative humidity of the atmosphere that surrounds the product. Water adsorption processes are not totally reversible. Initially, the water is absorbed, forming a monolayer in the product's surface, followed by a multilayer of adsorption. In Figure 9.2 a typical sorption isotherm for a food product is shown. It can be observed that there is an adsorption curve of adsorption and another of desorption, corresponding to humidification and drying processes of the product, respectively. This phenomenon is denominated hysteresis. Hysteresis in food can be distinguished as three types: (1) foods with high content of sugar and pectins, in which this phenomenon is said to be in the low humidity content region; (2) foods with high protein content, in those in which the hysteresis begins at high water activity values in the region of capillary condensation, extending until it reaches null water activity; and (3) foods with starch, where a large hysteresis cycle is presented with a maximum water activity value of 0.70 (in the region of capillary condensation). Temperature can affect the hysteresis phenomenon, although it depends on the food type, since in some foods it can be eliminated by increasing the temperature, while in other cases hysteresis is independent of temperature (Kapsalis, 1987). The interaction type between the water and the food matrix depends on the water activity value. Thus, the characteristics of bound water are

Water bound strongly	Water activity < 0.3
Water bound moderately	Water activity 0.3 at 0.7
Water bound without cohesion	Water activity > 0.7
Free water	Water activity = 1

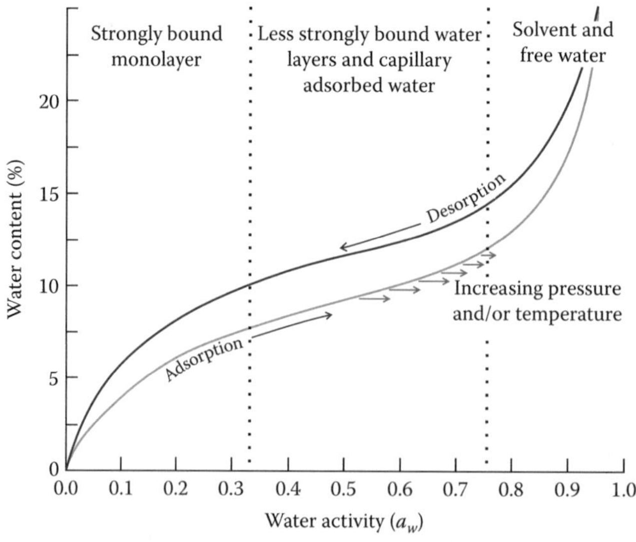

FIGURE 9.2 Typical sorption isotherm for a food.

Any sorption process for a fluid component (gas or liquid) on a solid surface is distinguished by the fact that the sorbate or solute is retained, while the solid that retains the solute is denominated adsorbent. The retention of the solute on the solid surface can be due to physical forces (van der Waals) or because of chemical reactions at certain points of the solid surface. Thus, one can speak of physisorption and chemisorption. In physisorption the sorption process rate is higher, and, generally, it does not present activation energy, although small activation energy sometimes appears due to the competition among molecules interacting at the active points of the adsorbent. In relation to the adsorbent, it is important to point out that the greater the solute retention the higher specific surface.

In sorption processes, once the solid is in equilibrium, there is a relationship between the solute content in the solid phase and the fluid phase. The concentration in the solid phase is expressed as a quantity of sorbate by adsorbent mass unit, while in the fluid phase it is the quantity of solute by fluid unit. For food sorption isotherms, the sorbate is water, while in the solid phase the concentration is expressed as equilibrium humidity in dry base (w_e); in the fluid phase, the air that surrounds the food and its relative humidity is expressed as per unit that coincides with the water activity.

The water contained in a food is a decisive factor in the inhibition or propagation of reactions, so there are many chemical, enzymatic, and microbiological reactions that influence final food quality. The different ways in which water correlates with the food matrix depend on the condition of the parameters of storage of a certain product, as well as their commercial life and the optimum conditions of processing. One way of correlating water availability with food humidity is the adsorption and desorption isotherm, which allows analysis of the relationships among the diverse components of the food with water.

The customary way of preserving food quality is to reduce water activity until it reaches low enough levels to inhibit microbial growth; thus, to obtain the sorption isotherm, it is essential to establish the limits of humidity starting with microbial growth and other types of reactions that are inhibited in the food during the storage.

Water adsorption processes are not totally reversible. Initially, water is adsorbed, forming a monolayer on the product surface, followed by a multilayer adsorption.

The sorption isotherms can be obtained in an experimental way, although this process usually requires an extensive amount of time. Therefore, it is important to obtain models that allow calculation of these sorption isotherms, in order to efficiently study the relationships between the water and substrate as a behavior of a certain food under specifically defined conditions in processing or storage. The different models can be classified into two categories: general and specific. The general models are based on theoretical considerations of physical or chemical type to predict the behavior of different types of foods in a wide range of variation of water activity.

The specific models are based on the empiric observation of the behavior of a certain food, and these models can only be applied to this food type and in the range of water activity obtained experimentally. In the literature, more than 200 equations of isotherms for biological materials have been proposed (van der Berg and Bruin, 1981). Some of these models are based on the Brunauer, Emmett, and Teller (BET) equation; however, others are empiric equations of two or three parameters that are useful in the prediction of water adsorption properties in foods, although they provide a little information about the interactions between water and other food components. Due to the great number of existent equations, only a few of these models will be presented in this chapter, and these examples are most frequently currently being used.

9.6.1 Adsorption Molecular Model

The prediction of water sorption isotherms is based on the ways in which the water is retained on a solid surface. Depending on how the adsorption is assumed to be either in monolayer or in multilayer, different models are obtained.

9.6.1.1 Langmuir Model

The first published model of water adsorption over a solid product goes back to 1918 (Langmuir, 1918), based on the physisorption of water on a solid surface. This model retains the concept of monolayer humidity (X_m), in which the monolayer or monomolecular humidity corresponds to the adsorption of the first layer of water molecules on the surface. The mathematical expression of this model is

$$\frac{a_w}{X_e} = \frac{k}{bX_m} + \frac{a_w}{X_m} \tag{9.38}$$

where

X_e is the equilibrium water content
X_m is the monolayer humidity content
k is the inverse of the vapor pressure for pure water at the system temperature
b is a constant

This equation has not been satisfactory for application in foods because the sorption heat is not constant in the whole surface (Barbosa-Cánovas and Vega-Mercado, 1996).

9.6.1.2 BET Model

This model is probably the one most frequently used in the characterization of water sorption in foods, and it is based on physisorption, under the name of the Brunauer, Emmett, and Teller (BET) model (1938). In the development of this model, it is assumed that the adsorbent surface is uniform and not porous and that there are no lateral interactions among the adsorbed molecules. Solute molecules are adsorbed on the surface in successive, complete layers or not, which are in equilibrium and with the molecules of the fluid. The retention rate on the first layer is equal to the desorption rate of the second layer. Also, it is assumed that the adsorbent–sorbate interaction energy of all the molecules on the first layer is identical, and the interaction energy in the other layers is equal to that among the molecules in the pure compound in the liquid phase (condensation heat). In fact, these assumptions are not completely met. The surfaces are not homogeneous, since there are more active than nonactive points; the first molecules of water are adsorbed in these more active points, and as the interaction is stronger, it will give off a greater heat. Also, the multilayer can begin to be formed before the monolayer is completed, since the adsorbent–sorbate and sorbate–sorbate interactions cannot be very different from each other.
The mathematical equation of the BET model can be expressed as (Barbosa-Cánovas and Vega-Mercado, 1996)

$$\frac{X_e}{X_m} = \frac{Ca_w}{(1-a_w)\left[1+a_w(C-1)\right]} \tag{9.39}$$

$$C = K\exp\left(\frac{Q}{RT}\right) \tag{9.40}$$

where

X_e is the adsorbed water content
X_m is the adsorbed monolayer water value
a_w is the water activity
Q is the sorption heat
T is the temperature
K is a constant

This equation can be applied for water activity values between 0.1 and 0.5.

The BET equation can be rearranged and expressed as

$$\frac{a_w}{X(1-a_w)} = \frac{1}{CX_m} + \frac{(C-1)}{CX_m} a_w \tag{9.41}$$

Representing $a_w/X(1 - a_w)$ versus a_w, a slope of a straight line $(C - 1)/CX_m$ and an intercept $1/CX_m$ is obtained, which allows obtaining the C value and the value of the monolayer X_m. According to this model, the sorption heat Q is constant until the monolayer is covered, and then it falls suddenly.

9.6.2 Empirical and Theoretical Models of Sorption Isotherms

It has already been mentioned that there are infinite isotherm equations to describe the behavior of biological materials. Not all of the models relate the data correctly for different types of foods, and due to the different forms of isotherms, there are models that adjust better to a series of foods than to others. Thus, the best model to predict the form of protein isotherms is the Oswin model (1946) that is an empiric model developed starting from a mathematical series for sigmoidal curves. The White and Eyring model (1947) has a similar form to the Oswin model but allows extrapolation of the data for high values of water activity. According to Boquet et al. (1978), the best model to predict isotherms for meats is the Halsey model (1948). Chirife and Iglesias (1978) defined their model as the most appropriate for fruits with high sugar content. The BET model is one of the most frequently used models, although other models are presented later in this chapter that have also been applied to adjust and to describe experimental data in an appropriate way.

Any sorption model is adjusted in an appropriate way in the whole interval of water activity, because the water associated with the food matrix is associated with different mechanisms and in different regions of water activity. Thus, before modeling a certain sorption isotherm for a food, it is convenient to know the main characteristics of the food and the adsorption/desorption process for the purpose of applying models that better adjust to each situation.

9.6.2.1 Oswin Model

This model's mathematical expression is (Oswin, 1946)

$$X_e = A\left[\frac{a_w}{1-a_w}\right]^B \tag{9.42}$$

where A and B are constants. The data are adjusted well for experimental data in sigmoidal form.

9.6.2.2 Smith Model

This model's equation is expressed as (Smith, 1947)

$$X_e = a + b\ln(1-a_w) \tag{9.43}$$

where a and b are constants. This equation has been applied to predict the effect of solute components such as starch, casein, and soy flour in water activity. It is usually applied for water activity ranging from 0.3 to 0.5.

9.6.2.3 Halsey Model

Temperature is a variable that affects isotherm shape in a very significant way. This model keeps in mind the effect of the temperature, so its mathematical equation is (Halsey, 1948)

$$a_w = \exp\left[\frac{-A}{RT\left(X_e/X_m\right)^n}\right] \tag{9.44}$$

where
- A and n are constants of the model
- X_m is the monolayer humidity

9.6.2.4 Henderson Model

This model is an empiric equation developed in 1952 whose mathematical expression is

$$a_w = 1 - \exp\left(-kX_e^n\right) \tag{9.45}$$

where n and k are constants. With the purpose of keeping in mind the influence of the temperature, this equation can be expressed as:

$$a_w = 1 - \exp\left(-k'TX_e^n\right) \tag{9.46}$$

where k' is a new constant of the model. This model has been applied to adjust the experimental data of adsorption in frozen and dried banana and for pulp of frozen and dehydrated pineapple (Barbosa-Cánovas and Vega-Mercado, 1996).

9.6.2.5 Caurie Model

When considering the study of dehydrated foods as highly concentrated solutions, and keeping in mind that the maximum humidity content is of approximately of 22%, so that the dehydrated foods are stable, the following model can be established (Caurie, 1970):

$$X_e = \exp\left(a_w \ln r - \frac{1}{4.5X_S}\right) \tag{9.47}$$

where
- r is the model constant
- X_S is the security humidity content that provides maximum stability of the food during storage

This model usually gives good adjustments for a water activity ranging from 0 to 0.85.

9.6.2.6 Iglesias and Chirife Model

This empiric model adequately describes the behavior of water adsorption in fruits and products with high sugar contents. Mathematically, it is expressed according to the equation (Chirife and Iglesias, 1978)

$$\ln\left[X_e + \sqrt{X_e^2 + X_{0.5}}\right] = ba_w + p \tag{9.48}$$

where
- a and b are the model constants
- $X_{0.5}$ is the water content in equilibrium to a water activity equal to 0.5

9.6.2.7 Guggenheim–Anderson–de Boer Model

This is a model derived from the BET model, where the modified properties of the adsorbed water in multilayers consider the modified interactive properties of the multi-region of the adsorbate molecules with the solvent. In addition to the parameters of the BET equation, a new parameter K appears that is a correction factor related with the multilayer sorption heat. This model is expressed as (van der Berg and Bruin, 1981)

$$\frac{X_e}{X_m} = \frac{CKa_w}{(1-Ka_w)\left[1+(C-1)Ka_w\right]} \tag{9.49}$$

$$C = c\exp\left(\frac{H_p - H_m}{RT}\right) \tag{9.50}$$

$$K = k\exp\left(\frac{H_p - H_n}{RT}\right) \tag{9.51}$$

where

C is the Guggenheim constant, characteristic of the product and related with the monolayer sorption heat
K is a correlation factor related to the multilayer sorption heat
c and k are entropic accommodation factors
H_p is the evaporation molar enthalpy of the liquid water
H_m is the monolayer sorption enthalpy
H_n is the multilayer sorption enthalpy

This model is one of the few that better fits the data of sorption isotherms for most foods (Kapsalis, 1987; Barbosa-Cánovas and Vega-Mercado, 1996).

PROBLEMS

9.1 Calculate the water activity of a sugary solution that contains 25% in weight of sucrose ($C_{12}H_{22}O_{11}$).

Molecular weight of sucrose is 342 g/mol. The molecular weight of water is 18 g/mol.

Sucrose molar fraction:

$$x_S = \frac{(25/342)}{(25/342)+(75/18)} = 0.017$$

Water molar fraction:

$$x_w = \frac{(75/18)}{(25/342)+(75/18)} = 0.983$$

Considering an ideal solution, the activity coefficient is equal to 1, Raoult's law can applied, and the water activity is equal to the water molar fraction; therefore, $a_w = 0.983$.

If the behavior is nonideal, Norrish's equation (Equation 9.12) for the sucrose gives a value of $K = 6.47$ (Table 9.2):

$$a_w = x_w \exp\left(-Kx_S^2\right) = (0.983)\exp\left[-(6.47)(0.017)^2\right] = 0.981$$

9.2 A cured sausage has a humidity of 37%, a 5% salt content. Calculate the water activity for this sausage.
NaCl molecular weight is 58.4 g/mol. The molecular weight of water is 18 g/mol.
Molal solute (NaCl) concentration:

$$m_S = \frac{(5/58.4)\,\text{mol NaCl}}{37\times10^{-3}\,\text{kg H}_2\text{O}} = 2.314\,\text{mol NaCl/kg H}_2\text{O}$$

Since in this food NaCl is present as a solute, it is an electrolyte, and for the calculation the osmotic coefficient of Bromley's equation will be used (Equation 9.17). For NaCl, from Table 9.3, the parameter $B = 0.0574$ is obtained. Also, the parameters F_1 and F_2 should be obtained, as well as the ionic strength I.
The ionic strength is calculated through Equation 9.18:

$$I = \frac{1}{2}\left[(2.314)(1)^2 + (2.314)(-1)^2\right] = 2.314$$

Parameter F_1 is calculated from Equation 9.19, keeping in mind that $A_\gamma = 0.511$:

$$F_1 = |(1)(-1)|(0.511)\left[\frac{1+2\left(1+2.314^{1/2}\right)\ln\left(1+2.314^{1/2}\right)-\left(1+2.314^{1/2}\right)}{2.314\left(1+2.314^{1/2}\right)}\right] = 0.1876$$

To calculate F_2 Equation 9.20 is used, where the a parameter value is obtained from Equation 9.21:

$$a = \frac{1.5}{|(1)(-1)|} = 1.5$$

Then

$$F_2 = |(1)(-1)|\left[\frac{(1+2(1.5)(2.314))}{(2.314)(1+(1.5)(2.314))^2} - \frac{\ln(1+(1.5)(2.314))}{(1.5)^2(2.314)}\right] = -0.1160$$

The obtained data are substituted in Equation 9.17:

$$\Phi = 1 + 2.303\left[(0.1876) + (0.06 + 0.6(0.0574))(-0.116) + 0.5(0.0574)(2.314)\right] = 1.560$$

Water activity is obtained from Equation 9.16:

$$a_w = \exp\left(-0.01802\Phi\sum_i \nu m_i\right) = \exp\left[-0.01802(1.560)(2)(2.314)\right] = 0.878$$

9.3 In the preparation of apple marmalade, apples are crushed and sucrose is added, followed by a cooking process in which some of the water is eliminated, so the final product contains 37% of water, 22% sucrose, 14% glucose, and 20% fructose. Calculate the water activity of this marmalade.

It is considered that the reducing sugars are a mixture of glucose and fructose, their molecular weight being 180 g/mol. The sucrose molecular weight is 342 g/mol. The water molecular weight is 18 g/mol.

In order to calculate the activity, Ross's equation will be used, in which it is considered that the sucrose and the reducing sugars form a solution with water in an independent way.

- Sucrose–water
  Sucrose molar fraction:

$$x_S = \frac{(22/342)}{(22/342)+(37/18)} = 0.0303$$

  Water molar fraction:

$$x_w = \frac{(37/18)}{(22/342)+(37/18)} = 0.9697$$

  For water activity of the system sucrose–water, we will use Norrish's equation, where parameter K for sucrose is 6.47 (Table 9.2):

$$a_w = x_w \exp\left(-Kx_S^2\right) = (0.9697)\exp\left[-(6.47)(0.0303)^2\right] = 0.964$$

- Reducing sugars–water
  Reducing sugars molar fraction:

$$x_S = \frac{(34/180)}{(34/180)+(37/18)} = 0.0842$$

  Water molar fraction:

$$x_w = \frac{(37/18)}{(34/180)+(37/18)} = 0.9158$$

  Parameter K for glucose is 2.25 (Table 9.2):

$$a_w = x_w \exp\left(-Kx_S^2\right) = (0.9158)\exp\left[-(2.25)(0.0842)^2\right] = 0.901$$

  Applying Ross' equation (Equation 9.26), the water activity for apple marmalade is obtained:

$$a_w = \prod_i (a_w)_i = (a_w)_1 (a_w)_2 = (0.964)(0.901) = 0.869$$

9.4 It is desired to obtain the sorption isotherm of mushrooms, where samples of these mushrooms were placed in chambers with a controlled relative humidity. Once equilibrium was reached, the samples were weighed and their humidity content was determined. The data obtained for the variation of water activity with the equilibrium humidity are presented in Table P.9.1. Model the sorption isotherm by means of the *n* of the BET equation obtaining the parameters for this model.

TABLE P.9.1
Variation of Water Activity with the Equilibrium Humidity

a_w	X_e (g H_2O/g Dry Solid)
0.115	0.0571
0.330	0.0939
0.446	0.1191
0.544	0.1527
0.766	0.3001

If the BET model is used, Equation 9.41 will be applied, and starting from the problem description, the following Table P.9.2 can be built:

TABLE P.9.2
Values of the First Member of Equation 9.41 Depending on a_w

a_w	$\frac{a_w}{X(1-a_w)}$
0.115	2.280
0.330	5.240
0.446	6.759
0.544	7.813
0.766	10.908

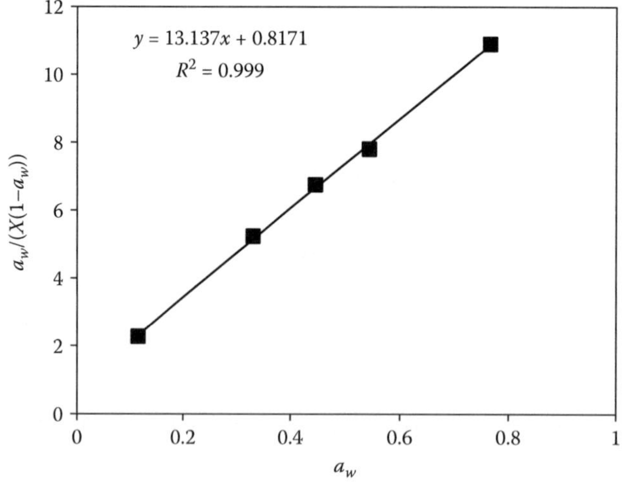

FIGURE P.9.1 Adjustment of the experimental data to the Equation 9.41.

Representing $a_w/X(1 - a_w)$ versus a_w, a slope of a straight line $(C - 1)/CX_m$ and an interception of $1/CX_m$ is obtained as shown in Figure P.9.1.

Therefore, it is obtained that

$$\text{Slope:} \qquad \frac{(C-1)}{CX_m} = 13.137$$

$$\text{Intercept:} \qquad \frac{1}{CX_m} = 0.8171$$

Operating the parameters for the BET model, the following is obtained: $C = 17.08$; monolayer humidity value $X_m = 0.072$ g H_2O/g dry solid.

10 Mechanical Properties
Structural and Geometric Properties

The structure, the physical state, and the response to the stress applied to food materials are the basis of mechanical properties. Within this group of properties, two groups can be distinguished, one that includes structural and geometric properties and another that includes properties related to stress and deformations; this group constitutes the rheological and textural properties of foods. In this chapter the properties related to structure and geometry will be presented, while the rheological and textural properties will be examined in Chapter 13.

10.1 DENSITY

Density is defined as the relation between the mass of a given sample and its volume. The unit for density in the International System (SI) is kg/m^3. Due to the characteristics of foods, different types of density can be distinguished. Thus, the following definitions can be given:

True density (ρ). It refers to a substance that does not have pores in its structure.
Particle density (ρ_p). This is the density of a particle that has not been modified. The particle can contain internal pores, and the solid material does not occupy all of the volume.
Apparent density (ρ_{app}). This is the density of a sample constituted by particles that can have closed pores and/or is connected externally to the atmosphere.
Bulk density (ρ_B). This is the density of a material when it is packed or stacked in bulk. This term is frequently used to describe the density of a particulate system.

Different expressions for the calculation of food density can be found in the literature. Thus, for fruit juices, density can be expressed as a function of the refraction index according to the expression (Riedel, 1949)

$$\rho = \frac{s^2 - 1}{s + 2} \frac{64.2}{0.206} 16.0185 \tag{10.1}$$

where
ρ is the density expressed in kg/m^3
s is the refraction index

TABLE 10.1
Parameters of Equation 10.4

Fruit Juice	Temperature Range (°C)	Concentration Range (°Brix)	*a*	*b*	*c*	*d*
Apple	5–70	10–71	989.98	0.5050	5.1709	0.0308
Pear	5–70	10–71	1011.30	0.5476	3.713	0.0174
Peach	0–80	10–60	1006.56	0.5155	4.1951	0.0135
Orange	0–80	10–60	1025.42	0.3289	3.2819	0.0178

There are equations in which density is expressed as a function of temperature and soluble solids content. For clarified apple juices, Constenla et al. (1989) presented the following equation:

$$\rho = 0.82780 + 0.34708\exp(0.01X) - 5.479\times10^{-4}\,T \tag{10.2}$$

Where
density is expressed in g/cm^3
X is the concentration in °Brix
T is the absolute temperature

This expression can be applied in the 20°C–80°C temperature range and in the 12°Brix–68.5°Brix range. These same authors expressed the density of these juices as a function of °Brix and density of water:

$$\rho = \frac{\rho_{Water}}{0.992417 - 3.7391\times10^{-3}X} \tag{10.3}$$

However, for different clarified fruit juices, the following polynomial equation can be used:

$$\rho = a - bT + cC + dC^2 \tag{10.4}$$

where
density is expressed in kg/m^3
C is expressed in °Brix
T is expressed in °C

Table 10.1 shows the values of these parameters for clarified apple, pear, peach, and orange juices (Aguado and Ibarz, 1988; Ibarz and Miguelsanz, 1989; Ramos and Ibarz, 1998), for different temperature and soluble solids content range.

Alvarado and Romero (1989) presented the following expression for different juices, for temperatures from 20°C to 40°C and for concentrations from 5°Brix to 30°Brix:

$$\rho = 1002 + 4.61C - 0.460T + 7.001\times10^{-3}T^2 + 9.175\times10^{-5}T^3 \tag{10.5}$$

Where
density is expressed in kg/m^3
C is expressed in °Brix
T is expressed in °C

For sucrose solutions with concentrations between 6°Brix and 65°Brix and a temperature of 20°C, Kimball (1986) reported the following equation:

$$\rho = 0.524484 \exp\left[\frac{(C + 330.872)^2}{170,435}\right] \tag{10.6}$$

Where
density is expressed in g/cm^3
C is expressed in °Brix

Manohar et al. (1991) presented a second-order polynomial equation as a function of the total soluble solids content for tamarind juices:

$$\rho = 1000 + 4.092C + 0.03136C^2 \tag{10.7}$$

where
density is obtained in kg/m^3
the concentration C is expressed in °Brix

Rambke and Konrad (1970) reported a second-order polynomial equation for milk as a function of the dry mass percentage:

$$\rho = a + bX_S + cX_S^2 \tag{10.8}$$

where
ρ is expressed in g/cm^3
X_s is the dry mass percentage

The coefficients of this equation for different temperatures are given in Table 10.2.

For temperatures higher than the boiling point, the equation of Berstch et al. (1982) can be used:

$$\rho = 1040.51 - 0.2655T - 2.307 \times 10^{-3}T^2 - (0.967 + 0.969 \times 10^{-2}T - 0.478 \times 10^{-4}T^2)f \tag{10.9}$$

where
ρ is expressed in kg/m^3
T is temperature in °C with its range from 65°C to 140°C
f is the fat content in % (w/w), for values between 0.02% and 15.5%

TABLE 10.2
Values of the Parameters of Equation 10.8

	Skim Milk			Whole Milk (c = 0)	
T (°C)	a	$b \times 10^3$	$c \times 10^5$	a	$b \times 10^3$
5	1.0000	3.616	1.827	1.0010	2.55
20	0.9982	3.519	1.782	1.0080	2.09
35	0.9941	3.504	1.664	1.0137	1.66
50	0.9881	3.568	1.366	0.9953	2.11
60	0.9806	3.601	1.308		

TABLE 10.3
Equations to Estimate Density of Food Components

		Temperature Functions[a]
ρ (kg/m^3)	Carbohydrate	$\rho = 1.5991 \times 10^3 - 0.31046T$
	Ash	$\rho = 2.4238 \times 10^3 - 0.28063T$
	Fiber	$\rho = 1.3115 \times 10^3 - 0.36589T$
	Fat	$\rho = 9.2559 \times 10^2 - 0.41757T$
	Protein	$\rho = 1.3299 \times 10^3 - 0.51840T$
	Water	$\rho = 997.18 + 3.1439 \times 10^{-3}T - 3.7574 \times 10^{-3}T^2$
	Ice	$\rho = 916.89 - 0.13071T$

Source: Choi, Y. and Okos, M.R., Thermal properties of liquid foods: Review, in *Physical and Chemical Properties of Foods*, M.R. Okos, ed., American Society of Agricultural Engineers, St. Joseph, MI, 1986.

[a] T in °C

Andrianov et al. (1968) reported the following equation for cream in the 40°C–80°C range and fat content between 30% and 83%:

$$\rho = 1.0435 - 1.17 \times 10^{-5} X_G - (0.52 \times 10^{-3} + 1.6 \times 10^{-8} X_G)T \tag{10.10}$$

Where
density is expressed in g/cm^3
temperature is expressed in °C
the fat content X_G is the mass fraction

Choi and Okos (1986) suggested an expression as a function of the density of the components of the product:

$$\rho = \frac{1}{\sum_i X_i^m / \rho_i} \tag{10.11}$$

in which X_i^m is the mass fraction of the component i and ρ_i its density.

Table 10.3 shows the expressions that allow calculation of the densities of the pure components as a function of temperature.

10.2 POROSITY

Porosity (ε) is defined as the fraction of void space or air in a given sample:

$$\varepsilon = \frac{\text{Void volume}}{\text{Total volume}} \tag{10.12}$$

Porosity can also be defined in different ways:

Apparent porosity (ε_{app}). This is the rate between the space occupied by the air and the total volume:

$$\varepsilon_{app} = \frac{\rho_s - \rho_{app}}{\rho_s} \tag{10.13}$$

Porosity of open pores (ε_{op}). This is the rate between the volume of pores connected to the atmosphere and the total volume:

$$\varepsilon_{op} = \frac{\rho_p - \rho_{app}}{\rho_p} \tag{10.14}$$

Porosity of closed pores (ε_{cp}). This is the rate between the volume of pores not connected to the atmosphere and the total volume. It is the difference between the apparent porosity and that of open pores:

$$\varepsilon_{cp} = \varepsilon_{app} - \varepsilon_{op} \tag{10.15}$$

Bulk porosity (ε_b). This is referred to as a packed sample that includes the external empty space the individual material:

$$\varepsilon_B = \frac{\rho_{app} - \rho_B}{\rho_{app}} \tag{10.16}$$

Total porosity (ε). This is the fraction of total holes of a material packed or bulk stacked.

$$\varepsilon_{Tp} = \varepsilon_{app} + \varepsilon_B \tag{10.17}$$

10.3 SIZE AND SHAPE

Size is a physical characteristic of the foods that is used to evaluate quality, as well as classification and the screening process for separating solid materials. In powdered foods, size is an important property since it is related to instant properties as in the case of powdered milk and dried coffee.

Shape is another physical attribute that, as in the case of size, is also used in the screening of solids, for example, in the classification and evaluation of the quality of foods. The main parameter used in such evaluation is called sphericity. Sphericity (ϕ) is defined as the ratio between the volume of a solid and the volume of a sphere that has a diameter similar to the diameter greater than the particle. It is assumed that the larger diameter of a particle is the one corresponding to a circumscribed sphere to the particle. In the case of spherical particles, the sphericity is equal to 1; however, for other types of particles, their value is different than the unit.

Another important attribute is the surface area or specific surface area of a particle, which is defined as the ratio between the surface and the volume of the particle:

$$a_{S0} = \frac{\text{Particle surface}}{\text{Particle volume}} \text{ expressed in m}^{-1} \tag{10.18}$$

For a spherical particle, whose diameter is d_r, the specific surface area is

$$a_{S0} = \frac{\pi d_r^2}{(\pi/6)d_r^3} = \frac{6}{d_r} \tag{10.19}$$

For non-spherical particles, the equivalent diameter (d_p) is defined as the diameter of a sphere whose relationship between the superficial area and its volume is the same as the one owned by the particle.

TABLE 10.4
Sphericity of Particles

Form of the Particle	Sphericity (ϕ)
Sphere	1
Cube	0.81
Cylinder	
$h = d$	0.87
$h = 5d$	0.70
$h = 10d$	0.58
Disk	
$h = d/3$	0.76
$h = d/6$	0.60
$h = d/10$	0.47
Wheat	0.85
Sunflower seeds	0.51–0.54

Source: Levenspiel, O., *Flujo de Fluidos. Intercambio de Calor*, Reverté, Barcelona, Spain, 1993.

The relationship between the equivalent diameter of the particle and the diameter of the sphere is denominated sphericity:

$$d_p = \phi d_r \tag{10.20}$$

In this way, the sphericity for an irregular particle can be defined as

$$\phi = \frac{\text{Specific surface area for a sphere having the same volume of the particle}}{\text{Specific surface area of the particle}} \tag{10.21}$$

Table 10.4 presents sphericity values for different types of particles.

For a bed of particles, the specific surface area is defined as the area of the bed surface exposed to the fluid by unit of bed volume:

$$a_S = \frac{\text{Area exposed to the fluid}}{\text{Bed volume}} \text{ expressed in m}^{-1} \tag{10.22}$$

Due to the bed's porosity, the specific surface of the bed and of the particle do not coincide; thus, they are related according to the equation

$$a_S = a_{S0}(1-\varepsilon) \tag{10.23}$$

PROBLEMS

10.1 Determine the density, at 25°C, of a food product that has been chemically analyzed and whose weight composition is 77% water, 19% carbohydrate, 3% protein, 0.2% fat, and 0.8% ash.

The equation of Choi and Okos is used. Therefore, the density of each component at 25°C was previously calculated, using the equations shown in Table 10.3.

The mass and density of each component are presented in the following table:

Component	X_i^m	ρ_i (kg/m³)
Water	0.77	994.91
Carbohydrate	0.19	1591.34
Protein	0.03	1316.94
Fat	0.002	915.15
Ash	0.008	2416.78

Density obtained from Equation 10.11:

$$\rho = \frac{1}{\sum_i \left(X_i^m / \rho_i \right)} = 1085 \text{ kg/m}^3$$

10.2 In a drying installation of cereals, hot air is injected into cylindrical tubes that contain cereal grains. Each cereal grain can be considered to be formed by a cylindrical body of 0.5 cm diameter and 0.75 cm of height, limited by two semispheres of the same diameter. The tubes that contain the cereal grains possess a diameter of 7 cm. Each tube meter contains 10,000 cereal grains. Calculate the porosity and the specific surface of the particle and of the bed.

Volume for 1 m of tube

$$V = \frac{\pi}{4} D^2 H = \frac{\pi}{4} (7 \text{ cm})^2 (100 \text{ cm}) = 3848.5 \text{ cm}^3$$

Volume of cereal grain

- Cylindrical part

$$V_c = \frac{\pi}{4} d^2 h = \frac{\pi}{4} (0.5 \text{ cm})^2 (0.75 \text{ cm}) = 0.1473 \text{ cm}^3$$

- Spherical part

$$V_e = \frac{\pi}{6} d^3 \left(\frac{d}{2} \right)^3 = \frac{\pi}{6} (0.5)^3 = 0.0655 \text{ cm}^3$$

- Grain volume

$$V_g = (0.1473)(0.0655) = 0.2128 \text{ cm}^3$$

Area of the cereal grain

- Cylindrical part

$$A_c = \pi d h = \pi (0.5 \text{ cm})(0.75 \text{ cm}) = 1.1781 \text{ cm}^2$$

- Spherical part

$$A_e = \pi d^2 = \pi (0.5)^2 = 0.7854 \text{ cm}^2$$

- Grain area

$$A_g = (1.1781)(0.7854) = 1.9635\ \text{cm}^2$$

Porosity is calculated from Equation 10.12:

$$\varepsilon = \frac{\text{Void volume}}{\text{Total volume}} = \frac{\text{Total volume} - \text{Grain volume}}{\text{Total volume}}$$

$$= 1 - \frac{\text{Grain volume}}{\text{Total volume}} = 1 - \frac{(10{,}000)(0.2128\ \text{cm}^3)}{(3{,}848.5\ \text{cm}^3)} = 0.553$$

The specific surface of the particle is calculated using Equation 10.18:

$$a_{S0} = \frac{\text{Particle surface}}{\text{Particle volume}} = \frac{(1.9635\ \text{cm}^2)}{(0.2128\ \text{cm}^3)} = 9.23\,\text{cm}^{-1} = 923\,\text{m}^{-1}$$

The bed specific surface is calculated using Equation 10.23:

$$a_S = a_{S0}(1-\varepsilon) = (9.23\,\text{cm}^{-1})(1-0.553) = 4.13\,\text{cm}^{-1} = 413\,\text{m}^{-1}$$

11 Thermal Properties of Foods

11.1 THERMAL CONDUCTIVITY

In heat transfer by conduction, under steady-state conditions, the flow of heat transmitted ($\dot{Q}$) through a solid is directly proportional to the transmission area (A) and to the temperature change (ΔT) and inversely proportional to the thickness of the solid (e). The proportionality constant in this relationship is called *thermal conductivity*:

$$\dot{Q} = k\frac{A\Delta T}{e}$$

Heat conduction under steady state has been used in different experiments to calculate the thermal conductivity of food, although experiments under nonsteady state can also be used. Either way, mathematical relationships that allow calculation of the thermal conductivity of a given food are sought as a function of temperature and composition.

Riedel (1949) developed an equation that allows calculation of the thermal conductivity of sugar solutions, fruit juices, and milk:

$$k = \left(326.8 + 1.0412T - 0.00337T^2\right)\left(0.44 + 0.54X^m_{WATER}\right)1.73\times10^{-3} \tag{11.1}$$

where

k is expressed in J/(s m °C)

T is expressed in °C

X^m_{WATER} is the mass fraction of water

This equation is valid for a temperature range between 0°C and 180°C.

Sweat (1974) gives the following equation for different fruits and vegetables:

$$k = 0.148 + 0.493X^m_{WATER} \tag{11.2}$$

which is valid for water contents higher than 60%, although it cannot be used with low-density foods or with foods that have pores (e.g., apples).

In the case of milk, Fernández-Martín (1982) gave a second-order polynomial expression with respect to temperature:

$$k = A + BT + CT^2 \tag{11.3}$$

in which the parameters A, B, and C are a function of the fat and nonfat content of milk.

An equation that allows us to obtain the thermal conductivity of cream (Gromov, 1974) is

$$k = \frac{[411.6 - 4.26\,(f - 10)] \times 10^{-6}}{1 - 0.0041\,(T - 30)}\rho \tag{11.4}$$

where

thermal conductivity is expressed in kcal/(h m °C)
f is the fat percentage, between 10% and 60%
ρ is the density of the sample at the corresponding temperature and composition, expressed in kg/m³
T is the temperature in °C in the 30°C–70°C range

Also, Fernández-Martín and Montes (1977) gave an expression for cream:

$$k = \left[12.63 + 0.051\,T - 0.000175T^2\right]\left[1 - (0.843 + 0.0019T)X_G^V\right] \times 10^{-4} \tag{11.5}$$

where

the thermal conductivity is expressed in cal/(s cm °C)
temperature is expressed in °C in the 0°C–80°C range
X_G^V is the volumetric fraction of the fat phase, for values lower than 0.52

This equation is applicable to a fat content of between 0.1% and 40%.

If the food composition is known, it is possible to find its thermal conductivity from the equation

$$k = \sum_i \left(k_i X_i^V\right) \tag{11.6}$$

where

k_i is the thermal conductivity of the component i
X_i^V is the volumetric fraction of such component

The volumetric fraction of the component i is given by the expression

$$X_i^V = \frac{(X_i^m/\rho)\rho_i}{\sum_i \left(X_i^m/\rho_i\right)} \tag{11.7}$$

where

X_i^m is the mass fraction of the component i
ρ_i is the density of i

The density of each component is calculated using the equations shown in Table 10.3.

Table 11.1 presents the thermal conductivity values of some foods. Table 11.2 shows the thermal conductivity of the main pure components of foods, while the conductivity of water and ice as a function of temperature is given in Table 11.3. All of these tables are located at the end of this chapter.

TABLE 11.1
Thermal Conductivity of Some Foods

Product	Water Content (%)	Temperature (°C)	Thermal Conductivity (J/s m °C)
Oil			
Olive	15	0.189	
	—	100	0.163
Soybean	13.2	7–10	0.069
Vegetable and animal	—	4–187	0.169
Sugars	—	29–62	0.087–0.22
Cod	83	2.8	0.544
Meats			
Pork			
Perpendicular to the fibers	75.1	6	0.488
		60	0.54
Parallel to the fibers	75.9	4	0.443
		61	0.489
Fatty meat	—	25	0.152
Lamb			
Perpendicular to the fibers	71.8	5	0.45
		61	0.478
Parallel to the fibers	71.0	5	0.415
		61	0.422
Veal			
Perpendicular to the fibers	75	6	0.476
		62	0.489
Parallel to the fibers	75	5	0.441
		60	0.452
Beef			
Freeze-dried			
1000 mm Hg	—	0	0.065
0.001 mm Hg	—	0	0.035
Lean			
Perpendicular to the fibers	78.9	7	0.476
	78.9	62	0.485
Parallel to the fibers	78.7	8	0.431
	78.7	61	0.447
Fatty	—	24–38	0.19
Strawberries	—	−14 to 25	0.675
Peas	—	3–17	0.312

TABLE 11.2
Equations to Calculate Thermal Properties

Thermal Property	Component	Equation as a Function of Temperature
k (W/m °C)	Carbohydrate	$k = 0.20141 + 1.3874 \times 10^{-3}\,T - 4.3312 \times 10^{-6}\,T^2$
	Ash	$k = 0.32962 + 1.4011 \times 10^{-3}\,T - 2.9069 \times 10^{-6}\,T^2$
	Fiber	$k = 0.18331 + 1.2497 \times 10^{-3}\,T - 3.1683 \times 10^{-6}\,T^2$
	Fat	$k = 0.18071 + 2.7604 \times 10^{-3}\,T - 1.7749 \times 10^{-7}\,T^2$
	Protein	$k = 0.17881 + 1.1958 \times 10^{-3}\,T - 2.7178 \times 10^{-6}\,T^2$
$\hat{C}_P$ (kJ/kg °C)	Carbohydrate	$\hat{C}_P = 1.5488 + 1.9625 \times 10^{-3}\,T - 5.9399 \times 10^{-6}\,T^2$
	Ash	$\hat{C}_P = 1.0926 + 1.8896 \times 10^{-3}\,T - 3.6817 \times 10^{-6}\,T^2$
	Fiber	$\hat{C}_P = 1.8459 + 1.8306 \times 10^{-3}\,T - 4.6509 \times 10^{-6}\,T^2$
	Fat	$\hat{C}_P = 1.9842 + 1.4733 \times 10^{-3}\,T - 4.8008 \times 10^{-6}\,T^2$
	Protein	$\hat{C}_P = 2.0082 + 1.2089 \times 10^{-3}\,T - 1.3129 \times 10^{-6}\,T^2$
$\alpha \times 10^6$ (m²/s)	Carbohydrate	$\alpha = 8.0842 \times 10^{-2} + 5.3052 \times 10^{-4}\,T - 2.3218 \times 10^{-6}\,T^2$
	Ash	$\alpha = 1.2461 \times 10^{-1} + 3.7321 \times 10^{-4}\,T - 1.2244 \times 10^{-6}\,T^2$
	Fiber	$\alpha = 7.3976 \times 10^{-2} + 5.1902 \times 10^{-4}\,T - 2.2202 \times 10^{-6}\,T^2$
	Fat	$\alpha = 9.8777 \times 10^{-2} + 1.2569 \times 10^{-4}\,T - 3.8286 \times 10^{-8}\,T^2$
	Protein	$\alpha = 6.8714 \times 10^{-2} + 4.7578 \times 10^{-4}\,T - 1.4646 \times 10^{-6}\,T^2$

Source: Choi, Y. and Okos, M.R., Thermal properties of liquid foods: Review, in *Physical and Chemical Properties of Foods*, M.R. Okos, ed., American Society of Agricultural Engineers, St. Joseph, MI, 1986.

TABLE 11.3
Equations to Calculate Thermal Properties of Water and Ice

	Temperature Functions[a]	
Water	$k_A = 0.57109 + 1.7625 \times 10^{-3}\,T - 6.7036 \times 10^{-6}\,T^2$	(W/m °C)
	$\alpha_A = [0.13168 + 6.2477 \times 10^{-4}\,T - 2.4022 \times 10^{-6}\,T^2] \times 10^{-6}$	(m²/s)
	$\hat{C}_{pA1} = 4.0817 - 5.3062 \times 10^{-3}\,T + 9.9516 \times 10^{-4}\,T^2$	(kJ/kg °C)
	$\hat{C}_{pA2} = 4.1762 - 9.0864 \times 10^{-5}\,T + 5.4731 \times 10^{-6}\,T^2$	(kJ/kg °C)
Ice	$k_H = 2.2196 - 6.2489 \times 10^{-3}\,T + 1.0154 \times 10^{-4}\,T^2$	(W/m °C)
	$\alpha_H = [1.1756 - 6.0833 \times 10^{-3}\,T + 9.5037 \times 10^{-5}\,T^2] \times 10^{-6}$	(m²/s)
	$\hat{C}_{pH} = 2.0623 + 6.0769 \times 10^{-3}\,T$	(kJ/kg °C)

Source: Choi, Y. and Okos, M.R., Thermal properties of liquid foods: Review, in *Physical and Chemical Properties of Foods*, M.R. Okos, ed., American Society of Agricultural Engineers, St. Joseph, MI, 1986.

[a] $\hat{C}_{pA1}$, for a temperature range between –40°C and 0°C; $\hat{C}_{pA2}$, for a temperature range between 0°C and 150°C.

11.2 SPECIFIC HEAT

Specific heat is defined as the energy needed to increase by one degree the temperature of one mass unit.

For foods with high water content, above the freezing point, the following equation can be used (Siebel, 1982):

$$\hat{C}_P = 0.837 + 3.349 X^m_{WATER} \tag{11.8}$$

where
$\hat{C}_P$ is expressed in kJ/(kg °C)
X^m_{WATER} is the mass fraction of the water in food

An equation given by Charm (1971) is

$$\hat{C}_P = 2.309 X^m_G + 1.256 X^m_s + 4.187 X^m_{WATER} \tag{11.9}$$

in which X^m_G and X^m_s are the mass fractions of fat and solids, respectively.

For milk, at temperatures higher than the final point of fusion of milk fat, the following expression can be used (Fernández-Martín, 1972):

$$\hat{C}_P = X^m_{WATER} + (0.238 + 0.0027T) X^m_{ST} \tag{11.10}$$

where
the specific heat is expressed in kcal/(kg °C)
temperature T is expressed in °C in a range from 40°C to 80°C
X^m_{WATER} and X^m_{ST} are the mass fractions of water and total solids, respectively

Gromov (1979) gave the next equation for cream:

$$\hat{C}_P = 4.187 X^m_{WATER} + (16.8T - 3.242)\left(1 - X^m_{WATER}\right) \tag{11.11}$$

expressing the specific heat in J/(kg K); temperature T in Kelvin, for the 272–353 K range; and the fat content between 9% and 40%.

Manohar et al. (1991) gave the following equation for tamarind juices:

$$\hat{C}_P = 4.18 + (6.839 \times 10^{-5} T - 0.0503) C \tag{11.12}$$

where
the specific heat is expressed in kJ/(kg K) if the temperature is given in Kelvin
C is the soluble solids content expressed in °Brix

Choi and Okos (1986) proposed an equation for the case in which the composition of the product is known:

$$\hat{C}_P = \sum_i \left(\hat{C}_{Pi} X^m_i\right) \tag{11.13}$$

where
$\hat{C}_{Pi}$ is the specific heat of the component i
X^m_i is the mass fraction of the component i

TABLE 11.4
Specific Heat for Some Foods

Product	Water (%)	Specific Heat (kJ/kg K)
Meats		
Bacon	49.9	2.01
Beef		
Lean beef	71.7	3.433
Roast beef	60.0	3.056
Hamburger	68.3	3.520
Veal	68.0	3.223
Prawns	66.2	3.014
Eggs		
Yolk	49.0	2.810
Milk		
Pasteurized, whole	87.0	3.852
Skim	90.5	3.977–4.019
Butter	15.5	2.051–2.135
Apples (raw)	84.4	3.726–4.019
Cucumbers	96.1	4.103
Potatoes	79.8	3.517
	75.0	3.517
Fish	80.0	3.600
Fresh	76.0	3.600
Cheese (fresh)	65.0	3.265
Sardines	57.4	3.014
Carrots (fresh)	88.2	3.810–3.935

Source: Reidy, G.A., Thermal properties of foods and methods of their determination, MS thesis, Food Science Department, Michigan State University, East Lansing, MI, 1968.

The specific heat values of different foods are listed in Table 11.4. Table 11.2 also presents expressions for the calculation of the specific heat of pure components as a function of temperature, while in Table 11.3 equations that allow calculation of the specific heat of water and ice as a function of temperature are given.

11.3 THERMAL DIFFUSIVITY

A widely used property in calculations of heat transfer by conduction is the thermal diffusivity that is defined according to the expression

$$\alpha = \frac{k}{\rho \hat{C}_P} \tag{11.14}$$

The value of the thermal diffusivity of a given food can be calculated if the thermal conductivity, density, and specific heat are known.

However, there are mathematical expressions that allow calculation of the thermal diffusivity according to water content. Thus, Martens (1980) reported the following equation:

$$\alpha = 5.7363 \times 10^{-8} X_{WATER}^{m} + 2.8 \times 10^{-10} T \tag{11.15}$$

where
α is the thermal diffusivity in m^2/s
X_{WATER}^{m} is the water mass fraction
T is the temperature in Kelvin

On the other hand, Dickerson (1969) presented an expression in which the food's thermal diffusivity is only a function of the water content and its thermal diffusivity:

$$\alpha = 8.8 \times 10^{-8} \left(1 - X_{WATER}^{m}\right) + \alpha_{WATER} X_{WATER}^{m} \tag{11.16}$$

Choi and Okos (1986) expressed thermal diffusivity as a function of the components, similar to other thermal properties:

$$\alpha = \sum_{i} \left(\alpha_i X_i^V\right) \tag{11.17}$$

where
α_i is the thermal diffusivity of the component i
X_i^V is the volumetric fraction of such component

Table 11.5 presents thermal diffusivity values for some foods. Tables 11.2 and 11.3 show the expressions that allow calculation of the thermal diffusivities of pure components.

PROBLEMS

11.1 Determine the thermal conductivity, specific heat, and thermal diffusivity, at 25°C, of a food product that has been chemically analyzed and whose weight composition is 77% water, 19% carbohydrate, 3% protein, 0.2% fat, and 0.8% ash.

The method of Choi and Okos (1986) is used. Therefore, the thermal properties of each component at 25°C are previously calculated. The following table contains the results obtained.

Component	ρ_i (kg/m³)	k_i (W/m °C)	$\hat{C}_{Pi}$ (kJ/kg °C)	$\alpha_i \times 10^7$ (m²/s)
Water	994.91	0.6110	4.1773	1.458
Carbohydrate	1591.34	0.2334	1.5942	0.927
Protein	1316.94	0.2070	2.0376	0.797
Fat	915.15	0.2496	2.0180	1.019
Ash	2416.78	0.3628	1.1375	1.332

TABLE 11.5
Thermal Diffusivity for Some Foods

Product	Water (%)	Temperature[a] (°C)	Thermal Diffusivity × 10^5 (m^2/s)
Fruits, vegetables			
Avocado (pulp)	—	24 (0)	1.24
Seed	—	24 (0)	1.29
Whole	—	41 (0)	1.54
Sweet potato	—	35	1.06
	—	55	1.39
	—	70	1.91
Cherries (pulp)	—	30 (0)	1.32
Squash	—	47 (0)	1.71
Strawberries (pulp)	92	5	1.27
Beans (purée)	—	26–122	1.80
Peas (purée)	—	26–128	1.82
String beans (cooked)	—	4–122	1.68
Limes	—	40 (0)	1.07
Apples	85	0–30	1.37
Applesauce	37	5	1.05
	37	65	1.12
	80	5	1.22
	80	65	1.40
	—	26–129	1.67
Peach	—	27 (4)	1.39
Turnip	—	48 (0)	1.34
Potato			
Pulp	—	25	1.70
Mashed (cooked)	78	5	1.23
Banana (pulp)	76	5	1.18
	76	65	1.42
Grapefruit (pulp)	88.8	—	1.27
Exocarp	72.2	—	1.09
Beet	—	14 (60)	1.26
Tomato (pulp)	—	4.26	1.48
Fishes and meats			
Cod	81	5	1.22
	81	65	1.42
Halibut	76	40–65	1.47
Salted meat	65	5	1.32
	65	65	1.18
Ham (smoked)	64	5	1.18
	64	40–65	1.38
Beef			
Loin[b]	66	40–65	1.23
Round	71	40–65	1.33
Tongue	68	40–65	1.32
Water	—	30	1.48
	—	65	1.60
Ice	—	0	11.82

Source: Singh, R.P., *Food Technol.*, 36(2), 87, 1982.

[a] The first temperature is the initial one, and that in parentheses is the one of the surroundings.

[b] Data are applicable if the juices exuded during storage remain in foods.

The volumetric fraction of each component is calculated by means of Equation 11.7. The mass and volumetric fractions of each component are presented next.

Component	X_i^m	X_i^V
Water	0.77	0.8398
Carbohydrate	0.19	0.1296
Protein	0.03	0.0247
Fat	0.002	0.0024
Ash	0.008	0.0036

Thermal conductivity, obtained from Equation 11.6

$$k = \sum\left(k_i X_i^V\right) = 0.55\ \text{W/(m °C)} = 5.5\times10^{-4}\,\text{kJ/(s m °C)}$$

- Specific heat, obtained from Equation 11.13

$$\hat{C}_P = \sum_i\left(\hat{C}_{Pi} X_i^m\right) = 3.594\ \text{kJ/(kg °C)}$$

- Thermal diffusivity, obtained from Equation 11.17

$$\alpha = \sum_i\left(\alpha_i X_i^V\right) = 1.37\times10^{-7}\ \text{m}^2/\text{s}$$

It can also be calculated by Equation 11.14:

$$\alpha = \frac{k}{\rho\hat{C}_P} = 1.41\times10^{-7}\ \text{m}^2/\text{s}$$

Result:

$$k = 0.50\ \text{W/(m °C)}$$

$$\hat{C}_P = 3.94\ \text{kJ/(kg °C)}$$

$$\alpha = 1.37\times10^{-7}\ \text{m}^2/\text{s}$$

12 Optical Properties of Foods

12.1 COLOR

The first perception that consumers have of food is its optical appearance. Color is an attribute of food that significantly influences the acceptance of the product. Also, it is a parameter that can be correlated with a number of properties of a given food such as maturity, degree of processing, and type of spoilage. For example, the evolution of the maturity of a fruit can be monitored by its color changes.

To better understand optical properties, it is necessary to introduce means to characterize and measure color. "Color is the visual perceptual property corresponding in humans to the categories called *red*, *blue*, *yellow*, *green*, and others. Color derives from the spectrum of light (distribution of light power versus wavelength) interacting in the eye with the spectral sensitivities of the light receptors." At the same time, it is essential to define the concept of light and identify the electromagnetic radiations that could be sensitive to human eyes. The rays of white light of the sun can break down according to the wavelength in a scale that ranges from red to violet (Table 12.1), so that the human eye is sensitive to the different colors, as shown in Figure 12.1. A group of terms corresponding to the colored lights has been designated. The color is a physiological sensation that is characterized in three ways: the nature of the object, the light that is reflected in the object itself and that is captured by our eyes, and the message received by the eye that is communicated to the brain.

The objective specification of a color is not that simple, and this is why it is necessary to rely on some specific metrics. The light can be determined by the wavelength, which allows classification along the spectrum, so the spectral bands can be denominated according to their dominant color. When a spectral band is very narrow, it is said to be monochromatic, although a light is considered monochromatic when it is formed by only one wavelength. Pure colors are those that are associated with monochromatic lights. The bodies can be classified in different colors; for example, a surface considered as white is one that diffuses the light in all directions without absorbing the visible radiations that it receives. A surface is considered black when it absorbs the entire incidental radiation, while gray bodies are those that diffuse or partially transmit in the same way all of the radiation they receive. All surfaces that are not white, black, or gray are considered as colored surfaces.

12.2 CHARACTERISTICS OF COLOR

To describe the color of a light source or an object, three essential characteristics are needed: value, hue, and chroma (saturation).

Value (lightness or lightness factor) describes overall intensity to how light or dark a color is. It is the only dimension of color that may exist by itself. When we describe a color as "light" or "dark," we are discussing its value or "brightness." This property of color tells us how light or dark a color is based on how close it is to white. Hue is the most obvious characteristic of a color. There are really an infinite number of possible hues. A full range of hues exists, for example, between red and yellow. In the middle of that range are all the orange hues. Similarly, there is a range of hues between any other two hues. The third characteristic is denominated the purity factor, which indicates whether the color is more or less similar to the corresponding pure color. The term corresponding to purity that characterizes the sensorial response of a subject is saturation. The dominant wavelength and the purity factor is the chromaticity of the light. The term used for the mean group of the tonality and the saturation is *chroma*.

TABLE 12.1
Color Classification according to Wavelength

Color	Wavelength Interval (nm)
Violet	397–424
Navy blue	424–455
Blue	455–492
Green	492–575
Yellow	575–585
Orange	585–647
Red	647–723

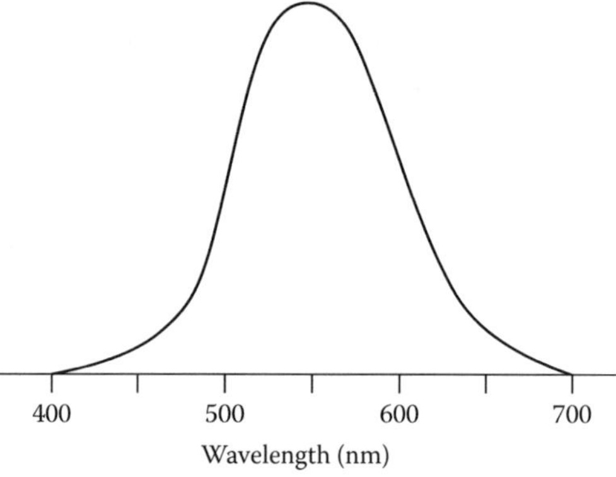

FIGURE 12.1 Sensibility curve of human eye.

The order of magnitude of the factor of brightness and the purity factor of a body can be expressed by means of a specific adjective; in this way, if the color of a surface is clear and saturated, it is denominated bright; when it is dark and saturated, it is denominated deep; if it is clear and light at the same time, it is denominated pale.

12.3 COLOR PERCEPTION

The eye is the organ that transmits to the brain, through the retina, the received sensations of the environment. The eye is an integrative organ, which, whether it receives violet light or an overlapping of blue and red light, will see only a violet color. The human eye is not able to separate the pure colors of a mixture. However, it is very sensitive and it is able to detect small differences among mixtures, depending on the spectral area. In Figure 12.2 the average curves of the sensitive

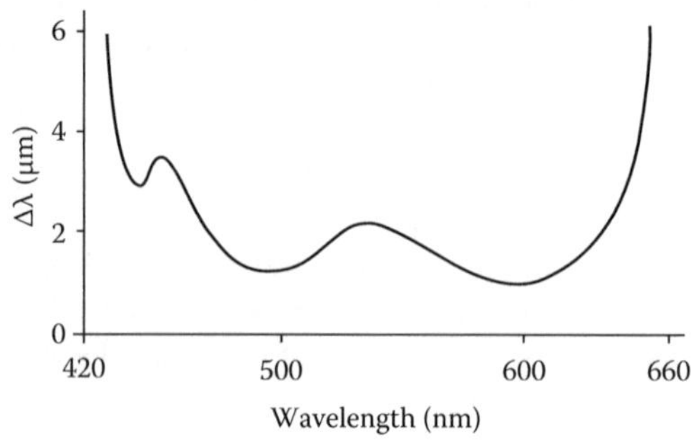

FIGURE 12.2 Differentiation of colors according to wavelength.

variations as a function of the wavelength are shown. According to this figure the discrimination of violet and red tones is easier, while it is more difficult to appreciate the difference of tones in the intervals of yellow. However, the saturation or grade of purity is easier to evaluate for yellow and green, while being much less so for red, but even easier for violets.

The eye can distinguish a great number of colors very easily, although it seems impossible that multiple receiving systems that distinguish the infinity of existent colors in nature can exist. Therefore, Young (1802) formulated the trichromatic theory that was later completed by Helmholtz. This theory basically states that all chromatic sensations can decrease to three basic sensations of color. The perception of color is the result of the combined action of three receiving mechanisms that possess different spectral sensibilities, being the basic or primary colors, blue, red, and green. The mixture of the three colors can give rise to an infinite quantity of color shades. The difference among the primary colors, according to the laws of physics and chemistry, is significant. From the point of view of physics theory, the primary colors are the three mentioned earlier; however, according to the chemical theory, colors such as cyan, yellow, and magenta are also considered as primary. Cyan is a clear blue color that is also called blue turquoise, with a wavelength of approximately 500 nm. The color magenta is a type of red that tends to violet.

In the human eye there are two types of light-sensitive cells, called photoreceivers, which are the rods and the cones; the cones are "in charge" of contributing the sensation of color. The curve of day vision is perceived by the cones, presenting a maximum wavelength of approximately 550 nm, decreasing toward both ends in a symmetrical way, in the range of 400–700 nm. The rods are sensitive to scarce illumination, corresponding to night vision, presenting a maximum at 500 nm. There are three cone types that present a maximum sensibility to the three primary colors. The receiving cones for green and red present a curve of similar sensibility, while the sensibility of the receivers of blue is the 20th part of the other two colors. The sensation of the perceived color is defined as the solution to each of the sensibility curves to the spectrum that emits the object, obtaining three different answers that correspond to each color. It is possible that two objects that emit different spectra will produce the same color sensation.

The formation of any color can be obtained by the addition of red, green, and blue colors or the subtraction of the colors cyan, magenta, and yellow. The models associated with the theory of color like RGB or CMY are known like this, due to the initials of the corresponding colors in English. In this way, the additive model RGB is based on that fact that all the possible colors can be created by the mixture (additive synthesis) of the projection of the three primary colors (red, green, and blue), and when none of these colors exists, black is perceived (Figure 12.3). The subtractive model CMY is based on the idea that any color can be formed from the subtractive mixture of the three primary pigments (cyan, magenta, and yellow); in this model, black is created by a mixture of all colors, and the target is the absence of any color (Figure 12.4). This subtractive synthesis explains

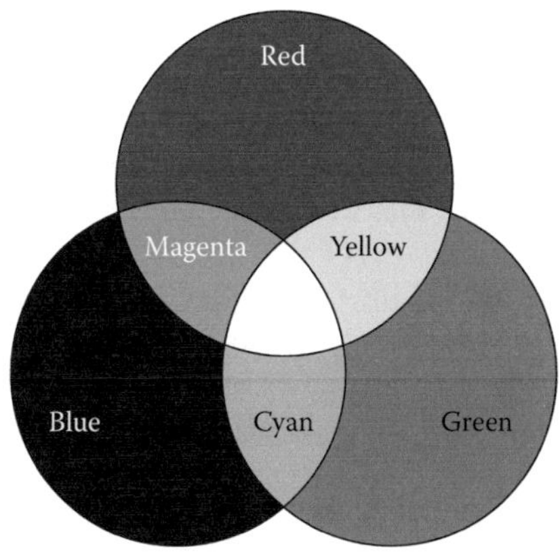

FIGURE 12.3 RGB color model.

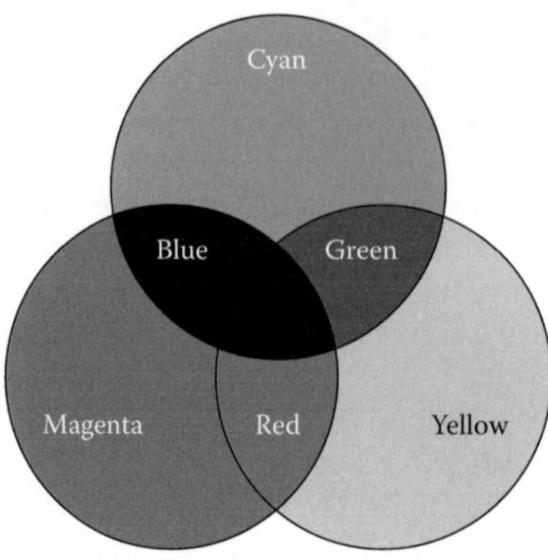

FIGURE 12.4 CMY color model.

the theory of the mixture of paintings, tints, inks, and natural colorants to create colors that absorb given wavelengths and reflect others. The color that seems to be present in a given object depends on what zones of the electromagnetic spectrum this color is reflected by the object or which spectrum zones are not absorbed.

12.4 COLORIMETRY

Any color can be defined by means of three physical attributes:

1. Hue: defined by a wavelength attributed arbitrarily to each color, as shown in Figure 12.1
2. Grade of purity, in reference to the target
3. Brightness factor or the quantity of transmitted light that classifies the colors as clear or dark

This classification leads to a great number of colors and shades that can be perceived by the eye. However, this perception is subjective, and ambiguities can be presented in the fixation of a certain color of a body. Colorimetry allows the determination of the factors that calculate color, so that it can be fixed without ambiguity. It might be necessary to show the spectral curve, in order to analyze with more detail the colors of a body or a luminous source. The instruments used for this purpose are denominated spectrophotocolorimeters.

When measuring the color of an object, it is necessary to define the source type or illuminant used. The most usual forms for illumination are the natural light of day or the simulated (artificial) light of day. The natural light of day is considered as illuminant, whenever it is not direct light, that is, it will be a light received by a window facing to the north that does not directly receive sunlight. However, this type of light can vary, since it depends on the localization and the hour of the day and the year. Therefore, for the measurements to be more reliable, illuminants are usually used to simulate the natural light of the day artificially, since they are stable light sources and they can be normalized. The *Commission Internationale de l'Eclairage* (CIE) has defined three illuminant types: illuminant A simulates the typical light of an incandescent lamp, illuminant type B simulates the light of the sun, while the illuminant type C represents the total sky average light of the day. Illuminants A and C are defined for temperature emitted in a given color. When heating a black body until it reaches incandescence, the emitted radiation depends on temperature. Red is emitted in low temperatures, while as the temperature increases, it moves toward the blue tonalities.

Illuminant A is related to the light that emits a wolfram filament lamp and it is equivalent to the light that emits a black body at 2855 K. Illuminant B refers to the light of noon, with a color temperature 4874 K. Illuminant C is related to the light of the day in the northern hemisphere without

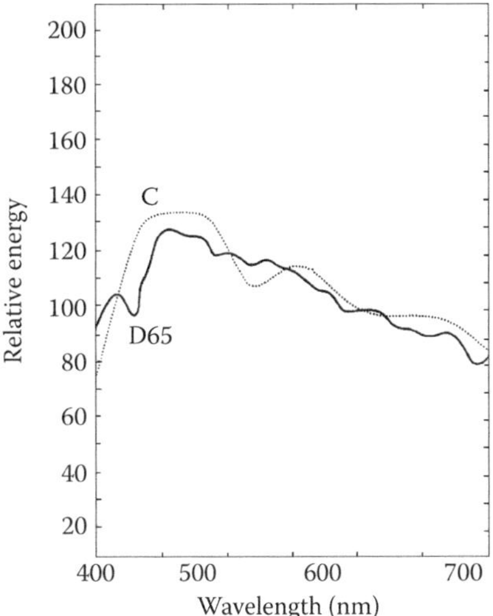

FIGURE 12.5 Illuminant emission spectra.

direct radiation, with a color temperature 6774 K. Illuminant C has been used very frequently, although the CIE has proposed the use of a new illuminant, D65, with a color temperature of 6500 K. In Figure 12.5 the emission spectra of illuminants A, C, and D65 are shown.

The object itself also influences the perception of color, since its characteristics will be those that determine the quantity of energy that is absorbed, transmitted, or reflected in relation to that which is received from the illuminant, and this perception causes the radiation type that emerges from the surface of the object with its defined colorimetric characteristics.

It is also important to normalize the conditions emitted by the body and received by the observer, since the perceived color can differ because the conditions that the body observes are also going to be measured. However, different systems are utilized to measure color, and some of those systems are based on the comparison of the object with well-defined standards. In these cases, to carry out a correct evaluation of the color, the comparison of the object color with that of the standards should be carried out under normalized conditions, as well as the observation of the illumination. In general, these observations are carried out in an illumination chamber that contains different illuminants; the measurements should be carried out so that the illumination is perpendicular to the plane of the object and the observation at 45°. This type of color evaluation can be subjective, since it depends on the sensation that is perceived by the observer. To avoid possible observation defects, other evaluation systems may be used. In these cases the exact conditions of evaluation consist of illuminating the object with an angle of 45°, with the observer being in a perpendicular position. The observer is placed at 45.7 cm (18 in.) from the object, observing it through a gap with a vision angle of 2°, although a vision angle of 10° is also sometimes used.

Because the evaluation of the color of an object is perceived by an observer's direct observation, it can have a certain subjective character; instruments such as spectrophotometers and colorimeters can be used for this purpose.

12.5 COLOR SYSTEMS

To describe the colors of objects, a series of systems has been developed. Thus, a system for notation has been defined that allows fixing the color of the object by means of certain coordinates that are 3D dispositions of color agreement with the object's appearance.

12.5.1 CIE Color System

In order to normalize the measurement of colors, the CIE developed a system that was based on the theory of trichromatic perception that helped to obtain a more objective measurement of the color of a sample. For this reason, the illumination source, the observation conditions, the observer's curves, and the appropriate mathematical units must be described. Previously, the illuminant types and the observer's conditions have been defined. In relation to the standard observer, the CIE describes in a numeric way the colors perceived by the eye. The quantification of these colors is carried out through some standard curves obtained by comparison of the monochromatic lights of the visible spectrum with mixtures of the three primary colors. These are defined as mathematical functions *X*, *Y*, and *Z* assigned to red, green, and blue light, respectively, and they are known as tristimulus coordinates, representing the quantity of each light required to match a certain color.

The response of the wavelengths in front of the eye is given in a normalized form, as shown in Figure 12.6, representing the CIE standard curves $\bar{x}$, $\bar{y}$, $\bar{z}$ of the observer.

The tristimulus coordinates can be calculated from the spectra of the sample (*R*), the illuminant (*E*), and the observer's curves integrating between the visible wavelengths and the following equations:

$$X = \int_{380}^{700} RE\bar{x}\,dx \tag{12.1}$$

$$Y = \int_{380}^{700} RE\bar{y}\,dy \tag{12.2}$$

$$Z = \int_{380}^{700} RE\bar{z}\,dz \tag{12.3}$$

From these three parameters or tristimulus coordinates, the CIE chose the *Y* value in such a way that coincides with the standard observer's curve, directly indicating the value of the brightness.

In addition to these tristimulus coordinates, chromaticity coordinates or chromaticity coefficients can be defined as the unitary fraction, according to the expressions

$$x = \frac{X}{X+Y+Z} \tag{12.4}$$

$$y = \frac{Y}{X+Y+Z} \tag{12.5}$$

$$z = \frac{Z}{X+Y+Z} \tag{12.6}$$

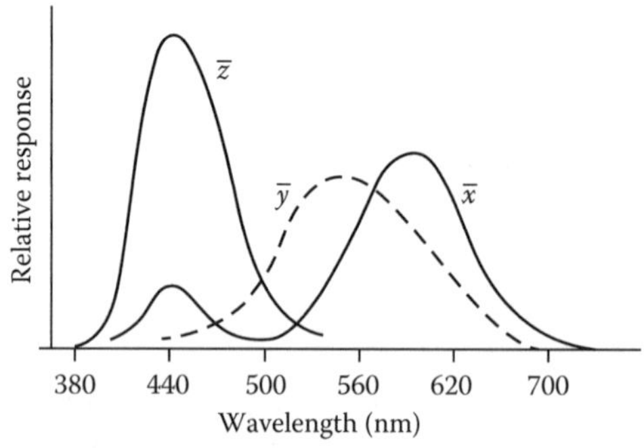

FIGURE 12.6 Tristimulus values for standard observer.

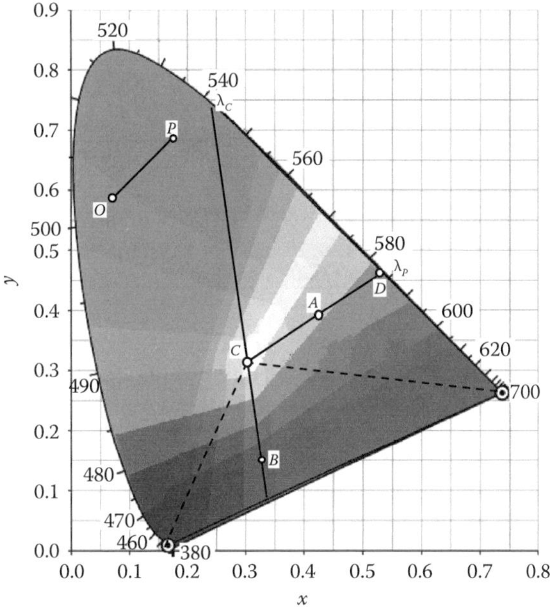

FIGURE 12.7 Chromatic diagram of CIE system.

It can be seen that the summation of the three chromatic coordinates is the unit. That is why if two of these coordinates are known, the third coordinate is obtained by subtracting from the unit the sum of the other two. Generally, to define the color of a sample, coordinate values of x and y are usually taken first. The third coordinate takes the value of the tristimulus Y that defines the brightness of the sample. When representing the values of y before x, the denominated chromatic diagram is obtained (Figure 12.7). The color of a sample will be given by the values of x and y, which determine its chromaticity, while the brightness Y is a value that is in the perpendicular straight line of the chromatic plane, where the white and black and gray scales are represented. For illuminant C, the coordinates in this diagram are $x_C = 0.3101$, $y_C = 0.3163$, while their tristimulus coordinates are $X_n = 98.4$, $Y_n = 100$, $Z_n = 118.12$.

For illuminant D65, the coordinates are $x_{D65} = 0.3127$, $y_{D65} = 0.3290$; their tristimulus coordinates being $X_n = 95.02$, $Y_n = 100$, $Z_n = 108.80$.

For a given color, represented by point A and possessing coordinates x, y, it can be associated to being a predominant wavelength that one obtains when tracing the straight line that passes through the point of the illuminant C and the point A representative of the color and prolonging the line until it cuts into the outlying line of the diagram, where the predominant wavelength predominant value will be obtained (P). For the points located in the triangle formed by the illuminant and the ends of the wavelengths 400 and 700 nm, as the point B, the predominant wavelength is not obtained in the straight line 400–700, but it is obtained in the back continuation until it cuts the outlying line curve, obtaining the complementary wavelength (C).

Purple colors are located in the chromatic diagram on the previously defined triangle, and they do not possess a predominant wavelength, but rather they possess a complementary wavelength. The sensorial perception of the saturation is the colorimetric purity, which for a point A can be defined as the relationship CA/CD. In the same chromatic diagram, around the illuminant, there is an area of neuter or achromatic colors, which lacks chromaticity. When it is desired to carry out the mixture of two represented colors, for example, for points A and B, the resulting color will be on the straight line that passes through these two points. It is also important to point out that each color has its complement and that in the trichromatic diagram they are on the straight line that goes by the point of the illuminant. The idea of complementary colors can be very useful in the food industry, because displaying a food with a certain color next to a food with a complementary color can allow it to stand out.

12.5.2 Munsell Color System

In this system, each color can be described by means of three parameters: hue H, the luminosity or value V, or chroma or saturation C, as shown in Figure 12.8. Brightness is represented in the perpendicular axis to the circle, indicating the scale of gray, with a scale that goes from black (0) to white (10). Brightness can also be also denominated as clarity, showing a relationship between the reflected and absorbed light.

Hue is the quality by which a color is distinguished from another color, with this attribute corresponding to its own color, such as blue, red, orange, or any other color. Hue is defined on the circumference, where the colors are identified, being differentiated between five main tones: red (R), yellow (Y), green (G), blue (B), and purple (P) (Figure 12.8b). The distance between each tone is divided in a scale from 1 to 10. At the middle point, there are intermediate colors, represented as red–yellow (RY), yellow–green (YG), green–blue (GB), purple–blue (PB), and red–purple (RP).

Chroma is given by the radial distance from the central axis, and it is the quality that distinguishes a pure hue.

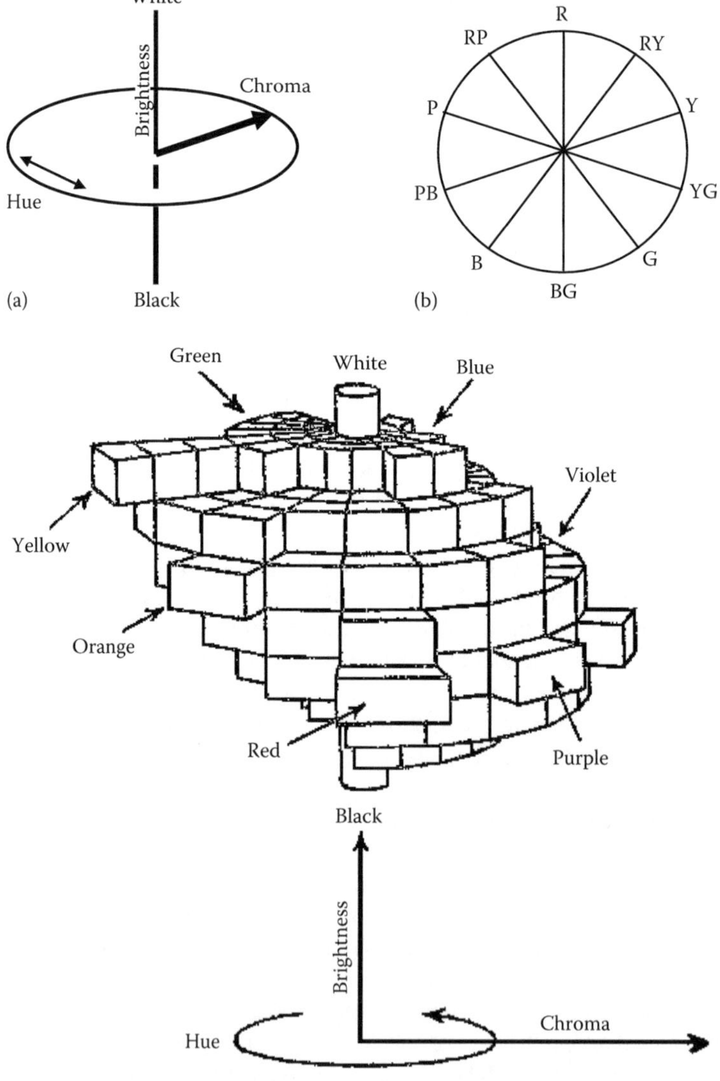

FIGURE 12.8 Munsell's color representation. (a) 3-D Munsell color system and (b) plane of hue.

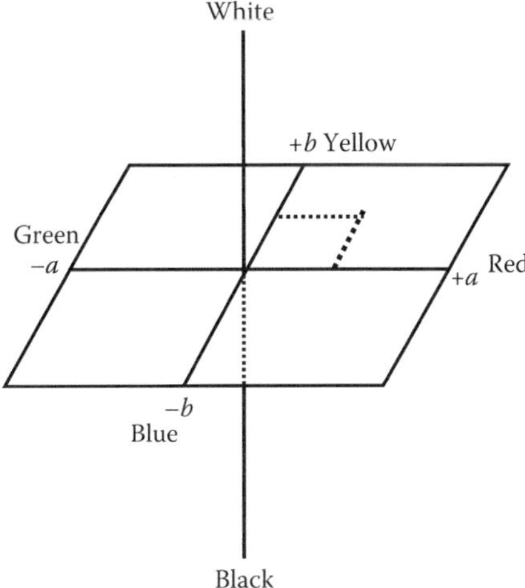

FIGURE 12.9 Coordinates in the Hunter color system.

To define a color according to the Munsell system, a notation such as 4 R 5/9 is used, indicating that it is a red color with a hue 4, brightness 5, and a chroma of 9.

12.5.3 Hunter *Lab*–Munsell Color System

In this system, the colors are defined by means of three coordinates, L, a, and b, as shown in Figure 12.9. The first parameter L is the brightness, while a and b define the chromaticity of the sample. On a Cartesian plane, the parameter a defines the red–green component, while b defines the yellow–blue component. The positive values of a represent the red color and the negative values represent the green. In the case of b, the positive values are for the yellow, while the negatives are for the blue. The brightness is represented on a perpendicular straight line to the plane a–b, corresponding to the white positive value of L, while the black corresponds to a negative value. These values of L for the Hunter system are directly comparable with Y from the CIE system and with the value of the Munsell system. The saturation is determined by the distance to the geometric center, so that the values that are far away correspond to colors of more saturation.

The values L, a, and b can be calculated starting from the tristimulus values X, Y, Z according to the following expressions:

$$L = 10\sqrt{Y} \tag{12.7}$$

$$a = \frac{17.5(1.02X - Y)}{\sqrt{Y}} \tag{12.8}$$

$$b = \frac{7.0(Y - 0.847Z)}{\sqrt{Y}} \tag{12.9}$$

12.5.4 *Lovibond* System

The Munsell color space is based on a method of addition of colors, while the Lovibond system is based on an opposite principle of subtracting colors. Therefore, starting from white and with adapted filters of red, yellow, and blue, the colors are subtracted from the initial white subtracting the initial white until the color of the sample is matched. The equalization of the color of the sample with the filters should

TABLE 12.2
Lovibond Measurements of Different Colors

Color	Lovibond Measurement
Light yellow	2.0–3.0
Medium yellow	3.0–4.5
Deep straw/gold	4.5–6.0
Deep gold	6.0–7.5
Light amber	7.5–9.0
Copper	9.0 – 11.0
Red/brown	11.0–14.0
Light brown	14.0–17.0
Medium brown	17.0–20.0
Light black	20.0–25.0
Black	>25

be carried out under defined conditions of illumination and of the observer, measuring the color with transmitted light. In Table 12.2 the measurements are given for the Lovibond scale for different colors. This is a system for measuring color that is applied in the brewery, oil, and honey industries.

12.5.5 *CIELAB* System

This system proposed by the CIE is a chromatic space that is defined in rectangular coordinates (L^*, a^*, b^*) together with another space that is defined in cylindrical coordinates (L^*, H^*, C^*). The coordinate L^* represents brightness, and it is the difference between white ($L^* = 100$) and black ($L^* = 0$). The coordinate a^* represents the difference between green ($-a^*$) and red ($+a^*$), while the coordinate b^* represents the difference between blue ($-b^*$) and yellow ($+b^*$). The knowledge of these three coordinates allows us to correctly describe the color of any sample and to locate it in the color space.

The CIELAB coordinates can be determined starting from the tristimulus values from the following equations:

$$L^* = 116\sqrt[3]{\frac{Y}{Y_n}} - 16 \tag{12.10}$$

$$a^* = 500\left(\sqrt[3]{\frac{X}{X_n}} - \sqrt[3]{\frac{Y}{Y_n}}\right) \tag{12.11}$$

$$b^* = 200\left(\sqrt[3]{\frac{Y}{Y_n}} - \sqrt[3]{\frac{Z}{Z_n}}\right) \tag{12.12}$$

with $Y/Y_n > 0.01$.

In these equations, X, Y, Z are the tristimulus values of the sample, while X_n, Y_n, Z_n are the tristimulus values corresponding to the used illuminant.

In many cases, it is important to compare the color of different samples to each other, or in relation to a standard color. Thus, the color difference between samples is defined according to the expression

$$\Delta E = \sqrt{(\Delta L^*)^2 + (\Delta a^*)^2 + (\Delta b^*)^2} \tag{12.13}$$

where Δ operator represents the increment of each of the parameters among the two compared samples.

To locate the color of a sample in the CIELAB color space, cylindrical coordinates can also be used. In this case, the coordinates that are used are brightness (L^*), chroma (saturation) (C^*), and hue (H^*). Brightness is defined as a rectangular coordinate, while chroma is the coordinate that represents the perpendicular distance from the axis of brightness to the considered point, and hue is the angle expressed in degrees determined by the straight line that bounds the origin of the coordinate with the chromatic point and the axis $+a^*$. Hue is 0° for the axis $+a^*$, 90° for the axis $+b^*$, 180° for the axis $-a^*$, 270° for the axis $-b^*$, and 360° = 0°, when returning to the axis $+a^*$. The mathematical expressions that allow the calculation of chroma and hue values are

$$C^* = \sqrt{(a^*)^2 + (b^*)^2} \tag{12.14}$$

$$H^* = \arctan\left(\frac{b^*}{a^*}\right) \tag{12.15}$$

Also, the chroma and hue differences between samples can be defined, although in this case the mathematical expressions that allow obtaining these values are

$$\Delta C^* = C^*_{sample} - C^*_{reference} \tag{12.16}$$

$$\Delta H^* = \sqrt{(\Delta a^*)^2 + (\Delta b^*)^2 - (\Delta C^*)^2} \tag{12.17}$$

PROBLEMS

12.1 There are 100 g of a colorant A and 200 g of another colorant B. The tristimulus coordinates of these two samples have been determined; the results are shown in the following table, as well as those of the illuminant C for observer 2°.

	X	Y	Z
Colorant A	50.0	58.5	20.4
Colorant B	74.5	35.0	6.2
Illuminant C	98.4	100.0	118.12

Calculate the initial colorants and the resulting mixture:

a. Trichromatic coordinates
b. Dominant and complementary wavelength
c. Brightness, hue, and purity

a. Trichromatic coordinates are calculated from Equations 12.4 through 12.6:

Colorant A

$$x_A = \frac{X}{X+Y+Z} = \frac{50}{50+58.5+20.4} = 0.388$$

$$y_A = \frac{Y}{X+Y+Z} = \frac{58.5}{50+58.5+20.4} = 0.454$$

$$z_A = \frac{Z}{X+Y+Z} = \frac{20.4}{50+58.5+20.4} = 0.158$$

Colorant *B*

$$x_B = \frac{74.5}{74.5 + 35.0 + 6.2} = 0.644$$

$$y_B = \frac{35}{74.5 + 35.0 + 6.2} = 0.303$$

$$z_B = \frac{6.2}{74.5 + 35.0 + 6.2} = 0.054$$

The distance of the two colorant representative points will be

$$d_{AB} = \sqrt{(x_A - x_B)^2 + (y_A - y_B)^2} = \sqrt{(0.3882 - 0.6439)^2 + (0.4542 - 0.3025)^2}$$

$$d_{AB} = 0.298$$

In the chromatic diagram (Figure P.12.1), it is possible to locate the points corresponding to the two colorants. The representative point of mixture (*M*) is in the straight line that bounds points *A* and *B*. The lever rule can be applied, then

$$100 \times d_{AM} = 200 \times (d_{AB} - d_{AM})$$

On the diagram the distance is measured between points *A* and *B*, obtaining $d_{AB} = 70$ mm.

When applying the lever rule, the distances between points *A* and *B* and the representative point of the mixture are obtained: $d_{AM} = 46.77$ mm and $d_{BM} = 23.3$ mm. This allows fixing the mixture point on the chromatic diagram, obtaining the following coordinates:

$$x_M = 0.55; \quad y_M = 0.35; \quad z_M = 1 - (0.55 + 0.35) = 0.10$$

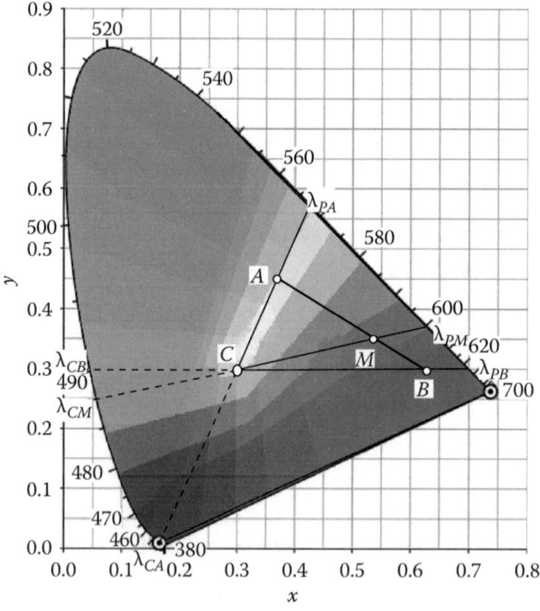

FIGURE P.12.1 Chromatic diagram. Location of points *A*, *B*, and *M*.

TABLE P.12.1
Dominant and Complementary Wavelengths

Sample	$\lambda_{predominant}$ (nm)	$\lambda_{complementary}$ (nm)
A	568	465
B	620	492
M	599	488

b. The dominant and complementary wavelengths.

Initially, the chromatic coordinates of the illuminant have been determined:

$$x_C = \frac{X}{X+Y+Z} = \frac{98.4}{98.4+100+118.12} = 0.311$$

$$y_C = \frac{Y}{X+Y+Z} = \frac{100}{98.4+100+118.12} = 0.316$$

To obtain the dominant wavelength, point *C* of the illuminant is located in the chromatic diagram, and a straight line is traced passing through this point and those corresponding to the two colors, obtaining the outlying curve of the predominant wavelength values. The complementary wavelengths are obtained when prolonging these lines in opposite directions and their intersection with the chromatic curve will give the value of these wavelengths. In Table P.12.1 the obtained values for the predominant and complementary wavelengths of the three considered points are shown.

c. Brightness, hue, and purity.

Brightness takes the *Y* tristimulus value for samples *A* and *B*; these values are those given by the problem:

$$Y_A = 58.5; \quad Y_B = 35.0$$

For the mixture, the value of the brightness is obtained from the following expression:

$$Y_M = \frac{m_A Y_A + m_B Y_B}{m_A + m_B} = \frac{(100)(58.5)+(200)(35)}{100+200} = 42.83$$

The hue gives the predominant wavelength, therefore the hue of the three samples is

$$\lambda_{PA} = 568; \quad \lambda_{PB} = 620; \quad \lambda_{PM} = 599$$

The purity is obtained from the relationship between the distance from the illuminant to the point of the sample and the distance of the illuminant until the representative point of the predominant wavelength is found. The distances of the illuminant to the different points are

$$d_{C-A} = \sqrt{(0.388-0.311)^2 + (0.454-0.316)^2} = 0.158$$

$$d_{C-PA} = \sqrt{(0.44-0.311)^2 + (0.56-0.316)^2} = 0.276$$

$$d_{C-B} = \sqrt{(0.644-0.311)^2 + (0.303-0.316)^2} = 0.333$$

$$d_{C-PB} = \sqrt{(0.70-0.311)^2 + (0.30-0.316)^2} = 0.389$$

$$d_{C-M} = \sqrt{(0.55-0.311)^2 + (0.35-0.316)^2} = 0.241$$

$$d_{C-PB} = \sqrt{(0.63-0.311)^2 + (0.37-0.316)^2} = 0.324$$

$$\text{Purity of } A = \frac{d_{C-A}}{d_{C-PA}} = \frac{0.158}{0.276} = 0.5724\,(57.24\%)$$

$$\text{Purity of } B = \frac{d_{C-B}}{d_{C-PB}} = \frac{0.333}{0.389} = 0.8560\,(85.60\%)$$

$$\text{Purity of } M = \frac{d_{C-M}}{d_{C-PM}} = \frac{0.241}{0.324} = 0.7438\,(74.38\%)$$

12.2 Many solid foods absorb water during storage; this usually affects their colorimetric properties. At the end of a drying process, the following tristimulus coordinates are obtained: $X = 70$, $Y = 65$, $Z = 48$. This food is stored in a cellar that presents a relative humidity of 95%, and after 3 months these coordinates are measured again, obtaining the following values: $X = 59$, $Y = 49$, $Z = 23$. In the colorimetric determinations, the illuminant C has been used ($X_n = 98.4$; $Y_n = 100$; $Z_n = 118.12$). Determine the trichromatic coordinates, the brightness, the purity, and the dominant and complementary wavelength. Also, calculate the color difference.

The trichromatic coordinates are calculated from Equations 12.4 through 12.6:

$$x_A = \frac{X}{X+Y+Z}, \quad y_A = \frac{Y}{X+Y+Z}, \quad z_A = \frac{Z}{X+Y+Z}$$

The chromatic coordinates of the initial and final sample, as well as of the illuminant, are shown in Table P.12.2.

TABLE P.12.2
Trichromatic Coordinates for Samples and Illuminant

	$M_{initial}$	M_{final}	$C_{illuminant}$
x	0.383	0.450	0.311
y	0.355	0.374	0.316
z	0.262	0.176	0.373
$\lambda_{predominant}$	586	589	
$\lambda_{complementary}$	484	486	

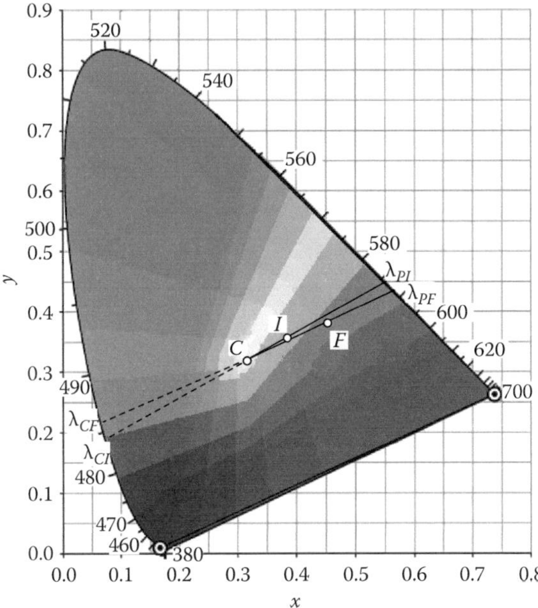

FIGURE P.12.2 Chromatic dominant diagram and complementary wavelength.

The predominant and complementary wavelengths are obtained from the representative points of each sample and of the illuminant, as shown in Figure P.12.2.

The brightness of the samples coincides with the value of the tristimulus coordinate Y; for the initial sample, the value of the brightness is $Y_I = 65$, and for the stored sample, $Y_F = 49$.

Purity is obtained from the relationship between the distance from the illuminant to the point of the sample and the distance of the illuminant to the representative point of the predominant wavelength is found. The distances of the illuminant to the different points are

$$d_{C-AI} = \sqrt{(0.383 - 0.311)^2 + (0.355 - 0.316)^2} = 0.082$$

$$d_{C-PI} = \sqrt{(0.55 - 0.311)^2 + (0.45 - 0.316)^2} = 0.274$$

$$d_{C-F} = \sqrt{(0.45 - 0.311)^2 + (0.374 - 0.316)^2} = 0.151$$

$$d_{C-PB} = \sqrt{(0.565 - 0.311)^2 + (0.435 - 0.316)^2} = 0.280$$

$$\text{Purity for the initial sample } I = \frac{d_{C-I}}{d_{C-PI}} = \frac{0.082}{0.274} = 0.2993\ (22.93\%)$$

$$\text{Purity for the final } F = \frac{d_{C-F}}{d_{C-PF}} = \frac{0.151}{0.280} = 0.5393\ (53.93\%)$$

For the calculation of the color difference, it is necessary to first calculate the values of the CIELAB coordinates. Thus, Equations 12.10 through 12.12 are applied:
CIELAB coordinates for the initial sample

$$L^* = 116\sqrt[3]{\frac{Y}{Y_n}} - 16 = 116\sqrt[3]{\frac{65}{100}} - 16 = 84.48$$

$$a^* = 500\left(\sqrt[3]{\frac{X}{X_n}} - \sqrt[3]{\frac{Y}{Y_n}}\right) = 500\left(\sqrt[3]{\frac{70}{98.4}} - \sqrt[3]{\frac{65}{100}}\right) = 13.23$$

$$b^* = 200\left(\sqrt[3]{\frac{Y}{Y_n}} - \sqrt[3]{\frac{Z}{Z_n}}\right) = 200\left(\sqrt[3]{\frac{65}{100}} - \sqrt[3]{\frac{48}{118.12}}\right) = 25.11$$

CIELAB coordinates for the final sample

$$L^* = 116\sqrt[3]{\frac{Y}{Y_n}} - 16 = 116\sqrt[3]{\frac{49}{100}} - 16 = 75.45$$

$$a^* = 500\left(\sqrt[3]{\frac{X}{X_n}} - \sqrt[3]{\frac{Y}{Y_n}}\right) = 500\left(\sqrt[3]{\frac{59}{98.4}} - \sqrt[3]{\frac{49}{100}}\right) = 27.43$$

$$b^* = 200\left(\sqrt[3]{\frac{Y}{Y_n}} - \sqrt[3]{\frac{Z}{Z_n}}\right) = 200\left(\sqrt[3]{\frac{49}{100}} - \sqrt[3]{\frac{23}{118.12}}\right) = 41.75$$

The color difference can be obtained from Equation 12.13:

$$\Delta E = \sqrt{(\Delta L^*)^2 + (\Delta a^*)^2 + (\Delta b^*)^2} = \sqrt{(84.48 - 75.45)^2 + (13.23 - 27.43)^2 + (25.11 - 41.75)^2}$$

$$= 23.67$$

12.3 An industry that produces concentrated fruit juices has performed a study of acceptability of its product colors. Then, the samples have been compared with a standard sample whose tristimulus coordinates are $X = 31.0$, $Y = 14.55$, $Z = 4.15$, obtaining that the sample with the greatest color difference has the following tristimulus values: $X = 29.8$, $Y = 12.95$, $Z = 3.88$. The measurements of the color have been conducted with observer #2, using the illuminant D_{65} ($X = 95.02$, $Y = 100$, $Z = 108.80$):

a. Calculate in CIELAB terms the color tolerance.
b. A study of color degradation of a juice sample has been analyzed. Therefore, glass containers have been filled with juice, and the containers have been stored at 25°C for one year; the color of these samples has been measured at weekly intervals. The evolution of the color difference has been fitted to a first-order kinetics, obtaining the following kinetic equation: $\Delta E = 0.15t$, where t is the storage time expressed in months.

The colorimetric control of two juice samples has been elaborated in the factory, obtaining the following values:

	L*	**a***	**b***
Sample 1	45.20	79.15	41.76
Sample 2	44.06	77.98	42.15

Calculate the commercial life of the two samples.

The color tolerance is obtained by noting the difference of existing color between the juices, which presents a greater difference in relation to the standard sample.

In order to obtain the color difference, it is necessary to first calculate the values of the CIELAB coordinates. Thus, Equations 12.10 through 12.12 are applied:

CIELAB coordinates for the standard sample are

$$L^* = 116\sqrt[3]{\frac{Y}{Y_n}} - 16 = 116\sqrt[3]{\frac{14.55}{100}} - 16 = 45.01$$

$$a^* = 500\left(\sqrt[3]{\frac{X}{X_n}} - \sqrt[3]{\frac{Y}{Y_n}}\right) = 500\left(\sqrt[3]{\frac{31}{95.02}} - \sqrt[3]{\frac{14.55}{100}}\right) = 81.23$$

$$b^* = 200\left(\sqrt[3]{\frac{Y}{Y_n}} - \sqrt[3]{\frac{Z}{Z_n}}\right) = 200\left(\sqrt[3]{\frac{14.55}{100}} - \sqrt[3]{\frac{4.15}{108.80}}\right) = 37.87$$

CIELAB coordinates for the sample with the greatest color difference.

$$L^* = 116\sqrt[3]{\frac{Y}{Y_n}} - 16 = 116\sqrt[3]{\frac{12.95}{100}} - 16 = 42.69$$

$$a^* = 500\left(\sqrt[3]{\frac{X}{X_n}} - \sqrt[3]{\frac{Y}{Y_n}}\right) = 500\left(\sqrt[3]{\frac{29.8}{95.02}} - \sqrt[3]{\frac{12.95}{100}}\right) = 86.74$$

$$b^* = 200\left(\sqrt[3]{\frac{Y}{Y_n}} - \sqrt[3]{\frac{Z}{Z_n}}\right) = 200\left(\sqrt[3]{\frac{12.95}{100}} - \sqrt[3]{\frac{3.88}{108.80}}\right) = 35.35$$

The allowed maximum color difference is

$$\Delta E = \sqrt{(\Delta L^*)^2 + (\Delta a^*)^2 + (\Delta b^*)^2} = \sqrt{(45.01 - 42.69)^2 + (81.23 - 86.74)^2 + (37.87 - 35.35)^2}$$

$$= 6.49$$

The color difference regarding the standard sample M1 is

$$\Delta E = \sqrt{(\Delta L^*)^2 + (\Delta a^*)^2 + (\Delta b^*)^2} = \sqrt{(45.01 - 45.20)^2 + (81.23 - 79.15)^2 + (37.87 - 41.76)^2}$$

$$= 4.41$$

The color difference regarding the standard sample M2 is

$$\Delta E = \sqrt{(\Delta L^*)^2 + (\Delta a^*)^2 + (\Delta b^*)^2} = \sqrt{(45.01 - 44.06)^2 + (81.23 - 77.98)^2 + (37.87 - 42.15)^2}$$

$$= 5.50$$

The two samples present a color difference smaller than the maximum allowed; therefore, the commercial lifetime of each one of the samples will be

$$\text{Sample 1: } t_{commercial} = \frac{6.49 - 4.41}{0.15} \approx 14 \text{ months}$$

$$\text{Sample 2: } t_{commercial} = \frac{6.49 - 5.50}{0.15} \approx 6.6 \text{ months}$$

13 Rheology of Food Products

13.1 INTRODUCTION

Rheology is the science that studies the flow and deformations of solids and fluids under the influence of mechanical forces. The rheological measurements of a product in different manufacturing stages can be useful for quality control purposes. The microstructure of a product can also be correlated with its rheological behavior, allowing the development of new materials. Rheometry permits attainment of rheological equations that are applied in engineering processes, particularly unit operations that involve heat and momentum transfer. Finally, knowing consumer preferences, it is possible to design a product that fulfills these preferences since rheological properties are well correlated with sensorial attributes.

Food industries frequently work with products that are in a liquid phase in all or some of the manufacturing stages (e.g., concentration, evaporation, pasteurization, pumping). Good design of each type of equipment is essential for optimum processing. In the design of every process, it is necessary to know, among other things, the physical characteristics of the flowing components. One of those characteristics is the rheological behavior of the fluid to be processed in order to avoid oversized pumps, pipes, evaporators, etc., that can have negative impact on the costs of the overall process.

Viscosity is used in the calculation of parameters of momentum and energy transport phenomena, as well as in the quality control of some products. For this reason, the rheological constants of processed fluids intervene in the equations of the mathematical model setup for the diverse operations that form a particular process. Such rheological constants should be determined, generally, by means of experimentation in each case. Thus, the rheological characterization of the different fluid streams is very important, as is the deduction of the equations that allow direct calculation of the rheological constants as a function of the considered food and the operation variables.

Foods in the liquid phase used in the industrial process or by the consumer include pastes, purées, soft drinks, egg products, milk products, natural fruit juices, vegetable concentrates, and sauces, among others.

13.2 STRESS AND DEFORMATION

When a force is applied to a body, the observed response is different, depending on the material. Thus, when the force is applied to an elastic solid, it deforms, but when the force stops acting, the solid recovers its initial shape. However, if the product is a Newtonian fluid, when the force stops acting, the fluid continues flowing.

The stress (σ) applied to a material is defined as the force (F) acting per area unit (A). The stress can be normal and tangential. Within the normal stresses are those of tension and compression. Consider a body, as shown in Figure 13.1. It can be observed that three stresses can be applied on each side of the body, one for each direction. Therefore, there will be nine components of stress, and each one will be named with two subindexes. The first subindex refers to the section on which it is applied, and the second subindex refers to the direction. Thus, σ_{ij} is a stress applied on section i in the j direction. The nine components of the stress define the stress *vector:*

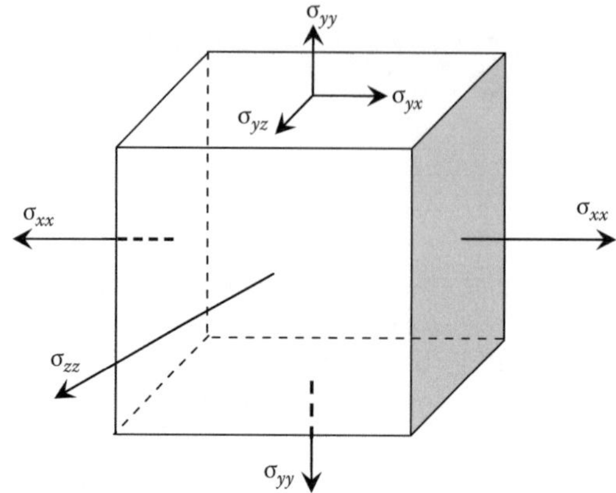

FIGURE 13.1 Stresses on a material.

$$\sigma_{ij} = \begin{pmatrix} \sigma_{11} & \sigma_{12} & \sigma_{13} \\ \sigma_{21} & \sigma_{22} & \sigma_{23} \\ \sigma_{31} & \sigma_{32} & \sigma_{33} \end{pmatrix} \tag{13.1}$$

The stress *vector* is symmetric, which implies that $\sigma_{ij} = \sigma_{ji}$. Therefore, out of the nine components of the stress *vector*, only six are independent.

The tangential stresses are those in which $i \neq j$, while in the normal stresses $i = j$. The normal tension stresses are positive ($\sigma_{ii} > 0$), while those of compression are negative ($\sigma_{ii} < 0$).

All stresses applied to a material produce a deformation. Deformations can be angular or longitudinal, depending on the type of stress applied. Normal stresses cause longitudinal deformations that may be lengthening or shortening, depending on whether a tension or compression stress is applied. Tangential stresses produce angular deformations.

If normal tension stress is applied to a bar with initial length L_0, it produces a lengthening of the bar in such a way that the final length is $L = L_0 + \Delta L$ (Figure 13.2). The deformation produced on this bar can be expressed as Cauchy's deformation:

$$\varepsilon_c = \frac{L - L_0}{L_0} = \frac{L}{L_0} - 1 \tag{13.2}$$

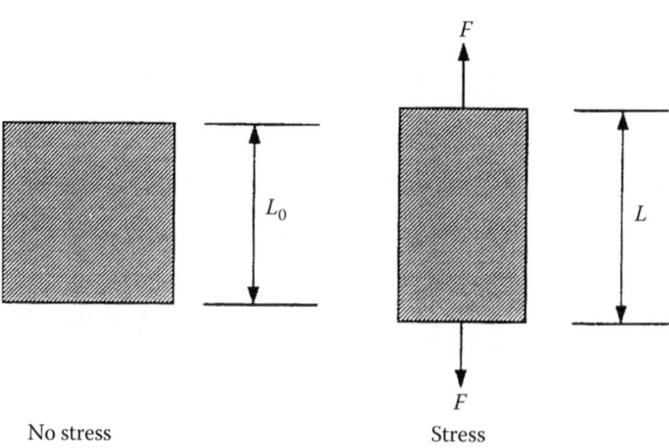

FIGURE 13.2 Deformation due to normal stress.

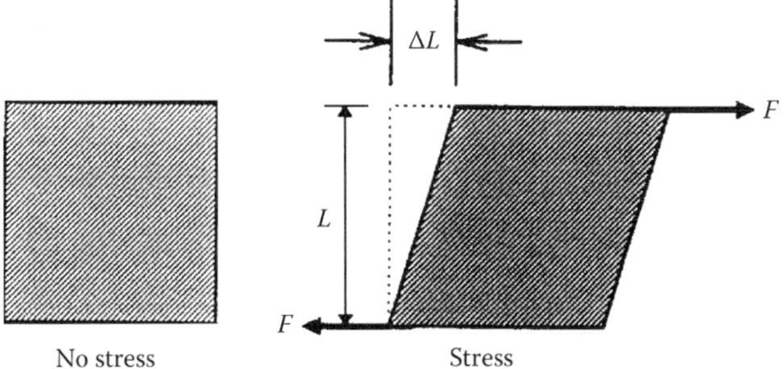

FIGURE 13.3 Shear deformation.

Another way to represent this would be to use Hencky's deformation equation:

$$\varepsilon_H = \ln\left(\frac{L}{L_0}\right) \tag{13.3}$$

For large deformations, Hencky's deformation definition is preferred over Cauchy's deformation.

Another type of deformation is the simple shear. If a tangential stress is applied to the upper side of a rectangular material, a deformation will occur, as shown in Figure 13.3, where the bottom side remains motionless. The angle of shear γ is calculated according to the following equation:

$$\tan\gamma = \frac{\Delta L}{L} \tag{13.4}$$

In the case of small deformations, the angle of shear in radians coincides with its tangent: $\tan\gamma = \gamma$, called shear strain.

13.3 ELASTIC SOLIDS AND NEWTONIAN FLUIDS

If tangential stress (σ_{ij}) is applied to a solid material, it is complied that the relation between the applied stress and the shear deformation (γ) produced is proportional:

$$\sigma_{ij} = G \cdot \gamma \tag{13.5}$$

then, the proportionality constant G is called shear or rigidity module. This type of material is called Hooke's solid, which is linearly elastic and does not flow. The stress remains constant until the deformation is eliminated; then, the material returns to its initial form.

The behavior of the solid of Hooke can also be studied by applying a normal stress (σ_{ii}) that produces a length change. The normal stress applied is directly proportional to the deformation of Cauchy produced:

$$\sigma_{ii} = E \cdot \varepsilon_C \tag{13.6}$$

where the proportionality constant E is denominated Young's or elasticity module.

The *elastic* solids are those that are deformed when subjected to tension forces, recovering their initial shape as such forces start to disappear. Solids that do not recover their original shape when the forces stop acting are known as *plastic* solids. *Elastoplastic* solids are those in between these two types; that is, if the applied stresses are lower than a certain value, they behave as elastic solids, but once they are over this stress value, they can no longer recover their initial shape, behaving then as plastic solids (Figure 13.4).

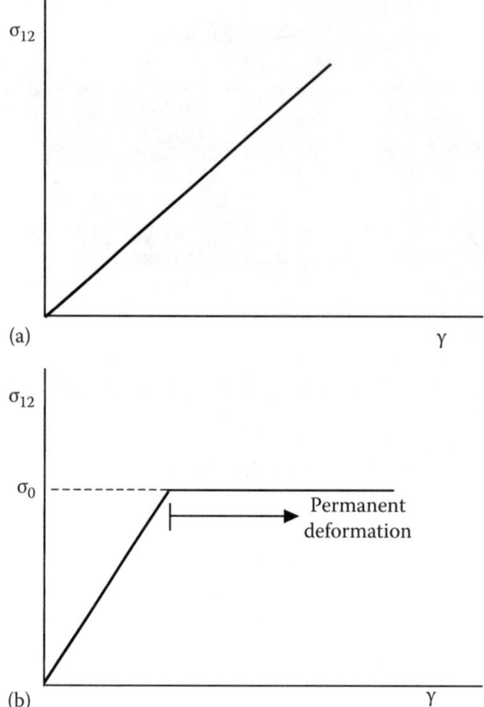

FIGURE 13.4 Deformation curves of solids: (a) elastic solid and (b) elastoplastic solid.

To study the ideal behavior of fluids, suppose a sample contained between two parallel slabs separated by a distance h, as shown in Figure 13.5, in which the lower plate remains fixed, while the upper plate moves at a constant velocity v_P. At a given time, this velocity is defined as

$$v_P = \frac{\delta x}{\delta t}$$

It is necessary to exert a force per unit area (σ_{12}) on the upper plate to retain this velocity. A velocity profile then develops along the height of the fluid, as shown in Figure 13.5. This type of flow is called simple shear. Shear rate ($\dot{\gamma}$) is defined as

$$\dot{\gamma} = \frac{dv_P}{dy} = \frac{d}{dy}\left(\frac{\delta x}{\delta t}\right) = \frac{d}{dt}\left(\frac{\delta x}{\delta y}\right) = \frac{d\gamma}{dt} \tag{13.7}$$

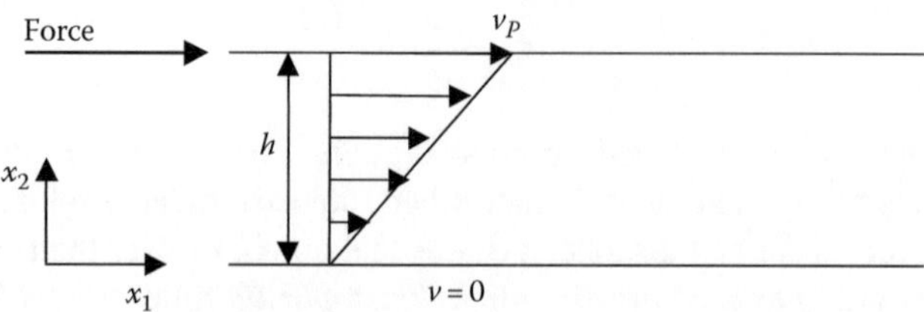

FIGURE 13.5 Velocity profile between parallel slabs.

The tangential stress applied σ_{12} and the rate of shear $\dot{\gamma}$ are directly proportional, and the proportionality constant is the viscosity (η):

$$\sigma_{12} = \eta \cdot \dot{\gamma} \tag{13.8}$$

This expression is known as Newton's law of viscosity.

In the same way, as solids that obey Hooke's law of deformation are considered as ideal elastic solids, fluids that follow Newton's law of viscosity are considered ideal viscous fluids.

13.4 VISCOMETRIC FUNCTIONS

In order to thoroughly study the flow of fluids, it is necessary to correlate the shear stresses and the rates of shear in three dimensions. The shear stress, as well as the shear rate, must be considered as *tensors*, each of which has nine components. However, for steady state and flow in simple shear, the stress *tensor* is reduced to

$$\sigma_{ij} = \begin{pmatrix} \sigma_{11} & \sigma_{12} & 0 \\ \sigma_{21} & \sigma_{22} & 0 \\ 0 & 0 & \sigma_{33} \end{pmatrix} \tag{13.9}$$

Simple shear flow is also called viscometric flow, and this is the type of flow that occurs in the rotational flow between concentric cylinders, axial flow in a pipe, rotational flow between parallel plates, and between plate and cone.

Only three functions are required to describe the behavior of a fluid's simple shear flow. These are the so-called viscometric functions, including the viscosity function $\eta(\dot{\gamma})$ and the first and second functions of normal stress, $\psi_1(\dot{\gamma})$ and $\psi_2(\dot{\gamma})$. These functions are defined as

$$\eta(\dot{\gamma}) = \phi_1(\dot{\gamma}) = \frac{\sigma_{12}}{\dot{\gamma}} \tag{13.10}$$

$$\psi_1(\dot{\gamma}) = \phi_2(\dot{\gamma}) = \frac{\sigma_{11} - \sigma_{22}}{(\dot{\gamma})^2} \tag{13.11}$$

$$\psi_2(\dot{\gamma}) = \phi_3(\dot{\gamma}) = \frac{\sigma_{22} - \sigma_{33}}{(\dot{\gamma})^2} \tag{13.12}$$

In some cases, the first and second differences of normal stresses are defined as

$$N_1 = \sigma_{11} - \sigma_{22} \tag{13.13}$$

$$N_2 = \sigma_{22} - \sigma_{33} \tag{13.14}$$

Generally, N_1 is much larger than N_2, and since the measurement of this last function is difficult, it can be supposed that its value is null.

13.5 RHEOLOGICAL CLASSIFICATION OF FLUID FOODS

The distinction between liquids and solids was very clear in classical mechanics, and separate physical laws were promulgated to describe their behavior; solids were represented by Hooke's law and liquids by Newton's law. However, there exist a variety of products whose behavior under flow lies between these two extremes. Such is the case for a great quantity of foods. To optimize the use of these foods in the industry, rheological characterization is required.

In a general way, a primary distinction can be made between foods with Newtonian and non-Newtonian behavior, depending on whether their rheological behavior can or cannot be described by Newton's law of viscosity. Also, the behavior of some foods depends on the time that the shear stress acted on them. Those fluids whose behavior is a function only of shear stress are called time independent, and their viscosity, at a given temperature, depends on the rate of shear. Time-dependent fluids are those in which the viscosity depends not only on the velocity gradient but also on the time in which such gradients act. Finally, there are foods that exhibit behaviors of both viscous fluids and elastic solids, that is, viscoelastic fluids.

The classification of fluid foods can be performed using the viscometric functions defined in the previous section. Thus, for Newtonian fluids the viscosity function is constant, and its value is the Newtonian viscosity ($\eta(\dot{\gamma}) = \eta =$ constant). In non-Newtonian fluids, this function is not a constant and may be time independent or dependent, which allows distinguishing between the time-independent and time-dependent non-Newtonian fluids.

In non-Newtonian fluids, we cannot talk about viscosity, since the relationship between the applied shear stress and the rate of shear is not constant. The viscosity function is called apparent viscosity and is a function of the rate of shear:

$$\eta_a = \frac{\sigma_{12}}{\dot{\gamma}} = \eta(\dot{\gamma}) \neq \text{constant} \tag{13.15}$$

The first normal stress difference (N_1) can be applied in the study of the viscoelastic behavior of the fluids that are represented.

In this way, a classification of fluid foods can be made according to the following scheme:

1. Newtonian flow
2. Non-Newtonian flow
   a. Time-independent behavior
      i. Plastic fluids
      ii. Pseudoplastic fluids
      iii. Dilatant fluids
   b. Time-dependent behavior
      i. Thixotropic fluids
      ii. Antithixotropic or rheopectic fluids

13.6 NEWTONIAN FLOW

The viscous flow of a Newtonian fluid implies a nonrecoverable deformation. This behavior is illustrated in Figure 13.5, in which a fluid is contained between two parallel plates. The upper plate moves at a velocity v, in relation to the bottom plate. The velocity results from the application of a stress force F per unit area (it is assumed that the plates have an infinite extension or that the edge effects are negligible). The layers of fluid in contact with the plates are considered to move at the same velocity as the surface with which they are in contact, assuming that no sliding of the walls takes place. The fluid behaves as a series of parallel layers, or slabs, whose velocities are proportional to the distance from the lower plate. Thus, for a Newtonian fluid, the shear stress is directly proportional to the rate at which velocity changes over distance:

$$\sigma = \frac{F}{A} = \eta \frac{dv}{dy} \tag{13.8a}$$

where η is the viscosity coefficient, although generally it is simply called viscosity. It can be said that for such fluids the velocity gradient is equal to the rate of shear, resulting in the viscosity equation that is more frequently used as

$$\sigma = \eta \dot{\gamma} \tag{13.8b}$$

in which $\dot{\gamma}$ is the rate of shear.

Simple liquids, true solutions, solvents of low molecular weight, diluted macromolecular dispersions, polymer solutions that do not interact, and pastes with low solid content exhibit ideal Newtonian behavior.

These flow characteristics are exhibited by most beverages, including tea, coffee, beer, wines, and soft drinks. Sugar solutions are also included. Many researchers have studied the viscosity of sucrose solutions, since they are often used to calibrate viscometers (Muller, 1973) (see Table 13.1). The viscosity of solutions of sugar mixtures can be easily estimated, as it can be approximated by addition.

Milk, which is an aqueous emulsion of 0.0015–0.001 mm diameter fat globules of butter and contains around 87% water, 4% fat, 5% sugar (mainly lactose), and 3% protein (mainly casein), is a Newtonian liquid. Fernández-Martín (1972) pointed out that the viscosity of milk depends on temperature, concentration, and physical state of fat and protein, which in turn is affected by thermal and mechanical treatments. He also found that concentrated milks are non-Newtonian liquids and that concentrated milk presents a slight dependence on shear. Skim milk is less viscous than whole milk, since its viscosity increases with fat content. The viscosity also increases as nonfat solids increase, but neither relation is simple. The viscosity of milk decreases as temperature increases, as with most liquids.

TABLE 13.1
Viscosity Coefficients of Sucrose Solutions at 20°C

Sucrose%	g/100 g Water	Viscosity (mPa s)
20	25.0	2.0
25	33.2	2.5
30	42.9	3.2
35	53.8	4.4
40	66.7	6.2
45	81.8	9.5
50	100.0	15.5
55	122.2	28.3
60	150.0	58.9
65	185.7	148.2
70	233.3	485.0
75	300.0	2344.0

Source: Muller, H.G., *An Introduction to Food Rheology*, Crane, Russak & Company, Inc., New York, 1973.

Oils are normally Newtonian, but at very high values of shear rate, they show a different behavior (this behavior will be referred as pseudoplasticity). This may be due to the alignment of unit cells at high shear stresses, which can cause a decrease in internal friction (Muller, 1973). All oils have a high viscosity due to their molecular structure consisting of long chains. The greater the length of the carbon chain of fatty acids, the greater the viscosity. Polymerized oils have a greater viscosity than nonpolymerized oils. The viscosity of oils also increases with the saturation of the double bonds of carbon atoms.

Generally, it seems that greater molecular interaction results in a greater viscosity. The main constituent of castor oil is ricinoleic acid, which contains 18 carbon atoms with a hydroxyl group position 12. The hydroxyl groups form hydrogen bonds, and for this reason, the viscosity of castor oil is higher than the viscosity of other similar oils.

Some fruit juices also exhibit a Newtonian flow, such as apple juice with pectin and solids up to 30°Brix and must up to 50°Brix, and filtered orange juice of 10°Brix and 18°Brix (Saravacos, 1970). This behavior was identified in the 20°C–70°C temperature range. Juices of different clarified fruits from which pectinases were removed, such as apple, pear, and peach, presented Newtonian behavior as well (Rao et al., 1984; Ibarz et al., 1987, 1989, 1992a). A list of products that present this behavior can be found in a review paper by Barbosa-Cánovas et al. (1993).

Another important type of food that presents this behavior is syrups such as honey, cereal syrups, and mixtures of sucrose and molasses.

Pryce-Jones (1953) showed that with the exception of products based on heather (*Calluna vulgaris*) from England, "manuka" (*Leptospermum scoparium*) from New Zealand, and *Eucalyptus ficifolia* from South Africa, most syrups are Newtonian liquids (Rao, 1977).

Munro (1943) showed that the effect of temperature on the viscosity of honey presents three different states. The greatest decrease in viscosity occurs with cold honey heated to room temperature; later, heating reduces the velocity of viscosity decrease, and heating above 30°C has a low practical value.

13.7 NON-NEWTONIAN FLOW

13.7.1 Time-Independent Flow

Although most of the simple gases and liquids experimentally behave as Newtonian fluids in the laminar flow region, various systems, including emulsions, suspensions, solutions of long molecules, and fluids of macromolecular mass, approximate Newtonian behavior only at very low shear stresses and rates of shear. At higher shear levels, such systems may deviate from ideal Newtonian behavior in one or several ways, as illustrated in Figure 13.6. Here, shear stress is schematically shown as a function of the rate of shear. The straight line *a*, which passes through the origin and has a constant slope, represents an ideal Newtonian fluid. Curve *c*, which also passes through the origin, represents a pseudoplastic fluid or shear thinning fluid. For such materials, the viscosity coefficient is not constant, as in Equation 13.8, but a function of the rate of shear. The viscosity coefficient takes a value at each time, and this quantity is known as apparent viscosity.

In a similar way, curve *b* passes through the origin and tends to a straight line near the origin, but in contrast to pseudoplastic fluids, the curve is concave toward the axis of shear stress as the rate of shear increases. Such materials are called dilatant fluids, or shear thickening, and are generally limited to concentrated suspensions or aqueous pastes.

Another important phenomenon linked to flow is the existence of a threshold value. Certain materials can flow under enough shearing but do not flow if the shear stress is smaller than a particular value. This is called threshold value or yield stress (Van Wazer et al., 1963). This type of behavior is shown by different materials such as frozen tart and whipped egg yolk, among others. Once the yield stress is exceeded, the rate of shear is proportional to the shear stress, as in the case of Newtonian fluids. A fluid that exhibits this behavior is denominated as a plastic substance or

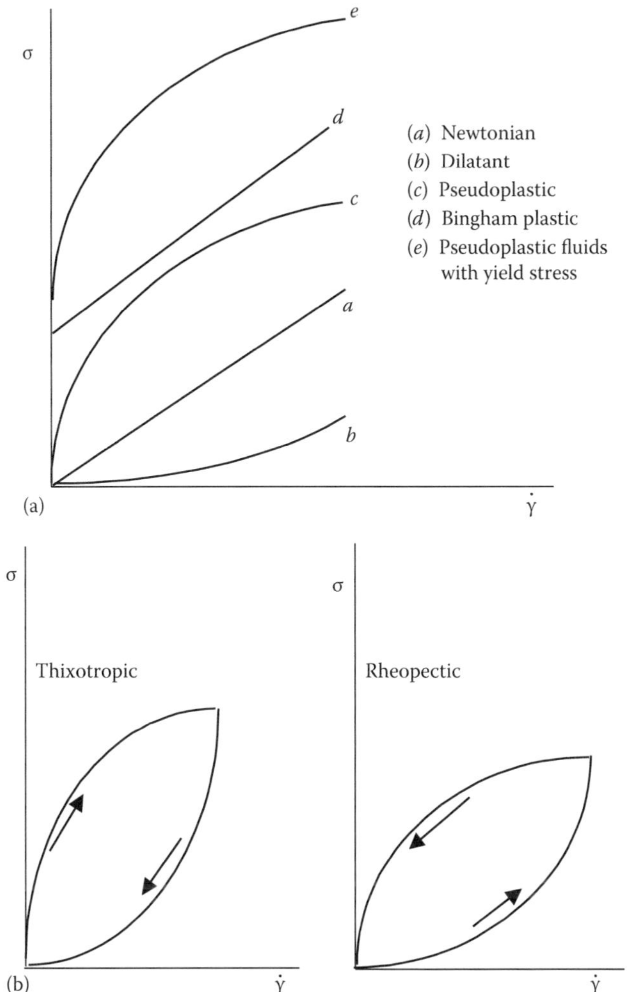

FIGURE 13.6 Rheograms of fluids: (a) time-independent Newtonian and non-Newtonian and (b) time-dependent non-Newtonian.

Bingham's body (see curve *d*, Figure 13.6). Materials that exhibit yield stress can also present non-linearity in the ratio obtained when dividing the shear stress by the shear rate.

The equation for this ratio is called apparent plastic viscosity, η_{pl}, and is given by

$$\eta_{pl} = \frac{\sigma - \sigma_0}{\dot{\gamma}} \tag{13.16}$$

where σ_0 is the yield stress. The fluids that exhibit nonlinear flow once the yield stress is exceeded are called pseudoplastic fluids with yield stress (curve *e*, Figure 13.6).

All of these types of flow behavior are referred to as "steady state," or independent time flow (such that equilibrium is reached in a relatively short time during the experiment). The Newtonian, pseudoplastic, and dilatant fluids do not have a yield stress; therefore, they can be called liquids. However, plastic substances exhibit the properties of liquids at shear stress values higher than yield stress, so they can be classified either as liquids or as solids.

To express the flow in a quantitative way and adjust the experimental data, in general, the more applied model is a power equation or Herschel–Bulkley model:

$$\sigma = k\dot{\gamma}^n + c \tag{13.17}$$

This equation includes shear stress (σ), yield stress (c), consistency index (k), shear rate ($\dot{\gamma}$), and index of flow behavior (n) and can represent Newtonian properties of Bingham's body, pseudoplastic or dilatant, depending on the value of the constants.

The dependence with respect to the shear rate points out the existence of a structure within the system. The pseudoplastic behavior indicates the continuous breakage or reorganization of structures, giving as a result an increase of resistance when applying a force (Rha, 1978). The pseudoplastic behavior is due to the presence of

1. Compounds with high molecular weight or long particles at low concentrations
2. High interaction among particles, causing their aggregation or association by secondary bonds
3. Large axial relation and asymmetry of the particles, requiring their orientation along the stream lines
4. Variation of shape and size of the particles, allowing them to pile up
5. Nonrigid or flexible particles that can suffer a change in their geometry or shape

The shear thickening behavior can be explained by the presence of tightly packed particles of different shapes and sizes, making the flow more difficult as pressure increases. With an increase of the rate of shear, the long and flexible particles can stretch, contributing to dilation (Rha, 1978).

A number of liquid foods exhibit a pseudoplastic behavior to flow. Scientists who have conducted investigations on this topic include Saravacos (1970), apple, grape, and commercial orange juice; Prentice (1968), cream; Cornford et al. (1969), melted samples of frozen whole eggs; and Tung et al. (1970), unmixed egg whites.

In general, fruit and vegetable purées are pseudoplastic fluids. The consistency of these products is an important quality parameter that is usually measured at only one point, using instruments such as the Adams and Bostwick consistometers, the Stormer viscometer, and the flow pipette (Rao, 1977). The consistency of tomato purées depends on the preparation method and the mechanical method to which samples have been subjected, as well as the variety and maturity of the tomatoes.

Harper and El Sahrigi (1965) studied the rheological behavior of tomato concentrates within the total solid range of 5.8%–30.0% in weight. The power model adequately describes the flow behavior of these concentrates.

Watson (1968) used the power model to characterize the rheological behavior of purées and concentrates of green and ripe apricots.

The rheological properties of other fruit purées have been determined, and in each case the power model has been used to describe the flow behavior. The products studied are apple sauce (Charm, 1960; Saravacos, 1968), banana purée (Charm, 1960), peach purée (Saravacos and Moyer, 1967; Saravacos, 1968), pear purée (Harper and Leberman, 1962; Saravacos, 1968), and sweet potato (Rao et al., 1975).

It has been found that ketchup and French mustard obey the power law of a yield stress (Higgs and Norrington, 1971). A few products exhibit a dilatant behavior such as that presented by the *E. ficifolia*, *Eucalyptus eugenioides*, *Eucalyptus corymbosa*, and *Opuntia engelmannii* syrups.

From the industrial viewpoint and also in many design equations for pumping installations, it is necessary to know the value of viscosity, which in the case of non-Newtonian fluids will be apparent viscosity. In most of non-Newtonian food fluids, apparent viscosity decreases with the increase of shear rate, and that is why some authors have developed mathematical expressions fitting the evolution of viscosity with shear rate. Some of these are Cross (1965), Powell-Eyring (Eyring, 1936; Ree et al., 1958; Christiansen et al., 1955), Carreau–Yasuda (Carreau, 1972; Yasuda, 1979), or Tscheuschner (1994) (Mezger, 2006) models. Nevertheless, these models are mathematically complex, with four or five parameters related in different ways. Falguera and Ibarz (2010) proposed an intuitive equation to describe apparent viscosity change with shear rate of non-Newtonian fluids, with only three parameters that can also be easily interpreted. At high shear rates, apparent viscosity reaches a constant value (η_∞), and if one considers that initial shear rate tends to zero, its value is η_0; the apparent viscosity for a certain shear rate can be expressed as

$$\eta_{app} = \eta_\infty + (\eta_0 - \eta_\infty)\lambda \tag{13.18}$$

where λ is a parameter that is a function of shear rate. It can be assumed that λ decreases with shear rate according to a power or exponential equation, obtaining the following equation:

$$\eta_{app} = \eta_\infty + (\eta_0 - \eta_\infty)(\dot{\gamma})^{-k} \tag{13.19a}$$

$$\eta_{app} = \eta_\infty + (\eta_0 - \eta_\infty)\exp(-k\dot{\gamma}) \tag{13.19b}$$

13.7.2 Time-Dependent Flow

Some materials exhibit time-dependent flow characteristics. Thus, as flow time under constant condition increases, such fluids may develop an increase or decrease in viscosity. An increase in viscosity is called rheopecticity, whereas a decrease in viscosity is known as thixotropy, and both are attributed to the continuous change of the structure of the material, which may be reversible or irreversible.

The factors that contribute to thixotropy also contribute to pseudoplasticity, and the factors that cause rheopecticity also cause shear thickening. Thixotropy is due to the dependence on time, similar to the dependence on shear, and it results from the structural reorganization in a decrease of the resistance to flow. Rheopectic behavior implies the formation or reorganization of the structure that brings with it an increase of the resistance to flow.

In other words, the phenomenological description of flow characteristics cannot be complete unless time is included, so, in the general case, an axis for time should be added to the flow curves.

Many researchers (Moore, 1959; Ree and Eyring, 1958; Peter, 1964; Cheng and Evans, 1965; Harris, 1967; Frederickson, 1970; Ritter and Govier, 1970; Joye and Poehlein, 1971; Lee and Brodkey, 1971; Mylins and Reher, 1972; Petrellis and Flumerfelt, 1973; Carleton et al., 1974; Lin, 1975; Zitny et al., 1978; Kemblowski and Petera, 1980; Barbosa-Cánovas and Peleg, 1983) have attempted the difficult task of formulating quantitative relationships between shear stress, rate of shear, and time.

Cheng and Evans (1965) and Petrellis and Flumerfelt (1973) modified the equation of Herschel–Bulkley with the objective of including a structural parameter that takes into account the effects of time dependence as

$$\sigma = \lambda[c + k\dot{\gamma}^n] \tag{13.20}$$

where λ is a structural parameter of time, whose value oscillates between one for time zero up to an equilibrium limit value, λ_e, that is smaller than one.

According to Petrellis and Flumerfelt (1973), the decrease in the value of the structural parameter λ with time is supposed to obey a kinetic equation of the second order:

$$\frac{d\lambda}{dt} = -k_1(\lambda - \lambda_e)^2 \quad \text{for } \lambda > \lambda e \tag{13.21}$$

in which the velocity constant, k_1, is a function of the rate of shear, which should be determined in an experimental way.

The determination of the kinetic constant k_1 of Equation 13.12 as a function of the rate of shear is difficult because the structural parameter λ cannot be obtained in an explicit form from experimental measurements. To overcome this difficulty, the instant and equilibrium structural parameters, λ and λ_e, are expressed in terms of the apparent viscosity:

$$\frac{d\eta}{dt} = -a_1(\eta - \eta_e)^2 \tag{13.22}$$

where

$$a_1\dot{\gamma} = \frac{k_1}{c} + k_1\dot{\gamma}^n \tag{13.23}$$

Integrating Equation 13.22 at a constant rate of shear, with limit conditions,

- For $t = 0 \quad \eta = \eta_0$
- For $t = t \quad \eta = \eta$

it is obtained that

$$\frac{1}{\eta - \eta_e} = \frac{1}{\eta_0 - \eta_e} + a_1 \cdot t \tag{13.24}$$

For a given rate of shear, when plotting $1/(\eta - \eta_e)$ against time, a straight line with slope a_1 is obtained. If the same procedure is repeated using different shear rates, a relation between a_1 and $\dot{\gamma}$ can be established, and k_1 is obtained from Equation 13.23. Tiu and Boger (1974) employed the kinetic–rheological model to characterize the thixotropic behavior of a mayonnaise sample.

Another kinetic model is the one given by Figoni and Shoemaker (1983), where it is assumed that the decrease in shear stress is an addition of kinetic functions of the first order:

$$\sigma - \sigma_e = \Sigma(\sigma_{0,i} - \sigma_{e,i})\exp(-k_i t) \tag{13.25}$$

where

σ_e is the equilibrium shear stress
σ_0 is the initial time shear stress
k_i is the kinetic constants of structural degradation

Tung et al. (1970) used the mathematical model of Weltmann (1943):

$$\sigma = A_1 - B_1 \log\ t \tag{13.26}$$

Another model is that of Hahn et al. (1959):

$$\log(\sigma - \sigma_e) = A_2 - B_2\ t \tag{13.27}$$

that describes the thixotropic behavior of egg white, fresh, old, and irradiated with gamma rays. In these equations, σ is the shear stress, σ_e is the equilibrium shear stress, and the coefficients B_1, B_2, A_1, and A_2 denote the initial shear stress. Longree et al. (1966) discussed the rheological properties of a set of milk cream systems based on milk–egg–starch mixtures. The products exhibited a thixotropic behavior and were highly non-Newtonian. Neither the power equation nor that of Casson (1959) yielded adequate results, so other equations were tested. It was found that some data fitted the equation:

$$\log(\eta - \eta_e) = -ct \tag{13.28}$$

where

η is the viscosity for a time t
η_e is the viscosity at equilibrium
c is an empirical constant

Higgs and Norrington (1971) studied the thixotropic behavior of sugared condensed milk. This product behaves as a Newtonian fluid at temperatures between 40°C and 55°C, and slightly non-Newtonian behavior is observed at lower temperatures; however, it presented a thixotropic behavior at all temperatures studied. The time coefficient for the thixotropic break B is given by

$$B = \frac{m_1 - m_2}{\ln(t_2/t_1)} \tag{13.29}$$

where m_1 and m_2 are the slopes of the return curves measured at the end of times t_1 and t_2, respectively, and the thixotropic break coefficient due to the increase of the rate of shear M is given by

$$M = \frac{m_1 - m_2}{\ln(N_2/N_1)} \tag{13.30}$$

where N_1 and N_2 are the angular velocities. The parameters of the power law and the thixotropic coefficients for sugared condensed milk are presented in Table 13.2.

In relaxation studies at a fixed shear rate, many foods initially show a growing shear stress. This phenomenon is known as shear stress overshoot.

Dickie and Kokini (1981) studied the shear stress overshoot in typical food products using the model of Bird–Leider:

$$\sigma_{rz} = m(\dot{\gamma})^n \left[1 + (b\dot{\gamma}t - 1)\exp\left(\frac{-t}{an\lambda}\right)\right] \tag{13.31}$$

where

m and n are the parameters of the power law
$\dot{\gamma}$ is the sudden imposed shear rate
t is the time
a and b are the fitting parameters
λ is a time constant

The constant of time in this equation is calculated in function of the viscosity and the primary coefficient of normal stresses (ψ_1') defined according to

$$\eta = m(\dot{\gamma})^{n-1} \tag{13.32}$$

$$\psi_1' = \tau_{11} - \tau_{22} = m'(\dot{\gamma})^{n'-2} \tag{13.33}$$

$$\lambda = \left(\frac{m'}{2m}\right)^{1/(m'-n)} \tag{13.34}$$

TABLE 13.2
Parameters of the Power Law and Thixotropic Coefficients for Sugared Condensed Milk

Temperature (°C)	k (Pa s^n)	n	B (Pa s)	M (Pa s)
25	3.6	0.834	4.08	18.2
40	0.818	1.0	1.20	8.65
55	0.479	1.0	0.529	1.88

TABLE 13.3
Rheological Parameters of Foods

Food	m (Pa s^n)	n	m' (Pa s$^{n'}$)	n'	λ (s)
Apple marmalade	222.90	0.145	156.03	0.566	8.21×10^{-2}
Canned candied products	355.84	0.117	816.11	0.244	2.90×10^{0}
Honey	15.39	0.989			
Ketchup	29.10	0.136	39.47	0.258	4.70×10^{-2}
"Acalia" cream	563.10	0.379	185.45	0.127	1.27×10^{3}
Mayonnaise	100.13	0.131	256.40	0.048	2.51×10^{-1}
Mustard	35.05	0.196	65.69	0.136	2.90×10^{0}
Peanut butter	501.13	0.065	3785.00	0.175	1.86×10^{5}
Butter bar	199.28	0.085	3403.00	0.398	1.06×10^{3}
Margarine bar	297.58	0.074	3010.13	0.299	1.34×10^{3}
Pressed margarine	8.68	0.124	15.70	0.168	9.93×10^{-2}
Margarine in tube	106.68	0.077	177.20	0.353	5.16×10^{-1}
Whipped butter	312.30	0.057	110.76	0.476	1.61×10^{-2}
Whipped cheese cream	422.30	0.058	363.70	0.418	8.60×10^{-2}
Whipped dessert garnish	35.98	0.120	138.00	0.309	3.09×10^{1}

Source: Kokini, J.L., Rheological properties of foods, in *Handbook of Food Engineering*, D.R. Heldman and D.B. Lund, eds., Marcel Dekker, Inc., New York, 1992, p. 1.

The foods used to study this model were ketchup, mustard, mayonnaise, apple marmalade, unsalted butter, margarine, and ice cream, among others. Kokini (1992) gave the parameters m, n, m', and n' for 15 typical food products (Table 13.3).

A great advantage of this equation is that for long times it converges to the power law. In addition, it perfectly describes the shear stress overshoot peaks. However, it does not adequately describe the decreasing part of the relaxation curve. For this reason, Mason et al. (1982) modified the model of Leider and Bird by adding several relaxation terms, obtaining the equation:

$$\sigma_{yx} = m(\dot{\gamma})^n \left[1 + (b_0\dot{\gamma}t - 1)\frac{\sum b_i \exp(-t/\lambda_i)}{\sum b_i} \right] \tag{13.35}$$

where

m and n are the parameters of the power law
$\dot{\gamma}$ is the shear rate
t is the time
λ_i is the constants of time
b_0 and b_i are constants

13.8 VISCOELASTICITY

Some semiliquid products present, in a joint form, properties of viscous fluid and elastic solid and are called viscoelastic. Knowing the viscoelastic properties is very useful in the design and prediction of the stability of stored samples. The typical example to describe a fluid with viscoelastic properties is the mixture formed by water and flour. If a strip is separated from the mass and pulled by the ends, it will be extended and seem to flow like a viscous liquid. If the ends are left free, the strip will shrink like a piece of soft rubber, although the recovery will be only partial and will not acquire its initial length, so that it is not perfectly elastic. This experiment shows that the mass

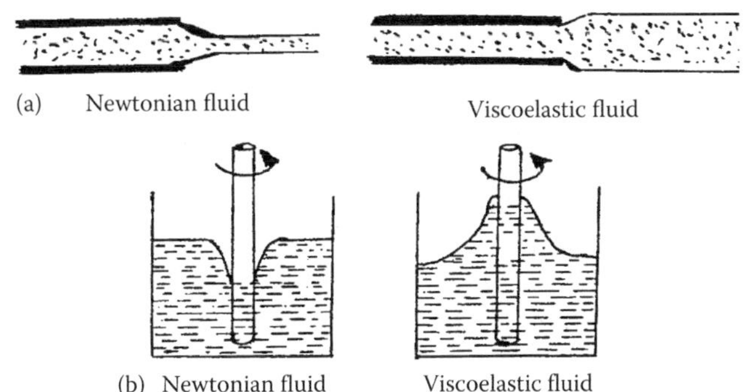

FIGURE 13.7 Nonlinear viscoelasticity: (a) Barus or Merrington effect and (b) Weissenberg effect.

simultaneously displays the properties associated with a viscous liquid and the elastic characteristics of a solid, that is, the dough is viscoelastic.

Viscoelastic materials present a characteristic behavior that makes them sharply different from other fluids. Thus, if a Newtonian fluid is allowed to spurt through a pipe, the diameter of the spurt contracts as it exits the pipe, while if it is a viscoelastic fluid, a marked widening of the spurt is produced (Figure 13.7a). This phenomenon is known as the Barus effect or jet swell. In addition, these fluids are able to form a tubeless siphon when they are removed from a tank by means of an upward pipe.

Another characteristic of these fluids is the Weissenberg effect. This effect is observed when a fluid in a container is agitated with a small rod. If the fluid is Newtonian or pseudoplastic, a free vortex with a surface profile, as shown in Figure 13.7b, is formed. In the case of viscoelastic fluids, the fluid tends to climb up the small rod.

To describe the viscoelastic behavior, some equations are given, in which the shear stress is a function of the viscous flow and of the elastic deformation. Among these equations, those of Voigt and Maxwell are the most used, depending on whether the behavior is more similar to an elastic solid or a viscous fluid.

The mathematical expression of Voigt's model is

$$\sigma = G\gamma + \eta\dot{\gamma} \tag{13.36}$$

When the velocity gradient tends to zero, the body behaves as an elastic solid. The quotient between the viscosity and the module G is called retard time of the solid:

$$t_E = \frac{\eta}{G} \tag{13.37}$$

and represents the time needed by the shear rate drop to half of its initial value, when performing at constant stress.

Maxwell's equation is given by

$$\sigma = \frac{\eta}{G}\frac{d\sigma}{dt} + \eta\dot{\gamma} \tag{13.38}$$

The relation between the viscosity and the module G is called relaxation time of the fluid:

$$t_R = \frac{\eta}{G} \tag{13.39}$$

and it represents the time needed by the stress that results from constant deformation to drop to half of its value.

Generally, the viscoelasticity presented by foods is nonlinear and is difficult to characterize; therefore, experimental conditions under which the relationship among the variables of deformation, tension, and time can be obtained should be defined.

The viscoelastic behavior of a sample can be studied and characterized in different ways. One way is to study the evolution of the shear stress with time, at a fixed rate of shear; it is possible to perform a comparative analysis of different samples from the obtained curves (Elliot and Green, 1972; Elliot and Ganz, 1977; Fiszman et al., 1984).

Viscoelastic materials also present normal stresses (the difference being with the former normal stresses that have been used) in the characterization of the viscoelasticity. Opposite to what occurred with shear stresses, the primary normal stresses do not present overshoot in most semisolid foods.

As indicated by Kokini and Plutchok (1987), the coefficients of the primary normal stresses can be defined under steady state according to the expression:

$$\Psi_1 = \frac{\sigma_{11} - \sigma_{22}}{\dot{\gamma}^2} \tag{13.40}$$

When plotting the coefficient of primary normal stresses under steady state against the shear rate for different semisolid foods using a logarithmic double scale, straight lines are obtained, indicating a behavior in accordance with the power law.

Oscillating tests are also used, in which a shear rate is applied to the sample in an oscillating continuous form, which causes a sinusoidal wave of stresses to appear. For elastic solids, this wave is in phase with the wave of the applied deformation, while for a perfect viscous fluid, there is a 90° phase out. For viscoelastic materials, the value of the phase out angle is between 0° and 90°.

Two rheological properties have been defined, the rigidity or storage modulus (G'), which represents the elastic part of the material, and the loss modulus (G''), which represents the viscous character. If $\gamma°$ and $\tau°$ are, respectively, the amplitudes of the deformation and stress waves and ε is the phase out angle, the G' and G'' moduli are defined by

$$G' = \left(\frac{\tau°}{\gamma°}\right)\cos\varepsilon \tag{13.41}$$

$$G'' = \left(\frac{\tau°}{\gamma°}\right)\sin\varepsilon \tag{13.42}$$

The complex viscosity for fluid systems is defined as

$$\eta^* = \left[(\eta')^2 + (\eta'')^2\right]^{1/2} \tag{13.43}$$

in which η' is the viscous component in phase between the stress and the rate of shear, while η'' is the elastic or phase out component. These viscosity functions are defined by

$$\eta' = \frac{G''}{\omega} \tag{13.44}$$

$$\eta'' = \frac{G'}{\omega} \tag{13.45}$$

This type of correlation has been used to characterize hydrocolloid solutions (Morris and Ross-Murphy, 1981). Also, in some cases, there are expressions that correlate the different shear properties and viscous functions under steady state with those of the oscillating states (Kokini and Plutchok, 1987).

It is interesting to have a constitutive equation that can adequately describe the real behavior of samples. One such model is Maxwell's, which combines the viscosity equation of Newton and the elasticity equation of Hooke:

$$\sigma + \lambda\dot{\sigma} = \eta\gamma \tag{13.46}$$

where λ is the relaxation time defined as the quotient between the Newtonian viscosity (η) and the elasticity modulus (G).

In stress relaxation experiments, the sample is subjected to a fix deformation for a period of time and the variation of stress with time is measured. The relaxation measurement is a simple method to obtain a qualitative description of the elastic behavior of a sample. The relaxation time is calculated from the relaxation experienced by the sample after a shear stress has been applied. In some cases, one relaxation time is not enough to describe the rheological behavior of the sample, and it is necessary to use the generalized Maxwell's model, which includes several elements of such a model.

In general, the viscoelasticity presented by foods is nonlinear, so Maxwell's model cannot be applied. For this reason, other constitutive models have been sought, such as the model of Bird–Carreau. This is an integral constitutive model based on Carreau's constitutive theory of molecular networks, in which the whole deformation history of the material is incorporated. In order to apply Bird–Carreau's model, it is necessary to know the values of the limit viscosity η_0 at rate of shear zero, as well as the time constants γ_1 and γ_2 and the constants α_1 and α_2. It is necessary to perform shear tests under steady state and oscillating tests to calculate these elements. Figure 13.8 shows the graphs obtained in this type of experiment that allow the determination of the Bird–Carreau constants (Bird et al., 1977).

From the shear experiment under steady state (Figure 13.8a), η_0 is obtained by extrapolation to $\gamma = 0$; the time constant λ_1 is the value of the inverse of the rate of shear at the intersection point of the straight line with value η_0 and the tangent to the curve in the non-Newtonian zone, while α_1 is obtained from the slope of the curve $\log \eta - \log \gamma$ in the non-Newtonian zone (Kokini, 1992).

From the oscillating experiments, it is possible to obtain graphs of η' and η''/ω against the angular velocity ω in double logarithmic coordinates (Figure 13.8b and c). The time constant λ_2 is obtained as the inverse of the value of the angular velocity ω at the intersection point of the straight line η_0 and the tangent to the curve of the non-Newtonian zone, while α_2 is obtained from the slope of this curve.

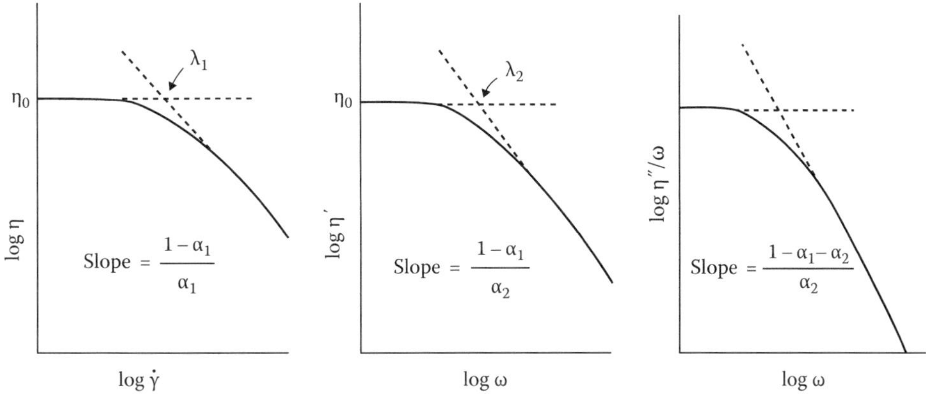

FIGURE 13.8 Determination of the Bird–Carreau constants λ_1, λ_2, x_1, and x_2.

Once these constants are known, it is possible to predict the values of η, η′, and η″ (Bird et al., 1977). Thus, the Bird–Carreau model that allows obtaining of the viscosity value η is

$$\eta = \sum_{p=1}^{\infty} \frac{\eta_P}{1+(\lambda_{1P}\dot{\gamma})^2} \tag{13.47}$$

where

$$\lambda_{1P} = \lambda_1 \left(\frac{2}{p+1}\right)^{\alpha_1} \tag{13.48}$$

$$\eta_P = \eta_0 \frac{\lambda_{1P}}{\sum \lambda_{1P}} \tag{13.49}$$

For high rates of shear, Equation 13.47 can approximate to the expression:

$$\eta = \frac{\pi\eta_0}{Z(\alpha_1)-1}\left[\frac{(2\alpha_1\lambda_1\dot{\gamma})^{(1-\alpha_1)/\alpha_1}}{2\alpha_1 sen(((1+\alpha_1)/2\alpha_1)\pi)}\right] \tag{13.50}$$

in which $Z(\alpha_1)$ is the Riemann zeta function

$$Z(\alpha_1) = \sum_{k=1}^{\infty} k^{-\alpha_1} \tag{13.51}$$

Expressions that allow calculation of η′ and η″, according to the constitutive model of Bird–Carreau, can be found in the literature (Bird et al., 1977; Kokini, 1992).

Creep and recovery tests have been applied to cream and ice cream (Sherman, 1966). In this case, a constant shear stress is applied during a determined time and the variation of the deformation with time is studied (Figure 13.9). The results are expressed in terms of $J(t)$ against time. The function $J(t)$ is the compliance, which is a relation between the deformation produced and the shear stress applied:

$$J(t) = \frac{\gamma}{\sigma} \tag{13.52}$$

For ice cream and frozen products, the creep compliance against time can be described according to the equation:

$$J(t) = J_0 + J_1\left[1-\exp\left(-\frac{t}{\theta_1}\right)\right] + J_2\left[1-\exp\left(-\frac{t}{\theta_2}\right)\right] + \frac{1}{\eta_N} \tag{13.53}$$

where

$J_0 = 1/E_0$ is the instant elastic compliance
E_0 is the instant elastic module
$J_1 = 1/E_1$ and $J_2 = 1/E_2$ are the compliance associated to the retarded elastic behavior
$\theta_1 = \eta_1/E_1$ and $\theta_2 = \eta_2/E_2$ are the retarded times associated to the retarded elasticity
E_1 and E_2 are the retarded elastic modules associated to the retarded elasticity
η_1 and η_2 are the viscosity components associated to retarded elasticity
η_N is thc viscosity associated to the Newtonian flow

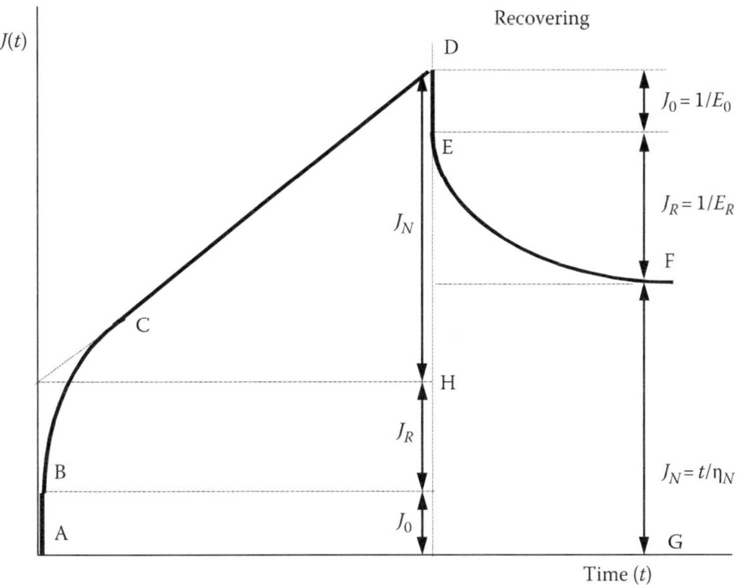

FIGURE 13.9 Curve of a "creep" model. Creep zone A–B: instantaneous elastic creep. B–C: retarded elastic creep. C–D: Newtonian fluid. Recovering zone: D–E: instantaneous elastic recovering. E–F: retarded elastic recovering. F–G: nonrecoverable deformation.

Likewise, it is possible to characterize the viscoelasticity of the samples by studying the development of the shear stress under unsteady state (shear stress overshoot), applying the model of Bird–Leider or the modified model, as stated in the last section.

Another method widely used in the viscoelastic characterization of foods is the use of mechanical models. Thus, for example, mixed and frozen products behave as two bodies of Kelvin–Voigt connected in series, while the behavior of melted ice cream can be represented as a unique body of Kelvin–Voigt.

13.9 EFFECT OF TEMPERATURE

During manufacture, storage, transport, sale, and consumption, fluid foods are subjected to continuous variations of temperature. For this reason, it is important to know the rheological properties of the products as a function of temperature.

In the case of Newtonian fluids, the expression that correlates viscosity and temperature is an Arrhenius-type equation. However, for non-Newtonian fluids, the apparent viscosity is usually related to a fixed rate of shear, rather than viscosity (Vitali et al., 1974; Moresi and Spinosi, 1984; Rao et al., 1984):

$$\eta_A = \eta_\infty \exp\left(\frac{E_a}{RT}\right) \quad (13.54)$$

where

E_a is the flow activation energy
η_∞ is a constant called viscosity of infinite deformation
R is the gas constant
T is temperature in degrees Kelvin

For non-Newtonian fluids, consistency index is also usually used, instead of the apparent viscosity (Harper and El Sahrigi, 1965; Vitali and Rao, 1984a).

Equation 13.54 is the most used for all types of food fluids, although there are others such as the one used by Sáenz and Costell (1986):

$$Y = Y_0 \exp(-BT) \tag{13.55}$$

where

Y is the viscosity or the yield stress
T is the temperature expressed in °C
B is a constant

In the case of kiwi juice (Ibarz et al., 1991), a linear equation has been applied to describe the variation of the yield stress with temperature.

Temperature can affect the different rheological parameters such as viscosity, consistency index, flow behavior index, and yield stress. Generally, the observed effect is as follows:

- Viscosity and consistency indexes decrease as temperature increases.
- Flow behavior index is not usually affected by the temperature variation (Mizrahi and Berk, 1972; Crandall et al., 1982; Sáenz and Costell, 1986). However, in some cases, it has been observed that a temperature increase could produce an increase in the flow behavior index (Ibarz and Pagán, 1987), changing from pseudoplastic to Newtonian behavior.
- The yield stress can also vary with temperature in such a way that when temperature increases, yield stress value decreases. Thus, for lemon juices (Sáenz and Costell, 1986), this variation in the yield stress can cause a change in the behavior of juices, from pseudoplastic to Newtonian.

13.10 EFFECT OF CONCENTRATION ON VISCOSITY

13.10.1 Structural Theories of Viscosity

Bondi (1956) revised the theories of the viscosity of liquids. According to the kinetic theory, the viscosity of Newtonian liquids can be obtained by the following equation (after numerous approximations):

$$\eta = \left[0.48\left(\frac{r_1}{v}\right) m\Phi_a(r_1)\right]^{1/2} \exp\left[\frac{\Phi_a(r_1)}{kT}\right] \tag{13.56}$$

where

r_1 is the interatomic distance characteristic of the liquid density
Φ_a is the attraction component of the interaction energy between a pair of isolated molecules
m is the molecular mass
v is the molar volume
k is Boltzmann's constant
T is the temperature

The advantage of Equation 13.56, and similar expressions, is that it relates the viscosity with molecular parameters, especially with the interaction energy between two isolated molecules. However, the mathematical problems when applying the basic molecular theory to the liquid viscosity are so difficult that the final equations derived at the present time do not have a practical value for rheologists (Van Wazer et al., 1963).

Another approach to the theoretical comprehension of the viscosity of liquids has been developed by Eyring (1936) based on the "absolute" theory of velocity processes. This theory describes the elementary processes that govern the kinetics of chemical reactions. This approach supposes that the liquid is an imperfect molecular reticule, with a number of vacant spaces of the reticule called "holes." According to Eyring, the equation for viscosity can be expressed as

$$\eta = \frac{hN}{v}\exp\left(\frac{\Delta F}{RT}\right) \tag{13.57}$$

where
ΔF is the free activation energy
h is Planck's constant
N is Avogadro's number
v is the molar volume

Brunner (1949) modified this equation to obtain

$$\eta = \left(\frac{\delta}{v_L}\right)(2\pi mkT)^{1/2}\exp\left(\frac{\Delta F}{RT}\right) \tag{13.58}$$

where
δ is the width of the potential barrier
m is the reduced mass of the activated complex
v_L is the molecular volume of the liquid

13.10.2 Viscosity of Solutions

The viscosity of diluted solutions has received considerable theoretical treatment. When talking about viscosity data at high dilution, different functions have been used. The viscosity rate (also known as relative viscosity)

$$\eta_r = \frac{\eta}{\eta_0} \tag{13.59}$$

where
η is the viscosity of the solution of concentration c
η_0 is the viscosity of the pure solvent

Specific viscosity

$$\eta_{sp} = \eta_r - 1 = \frac{\eta - \eta_0}{\eta_0} \tag{13.60}$$

Viscosity number (reduced viscosity)

$$\eta_{red} = \frac{\eta_{sp}}{C} \tag{13.61}$$

Inherent viscosity (logarithm viscosity number)

$$\eta_{inh} = \ln(\eta_r)/C \tag{13.62}$$

Intrinsic viscosity (limit viscosity number)

$$[\eta] = \left[\frac{\eta_{SP}}{C}\right]_{C\to 0} = \left[\left(\frac{\ln \eta_r}{C}\right)\right]_{C\to 0} \tag{13.63}$$

According to Einstein (1911), the limit viscosity number, in the case of suspension of rigid spheres, is 2.5 per volumetric concentration (C_v) measured in cm^3 of spheres per cm^3 of total volume.

For low concentrations, the viscosity number is given by the expression:

$$\eta_{red} = 2.5 + 14.1C_v \tag{13.64}$$

Einstein's viscosity relationship has been applied to different solutions and suspensions. For sugar solutions, the first for which this relationship was tested, the viscosity number experimentally determined was from one to one and a half times greater than the value of 2.5 obtained when assuming that the dissolved sugar was in the form of unsolvated spherical molecules (Van Wazer et al., 1963).

The viscosity data as a function of concentration are extrapolated up to infinite dilution by means of Huggins' equation (1942):

$$\frac{\eta_{sp}}{C} = [\eta] + k'[\eta]^2 C \tag{13.65}$$

where k' is a constant for a series of polymers of different molecular weight in a given solvent. An alternative definition of the intrinsic viscosity leads to the following equation (Kraemer, 1938):

$$\frac{\ln \eta_r}{C} = [\eta] + k''[\eta]^2 C \tag{13.66}$$

where $k' - k'' = 1/2$.

At intrinsic viscosity higher than approximately 2, and even lower in some cases, there can be a noticeable dependence of viscosity on the shear velocity in the viscometer. This dependence is not eliminated by extrapolation to infinite dilution, since the measure as a function of the shear velocity and extrapolation at zero rate of shear are also necessary (Zimm and Grothers, 1962).

Theories about the friction properties of polymer molecules in solution show that the intrinsic viscosity is proportional to the effective hydrodynamic volume of the molecule in solution divided by its molecular weight (M). The effective volume is proportional to the cube of a linear dimension of the random coiling of the chain. If $(r)^{1/2}$ is the chosen dimension,

$$[\eta] = \frac{\delta (r^2)^{3/2}}{M} \tag{13.67}$$

where δ is a universal constant (Billmeyer, 1971).

The effect exerted by concentration on a homogeneous system is to increase the viscosity or consistency index. Two types of correlation can be found in the literature (Harper and

El Sahrigi, 1965; Rao et al., 1984), according to a power model and an exponential model, in agreement with the following equations:

$$Y = K_1(C)^{A_1} \tag{13.68}$$

$$Y = K_2 \exp(A_2 C) \tag{13.69}$$

where
- Y can be the viscosity or the consistency index
- C is the concentration of some component of the sample

Most of the existing information refers to fruit-derived products, in which the effect of the soluble solids, pectin content, total solids, etc., was studied. From the two last equations, the former usually yields good results in purée-type foods, while the latter is successfully applied to concentrated fruit juices and pastes (Rao and Rizvi, 1986).

Besides viscosity and consistency index, concentration can also affect other parameters. Thus, the yield stress increases as concentration increases. For kiwi juices (Ibarz et al., 1991), as concentration decreases, the yield stress disappears, changing from a plastic to a pseudoplastic behavior.

In the case of flow behavior index, there are studies showing that concentration does not affect it (Sáenz and Costell, 1986), while in other studies, an increase in concentration decreased the value of the flow behavior index (Mizrahi and Berk, 1972).

13.10.3 Combined Effect Temperature–Concentration

From an engineering point of view, it is interesting to find a unique expression that correlates the effect of temperature and concentration on viscosity.

The equations generally used are

$$\eta_a = \alpha_1 (C)^{\beta_1} \exp\left(\frac{E_a}{RT}\right) \tag{13.70}$$

$$\eta_a = \alpha_2 \exp\left(\beta_2 C + \frac{E_a}{RT}\right) \tag{13.71}$$

where
- η_a is the viscosity for Newtonian fluids or either the apparent viscosity or consistency index for non-Newtonian fluids
- The parameters α_i and β_i are constants
- C is the concentration
- T is the absolute temperature

Generally, this type of equation is valid in the variable range for which they have been determined (Vitali and Rao, 1984; Rao and Rizvi, 1986). These equations have been applied to describe the combined effect of concentration and temperature in different food products (Rao et al., 1984a; Vitali and Rao, 1984a; Ibarz and Sintes, 1989; Ibarz et al., 1989, 1992a,b; Castaldo et al., 1990).

13.11 MECHANICAL MODELS

The rheological behavior of different substances can be described by means of mechanical models. The results obtained with mechanical models can be expressed by means of stress–strain and time diagrams that allow obtaining of equations of rheological application. The models used are those of spring and dashpot, as well as combinations of both.

13.11.1 Hooke's Model

Hooke's mechanical model (spring) serves to describe the behavior of ideal elastic solids. This type of model is represented in Figure 13.10, as well as the stress versus strain curve occurring in the spring.

13.11.2 Newton's Model

The mechanical model of Newtonian fluids is represented by a dashpot with fluid, as shown in Figure 13.11. A dashpot is a container filled with fluid in which a piston can freely move up and down. In this model the only resistance that impedes the movement of the piston is the viscosity of the liquid.

13.11.3 Kelvin's Model

Kelvin's mechanical model describes the rheological behavior of ideal viscoelastic materials and corresponds to the Voigt equation (Equation 13.36). This model considers a spring and a dashpot in parallel, that is, their movement occurs at the same time and at the same velocity. The two elements, spring and dashpot, undergo the same deformation, but not the same stress. This model is represented in Figure 13.12.

In the substances or Kelvin's bodies, when the stress stops acting, the strain is cancelled but in a retarded form, according to the following expression:

$$\varepsilon = \varepsilon_{max} \exp\left(\frac{-t}{t_E}\right) \tag{13.72}$$

where

ε is the relative strain
t is the time
t_E is the retardation time of the solid

This indicates that there exists elasticity but not flow. The delay to return to the initial state is due to the viscosity of the dashpot.

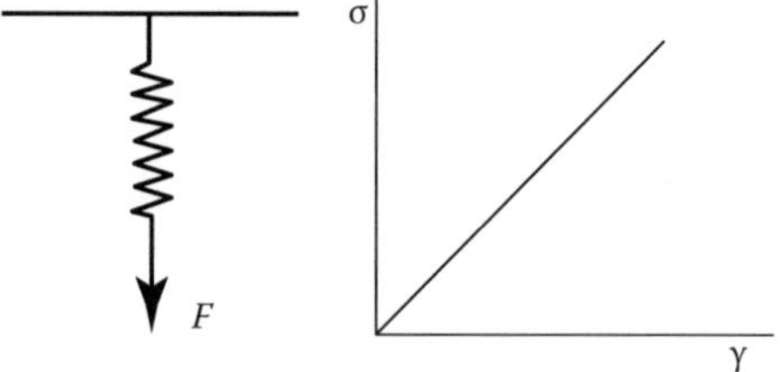

FIGURE 13.10 Hooke's mechanical model.

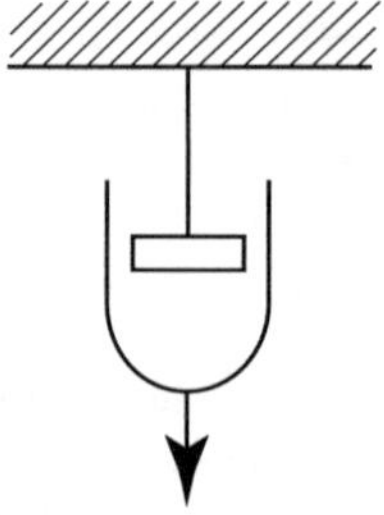

FIGURE 13.11 Mechanical model of a Newtonian fluid.

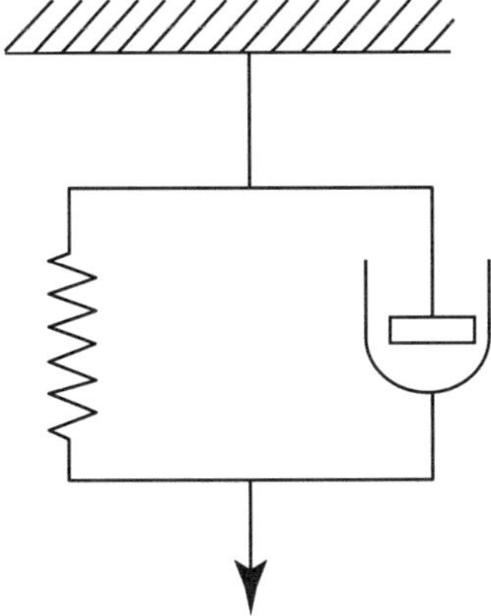

FIGURE 13.12 Mechanical model of Kelvin's body.

13.11.4 Maxwell's Model

This model considers a spring and a dashpot in series, corresponding to the behavior of ideal visco-elastic fluids. Both elements undergo the same stress, but not the same deformation. This model is shown in Figure 13.13. The change of deformation with time is linear at first because initially only the spring is deformed. Then the deformation of the dashpot begins, and once the spring is no longer able to deform, only the dashpot controls the deformation.

If in a Maxwell's body on which a shear stress σ_M has been applied, at a given instant, t_i, the achieved deformation is kept constant, the shear stress slowly ceases because of the viscosity of the dashpot, and it can be expressed according to the equation:

$$\sigma = \sigma_M \exp\left(\frac{-t}{t_R}\right) \tag{13.73}$$

where

σ_M is the shear stress applied
t_R is the relaxation time of the fluid

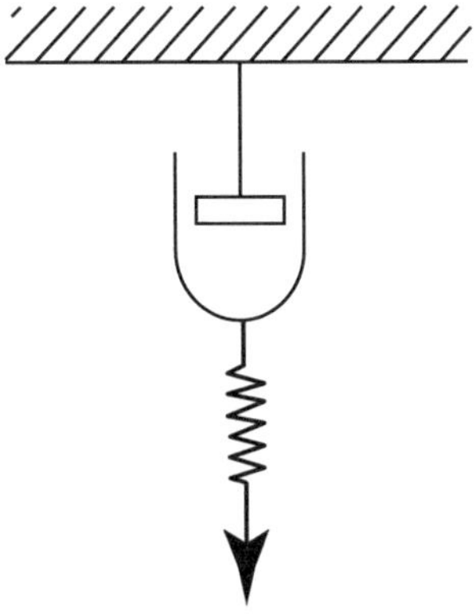

FIGURE 13.13 Maxwell's mechanical model.

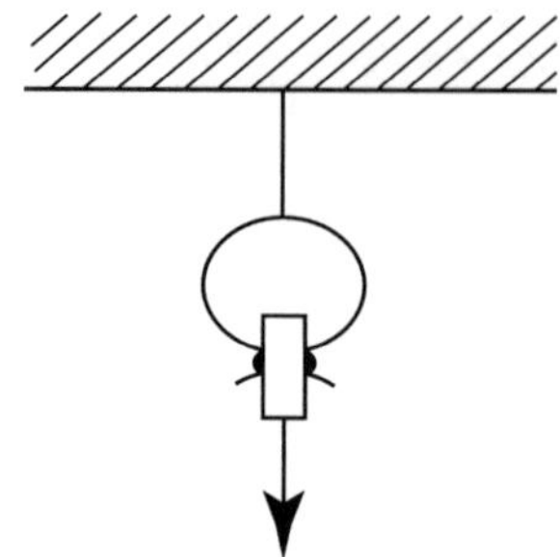

FIGURE 13.14 Saint-Venant's mechanical model.

This relaxation phenomenon occurs whenever a substance has plastic properties. An ideal solid body does not present relaxation, whereas a perfect viscous fluid presents an instantaneous relaxation. Every solid body that presents this relaxation phenomenon proves that it is also viscous.

13.11.5 Saint-Venant's Model

This model is called a glide, which is a type of tweezer holding a piece of material (Figure 13.14). If the applied shear stress is greater than the friction between the glide and the material, it will slide; otherwise, it will stay fixed.

13.11.6 Mechanical Model of Bingham's Body

The mechanical model that describes the rheological behavior of Bingham's plastics is a combination of a spring and a glide in series and a dashpot in parallel (Figure 13.15).

In addition to these mechanical models, there are others that are more complicated but that are a combination of the three basic models, that is, spring, dashpot, and glide. Thus, Figure 13.16 presents the mechanical model of Burgers, which describes the viscoelastic behavior of certain products.

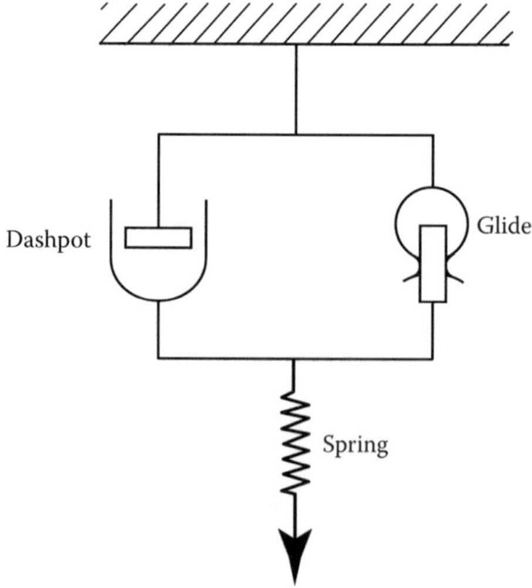

FIGURE 13.15 Bingham's mechanical model.

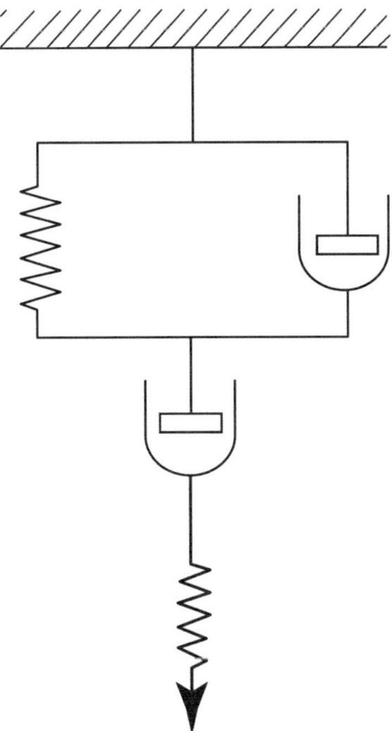

FIGURE 13.16 Burgers' mechanical model.

13.12 RHEOLOGICAL MEASURES IN SEMILIQUID FOODS

Scott-Blair (1958) classified the instruments for studying texture, while Rao (1980) classified the instruments for measuring the flow properties of fluid foods into three categories: (1) fundamental, (2) empirical, and (3) imitative. Fundamental experiments define physical properties and are independent of instruments. Empirical experiments measure parameters that are not clearly defined, but the parameters have been proved to be useful from past experience. Imitative experiments measure properties under similar conditions to those found in practice (White, 1970).

13.12.1 Fundamental Methods

Different instruments have been used to measure flow properties using fundamental methods. Oka (1960), Van Wazer et al. (1963), Sherman (1970), Rao (1977), and Shoemaker et al. (1987) have described several of the commercial instruments and their fundamental equations.

These fundamental methods can be classified according to the geometry used: capillary, Couette (concentric cylinder), plate and cone, parallel plates, back extrusion, or flow by compression.

Three requirements are common to the cited geometries. These are (1) laminar flow of the liquid, (2) isothermal operation, and (3) no sliding in the solid–fluid interface (Van Wazer et al., 1963).

13.12.1.1 Rotational Viscometers

A rotating body immersed in a liquid experiences a viscous drag or retarded force. The quantity of viscous drag is a function of the velocity of the rotation of the body. Using velocity equations, no difference is obtained whether it is the body or the container that rotates.

Rotational viscometers allow for continuous measurements at a given rate of shear or shear stress for long periods of time, permitting the determination of whether there is time dependence. These attributes are not typical of most capillary viscometers or of other types of viscometers. For these reasons, rotational viscometers are the most used type of instrument for rheological determinations.

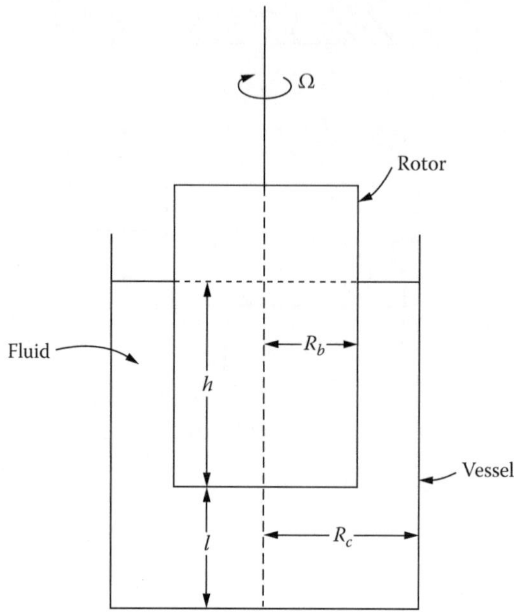

FIGURE 13.17 Viscometer of concentric cylinders.

13.12.1.1.1 Viscometers of Concentric Cylinders

Although rotational viscometers use a vessel and rotors in the form of spheres, disks, cones, and other particular forms, the most common type is that of concentric coaxial cylinders, schematically shown in Figure 13.17. One cylinder of radius R_b is suspended in a fluid sample in a container of radius R_c. The liquid covers the internal cylinder up to height h. The bottom part of the internal cylinder, or rotor, is separated from the bottom of the vessel by a distance l.

Complete descriptions of this type of flow have been given by Oka (1960), Van Wazer (1963), Reiner (1971), and Walters (1975), among others.

This is a rheological flow that has been used to characterize the shear behavior of non-Newtonian fluids. This configuration has also been used to characterize differences of normal stresses (Schowalter, 1978). According to Van Wazer et al. (1963), to obtain the fundamental equations, the following assumptions should be made:

1. The liquid is incompressible.
2. The movement of the liquid is under laminar regime.
3. There is a steady velocity that is only a function of the radius; it is supposed that the radial and axial flows are equal to zero (neglecting the centrifuge forces).
4. There is steady movement. All the derivatives with respect to time of the continuity and movement equations are zero.
5. There is no relative movement between the surface of the cylinders and the fluid in contact with the cylinders, that is, there is no sliding.
6. The movement is 2D (neglecting the final and edge effects).
7. The system is isothermal.

Couette's flow is an example of simple shear flow with a velocity field in which components as cylindrical coordinates are

$$v(r) = 0 \tag{13.74a}$$

$$v(\theta) = r\omega(r) \tag{13.74b}$$

$$v(z) = 0 \tag{13.74c}$$

The only nonnull velocity component is the one in the direction θ, $r\omega(r)$, in which $\omega(r)$ is the angular velocity.

According to the velocity field, the stress components are given by

$$\sigma_{rz} = \sigma_{\theta z} = 0 \tag{13.75}$$

$$\sigma_{r\theta} = \sigma(\dot{\gamma}) \tag{13.76}$$

$$\sigma_{rr} - \sigma_{zz} = \psi_1(\dot{\gamma}) \tag{13.77}$$

$$\sigma_{\theta\theta} - \sigma_{zz} = \psi_2(\dot{\gamma}) \tag{13.78}$$

where

σ_{rr}, $\sigma_{\theta\theta}$, σ_{zz}, $\sigma_{r\theta}$, σ_{rz}, and $\sigma_{\theta z}$ are the components of the stress tensor
$\psi_1(\dot{\gamma})$ and $\psi_2(\dot{\gamma})$ are the components of the function of normal stresses

It is assumed that the flow occurs between two infinite coaxial cylinders with radii R_i and R_0 ($R_i < R_0$). The internal cylinder rotates with certain angular velocity, while the external remains fixed. If the torque per height unit (T') exerted on the fluid in the interior of the cylindrical surface with constant r is calculated, the following relation is obtained:

$$T' = \sigma_{r\theta}(2\pi r)r = 2\pi r^2 \sigma_{r\theta} \tag{13.79}$$

According to Cauchy's first law of movement, it is obtained that

$$\sigma_{r\theta} = \frac{\mu}{r^2} \tag{13.80}$$

where μ is a constant. Equalizing Equations 13.76 and 13.79,

$$\sigma_{r\theta} = \frac{T'}{2\pi r^2} = \sigma(\dot{\gamma}) \tag{13.81}$$

If we assume that the fluid obeys the power law,

$$\sigma = m\dot{\gamma}^n = m\left[\frac{dv}{dr} - \frac{v}{r}\right]^n \tag{13.82}$$

and integrating between the ratio of the internal and external cylinders, the following expression is obtained:

$$\int_0^{\omega_i} d\omega = \left[\frac{T'}{2\pi m}\right]^{1/n} \int_{R_0}^{R_i} \frac{dr}{\left(r^{(n+2)/n}\right)} \tag{13.83}$$

where ω_i is the angular velocity at which the internal cylinders spin. Equation 13.83 can be integrated as follows:

$$\omega_i = 2\pi N' = \frac{n}{2}\left[\frac{T'}{2\pi m}\right]^{1/n}\left[\frac{1}{R_i^{2/n}} - \frac{1}{R_0^{2/n}}\right] \tag{13.84}$$

where N' is the number of revolutions per unit time at which the internal cylinder spins. The expression for Newtonian fluids is directly obtained for $n = 1$, from which Margules' equation for Newtonian viscosity is obtained:

$$\eta = \frac{T'}{4\pi\omega_i}\left[\frac{1}{R_i^2} - \frac{1}{R_0^2}\right] \tag{13.85}$$

When the measured fluid shows a Bingham's plastic behavior, the following expression is obtained:

$$\omega_i = 2\pi N' = \frac{1}{m}\frac{T'}{4\pi}\left[\frac{1}{R_i^2} - \frac{1}{R_0^2}\right] - \frac{\tau_0}{m}\ln\left(\frac{R_0}{R_i}\right) \tag{13.86}$$

13.12.1.1.2 Plate–Plate and Cone–Plate Viscometers

Other types of rotational viscometers consist of plate–plate and of cone–plate. As indicated, the plate and cone viscometer consists of a flat plate and a cone (Figure 13.18). The angle of the cone θ_0 is about 3° or less.

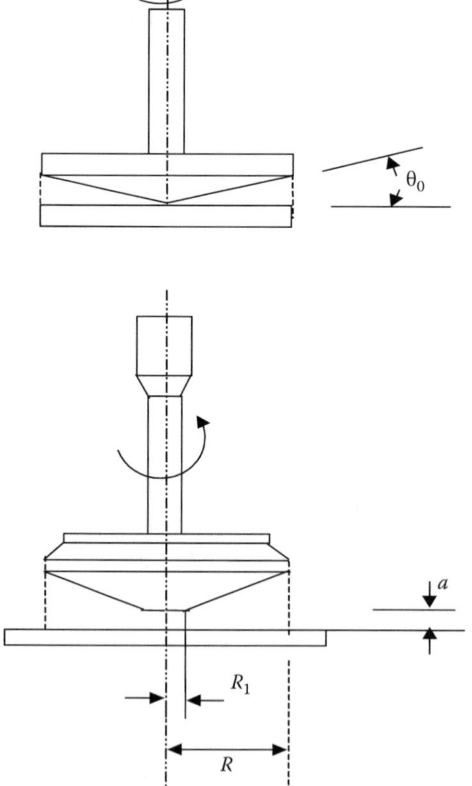

FIGURE 13.18 Scheme of cone–plate geometry.

According to Walters (1975), it is assumed that

1. Inertia effects are insignificant
2. The angle of the cone is very small
3. The cone and the plate have the same radius a
4. The free surface of the liquid is part of a sphere of radius a and is centered in the cone's vortex
5. The simple shear flow under steady state is continuous up to the free surface
6. Surface tension forces are insignificant, and the torque measurement (T) in the fixed plate, as a function of the angular velocity (Ω_1) of the cone, can be used to determine the rate of shear, which depends on the apparent viscosity, according to the following equations:

$$\dot{\gamma} = \Omega_1 \frac{a}{h} \quad (13.87)$$

$$\sigma = \frac{3T}{2\pi a^3}\left[1 + \frac{1}{3}\frac{d \ln T}{d\dot{\gamma} a}\right] \quad (13.88)$$

where
a is the radius
h is the distance between the plate and the cone

If it is a parallel plate system, the same equations are used to calculate the rate of shear and the shear stress. The velocity field is independent of material's properties for the flow as well as for the flow's torsion of the plate and cone.

Considerable attention has been paid to the situation of combined steady and oscillating flow since the mid-1960s. Such a situation is interesting in itself, since a rheometric flow can be interpreted relatively simply in terms of certain well-defined functions of the material. It also supplies a much more critical experiment than any rheological equation proposed separately in shear under steady and/or oscillating state (Walters, 1975; Kokini and Plutchok, 1987).

13.12.1.1.3 Error Sources

The basic equations obtained in the last section for the determination of the relation between σ and γ are based on the existence of a simple flow through the liquid, so numerous assumptions should be complied with to obtain the desired type of flow. An important consideration, therefore, is that these assumptions are valid in practical situations with the limitations that they impose on the operation conditions. Possible error sources affecting the interpretation of the experimental results in the Couette flow are reviewed in this section.

13.12.1.1.3.1 Viscous Heating When a liquid is sheared, part of the work applied is dissipated as heat. The heating induced by the shearing causes an unavoidable increase of the liquid's temperature, which constitutes a source of error that could be particularly important in the case of highly viscous liquids and/or at high shear value (Walters, 1975). The symbol β is used to describe the dependence on temperature of the material's properties ($\beta = 0$ represents materials whose properties are not affected by temperature changes, and high values of β represent materials whose properties are sensitive to temperature).

It is possible to identify β as the viscosity–temperature relation parameter. The parameter that defines whether the heating induced by shearing is important in a practical situation is the product

$\beta B(r)$, with $B(r)$ being Brinkman's number. For a Couette's flat flow with a separation between cylinders (gap) of width h, $B(r)$ is given by

$$B(r) = \frac{\sigma^2 h^2}{\eta K_T T_0} r^2 \tag{13.89}$$

where

K_T is the thermal conductivity
T_0 is a reference temperature (K)

It can be deduced from this equation that the heating effects should be significant in the higher viscosity and rate of shear ranges. A way to minimize such effects is to work with a narrow gap between cylinders.

13.12.1.1.3.2 Imperfections of the Instrument Reliable viscometric measurements require much precision in the geometry of the instrument as well as in the alignment of the surfaces and problems related to it (Walters, 1975).

Having perfect alignment in the case of narrow gap rheometry is a difficult task, and for highly viscous liquids, a slight slanting can cause a great positive pressure on the region of convergent flow and a great negative pressure on the region of divergent flow overimposed to the pressure resulting from the effect of normal stresses (Greensmith and Rivlin, 1953). This effect can be eliminated by changing the direction of movement and considering the average pressure during movement in both directions (McKennell, 1960).

Another source of error is the small axial movement of the rotating device during the experiment caused by imperfections of the connection (Adams and Lodge, 1964).

13.12.1.1.3.3 End and Edge Effects Different authors such as Oka (1960), Van Wazer (1963), and Walters (1975) have highlighted this type of error. This error is due to the fact that the devices of the instrument have finite dimensions. Consequently, in general, there always exists a viscous drag due to the stress at the bottom of the surface of the internal cylinder. Also, the distribution of stress on the cylindrical surface differs from the distribution on cylinders of infinite length, since the flow is affected by the end of the surface.

The end effect can be considered as being equivalent to an increase in the effective immersion length from h to $(h + \Delta h)$ (Oka, 1960). Margules' equation for a Newtonian flow becomes

$$\Omega = \left[\frac{T'h}{4\,\pi\eta(h+\Delta h)}\right]\left(\frac{1}{R_i^2} - \frac{1}{R_0^{\,2}}\right) \tag{13.90}$$

where Δh, the end correction, is in general a function of R_i, R_0, h, and the ending hollow l. The end correction (Δh) is normally obtained by experimentation when plotting T'/Ω versus h.

Oka (1960) deduced a relation for the end effect that satisfies Navier–Stokes' equation. The final expression is

$$\frac{\Delta h}{R_i} = \frac{1}{8}\frac{R_i}{l}\left[1 - \left(\frac{R_i}{R_0}\right)^2\right]\left[1 + 4\frac{1}{R_i}\sum_1^{\infty} A_n I_2\left(\frac{n\pi a}{1}\right) + \frac{8}{\pi}\frac{1}{R_i}\sum_1^{\infty} B_n \frac{senh(K_n h)}{K_n R_i}\right] \tag{13.91}$$

where

I_2 is the second-order modified Bessel function
K is the nth positive root of the equation $r = (KR_0) = 0$
A_n and B_n are functions of the dimensionless parameters R_i/R_0, h/R_0, and l/R_0

In this equation, the first term of the second brackets is only due to the end effect of the internal cylinder, the second form is due to the edge effect at the bottom, and the third term is due to the end effect and to the free surface of the fluid in the gap. It is evident that when l is very large compared to R_i and when R_i/R_0 is close to the unit, the term $(\Delta h/R_i)$ is very small. Based on these results, the influence of sphere and end effects may be and probably should be reduced, if the hollow between the cylinders is as narrow as possible.

13.12.1.1.3.4 Taylor's Vortex The presence of Taylor's vortexes is a problem that occurs when a sample is analyzed with a coaxial viscometer in which the internal cylinder rotates. The condition for this instability is given by Taylor's number (Schlichting, 1955):

$$T_a = \frac{v_i}{\nu}\left(\frac{d}{R_i}\right)^{1/2} \geq 41.3 \tag{13.92}$$

where

R_i is the radio of the internal cylinder
d is the hollow between cylinders
v_i is the peripheral velocity of the internal cylinder
ν is the kinematic viscosity

Three Taylor's numbers can be differentiated for three flow regimes:

$T_a < 41.3$ = Couette laminar
$41.3 < T_a < 400$ = laminar, with Taylor's vortexes
$T_a > 400$ = turbulent

13.12.1.1.3.5 Miscellaneous Sources of Error Other sources of error are

1. Inertia effects (this correction depends on $w(r)$, which in turn depends on the apparent viscosity function)
2. Inherent errors to the interpretation of the experimental results
3. Instabilities due to the turbulent flow
4. Homogeneity of the tested sample
5. Stability of the tested sample
6. Sliding

13.12.1.2 Oscillating Flow

Oscillating experiments can be carried out using the following geometries: parallel plates, plate and cone, and the Couette system of concentric cylinders. This type of experiment is used to study the viscoelasticity of foods. To do that, a simple harmonic movement of small amplitude is applied to the fixed parts of the previous measurement systems, which induces an oscillation in the rotor that is affected by the viscous resistance and the elastic force of the sample contained in the viscometer. The oscillating phase of the rotor is negatively phased out, and the induced amplitude also differs from the applied one.

It is possible to calculate η' and G' (Rao, 1986) from the expressions:

$$\eta' = \frac{-s(\theta_1/\theta_2)sinc}{\left[(\theta_1/\theta_2)^2 - 2(\theta_1/\theta_2)\cos c + 1\right]} \tag{13.93}$$

$$G' = \frac{\omega s(\theta_1/\theta_2)[\cos c - (\theta_1/\theta_2)]}{[(\theta_1/\theta_2)^2 - 2(\theta_1/\theta_2)\cos c + 1]} \tag{13.94}$$

The variables of these expressions are defined in different ways according to the measurement system used.

For the Couette system of concentric cylinders, the variables are defined as

$$s = \frac{\left(R_0^2 - R_i^2\right)(K - I\omega^2)}{4\pi L R_i^2 R_0^2 \omega} \tag{13.95}$$

In the expressions earlier, θ_1 is the angular amplitude of the internal cylinder and θ_2 of the external cylinder, L is the height of the sample contained between the cylinders, I is the inertia moment of the internal cylinder with respect to its axis, ω is the frequency, c is the phase out, and K is the restoration constant of the torsion pair.

For parallel plate systems, in which the lower plate is subjected to oscillation, the variable s is calculated by the following equation:

$$s = \frac{2h(K - I\omega^2)}{\pi a^2 \omega} \tag{13.96}$$

where

- h is the distance between plates to their radii
- I is the moment of inertia of the upper plate with respect to its axis
- θ_1 is the angular amplitude of the upper plate
- θ_2 of the lower plate
- c is the upper plate's phase out

In cone and plate systems in which the plate rotates, s is calculated using the following equation:

$$s = \frac{3\theta_0(K - I\omega^2)}{2\pi a^2 \omega} \tag{13.97}$$

where

- θ_0 is the cone's angle
- θ_1 is the angular amplitude of the cone
- θ_2 of the plate
- a is the radius of the plate
- c is the cone's phase out

13.12.1.3 Capillary Flow

When a liquid flows through a pipe, it forms a velocity gradient and a shearing effect occurs. Methods have been developed to measure the flow properties of fluids by using capillary tubes through which a fluid is forced to flow due to an applied pressure or to hydrostatic pressure. If the volumetric flux, tube dimensions, and applied pressure are known, curves of flow can be plotted and the apparent values of the viscosity can be calculated. Certain assumptions should be made to develop general equations that allow calculating the rates of shear and shear stresses for a specific point in the tube.

Rabinowitsch (Van Wazer et al., 1963) developed a general equation to calculate the rates of shear. Such an equation, which is valid for non-Newtonian as well as for Newtonian fluids, is

$$\dot{\gamma} = \frac{3 + b}{4(4q/\pi R^3)} \tag{13.98}$$

in which

$$b = \frac{d\log(4q/\pi R^3)}{d\log(\Delta PR/2l)} \tag{13.99}$$

where

q is the volumetric flow rate through the capillary of length l and radius R
P is the applied pressure

The value of term b can be calculated when plotting $(4q/R^3)$ versus $(\Delta PR/2l)$ in double logarithmic coordinates, b being the slope of the straight line built in such a way. For Newtonian liquids, the slope of the straight line is one and the equation reduces to

$$\dot{\gamma} = \frac{4q}{R^3} \tag{13.100}$$

When the slope of the straight line deviates from l, the fluid does not exhibit a Newtonian character, so the global equation should be used. The flow behavior of a variety of food suspensions has been studied using a capillary tube, including applesauce, baby foods, and tomato purée (Charm, 1960; Saravacos, 1968; Rao et al., 1974).

13.12.1.4 Back Extrusion Viscometry

A way to characterize the rheological behavior of non-Newtonian fluids independent of time is by means of back extrusion tests. Back extrusion is produced when a small rod is submerged in a cylindrical vessel that contains the sample to be tested (Figure 13.19). As the small rod penetrates, a sample movement is produced but in the opposite direction to the rod penetration, hence the name "back extrusion."

Osorio and Steffe (1987) applied this type of experiment for the determination of rheological parameters of fluids that obey the power law. Figure 13.20 shows the graphic obtained in a typical back extrusion experiment, in which the force applied to the small rod is each time greater, so that the length of the small rod that penetrates the sample increases. The penetration stops at a force F_T, appearing as an equilibrium force F_{Te}.

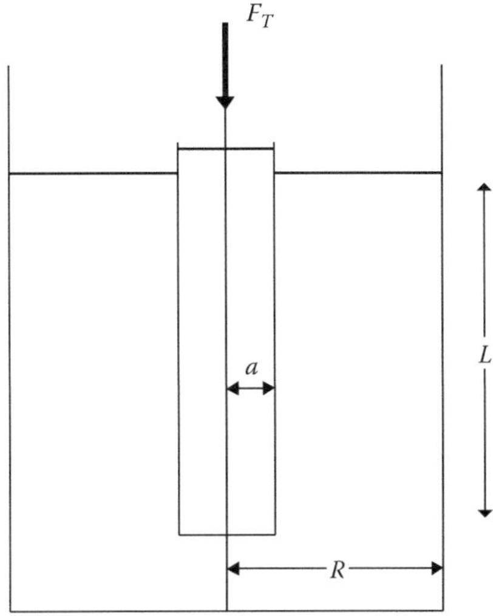

FIGURE 13.19 Position of a small rod and cylindrical vessel in a back extrusion test.

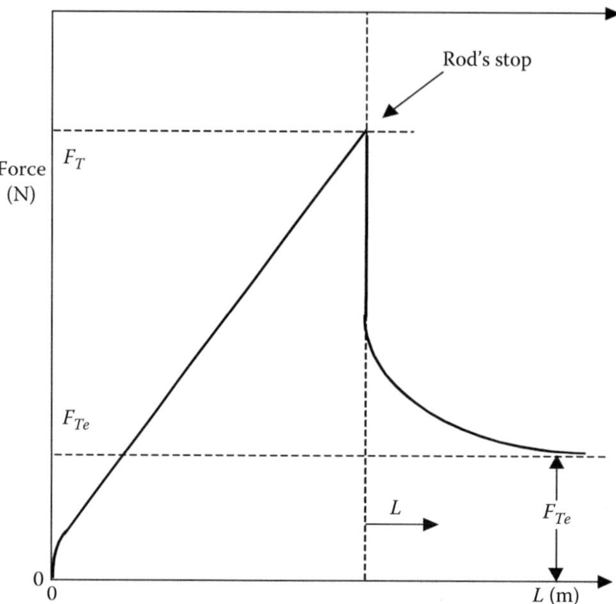

FIGURE 13.20 Typical data of back extrusion for a fluid of the power law.

When calculating the flow behavior index (n), it is necessary to perform two experiments, obtaining its value from the following expression:

$$n = \frac{\ln[(F_{b_2}/F_{b_1})(L_1/L_2)]}{\ln(v_1/v_2)} \tag{13.101}$$

where

the subindexes 1 and 2 refer to the two experiments

v is the velocity of the small rod

L is the length of the small rod submerged in the sample

F_b is the force corrected with the buoyancy force, defined by

$$F_b = F_T - \rho g L \pi a^2 \tag{13.102}$$

where

a is the radius of the small rod

ρ is the density of the sample

g is the gravitational acceleration

F_T is the force just before the small rod stops

To calculate the consistency index, the following equation is used:

$$k = \frac{RF_bK}{2\Gamma^2\pi LRa}\left(\frac{\Phi R}{vK^2}\right) \tag{13.103}$$

where

R is the radius of the external vessel

K is the relation between the radii of the small rod and the vessel ($K = a/R$)

Φ is the dimensionless flow velocity

Γ is the dimensionless radius when the stress is null

Φ and Γ are functions of K and of the flow behavior index, whose values are given in graphical form or tabulated (Osorio and Steffe, 1987; Steffe and Osorio, 1987).

This type of characterization can also be applied to fluids with Newtonian behavior, assuming that $n = 1$. In the same way, for Herschel–Bulkley's and Bingham's fluids, their corresponding rheological parameters can be obtained (Osorio, 1985; Osorio and Steffe, 1985). These fluids present a flow threshold that is possible to determine using this type of experiment.

It is also possible to apply this technique to the rheological characterization of fluids that present dependence on time; thus, it has been applied to the rheological study of children's food consisting of macaroni and cheese (Steffe and Osorio, 1987).

13.12.1.5 Squeezing Flow Viscometry

When a fluid material is compressed between two parallel plates, the compressed material flows in a perpendicular direction to the compression force exerted. This is the basis of squeezing flow viscometry and can be used in the rheological characterization of fluids that present slippage problems (Campanella and Peleg, 1987a,b).

There are four types of tests that allow adequate rheological characterization, which are at constant volume or constant area of the sample, combined with the application of constant force or rate of shear. Figure 13.21 presents the schemes of these four types of tests. If it is desired to obtain the parameters of the power law for a determined sample by means of this technique, it is necessary to be sure that the sample does not present viscoelasticity. Next, it is explained how to calculate the consistency index (k) and the flow behavior index (n) for squeezing flow tests of a sample of constant area introduced between two circular plates of radius R.

For a constant rate of shear (V_d), there is a variation of the sample height $H(t)$ and of the force $F(t)$ with time that can be related according to the following expression (Campanella and Peleg, 1987a):

$$F(t) = \frac{3^{(n+1)/2}\pi R^2 k (V_d)^n}{[H(t)]^2} \tag{13.104}$$

from which it is possible to obtain the values of k and n, through a nonlinear regression, by fitting the experimental data of the variation of $F(t)$ with $H(t)$.

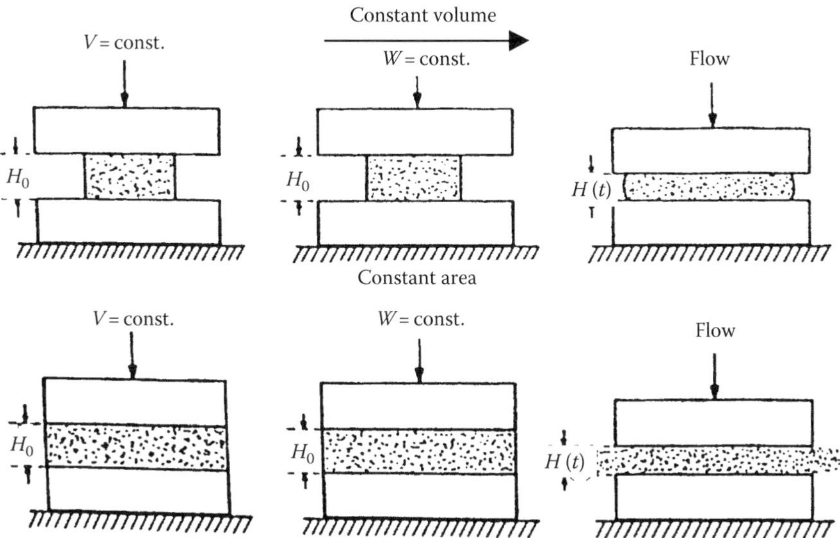

FIGURE 13.21 Types of squeezing flow viscometry tests.

These parameters can also be determined by an experiment at constant force. If W is the applied force and H_0 is the initial height of the sample, the variation of $H(t)$ with time is given by the expression (Campanella and Peleg, 1987a):

$$\ln\left[\frac{H(t)}{H_0}\right] = -St \quad (13.105)$$

in which

$$S = \left[\frac{W}{3^{(n+1)/2}\pi R^2 k}\right]^{1/n} \quad (13.106)$$

To determine the consistency index k and the flow behavior index n, it is necessary to perform at least two experiments, since the slope S of Equation 13.105 contains the two parameters to be determined. As will be shown later, squeezing flow viscometry can also be used to determine the yield stress in those products that exhibit it.

13.12.2 Empirical Methods

Many instruments have been developed to characterize the consistency of purees (Rao, 1980).

Rotational viscometers with spindles for which mathematical analysis is difficult have been used in empirical tests. The spindles can be needles with protuberances and fins. Kramer and Twigg (1962), Aronson and Nelson (1964), and Rao (1980) describe some of these methods used in the food industry.

13.12.2.1 Adams Consistometer

These instruments measure the consistency of foods based on the degree of extension or flow of the product in all directions in a determined time. This type of instrument has been used to measure tomato, pumpkin, cereal cream, and bean paste products (Lana and Tischer, 1951; Davis et al., 1954; Mason and Wiley, 1958).

13.12.2.2 Bostwick Consistometer

This instrument measures the consistency of viscous materials by measuring the distance that a material flows in all directions in a determined time. It has been used for ketchup (USDA, 1953), preserves and canned foods (Davis et al., 1954), and milk pudding (Rutgus, 1958).

13.12.2.3 Tube Flow Viscometer

The tube flow viscometer is used to measure the time needed by a determined quantity of fluid to cross a tube or capillary. This type of instrument is especially good for highly mobile materials, as is the Ostwald viscometer. Davis et al. (1954) used this type of viscometer to measure the consistency of tomato paste. They showed that the data obtained with the Adams and the Bostwick consistometers were linearly related and that with use of the efflux viscometers, different parameters to those from the consistometer are obtained.

13.12.3 Imitative Methods

These methods, in special cases, measure rheological properties under symmetry conditions similar to those presented in practice. This group includes (1) the Brabender Visco-Amylo-Graph, which is used in the evaluation of the consistency of flours, starches, and gums (Bhattacharya and Sowbhagya, 1978). The sample is placed in a container and the torque exerted on a rotational spade is measured. The temperature of the sample is raised or lowered at constant velocity and the results are recorded on a mobile registration paper. The Messtometer consists of a flow bridge in which the conduits are arranged in a way such that a differential pressure between two reference points is produced. The differential pressure in the flow bridge is given as a function of consistency (Eolkin, 1957).

13.12.4 Obtaining the Rheological Parameters

13.12.4.1 Capillary Viscometer

Figure 13.22 shows a typical assembly to obtain the rheological flow behavior parameters for a fluid. This viscometer type is a capillary tube and is also called an extrusion rheometer. The system consists of a closed reservoir that contains the fluid at a height h. At the bottom of the deposit, a capillary tube is connected to longitude L. By means of an appropriate mechanism, the deposit can be pressurized in such a way that a pressure drop $(-\Delta P)$ can be obtained between the reservoir head and the exit of the capillary tube.

When applying a balance of mechanical energy between the fluid surface in the reservoir and the tube exit hole (points 1 and 2), it is obtained that

$$g(z_2 - z_1) + \frac{\Delta P}{\rho} + \hat{E}_f = 0 \tag{13.107}$$

where

z is the height of the considered point

ρ the density of the fluid

ΔP is the pressure drop among the exit point 2 and the initial point 1

$\hat{E}_f$ being the mechanical energy losses by a mass unit that experiences the fluid for friction with the walls

According to Figure 13.22, $z_2 = 0$, while $z_1 = h + L$, the mechanical energy losses that the fluid experiences will be obtained when reordering Equation 13.107:

$$\hat{E}_f = g(h+L) + \frac{(-\Delta P)}{\rho} \tag{13.108}$$

Therefore, for a certain experiment, it is possible to obtain the mechanical energy losses of the fluid when passing through the capillary.

The mechanical energy losses can be calculated by means of the Fanning equation, in which the friction factor takes place; this factor is defined as the relationship between the shear stress at the

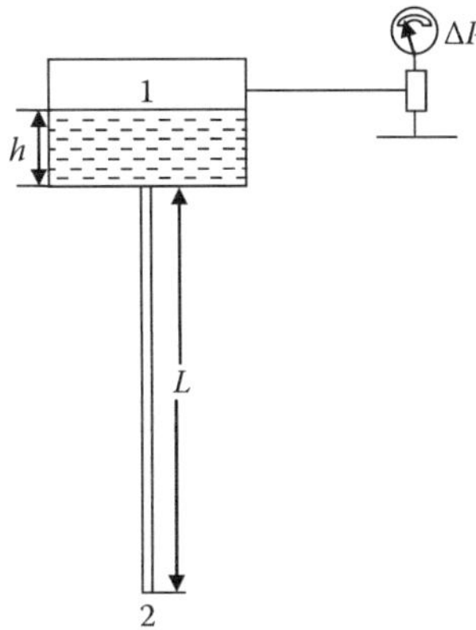

FIGURE 13.22 Capillary viscometer system.

wall and the kinetic energy for unit of mass. Combining these equations, it is possible obtaining the shear stress value at the capillary wall:

$$\sigma_W = \frac{\rho d\hat{E}_f}{4L} \tag{13.109}$$

Also, the shear rate at the wall can be obtained by the expression (Skelland, 1967):

$$\dot{\gamma}_W = \frac{8v}{d}\left(\frac{3n'+1}{4n'}\right) \tag{13.110}$$

where

- d is the diameter of the tube
- v is the lineal speed of the fluid that is obtained from the volumetric flow rate and of the transversal section of the capillary tube

The value of the parameter n' is given by

$$n' = \frac{d(\log \sigma_W)}{d(\log(8v/d))} \tag{13.111}$$

When graphically representing the shear stress values at the wall (σ_W), calculated from Equation 13.109, versus the values of $8v/d$, in logarithmic double coordinates, a straight line is obtained whose slope is the value of the parameter n'.

In the case that the fluid flow behavior is described by means of the power law equation ($\sigma = k\dot{\gamma}^n$), the value of the flow behavior index n coincides with n' ($n = n'$).

When substituting the shear rate value (Equation 13.110) in the power law equation, expressed in logarithmic form, the following expression is obtained:

$$\log \sigma_W = \log\left[k\left(\frac{3n+1}{4n}\right)^n\right] + n\log\left(\frac{8v}{d}\right) \tag{13.112}$$

It is observed that if the shear stress values at the wall (σ_W) are represented versus ($8v/d$), in logarithmic double coordinates, a straight line is obtained whose slope is the n value, while the origin ordinate in the origin is $\log[k((3n + 1)/4n)^n]$; from this value and the one obtained for n, it is possible to find the consistency coefficient value k.

13.12.4.2 Concentric Cylinder Viscometer

The rotational concentric cylinder viscometer is a good tool to determine fluid flow behavior. This type of viscometer consists of a cylinder (rotor), radius R_1, and height L that rotates inside a container that contains the fluid whose internal radius is R_2. When the rotor rotates at a certain rotational speed (N), a torque is created (T) that is able to carry out the measurement of both variables using the appropriate instruments. The turn of the rotor causes a shear stress at the rotor wall (σ_W) that is related to the created torque. The torque at the shear surface, with regard to the rotational axis, is the product of the force and the rotor radius:

$$T = \sigma_W(2\pi R_1 L)R_1 \tag{13.113}$$

From this equation, it is possible to obtain the shear stress:

$$\sigma_W = \frac{T}{2\pi R_1^2 L} \tag{13.114}$$

In the case that the tank containing the sample possesses a radius $R_2 \approx R_1$, the shear rate at the wall ($\dot{\gamma}_W$) can be calculated from the rotor speed by means of the equation:

$$\dot{\gamma}_W = \frac{2\pi R_1 N}{R_2 - R_1} \tag{13.115}$$

When the container radius is much larger than the rotor radius ($R_2 >>> R_1$), meaning that the rotor is submerged in a container whose radius is considerably greater than that corresponding to the rotor, the shear rate at the wall is obtained by the following equation:

$$\dot{\gamma}_W = 4\pi N \frac{d(\log N)}{d(\log T)} \tag{13.116}$$

In this case, $d(\log N)/d(\log T)$ represents the slope of the straight line that one obtains when representing N versus T, in logarithmic double coordinates.

In order to obtain an equation that describes the fluid flow behavior, it is necessary to perform a series of experiments, varying the rotor speed and obtaining its corresponding torque. With these experimental data, applying Equation 13.114, it is possible to obtain the shear stress values, while with Equation 13.115 or 13.116, it will be possible to calculate the value corresponding to shear rate, depending on the type of viscometer used. The value of $\sigma_W - \dot{\gamma}_W$ is fitted by means of regression methods to the different rheological models, and in this way, it will be possible to obtain an equation that better describes the fluid flow behavior (Table 13.4).

TABLE 13.4
Rheological Models for Time-Independent Viscous Foods

Denomination	Equation	Notes
Newton's law	$\sigma = \eta\dot{\gamma}$	One parameter
Bingham's model	$\sigma = \eta_{pl}\dot{\gamma} + c$	Two parameters
Ostwald–de Waele	$\sigma = \eta(\dot{\gamma})^n$	Two parameters
Nutting's model or power law		
Herschel–Buckley	$\sigma = \eta(\dot{\gamma})^n + c$	Three parameters
Modified power law		
Casson's model	$\sigma^{0.5} = k_1 + k_2(\dot{\gamma})^{0.5}$	Three parameters
Modified Casson's equation	$\sigma^{0.5} = k_1' + k_2'(\dot{\gamma})^m$	Three parameters
Elson's equation	$\sigma = \mu\dot{\gamma} + Bsenh^{-1} + \sigma_0$	Three parameters
Vocadlo's model	$\sigma = \left(\sigma_0^{1/n} + k\dot{\gamma}\right)^n$	Three parameters
Shangraw's model	$\sigma = a\dot{\gamma} + b(1 - \exp(-c\dot{\gamma}))$	Two parameters
Generalized model	$\dot{\gamma} = \frac{1}{\eta_0}\left[\frac{1+(\tau_{rz}/\tau_m)^{\alpha-1}}{1+(\tau_{rz}/\tau_m)^{\alpha-1}(\eta_\infty/\eta_0)}\right]$	Four parameters
Sutterby's model	$\tau_{rz} = -\mu_0\left[\frac{arcsen\beta\dot{\gamma}}{\beta\dot{\gamma}}\right]^\alpha\dot{\gamma}$	Three parameters
Springs truncated power law	$\tau_{rz} = -\mu_0\left[\frac{\dot{\gamma}}{\dot{\gamma}_0}\right]^{n-1}\dot{\gamma}$	Three parameters
Williamson's model	$\tau = A\frac{\dot{\gamma}}{B+\dot{\gamma}} + \mu_\infty\dot{\gamma}$	Three parameters
Sisko's model	$\tau = A\dot{\gamma} + B\dot{\gamma}^n$	Three parameters

PROBLEMS

13.1 The flow behavior of a 47.3°Brix clarified kiwi juice, with a given pectin content, was determined, obtaining that the best model to describe such behavior is the power equation. The rheological constants obtained, at different temperatures, are indicated in the following table:

T (°C)	K (mPa s^n)	n
4	2780	0.68
10	2287	0.68
15	1740	0.68
20	1247	0.71
25	1146	0.68
30	859	0.71
35	678	0.73
40	654	0.71
45	557	0.73
50	515	0.73
55	467	0.74
60	404	0.75
65	402	0.74

a. Determine the flow activation energy in kJ/mol.
b. Calculate the apparent viscosity of the kiwi juice with 47.3°Brix at 37°C, for a shear rate of 100 s^{-1}.
   a. The variation of viscosity with temperature can be correlated by an Arrhenius-type equation, and in the case of non-Newtonian fluids, the consistency index is used instead of the viscosity:

$$K = K_0 \exp\left(\frac{E_a}{RT}\right)$$

   When representing ln K against $1/T$, a straight line is obtained with an intercept origin ordinate of ln(K_0) and a slope of E_a/R. Carrying out this fitting with the data in the table, it is obtained that

$$K_0 = 3.5\times10^{-2} \text{ mPa s}^a, \frac{E_a}{R} = 3097 \text{ k}$$

   so the flow activation energy has a value of 25.75 kJ/mol.
   b. For a pseudoplastic fluid, the apparent viscosity is given by

$$\eta_a = K(\dot{\gamma})^{n-1}$$

   For 27°C, it can be taken that the flow behavior index is n = 0.72. The consistency index can be obtained from the Arrhenius equation, with the K_0 and E_a values obtained in the last section, for
   T = 310 K. With these data, it is obtained that K = 763.5 mPa s^a.
   The apparent viscosity for a shear rate of 100 s^{-1} is η_a = 210 mPa s.

13.2 Companies that process clarified juices without pectin usually concentrate them up to a soluble solid content close to 70°Brix, in a multiple evaporation system. The juice exits the evaporation stage at 60°C, and it should be cooled down to a storage temperature of 5°C, by means of a plate heat exchanger followed by one in a spiral shape. The plate heat exchanger only allows fluids with viscosity lower than 1500 mPa s to pass through.

The variation of the viscosity with temperature for a clarified peach juice without pectin of 69°Brix can be expressed by means of the following equation:

$$\eta = 7.76 \times 10^{-11} \exp\left(\frac{6690}{T}\right)$$

where

η is the viscosity in Pa s
T is the absolute temperature

a. Calculate the flow activation energy.
b. What is the minimum temperature to which a 69°Brix concentrate could be cooled using the plate heat exchanger?
   a. The variation of the viscosity with temperature can be correlated according to an Arrhenius-type equation. When comparing the Arrhenius equation with that in the problem's statement, it is possible to obtain that

$$\eta = 7.76 \times 10^{-11} \exp\left(\frac{6690}{T}\right)$$

   So $E_a/R = 6690$ K, an equation that allows a flow activation energy of $E_a = 55.62$ kJ/mol to be obtained.
   b. Since the maximum viscosity permitted by the plate heat exchanger is 1500 mPa s, the exit temperature of the fluid, once cooled, should be such as to have a fluid viscosity of precisely 1.5 Pa s at this temperature.

   The minimum exit temperature is obtained when substituting the viscosity value (1.5 Pa s) in the Arrhenius equation:

$$1.5 = 7.76 \times 10^{-11} \exp\left(\frac{6690}{T}\right)$$

   and rearranging, it is obtained that $T = 282.5$ K = 9.5°C.

13.3 The influence of the soluble solid content on the rheological behavior of a clarified pear juice without pectin was studied. With this purpose, a 70°Brix concentrated industrial juice was assayed, and by means of dilution using distilled water, juices with concentrations within the 30°Brix–70°Brix range were obtained. It was found that at 25°C, all of the juices presented a Newtonian behavior, obtaining the following viscosity of each sample:

C (°Brix)	30	40	45	50	55	60	65	70
η (mPa s)	3	5	8	13	19	41	74	233

Obtain an expression that describes the influence of the soluble solid content on viscosity.

In one of the stages of the industrial process, the pear juice should circulate through a pipe, having available a centrifuge pump that can propel fluids with a maximum viscosity of 100 mPa s.

Would this pump be useful to propel a 68°Brix juice at 25°C? What is the maximum concentration that this pump can propel?

The variation of the viscosity with the soluble solid content can be correlated by either of the following expressions:

$$\eta = K_1 \exp(a_1 C) \quad \text{or} \quad \eta = K_2 (C)^{a_2}$$

These equations can be linearized if they are taken in a logarithmic form, and with the data in the table, it is possible to find the different constants. From the fitting, it is obtained that

$$\eta = 7.9 \times 10^{-5} \exp(0.106C) \qquad r = 0.975$$

$$\eta = 9.3 \times 10^{-11} (C)^{4.89} \qquad r = 0.940$$

in which the viscosity is given in Pa s, if the soluble solid content is expressed in °Brix. It seems that the exponential model fits better, since its regression coefficient is higher.

If the exponential equation is taken, it is obtained, for a 68°Brix concentration, that the viscosity of such juice is 106.7 mPa s. Since this viscosity is higher than 100 mPa s, the pump could not propel the 68°Brix juice.

The 100 mPa s viscosity corresponds to a 67.4°Brix juice, according to the exponential equation. For this reason, the available pump could only propel juices with a soluble solid content lower than 67.4°Brix.

13.4 The rheological flow behavior of a clarified raspberry juice was studied. From a 41°Brix concentrated juice with a pectin content of 0.5 g galacturonic acid/kg juice, different samples up to 15°Brix have been prepared by dilution with distilled water. The rheological behavior of these samples has been studied within the 5°C–60°C temperature range. It was obtained that the power law is the model that best describes such behavior, obtaining that the consistency index and the flow behavior index vary with temperature and soluble solid content according to the expressions:

$$K = 1.198 \times 10^{-10} \exp\left(\frac{4560}{T} + 0.196C\right)$$

$$n = 1.123 - 8.52 \times 10^{-3}\, C$$

where

K is in Pa s^n
T is in Kelvin
C is in °Brix

What is the value of the flow activation energy, as expressed in kcal/mol and kJ/mol?

An industry that concentrates clarified raspberry juices needs to know the viscosity of a 27°Brix juice that should circulate at 50°C through a stainless steel conduit. If the velocity gradient exerted on such juice along the conduit is 100 s^{-1}, what is the viscosity in mPa s?

According to the expressions that describe the temperature–concentration combined effect,

$$K = K_1 \exp\left(\frac{E_a}{RT} + K_2 C\right)$$

indicating that $E_a = 4560R$

$$E_a = 4560 \text{ K } 1.987 \times 10^{-3} \text{ kcal/(mol K)} = 9.06 \text{ kcal/mol}$$

$$E_a = 4560 \text{ K } 8.314 \times 10^{-3} \text{ kJ/(mol K)} = 37.91 \text{ kJ/mol}$$

For 27°Brix raspberry juices, the flow behavior index at 50°C (323 K) is $n = 0.893$, while the consistency index will be $K = 32.2$ mPa s^n.

The apparent viscosity of a fluid that obeys the power law is $\eta_a = K(\dot{\gamma})^{n-1}$. When substituting the obtained data for the consistency and flow behavior indexes in the exponential equation and for a rate of shear of 100 s^{-1}, an apparent viscosity of 19.7 mPa s is obtained.

13.5 In order to obtain the flow behavior parameters of a non-Newtonian fluid, whose density is 1100 kg/m^3, a capillary device has been used with a tube of internal diameter 1.25 mm and a height of 20 cm. Several different experiments have been conducted, varying the pressure applied in the sample tank and having obtained different fluid flow rates through the capillary for different pressure drops. In the following table, the obtained results are shown.

$(-\Delta P)$ (kPa)	2300	1900	1300	850	300	50
q (L/h)	1.2	1.0	0.75	0.49	0.23	0.06

The shear stress at the capillary wall can be obtained from the following equation:

$$\sigma_W = \frac{\rho d \hat{E}_f}{4L}$$

From a mechanical energy balance applied to a capillary tube, it is possible to obtain the mechanical energy losses:

$$\hat{E}_f = gL + \frac{(-\Delta P)}{\rho}$$

where
L is the capillary height
$(-\Delta P)$ is the pressure drop

The average velocity for the fluid is obtained from the volumetric flow rate:

$$v = \frac{4q}{\pi d^2}$$

Given the data of the problem, it is possible to calculate the mechanical energy losses, the shear stress at the wall, and the average velocity for each of the experiments shown in the table:

$-\Delta P$ (Pa)	q (m^3/s)	$\hat{E}_f$ (J/kg)	v (m/s)	σ_W (Pa)	$8v/d$ (s^{-1})
2,300,000	3.33E–07	2092.87	0.272	3597.1	1738.40
1,900,000	2.78E–07	1729.23	0.226	2972.1	1448.66
1,300,000	2.08E–07	1183.78	0.170	2034.6	1086.50
850,000	1.36E–07	774.69	0.111	1331.5	709.85
300,000	6.39E–08	274.69	0.052	472.1	333.19
50,000	1.67E–08	47.41	0.014	81.5	86.92

The flow behavior index can be calculated by graphically representing the shear stress values at the wall (σ_W) versus the values of $8v/d$, in logarithmic double coordinates; a straight line is obtained whose slope is the value for the parameter n'. For power law fluids, $n' = n$.

The adjustment of the obtained data allows obtaining the following values:

Slope: $n = n' = 0.787$

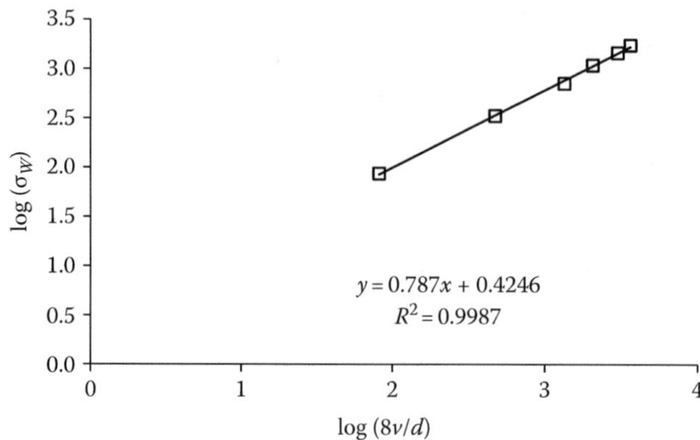

FIGURE P.13.5 Adjustments for obtaining rheological parameters.

Ordinate at origin

$$\log\left[k\left(\frac{3n+1}{4n}\right)^n\right] = 0.4246$$

From this latter value, it is possible to obtain the flow consistency index (Figure P.13.5):

$$k = 2.525 \text{ Pa s}^n$$

The rheological behavior for this fluid can be described as a power law fluid, whose flow behavior index $n = 0.787$ and consistency index $k = 2.525$ Pa s^n.

13.6 A company produces a concentrated peach puree and they want to pump it from the evaporator exit to its packing point. To determine its rheological behavior, a concentric cylinder viscometer is used, whose rotor possesses a radius of 5 cm and a height of 10 cm, while the container possesses an internal radius of 5.1 cm. The viscometer is shear stress controlled, and it has been observed that the rotor does not rotate if the applied torque is lower than 0.05 N m, while if the torque is 1.0 N m, the rotor rotates at a speed of 10 rpm. Determine the equation that best describes the derived peach flow behavior.

Because for torque values inferior to 0.5 N m the rotor does not work, this fluid type presents a shear yield that is calculated by means of the following equation:

$$\sigma_0 = \frac{T}{2\pi R_1^2 L} = \frac{0.05 \text{ N m}}{2\pi(0.05 \text{ m})^2(0.1 \text{ m})} = 31.8 \text{ Pa}$$

Since there is a unique value for the torque associated with the corresponding rotating velocity, it is a fluid that behaves as a Bingham plastic: $\sigma = \sigma_0 + \eta'\dot{\gamma}^n$.

The values of shear stress and shear rate are calculated:

$$\sigma = \frac{T}{2\pi R_1^2 L} = \frac{1.0 \text{ N m}}{2\pi(0.05 \text{ m})^2(0.1 \text{ m})} = 636.6 \text{ Pa}$$

$$\dot{\gamma}_W = \frac{2\pi R_1 N}{R_2 - R_1} = \frac{2\pi(0.05 \text{ m})(10/60 \text{ s}^{-1})}{0.05 - 0.051} = 52.4 \text{ s}^{-1}$$

The plastic viscosity is obtained from the following expression:

$$\eta' = \frac{\sigma - \sigma_0}{\dot{\gamma}} = \frac{(636.6 - 31.8)\ \text{Pa}}{52.4\ \text{s}^{-1}} = 11.5\ \text{Pa s}$$

13.7 A laboratory has a viscometer that consists of a rotor of 2.9 radius cm and of height 4 cm that is submerged in a container whose radius is very superior to that of the rotor. This viscometer is used to determine the rheological characteristics of tomato paste. For this purpose, the rotating velocity of the rotor is varied to measure the created torque. The data obtained are presented in the following table:

N (rpm)	60	120	210	350	420	490
T (N m)	0.21	0.315	0.45	0.48	0.62	0.68

The shear stress for all experiment is calculated from following equation: $\sigma_W = T/(2\pi R_1^2 L)$.

As the radius of the container that contains the sample is greater than the radius of the rotor, the shear rate corresponding to the shear stress is determined by means of the following equation: $\dot{\gamma}_W = 4\pi N\, d(\log N)/d(\log T)$.

For this calculation, it is necessary to determine the value of the straight line slope that is obtained when representing the rotation velocity of the rotor versus the corresponding torque. From the problem data, the rotation speed N is calculated in s^{-1} and it is represented versus the torque (Figure P.13.7.1).

The experimental data are fitted, obtaining a value of the slope of 15.28. This value is used in the calculation of the shear rate. In Table P.13.7, the different variable data are shown.

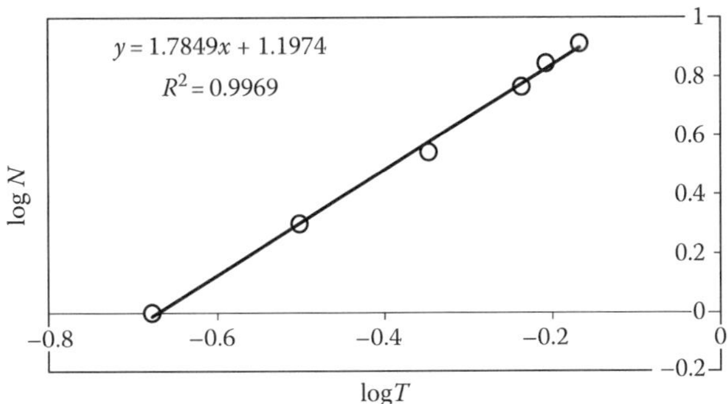

FIGURE P.13.7.1 Variation of torque with rotation velocity (double logarithmic scale).

TABLE P.13.7
Values for the Different Variables

N (rpm)	N (s^{-1})	T (N m)	σ_W (Pa)	$\dot{\gamma}$ (s^{-1})
60	1.00	0.21	994	22.4
120	2.00	0.315	1490	44.9
210	3.50	0.45	2129	78.5
350	5.83	0.58	2744	130.8
420	7.00	0.62	2933	157.0
490	8.17	0.68	3217	183.2

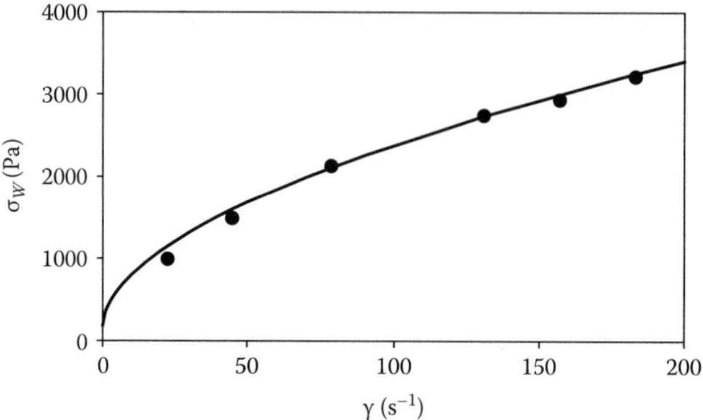

FIGURE P.13.7.2 Data fitted to the Herschel–Bulkley equation.

The values of the shear stress vary with the shear rate by means of a nonlinear regression of different equations, obtaining that the one that describes best the evaluation of the data is the Herschel–Bulkley equation. Figure P.13.7.2 shows the dependence of the fitted data, as well as the fitted line.

This adjustment allows to obtain a yield stress value σ_0 = 180 Pa, a consistency index k = 175 Pa s^n, and a flow behavior index n = 0.55.

PROPOSED PROBLEMS

13.1 A dairy industry is concentrating milk in a two-effect evaporator, obtaining as final product a concentrated milk that is at 80°C to the exit of the second effect. Since the high temperatures negatively affect the quality of the product, it is desired to reduce the temperature to 4°C. To do so, the company has installed a plate heat exchanger that uses glycol–water as coolant–refrigerant. The distance between the plates is 4 mm, which makes the viscosity of the fluid greater than 0.8 Pa s. It is not possible that concentrated milk can circulate through the heat exchanger. In laboratory experiments, it has been obtained that the viscosity of the concentrated milk varies with the temperature according to an Arrhenius equation type, with a flow activation energy of 80 kJ/mol and a viscosity of infinite deformation of 2×10^{-8} Pa s. Calculate the lowest temperature to cool down the condensed milk so that it can circulate in the heat plate exchanger.

13.2 A company wishes to elaborate and market a food puree, and it has initially built a small pilot plant. To pump the puree, it is necessary to know its rheological properties. For this purpose, a concentric cylinder viscometer has been used whose internal and external diameters are 4.95 and 5.05 cm, respectively, its height being 10 cm. It has been observed that when the torque applied is of 0.628 N m, the rotor rotates to 5 rpm; however, when the torque is lower than 0.314 N m, the cylinder does not rotate. Calculate the rheological parameters of the fluid.

13.3 In order to determine the rheological flow behavior parameters of an apple puree, an extrusion viscometer system has been used. The reservoir of this device contains the puree at a height of 10 cm, and the extrusion tube has a longitude of 1 m and a diameter of 1.2 cm. The apple puree possesses a density of 1200 kg/m^3. In an experimental series, the exit flow puree through the tube has been measured as a function of the pressure drop. The following table shows the obtained results:

$(-\Delta P)$ (Pa)	19,500	27,500	34,800	43,800
$q \times 10^3$ (L/s)	0.2	1.0	2.6	8.3

13.4 A company that produces a derived nutritious mash does not know the rheological behavior of this product. With the purpose of determining the rheological parameters, a series of experiments is performed with a concentric cylinder viscometer. The cylinder rotor has a diameter of 1.9 cm and a height of 4 cm. This rotor is contained in a tank whose diameter is greater than its own rotor. The following results have been obtained:

$T \times 10^6$ (N m)	42	63	107	152
$N \times 10^3$ (s^{-1})	0.1	0.2	0.5	1.0

14 Electrical Properties of Foods

14.1 INTRODUCTION

In many food processing operations, electricity is a form of energy that plays an important role, since electric power is used for different purposes, such as in the generation of processing vapor, in obtaining hot water and hot air, among others. Electricity can also be applied to produce heat directly into the food, such as in the case of ohmic processing. In processes where it is necessary to pump a liquid or to create air circulation, it is necessary to use pumps and fans or compressors that use electric power. Thus, in all operations that consume electric power, it is important to determine its consumption, which provides an idea of the energy requirement for a given process.

The electric properties of foods intervene in certain processing operations, and it is necessary to know them or to know how to evaluate them. The passage of an electrical current through a resistance generates heat; this is why it is necessary to know the resistance and the conductance of the food together with the dielectric properties. These last properties are those that intervene and control the interaction mechanisms between the foods and electromagnetic fields.

As a general rule, a body can have positive or negative charges, so if two bodies have charges of the same sign, they are repelled, and if they are of contrary signs, they are attracted to each other. The electric charge of the bodies, in the International System, is measured in coulombs (C). A material with a charge has an excess or deficiency of electrons. In the first case, it is said that the body has a negative charge, while in the second case, it has a positive charge.

When two points with different charges are connected by a material that allows the motion of electrons, an electric current takes place from the points with negative charge to the positive one. If there is a flow of electrons, it is necessary for the material to be conductive; thus, between both points, there exists a potential difference, due exactly to the charge difference. The electric current is due to the flow of electrons, and the current intensity depends on the flow speed of these electrons. It is considered that the electric current flows from the positive end to the negative end, in the opposite direction of the electron flow.

The current intensity (I) is measured as the charge by unit of time, so it is expressed as

$$I = \frac{Q}{t} \tag{14.1}$$

where

I is the current intensity, expressed in amperes (A)
Q is the charge, expressed in coulombs (C)
t is the time in seconds (s)

The current of 1 A corresponds to an electron flow of 6.26×10^{18} s^{-1}. Also, the charge of 1 C could be defined as the quantity of charge transported by a current of 1 A in 1 s.

Continuous electric current occurs when the electron flow takes place in only one direction, while in alternating current, the electron flow takes place in both directions and, in general, varies with time following a sinusoidal equation.

The charge difference between two points is the potential difference that is measured in volts (V), and this is the driving force of the electric current.

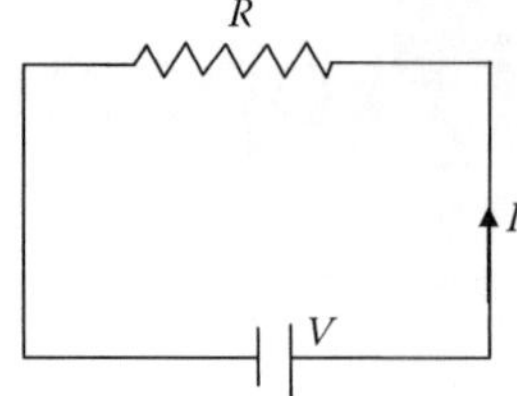

FIGURE 14.1 Electrical circuit.

14.2 ELECTRIC RESISTANCE AND CONDUCTIVITY

If we consider the electric circuit shown in Figure 14.1, where electric current flows through a conductor due to the potential difference created by an electromotive force (EMF) or voltage, there is a relationship between the voltage and the electric current as it passes through the conductor. This relationship is known as resistance of the conductor, and this is expressed according to the equation

$$R = \frac{V}{I} \tag{14.2a}$$

where
- V is the potential difference in volts
- I is the intensity in amperes
- R is the resistance expressed in ohm

The unit of resistance is ohm (Ω), and it can be defined as the resistance that a conductor offers to allow passage of a current of an ampere when there is a voltage of 1 V. The previous equation can be reordered as

$$I = \frac{V}{R} \tag{14.2b}$$

which is known as Ohm's law, and that indicates that the electric current passing through a conductor is directly proportional to a driving force, the potential difference, and inversely proportional to the resistance that the conductor offers for the current to pass.

Good conductor materials are those that possess low values of electric resistance, while insulating materials possess high values of resistance. Semiconductor materials possess values of electric resistance between both ends.

For many electric circuits, the current passes through conductive wires, so the electric resistance is directly proportional to the length of the conductor and inversely proportional to the cross section. The constant of proportionality is known as *resistivity*:

$$R = \rho \frac{L}{(\pi/4)d^2} \tag{14.3}$$

where
- L is the length of the piece of the material
- d is the diameter, both expressed in m
- ρ is the material resistivity

The resistivity units are of ohm per meter (Ω m). For conductive materials, as in the case of metals, resistivity presents values in the order of 10^{-7}–10^{-8} Ω m, while for insulating materials such as glass and mica, the value is in the order of 10^{10}–10^{12} Ω m. For foods such as potatoes and apples, the values are in the order of 30–100 Ω m (Mohsenin, 1984; Lewis, 1990). Resistivity is a parameter that can be used to determine the quality of products, as well as the content of sugar of some fruits (Mohsenin, 1984).

The inverse of resistivity is the *electric conductivity* (σ) or specific conductance that is the capacity that a body possesses to allow passage of electric current:

$$\sigma = \frac{1}{\rho} = \frac{L}{R(\pi/4)d^2} \tag{14.4}$$

The units of the electric conductivity are Ω^{-1} m^{-1}. The inverse of the ohm (Ω^{-1}) is siemens (S) or mho, which is why in the International System, the units of electric conductivity are of siemens by meter (S/m). Electric conductivity or specific conductance must not be confused with conductance (G), which is the inverse of resistance: $G = 1/R$.

Electric conductivity in liquid materials is closely bound to the presence of dissolved salts that leads to solutions that contain ions (anions and cations) that are able to conduct electricity. These types of substances are called electrolytes. Food solutions that contain ionic particles with charge possess this quality of conducting electricity, so as the quantity of dissolved ions increases, the conductivity of the solution increases. Electric conductivity is a property that is used to determine the quality of some foods, such as honey. Measures of conductivity in wines have been used in the process of removing tartrates that can create precipitation problems in white wines when they are stored at low temperatures. Also, in cation elimination processes such as that carried out in whey, it can be used to monitor its cation content. In fermentation processes, the acidity increase causes an increase in conductivity.

In crystalline solid materials, the valence bonds can be superimposed, which leads to a cloud of free electrons. When subjecting this type of solids to an electric field, the free electrons of the cloud are those that cause the electric current. These types of materials are called metallic electric conductors.

14.3 ELECTRIC ENERGY

When an electric potential difference is applied to a conductive material, power takes place in the form of electric energy that passes through the material, but it involves a certain quantity of energy that dissipates in the form of heat. The energy flow that takes place can be evaluated by the following equation:

$$\dot{Q} = \frac{E}{t} = VI \tag{14.5a}$$

where

$\dot{Q}$ is the generated heat flow rate, expressed in watts (W)
E is the generated electric power in joules (J)
V is the applied difference of electric potential in volts (V)
I is the current intensity in amperes (A)
t is the time in seconds (s)

It can be observed that the generated heat flow is a measurement of the developed electric power.

To measure the quantity of consumed electric power, the kilowatt-hour is used (kWh), which is the product of the power and time. The equivalence in energy units is

$$1\ \text{kWh} = \left(10^3\ \text{J/s}\right)\left(3600\ \text{s}\right) = 3.6 \times 10^6\ \text{J} = 3.6\ \text{MJ}$$

The flow of heat generated can also be obtained when introducing Equation 14.5a and the expression of Ohm's law (Equation 14.2):

$$\dot{Q} = \frac{V^2}{R} = I^2 R \tag{14.5b}$$

14.4 ALTERNATING CURRENT

In general, for industrial and domestic purposes, alternating current is usually used. Alternating current is described as a sinusoidal wave that can be generated when a coil rotates inside a magnetic field. As it is a sinusoidal wave, the voltage intensity acquires some maximum values that are presented only in a short period of time in practice, and that is why it is more convenient to use the effective currents or effective voltages whose values are as a function of the maxima:

$$V = \frac{V_{max}}{\sqrt{2}} \tag{14.6}$$

$$I = \frac{I_{max}}{\sqrt{2}} \tag{14.7}$$

The alternating current is measured in cycles per second denominated in hertz (Hz). In the European Union, the frequency of the alternating current is 50 Hz, while in the United States, it is 60 Hz.

In Figure 14.2 a typical circuit of alternating current is shown, in which in addition to the source of current generation, other elements appear such as a resistance, a capacitor, and a coil. In this type of circuits in series, the electric current that crosses each of the components is the same, but the voltage is different. The resistance of the circuit behaves the same way the continuous current case, where the current and the voltage are in phase. However, the presence of capacitors and coils causes the current and the voltage to be out of phase.

When an electric current passes through a coil, a magnetic field is created that induces an opposing current, and there is a resistance to the current flow, which is called inductive reactance. The property that originates this effect is known as autoinductance (L). In the coils or inductors, this is defined through the equation

$$V_L = L\frac{\Delta I}{\Delta t} \tag{14.8}$$

where

V_L is the potential difference between the coil terminals
I is the current intensity
t is the time

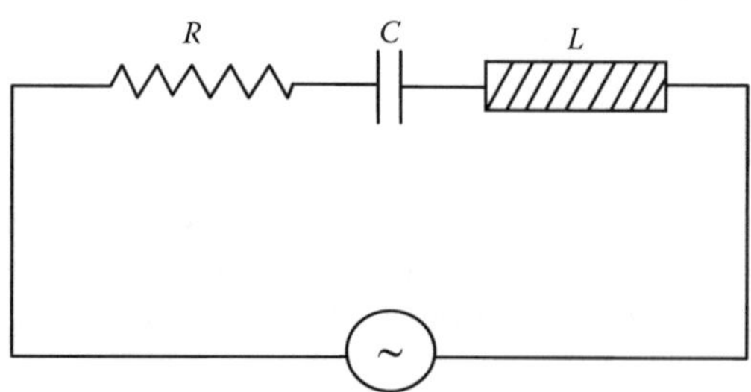

FIGURE 14.2 Alternating current circuit with a resistance, capacitor, and inductor.

In the International System, the inductance is expressed in Henry (H) that is the inductance when the induced voltage of 1 V changes to the speed of 1 A/s.

The inductive reactance (X_L) is calculated starting from the value of the coil inductance and of the frequency of the current change according to the equation

$$X_L = 2\pi f L \tag{14.9}$$

where f is the frequency, so the current intensity will be $I = V/X_L$.

The coils produce an out of phase between the maximum voltage and current, where the voltage leads the current, being the phase angle 90°.

The capacitors are devices that allow storing electric charge, and essentially, they consist of two parallel plates separated by an insulating material or dielectric. When the plates are bounded to the terminals of a battery, one of the plates acquires a positive charge, while the other one acquires a negative charge. The variable that relates the charge (Q) that acquires the plates and the potential difference (V) applied is called capacitance (C), then

$$C = \frac{Q}{V} \tag{14.10}$$

In the International System, the unit of capacity is the farad (F) that is defined as the capacitance that possesses a capacitor when it is charged with a load of 1 C and it acquires an electric potential difference of 1 V.

The capacitors allow the passage of the alternating current, but not the continuous current. Most capacitors are usually built with two metallic parallel plates separated by a material and rolled as a coil. If between the plates there is not a material but rather the space is occupied by a vacuum, the capacitance of this capacitor is a function of its geometric characteristics, so it is calculated by means of the following expression:

$$C = \varepsilon_0 \frac{A}{d} \tag{14.11}$$

where

A is the area of the plates, expressed in m^2

d is the distance between plates, expressed in m

ε_0 is the space permittivity whose value is 8.854×10^{-12} F/m

In the case that between plates there is a material, the capacitance of the capacitor is expressed as

$$C = \varepsilon' \varepsilon_0 \frac{A}{d} \tag{14.12}$$

where ε' is the relative permittivity or dielectric constant of the material. The presence of a material or dielectric between the capacitor plates allow to store a larger quantity of charge.

In electric circuits that contain capacitors, a resistance appears due to the capacitors that is called capacitive reactance (X_C) whose value can be obtained from the expression

$$X_c = \frac{1}{2\pi f C} \tag{14.13}$$

the current intensity will be $I = V/X_C$. In the capacitors, the current leads the voltage in 90°.

In circuits that contain a resistance, a coil, and a capacitor in series (Figure 14.2), the total resistance of the circuit is measured by means of the impedance (Z). Impedance is a variable that combines the effects of the resistance (R), the inductive reactance (X_L), and the capacitive reactance (X_C). The value of the impedance can be obtained from the expression

$$Z = \sqrt{\left[R^2 + \left(X_L - X_C\right)^2\right]} \tag{14.14}$$

The phase angle is obtained from the expression

$$\phi = \tan^{-1}\left(\frac{X_L - X_C}{R}\right) \tag{14.15}$$

14.5 DIELECTRIC PROPERTIES

Due to the application of microwaves and radiofrequency waves in the thermal treatments of foods, it is very important to know how different types of foods behave when they are treated with these technologies. For conventional thermal processing, the energy is transmitted due to temperature gradients; however, in microwave heating, the heating is due to the transformation of electromagnetic energy in thermal energy with direct interaction of incident radiation with the molecules. The radiation can penetrate into the material, where heat is being generated in the entire volume of the product. When the radiation impacts on the surface of a product, a part of the incident radiation can be absorbed, transmitted, or reflected. The properties that govern the proportion of each of these three categories are known as dielectric properties of the material.

Complex permittivity is defined as a fundamental electric property of the material (ε^*) and is mathematically expressed as

$$\varepsilon^* = \varepsilon' - j\varepsilon'' \tag{14.16}$$

where

ε' is the relative dielectric constant
ε'' is the loss relative dielectric factor

These two properties play an important role in the determination of the interaction between the incident microwaves and the impact of these waves in food.

The relative permittivity or dielectric constant is a macroscopic property that indicates how it can affect an electric field for a given material or how the material can affect an electric field. The value of the relative dielectric constant is the relationship of the permittivity value of the material and the space permittivity:

$$\varepsilon' = \frac{\varepsilon}{\varepsilon_0} \tag{14.17}$$

The absolute permittivity in the vacuum space (ε_0) is a function of light speed (c) and of the magnetic constant (μ_0), therefore

$$\varepsilon_0 \mu_0 c = 1 \tag{14.18}$$

The value of μ_0 is 1.26×10^{-6} H/m, while that of ε_0 is of approximately 8.854×10^{-12} F/m. The permittivity values in other materials are higher than the value corresponding to the space.

The relative loss dielectric factor ε'' is essential to evaluate the quantity of energy that will dissipate in the form of heat when the material is subjected to an electric field. The heat flow generated by unit of volume ($\tilde{q}_G$) in a given point of the food during heating waves is given by Decareau (1992):

$$\tilde{q}_G = 2\pi\varepsilon_0\varepsilon'' fE^2 \tag{14.19}$$

where

$\tilde{q}_G$ is expressed in W/(s m^3)
f is the frequency in Hz
E is the intensity of electric field in V/m

As can be observed, heat generation is directly proportional to the value of ε''.

Loss angle (δ) is defined as the relationship between the loss relative dielectric constant and the relative dielectric constant:

$$\tan\delta = \frac{\varepsilon''}{\varepsilon'} \tag{14.20}$$

The loss angle is equal to 90° minus the phase angle. The phase angle is the angle where the current intensity leads the voltage when in some circuits there is a dielectric material between the capacitor plates. If there are no dielectrics, the lead angle would be 90°.

For some dielectric materials, as the dielectric constant gets higher, it will reach the storage capacity. Therefore, this constant measures the capacity that a material has to store the electromagnetic energy received. Also, as the value of the dielectric constant becomes higher, the heat generation will grow as well.

In the microwave heating process, the absorption mechanisms of electromagnetic energy in the material are due mainly to two basic mechanisms: ionic interaction and bipolar rotation. In the case of foods that contain ions, the presence of an electric field causes migration of the present ions, so there will be motion in contrary directions, depending on the sign of the ion charge; this motion causes friction between the ions that give rise to frictional heating. On the other hand, molecules that possess separate opposite charges are known as polar molecules or dipoles. The most outstanding case is that of the molecule of water that possesses a distributed charge in an asymmetric way. When this type of molecule is under the effect of an electric field, it spreads to rotate and join with the electric field. In this alignment process, there are friction forces among the dipoles that give place to more heating.

Water is one of the main components of foods, although its content in different foods varies greatly. Therefore, dielectric properties should differ depending not only on the type of food but also on their water content. Values of the dielectric constant depend on the incident wavelength, and in the literature, values of the most used frequencies can be found for microwave treatments that are 915 and 2450 MHz (Venkatesh and Raghavan, 2004). Relative dielectric constants for different foods are presented in Table 14.1, for a temperature of 25°C (Sahin and Sumnu, 2006). In this table, it is observed that the smallest dielectric constant values correspond to oil, and this is due to oils' nonpolar characteristics. On the contrary, the biggest value in the dielectric constant belongs to the water, and products with high water content also have high values. For the loss dielectric constant, it can be observed that ham is the product that presents the biggest value, due to the presence of salt that promotes strong ionic interactions.

Water content is one of the characteristics that greatly influence the properties of dielectrics, due to the polarity of the water molecule that absorbs the energy of the waves through the

TABLE 14.1
Dielectric Constant Values for Different Foods

	ε'	ε''
Distilled water	77	10
5% salt solution	76	23
Carrot	73	20
Pear	70	15
Potato purée	69	19
Banana	66	18
Corn	60	16
Nonfat cheese	60	22
Beef	59	17
Poultry	58	17
Ham	57	33
Potato	50	15
Garlic	46	18
Dough bread	21	10
Cooked noodles	18	4.5
Bread	3.5	0.5
Oil	2	0.01

mechanism of dipolar rotation. For free water, dielectric properties can be predicted by means of Debye's models for polar solvents (Mudgett, 1986; Sahin and Sumnu, 2006):

$$\varepsilon' = \frac{(\varepsilon_S - \varepsilon_\infty)}{1+(\lambda_S/\lambda)^2} + \varepsilon_\infty \tag{14.21}$$

$$\varepsilon'' = \frac{(\varepsilon_S - \varepsilon_\infty)(\lambda_S/\lambda)}{1+(\lambda_S/\lambda)^2} \tag{14.22}$$

where
λ is the wavelength in the water
λ_S is the critical wavelength of the polar solvent
ε_S is the dielectric static constant (at low frequency)
ε_∞ is the dielectric optic constant (at high frequency)

These same equations can be used for the calculation of the dielectric properties for a polar solvent, although expressed in function of the frequency and of the denominated relaxation time (τ):

$$\varepsilon' = \frac{(\varepsilon_S - \varepsilon_\infty)}{1+(\omega\tau)^2} + \varepsilon_\infty \tag{14.23}$$

$$\varepsilon'' = \frac{(\varepsilon_S - \varepsilon_\infty)(\omega\tau)}{1+(\omega\tau)^2} \tag{14.24}$$

TABLE 14.2
Debye's Equation Parameters for Water

T (°C)	ε_S	ε_{00}	λ_S (cm)	Relaxation Time (10^{-12} s)	f_S (GHz)
0	87.9	5.7	3.3	17.67	9.007
10	83.9	5.5	2.4	12.68	12.552
20	80.2	5.6	1.8	9.36	17.004
30	76.6	5.2	1.4	7.28	21.862
40	73.2	3.9	1.1	5.82	27.346
50	69.9	4.0	0.9	4.75	33.506
60	66.7	4.2	0.8	4.01	39.69

Sources: Mudgett, R.E., *Food Technol.*, 40, 3, 1986; Venkatesh and Raghavan, 2004.

where τ is the relaxation time that is defined as the necessary time to obtain an alignment or maximum ordination state so that an irradiated material decreases at $1/e$ of their initial condition once the irradiation is suddenly removed. In this way, the relaxation time is expressed mathematically by means of the equation

$$\tau = \frac{1}{2\pi f_S}$$

where f_S is the relaxation frequency. The values of the parameters of these equations depend on the temperature; in Table 14.2, values of the different parameters are shown when the solvent is water (Venkatesh and Raghavan, 2004).

The form in which the water is inside the food matrix affects the values of the loss dielectric constant. Water in bounded form gives a value of this constant that varies with the water content of the food. However, if a critical value of the humidity is surpassed, the value of the loss dielectric constant increases notably with the increase of the humidity content of the product. When the water content of a food is low, the variation of dielectric properties with the humidity is also small; however, an increase in the water content increases the dielectric constant as well as the loss dielectric constant. In foods having a high water content, bound water barely affects their dielectric properties. Instead, these properties are primarily determined by constituents and overall water content. In the drying process, the values of the dielectric properties decrease as the humidity of the food decreases along the process.

Temperature is a variable that exerts great influence on the value of dielectric properties. In the case where water is in a bound form, an increase of temperature causes the dielectric properties values to increase; however, if the water is not bound, an increase of the temperature decreases the value of the dielectric properties. This can be observed in thawing processes. For a frozen food, at the beginning of the thawing process, the water is in solid form; therefore, when the temperature is increased, its dielectric properties will increase. At melting temperature, ice becomes liquid water, and if the temperature increases, it can be observed that the dielectric constant, as well as the loss, decreases with the increase of temperature. However, in those products that contains salts, as in the case of ham, it has been observed that once the melting temperature of ice is overcome, a temperature increase involves an increase of the dielectric properties (Bengtsson and Risman, 1971; Sahin and Sumnu, 2006). This is due to the fact that in this type of products, the two mechanisms of heat generation, the dipolar effect and the ionic effect, are conjugated. An increase of temperature causes the dielectric properties of the dipolar substances to decrease, while in the ionic substances, these properties increase with an increase of the temperature. With the loss of dielectric constant, at low temperatures, the dipolar effect is greater than the ionic effect, while at high temperatures, the ionic effect is greater than the dipolar effect; then, initially, the net value of this property experiences a decrease with the increase of temperature, to a minimum, from which starts to increase if the temperature keeps increasing.

In addition to salts, there are hydrocarbons in food, fat, and proteins that can affect the dielectric properties of food. In food systems, the main hydrocarbons are sugars and starches. For sugars, an increase in their content involves a decrease of the dielectric constant, because the water content is lower. However, for the loss dielectric constant, there is a critical concentration in glucose that affects this constant. When the temperature is greater than 40°C, an increase of sugar content causes an increase in the loss factor value; however, when the temperature is lower, the loss factor decreases with the concentration increase, because the sugar solution becomes saturated to a lower concentration. For starch, its influence in the dielectric properties depends on its solid state or its form of aqueous suspension. For powdered starch, the values of these constants increase with temperature, and they depend on the apparent density that they possess. Thus, the lower the apparent density of starch grains becomes, the lower is the loss dielectric constant value (Ndife et al., 1998). For starch suspensions, the dielectric properties decrease when temperature and concentration increase (Ndife et al., 1998; Sahin and Sumnu, 2004).

When increasing starch concentration, water percentage decreases, and this causes a decrease of the dielectric and loss constant. A phenomenon that could affect the values of the dielectric properties is the gelatinization of some food components such as starch. Thus, the dielectric constant for non-gelled starch is lower than in the case of gelled starch. This can be explained by the fact that the non-gelled starch possesses a greater quantity of bound water, which causes the dielectric constant value to be lower (Roebuck et al., 1972).

15 Physical and Chemical Properties of Food Powders

15.1 INTRODUCTION

Food powders represent a large fraction of the many food products available in the food industry, ranging from raw materials and ingredients, such as flours and spices, to processed products like instant coffee or powdered milk. Food powders can be distinguished not only by their composition and microstructure but also by particle size, size distribution, chemical and physical properties, and functionality. Historically, a number of unit operations have been developed and adopted for the production and handling of different food powders. Information on the physical properties, production, and functionality of food powders has been published, mainly through research and review articles, reports in trade magazines, and symposium presentation. Very recently, a few books dealing specifically with food powders have been published bringing additional attention to a subject that, in general, is not well covered in academic programs.

In powder technology, great attention has been paid to physical and chemical properties of powders. These properties are applicable in pharmaceutics, ceramic powders, metallurgy, detergent powders, and civil engineering, as well as in the food powder field and used by researchers to study the mechanisms of particle interactions and the evaluation of particles at a bulk level. Food powders participate in several food processing operations from raw material utilization, handling, and processing operations (e.g., blending, solubilization, fluidization, conveying) to storage operations (bulk disposal, packaging).

A powder is a complex form of solid material made up of a very large number of individuals, each different from its neighbor. The term "powders" will refer to food particulates in the size range of roughly 50–1000 μm (e.g., flour, spices, sugar, soup mixes, and instant beverages). Food powder properties can be classified into primary properties, that is, individual or inherent properties (particle density, particle porosity, shape, diameter, surface properties, hardness, or stickiness properties) and secondary or bulk properties (bulk density, bulk porosity, particle size distribution, moisture content). Secondary properties provide quantitative knowledge of a powder as a bulk.

Another classification of food powder properties can be into physical or chemical properties. Physical properties include particle shape, particle density and porosity, particle surface characteristics, hardness, particle diameter, and size. On the other hand, the term "chemical" in food materials is related with the food composition and its interaction with other substances like solvents or components within the food structure. In particular, instant properties and stickiness are composition related and characterize processes like dissolution and caking of powders, commonly found in different processing operations such as agglomeration, drying, mixing, and storing. Stickiness can be characterized with cohesion and cohesion properties such as tensile strength, angle of repose, and angle of internal friction that will be described.

This chapter covers definitions and determination methods of physical and chemical properties of food powders. Moreover, different applications of these properties and processing parameters will be discussed as well as recent developments in the field. It is the case to mention that many of the basic concepts presented are essential to understand the behavior of a particulate or particulate system under a given set of conditions. This basic understanding becomes essential at the time to get involved in food product, equipment and plant design, as well as in the identification of optimal processing conditions.

15.2 PHYSICAL PROPERTIES

The physical properties of powders, including those of food powders, are interdependent. A modification of particle size distribution or moisture content can result in simultaneous change in bulk density, flowability, and appearance. Physical properties provide a means to quantify the output particulate product of a certain drying, grinding, or size enlargement process and compare effects of composition, structural conformation, material origin, type of equipment used, or handling effects.

15.2.1 Density and Porosity

Density (ρ) is defined as the unit mass per unit volume measured in kg/m^3 in SI units and is of fundamental use for material property studies and industrial processes to adjust storage, processing, packaging, and distribution conditions. Density is given by the particle mass over its volume, and mass can be accurately determined even to a microgram with precise analytical balances. However, the way particle volume is measured will depend on the determination methods and its application and will define the type of density to be considered. There are three main types of densities: true density, apparent or particle density, and bulk density. Particularly, bulk density is one of the properties used as part of the specifications for a certain final product obtained from grinding or drying. Since other authors have used different names for the same condition, it is recommended that the definition of density be verified before using density data.

15.2.1.1 Particle True Density

True (or substance) density (ρ_s) represents the mass of the particle divided by its volume excluding all open and closed pores (Figure 15.1), that is, the density of the solid material of which the particle is made. Some metallic powders can present true densities of the order of 7000 kg/m^3, while most food particles have considerably lower true densities of about 1000–1500 kg/m^3. Most particles have similar true density (Table 15.1) due to similarity in its main biological, organic, and inorganic components.

15.2.1.2 Particle or Envelope Density

Apparent particle density (ρ_p) is the mass over the volume of a sample that has not been structurally modified. Volume includes internal pores not externally connected to the surrounding atmosphere

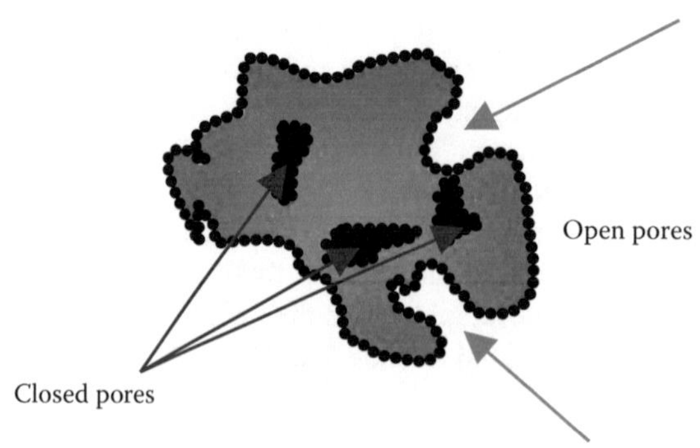

FIGURE 15.1 Volume definition depends on whether or not open and/or closed pores are accounted for.

TABLE 15.1
Densities of Common Food Powders

Powder	Particle/True Density, ρ_s (kg/m^3)	Bulk Density, ρ_b (kg/m^3)
Baby formula	—	400
Cellulose[a]	1270–1610	—
Citric acid[a]	1540	—
Cocoa powder, 10% fat	1450	350–400
Cocoa powder, 22% fat	1420	400–550
Coffee (ground and roasted)	—	310–400
Coffee (instant)	—	200–470
Coffee creamer	—	660
Corn meal	—	560
Egg (whole)	—	680
Fat[a]	900–950	—
Flour (corn)	1540	500–700
Flour (rye)	1450	450–700
Gelatin (ground)	—	680
Glucose[a]	1560	—
Microcrystalline cellulose	—	610
Instant dried whole milk	1300–1450	430–550
Instant dried skim milk	1200–1400	250–550
Oatmeal	—	510
Onion (powdered)	—	960
Protein (globular)[a]	1400	—
Rice, polished	1370–1390	700–800
Salt (granulated)	2160	950
Salt (powdered)	2160	280
Soy protein (precipitated)	—	800
Starch (corn)[a]	1500–1620	340–550
Starch (potato)[a]	1500–1650	650
Sugar (granulated)[a]	1590–1600	850–1050
Sugar (powdered)[a]	1590–1600	480
Wheat flour	1450–1490	550–650
Wheat (whole)	—	560
Whey	—	520
Yeast (active dry baker's)	—	820

Sources: Rao, 1992; Schubert, H., *J. Food Eng.*, 6, 22, 1987a; Schubert, H., *J. Food Eng.*, 6, 83, 1987b; Peleg, M., Physical characteristics of powders, in *Physical Properties of Foods*, M. Peleg and E.B. Bagley, Eds., Van Nostrand Reinhold/AVI, New York, 1983, pp. 293–324.

[a] Main components forming a particle together with water (1000 kg/m^3) and in some cases acting as a binding agent.

and excludes only the open pores. It is generally measured by gas or liquid displacement methods such as liquid or air pycnometry.

15.2.1.3 Bulk Density

The *bulk density* (ρ_b) of powders is determined by particle density, which in turn is determined by solid density and particle internal porosity, also by special arrangement of the particles in the container. Bulk density includes the volume of the solid and liquid materials and all pores closed or open to the surrounding atmosphere. Four different types of bulk density can be distinguished depending on their volume determination method:

- *Compact density* is determined after compressing the powder's bulk mass by mechanical pressure, vibration, and impact(s).
- *Tap density* results from a volume of powders after being tapped or vibrated under specific conditions and is the most useful to follow powder behavior during compaction.
- *Loose bulk density* is measured after a powder is freely poured into a container.
- *Aerated bulk density* is used for testing under fluidized conditions or pneumatic conveying applications when particles are separated from each other by a film of air, not being in direct contact with each other. In practice, it is the bulk density after the powder has been aerated, that is, the most loosely packed bulk density.

Table 15.1 lists typical densities for some food powders. It can be observed that inorganic salt presents a notably high particle density, while fat-rich powders present low densities. In general, bulk densities range 300–800 kg/m^3.

15.2.1.4 Particle Porosity

The volume fraction of air (or void space) over the total bed volume is indicated by porosity (ε) or voidage of the powder. Considering air density as ρ_a, bulk density can be expressed as a function of the solid density and porosity:

$$\rho_b = \rho_s(1-\varepsilon)+\rho_a\varepsilon \tag{15.1}$$

Air density is negligible with respect to the powder density. Thus, porosity expression can be calculated directly from the bulk and particle density of a given volume of powder mass:

$$\varepsilon = \frac{(\rho_s - \rho_b)}{\rho_s} \tag{15.2}$$

In general, food powders present a high porosity between 40% and 80% from internal, external, and interparticle pores. Bulk porosity can vary considerably due to mechanical compaction, difference in particle sizes (concentration of fines), moisture, and temperature. Furthermore, the porosity can be affected by the chemical nature of each constituent powder as well as from the process from which particles were originated. Changes in environmentally and time-dependant parameters such as moisture and temperature during storage can vary the interaction between particles therefore affecting its density, porosity, as well as the number of interparticle contact points. These changes are mostly due to volume reduction for increased adhesiveness and, in a lower extent, due to mass variations due to water sorption or evaporation or even due to phase changes with temperature (e.g., in fatty components).

15.2.1.5 Density Determination

A wide range of methodologies have been developed with different degrees of accuracy for particle and bulk density determination. Apparent particle density can be determined from fluid displacement methods or "pycnometry," using either a liquid or a gas (i.e., liquid pycnometry and air pycnometry).

15.2.1.5.1 Liquid Displacement Methods

For fine powders, a pycnometer bottle of 50 mL volume is normally employed (Figure 15.2), while coarse materials may require larger calibrated containers. The liquid should be a special solvent

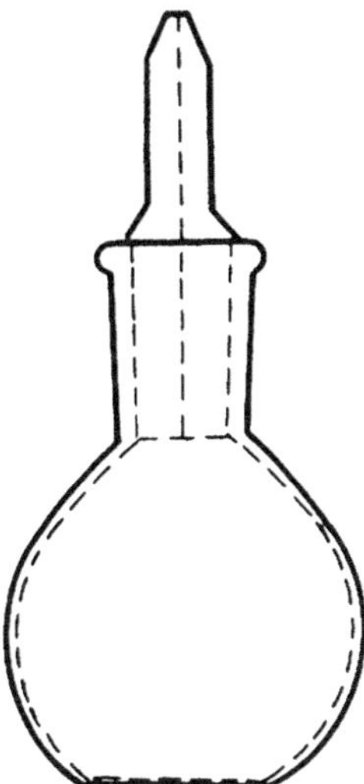

FIGURE 15.2 Conventional liquid pycnometer.

that does not dissolve, react, or penetrate the particulate food solid. The particle density ρ_s can be calculated from the net weight of dry powder divided by the net volume of the powder:

$$\rho_s = \frac{(m_s - m_0)\rho}{(m_l - m_0) - (m_{sl} - m_s)} \tag{15.3}$$

where

m_s is the weight of the bottle filled with the powder
m_0 is the weight of the empty bottle
ρ is the density of the liquid
m_l is the weight of the bottle filled with the liquid
m_{sl} is the weight of the bottle filled with both the solid and the liquid

A top-loading platform scale (Figure 15.3) is another liquid displacement method, which is lower in accuracy, and can be used for compressed or aggregated bulk powders of considerable size.

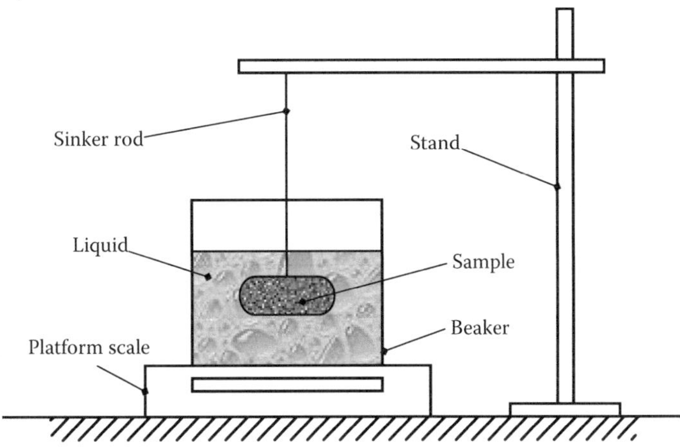

FIGURE 15.3 Top-loading platform scale for density determination of irregularly shaped objects.

This test is based on immersing the solid mass of powder and suspending it at the same time in a liquid contained in a beaker, which is on top of a scale. The volume of the solid V_s is equivalent to the mass of liquid displaced by its surface (Ma et al., 1997) and can be calculated by

$$V_s = \frac{m_{LCS} - m_{LC}}{\rho_L} \tag{15.4}$$

where

m_{LCS} is the weight of the container with liquid and submerged solid
m_{LC} is the weight of the container partially filled with liquid
ρ_L is the density of the liquid

15.2.1.5.2 Air Displacement Methods

Automatic pycnometers can calculate apparent density of particulate materials, either by having fixed-volume sample chambers of different sizes or by means of volume-filling inserts placed into the chamber. The system consists of two chambers, a pressure-measuring transducer, and regulating valves for atmospheric gas removal and replacement with helium gas. Sample volume is calculated from the observed pressure change that a gas undergoes when it expands from one chamber containing the sample into another chamber without the sample. Reproducibility of apparent density results is typically of ±0.01% when the sample volume fills the sample holder. This equipment can be applicable to several food powders, from coffee creamer to black pepper (Webb and Orr, 1997).

For instance, a special instrument based on two cylinders with a piston each (Figure 15.4), one used as reference and the other containing the powder sample, measures sample volume accurately. A differential pressure indicator is used to verify that both cylinders hold the same pressure before and after the test. The extra volume occupied by the solid sample is indicated by the displacement of the piston in the cylinder containing the sample, which is attached to a scale. The equipment can be calibrated in order to be read directly in cubic centimeters, usually with a digital counter.

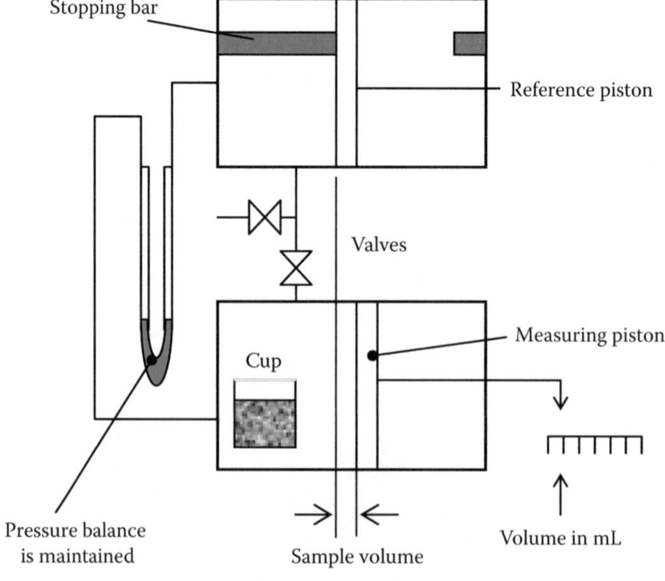

FIGURE 15.4 Air pycnometer. The reference piston reached the stop and the measuring piston is displaced a distance equivalent to the volume occupied by the sample.

After the test, the sample volume is divided into sample weight to give the apparent density. Volume occupied by open pores can be calculated, depending upon sample hydrophobic properties, by filling them either by wax impregnation or by adding water and then subtracting the envelope volume to the particle volume. By neglecting closed pores within each particle in the sample, this volume difference can give a measure of particle porosity.

The bed pressure method is based on passing gas through the powder bed in a laminar flow regime and measuring two different pressure drops. The Carman–Kozeny equation can be used to determine the particle density, ρ_p is the only unknown variable, which can be readily found by trial and error:

$$\frac{s_1}{s_2} = \left(\frac{\rho_{b1}}{\rho_{b2}}\right)\left[\frac{(\rho_p - \rho_{b2})}{(\rho_p - \rho_{b1})}\right]^3 \tag{15.5}$$

where

conditions 1 and 2 correspond to each set of measurements
s is the gradient of pressure drop
ρ_b is the powder bulk density

15.2.1.5.3 Porosity Measurements

Volume of the open pores could be determined with a mercury porosimeter. However, this is only suitable for coarse solids and requires costly equipment. The bed voidage method and the sand displacement method can be used alternatively. The bed voidage method (Abrahamsen and Geldart, 1980) consists of filling voids of a powder bed with a much finer powder. The fine powder is of known particle size and is used as control. The drawback of this method is to find control powders of the same shape that can yield the same voidage as the tested powder. The sand displacement method uses fine sand into which a known amount of coarse particles of the sample are mixed. The density of the sample is determined from the difference of the bulk density of the sand alone and that of the mixture. This method is sometimes used for density determinations of coarse bone particles, and it gives lower density than that of the solid bone as measured by pycnometry.

15.2.1.5.4 Measurements of Bulk Density

The aerated bulk density is measured by pouring powder through a vibrating sieve into a cup (Figure 15.5). A chute is attached to a mains-operated vibrator of variable amplitude and a stationary chute aligned with the center of a preweighed 100 mL cup. Vibration amplitude and sieve aperture size should be such that the time taken for the powder to fill the 100 mL cup is not less than 30 s. The Hosokawa powder tester incorporates automated tasks with increased reproducibility for quality control purposes. This instrument can also be used for tap bulk density determination by packing the powder by tapping, jolting, or vibrating the measuring vessel. After tapping, excess powder is scraped from the rim of the cup and the bulk density is determined by weighing the cup.

Another method for tap density determination is the tap density tester (or the Copley volumeter), which consists of a graduated cylinder and tapping mechanism (Figure 15.6). The advantage of the Copley over the Hosokawa tester is that a fixed mass of powder is used and there is no need to remove the container for weighing as the number of taps is progressively increased (Abdullah and Geldart, 1999). The tapping action is provided by a rotating cam driven by a 60 W motor with an output speed of 250 rpm. The tap density tester is described by norms from American Society for Testing and Materials Standards (ASTM) and gives standardized and repeatable results for measuring tapped or packed volumes of powders and granulated or flaked materials. These can have digital LED displays and user-selectable counter or timer operations, including dual nonrotating platform drive units and two graduated funnel-top cylinders, which generally are of 100 and 250 mL (Figure 15.6).

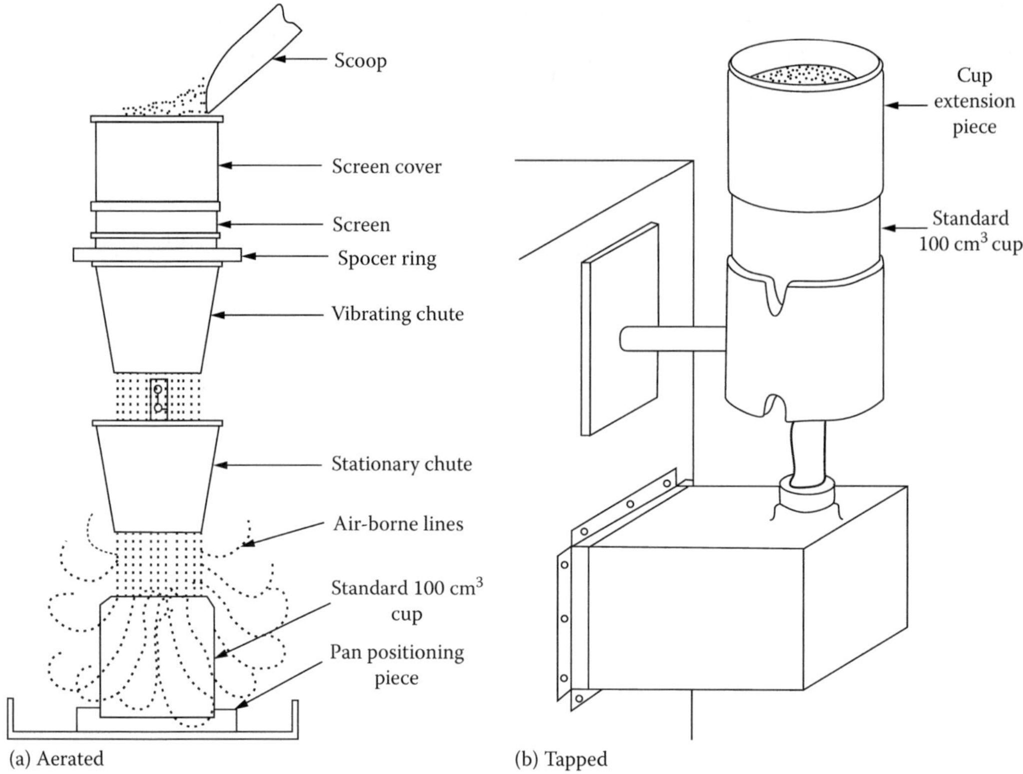

FIGURE 15.5 The Hosokawa powder tester for measuring the aerated and tapped bulk densities. (From Abdullah, E.C. and Geldart, D., *Powder Technol.*, 102, 151, 1999.)

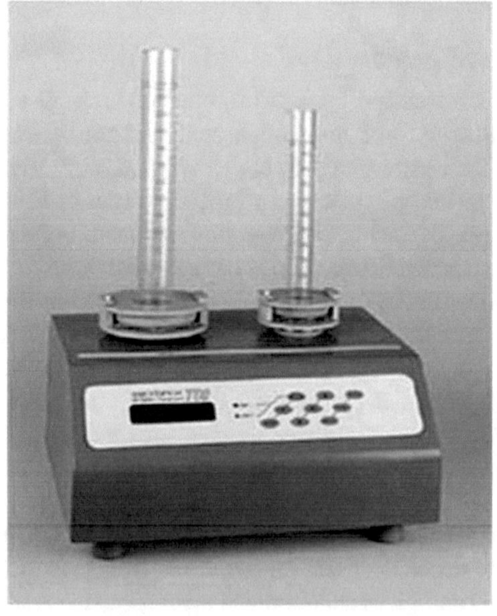

FIGURE 15.6 Tap density tester described by ASTM.

15.2.1.6 Ultimate Bulk Density

Yan et al. (2001) proposed the concept of "ultimate bulk density" while studying density changes in agglomerated food powders under high hydrostatic pressure. When the hydrostatic pressure was higher than a critical value (around 200 MPa for the powders used), the agglomerated food powders were compressed so densely that all of the agglomerates and primary particles were crushed and compressed together, leaving almost no open or closed pores. Because the final compressed bulk density is usually higher than the commonly used apparent "solid density," the bulk porosity (due to remaining pores under high pressure) will be a negative value without physical meaning. It is believed that the "ultimate bulk density" could be dependent on the product formulation, physical properties of product ingredients, and production conditions.

15.2.1.7 Hausner Ratio

Hausner ratio can be defined as the ratio of the tap bulk density to the loose bulk density. Hausner ratio can be used for fluidization studies (Geldart et al., 1984), to evaluate density changes with particle shape (Kostelnik and Beddow, 1970), to understand the influence of relative humidity upon process operations. Malavé-López et al. (1985) defined the ratio of asymptotic over initial bulk density by the relationship

$$\text{HR} = \frac{\rho_\infty}{\rho_o} \tag{15.6}$$

where

HR is the Hausner ratio
ρ_∞ is the asymptotic constant density after certain amounts of taps
ρ_o is the initial bulk density

A more practical equation widely used to evaluate flow properties can be given by the following equation, which calculates powder volume changes in a graduated cylinder after certain period of time or number of taps (Hayes, 1987):

$$\text{HR} = \frac{\rho_n}{\rho_o} = \frac{V_o}{V_n} \tag{15.7}$$

where

n is the number of taps provided to the sample
ρ_n and ρ_o are the tapped and loose bulk density
V_o and V_n are the loose and tapped volume, respectively

15.2.2 Particle Shape

Particle shape influences such properties as flowability of powder, packing, interaction with fluids, and covering powder of pigments and coatings. However, few quantitative works has been carried out on these relationships. There are two points of views regarding the assessment of particle shape. One is that the actual shape is unimportant and all that is required is a number for comparison purposes. The other is that it should be possible to regenerate the original particle shape from measurement data.

15.2.2.1 Shape Terms and Coefficients

Different porous structures yield different particle shapes as listed in Table 15.2. Qualitative terms may be used to give some indication of the particle shape and used for particle classification. The earliest method for describing the shape of particle outlines used length L, breadth or width B, and thickness T in expressions such as the *elongation* ratio (L/B) and the *flakiness* ratio (B/T) (Figure 15.7). However, one-number shape measurements could be ambiguous and the same single number could be obtained from more than one shape. A better definition is the term sphericity, Φ_s, defined by the relation

$$\Phi_s = \frac{6V_p}{x_p s_p} = \frac{6\alpha_V}{\alpha_s} \tag{15.8}$$

where

x_p is the equivalent particle diameter
s_p is the surface area of the particle
V_p is the particle volume
α_V and α_s are the volume and surface factors, respectively

TABLE 15.2
General Classification of Particle Shape

Shape Name	Shape Description
Acicular	Needle shape
Angular	Roughly polyhedral shape
Crystalline	Freely developed geometric shape in a fluid medium
Dentritic	Branched crystalline shape
Fibrous	Regular or irregular threadlike
Flaky	Platelike
Granular	Approximately equidimensional, irregular shape
Irregular	Lacking any symmetry
Modular	Rounded irregular shape
Spherical	Global shape

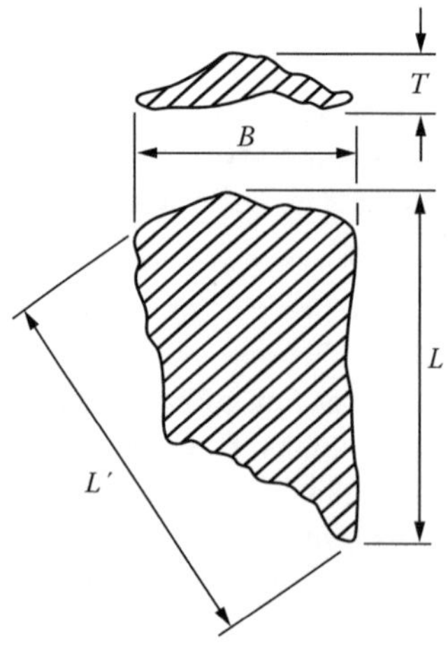

FIGURE 15.7 Heywood dimensions for flakiness and elongation ratio determination.

The numerical values of the surface factors are all dependent on the particle shape and the precise definition of the diameter (Parfitt and Sing, 1976). For spherical particles, Φ_s equals unity, while for many crushed materials, its value lies between 0.6 and 0.7.

15.2.2.2 Morphology Studies

Characterization of particle silhouettes for shape characterization can be done using polar coordinates of their peripheries using the center of gravity of the figure as origin (Figure 15.8). A truncated harmonic series represents the value of the radius vector R as it is rotated about the origin as a function of the angle of rotation θ:

$$R(\theta) = A_0 + \sum_{n=1}^{M} A_n \cos(n\theta - \phi_n) \tag{15.9}$$

where

ϕ_n is the phase angle of the nth harmonic

A_n is the amplitude

However, for fine detail and protuberances, a large number of terms must be used. An alternative approach is to represent any closed curve as a function of arc length by the accumulated change in direction of the curve. The outline is essentially described by taking a tangent around the shape and noting the change in the contour as a function of the angle θ. Computational programs have been designed to iterate and reproduce the equivalent harmonic closest to its shape (Jones, 1983). The surface of a particle in terms of roughness can be completed in detail with optical instruments for stereomicroscopy, scanning electron microscopy, or confocal scanning optical microscopy.

A mathematical description for modeling the silhouette of projected particle surface is based on boundary-line analysis by fractal mathematics (Mandelbrot, 1977, 1982). Fractal approach is applied to analyze 2D projections of particles by redefining its contour. A given step length can provide a close polygonal contour at different points of the particle projection that geometrically recreates a self-similar particle silhouette. The procedure is based on estimating the silhouette perimeter by adding the length of all the steps plus the exact length of the last side of the polygon (if different from the step length). The measurement of the perimeter of the silhouette is repeated by using several different step lengths and it gives different perimeters in each case. When measuring lengths of

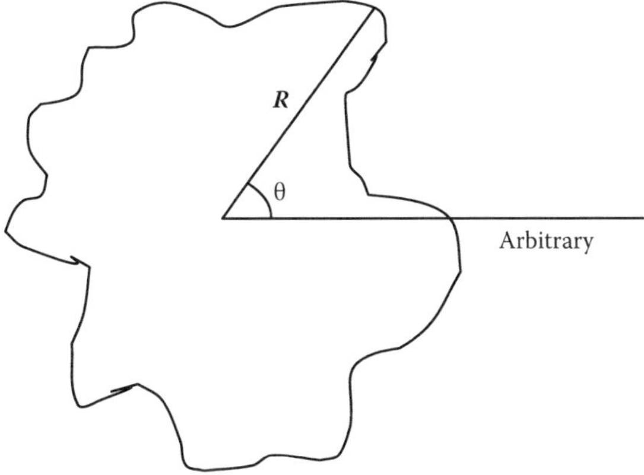

FIGURE 15.8 Representation of polar coordinates in an irregular silhouette.

irregular contours by stepping along them with a pair of dividers, as the step length s decreases, the estimated length L_s keeps increasing without limit. In such a case, L_s is related to s by

$$L_s = K(s)^{1-D} \tag{15.10}$$

where D is known as the fractal dimension, being different for jagged and for smoothly rounded contours. Fractal dimension provides useful information about particle shape, openness, and ruggedness in the form of single numerical numbers (Simons, 1996). Fractal characterization can be used as a tool to describe the eroding process in particulate materials during compaction (Barletta and Barbosa-Cánovas, 1993a). In fact, fractal dimension has been used as a sensitive attrition index, based on the fact that attrition causes changes in particle shape and surface, on the scale that the fractal approach is applicable (Peleg and Normand, 1985; Olivares-Francisco and Barbosa-Cánovas, 1990).

15.2.3 Strength Properties

There are a number of properties of particulate materials that determine particle breakage and resistance strength during mixing and handling operations, and great attention has been paid to the general behavior of powders under compression stress. Compression and compaction tests have been widely used in pharmaceutics, ceramic powders, metallurgy, and civil engineering, as well as in the food powder field and used by researchers to study of the mechanisms of particle interactions and the evaluation of particles at a bulk level. Many solid food materials, especially when dry, are brittle and fragile, showing a tendency to break down or disintegrate. Mechanical attrition of food powders usually occurs during handling or processing, when the particles are subjected to impact and frictional forces. Attrition represents a serious problem in most of the food processes where dry handling is involved, since it may cause undesirable results such as dust formation, health hazard, equipment damage, and material loss. Dust formation may be considered the worst of these aspects, as it may develop into a dust explosion hazard.

15.2.3.1 Hardness and Abrasiveness

The hardness of powders or granules is the degree of resistance of the surface of a particle to penetration by another body. It is related to the yield stress by considering the characteristics of the uniaxial stress–strain curve for several types of material failure (i.e., transition between elastic and plastic strains). Hardness can be determined as a qualitative property by using the Mohs hardness scale (Carr, 1976). In this scale, 10 selected minerals are listed in order of increasing hardness, by indicating qualitative resistance to plastic flow, so that a material of a given Mohs' number cannot scratch any substance of a higher number, but will scratch those of lower numbers.

A particle property related to hardness is the crushing strength (ASTM, 1986), which refers to the force required to crush a mass of dry powder or, conversely, the resistance of a mass of dry powder to withstand collapse from external compressive load. Bulk crushing strength can be evaluated by measuring either the amount of fines produced after compression of a fixed volume of particles at a predetermined pressure or the pressure required for producing a predetermined amount of fines.

Abrasiveness of bulk solids, that is, their ability to abrade or wear surfaces with which they come into contact, is considered a property closely related to the hardness of the material. The abrasiveness of food powders can be rated from the relative hardness of the particles and the surface with which they are in contact, using the Mohs hardness scale. From Mohs' hardness scale, materials can be generally rated as soft, medium hard, or hard, when they show values between 1–3, 3.5–5, and 5–10, respectively. The best way to assess abrasiveness is to use the actual bulk material and the contact surfaces in question. Abrasiveness and hardness are two major factors that govern the choice and design of different types of equipment, such as size reduction machines, air classifiers, mixers, and dryers. Many food materials are normally soft according to this criterion and, thus, the problems related to strength of materials normally faced in the food industry have to do with attrition and friability, rather than hardness and abrasion.

15.2.3.2 Friability and Attrition

Friability is the ability of breakdown of particles under impact or compressive forces. Impact tests can be used either on single particles or on a quantity of bulk solid. Impact methods measure powder's ability to resist high-rate loading. Different impact methods can be used in order to characterize powder strength, which include impacting powders with a falling mass, impact tests by ram, and pneumatic dropping of powders on a surface (Mohsenin, 1986; Hollman, 2001). Modern impact test equipments (e.g., universal testing machines [UTMs]) record the load on the specimen as a function of time and/or specimen deflection prior to fracture or particle breakage using electronic sensing instrumentation connected to a computer. Friability can be measured in terms of impact energy absorbed by the sample relative to the initial potential energy (gravitational or elastic) of the plunger. Impact tests can be used to determine food powder coating resistance.

Pneumatic dropping impact test was used for studying the influence of processing conditions (belt and pneumatic conveying among other operations) in NaCl crystals (Ghadiri et al., 1991). Furthermore, these tests can also be used for powder attrition testing by measuring the crushing strength necessary to produce a certain number of fines or, conversely, to evaluate the state of breakage (Couroyer et al., 2000).

Attrition is the deleterious particle breakdown, which increases the number of smaller particles by reducing particle size and affecting particle size distribution (Figure 15.9). Except for particle size reduction during comminution or grinding processes, attrition is undesirable in most processes. As a matter of fact, it is one of the most ubiquitous problems for a wide range of processing industries that deal with particulate solids. In food powders, it is more frequent in agglomerates, mainly because of their multiparticulate structure. Many food agglomerates possess brittle characteristics that make the product susceptible to vibrational, compressive, shear, or even convective forces applied to the particles during processing. Methods to determine extent of attrition are, namely, shear tests, vibration test, and the tumbler test.

Shear cells can be used to study attrition effects in particles under compression (Ghadiri and Ning, 1997). The direct shear cell usually consists of two compartments: one is placed on top of the other compartment, and one of them has a lid that covers the powder and acts like a piston. When a

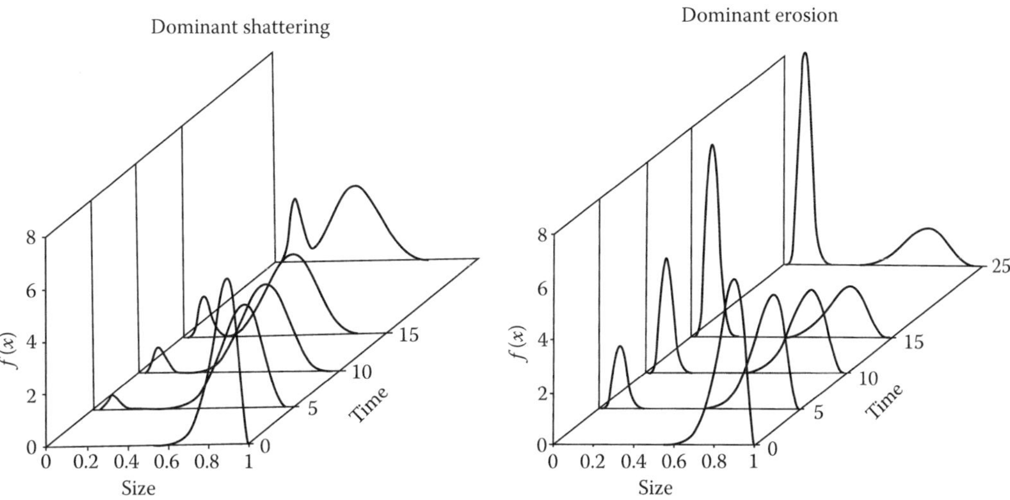

FIGURE 15.9 Changes in particle size distribution of disintegrating particulates at different tapping times. (From Popplewell, L.M. and Peleg, M., *Powder Technol.*, 58, 145, 1989.)

sample is put in the shear cell and compressed by a normal force from the lid, the base compartment can be placed in motion by a horizontal shearing force. Particle size distribution is compared before and after the experiment, and fines generated due to interlocking, frictional, and compaction forces are a measure of attrition.

Attrition of food agglomerates can be determined by using any form of resonance or a simple mechanical motion transmitted from a container to the particles in it. The tap density tester (Figure 15.6) can provide vertical vibration and measure the extent fine formation during handling and transportation. Research has been conducted for instant coffee, milk powders, and other agglomerated food powders (Malavé-López et al., 1985; Barletta et al., 1993a; Barletta and Barbosa-Cánovas, 1993b; Yan and Barbosa-Cánovas, 2001a). Tumbler test is another standard method (ASTM, 2002) used for agglomerates and determines the amount of fines formed from a presieved sample with homogeneous particle size after tumbling for a specified number of times. Fine formation is determined by separating the fines with a sieve of a specific aperture.

15.2.3.3 Compression Properties of Food Powders

The deformation mechanisms occurring during compression of fine and agglomerated foods depend on elastic and viscous flow, in addition to ductile yielding and brittle behavior that are common in pharmaceutical and food compaction processes (Barletta et al., 1993b). Some particulate materials are plastic and ductile if the stress is applied slowly but can be elastic or brittle if the stress is applied as by impact. Bulk density and particle size can vary with the rate with which a compressive stress is applied on a powder. Compressive stress and tensile stress are used to characterize cohesiveness between particles or within a certain powder cake as well as coating resistance in an encapsulated powder. A parameter that measures the change in bulk density with consolidating pressure acting on a powder bed is termed compressibility. Bulk density (in terms of apparent, compact, or tap density) and normal stress have been associated in empirical logarithmic or semilogarithmic relationships. Equation 15.11 is one of the most studied expressions in food powders, which relates the density fraction with the normal stress applied to a confined powder sample. It has been found valid up to the pressure range of 4.9 kPa with no expectation of particle yield or breakage (Peleg, 1978; Barbosa-Cánovas et al., 1987):

$$\frac{\rho_b - \rho_{b0}}{\rho_{b0}} = C_1 + C_2 \log \sigma \tag{15.11}$$

where

- ρ_b is the bulk density under compression stress σ
- ρ_{b0} is the powder's bulk density before compression
- C_1 and C_2 are characteristic constants of the powder; C_2 is known as the *compressibility index* representing the change of relative density with the applied stress

Results from different confined uniaxial compression tests have evaluated compressibility of the following food powders: fine salt, fine sucrose, cornstarch, baby formula, coffee creamer, soup mix and active baker's yeast instant agglomerated coffee, instant agglomerated low-fat (2%) milk, instant agglomerated nonfat milk, instant skim milk, ground coffee, ground corn, cornmeal, salt, sucrose, lactose, and flour (Kumar, 1973; Peleg and Mannheim, 1973; Moreyra and Peleg, 1980; Peleg, 1983; Konstance et al., 1995; Yan and Barbosa-Cánovas, 1997, 2000).

Particle size effect was evaluated in selected instant and non-instant materials (Molina et al., 1990; Yan and Barbosa-Cánovas, 1997, 2000; Onwulata et al., 1998), but no general tendency can be deducted due to differences in porosities and composition that provided different ductile or brittle behaviors and particle arrangement. On the other hand, compressibility has shown to increase with water activity in selected powders (Moreyra and Peleg, 1980) and in binary mixtures

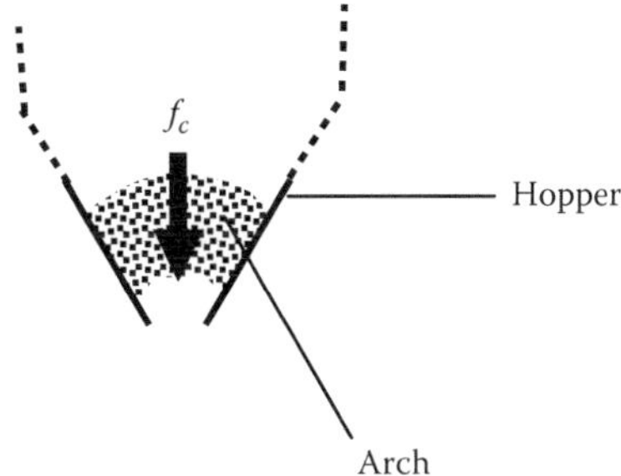

FIGURE 15.10 Unconfined yield stress f_c represented as the strength necessary to maintain a cohesive arch structure in a hopper.

(Barbosa-Cánovas et al., 1987) but tends to decrease in agglomerated powders (Yan and Barbosa-Cánovas, 2000). Other works present studies of compressibility in different cell geometries (Yan and Barbosa-Cánovas, 2000), using different anticaking agents (Hollenbach et al., 1982, Molina et al., 1990; Konstance et al., 1995; Onwulata et al., 1996) and characterizing flowability (Carr, 1965; Peleg, 1978; Ehlerrmann and Schubert, 1987). Compressibility can be used in feeder design to calculate loads that act on a feeder or gate and angle of wall friction to calculate the pressures acting perpendicular to the hopper wall. Furthermore, it can be used for quality control in order to determine the materials resistance to breakage from production to the consumer.

The unconfined yield stress f_c indicates the maximum compressive stress that a cohesive particle array is capable of sustaining at a particular porosity (Peleg, 1978; Mohsenin, 1986; Schubert, 1987a). It also represents the strength of the cohesive material at the surface of an arch (Figure 15.10), which is opposed by a lower stress induced by its own weight (Teunou et al., 1999). In flowability characterization, the unconfined yield stress refers to a situation in which the physical configuration of the system will allow the powder to flow before massive comminution of the particles occurs.

15.2.3.4 Compression Methods

Methods to determine compression behavior can be either static (dead load) or based on constantly increasing compression. Compression mechanisms can be approached from different tests such as uniaxial confined compression test, cubical triaxial tester, and the unconfined yield stress.

UTMs allow reading the maximum normal and shear forces, creep, and stress relaxation force–deformation graphs. Mechanical compressibility and breaking load under tension can be determined from stress–strain data by also using UTMs. The confined uniaxial compression test using a UTM (Figure 15.11) involves confining a bed of powder in a cylindrical cell and measuring the force applied to the flat-based piston as a function of the displacement of the piston, which is continuously

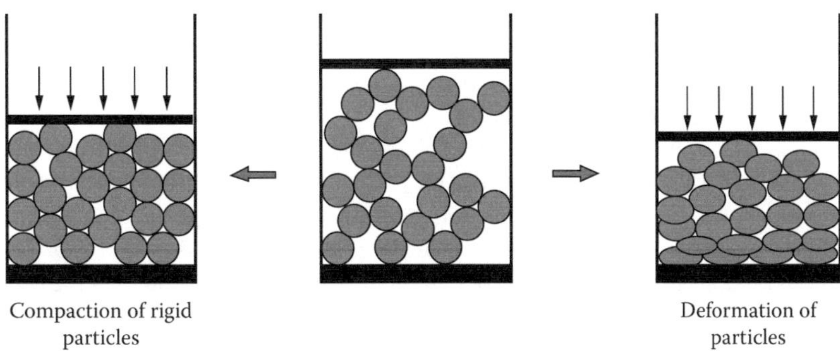

FIGURE 15.11 Confined uniaxial compression test—bed compression of rigid and deformable particles. (Adapted from Lu, W.-M. et al., *Powder Technol.*, 116, 1, 2001.)

monitored on an electronic recorder chart from a specific computer software package. Different vertical loads can be applied to a bulk solid sample of known mass and compression of the sample (Thomson, 1997).

This method can successfully evaluate particle attrition (Bemrose and Bridgwater, 1987), flowability (Peleg, 1977; Schubert, 1987a), compressibility (Malavé-López et al., 1985; Barbosa-Cánovas et al., 1987; Yan and Barbosa-Cánovas, 1997), and agglomerate strength (Adams et al., 1994). In particular, any agglomerate measurement will be affected by both the breakage properties of individual particles and the deformability of their assembly as a whole (Nuebel and Peleg, 1994).

A flexible boundary cubical triaxial test is another commonly used test for compression studies (Kamath et al., 1996; Li and Puri, 1996) and allows not only the three principal stresses to be applied independently but also volumetric deformation and deformations in three principal directions to be monitored constantly so that pressure in all three principal directions is the same. The cubical triaxial tester is useful to investigate anisotropy of cohesive and noncohesive powders and the effect of particle shape and sample deposition methods.

The unconfined yield test (Buma, 1971; Head, 1982) determines the unconfined yield stress of cohesive powder cakes. The load necessary to make the plug fail is defined as the unconfined yield stress. Unconfined yield stress values were obtained as an index of cohesion for whole milk powder and skim milk powder (Rennie et al., 1999). Dry whole milk was found to be more cohesive than skim milk with increasing temperature, indicating the influence of fat in the cohesive mechanism for whole milk. Rumpf (1961) and Pietsch (1969) have discussed about agglomerates strength and related compression mechanisms (Peleg and Hollenbach, 1984).

15.2.3.5 Compression Mechanisms during Uniaxial Compression Tests

The compressive mechanisms for fine food powders and agglomerates during uniaxial compression tests occur differently. In fine powders, compression process takes place, firstly, by particle movement by filling voids of the same or bigger size than that of the particles. The packing characteristics of particles or a high interparticulate friction between particles will prevent any further interparticulate movement (Nyström and Karehill, 1996). The second stage involves filling smaller voids by particles that are deformed either elastically (reversible deformation) and/or plastically (irreversible deformation) and eventually broken down (Kurup and Pilpel, 1978; Carstensen and Hou, 1985; Duberg and Nyström, 1986). Most of organic compounds exhibit consolidation properties, undergoing particle fragmentation during the initial loading, followed by elastic and/or plastic deformation at higher loads.

In agglomerates, bulk compression will happen in three distinct segments: (1) agglomerate particle rearrangement to fill the voids of the same or larger size than that of the agglomerates; (2) agglomerate deformation or brittle breakdown; (3) primary particle rearrangement, elastic and plastic deformation, and fracture (Mort et al., 1984; Nuebel and Peleg, 1994). These steps can be overlapped depending on whether it is an instant agglomerate obtained from spray or freeze drying (e.g., coffee, milk, instant juices) or a granulated material attached by binders (granulated or encapsulated powders). The main difference between bulk compression and individual compression of agglomerates is the bulk's cushioning effect among the particles that reduce the amount of fracture. This is common for brittle cellular solid foams.

Both compression mechanisms in fine and agglomerated powders are influenced by particle size and size distribution, particle shape, and surface properties. Potato starch and powdered milk demonstrated to be powders that will crackle during compression, that is, change in volume discontinuously (Gerritsen and Stemerding, 1980) because the material is packed in a loose state (has a considerable compressibility). If compressive forces are applied, these forces are transmitted at the contact points.

A decrease in compact porosity with increasing compression load is normally attributed to particle rearrangement elastic deformation, plastic deformation, and particle fragmentation. Scanning electron microscopy has been presented in the literature for the qualitative study of volume-reduction mechanisms (Figure 15.12).

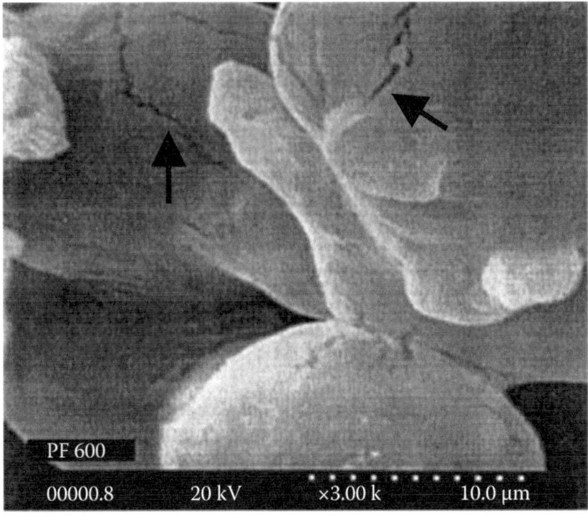

FIGURE 15.12 Fracture lines in a milk powder agglomerate after compression. (From Yan, H. and Barbosa-Cánovas, G.V., *Food Sci. Technol. Int.*, 3, 351, 2000.)

15.2.4 Surface Area

Another physical characteristic of food powders is the area of the surface of the solids, which can also be a measure of the powders porosity. The specific surface area of a powder is generally represented by the total particles contained in a unit mass of powder, that is, the internal and external surfaces that can be measured using various probes, such as gases and liquids (Chikazawa and Takei, 1997). If distributions of particle size and shape in a powder are known, the surface area of a powder $S_{specific}$ can be calculated from a relation of the particle surface and its volume:

$$S_{specific} = \frac{S}{V\rho} = \frac{3\sum_{i=1} R_i^2 N_i}{\rho \sum_{i=1} R_i^3 N_i} \tag{15.12}$$

where

R_i is the average particle radii
N_i is the number of particles
i is the size range

If particles where perfect spheres, $S_{specific}$ would have a simpler expression (Beddow, 1997):

$$S_{specific} = \frac{3}{\rho R} \tag{15.13}$$

Specific surface is important in applications such as powder coating, agglomeration, flow studies, and heat transfer studies, where the process is surface dependent. Surface-dependent phenomena such as permeametry and gas adsorption (chemisorption, physisorption) can be used for surface area measurement.

Permeametry is based on measuring the permeability of a packed bed of powder to a laminar gas flow, through pressure drop measurements. Evaluation of the resistance of fluid flow through a compact bed of powder measures the surface area of a solid volume of powder. The fluid mainly used is air since liquids can provide erroneous results due to adsorption and aggregation of fine particles.

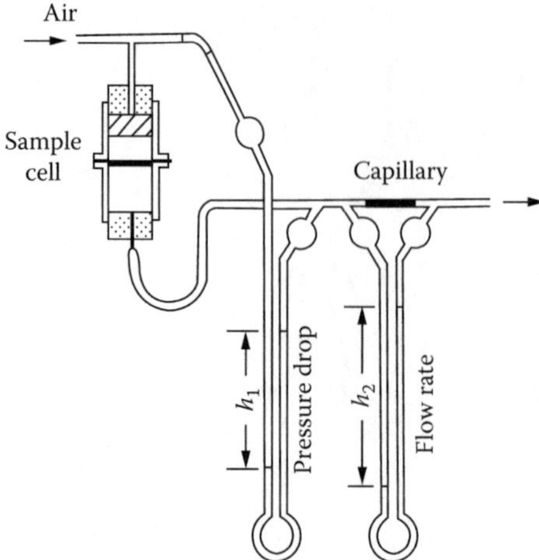

FIGURE 15.13 Lea and Nurse apparatus for permeametry measurements.

Permeametry is generally suitable for powders of average particle size between 0.2 and 50 μm, but it can be also used with coarser powders (up to 1000 μm average particle size) using a suitably scaled-up test equipment. With highly irregular particles, such as platelets or fibers, errors are mostly particle shape dependant.

Commercial instruments such as the Lea and Nurse apparatus compress the sample to a known porosity in the permeability cell at constant flow or constant pressure drop (Figure 15.13). Dry air, drawn by an aspirator or a pump, flows through the bed at a constant rate and then passes through a capillary that serves as a flow meter. Static pressure drop across the powder bed is measured with a manometer as static head h_1, while the flow rate is measured by means of the capillary flow meter, giving a reading h_2 on the second manometer. Both pressure drops are small compared with atmospheric pressure and, thus, the compressibility of the gas can be neglected. Permeability equipments are applicable for viscous flow and using the Carman–Kozeny (Carman, 1956) equation can be simplified to include the static head measurements h_1 and h_2, taking the following form:

$$S_w = \frac{14}{\rho_s(1-\varepsilon)}\sqrt{\frac{\varepsilon^3 A h_1}{cLh_2}} \tag{15.14}$$

where

A is the cross-sectional area of the bed

ρ_s is the solid's density

c is the flow meter conductance

Other constant flow instruments working on constant air flow are Fisher subsieve sizer, Blaine, and Arakawa-Shimadzu (Chikazawa and Takei, 1997). There also instruments to measure surface area by permeametry that operate on a variable flow mode (constant volume) such as the Griffin and George permeameter (oil suction), as well as the Reynolds and Branson apparatus (Parfitt and Sing, 1976).

Gas adsorption methods measure the surface area of powders from the amount of gas adsorbed onto the powder surface by determining the monolayer capacity and cross-sectional area of an adsorbate molecule. Nitrogen is most commonly used as operating gas, as well as organic adsorptives such as benzene and carbon tetrachloride. The sample adsorbs the adsorptive gas molecules and the vapor pressure decreases gradually until equilibrium attains. The adsorbed amount is determined from the difference between the introduced gas amount and the residual amount, which is measured by using helium gas that is not adsorbed.

15.3 CHEMICAL AND PHYSICOCHEMICAL PROPERTIES

Chemical reactions on food powders mainly occur due to the presence of free water within the particle or bulk. Temperature gradients can provide opportunities for water diffusion and interactions between particle proteinated and hydrophobic compounds to the formation of new products. However, the most relevant chemistry-related properties in food powder technology are related with its solubility during processing and consumption and caking capacity during storage. For instance, the rate at which dried foods pick up and absorb water has received attention from food drying technologists (Masters, 1985).

15.3.1 Instant Properties

The term "instant" is usually used in food industry to describe dispersion and dissolution properties of powders. Among different commercially available instant powders like milk, coffee, cocoa, baby foods, soups, sauces, and soft drinks, sugar mixtures can be found. The instantizing process "provides food powders with the 'instant' attributes so that they can dissolve or disperse more readily in aqueous liquids than when they are in their original powder forms" (Schubert, 1980). Instantizing is related to the formation of agglomerates, which have better instant properties than fine particles. Some definitions related with powder dispersion can are used to better describe the behavior of agglomerated powders in contact with water.

Instant properties of agglomerates are the most desirable properties of agglomeration processes and they can be measured by four dissolution properties (Schubert, 1987a) corresponding to four phases of dissolution, that is, wetting, sinking, dispersing, and solution. *Wettability* refers to the liquid penetration into a porous agglomerate system due to capillary action or the ability of agglomerates to be penetrated by the liquid and describes the capacity of the particles to absorb water on their surface (Pietsch, 1999). This property depends largely on particle size. Increasing particle size and/or agglomerating particles can reduce the incidence of clumping. The composition of the particle surface such as presence of free fat in the surface can affect wettability by reducing it. Surface-active agents, such as lecithin, improve wettability and provide instant abilities to fat-rich powders like cocoa.

Sinkability describes the ability of agglomerates to sink quickly below the water surface, which is controlled by the mass of the agglomerates (Pietsch, 1999). It depends mainly on the particle size and density, since larger and denser particles usually sink faster than finer and lighter ones. Particles with a lot of included air may exhibit poor sinkability because of their low density. *Dispersibility* is the ease with which the powder may be distributed as single particles over the surface and throughout the bulk of the reconstituting water. Dispersibility is reduced by clump formation and is improved when the sinkability is high. *Solubility* refers to the rate and extent to which the components of the powder particles dissolve in the water. It depends mainly on the chemical composition of the powder and its physical state.

15.3.1.1 Instant Property Evaluation

Wetting time is the most important parameter to measure instant properties. Evaluation procedures are set to a maximum allowable dissolution time when evaluating instant properties for quality control in different food powder products. Standard procedures to measure the instant properties define the specific solvent temperature, the liquid surface area, amount of material to dissolve, the method of depositing a certain amount of material onto the liquid surface, unassisted or predetermined mixing steps, and the timing procedure (Pietsch, 1999).

Different protocols have been standardized and determine food powder instant properties. Among them, the penetration speed test, the standard dynamic wetting test, and standard methods for dispersibility and wettability. The *penetration speed test* uses a test cell with a screen carrying a layer of agglomerates, which is retained with a Plexiglas cylinder from earlier. The test cell is put

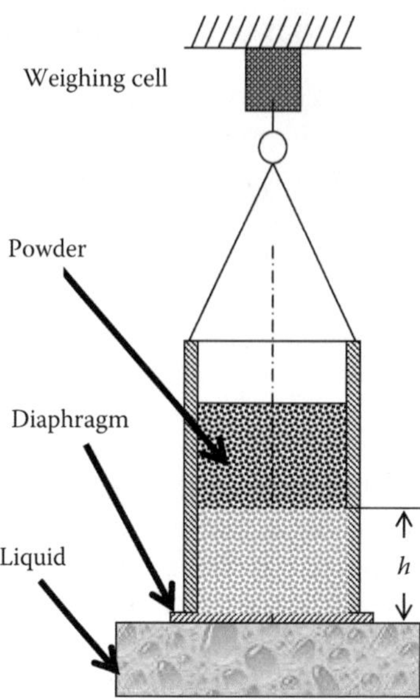

FIGURE 15.14 Dynamic wetting test—determination of the amount of liquid volume absorbed by capillary forces. (From Pietsch, W., *Chem. Eng. Prog.*, 8, 67, 1999.)

into water and penetration time is measured until the entire material bed is submerged (Pietsch, 1999). The *dynamic wetting test* uses a measuring cell attached to a weighing cell and in contact with the liquid by tilting the cell onto the liquid surface (Figure 15.14). The force measured by the weighing cell is proportional to the liquid volume absorbed due to capillary pressure and it is plotted against time (Schubert, 1980; Pietsch, 1999).

The International Dairy Federation (IDF) set the *dispersibility test* (Schubert, 1987a; Pietsch, 1999) using light transmission applied to disperse agglomerated food powders. Another instrument using the same light transmission mechanism measures wettability, dispersibility, and solubility in the same test and provide more information on the progress of wetting, dispersion, and dissolution. The *IDF standard method* (IDF, 1979) is specifically designed to determine the dispersibility of instant dried milk and is also a rapid routine method to determine wettability (i.e., wetting time).

15.3.2 Stickiness in Food Powders

Stickiness is the tendency to adhere to a contact surface and it may refer to particle–particle adhesion or bulk caking (Aguilera et al., 1995), adhesion to packages, containers, or to handling and processing equipment. As a matter of fact, stickiness is a prevalent problem that can cause lower product yield, operational problems, equipment wear, and fire hazards (Adhikari et al., 2001). The interaction of water with solids is the prime cause of stickiness and caking in low-moisture food powders.

Particularly, caking is a deleterious phenomenon by which amorphous food powders are transformed into a sticky undesirable material, resulting in loss of functionality and lowered quality (Aguilera et al., 1995). Water provides plasticity to food polymeric systems, reducing viscosity, and enhancing the molecular mobility of the system, which is linked to liquid and solid bridges formation and caking (Papadakis and Bahu, 1992; Peleg, 1993). Chemical caking is the most common type of caking mechanism, and it may be caused by chemical reactions in which a compound has been generated or modified, such as decomposition, hydration, dehydration, recrystallization, or sublimation. Plastic-flow caking occurs when the particles' yield value is exceeded and they stick together or merge into a single particulate form in amorphous materials such as tars, gels, lipids, or waxes.

TABLE 15.3
Cohesive Phenomena Contributing to Particle–Particle Attraction and Stickiness

Solid bridges
Liquid bridges
Intermolecular forces
Hydrogen bonding
Van der Waals
Electrostatic
Mechanical interlocking

Glass transition temperature can characterize the stickiness during powder storage (Roos and Karel, 1991). The presence of glassy low molecular weight materials such as glucose, fructose, and sucrose in spray- and freeze-dried fruit powders makes these powders particularly susceptible to stickiness due to their elevated hygroscopicity as well as chemical and conformational changes at increased temperatures. Table 15.3 describes different types of attractive particle–particle phenomena contributing to powder cohesion and stickiness, which will be described later.

15.3.2.1 Bridging

In general, interaction between particles is regulated by the relationship between the strength of the attractive (or repulsive) forces and gravitational forces. For all particles in the amorphous rubbery state (or above glass transition temperature), forces causing primary particles to stick together are the following (Schubert, 1981; Hartley et al., 1985): interparticle attraction forces (van der Waals or molecular forces and electrostatic forces), liquid bridges, and solid bridges. Interparticle forces are inversely related to the particle size (Buma, 1971; Rennie et al., 1999; Adhikari et al., 2001). Van der Waals and electrostatic attraction is not as high as the attraction force that comes from liquid bridges (Schubert, 1987a). A diagram showing the strength of agglomerate bond as a function of particle size is given in Figure 15.15.

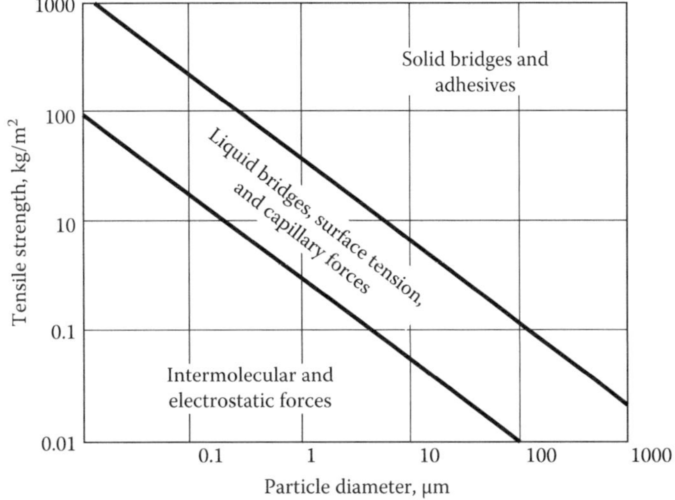

FIGURE 15.15 Strength of agglomerates and fine particle bonding as a function of particle size. (From Adhikari, B. et al., *Int. J. Prop.*, 4(1), 1, 2001.)

Van der Waals forces arise from electron motion among dipoles and act over very short distances within the material structure, becoming prevalent when the particle size is less than 1 μm (Hartley et al., 1985). *Electrostatic forces* are longer ranging forces that arise through surface differences on particles and are present when the material does not dissipate electrostatic charge.

Liquid bridges are related to chemical interactions between food particle components and result from the presence of bulk liquid (generally unbound water or melted lipids) between the individual particles. The forces of particle adhesion arise either from surface tension of the liquid/air system (as in the case of a liquid droplet) or from capillary pressure. According to Rumpf (1962), liquid bridges offer strong bonding that has practical significance in particulate aggregates having a size of 1 mm or less. The strength of a liquid bridge depends on factors that affect at the contact angles (e.g., the composition of the solid and the nature of the liquid solution) and factors that influence the radius of curvature (particle size, shape, interparticle distance, particle roughness, and the ratio of the amount of liquid to amount of solid in the agglomerate). The composition of the liquid of the bridge varies in different food materials. The "bridging potential" or "stickiness" is related to factors such as powder moisture, fat or low molecular weight sugar content, and shape of particles.

Solid bridges form as a result of sintering, solid diffusion, condensation, or chemical reaction and generally result from solidification at room temperature. The magnitude of the adhesion force depends on the diameter of the contact area and the strength of the bridge material (Loncin and Merson, 1979). Solid bridges are the structural bonds that hold individual particles in an agglomerate together or even in a powder cake (e.g., formed during fruit or coffee powder storage in a hopper).

15.3.2.2 Thermodynamic Adsorption

The adhesion of fluids or semisolid foods to different surfaces is not possible unless they have good spreadability to the adherend. Complete wetting of the surface will occur when the surface energy of the adherend is greater than that of the adhesive (Saunders et al., 1992). Low-energy materials absorb strongly to high-energy surfaces to lower the surface energy of the system.

The mechanism of thermodynamic adsorption is based on Dupre's energy equation (Michalski et al., 1997; Adhikari et al., 2001), which relates adhesive and adherend surface tensions of solid, liquid, and to the work of adhesion, W_s (J/m^2):

$$W_s = \gamma_s + \gamma_l \cdot \gamma_{sl} \tag{15.15}$$

where

γ_s (N/m) and γ_l (N/m) are the solid and liquid surface tensions
γ_{sl} (N/m) is the solid–liquid interfacial tension

Adhesion is attributable to electrodynamic intermolecular forces acting at the liquid–liquid, liquid–solid, and solid–solid interfaces, and that interfacial attraction between the adhesive and adherend can be expressed in terms of reversible work of adhesion that corresponds to material surface tensions. The work of adhesion is now recognized as a function of the reversible work of adhesion and the irreversible deformation of the substrate (Shanahan and Carre, 1995; Michalski et al., 1997).

15.3.2.3 Cohesion and Cohesion Properties

Powders can represent two types when referring to powder flowability: noncohesive powders and cohesive powders. Noncohesive (or "free flowing") powders are those powders in which interparticulate forces are negligible. Most powders are considered noncohesive only when they are dry and when their particle size is above 100–200 μm (Peleg, 1978; Teunou et al., 1999), then finer powders are susceptible to cohesion and their flowability is more difficult (Adhikari et al., 2001).

Powder cohesion occurs when interparticle forces play a significant role in the mechanical behavior of a powder bed. The dynamic behavior of powder seems to also be determined basically by interparticle forces and packing structure. Compressive and compaction behavior of powders is very important in evaluating flowability, since methods to determine flow properties like the angle of repose and angle of internal friction account for compressive and compacting mechanisms.

A standard for cohesive strength determination is the ASTM D 6128 (or Jenike method) where consolidating conditions found in the depth of a bin are reproduced in a shear cell. The shear cell is used to determine maximum shear force (or yield point) that a powder can undergo under a predetermined compression load. A plot (called *yield locus*) of maximum shear stress measured versus the corresponding normal consolidating stress gives a curve where two parameters can be obtained (Peleg, 1978): *cohesion C* and *angle of internal friction.*

Cohesion C is a measure of the attraction between particles and is due to the effect of internal forces within the bulk, which tend to prevent planar sliding of one internal surface of particles upon another. Cohesion has been proven proportional to *tensile stress* (Peleg, 1978) necessary to separate two bulk portions of a food powder sample. The angle of internal friction is a measure of the interaction between particles and is calculated from the slope of the yield locus. In free-flowing powders, it represents the friction between particles when flowing against each other. Therefore, it depends on their size, shape, roughness, and hardness.

The *static angle of repose* is defined as the angle at which a material will rest on a stationary heap and it is the angle θ formed by the heap slope and the horizontal when the powder is dropped on a platform. A higher angle will indicate a greater degree of cohesion and it is the actual flowability measurement applied by some laboratories in the food industry for quality control.

15.3.2.4 Test Methods

Various test methods have been adapted for characterization of food stickiness. Among them, shear cell method and glass transition methods have received the widest usage and provide a good degree of automation. Other methods such as the optical probe and the sticky point have received less attention. In particular, the optical probe is a novel method, which provides good quantification of stickiness, and may prove very useful for online monitoring of stickiness in powders as a function of moisture and temperature.

Measurement of cohesive and adhesive forces based on a *shear cell* has been widely used to characterize the degree of powder flowability by powder industries (Peleg, 1978; Teunou et al., 1999). A shear cell is a boxlike structure that is split in half horizontally and is equipped with a provision of applying various normal stresses (σ) and shear stresses (τ) on the top of it. The powder to be tested is preconsolidated with the maximum applicable consolidation load. If a sample is tested at different normal load, one can obtain a particular shear stress that causes failure (initiation of flow) and disruption of a cohesive structure generally formed in bins and hoppers.

Glass transition temperature of a system, predicts a bulk temperature and moisture matrix, which renders the product sticky. The differential scanning calorimeter (DSC) measures changes in specific heat capacity (C_p). Nuclear magnetic resonance method tracks the phase change by providing information of mobility of the entire molecule and the mobility of individual groups or region within a food structure (Blanshard, 1993). Mechanical–thermal methods measure the changes in loss modulus, elastic modulus, and their ratio as a function of temperature. A sharp change in these properties is indicative of glass transition temperature (Rahman, 1995). Up to now, glass transition temperature has been used to characterize low molecular weight carbohydrate systems (Adhikari et al., 2001).

The *optical probe* (Lockemann, 1999) measures the change in optical properties of a free-flowing powder when its flow pattern gets altered. A free-flowing sample in a glass test tube is immersed in temperature programmable oil bath and temperature is slowly raised while the test tube is rotated at a slow speed. An optical fiber sensor indicates the reflectance of light from the sample surface. The sticky-point temperature is when a sharp rise in reflectance is detected.

The *sticky-point* tester (Wallack and King, 1988) consists of a test tube immersed in a controlled temperature bath. Temperature is slowly raised and sample is stirred with a machine-driven impeller using double and curved blades imbedded in the sample. When the force experienced to drive the sample stirrer is maximized, the sticky-point temperature has been reached. Generally, this test fails to provide information regarding the cohesiveness status of a powder below the sticky-point temperature (Peleg, 1993). However, it is based on the softening of powders above their glass transition temperature when the glassy phase yields to a rubbery state because the plasticization of the particle surface accelerates once the glass transition temperature is exceeded (Peleg, 1993).

15.3.3 Water Activity and Glass Transition Temperature

The interaction of water with solids is the prime cause of stickiness and caking in low-moisture particulates. Particle surfaces bear adsorbed mono/multilayers of water or free water from capillary condensation. As a plasticizer, water can reduce surface microroughness of the particles that allows closer particle–particle contact thereby increasing forces of attraction between particles. This may be an important cause of caking of the particles in the presence of water (Iveson, 1997; Adhikari et al., 2001). In fact, an increase in water content will depress glass transition temperature (T_g) of particulates.

Generally, food powder surfaces are amorphous in nature and will undergo a phase change when they reach T_g. If amorphous foods are stored at a temperature higher than T_g, they will be transformed into a liquid-like, rubbery state that is associated with stickiness and caking (Roos and Karel, 1993; Adhikari et al., 2001). Caking of food powders in storage can be avoided simply by storing them below their glass transition temperature (Figure 15.16) provided that adsorption of moisture from the environment is avoided (Slade et al., 1993). Since the sticky-point temperature is 10°C–20°C higher than the glass transition temperature (Roos and Karel, 1993; Adhikari et al., 2001), stickiness and adhesion phenomena will not take place below the glass transition temperature.

Composition plays a fundamental role in stickiness in combination with factors such as surface relative humidity, temperature, and viscosity. The degree of contribution of food compositional factors on stickiness is presented in Table 15.4. The T_g of food materials such as sugars, starch, gluten, gelatin, hemicellulose, and elastin decreases rapidly to about 10°C or so when moisture content increases to 30% by mass (Atkins, 1987) and have glass transition temperatures ranging between 5°C and 100°C in their anhydrous state. Low molecular weight sugars are important players for stickiness and caking in food-based particulate systems due to their hygroscopicity and solubility.

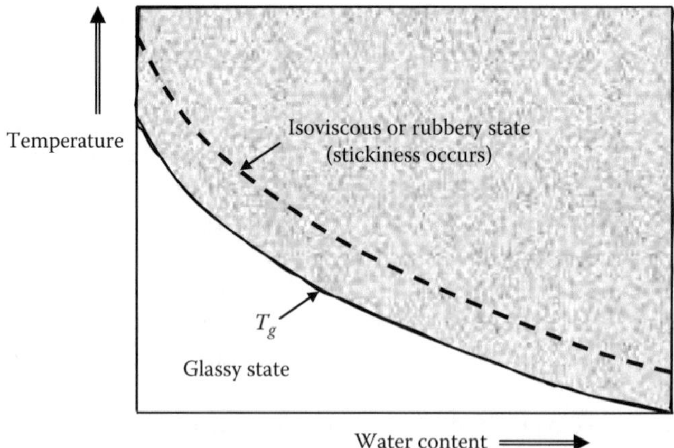

FIGURE 15.16 Schematic diagram of the relationship between glass transition temperature T_g, viscosity, water content, and occurrence of stickiness in food powders. (Adapted from Roos, Y.H., Time-dependent phenomena, in *Phase Transitions in Foods*, Academic Press, New York, 1995, pp. 193–245.)

TABLE 15.4
Factors Causing Stickiness and Their Relative Contribution

Factors	Relative Contribution to Stickiness
Protein	0
Polysaccharides	0
Fats	+
Low molecular sugars	++
Organic acids	++
Water/relative humidity	+++
Particle size distribution	+
Compression pressure	++
Temperature	+++
Viscosity	+++

Note: 0, base point (negligible contribution); +, high contribution; ++, higher contribution; +++, highest contribution.

Similarly, low molecular weight protein hydrolyzates presumably the amino acids were found to exhibit hygroscopicity and contribute to caking during storage of spray-dried fermented soy sauce powders (Hamano and Aoyama, 1974).

On the other hand, food powders with high-fat content might also induce powder caking if stored above the fat melting point. Above room temperature, liquefaction of fat provokes powder softening thereby increasing contact area and leading to the formation of liquid bridges (Adhikari et al., 2001).

15.4 APPLICATION OF COMPRESSION IN FOODS

The general behavior of powders under compressive stress or compaction due to mechanical motion fits into several processing operations where physical and chemical properties of food powders tend to change under preset temperature and relative humidity conditions. Food powder compression is involved in different industrial applications such as bulk size reduction, grinding, particle size enlargement, food tablet production, encapsulated material resistance, hopper and silo storage, mixing, and packaging. As mentioned before, handling operations are related to mechanical properties of materials (shear stress, tensile, and compressive stress) in that brittle agglomerated food powders can suffer the nondesirable effect of particle breakage or attrition.

Particle size reduction is a comminution process that includes operations such as crushing, grinding, and milling (e.g., milling of cereals, grinding of spices or coffee) where a food piece is deformed until breakage or failure. Breaking of hard materials can occur along cracks or defects in the structure during compression. Forces commonly used in food processes for particle size reduction are compressive, impact, attrition (or shear), and cutting. Particularly, crushing rolls use mainly compressive forces, hammer mills are based on impact, disk mills cause particle attrition through shear force application, and rotary knife cutters use cutting forces. In regard to particle shape and defects, a large piece with many defects can be broken with a small stress with very little deformation, while smaller pieces have fewer defects remaining and will need a higher breaking strength. For food materials of fibrous structures, shredding or cutting should be considered for the desired size reduction. Hard brittle materials like sugar crystals can be crushed, broken by impact, or ground by abrasion, while ductile cocoa is better reduced by impact (e.g., in a hammer mill).

TABLE 15.5
Instantizing Processes

Agglomeration	Nonagglomeration
Spray drying and agglomeration	Freeze drying
Rewetting agglomeration	Drum drying
Spray-bed dryer agglomeration	Additives (e.g., lecithin)
Press agglomeration	Fat removal
Thermal treatment	

Size enlargement operations include agglomeration (by compaction, granulation, tableting, briquetting, sintering) and powder encapsulation. These processes depend on material resistance properties as well as chemical properties like adhesiveness and stickiness. Agglomeration is a process aimed for controlling particle porosity and density and involves the aggregation of dispersed materials into materials with larger unit size held together by adhesive and/or cohesive forces (Bika et al., 2001).

Rewetting agglomeration (or instantizing process) is used in food processes partly to improve properties related to handling, while pressure agglomeration involves particulate confinement by compression into a mass that is then shaped and densified. Agglomerates formed by rewetting agglomeration develop lower strength levels primarily because they feature higher porosity from coalescence, while pressure agglomeration causes porosity to decrease, while density and strength increase. By agglomerating fine powders of about 100 μm in size into particles with the size of several millimeters, the wetting behavior of the particles is improved and lump formation can be avoided (Schubert, 1987a). Some instantizing processes are classified in Table 15.5. In rehydration operations, water aided by capillary forces penetrates into the narrow spaces between fine particles and lumps containing dry particles in the middle will be formed requiring strong mechanical stirring to be homogeneously dispersed or dissolved in the liquid (APV, 1989).

Agglomerate shaping takes place during pressure agglomeration by forcing the material through the holes. Chemical caking plays an important role in briquetting and tableting where compaction of food ingredients such as dextrose, gelatin, glucose, sucrose, lactose, or starch is pressed with food gums as binders.

Encapsulation is a process where a continuous thin coating is formed around solid particles (e.g., powdered sweeteners, vitamins, minerals, preservatives, antioxidants, cross-linking agents, leavening agents, colorants, and nutrients) in order to create a capsule wall (King, 1995; Risch, 1995). Encapsulation promotes easier handling of the core or interior material by preventing lumping; improving flowability, compression, and mixing properties; reducing core particle dustiness; and modifying particle density (Shahidi and Han, 1993). In particular, multiwall-structured capsules contain different concentric layers of the same or different composition and earn greater resistance to attrition during handling. Microencapsulation by spray drying of fats reduces adhesiveness (reducing clumping and caking) and enhances handling properties during storage transport and blending with nonfat ingredients.

Compaction, conveying, mixing, and metering among other types of handling of food powders can provoke attrition (Schubert, 1987a) bringing problems such as changes in bulk properties, segregation and, in some agglomerates, loss of instantaneity. Mixing mechanisms can be affected by the compaction properties such as mechanical interlocking, surface attraction, plastic welding (from high pressures between small contact areas), electrostatic attraction, and environmental factors such as ambient moisture and temperature fluctuations.

Compression and cohesion properties evaluation plays a key role in the study of stresses developed as a result of storage in high bins, hoppers, or silos. Mechanical compressibility can provide an idea of bulk density changes due to compacting pressure in stored powders (Thomson, 1997).

During storage, arches and ratholes are two of the most common flow obstruction problems caused by powdered material stored in hoppers and bins where cohesive strength causes stagnant material to bind together or interlock forming a narrow channel above the outlet. Since rathole material remains under storage, it can cake or degrade. Unconfined yield strength of the bulk solid is the main property that can be associated with arching and rathole formation.

15.5 RESEARCH UPDATE IN FOOD POWDER PROPERTIES

Among the latest research in food powder bulk properties, the study of electrostatic coating, use of padding materials, and ultimate bulk density can be highlighted.

Electrostatic coating properties. Electrostatic coating has been suggested as a method to improve powder coating efficiency of crackers, chips, and shredded cheese reducing powder waste and dust amounts and improving coating evenness (Ricks et al., 2002). Electrostatic coating is based on charging particles electrically so that they repel from each other and they seek for a grounded surface (to be coated). Flow properties (cohesion, angle of repose, particle density, and Hausner ratio, and flow index) were studied for selected food powders to predict transfer efficiency and dustiness. In this case, particle chargeability plays an important role in electrostatic transfer efficiency.

Padding materials. Some cushioning materials in a powder bed of agglomerates or on the container wall can absorb or reduce the mechanical or static impact thereby reducing their degree of attrition. Yan and Barbosa-Cánovas (2001b) studied the padding effect on agglomerated coffee and nonfat milk using pure polyurethane foam as padding material by confined uniaxial compression using a UTM. The padding effects were evaluated by using padding index that indicated if higher deformation was undergone by the padding material or the powder bed. The padding efficiency, related to powder attrition tendency, indicated the amount of powders retaining their original particle size when padding material was used. Force–deformation curves resulted smoother for instant coffee when padding was used, while curves for instant milk kept an exponential characteristic. The padding efficiency was better for lower-strength milk agglomerates than for harder-strength agglomerates such as coffee.

Ultimate bulk density. High hydrostatic pressure appears as a new possibility within the traditional powder compression methods where higher pressure provides higher bulk density up to a critical pressure that remains constant (ultimate bulk density). Yan et al. (2001) recently studied how bulk density of instant nonfat milk, spray-dried coffee, and freeze-dried coffee was affected by HHP processing times and particle size. The ultimate bulk density depends on the product formulation, physical properties of the ingredients, and its production conditions. For the same kind of agglomerates, even though they have different initial particle sizes, bulk densities, or water activities, their ultimate bulk densities are not significantly different under the same pressure. The ultimate bulk density concept could be a promising tool to detect composition variations due to a change in formulation of the product or change in the manufacturing conditions.

16 Heat Transfer by Conduction

16.1 FUNDAMENTAL EQUATIONS IN HEAT CONDUCTION

To analyze heat conduction in a given solid, where it is assumed heat transfer by convection and radiation is negligible, an energy balance should provide very meaningful results.

This balance provides an equation that is used to calculate the temperature profiles in the solid, as well as to obtain the heat flux that crosses it. The heat transfer per unit time, due to conduction, is related to the distribution of temperatures by Fourier's law.

The fundamental equation is obtained by carrying out an energy balance in a reference volume of the solid, according to the expression

{rate of thermal energy in} + {net rate of thermal energy generation} = {rate of thermal energy out} + {rate of accumulation of thermal energy}

The resulting expression of the equation depends on the geometry considered, and it is advantageous to derive different expressions according to the type of coordinates in which the system should be referenced.

16.1.1 Rectangular Coordinates

A volume of the solid is taken to perform the energy balance, as shown in Figure 16.1. It is supposed that there is only temperature variation along the x direction; this means that the solid's temperature is only a function of the coordinate x and a function of time. Also, it is considered that the characteristics of the material are constant in such a way that the thermal conductivity, k; the density, ρ; and the specific heat, $\hat{C}_P$, of the solid are constant.

With these assumptions, the energy balance applied to the volume of material in Figure 16.1 yields the following equation:

$$\rho\hat{C}_P\frac{\partial T}{\partial t} = k\frac{\partial^2 T}{\partial^2 x} + \tilde{q}_G$$

where

t is the time

T is the temperature

$\tilde{q}_G$ is the energy generation flow rate per unit volume

There are various terms in this general equation of heat transfer by conduction: the input and output of heat in volume of the solid considered, the generation, and the accumulation terms.

The input and output terms are governed by Fourier's equation. The generation term does not appear in all cases, but it does appear in cases in which chemical or nuclear reactions that generate heat occur in the considered volume or when there are electric currents applied to such volume that generate heat while traveling through it. When there is a variation in the temperature of the system, a variation in the internal energy will occur, so accumulation terms will appear in the fundamental equation of heat transfer by conduction. In the case where the temperature of the solid remains constant, there will not be an accumulation of energy, yielding to steady-state conditions.

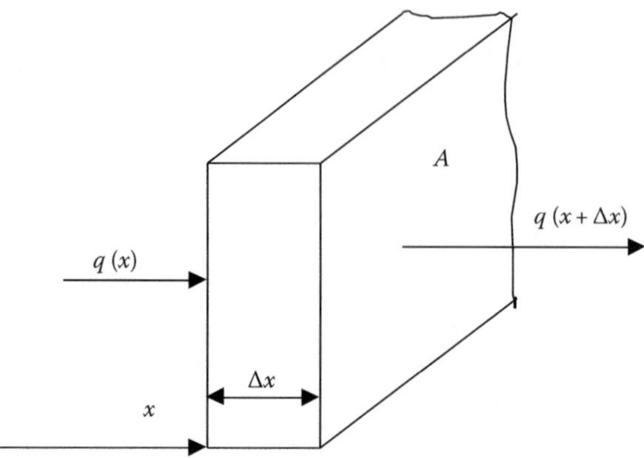

FIGURE 16.1 Volume of a solid in rectangular coordinates.

The heat balance is applied to a volume of solid, as in the example presented in Figure 16.1. The different terms of the energy balance are developed next.

- Input term
  The heat flow that enters the solid volume considered, according to Fourier's equation, is

$$\dot{Q}_e = q(x)A = -kA\frac{\partial T(x)}{\partial x}$$

  where
  A is the cross-sectional area through which the heat flows
  $q(x)$ is the heat flux
  $\dot{Q}_e$ is the heat flow that penetrates the solid through the area A in coordinate position x

- Output term
  Similar to the input term, the output term is

$$\dot{Q}_S = q(x+\Delta x)A = -kA\frac{\partial T(x+\Delta\Delta x)}{\partial x}$$

  where $\dot{Q}_S$ is the heat flow rate that exits of the solid through the A in position $(x + \Delta x)$.

- Generation term
  If the energy flow generated per unit of volume of solid is defined, $\tilde{q}_G$, the generation flow rate would be

$$\dot{Q}_G = \tilde{q}_G A\Delta x$$

- Accumulation term
  The volume of solid at the temperature T has an energy level given by

$$E = \rho A\Delta x\hat{C}_P T$$

If the solid is considered to be isotropic, then ρ and $\hat{C}_P$ are constant, the accumulation flow rate in the x direction is

$$\dot{Q}_A = \frac{\partial E}{\partial t} = \rho A \Delta x \hat{C}_P \frac{\partial T}{\partial t}$$

Therefore, the energy balance applied to the considered volume yields the expression

$$-kA\frac{\partial T(x)}{\partial x} + \tilde{q}_G A \Delta x = -kA\frac{\partial T(x+\Delta x)}{\partial x} + \rho A \Delta x \hat{C}_P \frac{\partial T}{\partial t}$$

If all of the terms of this expression are divided by the volume on which the balance $(A \cdot \Delta x)$ is applied, then rearranging this equation it is obtained that

$$\rho \hat{C}_P \frac{\partial T}{\partial t} = k\frac{(\partial T(x+\Delta x)/\partial t) - (\partial T(x)/\partial t)}{\Delta \Delta x} + \tilde{q}_G$$

If the limit of $\Delta x \rightarrow 0$ is taken, the following equation is obtained:

$$\rho \hat{C}_P \frac{\partial T}{\partial t} = k\frac{\partial^2 T}{\partial x^2} + \tilde{q}_G$$

This is the fundamental equation of heat transfer by conduction following the x direction.

This equation can be generalized to include the three directions x, y, and z; thus, the general equation of heat conduction is obtained:

$$\rho \hat{C}_P \frac{\partial T}{\partial t} = k\frac{\partial^2 T}{\partial^2 x} + \frac{\partial^2 T}{\partial^2 y} + \frac{\partial^2 T}{\partial^2 z} + \tilde{q}_G$$

If it is taken into account that the Laplacian operator ∇^2 is defined by

$$\nabla^2 = \frac{\partial^2}{\partial^2 x} + \frac{\partial^2}{\partial^2 y} + \frac{\partial^2}{\partial^2 z}$$

then the last equation can be expressed as

$$\rho \hat{C}_P \frac{\partial T}{\partial t} = k\nabla^2 T + \tilde{q}_G \quad (16.1)$$

16.1.2 Cylindrical Coordinates

The general equation of heat conduction for cylindrical coordinates is analogous to the one obtained for rectangular coordinates, with the exception that the Laplacian operator is expressed in a different way.

The general equation of heat conduction for cylindrical coordinates would be

$$\rho \hat{C}_P \frac{\partial T}{\partial t} = k\frac{1}{r}\frac{\partial}{\partial r}\left(r\frac{\partial T}{\partial r}\right) + \frac{1}{r^2}\frac{\partial^2 T}{\partial \phi^2} + \frac{\partial^2 T}{\partial z^2} + \tilde{q}_G \quad (16.2)$$

where r, ϕ, and z are the radial, angular, and axial coordinates, respectively, as can be seen in Figure 16.2.

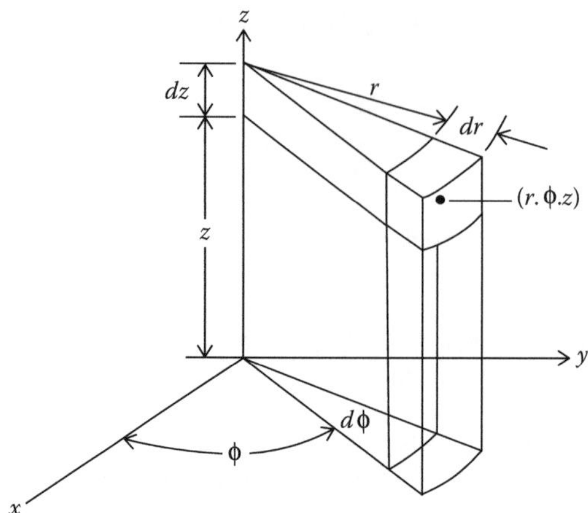

FIGURE 16.2 Cylindrical coordinates system.

16.1.3 Spherical Coordinates

Similar to the case of cylindrical coordinates, the general expression for heat conduction is obtained by expressing the Laplacian operator in spherical coordinates. If temperature is a function of the three coordinates and time, the resulting expression is

$$\rho\hat{C}_P\frac{\partial T}{\partial t} = k\frac{1}{r^2}\frac{\partial}{\partial r}\left(r^2\frac{\partial T}{\partial r}\right) + \frac{1}{r^2\text{sen}\theta}\frac{\partial}{\partial\theta}\left(\text{sen}\theta\frac{\partial T}{\partial\theta}\right) + \frac{1}{r^2\text{sen}^2\theta}\frac{\partial^2 T}{\partial\phi^2} + \tilde{q}_G \tag{16.3}$$

The meaning of each of the spherical coordinates can be seen in Figure 16.3.

16.2 HEAT CONDUCTION UNDER STEADY REGIME

In a case where no generation of energy in the system is considered, the term $\tilde{q}_G$ of the heat transfer equations disappears. If, additionally, heat conduction occurs under steady state, the accumulation term, which depends on time, is zero, so the general heat conduction equation is transformed into

$$\nabla^2 T = 0 \tag{16.4}$$

where ∇^2, the Laplacian operator, as seen before, has different expressions according to the type of coordinates used.

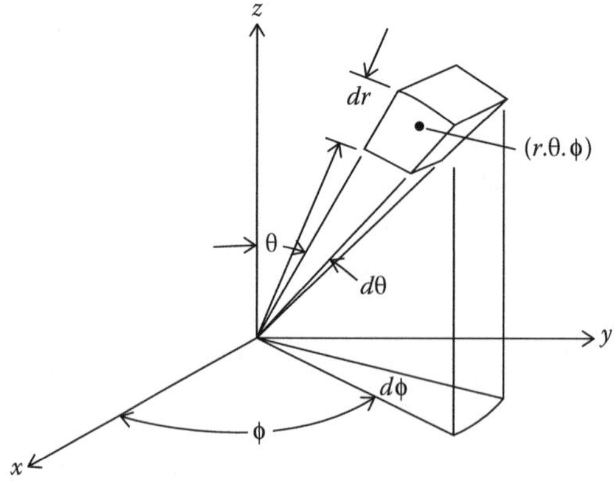

FIGURE 16.3 Spherical coordinates system.

This equation is called the Laplace equation and is widely used in various engineering areas since it appears in many problems related to the solution of mathematical models set up for different processes.

16.2.1 1D Heat Conduction

Heat conduction in only one direction will be studied in this section. This means that temperature changes will be analyzed only in one direction or coordinate. In the case of rectangular coordinates, it is considered that the temperature varies along the x coordinate, while for spherical and cylindrical coordinates, this temperature variation is assumed along the radial coordinate.

The temperature profile will be obtained along the desired direction in each case by integration of the Laplace equation, obtaining the integration constants by applying boundary conditions to the integrated equation. Once the temperature profile is obtained, Fourier's equation is applied to determine the heat transfer flow.

Among the different cases that can occur, those of heat transfer through slab, cylindrical, and spherical layers will be studied, since these are the most frequently used geometrical shapes.

16.2.1.1 Flat Wall

The most common way to decrease heat losses through a wall is to place an insulation layer, and it is important to know the width that such a layer should have. In this case, an infinite slab will be assumed, thus avoiding consideration of the end effects. Although this assumption is made, the results obtained can be applied to the case of finite slabs in a reliable way.

If a slab with infinite surface is assumed, with one face at temperature T_0 and the other at T_1, with $T_0 > T_1$, the heat flow goes from T_0 to T_1 (Figure 16.4).

The equation of Laplace for rectangular coordinates in one direction would be

$$\frac{\partial^2 T}{\partial x^2} = 0$$

Temperature is only a function of position x, so the partial derivatives can be substituted by total derivatives. The following is obtained when integrating this equation:

$$T = C_1 x + C_2$$

where C_1 and C_2 are integration constants whose values are obtained when applying the following boundary conditions:

- For $x = 0 \quad T = T_0$
- For $x = e \quad T = T_1$

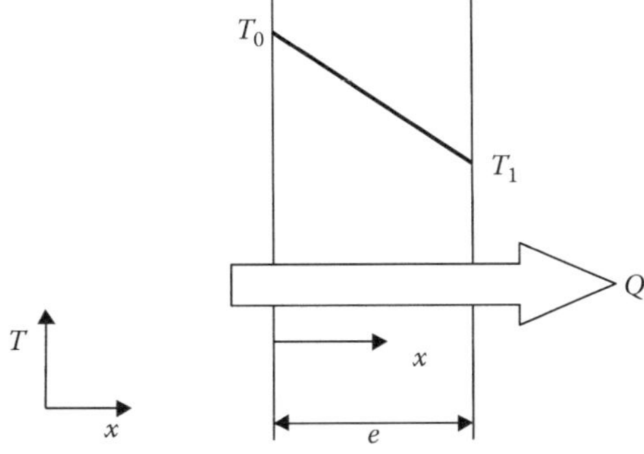

FIGURE 16.4 Temperature profile in a flat wall.

so

$$T = T_0 + \frac{T_1 - T_0}{e} x \tag{16.5}$$

Thus, temperature varies linearly with position. The temperature profile becomes flatter as the thermal conductivity of the solid becomes greater.

To calculate the heat flow rate that crosses such a slab, the heat flux is multiplied by the area of the slab:

$$\dot{Q} = qA = -kA\frac{dT}{dx}$$

The temperature profile dT/dx can be known, so

$$\dot{Q} = kA\frac{T_0 - T_1}{e} \tag{16.6}$$

In many cases it is better to express it as

$$\dot{Q} = \frac{T_0 - T_1}{(e/kA)} \tag{16.7}$$

This equation is analogous to Ohm's law, in which $(T_0 - T_1)$ represents the difference of the thermal potential and (e/kA) is the resistance to heat flow.

It is assumed that the medium is isotropic, so the thermal conductivity remains constant. In the case that the thermal conductivity varies linearly with temperature, the same type of equation can be used, but a value of conductivity that is the arithmetical mean between the values corresponding to the temperatures T_0 and T_1 must be used.

Usually, coatings are placed on walls to avoid heat loss. Generally, such coatings do not consist of one material only, but rather several coatings can be used, as shown in Figure 16.5.

For the ith resistance, the temperature between layer i and $i - 1$ will be T_{i-1}, and between layer i and $i + 1$ it will be T_i.

In the same way as for one layer, it can be demonstrated that for the ith layer, the temperature profile is defined by the equation

$$T(x_i) = T_{i-1} + \frac{T_i - T_{i-1}}{e_i} x_i \tag{16.8}$$

where

e_i is the thickness of layer i
x_i is the positional coordinate of layer i

Since there is no energy accumulation, the heat flow that crosses each layer is the same; hence

$$\dot{Q} = q_i A = \frac{T_0 - T_1}{(e_1/k_1 A)} = \frac{T_1 - T_2}{(e_2/k_2 A)} = \cdots = \frac{T_{N-1} - T_N}{(e_N/k_N A)}$$

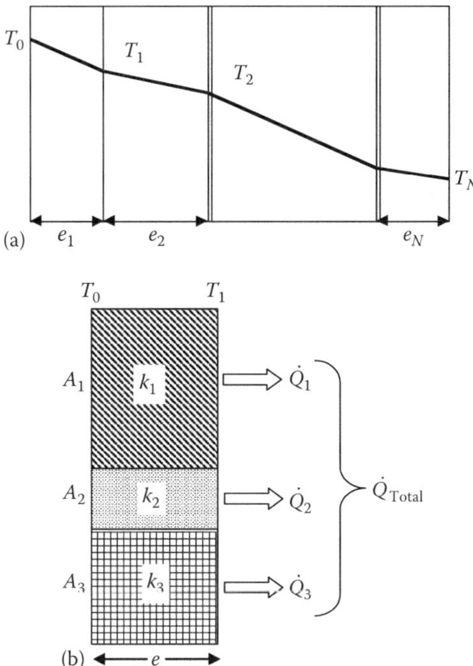

FIGURE 16.5 (a) Set of flat layers in series. (b) Heat transfer in a parallel slab.

Rearranging this equation according to the properties of the ratios,

$$\dot{Q} = \frac{T_0 - T_N}{\sum_{i=1}^{N} \left(e_i / k_i A\right)} \tag{16.9a}$$

Sometimes there are cases in which mixed layers having the same thickness exist, and then the heat transfer through the layers is considered as transfer through parallel resistances (Figure 16.5b). In this case, the plane layer is composed of three parallel resistances that possess the same thickness (e), and for a given section, the temperature is the same for all resistances. The global heat flow will be the sum of the flows that cross each one of the layers in parallel:

$$\dot{Q} = \dot{Q}_1 + \dot{Q}_2 + \dot{Q}_3 = \frac{T_0 - T_1}{\left(e/k_1 A_1\right)} + \frac{T_0 - T_1}{\left(e/k_2 A_2\right)} + \frac{T_0 - T_1}{\left(e/k_3 A_3\right)} \tag{16.9b}$$

16.2.1.2 Cylindrical Layer

The most common problem involving cylindrical bodies is related to pipes. They might transport fluids at temperatures higher or lower than room temperature, so they are coated with insulators to avoid heat transfer to the outside.

A hollow cylinder of radius r_0 covered with an insulator of thickness e is proposed to study this problem, as shown in Figure 16.6. The temperature of the fluid that circulates inside is T_0, while the temperature of the external medium is T_1.

The fundamental equation of heat conduction (Equation 16.2), if there is no energy generation, and under a steady state for the case of heat transfer along the radial coordinate r, becomes

$$\frac{1}{r}\frac{\partial}{\partial r}\left(r\frac{T}{\partial r}\right) = 0$$

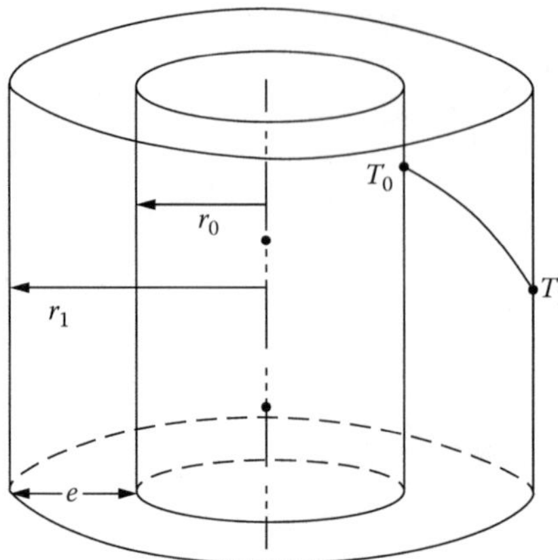

FIGURE 16.6 Temperature profile in a cylindrical layer.

or

$$r\frac{T}{\partial r} = C_1$$

where C_1 is a constant.

This equation can be expressed in total derivatives if temperature is a function only of the radial coordinate. The following equation is obtained after integration:

$$T = C_1 \ln r + C_2$$

where C_1 and C_2 are integration constants whose values are obtained by applying the following boundary conditions:

- For $r = r_0$ $T = T_0$
- For $r = r_1$ $T = T_1$

Hence,

$$T(r) = T_0 - (T_0 - T_1)\frac{\ln(r/r_0)}{\ln(r_1/r_0)} \tag{16.10}$$

Heat flow rate is obtained from Fourier's equation. If the cylinder has a length L, the cross-sectional area of the heat flux will be $A = 2\pi rL$, so such flow rate would be

$$\dot{Q} = qA = -kA\frac{dT}{dr} = -k2\pi rL\frac{dT}{dr}$$

In this equation, dT/dr is obtained by differentiating Equation 16.10, yielding

$$\dot{Q} = k2\pi rL\frac{T_0 - T_1}{\ln(r_1/r_0)} \tag{16.11}$$

If the numerator and the denominator of the second member are multiplied by the thickness $e = r_1 - r_0$, this equation becomes

$$\dot{Q} = \frac{T_0 - T_1}{(e/kA_{ml})} \tag{16.12}$$

where A_{ml} is the mean logarithmic area between the external and internal cylindrical surfaces:

$$A_{ml} = \frac{A_1 - A_0}{\ln\left(A_1/A_0\right)}$$

being $A_0 = 2\pi r_0 L$ and $A_1 = 2\pi r_1 L$.

If instead of only one coating there were N layers, the temperature profile of each layer i would be

$$T(r_i) = T_{i-1} - (T_{i-1} - T_i)\frac{\ln\left(r/r_{i-1}\right)}{\ln\left(r_i/r_{i-1}\right)} \tag{16.13}$$

The heat flux that crosses these layers is expressed as

$$\dot{Q} = \frac{T_0 - T_N}{\sum_{i=1}^{N} e_i/k_i(A_{ml})_i} \tag{16.14}$$

It can be observed that the equations obtained for heat flux are analogous to those obtained for the case of flat layers, with the difference being that for cylindrical geometry the area is the logarithmic average instead of the arithmetic one.

16.2.1.3 Spherical Layer

Consider a hollow sphere covered with one layer of a certain material with thickness $e = r_1 - r_0$, such as the one shown in Figure 16.7.

If temperature is a function only of the radial coordinate and if, also, there is no heat generation and the work is done under steady state, the heat conduction fundamental equation for spherical coordinates (Equation 16.3) becomes

$$\frac{1}{r^2}\frac{\partial}{\partial r}\left(r^2\frac{\partial T}{\partial r}\right) = 0$$

The following equation is obtained from the former:

$$r^2\frac{\partial T}{\partial r} = C_1$$

where C_1 is the integration constant.

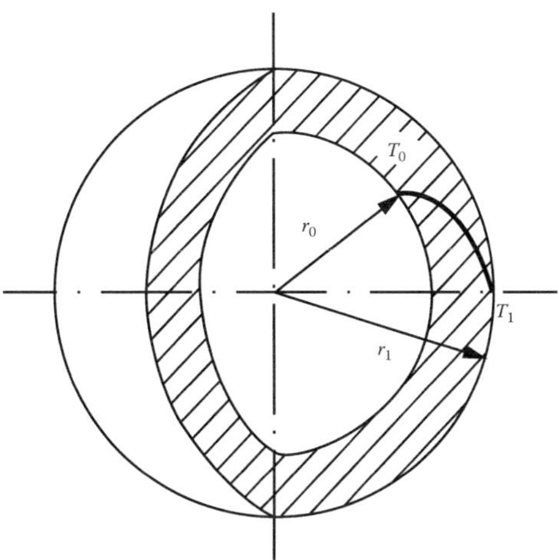

FIGURE 16.7 Thermal profile in a spherical layer.

Since temperature depends only on the radial coordinate, this equation can be expressed as a total derivative that, when integrated, becomes

$$T = C_2 - \frac{C_1}{r}$$

The integration constants of this equation are obtained by applying the following boundary conditions:

- For $r = r_0$ $T = T_0$
- For $r = r_1$ $T = T_1$

yielding a temperature profile according to the expression

$$T(r) = T_0 + (T_1 - T_0)\frac{r_1}{r_1 - r_0}\left(1 - \frac{r_0}{r}\right) \tag{16.15}$$

the heat flow rate will be

$$\dot{Q} = \frac{T_0 - T_1}{(e/kA_{mg})} \tag{16.16}$$

where
e is the layer's thickness
A_{mg} is the mean geometric area of the internal and external spherical areas

This equation was obtained similarly to the previous cases, substituting the expression of the temperature profile in Fourier's equation.

If instead of considering one layer surrounding the sphere, N layers are considered, the temperature profile in one of the intermediate layers i, will be

$$T(r_i) = T_{i-1} + (T_i - T_{i-1})\frac{r_i}{r_i - r_{i-1}}\left(1 - \frac{r_{i-1}}{r}\right) \tag{16.17}$$

The expression that allows calculation of the heat flow rate is

$$\dot{Q} = \frac{T_0 - T_N}{\sum_{i=1}^{N}(e_i/k_i A_{mgi})} \tag{16.18}$$

This equation is analogous to those obtained for other types of geometry studied, except that the area in the spherical geometry is the mean geometric area of the external and internal surfaces of the spherical layers.

16.2.2 2D Heat Conduction

Heat conduction in only one direction has been studied in previous sections. However, there are cases in which the problem cannot be reduced to 1D conduction.

If heat conduction occurs in two directions and there is no heat accumulation or generation, the following equation is obtained for rectangular coordinates:

$$\frac{\partial^2 T}{\partial x^2}+\frac{\partial^2 T}{\partial y^2}=0 \quad (16.19)$$

It can be observed that temperature depends on coordinates x and y.

This equation can be solved in different ways, although it is convenient to express it as finite differentials. Once expressed this way, analytical and numerical methods are applied to solve the equation. In addition to these methods, there are graphical and analogic methods that allow the solution of Laplace equation (Equation 16.4) in two directions.

In order to express Equation 16.19 in finite differentials, the solid in which an energy balance will be carried out is divided according to an increment of x and another of y (Figure 16.8), the two directions in which temperature varies.

The nodes resulting from the division are indicated according to direction x by subindex m and in direction y by subindex n.

The temperature around any node can be expressed as a function of the node. Taylor series expansion including up to the third term yields

$$T_{m+1,n}=T_{m,n}+\left(\frac{\partial T}{\partial x}\right)_{m,n}\Delta x+\frac{1}{2}\left(\frac{\partial^2 T}{\partial x^2}\right)_{m,n}(\Delta x)^2$$

$$T_{m-1,n}=T_{m,n}-\left(\frac{\partial T}{\partial x}\right)_{m,n}\Delta x+\frac{1}{2}\left(\frac{\partial^2 T}{\partial x^2}\right)_{m,n}(\Delta x)^2$$

$$T_{m,n+1}=T_{m,n}+\left(\frac{\partial T}{\partial y}\right)_{m,n}\Delta y+\frac{1}{2}\left(\frac{\partial^2 T}{\partial y^2}\right)_{m,n}(\Delta y)^2$$

$$T_{m,n-1}=T_{m,n}-\left(\frac{\partial T}{\partial y}\right)_{m,n}\Delta y+\frac{1}{2}\left(\frac{\partial^2 T}{\partial y^2}\right)_{m,n}(\Delta y)^2$$

The first and second derivatives of temperature at point (m, n) according to directions x and y can be obtained from this set of equations:

$$\left(\frac{\partial T}{\partial x}\right)_{m,n}=\frac{T_{m+1,n}-T_{m-1,n}}{2\Delta x}$$

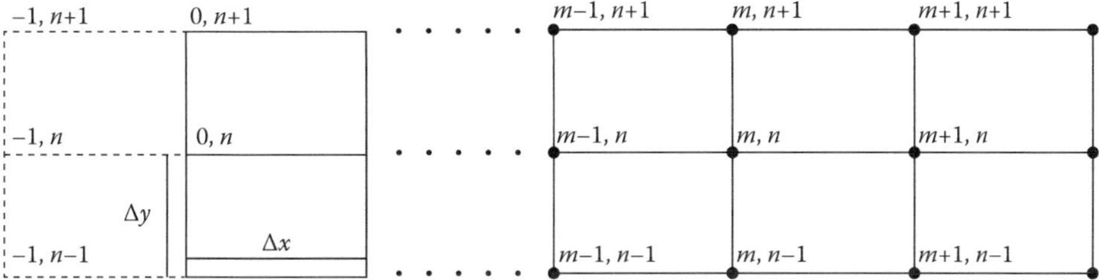

FIGURE 16.8 Notation for the numerical solution of 2D heat conduction.

$$\left(\frac{\partial T}{\partial y}\right)_{m,n} = \frac{T_{m,n+1} - T_{m,n-1}}{2\Delta y}$$

$$\left(\frac{\partial^2 T}{\partial x^2}\right)_{m,n} = \frac{T_{m+1,n} - 2T_{m,n} + T_{m-1,n}}{(\Delta x)^2}$$

$$\left(\frac{\partial^2 T}{\partial y^2}\right)_{m,n} = \frac{T_{m,n+1} - 2T_{m,n} + T_{m,n-1}}{(\Delta y)^2}$$

If these expressions are substituted in Equation 16.19, it is obtained that

$$\frac{T_{m+1,n} - 2T_{m,n} + T_{m-1,n}}{(\Delta x)^2} + \frac{T_{m,n+1} - 2T_{m,n} + T_{m,n-1}}{(\Delta y)^2} = 0$$

In the case that the division is carried out in such a way that the increments in the directions x and y are equal, $\Delta x = \Delta y$, this equation can be expressed as

$$T_{m,n} = \frac{1}{4}\left(T_{m+1,n} + T_{m-1,n} + T_{m,n+1} + T_{m,n-1}\right) \tag{16.20}$$

This equation indicates that temperature in one node is the arithmetic mean of the nodes around it.

The surrounding conditions can be expressed as an algebraic equation; thus, if the temperature at the wall of the solid is T_0, it can be expressed as

$$T_{0,n} = T_0 \quad \text{for } n = 0, 1, 2, \ldots, N$$

Also, since a maximum or minimum condition exists on the wall,

$$\left(\frac{\partial T}{\partial x}\right)_{0,n} = 0 \quad n = 0, 1, 2, \ldots, N$$

which expressed as finite differences becomes

$$T_{-1,n} = T_{1,n} \quad n = 0, 1, 2, \ldots, N$$

This equation indicates that there are points outside the system that have the same temperature as those symmetric with respect to the wall that belongs to the system.

Once these series of equations are stated, the problem should be easy to solve. Different methods can be used to do that. Next, two numerical methods to solve the equations are explained: Liebman's method and the relaxation method.

16.2.2.1 Liebman's Method

As obtained previously (Equation 16.20), the temperature at one node is the arithmetic mean of the temperatures around it.

This method distinguishes two types of alternating points (Figure 16.9), those marked with a circle and those marked with a cross. In order to find the temperature of the points, first the temperature

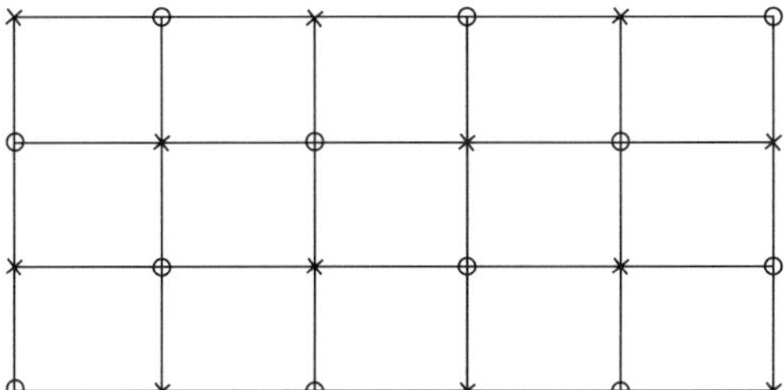

FIGURE 16.9 Division of a solid by Liebman's method.

of alternated points is given a value, for example, those marked with a circle (o). The temperatures corresponding to the points marked with a cross (x) can be calculated from the assumed temperatures using Equation 16.20. This process is repeated until the temperatures of two consecutive calculations coincide. At this point, the temperatures of each node are the temperatures we were looking for and the temperature profile is obtained.

16.2.2.2 Relaxation Method

In this method, the temperatures of all the nodes in the mesh are assumed. It is very difficult to guess correctly and comply with Equation 16.20 on the first attempt; however, it is possible to obtain a residual such as the following:

$$R = T_{m+1,n} + T_{m-1,n} + T_{m,n+1} + T_{m,n-1} - 4T_{m,n}$$

Once the values of the temperatures are supposed, the values of the residual of each node will be obtained and they should tend to zero, indicating that the supposed temperatures are the desired ones. Since this is difficult to do, temperatures should be successively corrected, according to the expression

$$\left(T_{m,n}\right)^{*} = T_{m,n} + \frac{R_{m,n}}{4}$$

The residuals of each point are calculated again using the corrected temperatures and the process is repeated until the residuals are negligible. In this case the temperatures at each point are the temperatures we were looking for.

Once the temperature of all the points is calculated using one of these methods, the temperature profile is known. It is possible to calculate the heat flow rate that crosses a given area from such a profile using the following expression:

$$\dot{Q} = \int_{S} k\nabla T \, dS \tag{16.21}$$

where dS is the cross-sectional area. Calculation of heat flow rate can be difficult, so graphical methods are used.

In short, the general problem in 2D heat conduction is the calculation of the temperature distribution, and once it is known, the calculation of the heat flow rate follows. Two numerical methods were shown to calculate the heat distribution or temperature profile, but other methods are explained elsewhere, such as graphical, analog, and matrix ones.

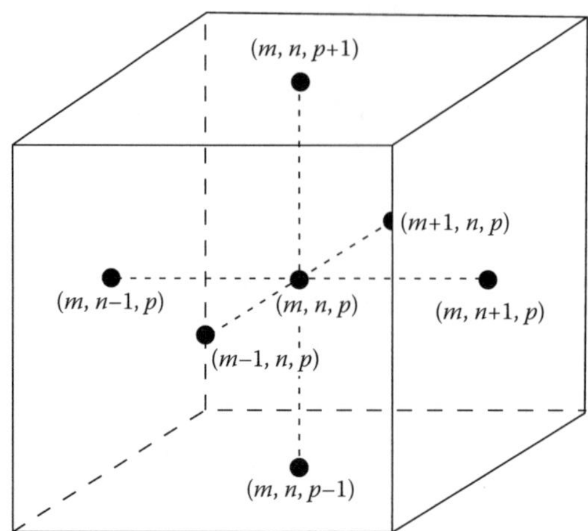

FIGURE 16.10 Division into slots for 3D heat conduction.

16.2.3 3D Heat Conduction

When heat conduction in a solid is such that three directions should be considered, the same techniques developed for the case of 2D conduction can be applied.

Consider a parallelepiped (Figure 16.10) through which heat is transmitted under steady state and also all the conditions surrounding it are known. When a heat balance expressed as finite differences is applied, the temperature of the point with coordinates (m, n, p) when the increments of the three space coordinates are equal $(\Delta x = \Delta y = \Delta z)$ is

$$T_{m,n,p} = \frac{1}{6}\left(T_{m+1,n,p} + T_{m-1,n,p} + T_{m,n+1,p} + T_{m,n-1,p} + T_{m,n,p+1} + T_{m,n,p-1}\right)$$

This equation indicates that the temperature of a node, for 3D problems without heat generation and under steady state, is the arithmetic mean of the temperature of the surrounding nodes. The procedure is analogous to the case of 2D heat conduction.

16.3 HEAT CONDUCTION UNDER UNSTEADY STATE

In all of the cases studied in Section 16.2, it was assumed that the temperature of any point of the solid remains constant with time. However, there are cases in which the temperature inside the solid, besides changing with position, also changes with time. Such is the case in freezing and thawing, in which it is desirable to know the time needed to reach a certain temperature at a given point of the solid or to know the temperature of such a point after a certain time. It should be taken into account that the process is developed under unsteady-state regime.

The following is obtained from the heat conduction fundamental equation in the case of rectangular coordinates (Equation 16.1):

$$\rho \hat{C}_P \frac{\partial T}{\partial t} = k\nabla^2 T + \tilde{q}_G \tag{16.1}$$

Taking into account that the thermal diffusivity α is defined by

$$\alpha = \frac{k}{\rho \hat{C}_P} \tag{16.22}$$

then Equation 16.1 is transformed into

$$\frac{\partial T}{\partial t} = \alpha \ \nabla^2 T + \tilde{q}_G \quad (16.23)$$

This equation is valid for any type of coordinates, but the Laplacian ∇^2 operator should be expressed in the adequate form in each case.

The mathematical treatments in the following sections will be conducted in rectangular coordinates, but it should be noted that for other types of coordinates the mathematical treatment is similar. In all cases presented next, it will be assumed that there is no heat generation.

16.3.1 1D Heat Conduction

If a semi-infinite solid is considered, as shown in Figure 16.11, in which the thickness in direction x is infinite, it can be supposed that the heat flow that crosses such a solid does so exclusively in the x direction. If inside the solid there is no heat generation, Equation 16.23 can be expressed as

$$\frac{\partial T}{\partial t} = \alpha \frac{\partial^2 T}{\partial x^2} \quad (16.24)$$

This equation is difficult to solve, but different methods such as analytical, numerical, or graphical can be applied. Some of these methods will be explained in the following sections.

16.3.1.1 Analytical Methods

The analytical solution of Equation 16.24 is obtained when the boundary conditions that allow us to integrate such an equation are specified.

The integration of this equation is possible only in some cases, such as semi-infinite solids. If at the beginning the solid has a temperature T_0 and at a given time the wall reaches a temperature T_e, it is considered that the temperature of a point far away from the wall continues to be T_0 at any instant. Therefore, the conditions to integrate Equation 16.24 are

- Initial condition: For $\forall x$ $t = 0$ $T = T_0$
- Boundary condition: For $\forall t$ $x \to \infty$ $T = T_0$
  For $\forall t$ $x = 0$ $T = T_e$

The boundary conditions assume that the wall is always at a constant temperature T_e; thus, it is an isothermal condition.

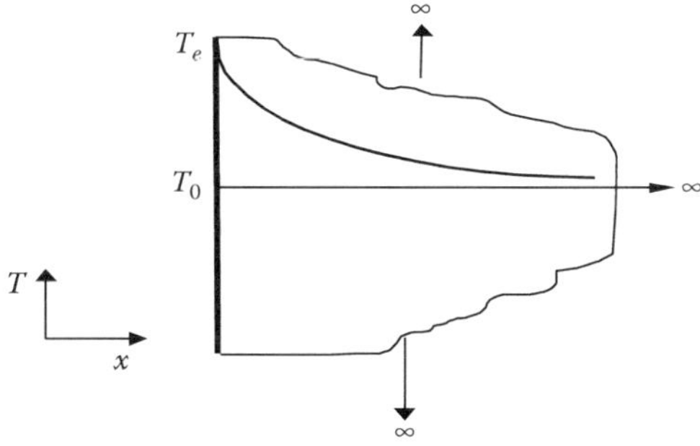

FIGURE 16.11 Semi-infinite solid.

The solution of Equation 16.24 is obtained under these boundary conditions, and it can be demonstrated that the equation that describes the temperature distribution in the slab is

$$\frac{T_e - T}{T_e - T_0} = erf\left(\frac{x}{2\sqrt{\alpha t}}\right) \quad (16.25)$$

In this equation *erf* is the Gauss error function, defined by

$$erf(\eta) = \frac{2}{\sqrt{\pi}} \int_0^{\eta} \exp(-\eta^2)\, d\eta \quad (16.26)$$

where

$$\eta = \frac{x}{2\sqrt{\alpha t}}$$

Gauss error function can be tabulated for the different values of position x. Such values are presented in Table 16.1 at the end of this chapter.

TABLE 16.1
Gauss Error Function

$$erf(\eta) = \frac{2}{\sqrt{\pi}} \int_0^{\eta} \exp(-\eta^2)\, d\eta$$

$$\eta = \frac{x}{2\sqrt{\alpha t}}$$

η	*erf* (η)	η	*erf* (η)
0.00	0.00000	1.00	0.84270
0.04	0.45110	1.10	0.88020
0.08	0.09008	1.20	0.91031
0.12	0.13476	1.30	0.93401
0.16	0.17901	1.40	0.95228
0.20	0.22270	1.50	0.96610
0.24	0.25670	1.60	0.97635
0.28	0.30788	1.70	0.98379
0.32	0.34913	1.80	0.98909
0.36	0.38933	1.90	0.99279
0.40	0.42839	2.00	0.99532
0.44	0.46622	2.10	0.99702
0.48	0.50275	2.20	0.99814
0.52	0.53790	2.30	0.99886
0.56	0.57162	2.40	0.99931
0.60	0.60386	2.50	0.99959
0.64	0.63459	2.60	0.99976
0.68	0.66278	2.70	0.99987
0.72	0.69143	2.80	0.99993
0.76	0.71754	2.90	0.99996
0.80	0.74210	3.00	0.99998
0.84	0.76514	3.20	0.999994
0.88	0.78669	3.40	0.999998
0.92	0.80677	3.60	1.00000
0.96	0.82542		

Another case in which it is possible to find an analytical solution to Equation 16.24 is when the semi-infinite solid has a simple geometry. Such is the case for infinite slabs with finite thickness or for cylinders with infinite height. It is also possible to find analytical solutions for spherical solids.

The case of heat transfer through an infinite slab with thickness $2x_0$ will be studied next. Initially, the slab is at a temperature T_0 and it is immersed in a fluid at a temperature T_e. In this case the surrounding condition is convection, since there is heat transfer from the fluid to the solid's wall by a convection mechanism, while within the solid the heat is transferred by conduction. Thus, it can be written that

$$h\left(T_e - T_P\right) = -k\left(\frac{\partial T}{\partial x}\right)_P \tag{16.27}$$

where
T_P is the temperature of the slab wall in contact with the fluid at a given time
h is the individual convective heat transfer coefficient between the fluid and the solid

The boundary conditions to obtain an analytical solution are

- Initial condition: For $\forall x$ $\quad t = 0 \quad T = T_0$
- Boundary condition: For $t \to \infty$ $\quad x = x_0 \quad T = T_e$

For $\forall t$ $\quad x = 0 \quad \dfrac{\partial T}{\partial x} = 0$

This last boundary condition is due to the fact that in the center of the slab ($x = 0$), an optimum maximum or minimum condition should exist, depending on whether the solid becomes cooler or hotter.

The analytical solution of the problem becomes easier if the variables are expressed in a dimensionless way. Hence, the following dimensionless temperature, time, and position variables are defined:

$$Y = \frac{T_e - T}{T_e - T_0} \tag{16.28a}$$

$$\tau = \left(\mathrm{Fo}\right) = \frac{\alpha t}{(x_0)^2} \tag{16.28b}$$

$$n = \frac{x}{x_0} \tag{16.28c}$$

It should be noted that the dimensionless time variable is an expression of Fourier's number.

The substitution of the dimensionless variables in Equation 16.24 yields

$$\frac{\partial T}{\partial t} = \frac{\partial^2 Y}{\partial n^2} \tag{16.29}$$

on the boundary conditions

- For $\forall n$ $\quad t = 0 \quad Y = 1$
- For $n = 1$ $\quad t \to \infty \quad Y = 0$
- For $n = 0$ $\quad \forall \tau \quad \dfrac{\partial Y}{\partial n} = 0$

Equation 16.29 can be solved by separation of variables, obtaining the solution

$$Y = 4\sum_{i=1}^{\infty} \frac{\text{sen}A_i}{2A_i + \text{sen}(2A_i)} \cos(A_i n)\exp(-A_i \tau) \tag{16.30}$$

In this expression A_i are the infinite solutions of the equation

$$A_i \, tg(A_i) = \frac{1}{m} \tag{16.31}$$

where $1/m$ is the *Biot number*, defined according to

$$(\text{Bi}) = \frac{1}{m} = \frac{hx_0}{k} \tag{16.32}$$

These solutions are valid for values of the Biot number between 0.1 and 40, for which there is a combined conduction–convection mechanism for heat transfer. In order to obtain the different values of A_i, it is necessary to carry out a calculation process. With the value of the Biot modulus, Equation 16.31 is solved, obtaining the A_i values, satisfying this equation. Depending on the i value, the A_i value is compressed between

$$0 < A_1 < \pi/2$$

$$\pi < A_2 < 3\pi/2$$

$$2\pi < A_3 < 5\pi/2$$

$$\ldots$$

When Fourier's number value is greater than 0.25 (Fo > 0.25), in the calculation of the dimensionless temperature, it is sufficient to use the first term of Equation 16.30.

If the Biot number value is lower than 0.1, the dimensionless temperature can be obtained from the expression

$$Y = \exp(-\text{Bi} \cdot \text{Fo}) \tag{16.33}$$

For the Biot number values greater than 40, Equation 16.30 transforms to

$$Y = 4\sum_{i=1}^{\infty} -\frac{(-1)^i}{(2i-1)\pi} \cos\left(\frac{2i-1}{2}\pi n\right) \exp\left(-\left(\frac{2i-1}{2}\pi\right)^2 \cdot \text{Fo}\right) \tag{16.34}$$

It is considered that this equation is accomplished for the Biot number values greater than 40 (Bi > 40), and in the case that Fourier's number is greater than 0.25, the solution only takes the first term of Equation 16.34.

This analytical solution was obtained for an infinite slab with thickness $2x_0$. Analogous solutions can be obtained for a cylinder of radius x_0 and infinite height, as well as for a sphere of radius x_0, although the solution of Equation 16.29 is more complex.

For a solid cylinder, the solution of the equation in dimensionless variables can be expressed as

$$Y = 2\sum_{i=1}^{\infty} \frac{J_1(A_i)}{J_0^2(A_i)+J_1^2(A_i)} J_0(A_i n)\exp\left(-A_i^2\text{Fo}\right) \tag{16.35}$$

where J_0 and J_1 are the first class and zero order Bessel functions, respectively. In this case, this expression is obtained:

$$\frac{A_i \cdot J_1(A_i)}{J_0(A_i)} = \text{Bi} \tag{16.36}$$

The Bessel functions are tabulated (see Appendix A), for different values of A_i. In this case, to obtain the Bessel functions and A_i values, it is necessary to obtain the values of these variables for each value of i. Depending on the ith value, the A_i value is compressed between

$$0 < A_1 < 2.405 < A_2 < 5.520 < A_3 < 8.654 < A_4 < 11.792 \ldots$$

At the same time, if the value of Fourier's number is greater than 0.25, it should only take the first term of Equation 16.35.

For a sphere, the equation in dimensionless variables also presents an analytical solution, in such a way that the dimensionless temperature can be calculated from the following equation:

$$Y = 4\sum_{i=1}^{\infty} \frac{\text{sen}A_i - A_i \cdot \cos(A_i)}{2A_i - \text{sen}(2A_i)} \frac{\text{sen}(A_i \cdot n)}{A_i \cdot n} \exp\left(-A_i^2 \cdot \text{Fo}\right) \tag{16.37}$$

where A_i are the infinite solutions for the equation

$$1 - \frac{A_i}{tg(A_i)} = \text{Bi} \tag{16.38}$$

The variation limits for A_i depend on the ith value, yielding

$$0 < A_1 < \pi < A_2 < 2\pi < A_3 < 3\pi < A_4 < 4\pi \ldots$$

For Fourier's numbers greater than 0.25, only the first term of Equation 16.37 should be taken.

The solution of the unsteady-state heat transfer equation (Equation 16.24 or 16.29) is the same as the solutions of the unsteady-state mass transfer equation given in Chapter 7, with a corresponding relationship between mass and heat transfer variables.

Such analytical solutions are usually represented as graphs (Figures 16.12 through 16.14), making the solution to different problems easier to be calculated. It should be noticed that these figures are the graphical solution of a dimensionless equation expressed as partial derivatives (Equation 16.29) under specific boundary conditions. For this reason, these graphs can be used for all problems expressed in this type of differential equations and boundary conditions.

In addition to these solutions, there is a numerical method that allows us to evaluate the temperature in the center of a solid. As noted previously, it is difficult to obtain an exact solution to this type

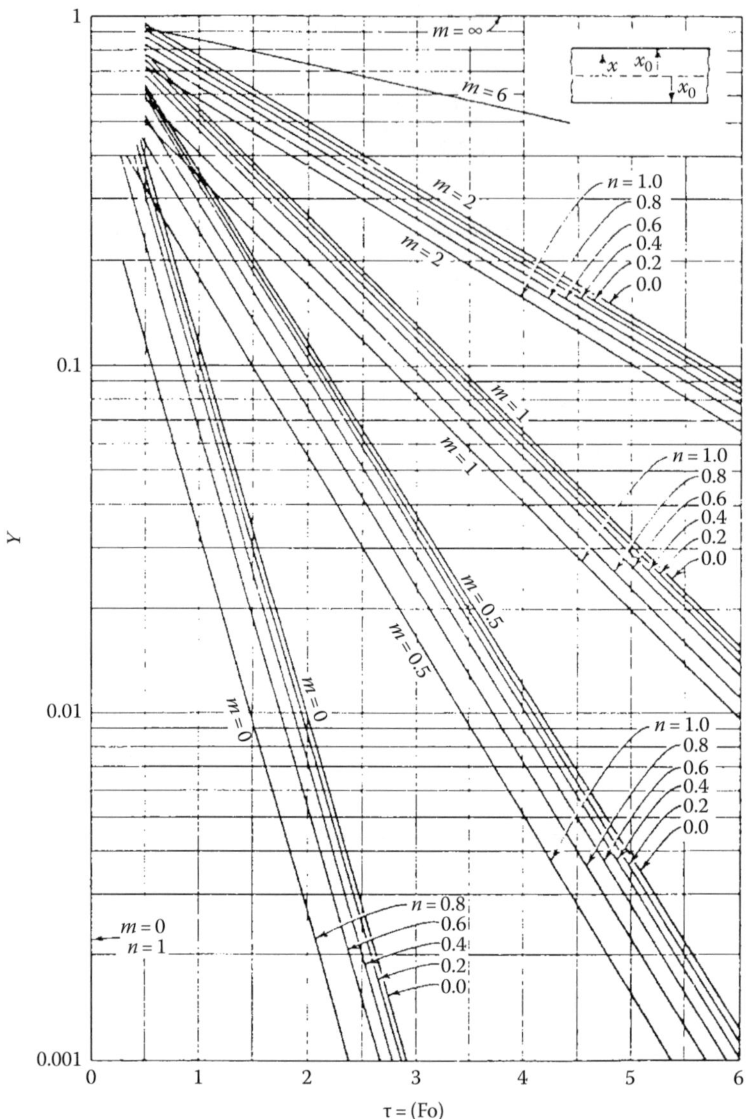

FIGURE 16.12 Dimensionless temperature as a function of time and position for an infinite slab.

of problem. However, for simple geometries, a series is obtained as a solution to the problem. For the specific case where the value of Fourier's number (Fo) is larger than 0.2, only the first term of the series is important. In this way, the solution to Equation 16.24 is a function of Fourier's number, according to the expression

$$Y = \frac{T_e - T_f}{T_e - T_0} = C_1 \exp\left(-\xi^2 \text{Fo}\right) \tag{16.39}$$

where T_f is the temperature at the center of the solid at time t, while T_e and T_0 are the temperatures of the external fluid and at a zero time of the solid, respectively. The parameters C_1 and ξ of this equation can be found in Table 16.2 and are a function of the value of the Biot

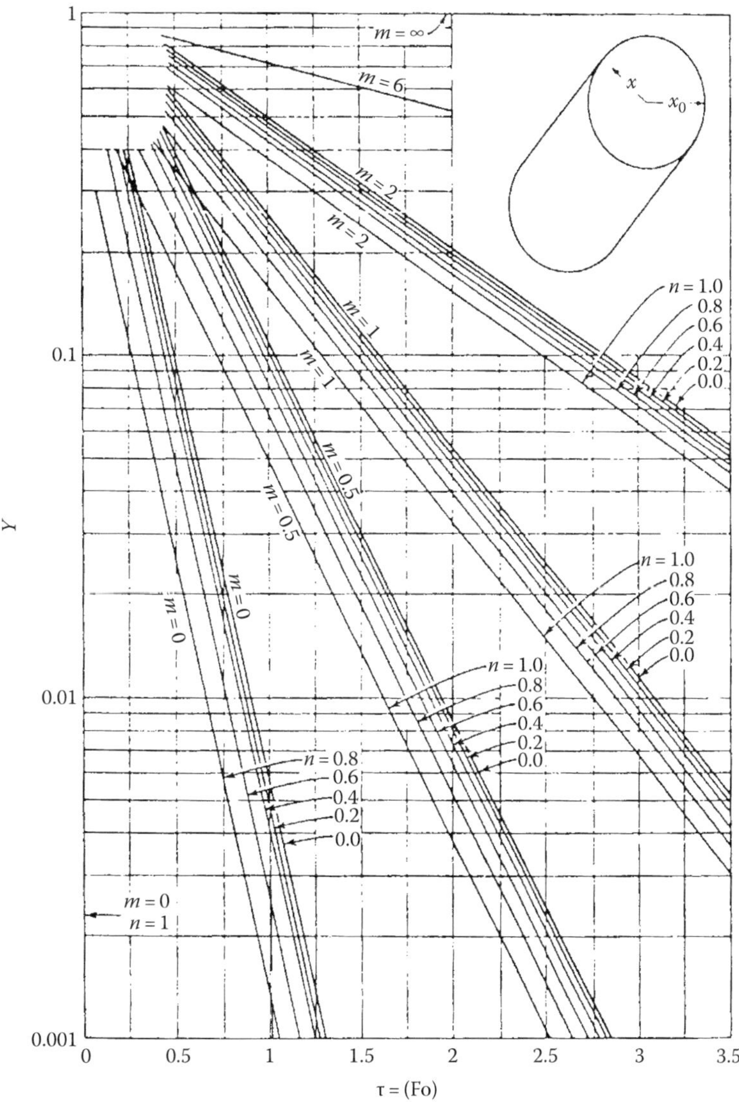

FIGURE 16.13 Dimensionless temperature as a function of time and position for a cylinder with infinite height.

number and the type of solid considered. It should be taken into account that Fourier's number is defined according to the expression

$$(\text{Fo}) = \frac{\alpha t}{L^2} \tag{16.40}$$

where

α is the thermal diffusivity
L is the characteristic length of the solid
t is the time

In those cases in which the value of the Biot number is lower than 0.1, convection controls the heat transfer process. Since convection is the controlling mechanism, the temperature of

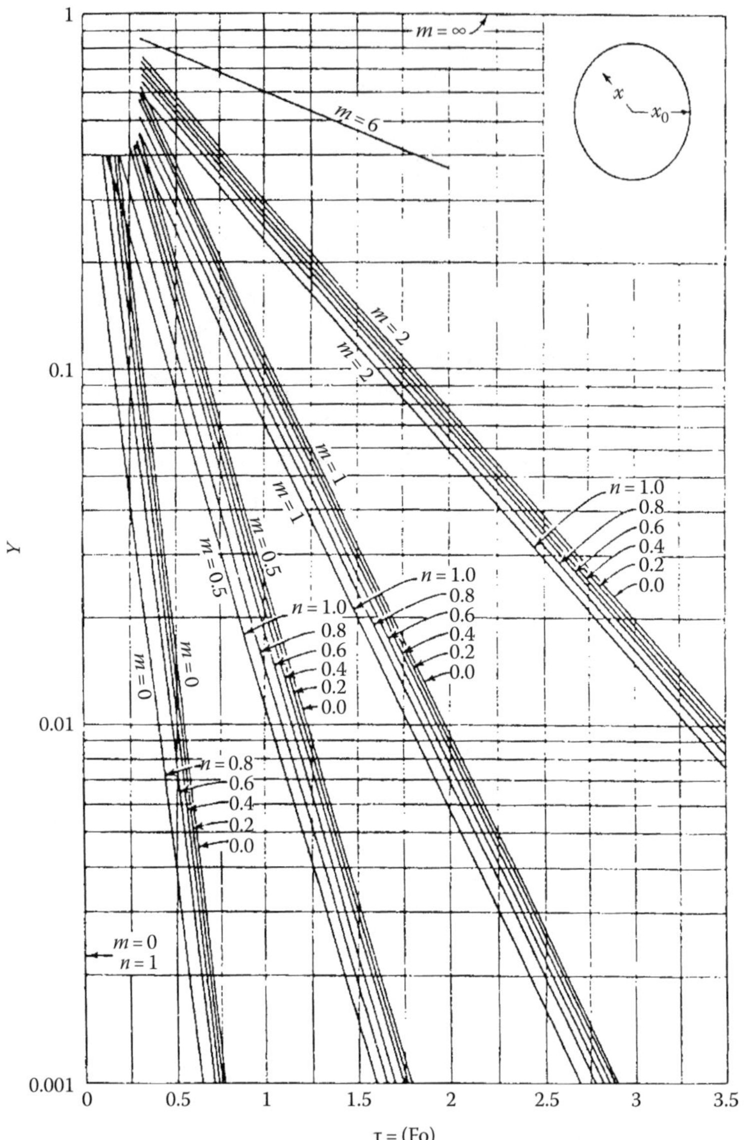

FIGURE 16.14 Dimensionless temperature as a function of time and position for a sphere.

the solid can be considered to be uniform. It is possible to obtain the following solution to the problem from an energy balance:

$$Y = \frac{T_e - T_f}{T_e - T_0} = \exp\left(-\frac{hAt}{m\hat{C}_P}\right) \tag{16.41a}$$

where

A is the surface area of the solid

$\hat{C}_P$ is its specific heat

m is the mass of the solid, while T_e and T_0 are the temperatures of the external fluid and initial of the solid

T_f is the temperature of the solid at time t

TABLE 16.2
Coefficients for the First Term of Solutions for 1D Heat Conduction under Unsteady State

	Flat Wall		Infinite Cylinder		Sphere	
Bi	ξ	C_1	ξ	C_1	ξ	C_1
0.01	0.0998	1.0017	0.1412	1.0025	0.1730	1.0030
0.02	0.1410	1.0033	0.1995	1.0050	0.2445	1.0060
0.03	0.1732	1.0049	0.2439	1.0075	0.2989	1.0090
0.04	0.1987	1.0066	0.2814	1.0099	0.3450	1.0120
0.05	0.2217	1.0082	0.3142	1.0124	0.3852	1.0149
0.06	0.2425	1.0098	0.3438	1.0148	0.4217	1.0179
0.07	0.2615	1.0114	0.3708	1.0173	0.4550	1.0209
0.08	0.2791	1.0130	0.3960	1.0197	0.4860	1.0239
0.09	0.2956	1.0145	0.4195	1.0222	0.5150	1.0268
0.10	0.3111	1.0160	0.4417	1.0246	0.5423	1.0298
0.15	0.3779	1.0237	0.5376	1.0365	0.6608	1.0445
0.20	0.4328	1.0311	0.6170	1.0483	0.7593	1.0592
0.25	0.4801	1.0382	0.6856	1.0598	0.8448	1.0737
0.30	0.5218	1.0450	0.7465	1.0712	0.9208	1.0880
0.40	0.5932	1.0580	0.8516	1.0932	1.0528	1.1164
0.50	0.6533	1.0701	0.9408	1.1143	1.1656	1.1441
0.60	0.7051	1.0814	1.0185	1.1346	1.2644	1.1713
0.70	0.7506	1.0919	1.0873	1.1539	1.3525	1.1978
0.80	0.7910	1.1016	1.1490	1.1725	1.4320	1.2236
0.90	0.8274	1.1107	1.2048	1.1902	1.5044	1.2488
1.0	0.8603	1.1191	1.2558	1.2071	1.5708	1.2732
2.0	1.0769	1.1795	1.5995	1.3384	2.0288	1.4793
3.0	1.1925	1.2102	1.7887	1.4191	2.2889	1.6227
4.0	1.2646	1.2287	1.9081	1.4698	2.4556	1.7201
5.0	1.3138	1.2402	1.9898	1.5029	2.5704	1.7870
6.0	1.3496	1.2479	2.0490	1.5253	2.6537	1.8338
7.0	1.3766	1.2532	2.0937	1.5411	2.7165	1.8674
8.0	1.3978	1.2570	2.1286	1.5526	2.7654	1.8921
9.0	1.4149	1.2598	2.1566	1.5611	2.8044	1.9106
10.0	1.4289	1.2620	2.1795	1.5677	2.8363	1.9249
20.0	1.4961	1.2699	2.2881	1.5919	2.9857	1.9781
30.0	1.5202	1.2717	2.3261	1.5973	3.0372	1.9898
40.0	1.5325	1.2723	2.3455	1.5993	3.0632	1.9942
50.0	1.5400	1.2727	2.3572	1.6002	3.0788	1.9962
100.0	1.5552	1.2731	2.3809	1.6015	3.1102	1.9990

If the expressions of the Biot and Fourier's numbers are taken into account (Equations 16.32 and 16.40), the last equation can be expressed in a dimensionless form as

$$Y = \frac{T_e - T_f}{T_e - T_0} = \exp\left[-(\mathrm{Bi})(\mathrm{Fo})\right] \tag{16.41b}$$

16.3.1.2 Numerical and Graphical Methods

Besides the analytical method obtained in the previous section for the solution of problems under unsteady state, numerical or graphical methods can be applied.

The numerical solution would be applied to simple cases such as slabs. To do this, the solid is divided into cells and it is observed how the temperature varies in each cell with time.

The equation for the heat balance under unsteady state, without heat generation, for rectangular coordinates in one direction only is

$$\frac{\partial T}{\partial t} = \alpha \frac{\partial^2 T}{\partial x^2} \tag{16.24}$$

This equation can be expressed as finite differences, so its partial derivatives are expressed as

$$\frac{\partial T}{\partial t} = \frac{T(x, t+\Delta t) - T(x, t)}{\Delta t} \tag{16.42}$$

$$\frac{\partial^2 T}{\partial x^2} = \frac{T(x+\Delta x, t) - 2T(x, t) + T(x-\Delta x, t)}{(\Delta x)^2} \tag{16.43}$$

Substitution of these derivatives in the last equation yields

$$T(x, t+\Delta t) - T(x, t) = \frac{\Delta t}{(\Delta x)^2}\alpha\left[T(x+\Delta x, t) - 2T(x, t) + T(x-\Delta x, t)\right]$$

If Fourier's number, which is defined by Equation 16.40,

$$(\text{Fo}) = \frac{\alpha \Delta t}{(\Delta x)^2}$$

is substituted in the last equation, an expression that gives the temperature of one cell in the solid for a given time ($t + \Delta t$) later than time t is obtained:

$$T(x, t+\Delta t) = \text{Fo}\left[T(x+\Delta x, t) + T(x-\Delta x, t) + \left(\frac{1}{\text{Fo}} - 2\right)T(x, t)\right] \tag{16.44}$$

The selection of the value of Fourier's number is limited, since for values of F_0 greater than ½, errors accumulate as the calculation process advances.

The selection of the value of F_0 fixes the time increase Δt since

$$\Delta t = \text{Fo}\frac{(\Delta x)^2}{\alpha} \tag{16.45}$$

In the case where a value of $F_0 = ½$ is selected, the following is obtained:

$$T(x, t+\Delta t) = \frac{1}{2}\left[T(x+\Delta x, t) + T(x-\Delta x, t)\right] \tag{16.46}$$

which points out that the temperature of a cell is the arithmetic mean of the temperatures of the adjacent cells in the last time period.

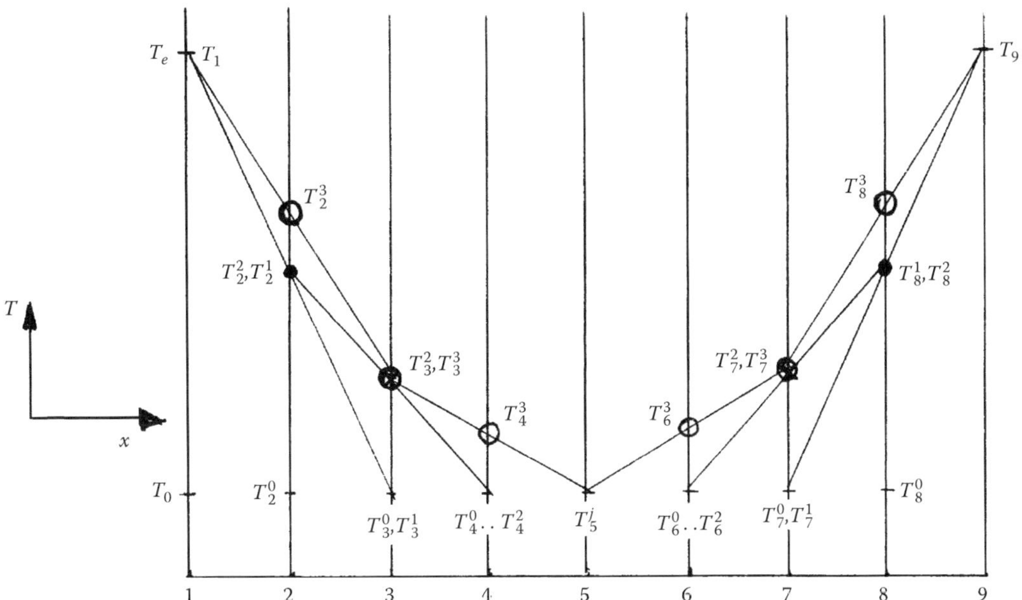

FIGURE 16.15 Division of a solid and evolution of the temperature profile.

If at the beginning the solid of Figure 16.15 was at temperature T_0 and the external faces were exposed to a temperature T_e, the way to operate would be explained as follows. The solid is divided into an odd number of cells (in this case 9), and the temperature profiles for the different time periods are

- For $t = 0$ $\quad T_1 = T_9 = T_e$

$$T_2 = T_3 = \cdots = T_8 = T_0$$

- For $t = 0 + \Delta t$ $\quad T_1 = T_9 = T_e$

$$T_2 = T_8 = (T_0 + T_e)/2$$

$$T_3 = T_4 = \cdots = T_7 = T_0$$

- For $t = 0 + 2\Delta t$ $\quad T_1 = T_9 = T_e$

$$T_2 = T_8 \text{ do not change}$$
$$T_3 = T_7 = (T_2 + T_0)/2$$
$$T_4 = T_5 = T_6 = T_0$$

and so on, until reaching the fixed Δt.

When Fourier's number is selected as Fo = ½, a simple graphical solution for the heat conduction problems under unsteady state is obtained. The graphical method based on this procedure is called Binder–Schmidt's method.

According to Equation 16.46, the temperature of a cell is the arithmetic mean between the temperatures of the adjacent cells at a previous time period, so the temperature of the intermediate cell can be obtained graphically by connecting the temperatures on their sides (Figure 16.16).

The Binder–Schmidt method assumes that the solid is divided into equal x increments (Δx) but can only be applied to 1D heat conduction problems. Figure 16.15 shows this graphical method developed for a slab divided into cells in such a way that there are only nine sections. The solid is initially at a temperature T_0 and both external faces are exposed to a temperature T_e. The variation

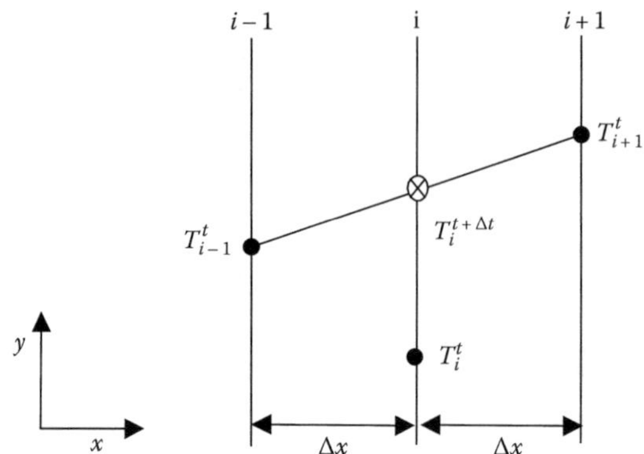

FIGURE 16.16 Graphical method of Binder–Schmidt.

of the temperature profile for the first four time increments can be observed in Figure 16.15. The temperature of each cell is indicated by T_i^j, in which the subindex *i* indicates the position of the plot and the superindex *j* indicates the time period.

16.3.2 2D and 3D Heat Conduction: Newman's Rule

Heat transfer by conduction under unsteady state has been studied in previous sections for infinite solids. However, in practice the slabs and cylinders do not have a length–thickness ratio large enough to be considered as solids of infinite dimensions.

The way in which this problem can be solved is simple. The method is called Newman's rule; Newman demonstrated that for a heating or cooling solid parallelepiped, the solution that describes the temperature variation as a function of time and of position can be expressed as

$$Y = Y_X Y_Y Y_Z \tag{16.47}$$

where

Y is the dimensionless temperature defined in Equation 16.28a

the variables Y_X, Y_Y, and Y_Z are the dimensionless values of temperatures according to directions *x*, *y*, and *z*, respectively, considering the other two directions as infinite

The assumption made is that the parallelepiped consists of three semi-infinite slabs of finite thickness that correspond to each of the three dimensions of the parallelepiped.

The case of a finite cylinder can be considered to be the intersection of an infinite length cylinder and a semi-infinite slab. In this way the solution is given by

$$Y = Y_C Y_L \tag{16.48}$$

where Y_C is the *a*-dimensional value of temperature for a cylinder of infinite length, while Y_L corresponds to a semi-infinite slab.

PROBLEMS

16.1 The heating fluid in a pasteurization process is heated to the processing temperature in an oven with a three-layer wall. The first wall is made of refractory bricks, the second wall of insulating bricks, and the third wall is a 6.3 mm steel veneer for mechanical protection. The temperature of a refractory brick in contact with the oven is 1371°C, while the external temperature of the steel veneer is 38°C. Calculate the thickness of the brick layers if the total heat loss through the oven wall is 1570 kJ/(h m^2).

Data: properties of the materials

Material	Maximum Temperature of Use (°C)	Thermal Conductivity (W/(m °C)) 38°C	1100°C
Refractory brick	1425	3.03	6.23
Insulating brick	1093	1.56	3.03
Steel	—	45.00	—

Since the maximum temperature the insulating brick can stand is 1093°C, this will be the temperature that one of the layers should have to minimize the total thickness. Therefore, T_1 = 1093°C.

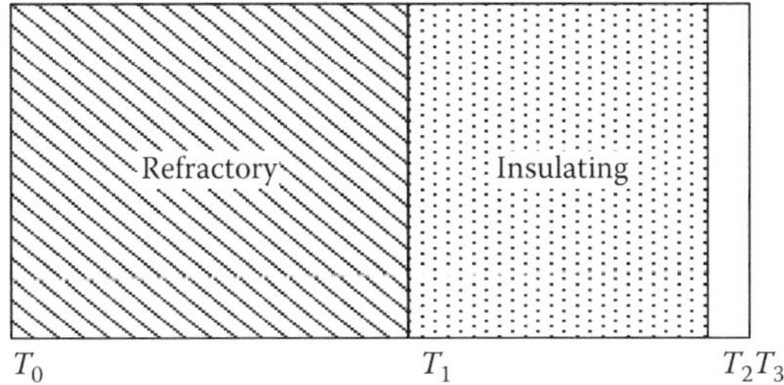

The temperature profile in a layer is given by Equation 16.8:

$$T(x_i) = T_{i-1} + \frac{T_i - T_{i-1}}{e_i} x_i$$

The heat flux is expressed according to Equation 16.6:

$$\frac{\dot{Q}}{A} = q = -k\frac{dT}{dx} = k\frac{T_{i-1} - T_i}{e_i}$$

Refractory brick

$$T = T_0 + \frac{T_1 - T_0}{e_1} x$$

$$q = -k_1 \frac{T_1 - T_0}{e_1} \Rightarrow + \frac{T_1 - T_0}{e_1} = \frac{-q}{k_1} = \frac{-15{,}750\ \text{W/(m}^2)}{6.23\ \text{W/(m °C)}} = -2{,}528\ °\text{C/m}$$

If

$$\left.\begin{aligned} T = T_1 = 1093°\text{C} \\ x = e_1 \end{aligned}\right\} \Rightarrow 1093 = 1371 - 2528 e_1 \Rightarrow e_1 \approx 0.11\ \text{m}$$

The thickness of the refractory brick layer is 0.11 m.

Insulating brick

$$T = T_1 + \frac{T_2 - T_1}{e_2} x$$

If the thermal conductivity k of a material varies linearly with temperature $k = a + bT$, it can be demonstrated that for any intermediate temperature,

$$k_m = \frac{1}{2}\left(k_{T1} + k_{T2}\right)$$

$$k_2 = \frac{1}{2}\left[\left(k_{insul}\right)_{1093°C} + \left(k_{insul}\right)_{38°C}\right] = 2.30 \text{ W/(m °C)}$$

so

$$q = -k_2 \frac{T_2 - T_1}{e_2} \Rightarrow \frac{T_2 - T_1}{e_2} = \frac{-q}{k_2} = \frac{-15{,}750 \text{ W/m}^2}{2.3 \text{ W/(m °C)}} = -6{,}847.8 \text{ °C/m}$$

Applying the limit condition,

$$T = T_2 \quad x = e_2$$

$$T_2 = 1093 - 6874.3 e_2$$

Temperature T_2 should be known to calculate the thickness e_2.

Steel veneer

$$T = T_2 + \frac{T_3 - T_2}{e_3} x$$

$$q = -k_3 \frac{T_3 - T_2}{e_3} \Rightarrow \frac{T_3 - T_2}{e_3} = \frac{-q}{k_3}$$

$$T_2 = T_3 + q \frac{e_3}{k_3}$$

$$T_2 = 38°\text{C} + 15{,}750 \text{ W/m}^2 \frac{0.0063 \text{ m}}{45 \text{ W/(m °C)}} = 40.2 \text{ °C}$$

$$\text{Hence, } 40.2 = 1093 - \left(6847.8\right) e_2 \Rightarrow e_2 = 0.154 \text{ m}$$

The thickness of the insulating layer is 0.154 m.

16.2 A new material is being tested to insulate a refrigerating chamber, and its thermal conductivity should be determined, so a hollow sphere of such material is built. An electric resistance of 15 W is placed in the center of the sphere, and the temperatures of the surfaces are measured with thermocouple, once the steady conditions are reached. Calculate (a) the thermal conductivity of the material and (b) the temperature in an intermediate point of the sphere wall.

Data: internal radius of the sphere, 3 cm; external radius, 8 cm

Temperatures: internal wall, 98°C; external wall, 85°C

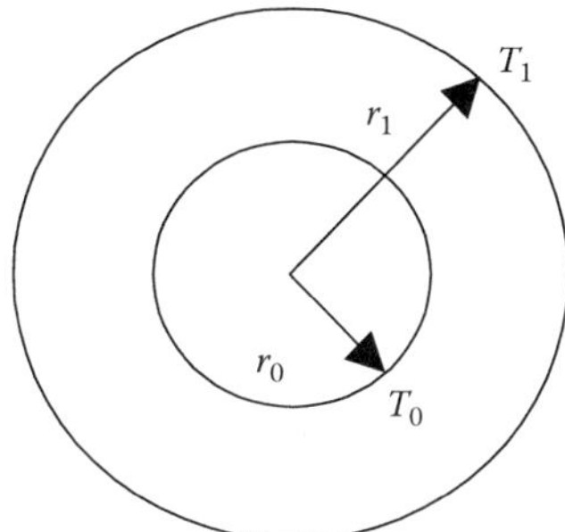

Under steady state, the temperature profile is given by Equation 16.15:

$$T = T_0 + (T_1 - T_0)\frac{r_1}{r_1 - r_0}\left(1 - \frac{r_0}{r}\right)$$

The heat flow rate will be constant and equal to

$$\dot{Q} = qA = -kA\frac{dT}{dr} = -k4\pi r^2\frac{dT}{dr} = 4\pi k(T_0 - T_1)\frac{r_1 r_0}{r_1 - r_0}$$

a. Thermal conductivity of material

$$k = \frac{\dot{Q}(r_1 - r_0)}{4\pi(T_0 - T_1)r_1 r_0}$$

$$\dot{Q} = 15\text{ W} = 15\text{ J/s}$$

$$r_1 = 0.08\text{ m};\ r_0 = 0.03\text{ m}$$

$$T_0 = 98°\text{C};\ T_1 = 85°\text{C}$$

$$k = \frac{(15\text{ W})(0.08 - 0.03)\text{ m}}{4\pi(98 - 85)°\text{C}(0.08\text{ m})(0.03\text{ m})} \approx 1.91\text{ W/(m °C)}$$

b. Temperature in an intermediate point

If it is assumed that

$$r_m = \frac{r_0 + r_1}{2} \quad r_m = 0.055\text{ m}$$

$$T_m = T_0 + (T_1 - T_0)\frac{r_1}{r_1 - r_0}\left(1 - \frac{r_0}{r_m}\right)$$

Substituting the data in the corresponding units,

$$r_0 = 0.03 \text{ m} \quad T_0 = 98°\text{C}$$

$$r_1 = 0.08 \text{ m} \quad T_1 = 85°\text{C}$$

$$r_m = 0.055 \text{ m}$$

$$T_m = 98 + (85 - 98)\frac{0.08}{0.08 - 0.03}\left(1 - \frac{0.03}{0.055}\right)$$

$$T_m = 98 - 9.5 = 88.5°\text{C}$$

16.3 Combustion smoke circulates through a square chimney of 45 m of height, so that the internal wall is at 300°C, while the external wall is at 30°C. If the dimensions of the chimney are those indicated in the attached figure, calculate the temperature at the center of the chimney wall.

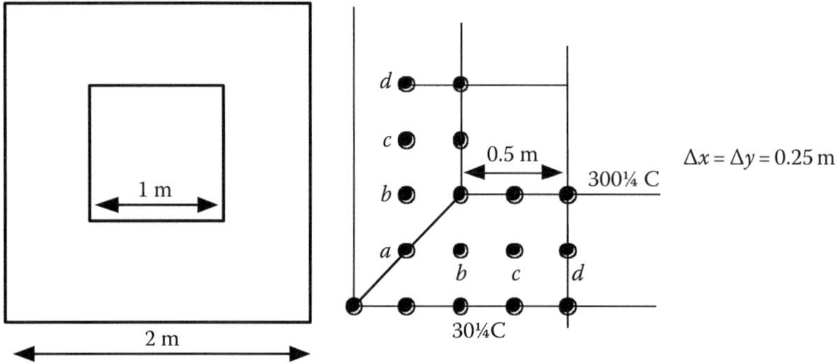

Due to the symmetry of the cross section of the chimney, the calculation of the temperature at the center points of the wall thickness is equal to calculating the corresponding temperature of the points indicated as *a*, *b*, *c*, and *d* in the fourth side of the chimney.

It can be observed that the grid is defined such that $\Delta x = \Delta y = 0.25$ m.

In this problem, since the points where temperature should be calculated are few, the general equation can be applied to obtain such temperatures (Equation 16.20):

$$T_{m,n} = \frac{1}{4}\left(T_{m+1,n} + T_{m\text{-}1,n} + T_{m,n+1} + T_{m,n-1}\right)$$

for the four points considered:

$$\left.\begin{aligned} 4T_a &= 30 + T_b + 30 + T_b \\ 4T_b &= 300 + 30 + T_a + T_c \\ 4T_c &= 300 + 30 + T_b + T_d \\ 4T_d &= 300 + 30 + T_c + T_c \end{aligned}\right\} \quad \left.\begin{aligned} 4T_a &= 60 + 2T_b \\ 4T_b &= 330 + T_a + T_c \\ 4T_c &= 330 + T_b + T_d \\ 4T_d &= 330 + 2T_c \end{aligned}\right\}$$

Therefore, a system of four equations with four unknowns is obtained, which when solved allows us to know the temperatures of the intermediate points of the wall:

$$T_a = 87°C \quad T_b = 144°C \quad T_c = 159°C \quad T_d = 162°C$$

Liebman's method

Initially, the temperature of two points T_a and T_c is assumed, and the temperature of the other two points (T_b and T_d) is calculated from the general equation. The temperature of the two first points is recalculated using the new temperatures, repeating the iterative process until the temperatures of two consecutive iterations coincide.

It is supposed that $T_a = T_c = 180°C$. The following table shows the results of different iterations.

Liebman's Method

Iteration	T_a (°C)	T_b (°C)	T_c (°C)	T_d (°C)
0	180		180	
1		172.5		172.5
2	101.25		168.75	
3		150		166.88
4	90		161.72	
5		145.43		163.36
6	87.72		159.70	
7		144.35		162.35
8	87.18		159.18	
9		144.09		162.09
10	87.05		159.05	
11		144.02		162.02
12	87.01		159.01	

Relaxation method

Initially, the temperature of all of the points is assumed to be known, and the value of the residual is calculated as

$$R_{m,n} = T_{m+1,n} + T_{m\ 1,n} + T_{m,n+1} + T_{m,n-1} - 4T_{m,n}$$

The value of these residuals should be zero. If not, the temperature is corrected at each point, using the expression

$$T^*_{m,n}\left(i+1\right) = T_{m,n}\left(i\right) + \frac{1}{4}R_{m,n}\left(i\right) \quad i = n° \text{ iteration}$$

The following table presents the results of the iterative process of this method:

Iteration		*a*	*b*	*c*	*d*
1	*T*	180	180	180	180
	R	−300	−30	−30	−30
2	*T*	105	172.5	172.5	172.5
	R	−15	−82.5	−15	−15
3	*T*	101.25	151.88	168.75	168.75
	R	−41.25	−7.5	−24.38	−7.5
4	*T*	90.94	150	162.66	166.88
	R	−3.76	−16.4	−3.77	−12.18
5	*T*	90	145.9	161.72	163.83
	R	−8.2	−1.88	−7.15	−1.88
6	*T*	87.95	145.43	159.93	163.36
	R	−0.94	−3.84	−0.93	−3.58
7	*T*	87.72	144.47	159.7	162.47
	R	−1.94	−0.46	−1.86	−0.48
8	*T*	87.24	144.36	159.24	162.35
	R	−0.24	−0.96	−0.25	−0.92
9	*T*	87.18	144.12	159.18	162.12
	R	−0.48	−0.12	−0.48	−0.24
10	*T*	87.05	144.09	159.06	162.06
	R	−0.02	−0.25	−0.09	−0.12
11	*T*	87.05	144.03	159.04	162.03
	R	−0.12	−0.035	−0.1	−0.04
12	*T*	87.02	144.02	159.02	162.02
	R	−0.04	−0.04	−0.04	−0.04

Note: *T*, temperature; *R*, residual

16.4 A solid vegetable body of large dimensions is at 22°C. One of the stages of the vegetable preservation process consists of cooking a vegetable in a tank with boiling water at 100°C. Calculate the temperature of a point placed at 15 mm under the surface after 10 min.

Data (properties of the vegetable): density, 700 kg/m^3; specific heat, 3.89 kJ/(kg °C); thermal conductivity, 0.40 W/(m °C)

This stage consists of a large solid heating under steady state. Because it is a solid of large dimensions, the problem is considered as heat conduction in semi-infinite solids:

$$\frac{\partial T}{\partial t} = \alpha \frac{\partial^2 T}{\partial x^2}$$

$$\alpha = \frac{k}{\rho \hat{C}_P} = \frac{0.40\ \text{J/(s m °C)}}{\left(700\ \text{kg/m}^3\right)\left(3890\ \text{J/(kg °C)}\right)} = 1.46 \times 10^{-7}\ \text{m}^2/\text{s}$$

Since this case involves a cooking stage in a boiling water tank, the convection coefficient can be considered high enough ($h \to \infty$) so as to suppose that the temperature at the surface of the solid is equal to the temperature of the water at the time when the vegetable is first immersed.

Since the tank contains boiling water ($h \to \infty$), it is assumed that $T_w \approx T_e$. The last differential equation can be integrated on the boundary conditions that allow us to obtain Equation 16.25:

$$\frac{T_e - T}{T_e - T_0} = erf\left(\frac{x}{2\sqrt{\alpha t}}\right)$$

where *erf* is the Gauss error function:

$$\frac{x}{2\sqrt{\alpha t}} = \frac{15\times10^{-3}\ \text{m}}{2\sqrt{\left(1.46\times10^{-7}\ \text{m}^2/\text{s}\right)\left(10\ \text{min}\right)\left(60\ \text{s}/1\ \text{min}\right)}} = 0.8013$$

The Gauss error function is interpolated between the values of the correspondent table (Table 16.1), obtaining

$$erf\left(0.8013\right) = 0.74286$$

$$\frac{T_e - T}{T_e - T_0} = 0.74286$$

Since $T_0 = 22°\text{C}$ and $T_e = 100°\text{C}$, it is obtained that $T \approx 42.1°\text{C}$.

16.5 A company that processes potatoes acquires them from a region where there is a variable climate. The commercial agent in charge of buying the potatoes found out that some days before harvesting, a strong cool wind came from the north and was blowing for 10 h with a temperature of −10°C. Potatoes deteriorate if their surface reaches 0°C. If the average soil depth of the potatoes is 10 cm, and at the beginning of the frost the soil was at a temperature of 5°C, what would be the advice that the commercial agent will give regarding the purchase of the potatoes?

Data (physical properties of soil): density, 1600 kg/m^3; specific heat, 3.976 kJ/(kg °C); thermal conductivity, 1 W/(m °C)

Thermal diffusivity

$$\alpha = \frac{k}{\rho\hat{C}_p} = \frac{1\ \text{J/(s m °C)}}{\left(1600\ \text{kg/m}^3\right)\left(3976\ \text{J/(kg °C)}\right)} = 1.57\times10^{-7}\ \text{m}^2/\text{s}$$

The heat transfer is carried out in a semi-infinite solid, complying that

$$\frac{T_e - T}{T_e - T_0} = erf\left(\frac{x}{2\sqrt{\alpha t}}\right)$$

$$\text{If } \eta = \frac{x}{2\sqrt{\alpha t}} = \frac{10\times10^{-2}\ \text{m}}{2\sqrt{\left(1.57\times10^{-7}\ \text{m}^2/\text{s}\right)\left(10\ \text{h}(3600\ \text{s}/1\text{h})\right)}} = 0.6651$$

The Gauss' error function takes a value of *erf* (0.6651) = 0.65534

$$\text{Hence, } \frac{T_e - T}{T_e - T_0} = 0.65534$$

Since $T_e = -10°\text{C}$ and $T_0 = 5°\text{C}$, it is obtained that $T = -0.2°\text{C}$, indicating that the potatoes will be affected and, therefore, it is advisable not to acquire them.

16.6 It is intended to process sausages in an autoclave. It can be considered that the sausage is equivalent to a cylinder of 30 cm in length and 10 cm in diameter. If the sausages are initially at 21°C and the temperature of the autoclave is kept at 116°C, calculate the temperature in the center of a sausage 2 h after its introduction into the autoclave.

Data: surface heat transfer coefficient in the autoclave to the surface of the sausage is 1220 W/(m² °C).

Sausage properties: density, 1070 kg/m³; specific heat, 3.35 kJ/(kg °C); thermal conductivity, 0.50 W/(m °C)

Since the sausage has a cylindrical shape, it is considered that it is formed from the intersection of a cylinder of infinite height of radius r_0 and an infinite slab of thickness $2x_0$:

$$\left.\begin{aligned} r_0 &= 5\text{ cm} \\ 2x_0 &= h = 30\text{ cm} \end{aligned}\right\} \quad \begin{aligned} r_0 &= 5\text{ cm} \\ x_0 &= 15\text{ cm} \end{aligned}$$

According to Newman's rule, $Y = Y_c\, Y_L$
in which

$$Y = \frac{T_e - T_A}{T_e - T_0} = \frac{116 - T_A}{116 - 21}$$

T_A being the temperature in the geometric center of the sausage, whose coordinates are $(r, x) = (0, 0)$.

Calculation of Y_C

$$n_c = \frac{r}{r_0} = \frac{0}{5} = 0$$

$$m_c = \frac{k}{hr_0} = \frac{0.50\text{ W/(m °C)}}{\left(1220\text{ W}/(\text{m}^2\text{ °C})\right)\left(5\times10^{-2}\text{ m}\right)} = 0.0082 \approx 0$$

$$\tau_c = \frac{kt}{\rho\hat{C}_p r_0^2} = \frac{\left(0.50\text{ W/(m °C)}\right)\left(7200\text{ s}\right)}{\left(1070\text{ kg/m}^3\right)\left(3350\text{ J(kg °C)}\right)\left(5\times10^{-2}\right)^2\text{ m}^2} = 0.4$$

From graph 16.13, $Y_c = 0.17$

Calculation of Y_L

$$n_L = \frac{x}{x_0} = \frac{0}{15} = 0$$

$$m_L = \frac{k}{hx_0} = \frac{0.50\text{ W/(m °C)}}{\left(1220\text{ W/(m}^2\text{ °C)}\right)\left(15\times10^{-2}\text{ m}\right)} = 0.0027 \approx 0$$

$$\tau_L = \frac{kt}{\rho\hat{C}_p x_0^2} = \frac{\left(0.50\text{ W/(m °C)}\right)\left(7200\text{ s}\right)}{\left(1070\text{ kg/m}^3\right)\left(3350\text{ J/(kg °C)}\right)\left(15\times10^{-2}\text{ m}\right)^2} = 0.0446$$

$Y_L \approx 0.98$ (from graph 16.12)

Therefore,

$$Y = Y_c Y_L = (0.17)(0.98) = 0.1666$$

$$Y = \frac{T_e - T_A}{T_e - T_0}$$

$$0.1666 = \frac{116 - T_A}{116 - 21}, \quad \text{hence } T_A = 100.2°\text{C}$$

16.7 It is desired to cook a piece of meat of 3 cm of thickness by placing it in a water bath at 100°C. Consider that the meat has a conductivity of 0.56 W/(m °C), a density of 1200 kg/m³, and a specific heat of 3.35 kJ/(kg °C). Initially, the meat has a temperature of 20°C and it is considered that it is cooked when the center reaches 70°C. Calculate the cooking time for (a) when the transfer convection heat coefficient is very high and (b) if this coefficient possesses a value of 30 W/(m² °C).

$$\text{The meat diffusivity is } \alpha = \frac{k}{\rho \hat{C}_p} = \frac{(0.56)}{(1200)(3350)} = 1.39 \times 10^{-7}\ \text{m}^2/\text{s}$$

For the center, the dimensionless position variable is $n = 0$:

a. For very high values of the heat transfer coefficient by convection, the Biot number is very high, in which case Equation 16.40 will be used. Since the module of Fourier's number is unknown, in first approach is to consider that its value is greater than 0.25, so that it will only take the first term of Equation 16.40:

$$Y = \frac{T_e - T}{T_e - T_0} = 4\frac{1}{\pi}\cos\left(\frac{\pi}{2}n\right)\exp\left(-\left(\frac{\pi}{2}\right)^2 \cdot \text{Fo}\right)$$

Substituting data in this equation,

$$\frac{100 - 70}{100 - 20} = 4\frac{1}{\pi}\cos\left(\frac{\pi}{2}\cdot 0\right)\exp\left(-\left(\frac{\pi}{2}\right)^2 \cdot \text{Fo}\right)$$

it is possible to obtain Fourier's number value Fo = 0.495. As this value is greater than 0.25, the previous supposition is correct. The cooking time is obtained from Equation 16.28a:

$$t = \frac{\text{Fo}(x_0)^2}{\alpha} = \frac{(0.495)(0.015)^2}{(1.39 \times 10^{-7})} = 801\ \text{s} = 13.35\ \text{min}$$

b. In this case the Biot number is

$$\text{Bi} = \frac{hx_0}{k} = \frac{(30)(0.015)}{(0.56)} = 0.8036$$

Assuming that Fourier's number is greater than 0.25, and applying Equation 16.30 for the first addition term,

$$Y = 4\frac{\sin A_1}{2A_1 + \sin(2A_1)}\cos(A_1 n)\exp(-A_1 \cdot \text{Fo})$$

The A_1 value must be accomplished (Equation 16.31):

$$A_1 \cdot \tan(A_1) = \frac{1}{m} = \text{Bi} = 0.8036$$

Solving this equation, it is obtained as $A_1 = 0.7924$.
This value is substituted in the previous equation, obtaining

$$\frac{100-70}{100-20} = 4\frac{\sin(0.794\ )}{2\times 0.7924 + \sin(2\times 0.7924)}\cos(0.7924\times 0)\exp(-0.7924 \cdot \text{Fo})$$

Solving this equation, Fourier's number is Fo = 1.36. As this value is greater than 0.25, the previous supposition is correct. The cooking time is obtained from Equation 16.28a:

$$t = \frac{\text{Fo}(x_0)^2}{\alpha} = \frac{(1.36)(0.015)^2}{(1.39\times 10^{-7})} = 2201\ \text{s} = 36.7\ \text{min}$$

16.8 A boiling water bath is used during the manufacture of sausages at the cooking stage. It can be considered that the sausages have a cylindrical shape of 5 cm diameter and 50 cm length. Initially, the sausage is at 22°C, and when it is immersed in the boiling water bath, its surface instantaneously reaches the temperature of the bath. Estimate the time that should elapse from when the sausage is introduced in the bath until the geometric center reaches 85°C.

Data (thermal properties of the sausage): thermal conductivity, 0.44 W/(m °C); density, 1260 kg/m^3; specific heat, 2.80 kJ/(kg °C)

Thermal diffusivity

$$\alpha = \frac{k}{\rho \hat{C}_P} = \frac{0.44\ \text{J}/(\text{s m °C})}{(1260\ \text{kg/m}^3)(2.80\times 10^3\ \text{J}/(\text{kg °C}))} = 1.247\times 10^{-7}\ \text{m}^2/\text{s}$$

The sausage can be considered as an object formed from the intersection of an infinite slab of 50 cm of thickness and a cylinder of 5 cm of diameter and infinite length. Newman's rule can be applied as

$$Y = Y_c Y_L = \frac{T_e - T}{T_e - T_0} = \frac{100-85}{100-22} = 0.1923$$

The time it takes to make the product $Y_c \cdot Y_L$ to be exactly 0.1923 has to be determined:

- Slab: $x_0 = 25$ cm

$$\tau_L = \frac{\alpha t}{(x_0)^2} = \frac{\left(1.247\times10^{-7}\ \text{m}^2/\text{s}\right)}{\left(0.25\ \text{m}\right)^2}\left(\frac{60\ \text{s}}{1\ \text{min}}t_L\ \text{min}\right) \approx 1.2\times10^{-4}\ t_L$$

$$n_L = \frac{x}{x_0} = 0$$

$$m_L = \frac{k}{hx_0} = 0 \quad \left(\text{since } h \to \infty\right)$$

- Cylinder: $r_0 = 2.5$ cm

$$\tau_c = \frac{\alpha t}{(r_0)^2} = \frac{1.247\times10^{-7}\ \text{m}^2/\text{s}}{\left(0.025\ \text{m}\right)^2}\left(\frac{60\ \text{s}}{1\ \text{min}}t_c\ \text{min}\right) \approx 0.012\,t_c$$

$$n_c = \frac{r}{r_0} = 0$$

$$m_L = \frac{k}{hr_0} = 0$$

τ_L and τ_c should be given in minutes.

The problem is solved by iteration, assuming a time and determining with graphs Y_c and Y_L, whose product yields 0.1923. The following table shows the values of the iterative process.

t (min)	τ_L	τ_c	Y_L	Y_c	Y
60	7.18×10^{-3}	0.718	≈1	0.0273	0.0273
30	3.59×10^{-3}	0.359	≈1	0.20	0.20
35	4.19×10^{-3}	0.419	≈1	0.152	0.152
32	3.84×10^{-3}	0.384	≈1	0.19	0.19

obtaining $t \approx 32$ min.

Since the height of the sausage is much greater than its diameter, it could be assumed to be a cylinder of infinite height, so $Y \approx Y_c = 0.1923$. It is obtained, graphically, that for this Y the value of the dimensionless time is $\tau_c = 0.012$.

From this value the following time is obtained: $t = 0.4/0.012 \approx 33$ min.

16.9 A piece of meat with the shape of a parallelepiped, of dimensions 1 m × 1 m × 6 cm, is submerged in a tank containing water at 3°C. The initial distribution of temperatures along the width of the meat piece is

Distance from the surface (cm)	0	1	2	3
Temperature (°C)	27	24	23	22

Determine the distribution of temperatures after 30 min.

Data and notes: the convective heat transfer coefficient from the piece of meat to water is high enough to assume that the temperature on the wall of the meat at the initial time is the arithmetic mean of temperature of the bath ant the meat's wall. After the first time period, assume that the wall reaches the temperature of the bath:

Thermal conductivity of meat: 0.56 W/(m °C)

Specific heat of meat: 3.35 kJ/(kg °C)

Density of meat: 1200 kg/m^3

The piece of meat will be considered as an infinite slab with a 6 cm thickness, so heat transfer is carried out through this thickness. The 1D heat balance is

$$\frac{\partial T}{\partial t} = \alpha \frac{\partial^2 T}{\partial x^2}$$

This equation is solved by finite differences, obtaining that the temperature at a point for a given time is a function of the temperature of the adjacent points during the last period, according to the expression

$$T(x, t+\Delta t) = (\text{Fo})\left[T(x+\Delta x, t) + T(x-\Delta x, t) + \left(\frac{1}{\text{Fo}} - 2\right)T(x,t)\right]$$

$$\text{where } \frac{1}{\text{Fo}} = \frac{(\Delta x)^2}{\alpha \Delta t}$$

The Binder–Schmidt method assumes (Fo) = ½ and $\Delta x = 0.01$ m; therefore, the time period Δt will be

$$\Delta t = \frac{(\Delta x)^2 \rho \hat{C}_p (\text{Fo})}{k} = \frac{(0.01\ \text{m})^2 (1200\ \text{kg/m}^3)(3350\ \text{J/(kg °C)})}{2\ (0.56\ \text{J/(s m °C)})} = 359\ \text{s}$$

This is $\Delta t = 359\ \text{s} \approx 0.1$ h.

Since it is desired to determine the temperature distribution at the end of 30 min, the number of time periods will be

$$n° \Delta t = \frac{0.5\ \text{h}}{\Delta t} = \frac{0.5}{0.1} = 5\ \text{periods}$$

$$\text{Since } \frac{1}{\text{Fo}} = \frac{1}{2T}(x, t+\Delta t) = \frac{1}{2}\left[T(x+\Delta x, t) + T(x-\Delta x, t)\right]$$

The following table presents the values of the temperatures of the different layers, as a function of time:

Time (h)	T_1 (°C)	T_2 (°C)	T_3 (°C)	T_4 (°C)	T_5 (°C)	T_6 (°C)	T_7 (°C)
Before cooling	27	24	23	22	23	24	27
0	15	24	23	22	23	24	15
0.1	3	19	23	23	23	19	3
0.2	3	13	21	23	21	13	3
0.3	3	12	18	21	18	12	3
0.4	3	10.5	16.5	18	16.5	10.5	3
0.5	3	9.75	14.25	16.5	14.25	9.75	3

Figure P.16.1 shows the plotting of the solid in different layers.

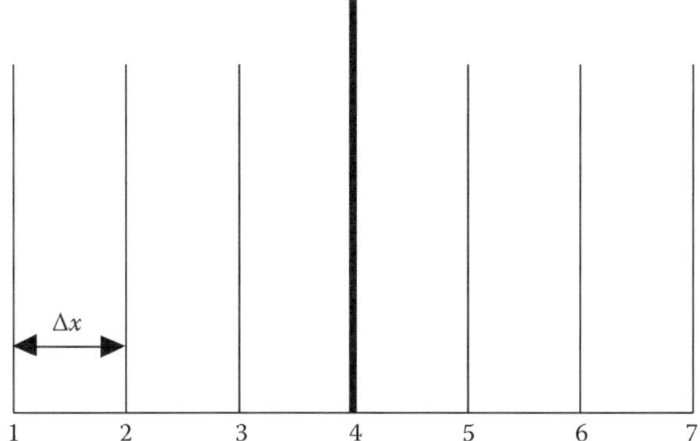

Figure P.16.1 Division of a solid into six parallel sections with $\Delta x = 1$ cm.

PROPOSED PROBLEMS

16.1 The wall of a bakery oven is built with refractory brick of 10 cm thickness whose thermal conductivity is 0.205 W/(m °C). In the wall there is a series of steel pieces whose area represents 1% of the total area of the internal wall. The steel thermal conductivity is 45 W/(m °C). Calculate (a) the heat flux losses through the oven wall, if the internal wall is at 230°C and the external at 25°C and (b) the total heat transmitted through the wall by the brick and the steel.

16.2 The wall of a cold store is composed of 20 cm of external brick, 5 cm of insulating brick, and an internal layer of cork of 8 cm. Inside the chamber the average temperature is 10°C, and the external brick temperature is 25°C. Calculate the heat flux of the wall. Also, determine the temperatures at the interfaces between the brick layers and that of the insulating brick and cork.

Data and notes: assume that the internal temperature of the cork layer coincides with the average internal temperature of the store. Thermal conductivities of the materials are as follows:

Brick 0.70 W/(m °C)
Insulating brick 0.25 W/(m °C)
Cork 0.043 W/(m °C)

16.3 A cold store has the following dimensions: 6 m × 5 m and 3 m of height. This store is isolated thermally at the floor and the roof. The lateral walls have 15 cm of thickness and they have been manufactured of concrete at 1 m of height and the rest until the roof by brick. To avoid energy losses, the internal wall has been covered with 5 cm of an insulating material, while in the external wall a steel sheet of 10 mm of thickness has been placed.

If inside the chamber the temperature is 2°C and the external air is at 30°C, calculate (a) the heat flow rate lost through the lateral walls and (b) the internal and external temperatures of the store wall.

Data: assume that the door has the same characteristics as the wall.

Thermal conductivity: concrete, 1.3 W/(m °C); brick, 0.75 W/(m °C); steel, 45 W/(m °C) insulating material, 0.035 W/(m °C).

16.4 A refrigerating fluid is flowing through a pipe of commercial steel of 2 in. of nominal diameter. This pipe is covered by a material of 0.5 cm of thickness that maintains the internal and external wall temperatures at 30°C and 25°C, respectively. Due to a change in atmospheric conditions, an ice layer begins to form that covers the pipe. What will the thickness of this layer be if the heat flow rate from the exterior toward the refrigerating fluid decreases to 99%?

Data: commercial steel pipe 2 in., $d_i = 52.5$ mm $d_e = 60.3$ mm.

Thermal conductivity: steel, 45 W/(m °C); material, 11 W/(m °C); ice, 2.2 W/(m °C).

16.5 A vegetable solid body of large dimensions at 2°C is inside a cold store, and it is desired to cook it in a water boiling bathroom at atmospheric pressure. Calculate the temperature of a point located 9 mm from the surface, after 5 min.

Data (properties of the vegetable body): density, 700 kg/m^3; thermal conductivity, 0.41 W/(m °C); specific heat, 3.9 kJ/(kg °C).

16.6 A vegetable solid body of large dimensions is at 80°C after a cooking process. For cooling it is placed in a cold chamber at 2°C, in which there are great air currents. Calculate the temperature of a point located 9 mm from the surface after 5 min.

Data (properties of the vegetable body): density, 700 kg/m^3; thermal conductivity, 0.41 W/(m °C); specific heat, 3.9 kJ/(kg °C).

16.7 A sudden cold wave, lasting for duration of 12 h, cools at atmospheric temperature up to −25°C. If the earth was initially at 5°C, at what depth should a water pipe be buried in order to prevent the danger of freezing? What would the penetration distance be under these conditions? For earth diffusivity, a value of 0.0011 m^2/h can be taken.

Note: the penetration distance is defined arbitrarily as the distance from the surface for which the temperature variation is 1% of the initial variation that the surface temperature.

16.8 An orange tree plantation produces oranges of 8 cm average diameter, and it is located in an area in which freezing temperatures are common. On a certain night, when the ambient temperature was 5°C, an abrupt temperature descent of −7°C, having a 2 h duration, occurred. When the superficial temperature of oranges arrives at −1°C, their structure is affected. Determine whether the abrupt temperature fall has affected the quality of the orange crop.

Data and notes: heat convection coefficient air-orange, 6 W/(m^2 °C); orange properties, density, 950 kg/m^3; thermal conductivity, 0.47 W/(m °C); specific heat, 3.85 kJ/(kg °C).

16.9 Before introducing apples to cold storage, they should be cooled to a temperature of 3°C in order to avoid problems when exposing the hot apples to the much lower temperature in the storage. It is assumed that the apples are initially at 25°C, they are spheres of 7 cm diameter and that the cooling is carried out by means of an airstream at −1°C, to a such velocity that the heat convection coefficient is 30 W/(m^2 °C). Calculate the necessary time to cool the apples so that its geometric center reaches 3°C.

Data (apple properties): density, 930 kg/m^3; thermal conductivity, 0.50 W/(m °C); specific heat, 3.6 kJ/(kg °C).

16.10 A can containing pea puree (18 cm length and 10 cm diameter) is at 80°C after being subjected to a pasteurization process. For cooling, an airflow at 20°C is used. If the air in the can is 16 W/(m^2 °C), determine (a) the temperature of a point located at 4.5 cm below the geometric center of the can and at 4 cm to the left of this center, after 4 h; (b) what time is necessary for this point to reach 30°C?; and (c) what should be the temperature of the cooling air for the same point to reach 50°C after 2 h?

Data (pea puree properties): density, 1100 kg/m^3; thermal conductivity, 0.80 W/(m °C); specific heat, 3.8 kJ/(kg °C).

16.11 In order to calculate the thermal conductivity of a fruit, it will be cut into a cylindrical form, 3 cm diameter and 6 cm height. Initially it is at 25°C, being placed inside a cold storage at 0°C. By means of a thermocouple, the temperature evolution in the geometric center can be monitored. After 45 min the temperature of this point is 1.5°C. Calculate the thermal conductivity of this fruit.

Data: heat convection coefficient fruit-air, 23 W/(m^2 °C); (fruit properties) specific heat, 3.56 kJ/(kg °C); density, 800 kg/m^3.

16.12 A cylindrical meat sausage is 6 cm diameter and 60 cm in length, at 30°C. It is desired to cook it in a water bath at 100°C. This treatment stops when the temperature in the point of smaller heating is 80°C. Assuming that the surface sausage temperature coincides with that of the bath, calculate the cooking time.

Once this cooking treatment is carried out, the sausage is placed in a dryer where the air temperature is 20°C. Determine the temperature of a point located at 1.5 cm from the central axis after 1 h.

Data (sausage properties): density, 1200 kg/m^3; thermal conductivity, 0.56 W/(m °C); specific heat, 3.35 kJ/(kg °C); heat convection coefficient sausage-air, 37 W/(m^2 °C).

17 Heat Transfer by Convection

17.1 INTRODUCTION

Convective heat transfer, often referred to simply as convection, is the transfer of heat from one place to another by the movement of fluids. Convection is usually the dominant form of heat transfer in liquids and gases.

If a fluid is in contact with a solid that is at a greater temperature, the fluid receives heat that is transferred within it by movement of the fluid particles. This movement causes heat transport by convection, and it can occur by means of natural or forced forms. The first case occurs when there is no mechanical agitation, and it is attributed to density differences at different points of the fluid caused by the effect of temperature. On the contrary, forced convection occurs when the movement of the fluid is produced mechanically using devices such as agitators and pumps, among others.

Heat transfer by convection is very important when studying the heat exchange between two fluids separated by a wall in such a way that one of them gives up heat to the other one, so that the first fluid cools while the second one heats up. The devices in which this heat transmission is performed are called heat exchangers.

17.2 HEAT-TRANSFER COEFFICIENTS

17.2.1 Individual Coefficients

If we consider a fluid that circulates through a solid conduit or around a solid surface, the heat transfer from the solid to the fluid (or vice versa) depends on the fluid–solid contact area and on the temperature difference. Thus, for a system such as the one shown in Figure 17.1, in which a solid of differential area dA at a temperature T_S is in contact with a fluid at temperature T_f, then $T_S > T_f$. There is heat transmission from the solid to the fluid in a way such that the heat flow will be proportional to dA and $(T_S - T_f)$, which is

$$dQ \propto dA\,(T_S - T_f)$$

A proportionality coefficient h is defined in such a way that

$$dQ = h\,dA\,(T_S - T_f) \tag{17.1}$$

This proportionality constant is called individual convective heat-transfer coefficient or film coefficient. This coefficient depends on the physical and dynamical properties of the fluid. Some relationships between these properties can be found by means of dimensional analysis, which when complemented with further experimentation will allow us to obtain equations to calculate such coefficients.

The individual convective heat-transfer coefficient h depends on the physical properties that affect flow (density, ρ, and viscosity, η), on its thermal properties (specific heat, $\hat{C}_P$, and thermal conductivity, k), on the characteristic length L of the transmission area (when the diameter of the cylindrical conduit is d), on gravity acceleration g, on the velocity v at which the fluid circulates, on the temperature difference between the solid and the fluid $(T_S - T_f)$, and on the cubic or volumetric expansion coefficient β (which in the case of an ideal gas coincides with the inverse of the absolute temperature).

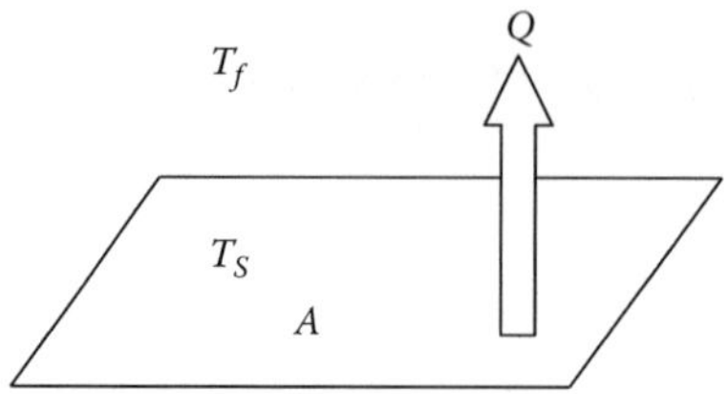

FIGURE 17.1 Heat transfer by convection.

A relationship among the different dimensionless groups is obtained by applying a dimensional analysis:

$$\text{Nu} = \phi[(\text{Re})(\text{Pr})(\text{Gr})] \tag{17.2}$$

in which the dimensionless groups are defined as follows:

Nusselt number

$$(\text{Nu}) = \frac{hd}{k} \tag{17.3}$$

Reynolds number

$$(\text{Re}) = \frac{\rho v d}{\eta} \tag{17.4}$$

Prandtl number

$$(\text{Pr}) = \frac{\hat{C}_P \eta}{k} \tag{17.5}$$

Grashof module

$$(\text{Gr}) = \frac{\beta g \Delta T d^3 \rho^2}{\eta^2} \tag{17.6}$$

Sometimes it is possible to express the relationship among dimensionless groups as a function of the Peclet and Graetz numbers, which are combinations of Reynolds and Prandtl numbers and are used in forced convection problems:

Peclet number

$$(\text{Pe}) = (\text{Re})(\text{Pr}) = \frac{\rho v d \hat{C}_P}{k} \tag{17.7}$$

Graetz number

$$(\text{Gz}) = (\text{Re})(\text{Pr})\left(\frac{d}{L}\right) = \frac{\rho v d^2 \hat{C}_P}{kL} \tag{17.8}$$

In natural convection, velocity depends on the flotation effects, so the Reynolds number can be omitted in Equation 17.2, whereas in forced convection, the Grashof number can be omitted. Thus,

Natural convection

$$\mathrm{Nu} = \phi[(\mathrm{Pr})(\mathrm{Gr})]$$

Forced convection

$$\mathrm{Nu} = \phi[(\mathrm{Re})(\mathrm{Pr})] = \phi(\mathrm{Gz})$$

It is interesting to point out that within a wide temperature and pressure range, for most gases, the Prandtl number is almost constant, so it can be omitted, yielding to simpler equations for the calculation of individual heat-transfer coefficients.

Once the relationships among the different dimensionless groups are found, final equations that allow the calculation of film coefficients can be obtained by experimentation. Although there are numerous expressions presented in the literature to calculate such individual coefficients, a series of expressions that are the most used in practice will be given next.

17.2.1.1 Natural Convection

As stated before, the Nusselt number is only a function of the Grashof and Prandtl numbers for free or natural convection. The equation that relates these three numbers is

$$\mathrm{Nu} = a(\mathrm{Gr}\cdot\mathrm{Pr})^b \tag{17.9}$$

in which the values of parameters a and b are dependent upon the system and working conditions.

Configuration	Gr · Pr	a	b
Vertical plates and cylinders			
Length > 1 m			
Laminar	$<10^4$	1.36	1/5
Laminar	$10^4 < \mathrm{Gr}\cdot\mathrm{Pr} < 10^9$	0.55	1/4
Turbulent	$>10^9$	0.13	1/3
Spheres and horizontal cylinders			
Diameter < 0.2 m			
Laminar	$10^3 < \mathrm{Gr}\cdot\mathrm{Pr} < 10^9$	0.53	1/4
Turbulent	$>10^9$	0.13	1/3
Horizontal plates (heated upward, cooled downward)			
Laminar	$10^5 < \mathrm{Gr}\cdot\mathrm{Pr} < 2 \times 10^7$	0.54	1/4
Turbulent (heated upward, cooled downward)	$2 \times 10^7 < \mathrm{Gr}\cdot\mathrm{Pr} < 3 \times 10^9$	0.55	1/3
Laminar	$3 \times 10^5 < \mathrm{Gr}\cdot\mathrm{Pr} < 3 \times 10^{10}$	0.27	1/4

In the case where the fluid is air in laminar flow, the equations to be used are

Horizontal walls

$$h = C(\Delta T)^{1/4} \tag{17.10}$$

Upward $C = 2.4$
Downward $C = 1.3$

Vertical walls

$$(L > 0.4 \text{ m}): h = 1.8(\Delta T)^{1/4} \tag{17.11}$$

Vertical walls

$$(L < 0.4\,\text{m}): \quad h = 1.4\left(\frac{\Delta T}{L}\right)^{1/4} \tag{17.12}$$

Horizontal and vertical pipes

$$h = 1.3\left(\frac{\Delta T}{d}\right)^{1/4} \tag{17.13}$$

In all these expressions, d and L should be given in meters to obtain the value of the film coefficient in J/(s m² °C).

17.2.1.2 Forced Convection

The properties of the fluid should be calculated at a mean global temperature in all of the expressions stated next:

17.2.1.2.1 Fluids inside Pipes

Flow in turbulent regime

One of the more used equations is the Dittus–Boelter equation:

$$\text{Nu} = 0.023(\text{Re})^{0.8}(\text{Pr})^{n} \tag{17.14}$$

$n = 0.4$ for fluids that heat up
$n = 0.3$ for fluids that cool down

This equation is valid for $\text{Re} > 10^4$ and $0.7 < \text{Pr} < 160$.

In some cases, it is convenient to use an expression that includes the number of Stanton (St):

$$(\text{St}) = \frac{h}{\hat{C}_P \rho v} \tag{17.15}$$

in such a way that the equation to use is

$$(\text{St})(\text{Pr})^{2/3} = 0.023(\text{Re})^{-0.2} \tag{17.16a}$$

The second member of this equation is the denominated factor j_H of Colburn.

For viscous fluids that move at $\text{Re} < 8000$ and $\text{Pr} < 10^4$, the equation of Sieder–Tate is commonly used:

$$\text{Nu} = 0.027(\text{Re})^{0.8}(\text{Pr})^{1/3}\left(\frac{\eta}{\eta_W}\right)^{0.14} \tag{17.17}$$

All of the properties of the fluid are calculated at a mean temperature, except η_w, which is calculated at the mean temperature of the wall.

Another equation that includes the viscosity term at the temperature of the wall is

$$(\text{Nu}) = 0.023(\text{Re})^{0.8}(\text{Pr})^{2/3}\left(\frac{\eta}{\eta_W}\right)^{0.14} \tag{17.18}$$

As indicated before, the value of the Prandtl number is usually constant in the case of gases circulating inside the conduits, so the equation that can be used is simpler:

$$(\text{Nu}) = 0.021(\text{Re})^{0.8} \tag{17.19}$$

Flow in transition regime
For $2100 < \text{Re} < 10^4$, the following expression is used:

$$(\text{Nu}) = 0.116[(\text{Re})^{2/3} - 125]\left[1+\left(\frac{d}{L}\right)^{2/3}\right]\left(\frac{\eta}{\eta_W}\right)^{0.14} \tag{17.20}$$

Flow in laminar regime
Heat transfer to a fluid, in a laminar regime ($\text{Re} < 2100$), takes place almost exclusively by conduction, and the distribution of velocity is also parabolic, so the equations to be used are different from those presented earlier. However, it is convenient to obtain a similar expression to those used for a turbulent regime, one of the most common being

$$(\text{Nu}) = 1.86\left[(\text{Re})(\text{Pr})\left(\frac{d}{L}\right)\right]^{1/3}\left(\frac{\eta}{\eta_W}\right)^{0.14} \tag{17.21}$$

or

$$(\text{Nu}) = 1.86(\text{Gz})^{1/3}\left(\frac{\eta}{\eta_W}\right)^{0.14} \tag{17.22}$$

17.2.1.2.2 Fluids Flowing on the Exterior of Solids

Some of the most common expressions will be presented next:

Flow in turbulent regime
For gases,

$$(\text{Nu}) = 0.26(\text{Re})^{0.6}(\text{Pr})^{0.3} \tag{17.23}$$

For liquids,

$$(\text{Nu}) = [0.35 + 0.47(\text{Re})^{0.52}](\text{Pr})^{0.3} \tag{17.24}$$

In the case of liquids that move by the annular space of concentric tubes, the equation of Davis can be used:

$$(\text{St}) = 0.029(\text{Re})^{-0.2}(\text{Pr})^{-2/3}\left(\frac{\eta}{\eta_W}\right)^{0.14}\left(\frac{d_0}{d_i}\right)^{0.15} \tag{17.25}$$

where d_0 and d_i are the external and internal diameters of the circular ring, respectively. Sometimes equations for flow inside tubes are used instead of Equation 17.25, exchanging the diameter of the pipe for the equivalent diameter:

Flow in the laminar regime
The following equation can be used for liquids in which $0.2 < \text{Re} < 200$:

$$(\text{Nu}) = 0.86(\text{Re})^{0.43}(\text{Pr})^{0.3} \tag{17.26}$$

For liquids that circulate at $\text{Re} > 200$ and gases at $0.1 < \text{Re} < 10^3$, the following expression is used:

$$(\text{Nu}) = [0.35 + 0.56(\text{Re})^{0.52}](\text{Pr})^{0.3} \tag{17.27}$$

If the Prandtl number is 0.74, the equation that can be used is

$$(\text{Nu}) = 0.24(\text{Re})^{0.6} \tag{17.28}$$

17.2.1.2.3 Heating or Cooling of Flat Surfaces

For $\text{Re} > 20{,}000$, considering that the characteristic length of the flat surface is in the same direction as the flow, the following expression is used:

$$(\text{Nu}) = 0.036(\text{Re})^{0.8}(\text{Pr})^{1/3} \tag{17.29}$$

For air over smooth slabs,

$$h = 5.7 + 3.9v \ (v > 5 \text{ m/s}) \tag{17.30a}$$

$$h = 7.4(v)^{0.8} \ (5 < v < 30 \text{ m/s}) \tag{17.30b}$$

where
v is the velocity of the fluid in m/s
h is obtained in J/(m² s °C)

17.2.1.3 Convection in Non-Newtonian Fluids

Many food fluids do not present Newtonian behavior, so for those fluids the equations given previously to calculate film coefficients cannot be used. Little data can be found in the literature to calculate such coefficients for non-Newtonian fluids. Some equations that can be used to do so are presented next:

1. Circulation in plug flow
   When a fluid circulates in plug flow through a pipe and for values of Graetz number higher than 500, the following expression is useful:

$$(\text{Nu}) = \frac{8}{\pi} + \frac{4}{\pi}(\text{Gz})^{1/2} \tag{17.31}$$

2. Circulation in turbulent regime
   When the flow behavior of the fluid can be described by the power law, the following equation can be used:

$$(\text{Nu}) = 1.75\left(\frac{3n+1}{4n}\right)^{1/3}(\text{Gz})^{1/3} \tag{17.32}$$

17.2.1.4 Convection in Condensing Fluids

The determination of individual convection coefficients for condensation fluids is essentially controlled by the condensed film deposited on the condensation surface. The condensed film will be different if the condensation process is carried out on vertical or horizontal surfaces:

1. *Vertical pipes and surfaces*
   If condensation is carried out on a vertical surface, the condensed film thickness increases with time. For Reynolds number values greater than 2100, the convection coefficient can be calculated from the equation:

$$\mathrm{Nu} = \frac{hL}{k} = 0.943\left(\frac{L^3\rho^2 g\lambda}{k\eta\Delta T}\right)^{1/4} \tag{17.33}$$

   where
   λ is the condensation latent heat
   ΔT is the difference temperatures between condensation vapor and cold fluid
   g is gravity constant
   L is the characteristic surface length (for pipes the external diameter)

2. *Horizontal pipes*
   The condensation on a horizontal pipe presents a condensed film that possesses a greater thickness in its bottom part. Generally, the vapor condensation is carried out in condensers that usually consist of the shell and pipe type, with several lines of pipes. The condensed liquid over a tube pours onto the tube of the bottom line; for this reason, in the determination of the heat convective coefficient, it is necessary to take into account the number of present lines. For Reynolds number values lower than 2100, an equation that can be used to determine film convection coefficient is

$$\mathrm{Nu} = \frac{hd}{k} = 0.73\left(\frac{d^3\rho^2 g\lambda}{N_t k\eta\Delta T}\right)^{1/4} \tag{17.34}$$

   where N_t is the pipe number on the same vertical.
   Also, in addition to the expressions given here, more can be found in the literature, and some of those will be presented in detail in the section that corresponds to film coefficients of shell and tube-heat exchangers.
   It is important to point out that in expressions in which the viscosity of the fluid at the temperature of the wall (η_W) appears, it is necessary to perform an iterative process to calculate the film coefficient. This is due to the fact that to obtain the value of η_W, it is necessary to previously know the value of the temperature on the exchanger wall, and to do that, the value of the film coefficient is needed.

17.2.2 Global Coefficients

The previous section presented methods to calculate the heat-transfer coefficients from a solid surface to a fluid or vice versa. However, there are cases in practice in which a fluid is cooled or heated by means of another fluid that cools down or heats up, and both fluids are separated by a solid surface. In this case, it is necessary to find a heat-transfer coefficient that allows the calculation of the heat transferred from one fluid to another through a solid wall.

Suppose a system (Figure 17.2) in which a hot fluid at temperature T transfers heat to another cool fluid at temperature t, through a solid surface. Heat transfer is carried out from the hot to the

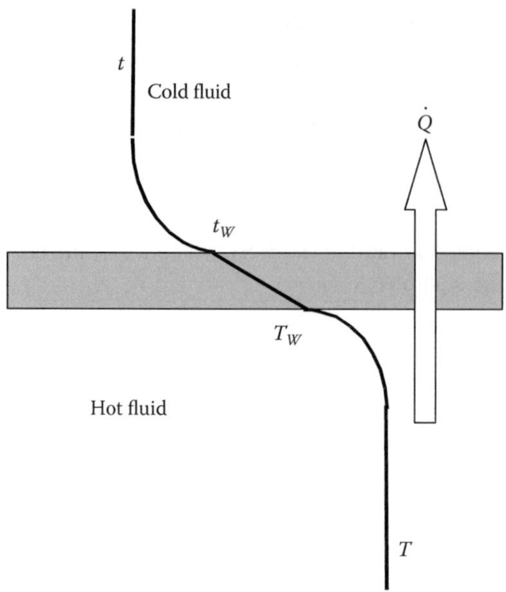

FIGURE 17.2 Heat exchange between fluids separated by a solid.

cold fluid. Initially, heat is transferred by convection from the fluid at temperature T to the surface at temperature T_W. Heat is transferred by conduction through the solid, so there is a temperature drop from T_W to t_W. Then heat is transferred by convection from the wall to the cold fluid at temperature t. The area of the solid surface in contact with the hot fluid is called A_H, while that in contact with the cold fluid will be A_F. If the surface is flat, these areas coincide and $A_H = A_C$. However, in general, these areas do not coincide. Next, the heat-transfer process described is studied:

Heat convection on the side of the hot fluid

$$\dot{Q} = h_h(T - T_W)A_h = \frac{T - T_W}{(1/h_h A_h)} \tag{17.35}$$

in which h_h is the individual heat-transfer coefficient in the hot fluid.

Heat conduction through the solid surface

$$\dot{Q} = \frac{k}{e}(T_W - t_W) = \frac{T_W - t_W}{(e/kA_{ml})} \tag{17.36}$$

where

k is the thermal conductivity of the solid
e is the thickness of the solid
A_{ml} is the logarithmic mean area, defined by

$$A_{ml} = \frac{A_c - A_h}{\ln(A_c/A_h)} \tag{17.37}$$

Heat convection on the side of the cold fluid

$$\dot{Q} = h_c(t_W - t)A_c = \frac{t_W - t}{(1/h_c A_c)} \tag{17.38}$$

in which h_c is the individual convective heat-transfer coefficient in the cold fluid.

The heat flux transferred is the same in each case, so

$$\dot{Q} = \frac{T - T_W}{(1/h_h A_h)} = \frac{T_W - t_W}{(e/kA_{ml})} = \frac{t_W - t}{(1/h_c A_c)} \tag{17.39}$$

Due to the properties of the ratios, the last equation can be expressed as the quotient between the sum of the numerators and of the denominators, in such a way that

$$\dot{Q} = \frac{T - t}{(1/U_h A_h)} = \frac{T - t}{(1/U_c A_c)} \tag{17.40}$$

where U_c and U_h are the heat-transfer global coefficients referring to the areas that are in contact with the hot and cold fluid, respectively. These global coefficients are defined by the expressions:

$$\frac{1}{U_h} = \frac{1}{h_h} + \frac{e}{k\,(A_{ml}/A_h)} + \frac{1}{h_c\,(A_c/A_h)} \tag{17.41a}$$

$$\frac{1}{U_c} = \frac{1}{h_h\,(A_h/A_c)} + \frac{e}{k(A_{ml}/A_c)} + \frac{1}{h_c} \tag{17.41b}$$

It can be observed that the expression $1/(UA)$ represents the global resistance to heat transfer from the hot to the cold fluid. This resistance is the sum of three resistances in series: those due to the hot fluid, to the solid wall, and to the cold fluid.

In fact, the calculation of this coefficient is more complicated, since the fluids can leave sediments on the solid surface. For this reason, two new resistances appear in the last equations, R_C and R_F, which are due to the deposits or sediments of the hot and cold fluids, respectively. So, the heat-transfer global coefficient, if the deposits or sediments are taken into account, can be expressed as

$$\frac{1}{U_D} = \frac{1}{U} + R_h + R_c \tag{17.42}$$

where

U is the global coefficient calculated by Equation 17.41

U_D is the heat-transfer global coefficient that includes the resistance due to deposits

Table 17.1 shows the typical values of fouling factors for water and other types of fluids.

TABLE 17.1
Fouling Factors

Product	*R* (m² °C/kW)	Product	*R* (m² °C/kW)
Water		*Liquids*	
Distilled	0.09	Brine	0.264
Sea	0.09	Organic	0.176
River	0.21	Fuel oil	1.056
Boiler	0.26	Tars	1.76
Very hard	0.58		
Gases			
Steam		Air	0.26–0.53
Good quality	0.052	Solvents' vapor	0.14
Bad quality	0.09		

The calculation of the heat-transfer global coefficients is performed as indicated, although approximated values of this coefficient can be found in the literature for different types of devices and operations (Kreith and Black, 1983).

The value of the coefficient on the internal side of the tubes is obtained in reference to the internal area, although sometimes it can be interesting to know the value of the heat-transfer coefficient in reference to the external area. If h_i and h_{ie} are the film coefficients of the fluid that circulates inside the tubing, referring to the internal and external area, respectively, then the following is complied with

$$h_{ie} = h_i \frac{d_e}{d_i} \frac{\phi_{te}}{\phi_{ti}}$$

where d_e and d_i are the external and internal diameters of the tubes, respectively.

The relation of viscosities defines the coefficient ϕ_t:

$$\phi_t = \frac{\eta}{\eta_W} \tag{17.43}$$

Therefore,

$$\frac{\phi_{ti}}{\phi_{te}} = \frac{\eta_{We}}{\eta_{Wi}}$$

where

η_{We} is the viscosity of the fluid at the external temperature of the wall of the tubes
η_{Wi} corresponds to the internal temperature

17.3 CONCENTRIC TUBE-HEAT EXCHANGERS

This is the simplest type of heat exchanger, consisting of two tubes of different diameters, arranged one inside the other. A fluid circulates inside the internal tube, while another fluid circulates through the circular ring. The hot or cold fluid can circulate indiscriminately through the interior tube or through the circular ring, although if the hot fluid circulates through the internal tube, the heat transfer is better.

These exchangers can operate in two ways: in parallel and in countercurrent (Figure 17.3). In the first case, the cold fluid as well as the hot fluid circulates in the same direction, while in countercurrent, the fluids circulate in opposite ways.

17.3.1 Design Characteristics

The setup of the mathematical model will be performed from the energy balances for the hot and cold fluid and through the exchange surface. This study will be made for the two operation methods mentioned previously.

17.3.1.1 Operation in Parallel

Suppose a system such as that shown in Figure 17.4, in which the fluid enters by section 1 and exits by section 2.

The circulation flows are w_h and w_c, and the specific temperature is $\hat{C}_P)_h$ and $\hat{C}_P)_c$ for the hot and cold fluids, respectively.

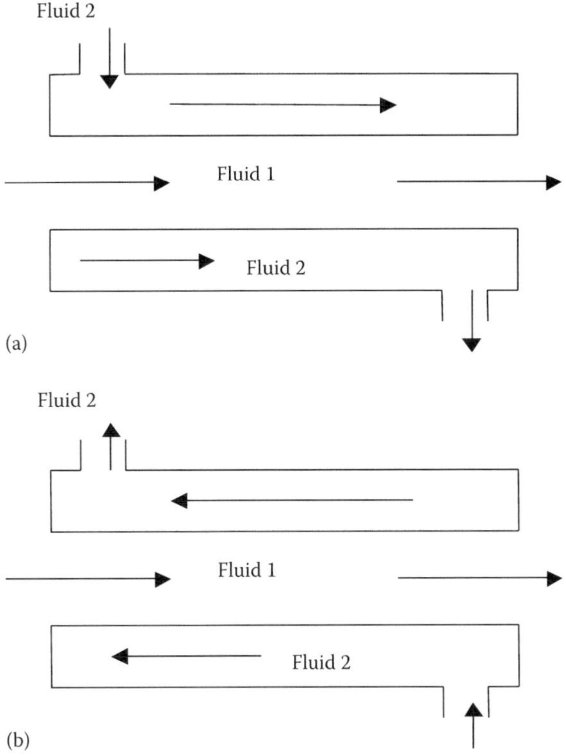

FIGURE 17.3 Concentric tube-heat exchanger: (a) parallel flow and (b) countercurrent flow.

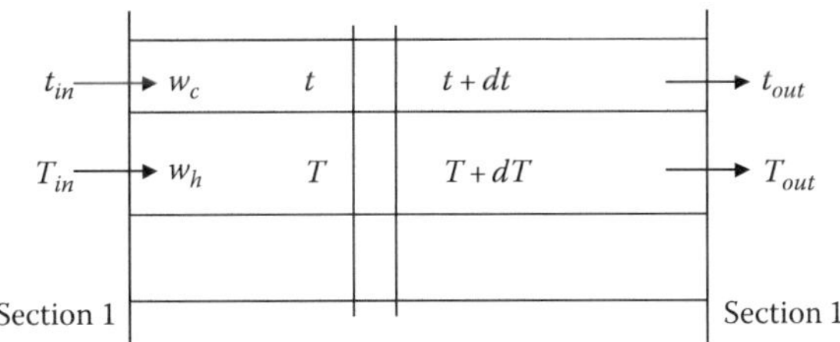

FIGURE 17.4 Concentric tube-heat exchanger; fluids flow.

The cold fluid captures the heat flux transferred by the hot fluid, and if there are no heat losses outside, then the following is complied with

$$\dot{Q} = w_h \hat{C}_P\Big)_h (T_{in} - T_{out}) = w_c \hat{C}_P\Big)_c (t_{out} - t_{in}) \tag{17.44}$$

where

w is the mass flow of each stream
T is the temperature of the hot fluid
t is the temperature of the cold fluid
$\hat{C}_P$ is the specific heat of the hot (h) and cold (c) fluid, respectively

Consider a surface differential (dA) of the heat exchanger (Figure 17.4). The temperatures of the fluids are T and t at the inlet of this element of the exchanger, with $(T + dT)$ and $(t + dt)$ being those at the outlet.

When an energy balance is performed on this part of the exchanger, the following is obtained for the hot fluid:

$$w_h\hat{C}_P\big)_h T = d\dot{Q} + w_h\hat{C}_P\big)_h (T + dT)$$

$$d\dot{Q} = U(T - t)dA$$

Hence,

$$d\dot{Q} = U(T - t)dA = -w_h\hat{C}_P\big)_h \, dT \tag{17.45}$$

In an analogous way, for the cold fluid,

$$d\dot{Q} = U(T - t)dA = w_c\hat{C}_P\big)_c \, dt \tag{17.46}$$

To calculate the exchange area, the following equations should be integrated and arranged as separated variables:

$$\frac{U}{w_h\hat{C}_P\big)_h} dA = \frac{-dT}{T - t} \tag{17.47}$$

$$\frac{U}{w_c\hat{C}_P\big)_c} dA = \frac{dt}{T - t} \tag{17.48}$$

The integration of these equations allows the calculation of the area of the exchanger:

$$A = \int dA = \int \frac{w\hat{C}_P\big)_h}{U} \frac{-dT}{T - t} \tag{17.49a}$$

$$A = \int dA = \int \frac{w_F\hat{C}_P\big)_h}{U} \frac{dt}{T - t} \tag{17.49b}$$

Integration is not as simple as it seems, since the heat-transfer global coefficient U, in general, is not constant, since its value depends on fluid and flow properties, which vary along the heat exchanger as the temperature changes.

The following expression can be obtained from the differential equations on separated variables obtained before:

$$\frac{-d(T - t)}{T - t} = -d\ln(T - t) = U\left(\frac{1}{w_h\hat{C}_P\big)_h} + \frac{1}{w_c\hat{C}_P\big)_c}\right)dA$$

If it is supposed that the heat-transfer global coefficient is *constant* (U = constant) and the difference between both fluids is defined as $\Delta T = T - t$, when integrating the whole area, it is obtained that

$$\ln\left(\frac{\Delta T_1}{\Delta T_2}\right) = UA\left(\frac{1}{w_h\hat{C}_P\big)_h} + \frac{1}{w_c\hat{C}_P\big)_c}\right)$$

The following expression is obtained from the global balance (Equation 17.44):

$$\left(\frac{1}{\left.w_h \hat{C}_P\right)_h} + \frac{1}{\left.w_c \hat{C}_P\right)_c} \right) = \left(\frac{\Delta T_1 - \Delta T_2}{\dot{Q}} \right)$$

so the heat flow rate transferred through the area of the exchanger is

$$\dot{Q} = UA(\Delta T)_{ml} \tag{17.50}$$

$(\Delta T)_{ml}$ being the logarithmic mean temperature difference:

$$(\Delta T)_{ml} = \frac{\Delta T_1 - \Delta T_2}{\ln(\Delta T_1 / \Delta T_2)} \tag{17.51}$$

If the heat-transfer global coefficient is not constant, the transmission area should be calculated by graphical or numerical integration of

$$A = -\int \frac{\left.w_h \hat{C}_P\right)_h dT}{U(T-t)} = \int \frac{\left.w_c \hat{C}_P\right)_c dt}{U(T-t)} \tag{17.52}$$

It should be taken into account that the relationship between the temperature of the fluids, t and T, is given by the heat balance between the inlet and outlet section and any other section in the exchanger:

$$\left.w_h \hat{C}_P\right)_h (T_{in} - T) = \left.w_c \hat{C}_P\right)_c (t - t_{in}) \tag{17.53}$$

However, it can be mathematically demonstrated that if the heat-transfer global coefficient, U, varies linearly with ΔT or with one of the temperatures T or t, the value of the heat flux transferred through the exchange area is

$$\dot{Q} = A(U\Delta T)_{mlc} \tag{17.54}$$

where $(U\Delta T)_{mlc}$ is the crossed logarithmic mean, defined by the expression

$$(U\Delta T)_{mlc} = \frac{U_2 \Delta T_1 - U_1 \Delta T_2}{\ln(U_2 \Delta T_1 / U_1 \Delta T_2)} \tag{17.55}$$

The calculation of the crossed logarithmic mean implies previous knowledge of the value of the heat-transfer global coefficient at the ends of the exchanger.

17.3.1.2 Countercurrent Operation

In this case, as indicated by its name, the fluids circulate in countercurrent (Figure 17.3). It can be demonstrated, in an analogous way to the parallel operation, that the equations obtained from the heat balance are

For a constant U,

$$\dot{Q} = UA(\Delta T)_{ml}$$

If U varies linearly with temperature,

$$\dot{Q} = A(U\Delta T)_{mlc}$$

For the same heat requirement, that is, the same values of inlet and outlet temperatures, the logarithmic mean of the temperatures is higher for countercurrent operations than for parallel operations. For this reason, the surface required to perform the desired heat transfer will be smaller in a countercurrent operation. It is important to point out that in the calculation of the logarithmic mean temperature, the temperature increments in each section vary with respect to the operation in parallel, since in this case it is the case that $\Delta T_1 = T_{in} - t_{out}$ and $\Delta T_2 = T_{out} - t_{in}$.

17.3.2 Calculation of Individual Coefficients

The different expressions presented in Section 17.2.1 are used to calculate individual or film coefficients, although one of the most used is the Dittus–Boelter equation (Equation 17.14).

In these equations, the diameter used to calculate the Reynolds and Nusselt numbers is the inside diameter of the interior tube. The equivalent diameter is used for the fluid that circulates by the annular space.

The *equivalent diameter* for a pipe of section other than circular is defined as four times the hydraulic radius, which is the quotient between the cross-sectional area and the wet perimeter:

$$D_e = 4r_H = 4\frac{\text{Cross-sectional area}}{\text{Wet perimeter}} \tag{17.56}$$

For a circular ring (Figure 17.5), the wet perimeter is considered to be only that part corresponding to the heat exchange area, that is, the perimeter "wet" by the heat transfer:

$$D_e = 4r_H = 4\frac{(\pi/4)\left(d_0^2 - d_e^2\right)}{\pi d_e}$$

Hence,

$$D_e = \frac{\left(d_0^2 - d_e^2\right)}{d_e} \tag{17.57}$$

where

d_0 is the internal diameter of the external tube
d_e is the external diameter of the internal tube

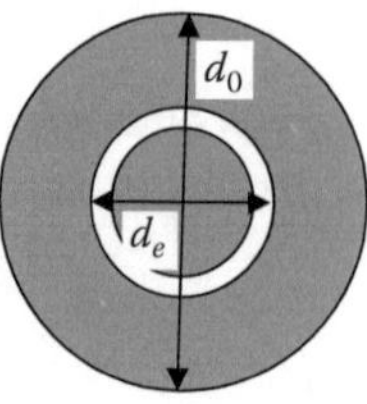

FIGURE 17.5 Cross section of a concentric tube-heat exchanger.

17.3.3 Calculation of Head Losses

Regarding the circulation of fluids inside pipes, head losses can be calculated from the Fanning equation. Diameter appears in this last expression, which in the case of the internal pipe coincides with its internal diameter. However, the equivalent diameter should be used in the case of the circular ring, and it varies from that obtained in the last section, since the wet perimeter is different:

$$D_e = 4r_H = 4\frac{(\pi/4)\left(d_0^2 - d_e^2\right)}{\pi\left(d_0 + d_e\right)} = d_0 - d_e \tag{17.58}$$

The friction factor can be calculated from the different equations that can be found in the literature. These equations vary depending on the circulation regime, being different according to the value of the Reynolds number.

17.4 SHELL AND TUBE-HEAT EXCHANGERS

In such cases in which it is necessary to have a large transfer surface to perform the heat transfer from one fluid to another, the use of a set of tubes contained in a shell is recommended (Figure 17.6). The set of tubes is called a bundle, and a supporting plate fastens the bundle. In this type of heat exchangers, one fluid circulates inside the tubes, and the other one circulates outside the tubes and inside the shell. Usually, this last fluid is forced to change direction due to the presence of supporting plates. Such plates or baffles can be perpendicular or horizontal to the tube bundle. In the first case, they are called *deflectors*, and different types of deflectors are shown in Figure 17.7 (orifice baffles, disk and doughnut baffles, and segmental baffles). The distance between baffles determines the velocity of the fluid, being in general 0.2 times the value of the diameter of the shell. Also, the percentage of the diameter of the shell not occupied by the deflector is called *bypassing*.

When the supporting plates placed inside the shell are slabs, parallel to the tubes, they divide the shell in a set of passes in such a way that the fluid that circulates outside the tubes changes direction

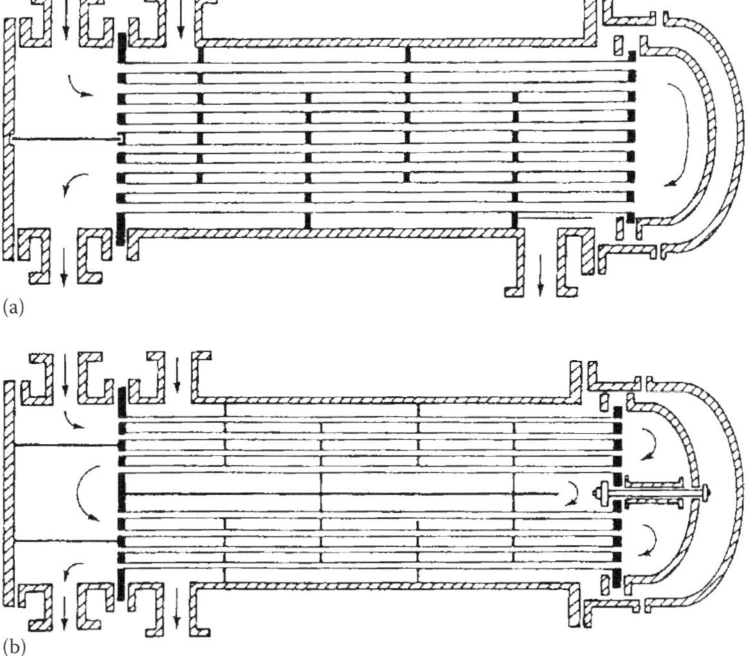

FIGURE 17.6 Shell and tube-heat exchanger: (a) 1/2 heat exchanger and (b) 2/4 heat exchanger.

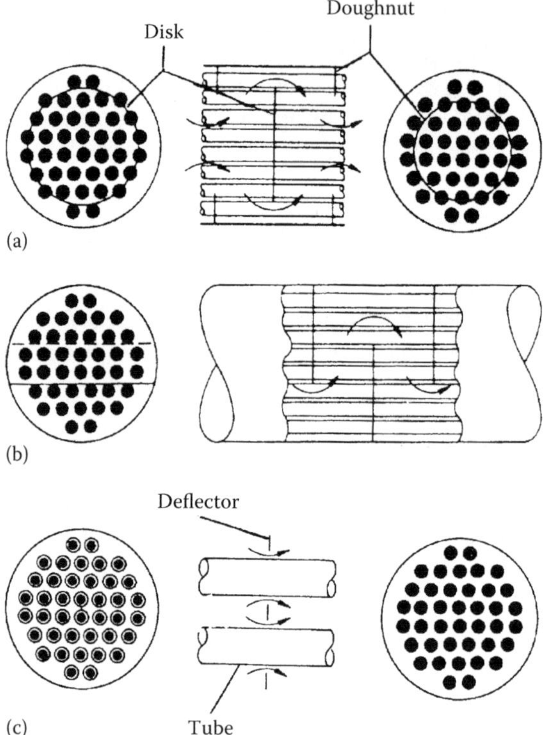

FIGURE 17.7 Types of baffles and shell and tube-heat exchangers: (a) with disk and doughnut baffles, (b) segmental baffles, and (c) orifice baffles. (Adapted from Kreith, F. and Black, W.Z., *La Transmisión del Calor. Principios Fundamentales*, Alhambra, Madrid, Spain, 1983.)

in two consecutive passes. In the same way, the directional change of the fluid that circulates inside the tube bundle is called a pass. Different types of shell and tube-heat exchangers with different fluid passes are represented in Figure 17.6.

On the other hand, the tubes can be arranged as triangles, squares, or rotated squares (Figure 17.8). The square arrangement allows an easier cleaning of the external part of the tubes. The triangular arrangement facilitates the turbulence of the fluid, so the heat-transfer coefficient increases, although the head losses are also increased.

17.4.1 Design Characteristics

Certain considerations regarding which fluids are advisable to be circulated inside and outside the tubes should be taken into account for the design of shell and tube-heat exchangers.

In general, to reduce heat losses to the exterior, the hot fluid should circulate inside the tube bundle. However, since it is easier to clean the inside of the tubes, it is preferred to circulate the dirtiest fluid or the one that leaves more deposits or sediments inside the tube bundle. If both fluids produce similar dirtiness, it is preferred to circulate the one with greater pressure inside the tubes, since, on the contrary, the cost of a pressured shell is high. For viscous fluids, it is preferred to circulate them inside the tubes, since, in general, this type of fluids deposits dirt. However, when they circulate in the laminar regime, it is preferred to circulate through the outside, since the baffles will help to increase the turbulence and, therefore, will increase the heat transfer. If the fluids are corrosive, they will be circulated inside the tubes, since if they circulate outside, the use of an anticorrosive will be needed to avoid the deterioration not only of the outside part of the tubes but also of the shell.

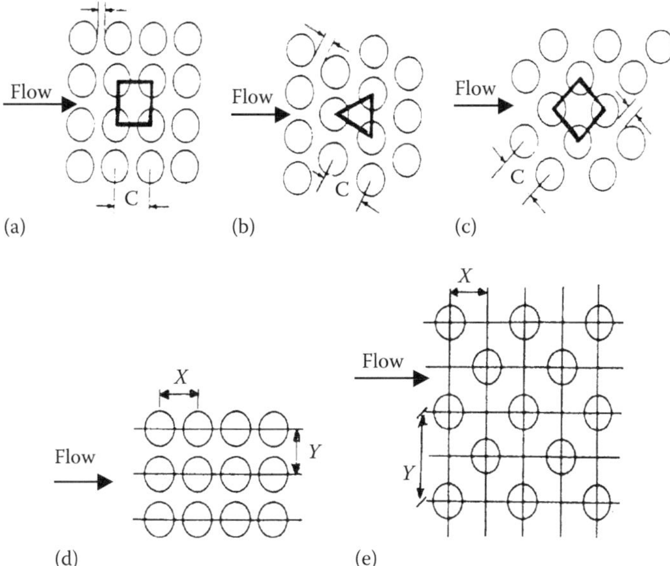

FIGURE 17.8 Tube bundle arrangement: (a) square arrangement, (b) triangular arrangement, (c) rotated square arrangement, (d) flow for square or line arrangement, and (e) flow for triangular or diamond arrangement.

Having explained these considerations, the setup of the mathematical model needed to design these types of heat exchangers will be presented next. The global heat-transfer balances for these heat exchangers are equal to those for the double tube-heat exchanger:

$$\dot{Q} = w_h \hat{C}_P\Big)_h (T_{in} - T_{out}) = w_c \hat{C}_P\Big)_c (t_{out} - t_{in}) \tag{17.44}$$

Regarding the equation of heat-transfer velocity through the exchange area, it is difficult to apply an expression such as that used for concentric tube-heat exchangers, since in this case the fluid that circulates through the shell experiences continuous direction changes, so parallel or countercurrent operations do not pertain. However, it is considered that the velocity equation is similar to the equation for double tube-heat exchangers, although the logarithmic mean temperature difference will be affected by a correction factor. Hence,

$$\dot{Q} = UA(\Delta T)_{ml} F \tag{17.59}$$

The logarithmic mean temperature difference is calculated as if the operation is performed in countercurrent:

$$(\Delta T)_{ml} = \frac{(T_{in} - t_{out}) - (T_{out} - t_{in})}{\ln\left[(T_{in} - t_{out})/(T_{out} - t_{in})\right]} \tag{17.60}$$

The factor F that corrects such logarithmic mean temperature difference is dimensionless and depends on the inlet and outlet temperature of the fluids and the type of heat exchanger.

17.4.2 Calculation of the True Logarithmic Mean Temperature

As mentioned before, the heat-transfer velocity equation is affected by a factor F, due to the geometry of the system. This factor corrects the logarithmic mean temperature increase $(\Delta T)_{ml}$, obtained at the end of the true value of this expression. A series of considerations related to the

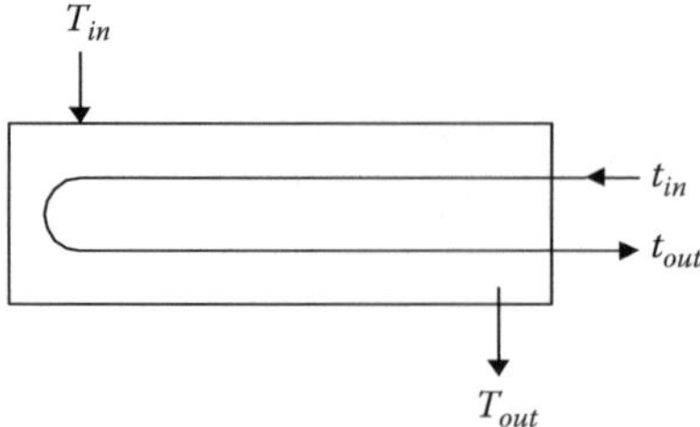

FIGURE 17.9 Shell and tube-heat exchanger 1/2.

functioning of this type of heat exchangers should be made for the calculation of this factor. The following assumptions are made:

- The temperature of the fluid in any cross section is considered to be the same.
- The heat-transfer coefficient is constant.
- The heat transfer is equal at each pass.
- The specific heat of each fluid is constant.
- The mass flow of each fluid is constant.
- There are no phase changes due to condensation or evaporation.
- Heat losses to the exterior are negligible.

The following dimensionless factor can be defined if the heat exchanger of Figure 17.9 is considered:

$$Z = \frac{T_{in} - T_{out}}{t_{out} - t_{in}} = \frac{w_c \hat{C}_P)_c}{w_h \hat{C}_P)_h} \tag{17.61}$$

It can be observed that it is a relationship between the heat capacities by hour, that is, the heat required to increase the flow per hour in 1°C for each fluid.

Also, a new dimensionless temperature factor can be defined according to the expression:

$$\varepsilon = \frac{t_{out} - t_{in}}{T_{in} - t_{in}} \tag{17.62}$$

This temperature relationship indicates the heating or cooling efficiency. This factor can vary from zero for the case of constant temperature of one of the fluids to the unit, when the inlet temperature of the hot fluid (T_{in}) coincides with the temperature of the cold fluid at the outlet (t_{out}).

The correction factor F is a function of these dimensionless parameters and of the type of heat exchanger: $F = F(Z, \varepsilon, \text{type})$. Figure 17.10 shows the graphical correlation for the different types of shell and tube-heat exchangers.

When applying the correction factors, it is not important whether the hot fluid circulates through the shell or the cold fluid circulates through the tube bundle. If the temperature of one of the fluids remains constant, the flow direction is not important, since in this case $F = 1$.

17.4.3 Calculation of Individual Coefficients

In order to calculate the heat-transfer global coefficient, film coefficients inside and outside the tubes should be known previously. The way to calculate individual coefficients will be explained next.

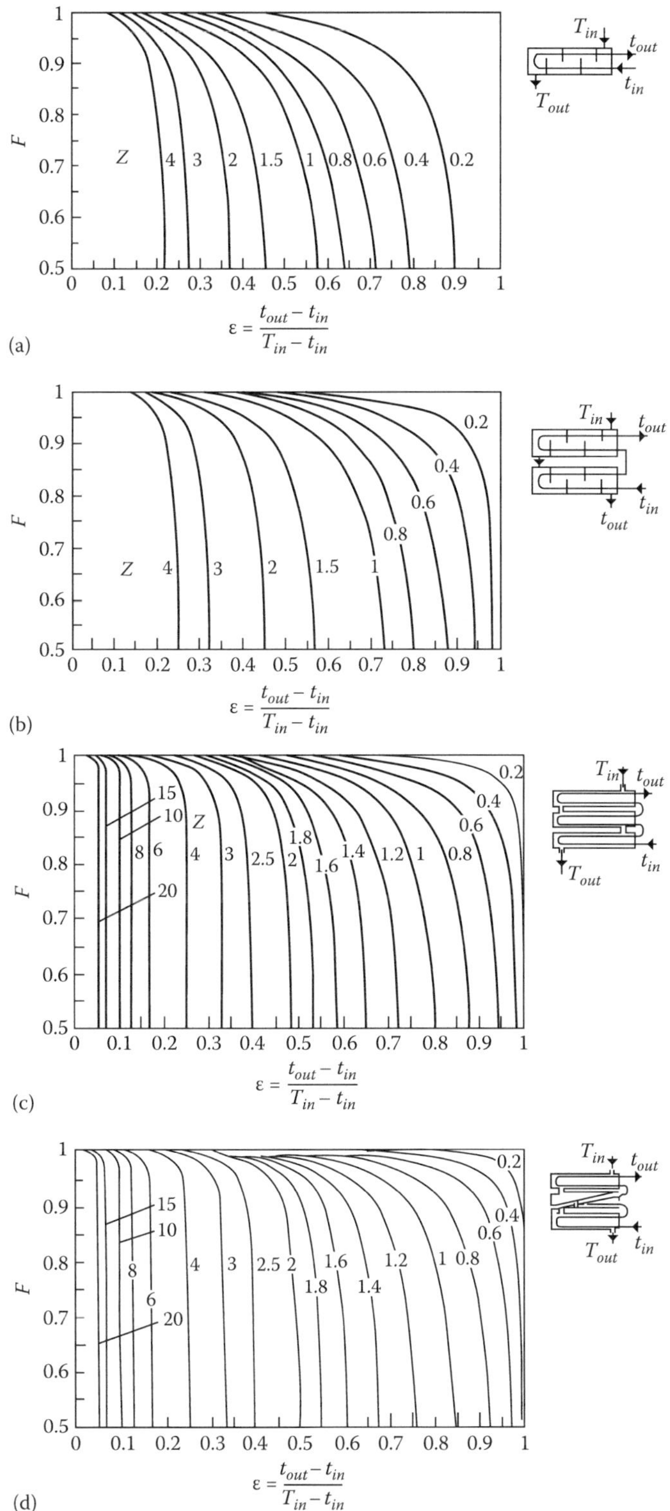

FIGURE 17.10 *F* factor for shell and tube exchangers: (a) one shell crossing, two or more tubes; (b) two shell crossings, four or more tubes; (c) three shell crossings; and (d) four shell crossings. (Adapted from Coulson, J.M. and Richardson, J.F., *Ingeniería Química*, Vols. I a VI, Reverté, Barcelona, Spain, 1979–1981.)

17.4.3.1 Coefficients for the Inside of the Tubes

The way to calculate the coefficient of the inside of the tubes was indicated in Section 17.2.1, and any of the expressions presented there can be used. One of the most used is that of the Sieder–Tate (Equation 17.17) or, also, the factor j_H of Colburn (Equation 17.16a):

$$j_H = (\text{Nu})(\text{Pr})^{-1/3}\left(\frac{\eta}{\eta_W}\right)^{-0.14} = 0.027(\text{Re})^{0.8} \tag{17.16b}$$

The Reynolds number appears in all these expressions, which is defined by

$$(\text{Re}) = \frac{\rho v d_i}{\eta} = \frac{G_t d_i}{\eta} \tag{17.63}$$

where

v is the linear circulation velocity of the fluid inside a tube
G_t is the mass flux of the fluid that circulates inside a tube

The value of G_t is calculated from the global mass flow rate of the fluid (w), from the cross section of a tube (a_t), from the total number of tubes in the bundle (N_t), and from their number of passes (n). Hence,

$$G_t = \frac{w}{N_t a_t / n} \tag{17.64}$$

17.4.3.2 Coefficients on the Side of the Shell

The calculation of the coefficient for the side of the shell is more complicated than in the last case, since it cannot be calculated using the expressions presented in other sections. This is due to different causes, one of them being that the fluid in the shell continuously changes direction. Also, turbulence that depends on the type of distribution of the tubes is created. Thus, in the case of a flow perpendicular to the tubes, the resistance of the square arrangement is lower than the resistance presented in the diamond arrangement (Figure 17.8). For this reason, it can be said that the value of the heat-transfer coefficient for the shell side depends on the Reynolds number, the type of deflectors, and the tube arrangement used.

Different methods to calculate the coefficient on the side of the shell can be found in the literature. One of these methods, suggested by Kern, uses the concept of the equivalent diameter for the shell, in the case of flow parallel to the tubes. The expression that allows the calculation of the equivalent diameter is different according to the arrangement of the tube bundle:

For square arrangements,

$$D_e = \frac{4\left[Y^2 - (\pi/4)d_e^2\right]}{\pi d_e} \tag{17.65a}$$

For triangle arrangements,

$$D_e = \frac{4\left[XY - (\pi/4)d_e^2\right]}{\pi d_e} \tag{17.65b}$$

where

d_e is the external diameter of the tubes
X and Y are defined by Figure 17.8

The equation that allows the calculation of the film coefficient is

$$(\mathrm{Nu}) = 0.36(\mathrm{Re})^{0.55}(\mathrm{Pr})^{1/3}\left(\frac{\eta}{\eta_W}\right)^{0.14} \quad (17.66)$$

in which the equivalent diameter defined by any of Equation 17.65 should be used in the Nusselt number. The Reynolds number is calculated according to the expression:

$$(\mathrm{Re}) = \frac{G_c D_e}{\eta} \quad (17.67)$$

G_c being the mass flux referred to the maximum area (A_c) for the transversal flow. The expression that allows the calculation of this area is

$$A_c = \frac{D_c B C'}{nY} \quad (17.68)$$

where

n is the number of passes that the fluid experiences in the shell
B is the separation between the deflector plates
D_c is the internal diameter of the shell
Y is the distance between the centers of the tubes
C' is the distance between two consecutive tubes ($C' = Y - d_e$)

The value of G_c is obtained from the mass flow rate w and the area A_c:

$$G_c = \frac{w}{A_c} \quad (17.69)$$

In the case of flow of gases through tube bundles, one of the most used equations is

$$(\mathrm{Nu}) = 0.33C_h(\mathrm{Re}_{max})^{0.5}(\mathrm{Pr})^{0.3} \quad (17.70)$$

in which C_h is a constant that depends on the tube arrangement and whose values for different arrangements are given in Table 17.2. In this equation, the maximum flow density value should be used to calculate the Reynolds number, in such a way that its value is obtained from Equation 17.69.

Different correlations that allow the calculation of heat-transfer coefficients for fluids that flow perpendicularly to the tube bundle can be found in the literature for different types of heat exchangers (Foust et al., 1960).

There are other expressions that allow the calculation of the coefficients for the fluid that circulates on the shell side, using the heat-transfer *factor* j_H, in such a way that

$$j_H = (\mathrm{Nu})(\mathrm{Pr})^{-1/3}\left(\frac{\eta}{\eta_w}\right)^{-0.14} \quad (17.71)$$

in which j_H is obtained from the empirical equations that take into account the type of circulation regime and influence of the bypassing of the deflectors and of the arrangement of the tube bundle.

TABLE 17.2
C_h and C_f Factors for Shell and Tube-Heat Exchangers

	$X = 1.25d_0$				$X = 1.5d_0$			
	Linear		Diamond		Linear		Diamond	
Re_{max}	C_h	C_f	C_h	C_f	C_h	C_f	C_h	C_f
				$Y = 1.25d_0$				
2,000	1.06	1.68	1.21	2.52	1.06	1.74	1.16	2.58
20,000	1.00	1.44	1.06	1.56	1.00	1.56	1.05	1.74
40,000	1.00	1.20	1.03	1.26	1.00	1.32	1.02	1.50
				$Y = 1.5d_0$				
2,000	0.95	0.79	1.17	1.80	0.95	0.97	1.15	1.80
20,000	0.96	0.84	1.04	1.10	0.96	0.96	1.02	1.16
40,000	0.96	0.74	0.99	0.88	0.96	0.85	0.98	0.96

Source: Coulson, J.M. and Richardson, J.F., *Ingeniería Química*, Vols. I a VI, Reverté, Barcelona, Spain, 1979–1981.

The most used expressions to calculate the factor j_H are explained next:
For turbulent regime ($2100 < \text{Re} < 10^6$),

$$j_H = a(\text{Re})^b \tag{17.72}$$

in which a and b are parameters that depend on the deflector and type of arrangement, respectively. The values of a for different values of deflector bypassing are as follows:

Deflector bypassing	5	25	35	45
a	0.31	0.35	0.30	0.27

The value of b will be
$b = 0.55$ for triangular and normal square arrangements
$b = 0.56$ for rotated square arrangement

For laminar regime ($\text{Re} < 2100$),

$$j_H = a(\text{Re})^{0.43} \tag{17.73}$$

where the parameter a is a function of the deflector bypassing:

Deflector bypassing	15	25	35	45
a	0.84	0.69	0.64	0.59

17.4.4 Calculation of Head Losses

Head losses of the fluid that circulates inside the tubes should be differentiated from those of the fluid that circulates outside.

17.4.4.1 Head Losses inside Tubes

When calculating the pressure drop inside the tubes, the pressure drop due to friction (ΔP_t) should be distinguished from the pressure drop due to changing from one pass to the next (ΔP_r).

The pressure drop due to the tubes can be calculated from the equation of Fanning:

$$\Delta P_t = (4f)\frac{L}{d_i}n\frac{G_t^2}{2\rho} \tag{17.74}$$

where

L is the length of the tubes
d_i is their internal diameter
G_t is the mass flux with which the fluid circulates through each tube
f is the friction factor that depends on the Reynolds number and the relative roughness, as well as on the circulation regime

The pressure drop for the direction change (180°) is calculated by the equation:

$$\Delta P_r = 4\frac{\rho v^2}{2} = 4\frac{G_t^2}{2\rho} \tag{17.75}$$

If there are n passes, the total head loss is

$$\Delta P = \Delta P_t + n\Delta P_r \tag{17.76}$$

17.4.4.2 Head Loss on the Shell Side

The equation that can be used to calculate head losses on the shell side is

$$\Delta P_c = \frac{f' G_c^2 D_c (N_c + 1)}{2\rho D_e} \tag{17.77}$$

where

D_c is the internal diameter of the shell
D_e is the equivalent diameter of the shell (Equation 17.65)
N_c is the number of baffles, so $(N_c + 1)$ is the number of times the fluid crosses the shell
If L is the length of the shell and B is the distance between baffles, then $N_c + 1 = L/B$.

The parameter f' is the friction coefficient, which depends on the Reynolds number, the type of tube arrangement, and the baffle bypassing. This friction factor can be calculated from graphs (Figure 17.11).

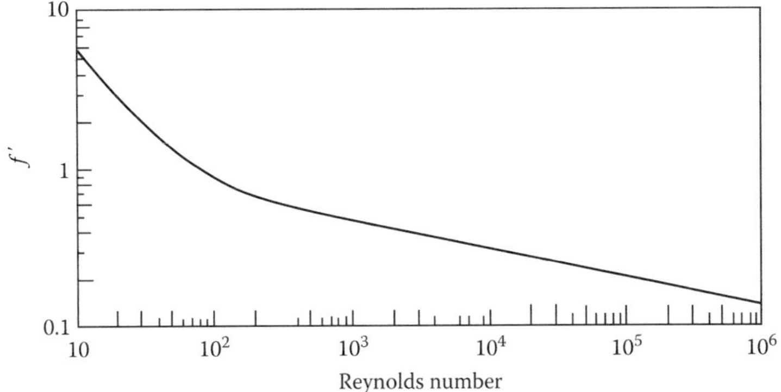

FIGURE 17.11 Friction factor for the shell fluid. (Adapted from Coulson, J.M. and Richardson, J.F., *Ingeniería Química*, Vols. I a VI, Reverté, Barcelona, Spain, 1979–1981.)

Head losses can also be calculated from the following equation:

$$\Delta P_c = C_f m \frac{\rho v_c^2}{6} \tag{17.78}$$

where

v is the circulation velocity of the fluid through the shell

m is the number of rows of tubes

C_f is a factor that depends on the type of tube arrangement, whose value can be obtained from data in Table 17.2

17.5 PLATE-TYPE HEAT EXCHANGERS

The plate-type heat exchangers are constituted by a series of corrugated plates, created by a simple stamp, and are drilled on the ends to allow or direct the flow of liquid to be heated or cooled. These plates are placed one in front of the other in such a way that fluids can circulate between them. Also, they are compressed by screws, so that they can support internal pressure. Between the plates and on their edges, rubber gaskets are placed to avoid fluid leakage or mixings. It should be noted that these rubber gaskets limit the areas in which these heat exchangers can be used. Once the set of plates is ready, the arrangement of the drilled holes on each plate directs the passage of fluids, as can be observed in Figure 17.12.

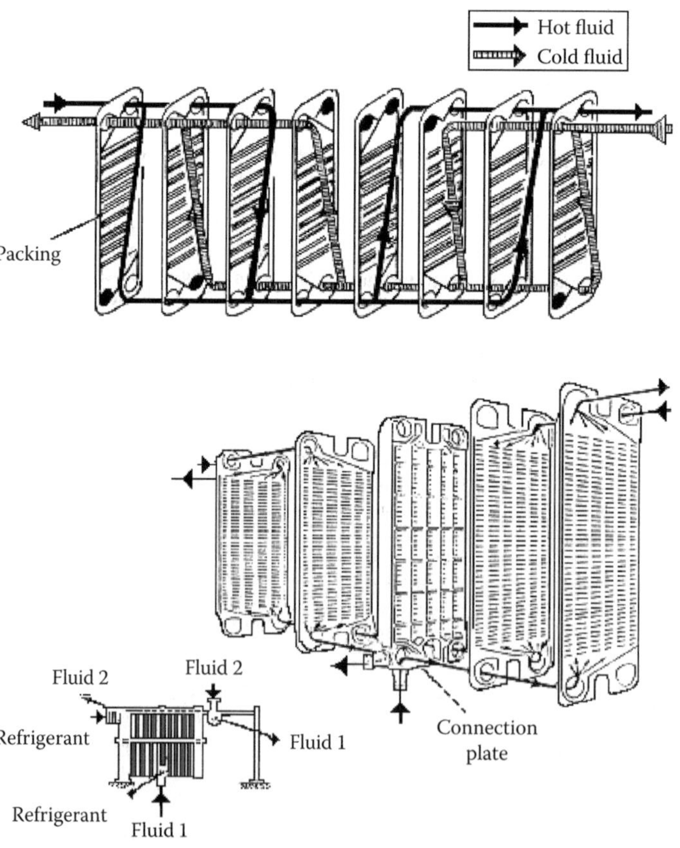

FIGURE 17.12 Plate-type heat exchanger.

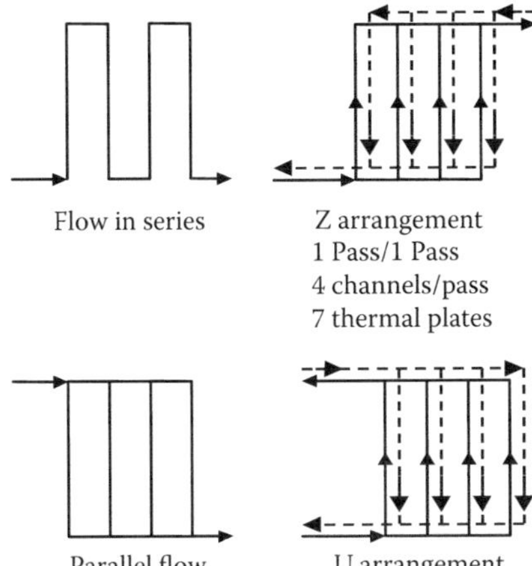

FIGURE 17.13 Fluid circulation for different arrangements.

According to how the plates are arranged, different forms for the pass of fluid can be obtained. Thus, plates can be combined in such a way that distributions such as those shown in Figure 17.13 can be formed, and these are briefly discussed next:

- Flow in series, in which a continuous stream changes its direction after each pass.
- Flow in parallel, in which the stream is divided into various substreams, and once passed through the plates, it converges again.
- Z arrangement, in which both streams flow in parallel, but the outlet is at the opposite side to the inlet.
- U arrangement, in which both streams flow in parallel, but the outlet and inlet points are on the same side of the heat exchanger.
- There are other types of arrangements, which are more complex. Some examples are shown in Figure 17.13.

The applications of these types of heat exchangers are very flexible, since when the arrangement of the plates or drilled holes is changed, the flow characteristics in such heat exchangers also change. In addition, it should be pointed out that it is possible to perform operations with various fluids in the same heat exchanger by inserting a connecting grid that allows the inlet or outlet of selected fluids, as can be observed in Figure 17.12. One example of this case is milk pasteurizers, in which milk can be gradually heated or cooled in one heat exchanger using hot or cold water streams and taking advantage of the heat carried by milk that should be cooled after heating. Other processes in which plate-type heat exchangers are used in a similar way as that described here can be found in the industry.

One of the problems of any heat exchanger is dirtiness, since the deposits formed on the walls decrease the transfer of heat. These deposits can prevent the heat exchanger from meeting the specifications for which it was designed if the heat exchanger is not adequately scaled. This can be a serious problem in a tubular heat exchanger, since it needs to be cleaned to function adequately again. However, plate-type heat exchangers have the advantage of being easily expanded by adding new plates, which allows a longer time between cleanings. Also, the dirtiness process in these heat exchangers is slower, since fluids circulate with high turbulence and the surfaces of plates are smoother, avoiding low-velocity zones.

In plate-type heat exchangers, turbulence is reached for values of Reynolds numbers from 10 to 500, since plate folds break the film stagnated on the heat-transfer surface. Corrugation of plates, in addition to holder together and keeping a constant separation between them, produces high turbulence in the fluid that circulates between the plates. On the other hand, corrugation allows a considerable increase in the transfer surface per plate. All of these factors contribute to high heat-transfer coefficients; under similar conditions, this coefficient is 10 times higher in a plate-type heat exchanger than the heat-transfer coefficient for a fluid circulation inside a tube of a conventional tube-heat exchanger. Thus, for the same exchange capacity, the transfer area is much lower in plate-type heat exchangers.

In spite of the advantages discussed earlier, plate-type heat exchangers have some limitations with respect to tube-heat exchangers. Thus, it is possible to build plate-type heat exchangers that can withstand 20 atm as maximum pressure, and in extreme cases, the gaskets can withstand a maximum of 260°C. In addition, the pressure drop in this type of heat exchangers is greater than in tube exchangers. These limitations define the application of the plate-type heat exchangers.

17.5.1 Design Characteristics

Consider a heat exchanger formed by four plates, like the one shown in Figure 17.14. It can be observed that one of the fluids circulates through a channel receiving heat from the other fluid through the plates.

This is a simple example involving plate-type heat exchangers. However, in general, they are constituted by a greater number of plates placed one next to the other, and fluids flow according to the arrangements shown in Figure 17.13 or with more complicated arrangements.

In any case, the basic equations used to set up the mathematical model are similar to those of other types of heat exchangers. Thus, the fundamental equations are

- Global energy balance. The cold fluid, increasing in this way its temperature, absorbs the heat flow given up by the hot fluid:

$$\dot{Q} = w_h\hat{C}_P\Big)_h (T_{in} - T_{out}) = w_c\hat{C}_P\Big)_c (t_{out} - t_{in}) \tag{17.79}$$

- Heat-transfer equation. The following equation can be applied to a system like the one shown in Figure 17.14:

$$\dot{Q} = U(2A_p)(\Delta T)_{ml} \tag{17.80}$$

where

U is the global heat-transfer coefficient

A_P is the area of one plate

$(\Delta T)_{ml}$ is the logarithmic mean temperature difference, taken as if the fluids circulate in countercurrent

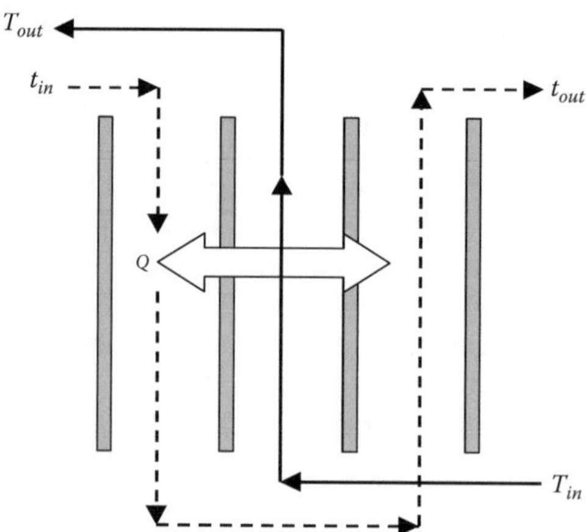

FIGURE 17.14 Basic unit of a plate-type heat exchanger.

For more complex systems, the velocity equation is similar to the previous one, but it includes a correction factor F for the logarithmic mean temperature difference, and its value depends on the passing system of the fluids and on the number of transfer units (NUT). Therefore,

$$\dot{Q} = UA_t(\Delta T)_{ml} F \tag{17.81}$$

In this equation, A_t is the total heat-transfer area of the exchanger.

The total area is the product of the area of one plate times the number of plates in which there is heat transfer:

$$A_t = N \cdot A_p \tag{17.82}$$

The plates through which heat is transferred are called *thermal plates*. It should be indicated that the plates at the end of the heat exchanger and the intermediate plates that distribute the fluids are not thermal, since in these plates there is no heat exchange between the fluids. Therefore, if there are N thermal plates, the number of channels by which fluids circulate is $N + 1$.

17.5.2 Number of Transfer Units

The NTU is defined as the relationship between the temperature increase experienced by the fluid that is being processed and the logarithmic mean temperature increase:

$$\text{NTU} = \frac{T_{in} - T_{out}}{(\Delta T)_{ml}} \tag{17.83}$$

From the velocity equation and from the total heat flow transferred (Equations 17.79 and 17.80), it is obtained that

$$\dot{Q} = U(2A_p)(\Delta T)_{ml} = w_h\hat{C}_P\big)_h (T_{in} - T_{out}) \tag{17.84}$$

Hence, the NTU can be expressed as

$$\text{NTU} = \frac{2A_pU}{w_h\hat{C}_P\big)_h} \tag{17.85}$$

Considering a global mean heat-transfer coefficient and the total area,

$$\text{NTU} = \frac{A_tU_m}{w_h\hat{C}_P\big)_h} \tag{17.86}$$

The NTU is also called execution factor, thermal length, or temperature ratio.

This factor is needed to evaluate the correction factor F of the logarithmic mean temperature difference, as will be seen later. Also, depending on the value of this factor, the type of plates that should be used in the heat exchanger for a given process can be chosen. Thus, for low values of NTU, short and wide plates should be used, which are characterized by having low heat-transfer coefficients and small pressure drops in each pass. On the contrary, for high NTU values, long and narrow plates with deep folds on their surfaces are used, and small hollows between plates are retained in order to have narrow channels.

Likewise, it should be pointed out that the plates in which folds are formed by channels that have obtuse angles yield high NTU values for each plate, while if the angle formed by the folds is acute, the value of NTU of the plate is low, thus offering smaller resistance to the flow of the fluid that passes between plates.

17.5.3 Calculation of the True Logarithmic Mean Temperature Difference

The true increase or logarithmic mean temperature difference is determined, in this type of heat exchangers, in a similar way to the case of shell and tube-heat exchangers. The logarithmic mean temperature increase is defined as if the operation were carried out in countercurrent (Equation 17.60).

The temperature increase should be corrected with a factor F, whose value can be graphically obtained once the NTU and the pass system of hot and cold fluids through the channels of the heat exchangers are known. F can be calculated from Figure 17.15, from the NTU and the type of passes of the fluids (Marriott, 1971; Raju and Chand, 1980).

In these heat exchangers, the correction factor can have high values, close to one, and in addition to the fact that they usually present high heat-transfer coefficients, very high performances can be obtained. In plate-type heat exchangers, the temperature difference between the hot stream at the inlet point (T_{in}) and the cold fluid at the outlet point (t_{out}) can be as small as 1°C, whereas for a shell and tube-heat exchanger, this difference is close to 5°C.

17.5.4 Calculation of Heat-Transfer Coefficients

There is an expression for plate-type heat exchangers that relates Nusselt, Reynolds, and Prandtl numbers according to the equation:

$$(\text{Nu}) = C(\text{Re})^x(\text{Pr})^y\left(\frac{\eta}{\eta_w}\right)^z$$

where C is a constant.

The values of the constant and exponents of this equation are experimentally determined, and they are only valid for each type of plate for which they were obtained. Also, they depend on the circulation regime. In these heat exchangers, it is considered that the fluid circulates in the laminar regime when the value of the Reynolds number is lower than 400, while for values higher than 400, the fluid circulates in the turbulent regime.

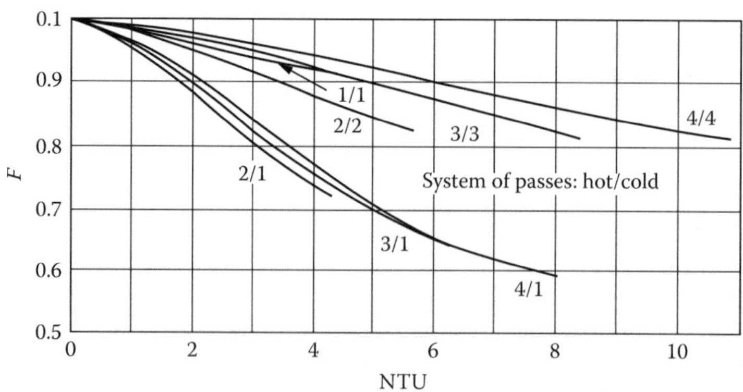

FIGURE 17.15 Correction factors F for different systems of passes. (Adapted from Marriott, J., *Chem. Eng.*, 5, 127, 1971.)

In the turbulent regime (Re > 400), these variables range between the following values (Marriott, 1971):

$C = 0.15–0.40$
$x = 0.65–0.85$
$y = 0.30–0.45$ (generally 0.333)
$z = 0.05–0.20$

The Reynolds number depends on the system of passes by which different fluids flow. Thus, for flow in series, in which each fluid passes through each channel as one stream, the Reynolds number is defined according to the equation:

$$(\text{Re}) = \frac{\rho v D_e}{\eta} = \frac{G D_e}{\eta} \tag{17.87a}$$

where
v is the linear circulation velocity of the fluid
G its correspondent global mass flux
D_e is the equivalent diameter

When the system of passes is in parallel flow, the streams of each fluid arc subdivided into sub-streams (Figure 17.13) that cross the channels between plates. The Reynolds number is given by

$$(\text{Re}) = \frac{(G/n)D_e}{\eta} \tag{17.87b}$$

In this equation, n is the number of channels of each type of fluid. If η_c is the number of channels for the hot fluid, the Reynolds number for this fluid is calculated by applying the last equation for $n = n_h$. For the cold fluid, the same equation is applied by using $n = n_c$, where η_c is the number of channels in which the stream of this fluid is divided.

One of the most used equations for the calculation of individual heat-transfer coefficients in a turbulent regime is

$$(\text{Nu}) = 0.374(\text{Re})^{2/3}(\text{Pr})^{1/3}\left(\frac{\eta}{\eta_W}\right)^{0.15} \tag{17.88}$$

An equation that is easy to apply is turbulent regime, and that equation uses the mean values of Reynolds and Prandtl numbers (Buonopane et al., 1963; Usher, 1970):

$$(\text{Nu}) = 0.2536(\text{Re})^{0.65}(\text{Pr})^{0.4} \tag{17.89}$$

The geometry of plates is not taken into account in all of these equations. However, there are some equations that include geometric factors, such as pass length (L) and the spacing between plates (b). One of these is that of Troupe et al. (1960):

$$(\text{Nu}) = (0.383 - 0.0505^{L/b})(\text{Re})^{0.65}(\text{Pr})^{0.4} \tag{17.90}$$

Other equations have been proposed for the calculation of film coefficients when working in a laminar regime (Re < 400), one of them being a variation of the Sieder–Tate equation (Marriott, 1971):

$$(\text{Nu}) = C\left(\text{Re}\,\text{Pr}\frac{D_e}{L}\right)^{1/3}\left(\frac{\eta}{\eta_W}\right)^{0.14} \tag{17.91}$$

where the value of the constant C ranges between 1.86 and 4.5, depending on the geometry of the system.

Another equation used in laminar regime is (Jackson and Troupe, 1964; Raju and Chand, 1980)

$$h = 0.742\hat{C}_P G(\text{Re})^{-0.62}(\text{Pr})^{-2/3}\left(\frac{\eta}{\eta_W}\right)^{0.14} \tag{17.92}$$

The equivalent diameter D_e that appears in the Reynolds and Nusselt numbers is defined as four times the hydraulic radius, which in turn is the ratio between the area through which the fluid flows between the plates and the wet perimeter:

$$D_e = 4r_H = 4\frac{ab}{2a} = 2b \tag{17.93}$$

where
a is the width of the plates
b is the distance between them

It is interesting to point out that for a tube-heat exchanger, the equivalent diameter coincides with the pipe diameter. Therefore, from the heat exchange standpoint, the plate-type heat exchanger will be equal to a tube-heat exchanger whose diameter is equal to double the thickness of the packing. If we take into account that the thickness is usually 2–5 mm, it turns out that the plate-type heat exchanger behaves the same as one with tubes of 4–10 mm diameter. In spite of the fact that a tube-heat exchanger with these characteristics would be ideal for the desired heat exchange, this result is very difficult to achieve, since mechanical difficulties arise during the building process.

For the calculation of individual coefficients, there exist graphs in which the values of such coefficients correlate with the so-called *specific pressure drop* (Figure 17.16).

The specific pressure drop (J) is defined as the ratio between the pressure drop and the NTU:

$$J = \frac{\Delta P}{\text{NTU}} \tag{17.94}$$

It can be seen in Figure 17.16 that in most plate-type heat exchangers, the variation of film coefficients due to specific pressure drop is similar. Curve A refers to special plates that have a spacing greater than usual, designed for operations in which the value of NTU is very low or for processes

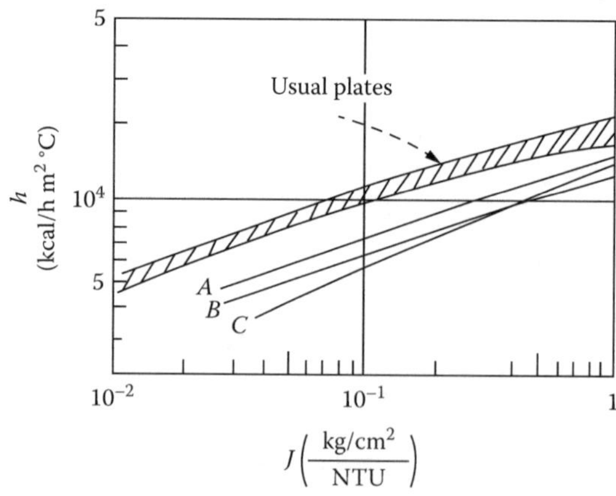

FIGURE 17.16 Variation of individual heat-transfer coefficients with the specific pressure drop. (Adapted from Marriott, J., *Chem. Eng.*, 5, 127, 1971.)

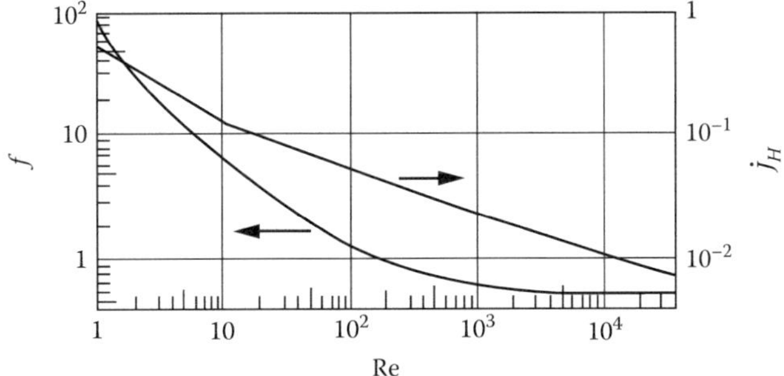

FIGURE 17.17 Variation of the friction factor f and j_H with the Reynolds number.

in which fluids with high solid content should be treated. Curve B is used for similar cases, but in which plates have fewer foldings per unit area and, therefore, fewer contact points. Finally, curve C is used for smooth plates with stamped waves that have little effect on fluid flow.

The film coefficients can also be obtained from the Colburn coefficient j_H for heat transfer. Figure 17.17 graphically shows the value of j_H as a function of the Reynolds number for one type of plate.

Once the individual coefficients are obtained, the global heat-transfer coefficient must be calculated according to the global equation:

$$\frac{1}{U} = \frac{1}{h_C} + \frac{e}{k_P} + \frac{1}{h_F} + R_h + R_c \tag{17.95}$$

where

e is the thickness of the plate wall
k_P is the thermal conductivity of the material of which the plate is made
R_h and R_c are the fouling factors of the hot and cold fluids, respectively

17.5.5 Calculation of Head Losses

It is extremely important to know the head or pressure losses experienced by the fluids when flowing through plate-type heat exchangers, since when such a drop is known, the type of plates for the heat exchanger can be chosen.

Variations of the Fanning equation can be used to calculate the pressure drop. One of these equations is

$$\Delta P = 2f\frac{G^2L}{gD_e\rho} \tag{17.96}$$

where

G is the mass flux of the fluid
L is the length of the plate or the distance that each fluid should cross when passing through the channel between two plates
g is the gravitational constant
f is the friction factor

The friction factors calculated from the flowing linear velocity of the fluids, in a turbulent regime, are 10–60 times higher in the flow through channels in a plate-type heat exchanger than in the flow through a tube for similar Reynolds numbers.

For the turbulent circulation regime, the friction factor can be calculated from the following equation (Cooper, 1974; Raju and Chand, 1980):

$$f = \frac{2.5}{(\text{Re})^{0.3}} \tag{17.97}$$

The friction factor depends on the type of plate. Figure 17.17 graphically shows the variation of the friction factor with the Reynolds number.

17.5.6 Design Procedure

Generally, in the design of heat exchangers, it is desirable to calculate the area of heat exchange or the number of thermal plates required to perform an operation in which the inlet and outlet temperatures of one of the fluids, the inlet temperature of the other fluid, the mass flow of both streams, the physical properties of both fluids, and the characteristics of the plates to use are known.

The design of a plate-type heat exchanger is complex, so in order to find a solution for the mathematical model, it is necessary to use computer programs. However, there are simpler methods that allow us to obtain a good approximation for the design of these heat exchangers, even in such cases in which the fluids that circulate behave as non-Newtonian.

A design method, based on the use of correction factors F of the logarithmic mean temperature difference, is explained next. A series of hypotheses should be made to solve the mathematical model stated. It will be assumed that the following conditions are complied with:

- Heat losses to the surroundings are negligible.
- No air bags are formed inside the heat exchanger.
- The global heat-transfer coefficient is constant along the heat exchanger.
- The temperature inside each channel only varies along the direction of flow.
- In parallel flow, the global stream is equally divided among all the channels.

These assumptions allow us to solve the problem setup at the beginning of the section, for fluids flowing either in series or in parallel.

Next, the steps to calculate the number of plates in the design of a desired heat exchanger will be explained:

Data: T_{in}, T_{out}, t_{in}, w_h, w_c
Physical properties of the fluids: k, ρ, η, $\hat{C}_P$ (obtained from tables, graphs, or equations)
Characteristics of the plates: k_p, L, a, b, e (given by the manufacturer)
Calculation steps

1. Calculation of the total heat flow transferred by the hot fluid and gained by the cold fluid, using Equation 17.79:

$$\dot{Q} = w_h \hat{C}_P\big)_h (T_{in} - T_{out})$$

2. Calculation of the outlet temperature of the cold fluid, from Equation 17.79:

$$t_{out} = t_{in} + \frac{\dot{Q}}{w_c \hat{C}_P\big)_c}$$

3. Calculation of the physical properties of the fluids at the mean temperatures, using tables, graphs, or appropriate equations.
4. Calculation of the logarithmic mean temperature difference $(\Delta T)_{ml}$.
5. Determination of the NTU, with Equation 17.83.
6. Calculation of the correction factor F of the logarithmic mean temperature using Figure 17.15.
7. Determination of the Reynolds number of each stream, using Equation 17.87, depending on the arrangement of the flow: series or parallel.
8. Calculation of the individual heat-transfer coefficients h, by Equations 17.88 through 17.92, depending on the circulation regime. They can also be determined from Figures 17.16 and 17.17.
9. Calculation of the global heat-transfer coefficients U, using Equation 17.95, once the film coefficients are known and the fouling factors have been estimated.
10. Calculation of the total heat-transfer area from Equation 17.81:

$$A_t = \frac{\dot{Q}}{U(\Delta T)_{ml} F}$$

11. Determination of the number of thermal plates from the total area and the area of each plate: $N = A_t/A_p$.

These steps are common for the calculation of the heat exchangers that work either in series or parallel. However, in the case of an operation in parallel, an iterative process for the calculation of the number of thermal plates should be carried out. This is due to the fact that in step 7, the number of channels of the hot (n_h) and cold (n_c) streams should be assumed in order to calculate the corresponding Reynolds number, which allows continuation of the following calculation steps until obtaining the number of thermal plates. Once this number N is known, the number of total channels will be known $n = N + 1$. In the case that N is odd, the number of channels is even, so the values of n_h and n_c are equal ($n_h = n_c = (N + 1)/2$). On the contrary, if N is even, the number of channels is odd, so the number of channels of a type is higher than the other type in one unit. Once the values of n_h and n_c are obtained from N, they are compared with those assumed in step 7. If the assumed and calculated values coincide, then the process is finished; otherwise, the process is repeated from step 7 to 11. The iterative process is repeated until such values coincide.

17.6 EXTENDED SURFACE HEAT EXCHANGERS

In many cases, heat exchangers whose surface has been extended by means of *fins* are used to improve the heat transfer. Fins can have different shapes (Figure 17.18), such as longitudinal, circular or spiral, or tang, among others. Fins can be of the same material as the tubes or of a different material, although in any case a good contact between the tube and the fin should be assured. Fins are widely used in cooling processes, especially when it is desired to eliminate heat by using fluids that are bad conductors, such as gases.

A fin will not always noticeably improve heat transfer, since the transfer depends on the resistance that more greatly affects such transfer. Thus, if the general expression relates the global heat-transfer coefficient in reference to the internal area with the different resistance, neglecting the possible fouling effects, then it is obtained that

$$\frac{1}{U_i A_i} = \frac{1}{h_e A_e} + \frac{e}{k A_{ml}} + \frac{1}{h_i A_i} \tag{17.98}$$

The left-hand side of this equation represents the global resistance to heat transfer and is the summation of three resistances, due to two fluids, plus the resistance offered by the metallic wall.

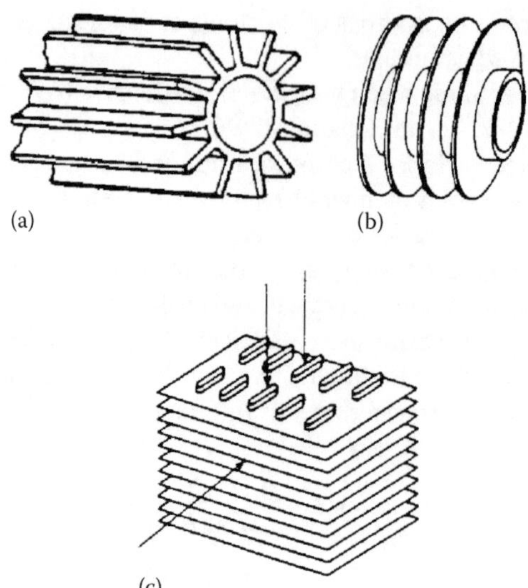

FIGURE 17.18 Different types of fins: (a) longitudinal, (b) transversal, and (c) tubes with continuous fins.

Generally, the resistance exerted by metal is not high when compared to the other two resistances, so it can be considered as negligible.

If Equation 17.98 is observed, it can be seen that A_e is the external area. Therefore, if fins are extended, then the value of $1/(h_eA_e)$ is decreased, so the global resistance also decreases in such a way that heat transfer is favored. This heat removal will be more or less important depending on the values acquired by the individual heat-transfer coefficients of the fluids. If the external coefficient h_e is very small compared to h_i, the external resistance will be much greater than the internal resistance, so if the external area A_e is increased, the global resistance decreases almost proportionally with the increase of the external area. On the contrary, if h_i is much smaller than h_e, the resistance of the internal fluid will be much greater than the external resistance; therefore, the increase in the external area will slightly influence the global resistance, and then it will be found that fins do not always noticeably increase heat transfer.

17.6.1 Mathematical Model

Among the different types of fins, only the problem of the longitudinal fins will be set up. This model is similar for the other types of fins, although the mathematical solution has not been performed in all cases because it is difficult to obtain.

Consider the fin shown in Figure 17.19. Its dimensions are indicated; T_W is the temperature at the base of the fin and is constant. An energy balance can be assumed according to the following:

- Heat transfer by conduction along the fins occurs only in the radial direction, so temperature is a function only of this coordinate.
- A fin dissipates heat by convection to the outside only along it and not by its edges. That is, it dissipates heat through the lateral area to the sides.
- The density of the heat flow driven to the outside at any point of the fin, with a temperature T, is expressed as $q = h_e\,(T - T_{in})$, where T_{in} is the temperature of the external fluid and h_e the individual convective heat-transfer coefficient within the external fluid.

The following equation is obtained when carrying out an energy balance in a differential element of the fin with a height Δx:

$$Mbq(x) - Mbq(x+\Delta x) - 2M\Delta x h_e (t - T_{in}) = 0 \qquad (17.99)$$

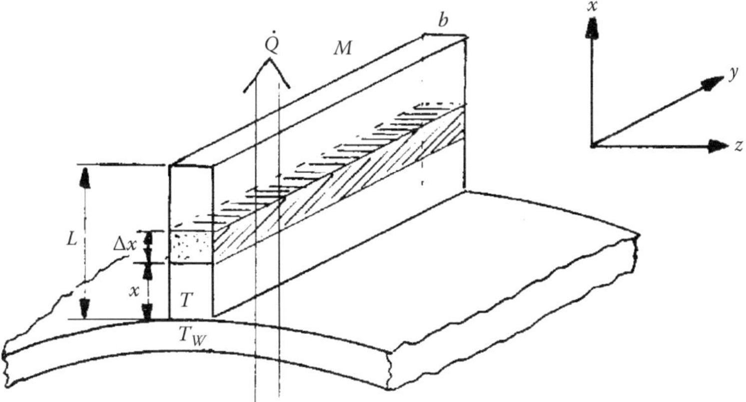

FIGURE 17.19 Longitudinal fin, geometric characteristics, and direction of the heat flow.

If the last expression is divided by the differential of the fin's volume ($Mb\Delta x$) and it is taken as the limit for $\Delta x \rightarrow 0$, the following equation results:

$$\frac{dq}{dx} = -\frac{2}{b} h_e (t - T_{in}) \tag{17.100}$$

Taking into account that the density of the heat flow, according to Fourier's law, is

$$q = -k\frac{dT}{dx}$$

then the following is obtained from Equation 17.100:

$$k\frac{d^2T}{dx^2} = \frac{2h_e}{b}(T - T_{in}) \tag{17.101}$$

This equation can be integrated on the boundary conditions:

$$\text{For } x = 0 \qquad T = T_W$$

$$\text{For } x = L \qquad \frac{dT}{dx} = 0$$

This last condition is a maximum or minimum since at the end of the fin the temperature will be maximum or minimum depending on whether it heats or cools.

The integration of the differential equation on the boundary conditions stated before yields the following expression:

$$\frac{T - T_{in}}{T_W - T_{in}} = \frac{\cosh[(L - x)(2h_e/kb)^{1/2}]}{\cosh[L(2h_e/kb)^{1/2}]} \tag{17.102}$$

17.6.2 Efficiency of a Fin

The efficiency of a fin is defined as the ratio between the heat dissipated by the surface of the fin and the heat that would be dissipated if the entire surface were at the temperature of the base (T_W):

$$\eta = \frac{\text{Heat dissipated by the fin's surface}}{\text{Heat that would be dissipated if the surface were at } T_w}$$

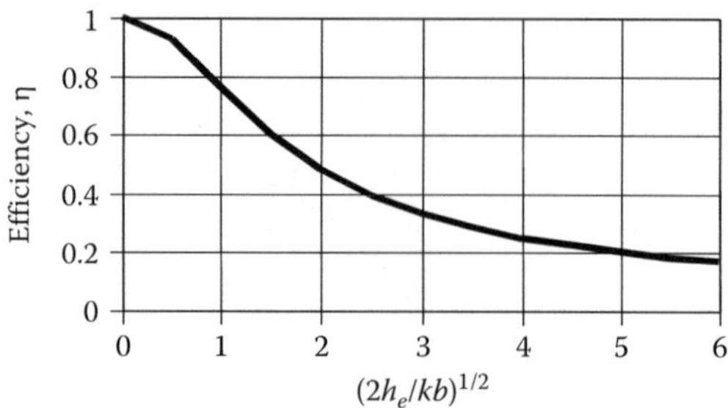

FIGURE 17.20 Efficiency of a longitudinal fin.

In the case presented in the last section for longitudinal fins, the expression that can be used to obtain the efficiency of a fin is

$$\eta = \frac{\int_0^L \int_0^M h_e (T - T_{in}) dy\, dx}{\int_0^L \int_0^M h_e (T_W - T_{in}) dy\, dx} \tag{17.103}$$

The integration of this expression yields the equation

$$\eta = \frac{\tanh\left[L\sqrt{2h_e/kb}\right]}{L\sqrt{2h_e/kb}} \tag{17.104}$$

This last equation can be expressed in graphical form (Figure 17.20), representing the efficiency against the dimensionless product $L(2h_e/kb)^{1/2}$. It can be observed that the value of this product depends on the ratio h_e/k in such a way that when it is small, the efficiency of the fin tends to 1. On the other hand, when the values of h_e are high, the efficiency decreases to very low values, so the use of fins is neither needed nor efficient, as indicated previously.

The increase of temperatures between a given section and the external fluid appears in the expression that defines the efficiency (Equation 17.103), in such way that $\Delta T = T - T_{in}$. The value of ΔT varies from the base of the fin up to its end, so the solution of the last integral should take into account this variation. This was considered when obtaining the last expression of efficiency. However, if an average of the difference between the temperature of the fin and the fluid is used $(\Delta T)_m = T - T_{in}$, where T_m is a mean temperature of the fin, the integration of the expression of the efficiency will be

$$\eta = \frac{\int_0^L \int_0^M h_e (T_m - T_e) dy\, dx}{\int_0^L \int_0^M h_e (T_W - T_e) dy\, dx} = \frac{T_m - T_{in}}{T_W - T_{in}} = \frac{(\Delta T)_m}{(\Delta T)_W} \tag{17.105}$$

where $(\Delta T)_W = T_W - T_{in}$.

This last equation points out that the efficiency of this type of fin can be expressed as a relationship between the average temperature increase of the fin and of the fluid and the difference between the temperature of the fin's base and the fluid.

The mathematical model and its solution, as well as the expressions of the temperature profile along the fins and its efficiency, performed here for longitudinal fins, can be similarly conducted for other types of fins.

$$L\sqrt{\frac{2h_e}{kb}}$$

17.6.3 Calculation of Extended Surface Heat Exchangers

The calculation of the global heat-transfer coefficient in heat exchangers with fins depends on the presence of fins, since the heat transferred from the metal surface to the fluid will be the addition of two flows, one through the area of the naked tube and the other through the surface of the fin. Therefore, the global coefficient referred to the internal area will be expressed as

$$U_i = \frac{1}{(A_i/h_e\,(\eta A_a + A_e)) + (eA_i/kA_{ml}) + (1/h_i)} \tag{17.106}$$

where
A_i represents the internal surface of the tube
A_a the fin's surface
A_e that of the naked tube
η is the efficiency of the fin

Obtaining of U_i is not so easy, since the calculation of h_e adds more difficulties than in the cases of heat exchangers made of naked tubes. This is due to the fact that the flow outside the fins is modified, so the film coefficient will be different from the coefficient of a smooth tube. The values of h_e for the case of tubes with fins should be experimentally obtained for the different types of fins. Generally, the manufacturers of this type of heat exchangers should provide these data.

17.7 SCRAPED-SURFACE HEAT EXCHANGERS

Scraped-surface heat exchangers consist of two concentric tubes in which, generally, a high viscosity fluid circulates inside the internal tube. The internal surface of this tube is scraped by a set of blades inserted in a central axis that spins at a certain number of revolutions. This type of heat exchanger is used in the food industry and is known as the Votator heat exchanger.

The design of this type of heat exchanger assumes that the heat transferred from the surface of the tube to the viscous fluid or vice versa is carried out by conduction under an unsteady state. It is considered that such heat transfer by conduction is similar to conduction in a semi-infinite solid. Thus, the equation that relates the temperature T of a point of the fluid at a distance x from the wall with time can be expressed as

$$\frac{T_W - T}{T_W - T_0} = erf\left(\frac{x}{2\sqrt{\alpha t}}\right) \tag{17.107}$$

where
T_W is the temperature of the wall
T_0 is the initial temperature of the considered point
erf is the so-called Gauss error function
α is the thermal diffusivity of the fluid

If there are P scraping blades inserted in the central axis, which spin at a frequency N, the time that a blade takes to pass at the same point of the fluid is $t_N = 1/(PN)$. Heat transfer by conduction within the viscous fluid occurs during this period.

The total heat transferred from the wall to the fluid during this period is

$$Q_T = h_i A (T_W - T_0) t_N \tag{17.108}$$

where
h_i is the individual convective heat-transfer coefficient for the viscous fluid
A is the heat-transfer area

On the other hand, if an infinitesimal time is considered, the density of the heat flow that can be expressed according to Fourier's equation, such that in the tube's wall, for $x = 0$, is

$$dq = -k\left(\frac{\partial T}{\partial x}\right)_{x=0}$$

If it is known that

$$\frac{d}{dx} erf\left(\frac{x}{2\sqrt{\alpha t}}\right) = \frac{1}{\sqrt{\pi\alpha t}} \exp\left(\frac{-x^2}{4\alpha t}\right)$$

the temperature gradient for $x = 0$ can be obtained from Equation 17.107:

$$\left(\frac{\partial T}{\partial x}\right)_{x=0} = \frac{T_W - T_0}{\sqrt{\pi\alpha t}}$$

Hence,

$$dq = k\frac{T_W - T_0}{\sqrt{\pi\alpha t}}$$

Since

$$dq = \frac{1}{A}\frac{dQ}{dt}$$

$$\frac{1}{A}\frac{dQ}{dt} = k\frac{T_W - T_0}{\sqrt{\pi\alpha t}}$$

This equation, with separable variables, can be integrated on the boundary condition:

$$\text{For } t = t_N \qquad Q = Q_T$$

obtaining

$$Q_T = A2k(T_W - T_0)\sqrt{\frac{t_N}{\pi\alpha}}$$

It is easy to deduce the following when comparing the last expression with Equation 17.108:

$$h_i = 2\sqrt{\frac{k^2}{\pi t_N \alpha}}$$

Since $\alpha = k/(\rho\hat{C}_P)$ and $t_N = 1/(PN)$, the individual coefficient for the viscous fluid is

$$h_i = 2\sqrt{\frac{k\rho\hat{C}_P PN}{\pi}} \tag{17.109}$$

There are other equations for the calculation of the individual coefficient on the internal side besides the last expression. Thus, for heat exchange between hot and cold water, there is an equation that relates the coefficient h_i with the spinning velocity of the rotor:

$$\log N = 8.36 \times 10^{-5}\, h_i + 0.164 \tag{17.110}$$

for values between 5 and 31.67 rps.

Another equation to calculate h_i is

$$(\mathrm{Nu}) = 4.9(\mathrm{Re})^{0.57}(\mathrm{Pr})^{0.47}\left(\frac{d_i N}{v}\right)^{0.17}\left(\frac{d_i}{L}\right)^{0.37} \tag{17.111}$$

for N values between 1.5 and 7.5 rps.

17.8 AGITATED VESSELS WITH JACKETS AND COILS

The calculation of this type of equipment requires previous knowledge of the individual coefficient of the fluid contained inside the vessel and of the fluid that circulates inside the coil or the jacket.

17.8.1 Individual Coefficient inside the Vessel

1. Helicoidal coils

   For the case of baffled agitators and for values of Reynolds number between 300 and 4×10^5, the following expression is used:

$$\frac{hD_T}{k} = 0.87\left(\frac{D_P^2 N\rho}{\eta}\right)^{0.62}\left(\frac{\hat{C}_P\eta}{k}\right)^{1/3}\left(\frac{\eta}{\eta_W}\right)^{0.14} \tag{17.112}$$

   In the case of flat-baffle disk turbine agitators and values of the Reynolds module between 400 and 2×10^5, the following equation is used:

$$\frac{hd_0}{k} = 0.17\left(\frac{D_P^2 N\rho}{\eta}\right)^{0.67}\left(\frac{\hat{C}_P\eta}{k}\right)^{0.37}\left(\frac{D_P}{D_T}\right)^{0.1}\left(\frac{d_0}{D_T}\right)^{0.5} \tag{17.113}$$

   The following variables are defined in these expressions:

   D_T is the internal diameter of the vessel

   D_P is the diameter of the agitator baffle

   d_0 is the external diameter of the coil tube

   N is the spinning velocity of the agitator

2. Jackets

   The following expression is used to calculate the coefficient of the fluid inside the agitated vessel:

$$\frac{hD_T}{k} = a\left(\frac{D_P^2 N\rho}{\eta}\right)^{b}\left(\frac{\hat{C}_P\eta}{k}\right)^{1/3}\left(\frac{\eta}{\eta_W}\right)^{m} \tag{17.114}$$

in which *a*, *b*, and *m* depend on the type of agitator used. The values of these parameters are given in the next table:

Agitator	a	b	m	Re
Baffles	0.36	2/3	0.21	$300–3 \times 10^5$
Helical	0.54	2/3	0.14	2000
Anchor	1	1/2	0.18	10–300
Anchor	0.36	2/3	0.18	$300–4 \times 10^4$
Flat-baffle disk turbine	0.54	2/3	0.14	$40–3 \times 10^5$
Helical coil	0.633	1/2	0.18	$8–10^5$

17.8.2 Individual Coefficient inside the Coil

The calculation of this individual coefficient is performed as in the case of pipes, and the value obtained should be multiplied by the value of the coefficient γ, defined by the equation:

$$\gamma = 1 + 3.5\frac{d_i}{D_S} \qquad (17.115)$$

where

d_i is the internal diameter of the coil tube
D_S is the diameter of the coil

17.8.3 Individual Coefficient in the Jacket

Generally, condensing vapor circulates through the jacket, the individual heat-transfer coefficients usually have high values, between 8000 and 9000 W/(m² °C).

The calculation of this coefficient can also be made, assuming that the jacket is a circular ring and using one of the equations given in Section 17.2.3.

17.9 HEAT EXCHANGE EFFICIENCY

The heat exchange efficiency is defined as the relation between the real heat flow and the maximum heat flow possible. This value could be obtained in a heat exchanger working in countercurrent and with an infinite area.

Assuming that there are no heat losses to the outside, two cases can occur for a heat exchanger with infinite area:

1. The first is $w_h\hat{C}_P)_h < w_c\hat{C}_P)_c$, where the outlet temperature of the hot fluid is equal to the temperature of the cold fluid at the inlet.
2. The second is $w_c\hat{C}_P)_c < w_h\hat{C}_P)_h$, where case the outlet temperature of the cold fluid coincides with the inlet temperature of the hot fluid.

According to these statements, it can be said that efficiency relates the heat flow of the processed fluid with the maximum possible flow, within the limit of not contradicting the second law of thermodynamics.

The expression that defines the efficiency is

$$\varepsilon = \frac{w_h\hat{C}_P\big)_h (T_{in} - T_{out})}{\left(w\hat{C}_P\right)_{min} (T_{in} - t_{in})} = \frac{w_c\hat{C}_P\big)_c (t_{out} - t_{in})}{\left(w\hat{C}_P\right)_{min} (T_{in} - t_{in})} \qquad (17.116)$$

in which $(w_h\hat{C}_P)_{min}$ is the lower value of $w_h\hat{C}_P)_h$ and $w_c\hat{C}_P)_c$.

If thermal capacity per unit time of any fluid is defined as the product of the mass flow rate times its specific heat, that is,

$$C = w\hat{C}_P \tag{17.117}$$

then Equation 17.116 can be expressed as

$$\varepsilon = \frac{C_h(T_{in} - T_{out})}{C_{min}(T_{in} - t_{in})} = \frac{C_c(t_{out} - t_{in})}{C_{min}(T_{in} - t_{in})} \tag{17.118}$$

In any case, if the efficiency of a determined heat exchanger is known, it is possible to know the heat flow transferred by knowing only the inlet temperatures of the hot and cold fluids:

$$\dot{Q} = \varepsilon C_{min}(T_{in} - t_{in}) \tag{17.119}$$

since it can be observed from Equation 17.116 that

$$\dot{Q} = C_h(T_{in} - T_{out}) = C_c(t_{out} - t_{in}) = \varepsilon C_{min}(T_{in} - t_{in}) \tag{17.120}$$

It can be seen that it is possible to calculate the heat flow transferred without knowing the outlet temperatures of the fluids, which are needed in conventional calculations of heat exchangers, by using Equation 17.120, since for the calculation of the heat flow transferred through the exchange area, the logarithmic mean temperature $(\Delta T)_{ml}$ should be calculated first.

Next, an example of the calculation system using the efficiency for a countercurrent system will be presented. Consider the heat flow transferred in a differential area dA for the system shown in Figure 17.21. Given an energy balance in this differential area and taking into account the heat-transfer velocity,

$$d\dot{Q} = C_h dT = C_c dt = U_m dA(T - t) \tag{17.121}$$

If the energy balance is performed between section 1 and any other, it is obtained that

$$C_h(T_{in} - T) = C_c(t_{out} - t)$$

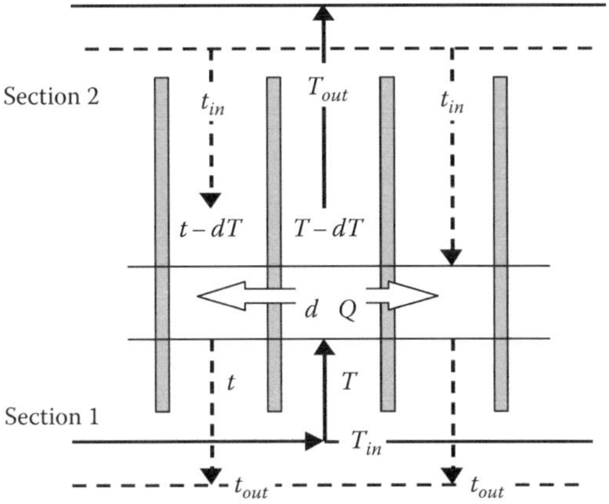

FIGURE 17.21 Heat transfer for the circulation of fluids in countercurrent.

Hence,

$$T - t = T_{in} - \frac{C_c}{C_h} t_{out} + \left(\frac{C_c}{C_h} - 1 \right) t \tag{17.122}$$

Substituting this expression in Equation 17.121 yields

$$\frac{U_m dA}{C_c} = \frac{dt}{T_{in} - (C_c/C_h)\, t_{out} + ((C_c/C_h) - 1)t}$$

This equation can be integrated on the boundary conditions:

$$\text{For } A = 0 \quad t = t_{out}$$

$$\text{For } A = A \quad t = t_{in}$$

obtaining

$$U_m A \left(\frac{C_c - C_h}{C_c C_h} \right) = \ln \left[\frac{T_{in} - t_{out}}{T_{in} - (C_c/C_h)\, t_{out} \; + \; ((C_c/C_h) - 1)\, t_{in}} \right]$$

or

$$-U_m A \left(\frac{C_c - C_h}{C_c C_h} \right) = \ln \left[\frac{(T_{in} - t_{in}) - (C_c/C_h)\,(t_{out} - t_{in})}{(T_{in} - t_{in}) - (t_{out} - t_{in})} \right]$$

Equation 17.118 can be expressed as

$$\frac{t_{out} - t_{in}}{T_{in} - t_{in}} = \varepsilon \frac{C_{min}}{C_c}$$

that substituted in the last equation allows us to obtain the efficiency value, according to the expression:

$$\varepsilon = \frac{1 - \exp(-U_m A(C_c - C_h/C_c C_h))}{(C_{min}/C_h) - (C_{min}/C_c) \exp(-U_m A(C_c - C_h/C_c C_h))} \tag{17.123}$$

Two cases can occur:

1. If $C_h < C_c$, then $C_{min} = C_h$

   $$C_{max} = C_c$$

   so

$$\varepsilon = \frac{1 - \exp(-(U_m A/C_h)(1 - (C_h/C_c)))}{1 - (C_h/C_c) \exp(-(U_m A/C_h)(1 - (C_h/C_c)))} \tag{17.124}$$

2. If $C_c < C_h$, then $C_{min} = C_c$

$$C_{max} = C_h$$

then

$$\varepsilon = \frac{1-\exp(-(U_m A/C_c)((C_c/C_h)-1))}{(C_c/C_h)-\exp(-(U_m A/C_c)((C_c/C_h)-1))}$$

$$= \frac{1-\exp(-(U_m A/C_c)(1-(C_c/C_h)))}{1-(C_c/C_h)\exp(-(U_m A/C_F)(1-(C_c/C_h)))} \tag{17.125}$$

Therefore, for both cases, efficiency can be expressed according to the following equation:

$$\varepsilon = \frac{1-\exp(-(U_m A/C_{min})(1-(C_{min}/C_{max})))}{1-(C_{min}/C_{max})\exp(-(U_m A/C_{min})(1-(C_{min}/C_{max})))} \tag{17.126}$$

This equation takes into account the definition of the number of heat-transfer units, defining the parameter β as the relationship between the minimum and maximum thermal capacities per unit time:

$$\text{NTU} = \frac{U_m A}{C_{min}} \tag{17.127}$$

$$\beta = \frac{C_{min}}{C_{max}} \tag{17.128}$$

In this way, the heat-transfer efficiency can be expressed as a function of these parameters, according to the equation:

$$\varepsilon = \frac{1-\exp[-\text{NTU}(1-\beta)]}{1-\beta\exp\ [-\text{NTU}(1-\beta)]} \tag{17.129}$$

For the case of flow in parallel, the treatment would be similar to the countercurrent flow. Therefore, the efficiency is expressed as

$$\varepsilon = \frac{1-\exp[-\text{NTU}(1+\beta)]}{1+\beta} \tag{17.130}$$

It is observed that for operations in parallel as well as in countercurrent, the efficiency is a function of two dimensionless parameters, that is, the number of heat-transfer units (NTU) and the ratio between the maximum and minimum thermal capacities per unit time (β).

These efficiency expressions, developed for parallel and countercurrent flow cases, can be obtained for any type of heat exchangers. The final equations are usually given in graphical form in which efficiency ε is plotted in the ordinates and the number of heat-transfer units (NTU) in the abscissas, β being the constant parameter of each curve (Figures 17.22 through 17.24). Different efficiency diagrams for different types and arrangements of plate-type heat exchangers can be found in the literature (Raju and Chand, 1980).

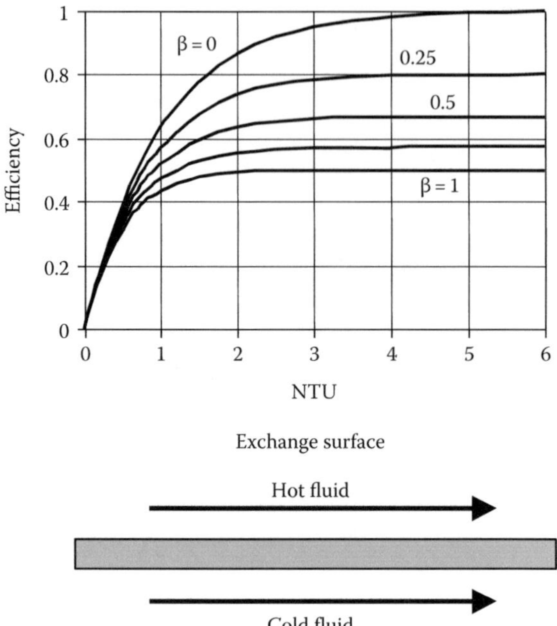

FIGURE 17.22 Efficiency diagram for parallel flow.

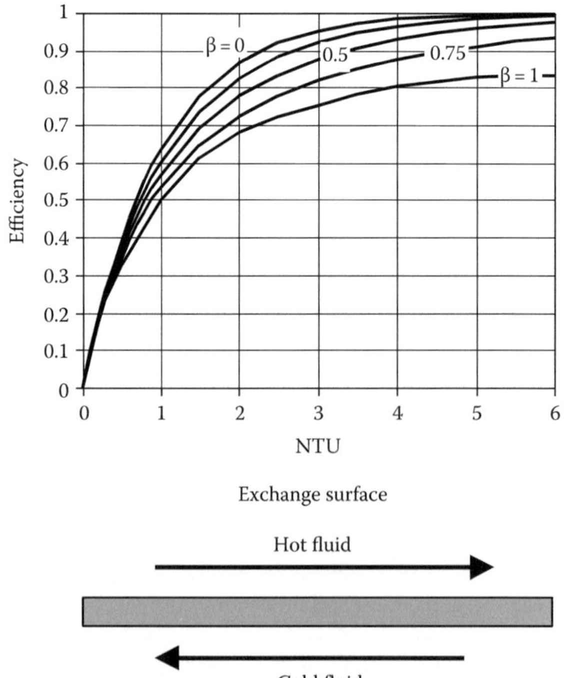

FIGURE 17.23 Efficiency diagram for countercurrent flow.

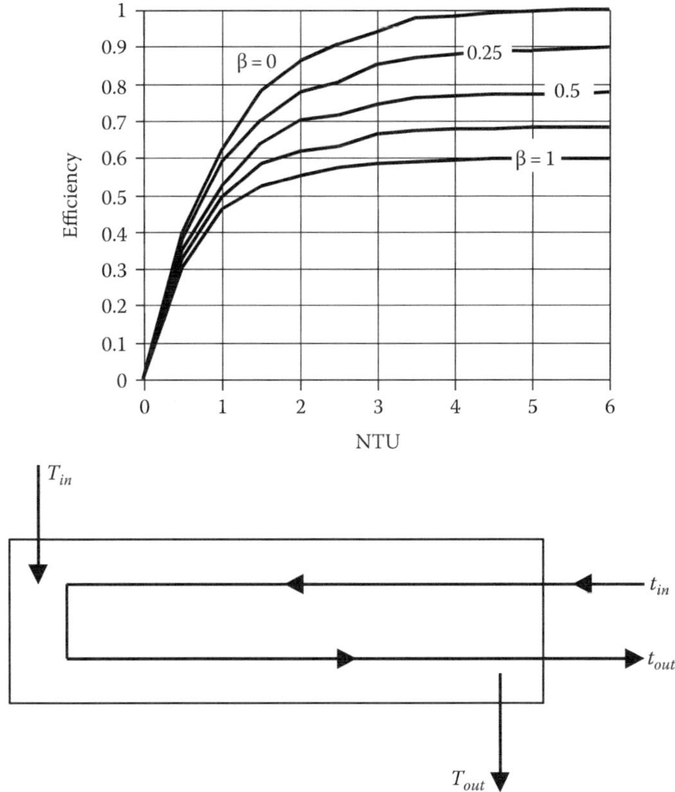

FIGURE 17.24 Efficiency diagram for shell and tube-heat exchanger.

PROBLEMS

17.1 It is desired to heat 12,000 kg/h of smashed tomatoes from 18°C to 75°C using a concentric tube-heat exchanger. Tomato circulates through a stainless steel tube AISI 304 standard of 2 in., while on the outside it condenses saturated steam at 105°C. If the resistance imposed by the film of condensate and the tube wall can be considered as negligible, calculate the length that the heat exchanger should have to perform the heating.

Data: Properties of the smashed tomatoes in the operation temperature range: Specific heat 3.98 kJ/(kg °C); thermal conductivity 0.5 W/(m °C); density 1033 kg/m^3. Viscosity varies with temperature according to the expression: $\eta = 1.75 \times 10^{-4} \exp(4000/T)$ mPa s, in which T is the absolute temperature.

Dimensions of the 2 in. steel tube: Internal diameter 5.25 cm; external diameter 6.03 cm.

Assume that the global heat-transfer coefficient varies linearly with temperature:

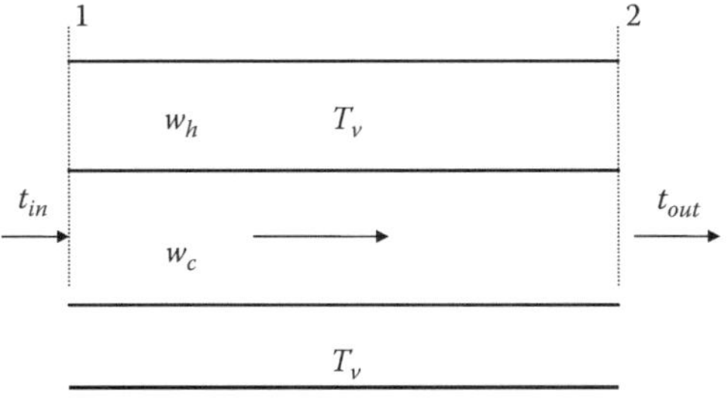

The condensation latent heat, obtained from saturated steam tables for $T_v = 105°C$, is

$$\lambda_v = 2242\,\text{kJ/kg}$$

The condensation heat of steam is applied to the tomato to increase its temperature from 18°C to 75°C:

$$w_h \lambda_v = w_c \hat{C}_p\Big)_c (t_{out} - t_{in})$$

$$w_h(2,242\,\text{kJ/kg}) = (12,000\,\text{kg/h})(3.98\,\text{kJ/kg °C})(75°\text{C} - 18°\text{C})$$

$$w_h = 1,214.2\,\text{kg/h}$$

According to the problem statement, the resistance offered to heat transfer by the condensed steam and by the wall can be considered as negligible, so $U = h_i$.

Since the global heat-transfer coefficient varies linearly with temperature, $U = a + b$, the heat flux that crosses the lateral section of the metal pipe is

$$\dot{Q} = A(U\Delta T)_{mlc} = \pi d_i L (U\Delta T)_{mlc}$$

$$(U\Delta T)_{mlc} = \frac{U_2 \Delta T_1 - U_1 \Delta T_2}{\ln(U_2 \Delta T_1 / U_1 \Delta T_2)}$$

$$U_1 = h_1; \quad \Delta T_1 = T_v - t_{in} = (105°\text{C} - 18°\text{C}) = 87°\text{C}$$

$$U_2 = h_2; \quad \Delta T_2 = T_v - t_{out} = (105°\text{C} - 75°\text{C}) = 30°\text{C}$$

The individual heat-transfer coefficients are calculated by using the equation of Sieder–Tate:

$$h = \frac{0.027k}{d} (\text{Re})^{0.8} (\text{Pr})^{0.33} \left(\frac{\eta}{\eta_W} \right)^{0.14}$$

Calculation of the flux of the tomato stream

$$G = \frac{w_t}{(\pi/4) d_i^2} = \frac{4(12,000\,\text{kg/h})}{\pi(5.25 \times 10^{-2})^2 \text{m}^2} \left(\frac{1\,\text{h}}{3,600\,\text{s}} \right) = 1,539.8\,\text{kg/m}^2\,\text{s}$$

The temperature of the metal wall will coincide with the temperature of steam condensation, since there is no resistance due to the metal wall and the condensation layer: $T_W = T_v = 105°C$.

The calculation of (Re), (Pr), h_1, and h_2 requires knowledge of the viscosity values at the corresponding temperatures. For this reason, the following expression is used:

$$\eta = 1.75 \times 10^{-4} \exp\left(\frac{4000}{T} \right)$$

The following table presents the values of (Re), (Pr), and η calculated by using the previous equations:

t (°C)	T (K)	η (mPa s)	Re	Pr	h_1 W/(m² °C)	h_2 W/(m² °C)
18	291	163.2	496	1298	$892L^{-1/3}$	—
75	348	17.2	4700	136.8	—	1283
105	378	6.9	—	—	—	—

It can be observed that at the inlet $(\text{Re})_1 = 496$, so the expression for laminar flow should be used to calculate h_1:

$$\text{Nu} = 1.86\left[(\text{Re})(\text{Pr})\left(\frac{d}{L}\right)\right]^{1/3}\left(\frac{\eta}{\eta_W}\right)^{0.14}$$

Then,

$$h_1 = \frac{1.86k}{d_i}[(\text{Re})(\text{Pr})(d_i)]^{1/3}\left(\frac{\eta}{\eta_W}\right)^{0.14} L^{-1/3}$$

For calculation of h_1, it is necessary to know the length of the heat exchanger, so the problem should be solved by iteration.

The following is obtained when substituting the values of the variables: $h_1 = 892L^{-1/3}$ W/(m² °C)

The calculation of h_2 can be performed by the equation of Sieder–Tate:

The length of the heat exchanger is calculated by the expression:

$$L = \frac{\dot{Q}}{\pi d_i (U\Delta T)_{mlc}}$$

where

$$\dot{Q} = w_h \lambda_v = 756.18\,\text{kW} \quad \text{and} \quad d_i = 5.25 \times 10^{-2}\,\text{m}$$

Substitution of these values in the expression of the heat exchanger length yields

$$L = \frac{160.62\ln(3.91L^{1/3})}{3.91 - L^{-1/3}}$$

Calculation of L is done by iteration, so an initial length is supposed and substituted in the right-hand term, obtaining the length that should coincide with the supposed one. The length of the heat exchanger is $L = 129$ m.

17.2 A water conduit, made of 2 in. steel tubing, crosses a 30 m room that is kept at 18°C. The temperature of the water at the inlet point is 15°C, and the pipe is insulated with a 6 cm thick cover made of fiber glass. Calculate

a. The minimal water flow to avoid freezing

b. The time needed by water to freeze inside the pipe in case water circulation stops, assuming that the global heat-transfer coefficient remains constant

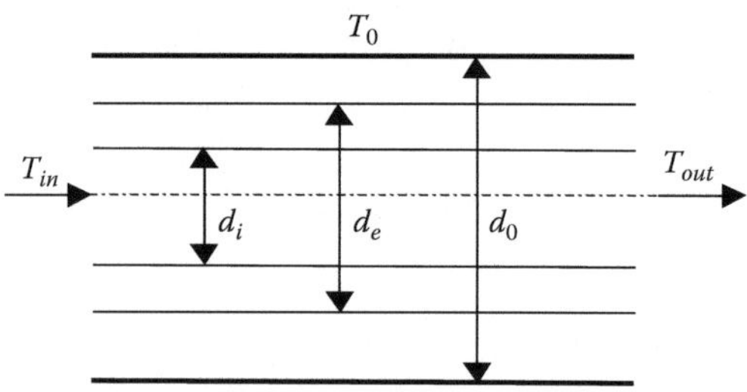

Data: Tube-air convection coefficient 12 W/(m^2 °C). Thermal conductivity of fiber glass 0.07 W/(m °C). Fusion latent heat of ice at 1 atm 335 kJ/kg. Thermal conductivity of water 0.60 W/(m °C).

Steel pipe: Thermal conductivity 45 W/(m °C)
Internal diameter 52.5 mm. External diameter 60.3 mm

$$d_i = 0.0525\,\text{m}$$

$$d_e = 0.0603\,\text{m}$$

$$d_0 = 0.1803\,\text{m}$$

Insulation thickness

$$e_a = 6\,\text{cm} = 0.06\,\text{m}$$

Pipe thickness

$$e_p = \frac{(d_e - d_i)}{2} = 3.9\,\text{mm} = 0.0039\,\text{m}$$

Water should have a temperature higher than 0°C at the outlet point to prevent freezing. The heat flux given by water is

$$\dot{Q} = w\hat{C}_p(T_{in} - T_{out})$$

where w is the maximum flow at which water circulates, T_e = 15°C, and $T_s \geq$ 0°C. Such heat flow rate is lost through the lateral wall according to the expression:

$$\dot{Q} = U_i A_i (\Delta T)_{ml}$$

where

U_i is the global heat-transfer coefficient referred to the internal area
A_i is the heat-transfer internal area
$(\Delta T)_{ml}$ is the logarithmic mean temperature difference

$$(\Delta T)_{ml} = \frac{(T_{in} - T_0) - (T_{out} - T_0)}{\ln((T_{in} - T_0)/(T_{out} - T_0))} = \frac{(15+18)-(0+18)}{\ln((15+18)/(0+18))} = 24.8°\text{C}$$

$$A_i = \pi d_i L = \pi(0.0525\,\text{m})(30\,\text{m}) = 4.948\,\text{m}^2$$

$$A_e = \pi d_e L = \pi(0.0603\,\text{m})(30\,\text{m}) = 5.683\,\text{m}^2$$

$$A_0 = \pi d_0 L = \pi(0.1803\,\text{m})(30\,\text{m}) = 16.993\,\text{m}^2$$

$$A_{ml_1} = \frac{A_e - A_i}{\ln(A_e/A_i)} = 5.307\,\text{m}^2; \quad A_{ml_2} = \frac{A_0 - A_e}{\ln(A_0/A_e)} = 10.326\,\text{m}^2$$

Calculation of U_i

$$\frac{1}{U_i} = \frac{1}{h_i} + \frac{e_p}{k_p(A_{ml_1}/A_i)} + \frac{e_a}{k_a(A_{ml_2}/A_i)} + \frac{1}{h_0(A_0/A_i)}$$

$$\frac{1}{U_i} = \frac{1}{h_i} + \frac{0.0039\,\text{m}}{(45\,\text{W/m}\,°\text{C})(5.307\,\text{m}^2/4.948\,\text{m}^2)} + \frac{0.06\,\text{m}}{(0.07\,\text{W/m}\,°\text{C})(10.326\,\text{m}^2/4.948\,\text{m}^2)}$$
$$+ \frac{1}{(12\,\text{W/m}^2\,°\text{C})(16.993\,\text{m}^2/4.948\,\text{m}^2)}$$

$$\frac{1}{U_i} = \frac{1}{h_i} + 0.435 \quad U_i = \frac{h_i}{1+0.43h_i}\,\text{W/m}^2\,°\text{C}$$

It is necessary to calculate the value of the convective coefficient for the inside of the pipe h_i, which can be done by using the equation of Dittus–Boelter for fluids that cool down: $h_i = 0.023(k/d_i)(\text{Re})^{0.8}(\text{Pr})^{0.3}$

$$(\text{Re}) = \frac{\rho v d_i}{\eta} = \frac{w4}{\pi d_i \eta} = 4\frac{(w\,\text{kg/h})(1\,\text{h}/3600\,\text{s})}{\pi(0.0525\,\text{m})(10^{-3}\,\text{Pa s})} = 6.74w$$

$$(\text{Pr}) = \frac{\hat{C}_p \eta}{k} = \frac{(4.185\,\text{kJ/kg}\,°\text{C})(10^{-3}\,\text{Pa s})}{(0.6\times10^{-3}\,\text{kJ/s m}\,°\text{C})} \approx 7$$

$$h_i = 0.023\frac{0.6\,\text{W/m}\,°\text{C}}{0.0525\,\text{m}}(6.74w)^{0.8}(7)^{0.3} = 2.17w^{0.8}$$

$$h_i = 2.17\,w^{0.8}\,\text{W/m}^2\,°\text{C}$$

From the energy balance,

$$w\hat{C}_p(T_{in} - T_{out}) = U_i A_i (\Delta T)_{ml}$$

$$(w\,\text{kg/h})(1\,\text{h}/3600\,\text{s})\,(4.185\,\text{kJ/kg}\,°\text{C})(15°\text{C}-0°\text{C})$$

$$=\left(\frac{h_i}{1+0.435h_i}10^{-3}\,\text{kJ/s}\,\text{m}^2\,°\text{C}\right)(4.948\,\text{m}^2)(24.8°\text{C})$$

Hence,

$$w=\frac{7.04h_i}{1+0.435h_i}\,\text{kg/h}$$

From the substitution of the expression of h_i as a function of the flow rate w, it is obtained that

$$w=\frac{15.27w^{0.8}}{1+0.94w^{0.8}}$$

This equation is solved by iteration, yielding w = 14.43 kg/h.

For this flow,

$$(\text{Re})=(6.74)(14.43)=97$$

Since the regime is laminar, the equation of Dittus–Boelter cannot be used to calculate h_i. Instead, another expression for laminar regime should be tried:

$$h_i=1.86\frac{k}{d_i}\left(\text{Re}\,\text{Pr}\frac{d_i}{L}\right)^{1/3}$$

$$h_i=1.86\frac{0.6\,\text{W/m}\,°\text{C}}{0.0525\,\text{m}}\left(6.74w7\frac{0.0525\,\text{m}}{30\,\text{m}}\right)^{1/3}$$

$$h_i\approx 9.26w^{1/3}\,\text{W/(m}^2\,°\text{C)}$$

The following equation is obtained when substituting the last expression in the energy balance:

$$w=\frac{65.19w^{1/3}}{1+4.03\,w^{1/3}}$$

After solving by iteration, w = 14.69 kg/h.

This means that if water circulates with a flow lower than 14.69 kg/h, it can freeze inside the pipe.

For w = 14.69 kg/h, it is obtained that

$$h_i\approx 22.7\,\text{W/(m}^2\,°\text{C)}$$

$$U_i\approx 2.1\,\text{W/(m}^2\,°\text{C)}$$

In case the circulation of water stops, then there will be a mass of water inside the tubing:

$$m = \rho V = \rho \frac{\pi}{4} d_i^2 L = \frac{\pi}{4}(10^3\ \text{kg/m}^3)(0.0525\,\text{m})^2(30\,\text{m})$$

$$m = 64.94\ \text{kg}$$

The time this mass of water needs to cool down to 0°C can be obtained from the energy balance under unsteady state. The heat output term is equal to the accumulation:

$$U_i A_i (T_i - T_0) = -m\hat{C}_p \frac{dT_i}{dt}$$

If this equation is integrated on the boundary conditions,

$$\text{For} \quad t = 0 \quad T_i = T_{i0}$$
$$t = t \quad T_i = T_C = 0°\text{C}$$

the following is obtained:

$$t = \frac{m\hat{C}_p}{U_i A_i} \ln\left(\frac{T_{i0} - T_0}{T_C - T_0}\right)$$

If initially it is assumed that

$$T_{i0} = \frac{(T_{in} + T_{out})}{2} = 7.5°\text{C}$$

then

$$t = \frac{(64.94\,\text{kg})(4.185\,\text{kJ/(kg°C)})}{2.1\times 10^{-3}\,\text{kJ/(s\,m}^2\,°\text{C)}(4.948\,\text{m}^2)} \ln\left(\frac{7.5+18}{0+18}\right) = 9110\,\text{s} = 2.53\,\text{h}$$

The heat flow rate that should be eliminated to freeze this water is

$$\dot{Q}_c = \frac{\lambda m}{t_c} = U_i A_i (T_C - T_0)$$

$$t_C = \frac{\lambda m}{U_i A_i (T_C - T_0)} = \frac{(335\,\text{kJ/kg})(64.94\,\text{kg})}{(2.1\times 10^{-3}\,\text{kJ/(s\,m}^2\,°\text{C)})(4.948\,\text{m}^2)(0°\text{C} + 18°\text{C})} = 116{,}315\,\text{s}$$

The time that water needs to freeze is $t_C = 32.31$ h.

17.3 A thermal insulated heat exchanger is made of two concentric steel tubes. The external diameter of the inner tube is 6 cm and has a wall thickness of 5 mm, while the internal diameter of the external tube is 10 cm. A fluid circulates through the internal tube with a flow of 1000 kg/h, and it is heated from 22°C to 78°C by circulating 100 kg/h of steam in countercurrent at 1 atm. Determine

a. Temperature of the heating fluid at the outlet point
b. Individual convective heat-transfer coefficient for the fluid being processed
c. Area of the heat-transfer surface
d. Length of the heat exchanger

Data: The resistance offered by the steel tube to heat transfer can be considered as negligible.

Fluid properties: Specific heat 4.185 kJ/(kg °C). Viscosity 1.2 mPa s. Density 1000 kg/m³. Thermal conductivity 0.56 W/(m °C).

Steam's latent heat 2256 kJ/kg

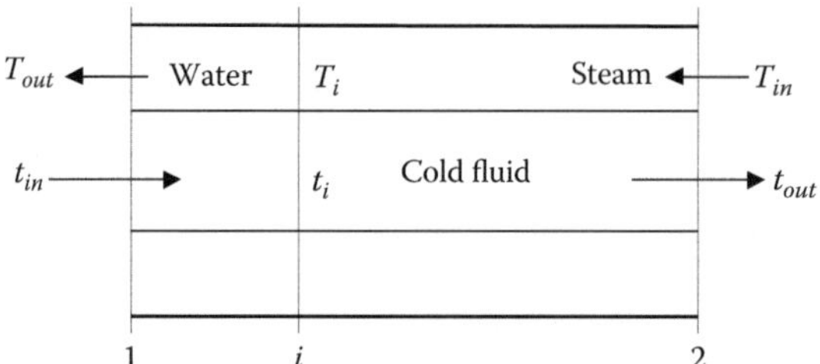

For the work conditions of the heat exchanger, it can be considered that for steam condensation, the individual convective heat-transfer coefficient is 870 W/(m² °C), while for liquid water at temperature higher than 75°C, this coefficient can be considered as equal to 580 W/(m² °C):

$$t_{in} = 22°\text{C} \quad t_{out} = 78°\text{C} \qquad T_{in} = 100°\text{C}$$
$$d_e = 6 \text{ cm} \quad d_i = d_e - 2e = 5 \text{ cm}$$

a. Steam's heat condensation

$$\dot{Q}_h = w_h\lambda = (100\,\text{kg/h})(2,256\,\text{kJ/kg})$$

$$\dot{Q}_h = 225,600\,\text{kJ/h}$$

Heat gained by the cold fluid

$$\dot{Q}_c = w_c\hat{C}_p(t_{out} - t_{in}) = (1,000)(4.185)(78 - 22)$$

$$\dot{Q}_c = 234,360\,\text{kJ/h}$$

Since $\dot{Q}_h \le \dot{Q}_c$, then the vapor condenses at an intermediate point, and from there on, it lowers its temperature, leaving the heat exchanger at a temperature lower than 100°C. The following is found at the point where the steam begins to condense:

$$w_h\lambda = w_c\left.\hat{C}_p\right)_c(t_{out} - t_i)$$

$$(100)(2256) = (1000)(4.185)(78 - t_i) \Rightarrow t_i = 24.1°\text{C}$$

When the fluid condenses, T_i = 100°C, and water gives heat up to a temperature T_s, finding that

$$w_h\left.\hat{C}_P\right)_h(T_i - T_{out}) = w_c\left.\hat{C}_P\right)_c(t_i - t_e)$$

$$(100)(4.185)(100 - T_{out}) = (1000)(4.185)(24.1 - 22) \Rightarrow T_{out} = 79°\text{C}$$

b. Calculation of h_i

$$G = \frac{w}{S} = \frac{1000\,\text{kg/h}}{(\pi/4)(0.05)^2\,\text{m}^2} \cdot \frac{1\,\text{h}}{3600\,\text{s}} \approx 141.5\,\text{kg/(m}^2\,\text{s)}$$

$$(\text{Re}) = \frac{\rho v d}{\eta} = \frac{G d_i}{\eta} = \frac{(141.5\,\text{kg/m}^2\,\text{s})(0.05\,\text{m})}{1.2\times 10^{-3}\,\text{Pa s}} \approx 5.9\times 10^3$$

$$(\text{Pr}) = \frac{\hat{C}_p \eta}{k} = \frac{(4.185\,\text{kJ/kg °C})(1.2\times 10^{-3}\,\text{Pa s})}{(0.56\times 10^{-3}\,\text{kJ/s m °C})} = 9$$

Applying the equation of Dittus–Boelter,

$$(\text{Nu}) = \frac{h\cdot d}{k} = 0.023(\text{Re})^{0.8}(\text{Pr})^{0.4}$$

$$\frac{h(0.05\,\text{m})}{0.56\,\text{W/m °C}} = 0.023(5.9\times 10^3)^{0.8}(9)^{0.4} \Rightarrow h_i \approx 645\,\text{W/(m}^2\,\text{°C)}$$

$$h_{ie} = h_i \cdot \frac{d_i}{d_e} = 645\left(\frac{5}{6}\right) = 537.5\,\text{W/(m}^2\,\text{°C)}$$

c. The heat exchanger is divided into two zones for the calculation of the area: condensation and cooling:
   - Steam condensation zone

$$\frac{1}{U_e} = \frac{1}{h_e} + \frac{e}{k(d_{ml}/d_e)} + \frac{1}{h_i(d_i/d_e)} = \frac{1}{h_e} + \frac{1}{h_{ie}} = \frac{h_{ie}+h_e}{h_e h_{ie}}$$

$$U_e = \frac{h_{ie}\cdot h_e}{h_{ie}+h_e} \quad U_e = \frac{(537.5)\cdot(870)}{537.5+870} \approx 332\,\text{W/(m}^2\,\text{°C)}$$

$$\dot{Q}_C = w_h\lambda = (100)(2,256) = 225,600\,\text{kJ/h}$$

$$\dot{Q}_C = U_e A'_e(\Delta T_{ml}) = 22,600\,\text{kJ/h}$$

$$\Delta T_{ml} = \frac{(T_i - t_i) - (T_{in} - t_{out})}{\ln((T_i - t_i)/(T_{in} - t_{out}))} = \frac{(100-24.1)-(100-79)}{\ln((100-24.1)/(100-79))} = 42.7\text{°C}$$

$$A'_e = \frac{(225,600\,\text{kJ/h})\left(1\,\text{h}/3,600\,\text{s}\right)}{(332\times 10^{-3}\,\text{kJ/s m}^2\,\text{°C})(42.7\text{°C})} \approx 4.42\,\text{m}^2$$

   - Zone of condensed water

$$\dot{Q}_e = w_h \hat{C}_{Pi}(T_i - T_{out}) = (100)(4.185)(100-79) = 8789\,\text{kJ/h}$$

$$U_e = \frac{(537.5)(580)}{(537.5)+(580)} = 279\,\text{W/(m}^2\,°\text{C)}$$

$$\Delta T_{ml} = \frac{(T_i - t_i)-(T_{out} - t_{in})}{\ln((T_i - t_i)/(T_{out} - t_{in}))} = \frac{(100-24.1)-(79-22)}{\ln((100-24.1)/(79-22))} \approx 66°\text{C}$$

$$A_e'' = \frac{(8789\,\text{kJ/h})(1\text{h}/3600\,\text{s})}{(279\times10^{-3}\,\text{kJ/s}\,\text{m}^2\,°\text{C})(66°\text{C})} \approx 0.13\,\text{m}^2$$

Total area

$$A_T = A_e' + A_e'' = 4.55\,\text{m}^2$$

d. The heat exchanger length can also be divided into two parts:

$$A_e = \pi d_e L$$

$$L' = \frac{A_e'}{\pi d_e} = \frac{(4.42)}{\pi(0.06)} = 23.45\,\text{m}$$

$$L'' = \frac{A_e''}{\pi d_e} = \frac{(0.13)}{\pi(0.06)} = 0.70\,\text{m} = 70\,\text{cm}$$

The total length of the heat exchanger is $L_T = L' + L'' \approx 24.15$ m.

17.4 It is desired to have a water stream of 4 kg/s at 90°C in an industrial facility where sausages are dried. The stream will be used to condition the drying air. A heat exchanger that has 2 tube passes (50 tubes per pass) per one shell is available to heat water up to 90°C. Oil circulating outside the tubes at 3.5 kg/s is used as heating fluid, and its temperature at the inlet point is 250°C.

If the inlet temperature of water is 18°C and the tubes by which it circulates are made of steel with an external diameter of 3 cm and an internal diameter of 2.85 cm in a diamond arrangement, determine

a. The heat-transfer coefficient for the inside of the tubes
b. The area of the heat-transfer surface
c. Length of the tubes

Data and notes: The resistance to heat flow offered by the surface of the tubes can be considered as negligible. The heat-transfer coefficient on the oil side is 580 W/(m^2 °C).
Properties of water: Density 1000 kg/m^3. Specific heat 4.185 kJ/(kg °C). Thermal conductivity 0.65 W/(m °C). Viscosity 1 mPa s.
Properties of oil: Density 850 kg/m^3. Specific heat 2.1 kJ/(kg °C). Viscosity 30 mPa s. Thermal conductivity 0.14 W/(m °C).

$$w_c = 4\,\text{kg/s} = 14{,}400\,\text{kg/h}$$

$$w_h = 3.5\,\text{kg/s} = 12{,}600\,\text{kg/h}$$

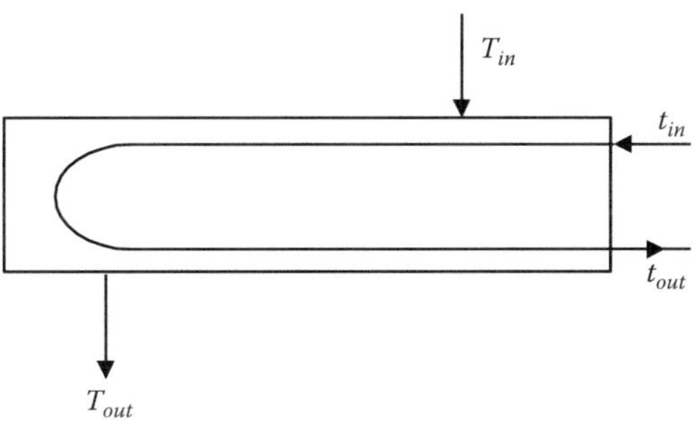

$$\dot{Q} = w\hat{C}_P(\Delta t) = U_e A_e \Delta T_{ml} F$$

$$\dot{Q} = (14{,}400\,\text{kg/h})(4.185\,\text{kJ/kg}\,°\text{C})(90°\text{C} - 18°\text{C}) = (12{,}600\,\text{kg/h})(2.1\,\text{kJ/kg}\,°\text{C})(250 - T_{out})$$

From which the flowing is obtained,

$$T_{out} = 85.4°\text{C}$$

$$\left.\begin{aligned} Z &= \frac{T_{in} - T_{out}}{t_{out} - t_{in}} = \frac{250 - 85.4}{90 - 18} \approx 2.3 \\ \varepsilon &= \frac{t_{out} - t_{in}}{T_{in} - t_{in}} = \frac{90 - 18}{250 - 18} = 0.31 \end{aligned}\right\} \xrightarrow{\text{Graph}} F \approx 0.83$$

$$d_{internal} = 2.85 \times 10^{-2}\,\text{m}$$

Cross-flow section per tube

$$S = \frac{\pi}{4} d_i^2 = 6.38 \times 10^{-4}\,\text{m}^2$$

$$G = \frac{w}{S_T} = \frac{4\,\text{kg/s}}{50(6.38 \times 10^{-4})\,\text{m}^2} \approx 125.4\,\text{kg/s}\,\text{m}^2$$

$$(\text{Re}) = \frac{Gd_i}{\eta} = \frac{(125.4)(2.85 \times 10^{-2})}{10^{-3}} = 3574 \Rightarrow \text{Turbulent flow}$$

$$(\text{Nu}) = \frac{hd_i}{k} = 0.027(\text{Re})^{0.8}(\text{Pr})^{1/3}\left(\frac{\eta}{\eta_W}\right)^{0.14}$$

It is taken that

$$\frac{\eta}{\eta_W} = 1$$

$$(\text{Pr}) = \frac{\hat{C}_p \eta}{k} = \frac{(4.185\,\text{kJ/kg}\,°\text{C})(10^{-3}\,\text{kg/m}\,\text{s})}{(0.65 \times 10^{-3}\,\text{kJ/s}\,\text{m}\,°\text{C})} = 6.44$$

Substitution in the last equation yields

$$\frac{h(2.85\times10^{-2}\,\text{m})}{0.65\,\text{W/m}\,°\text{C}} = 0.027(3.57\times10^{3})^{0.8}(6.44)^{1/3} \Rightarrow h_f = 796.5\,\text{W/m}^2\,°\text{C}$$

$$h_{t0} = h_t\frac{d_i}{d_e} = 756.7\,\text{W/m}^2\,°\text{C} \quad \frac{1}{U_e} = \frac{1}{h_c} + \frac{1}{h_{t0}}$$

$$\frac{1}{U_e} = \frac{1}{580} + \frac{1}{756.7} \Rightarrow U_e = 328.3\,\text{W/m}^2\,°\text{C}$$

The heat flow rate transferred through the exchange area is

$$\dot{Q} = U_e A_e (\Delta T)_{ml} F$$

The logarithmic mean temperature is

$$(\Delta T)_{ml} = \frac{(T_{in} - t_{out}) - (T_{out} - t_{in})}{\ln((T_{in} - t_{out})/(T_{out} - t_{in}))} = \frac{(250-90)-(85.4-18)}{\ln((250-90)/(85.4-18))} \approx 107°\text{C}$$

The heat flow rate exchanged by one of the fluids is

$$\dot{Q} = (14{,}400\,\text{kg/h})(4.185\,\text{kJ/kg}\,°\text{C})(90°\text{C} - 18°\text{C}) = 4{,}339{,}008\,\text{kJ/h}$$

Equaling the last result to the heat transferred through the exchange area yields

$$(4{,}339{,}008\,\text{kJ/h})\left(\frac{1\,\text{h}}{3{,}600\,\text{s}}\right) = (328.3\,\text{W/m}^2\,°\text{C})A_e(107\,°\text{C})(0.83)$$

Hence, the following is obtained:

$$A_e \approx 41.34\,\text{m}^2$$

Lateral area of a tube

$$\pi d_e L$$

$$A_e = 2\times50\times A_{tube} = 2\times50\pi d_e L \begin{cases} 2\,\text{passes} \\ 50\,\text{tubes/pass} \end{cases}$$

$$41.34\,\text{m}^2 = 2(50)\pi(0.03\,\text{m})L$$

Thus, the tube's length is $L = 4.39$ m.

If the equation of Dittus–Boelter is applied to the calculation of the individual coefficient h_t, the following is obtained:

$$h_t = 748.6\,\mathrm{W/(m^2\,°C)} \Rightarrow h_{t0} \approx 711.1\,\mathrm{W/(m^2\,°C)}$$

in which the global coefficient would be $U_e = 319.5$ W/(m² °C).

Hence, the following area and length of the tube would be obtained:

$$A_e = 42.48\,\mathrm{m^2} \Rightarrow L \approx 4.51\,\mathrm{m}$$

17.5 Tartrates present in white wine should be eliminated to avoid later precipitation problems in the bottle. For this reason, prior to bottling, the wine is cooled to 4°C to cause precipitation of tartrates, which are then separated from the wine. It is desired to cool 1000 kg/h of white wine from 16°C to 4°C in a wine cellar, using a ½ shell and tube-heat exchanger, with 18 tubes arranged in squares and with a distance between the centers equal to 1.5 times the diameter of the tubes. The shell has a diameter of 20 cm and is equipped with baffles that have a baffle cut of 15% and a baffle spacing of 25 cm. A 1200 kg/h stream of water with glycol at −6°C is used as refrigerating fluid, circulating outside the tubes. Due to the possible deposit of tartrates, the dirtiness factor should be considered as 0.00043 (m² °C)/W. Calculate

a. Global heat-transfer coefficient
b. Total heat-transfer area
c. Tube length

Data. Characteristics of the tubes: Internal diameter 10 mm, thickness 1.5 mm. Thermal conductivity 40 W/(m °C).

Consider that the properties of both fluids are similar and equal to those of water:

$$d_i = 0.01\,\mathrm{m}$$

$$d_e = 0.013\,\mathrm{m}$$

$$X = Y = 1.5d_e = 0.0195\,\mathrm{m}$$

$$C' = Y - d_e = 0.0065\,\mathrm{m}$$

Square arrangement

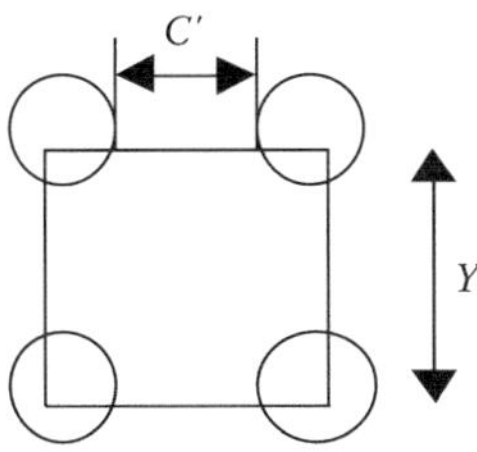

Energy balance

$$w_h\hat{C}_{ph}(T_{in} - T_{out}) = w_c\hat{C}_{pc}(t_{out} - t_{in})$$

$$(1000\,\mathrm{kg/h})(4.185\,\mathrm{kJ/kg\,°C})(16°\mathrm{C} - 4°\mathrm{C}) = (1200\,\mathrm{kg/h})(4.185\,\mathrm{kJ/kg\,°C})(t_s - (-6))°\mathrm{C}$$

Hence,

$$t_{out} = 4°\text{C}$$

Coefficient on the tube side

$$(\text{Nu}) = \frac{h_i d_i}{k} = 0.027(\text{Re})^{0.8}(\text{Pr})^{1/3}$$

$$G_t = \frac{4wn}{N_t \pi d_i^2} = \frac{4(1000\,\text{kg/h})2}{18\pi(0.01)^2\,\text{m}^2}\frac{1\,\text{h}}{3600\,\text{s}} \approx 393\,\text{kg/m}^2\,\text{s}$$

$$(\text{Re}) = \frac{G_t d_i}{\eta} = \frac{(393\,\text{kg/m}^2\,\text{s})(0.01\,\text{m})}{10^{-3}\,\text{Pa s}} = 3930$$

$$(\text{Pr}) = \frac{\hat{C}_p \eta}{k} = \frac{(4.185\,\text{kJ/kg °C})(10^{-3}\,\text{Pa s})}{(0.58 \times 10^{-3}\,\text{kJ/s m °C})} = 7.2$$

Therefore,

$$\frac{h_i(0.01\,\text{m})}{0.58\,\text{W/(m °C)}} = 0.027(3930)^{0.8}(7.2)^{1/3}$$

obtaining

$$h_i = 2270\,\text{W/m}^2\,°\text{C}$$

Coefficient of the shell

$$D_e = \frac{4\left[Y^2 - (\pi/4)d_e^2\right]}{\pi d_e} = \frac{4[0.0195^2 - (\pi/4)0.013^2]}{\pi 0.013} = 0.0242\,\text{m}$$

$$G_e = \frac{wnY}{D_c BC'} = \frac{(1200\,\text{kg/h})(1)(0.0195\,\text{m})}{(0.20\,\text{m})(0.25\,\text{m})(0.0065\,\text{m})}\frac{1\,\text{h}}{3600\,\text{s}} = 20\,\text{kg/m}^2\,\text{s}$$

$$(\text{Re}) = \frac{G_c D_e}{\eta} = \frac{(20\,\text{kg/m}^2\,\text{s})(0.0242\,\text{m})}{10^{-3}\,\text{Pa s}} = 484$$

$$(\text{Pr}) = 7.2$$

$$\frac{h_e D_e}{k} = 0.36(\text{Re})^{0.55}(\text{Pr})^{1/3}$$

$$\frac{h_e(0.0242\,\text{m})}{0.58\,\text{W/(m °C)}} = 0.36(484)^{0.55}(7.2)^{1/3}$$

obtaining

$$h_e = 499.3\,\text{W/m}^2\,°\text{C}$$

Global heat-transfer coefficient

$$\frac{1}{U_e} = \frac{1}{499.3} + \frac{1.5\times10^{-3}}{(40)(0.01134/0.013)} + \frac{1}{(2270)(0.01/0.013)} + 0.00043$$

$$d_{ml} = \frac{d_e - d_i}{\ln(d_e/d_i)} = \frac{(0.013-0.01)\,\text{m}}{\ln(0.013/0.01)} = 0.01134\,\text{m}$$

Hence,

$$U_e = 328\,\text{W/m}^2\,°\text{C}$$

Heat transfer through the exchange area

$$\dot{Q} = U_e A_e \Delta T_{ml} F$$

$$\Delta T_{ml} = \frac{(T_{in} - t_{out}) - (T_{out} - t_{in})}{\ln((T_{in} - t_{out})/(T_{out} - t_{in}))} = \frac{(16°\text{C} - 4°\text{C}) - (4°\text{C} - (-6°\text{C}))}{\ln(12/10)} \approx 11°\text{C}$$

Calculation of F

$$Z = \frac{T_{in} - T_{out}}{t_{out} - t_{in}} = \frac{16-4}{4-(-6)} = \frac{12}{10} = 1.2$$

$$\varepsilon = \frac{t_{out} - t_{in}}{T_{in} - t_{in}} = \frac{4-(-6)}{16-(-6)} = \frac{10}{22} \approx 0.455$$

The following is obtained for a 1/2 shell and tube-heat exchanger: $F \approx 0.81$.

Heat-transfer area

$$A_e = \frac{\dot{Q}}{U_e \Delta T_{ml} F} = \frac{\left.w_h \hat{C}_p\right)_h (T_{in} - T_{out})}{U_e \Delta T_{ml} F}$$

$$A_e = \frac{(1000\,\text{kg/h})(1\,\text{h}/3600\,\text{s})(4.185\,\text{kJ/kg}\,°\text{C})(16°\text{C} - 4°\text{C})}{(328\times10^{-3}\,\text{kJ/s}\,\text{m}^2\,°\text{C})(11°\text{C})(0.81)} = 4.773\,\text{m}^2$$

External heat exchange area

$$A_e = 4.773\,\text{m}^2$$

Length of tubes

$$A_e = N_t \pi d_e L$$

$$L = \frac{A_e}{N_t \pi d_e} = \frac{4.773\,\text{m}^2}{18\pi(0.013\,\text{m})} = 6.49\,\text{m}$$

17.6 A hot air stream at 95°C is required in a drying process. For this reason, 3500 kg/h of air at 15°C is circulated inside a tubular heat exchanger whose shell has a 0.5 m diameter, while in the outside combustion, gases entering the heat exchanger at 400°C and exiting at 100°C circulate. The heat exchanger has 1 shell pass and 2 tube passes, with a total of 400 tubes arranged in squares. The distance between the tube centers is two times the diameter of the tubes, and baffles with a 25% cut are placed every 25 cm. If the heat transferred by radiation is considered as negligible, calculate

a. Flow rate of combustion gases
b. Individual heat-transfer coefficients for both fluids
c. Total heat-transfer area
d. Length that the tubes should have

Data tube characteristics: Thermal conductivity 44 W/(m °C). Internal diameter 11 mm. External diameter 12.5 mm.

Consider that the properties of both fluids are similar. These values are presented in the following table:

T (°C)	k (W/m °C)	η (mPa s)	$\hat{C}_p$ (kJ/kg °C)
0	0.023	0.0180	0.996
250	0.046	0.0265	0.979
500	0.058	0.0350	0.963

Air mean temperature

$$t_m = \frac{t_{in} + t_{out}}{2} = \frac{15 + 95}{2} = 55°\text{C}$$

Combustion gas mean temperature

$$T_m = \frac{T_{in} + T_{out}}{2} = \frac{400 + 100}{2} = 250°\text{C}$$

Properties of fluids at mean temperatures

Fluid	T (°C)	k (W/m °C)	η (mPa s)	$\hat{C}_p$ (kJ/kg °C)
Air	55	0.028	0.0189	0.992
Gases	250	0.046	0.0265	0.979

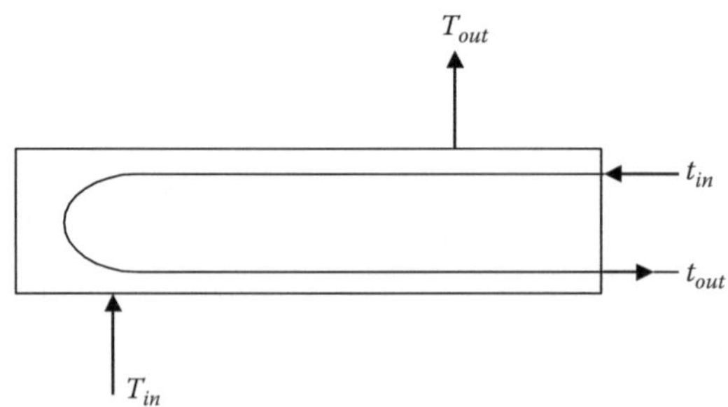

FIGURE P17.6A Tubes and shell heat exchanger 1/2.

a. Energy balance

$$w_h \hat{C}_p\Big)_h (T_{in} - T_{out}) = w_c \hat{C}_p\Big)_c (t_{out} - t_{in})$$

$$w_h(0.97\,\text{kJ/kg}\,°\text{C})(400°\text{C} - 100°\text{C}) = (3500\,\text{kg/h})(0.992\,\text{kJ/kg}\,°\text{C})(95°\text{C} - 15°\text{C})$$

Mass flow rate of combustion gases

$$w_h = 945.7\,\text{kg/h}$$

b. The heat-transfer *individual coefficients* would be calculated from the following expression:

$$(\text{Nu}) = 0.33C_h(\text{Re})^{0.6}(\text{Pr})^{0.3}$$

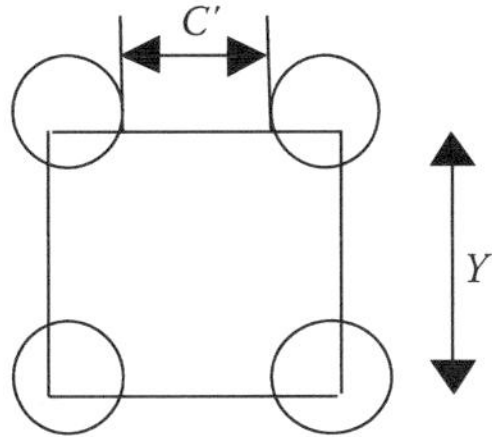

FIGURE P17.6B Arrangement of tubes.

For the combustion gases that circulate through the shell,
Maximum mass flux

$$G_h = \frac{w_h}{A_C}$$

$$A_C = \frac{D_C BC'}{nY} = \frac{D_C B(Y - d_0)}{nY}$$

$$Y = 2d_0 = 2 \times 12.5 \times 10^{-3}\,\text{m} = 0.025\,\text{m}$$

$$B = 25\,\text{cm} \quad D_C = 0.5\,\text{m} \quad n = 1\,\text{shell pass}$$

Hence,

$$A_C = 0.0625\,\text{m}^2$$

The mass flux is:

$$G_h = \frac{945.7\,\text{kg/h}}{0.0625\,\text{m}^2} = 15{,}131.2\,\text{kg/m}^2\,\text{h} = 4.20\,\text{kg/m}^2\,\text{s}$$

The equivalent diameter for a square arrangement is

$$D_e = \frac{4\left(Y^2 - (\pi/4)d_0^2\right)}{\pi\pi d_0} = 0.051\,\text{m}$$

Calculation of the Reynolds number

$$(\text{Re})_{max} = \frac{G_c D_e}{\eta} = \frac{(4.20\,\text{kg/m}^2\,\text{s})(0.051\,\text{m})}{0.0265 \times 10^{-3}\,\text{Pa s}} = 8083$$

Prandtl number

$$(\text{Pr}) = \frac{\hat{C}_p \eta}{k} = \frac{(0.979\,\text{kJ/kg °C})(0.0265 \times 10^{-3}\,\text{Pa s})}{(0.046 \times 10^{-3}\,\text{kJ/s m °C})} = 0.65$$

Calculation of the coefficient C_h

$$\left.\begin{aligned} Y &= 2d_0 \\ X = Y &= 2d_0 \end{aligned}\right\}$$

The values that correspond to $Y = X = 1.5d_0$ are taken from tables:

For	Re = 2,000	$C_h = 0.95$
	Re = 20,000	$C_h = 0.96$

The value $C_h = 0.95$ is taken.

The individual heat-transfer coefficient for the combustion gases can be obtained from the obtained values:

$$h_h = \frac{k}{D_e} 0.33\,(\text{Re}_{max})^{0.6} (\text{Pr})^{0.3} C_h$$

$$h_h = \frac{0.046\,\text{W/m °C}}{0.051\,\text{m}} 0.33(8083)^{0.6}(0.65)^{0.3}(0.95)$$

yielding

$$h_h = 54.9\,\text{W/m}^2\,\text{°C}$$

The calculation of the individual coefficients *for the air* that circulates inside the tubes is conducted using the following expression: (Nu) = 0.021(Re)$^{0.8}$.

If w_t is the air mass flow rate, the air flux that circulates inside the tube is

$$G_t = \frac{w_t}{N_t\, a_t / n} = \frac{wn4}{N_t \pi d_i^2}$$

where

$w_t = 3500$ kg/h mass flow rate
$n = 2$ tube passes
$N_t = 400$ total number of tubes
$a_t = (\pi/4) d_i^2$ transversal section of a tube

Substitution of data yields the mass flow density:

$$G_t = \frac{(3{,}500\,\text{kg/h})(2)(4)}{(400)\pi(11\times10^{-3})^2\,\text{m}^2} = 184{,}146.2\,\text{kg/h m}^2 = 55.15\,\text{kg/m}^2\,\text{s}$$

$$(\text{Re}) = \frac{G_t d_i}{\eta} = \frac{(51.15\,\text{kg/m}^2\,\text{s})(11\times10^{-2}\,\text{m})}{0.0189\times10^{-3}\,\text{Pa s}} = 29{,}971$$

$$(\text{Nu}) = \frac{h_t d_i}{k} = 0.021(29{,}971)^{0.8}$$

Therefore, the individual coefficient on the tube side can be obtained:

$$h_t = \frac{0.028\,\text{W/(m °C)}}{11\times10^{-3}\,\text{m}} 0.021(29{,}971)^{0.8}$$

$$h_t = 204\,\text{W/m}^2\,°\text{C}$$

c. Calculation of the total external area of the tubes

$$A_e = \frac{\dot{Q}}{U_e(\Delta T)_{ml}F}$$

$$d_{ml} = \frac{d_0 - d_i}{\ln(d_0/d_i)} = \frac{(12.5-11)\times10^{-3}}{\ln(12.5/11)}\text{m} = 11.73\times10^{-3}\,\text{m}$$

$$e = \frac{d_0 - d_i}{2} = \frac{(12.5-11)\times10^{-3}}{2}\text{m} = 7.5\times10^{-4}\,\text{m}$$

The global coefficient referred to the external area is calculated from the following equation:

$$\frac{1}{U_e} = \frac{1}{549\,\text{W/m}^2\,°\text{C}} + \frac{(7.5\times10^{-4}\,\text{m})(12.5\times10^{-3}\,\text{m})}{(44\,\text{W/m °C})(11.73\times10^{-3}\,\text{m})} + \frac{12.5\times10^{-3}\,\text{m}}{(204\,\text{W/m}^2\,°\text{C})(11\times10^{-3}\,\text{m})}$$

obtaining

$$U_e = 42\,\text{W/m}^2\,°\text{C}$$

Calculation of the factor F

$$F = F(Z,\ \varepsilon,\ type)$$

$$\left.\begin{aligned} Z &= \frac{T_{in}-T_{out}}{t_{out}-t_{in}} = \frac{400-100}{95-15} = 3.75 \\ \varepsilon &= \frac{t_{out}-t_{in}}{T_{in}-t_{in}} = \frac{95-15}{400-15} = 0.21 \end{aligned}\right\} \xrightarrow{\text{Graph}} F \approx 0.83$$

Calculation of $(\Delta T)_{ml}$

$$(\Delta T)_{ml} = \frac{(400-95)-(100-15)}{\ln((400-95)/(100-15))} = 172.2°\text{C}$$

Calculation of the total heat flow rate transferred

$$\dot{Q} = w_c \left.\hat{C}_p\right)_c (t_{out} - t_{in}) = (3500\,\text{kg/h})(0.992\,\text{kJ/kg °C})(95°\text{C} - 15°\text{C})$$

$$\dot{Q} = 277{,}760\,\text{kJ/h}$$

Total external area

$$A_e = \frac{\dot{Q}}{U_e(\Delta T)_{ml} F} = \frac{(277{,}760\,\text{kJ/h})(1\,\text{h}/3600\,\text{s})}{(42\times10^{-3}\,\text{kJ/s m}^2\,°\text{C})(172.2°\text{C})(0.83)} = 12.853\,\text{m}^2$$

d. Calculation of the tube length

$$A_e = \pi d_0 L N_t$$

where

L is the tube length
d_0 is the external diameter of the tubes
N_t equals 400 tubes

$$L = \frac{A_e}{\pi d_0 N_t} = \frac{12.853\,\text{m}^2}{\pi(12.5\times10^{-3}\,\text{m})(400)} = 0.818\,\text{m}$$

$L = 81.8$ cm

17.7 It is desired to refrigerate a must that is at 25°C, to send it to a jacketed tank where its fermentation will be carried out at a controlled temperature of 16°C. In order to perform this refrigeration, 20,000 kg/h of must is fed into a plate-type heat exchanger that operates in parallel, which handles a 2/1 hot fluid/cold fluid ratio, using fluid at 3°C as refrigerant fluid. If the temperature at which water leaves the heat exchanger is 14°C, calculate (a) the flow rate of refrigeration water, (b) the heat-transfer coefficients, and (c) the number of thermal plates.

Data: Characteristics of the plates: Stainless steel AISI 316. Dimensions 75 cm × 25 cm × 1.5 mm. Surface 0.165 m^2. Gasket thickness 5 mm.
Thermal conductivity 16 W/(m °C)

Assume that in the range of work temperature, the thermal properties of the fluids do not vary:

Water	Thermal conductivity 0.58 W/(m °C)	Viscosity 1 mPa s
	Density 1000 kg/m^3	Specific heat 4.185 kJ/(kg °C)
Grape must	Thermal conductivity 0.52 W/(m °C)	Viscosity 1.2 mPa s
	Density 1030 kg/m^3	Specific heat 4.06 kJ/(kg °C)

The following figure shows a schema of the plate-type heat exchanger and the dimensions of a channel between two plates:

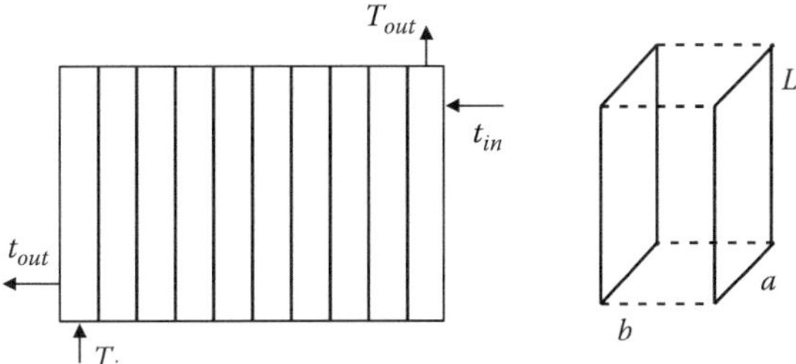

a. Energy balance

$$\dot{Q} = w_h \hat{C}_p\Big)_h (T_{in} - T_{out}) = w_c \hat{C}_p\Big)_c (t_{out} - t_{in})$$

$$(20,000\,\text{kg/h})(4.06\,\text{kJ/kg}\,°\text{C})(25°\text{C} - 16°\text{C}) = w_c\,(4.185\,\text{kJ/kg}\,°\text{C})(14°\text{C} - 3°\text{C})$$

$$w_c \approx 15,874.9\,\text{kg/h}$$

b. Data management

$$D_e = 4r_H = 2b = 2 \times 5\,\text{mm} = 10\,\text{mm} = 0.01\,\text{m}$$

$$S = a \cdot b = (0.25\,\text{m})(5 \times 10^{-3}\,\text{m}) = 1.25 \times 10^{-3}\,\text{m}^2$$

Flux of the fluids

$$G = \frac{w}{S}$$

$$G_h = \frac{20,000}{1.25 \times 10^{-3}} \cdot \frac{1}{3,600} = 4,444.44\,\text{kg/m}^2\,\text{s}$$

$$G_c = \frac{15,873}{1.25 \times 10^{-3}} \cdot \frac{1}{3,600} = 3,527.76\,\text{kg/m}^2\,\text{s}$$

Prandtl number

$$(\text{Pr}) = \frac{\hat{C}_p \eta}{k}$$

Hot fluid

$$\text{Pr})_h = 9.4$$

Cold fluid

$$\text{Pr})_c = 7.2$$

Reynolds number

$$(\text{Re}) = \frac{(G/n)\,D_e}{\eta\eta}$$

$$\text{Hot fluid} \qquad (\text{Re})_h = \frac{37{,}037}{n_h} \qquad \text{Turbulent if } n_f < 93$$

$$\text{Cold fluid} \qquad (\text{Re})_c = \frac{35{,}277.6}{n_c} \qquad \text{Turbulent if } n_f < 89$$

**Individual coefficients*

$$(\text{Nu}) = \frac{hD_e}{k} = 0.374(\text{Re})^{0.668}(\text{Pr})^{0.333}$$

Hot fluid

$$h_h = 51{,}546.6(n_c)^{-0.668}\ \text{W/(m}^2\,^\circ\text{C)}$$

Cold fluid

$$h_c = 40{,}935.1(n_f)^{-0.668}\ \text{W/(m}^2\,^\circ\text{C)}$$

Global coefficient

$$\frac{1}{U} = \frac{1}{h_h} + \frac{1}{h_c} + \frac{1.5\times 10^{-3}\ \text{m}}{16\ \text{W/(m}\,^\circ\text{C)}}$$

$$\Delta T_{ml} = \frac{(T_{in} - t_{out}) - (T_{out} - t_{in})}{\ln((T_{in} - t_{out})/(T_{out} - t_{in}))} = \frac{(25-14)-(16-3)}{\ln(11/13)} = 12^\circ\text{C}$$

$$\text{NTU} = \frac{T_{in} - T_{out}}{\Delta T_{ml}} = \frac{25-16}{12} = 0.75$$

From the graph, $F = F(\text{NTU}, \text{type of pass})$, obtaining $F \approx 0.96$.
From the thermal balance,

$$\dot{Q} = w_h \hat{C}_p\Big)_h (T_{in} - T_{out})$$

From the velocity equation,

$$\dot{Q} = UA_T \Delta T_{ml} F$$

The total area is obtained from these equations:

$$A_T = \frac{w_h \left.\hat{C}_p\right)_h (T_{in} - T_{out})}{U \Delta T_{ml} F}$$

Substitution of data yields

$$A_T = \frac{17{,}621.5}{U} \text{m}^2$$

The number of thermal plates is

$$N = \frac{A_T}{A_P}$$

It is necessary to perform an iterative calculation process:

It is supposed that the number of channels for both fluids (n_h and n_c) and the heat-transfer coefficients is calculated (h_h, h_c and U). Next, the total area is calculated, and with it, the number of thermal plates. The total number of channels: $n = n_h + n_c = N + 1$. If the number of channels obtained coincide with the one supposed, then the process ends. If that is not the case, the process should be performed again. Table P.17.7 presents the values of the different variables in each stage of the iterative calculation.

Heat-transfer coefficients

Grape must

$$h_c = 5703\ \text{W/m}^2\ \text{°C}$$

Water

$$h_h = 4529\ \text{W/m}^2\ \text{°C}$$

Global

$$U = 2043\ \text{W/m}^2\ \text{°C}$$

Number of thermal plates: $N = 53$

TABLE P.17.7
Variables of the Iterative Calculation Process of the Number of Thermal Plates

n_h	n_c	Re_h	Re_c	h_h	h_c	U	A_T	N	n_h	n_c
					(W/m² °C)		(m²)			
30	30	1235	1176	5315	4221	1928	9.140	56	29	28
28	28	1277	1260	5434	4420	1985	8.877	54	28	27
28	27	1323	1307	5566	4529	2024	8.706	53	27	27
27	27	1372	1307	5703	4529	2043	8.630	53	27	27

Operation in series

$$n_h = n_c = 1$$

$$(\text{Re})_c = 35{,}278 \qquad (\text{Pr})_c = 7.2$$

$$(\text{Re})_h = 37{,}037 \qquad (\text{Pr})_h = 9.4$$

$$h_c = 40{,}935\,\text{W/m}^2\,°\text{C} \qquad h_h = 51{,}547\,\text{W/m}^2\,°\text{C}$$

$$U = 7{,}269\,\text{W/m}^2\,°\text{C} \qquad A_T = 2.424\,\text{m}^2$$

Number of thermal plates: $N = 15$ thermal plates

17.8 A viscous food fluid is at 15°C, and it is desired to increase its temperature to 40°C to introduce it to a plate-type pasteurizer at a flow of 3000 kg/h. To carry out such heating, a scraped-surface concentric tube-heat exchanger, with four blades inserted in a central axis that rotates at 6 rpm, will be used. 10,000 kg/h of hot water that circulates by the annular section is introduced at 98°C. If the heat exchanger is perfectly insulated to avoid heat loss to the outside, calculate its length.

Data: The thickness of the central axis and of the blades can be considered as negligible.
Properties of the food fluid: Specific heat 3.35 kJ/(kg °C). Thermal conductivity 0.52 W/(m °C). Density 1100 kg/m³. Viscosity 1.6 Pa s.
Water properties: Specific heat 4.185 kJ/(kg °C). Viscosity 1 mPa s. Thermal conductivity 0.58 W/(m °C). Density 1000 kg/m³.
The tubes are made of stainless steel, and their thermal conductivity is 23 W/(m °C). The internal tube has an internal diameter of 22 cm and a wall thickness of 8 mm. The external tube has an internal diameter of 30 cm:

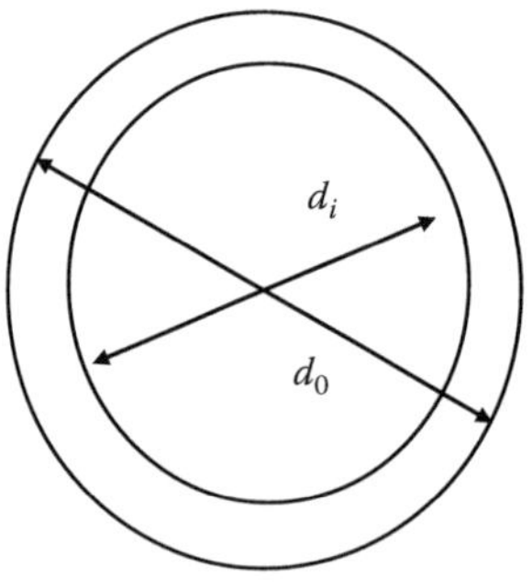

$$d_i = 22\,\text{cm} = 0.22\,\text{m}$$

$$d_e = d_i + 2e = 23.6\,\text{cm} = 0.236\,\text{m}$$

$$d_0 = 30\,\text{cm} = 0.3\,\text{m}$$

Internal tube section

$$S_i = \frac{\pi}{4} \cdot d_i^2 = 0.0380\,\text{m}^2$$

Annular section

$$S_a = \frac{\pi}{4}\left(d_0^2 - d_e^2\right) = 0.02694\,\text{m}^2$$

Equivalent diameter of the annular section

$$D_e = 4r_H = 4\cdot\frac{(\pi/4)\left(d_0^2 - d_e^2\right)}{\pi d_e} = \frac{d_0^2 - d_e^2}{d_e} = 0.1454\,\text{m}$$

Mass flux for water

$$G_h = \frac{w_h}{S_a} = \frac{10{,}000\,\text{kg/h}}{0.02694}\cdot\frac{1\,\text{h}}{3{,}600\,\text{s}} = 103.1\,\text{kg/m}^2\,\text{s}$$

The heat-transfer coefficient for water is calculated from the equation of Dittus–Boelter for fluids that cool down:

$$(\text{Nu}) = \frac{hD_e}{k} = 0.023\,(\text{Re})^{0.8}(\text{Pr})^{0.3}$$

Prandtl number

$$(\text{Pr}) = \frac{\hat{C}_p\eta}{k} = \frac{(4.185\,\text{kJ/kg}\,°\text{C})(10^{-3}\,\text{Pa}\,\text{s})}{(0.58\times10^{-3}\,\text{kJ/s}\,\text{m}\,°\text{C})} = 7.2$$

Reynolds number

$$(\text{Re}) = \frac{\rho v D_e}{\eta} = \frac{G_c D_e}{\eta} = \frac{(103.1\,\text{kg/m}^2\,\text{s})(0.1454\,\text{m})}{10^{-3}\,\text{Pa}\,\text{s}} \approx 1.5\times10^4$$

Substitution in the equation of Dittus–Boelter

$$h_e = \frac{(0.58\,\text{W/m}\,°\text{C})}{(0.1454\,\text{m})}0.023(1.5\times10^4)^{0.8}(7.2)^{0.3} \approx 363\,\text{W/m}^2\,°\text{C}$$

The individual heat-transfer coefficient for the food fluid is calculated from the following equation:

$$h_i = 2\left(\frac{k\rho\hat{C}_p PN}{\pi}\right)^{1/2}$$

$$h_i = 2\left(\frac{(0.52\,\text{W/m}\,°\text{C})(1100\,\text{kg/m}^3)(3.35\times10^3\,\text{J/kg}\,°\text{C})(4)(6\,\text{min}^{-1}(1\,\text{min}/60\,\text{s}))}{\pi}\right)^{1/2}$$

obtaining

$$h_i = 988\ \text{W/m}^2\ ^\circ\text{C}$$

Global heat-transfer coefficient referred to the internal area

$$\frac{1}{U_i} = \frac{1}{988\ \text{W/m}^2\ ^\circ\text{C}} + \frac{8\times10^{-3}\ \text{m}}{(23\ \text{W/m}\ ^\circ\text{C})(0.228\ \text{m}/0.22\ \text{m})} + \frac{1}{(363\ \text{W/m}^2\ ^\circ\text{C})\left(0.236\ \text{m}/0.22\ \text{m}\right)}$$

in which D_{ml} is

$$D_{ml} = \frac{d_e - d_i}{\ln(d_e/d_i)} = \frac{0.236-0.22}{\ln(0.236/0.22)} = 0.228\ \text{m}$$

yielding

$$U_i = 255\ \text{W/m}^2\ ^\circ\text{C}$$

From the energy balance,

$$(10{,}000\ \text{kg/h})(4.185\ \text{kJ/kg}\ ^\circ\text{C})(98 - T_{out})^\circ\text{C} = (3{,}000\ \text{kg/h})(3.35\ \text{kJ/kg}\ ^\circ\text{C})(40-15)^\circ\text{C}$$

resulting in a temperature of $T_{out} = 92^\circ$C.
From the velocity equation,

$$\dot{Q} = U_i A_i \Delta T_{ml} = U_i \pi\, d_i L \Delta T_{ml}$$

where

$$\dot{Q} = w_c \hat{C}_{pc}(t_{out} - t_{in}) = 251{,}250\ \text{kJ/h}$$

If the fluids circulate in *countercurrent*,

$$\Delta T_m = \frac{(T_{in}-t_{out})-(T_{out}-t_{in})}{\ln((T_{in}-t_{out})/(T_{out}-t_{in}))} = \frac{(98-40)-(92-15)}{\ln((98-40)/(92-15))} \approx 67^\circ\text{C}$$

$$A_i = \frac{(251{,}250\ \text{kJ/h})(1\ \text{h}/3{,}600\ \text{s})}{(255\times10^{-3}\ \text{kJ/s m}^2\ ^\circ\text{C})(67^\circ\text{C})} = 4.085\ \text{m}^2$$

the length of the heat exchanger being

$$L = \frac{4.085\ \text{m}^2}{\pi(0.22\ \text{m})} = 5.91\ \text{m}$$

If the fluids circulate in *parallel*,

$$\Delta T_{ml} = \frac{(T_{in} - t_{in}) - (T_{out} - t_{out})}{\ln((T_{in} - t_{in})/(T_{out} - t_{out}))} = \frac{(98-15)-(92-40)}{\ln((98-15)/(92-40))} \approx 66.3°\text{C}$$

$$A_i = \frac{(251{,}250\,\text{kJ/h})(1\,\text{h}/3{,}600\,\text{s})}{(255\times10^{-3}\,\text{kJ/s}\,\text{m}^2\,°\text{C})(66.3°\text{C})} = 4.128\,\text{m}^2$$

the length of the heat exchanger being

$$L = \frac{4.128\,\text{m}^2}{\pi(0.22\,\text{m})} = 5.97\,\text{m}$$

If the following equation is used to calculate h_i, the length L should be determined by iteration:

$$(\text{Nu}) = \frac{h_i d_i}{k} = 4.9(\text{Re})^{0.57}\,(\text{Pr})^{0.47}\left(\frac{d_i \cdot N}{v}\right)^{0.17}\left(\frac{d_i}{L}\right)^{0.37}$$

$$(\text{Re}) = \frac{G_f \cdot d_i}{\eta} = \frac{w_f d_i}{S_i \eta} = \frac{(3000\,\text{kg/h})(1\,\text{h}/3600\,\text{s})(0.22\,\text{m})}{(0.038\,\text{m}^2)(1.6\,\text{Pa}\,\text{s})} = 3$$

$$(\text{Pr}) = \frac{\hat{C}_p \eta}{k} = \frac{(3.35\,\text{kJ/kg}\,°\text{C})(1.6\,\text{Pa}\,\text{s})}{(0.52\times10^{-3}\,\text{kJ/s}\,\text{m}\,°\text{C})} = 10{,}308$$

$$\left(\frac{d_i N}{v}\right) = \frac{(0.22\,\text{m})(6\,\text{min}^{-1})}{0.02\,\text{m/s}}\left(\frac{1\,\text{min}}{60\,\text{s}}\right) \approx 1.104$$

The linear velocity of circulation velocity is

$$v = \frac{w_c}{\rho S} = \frac{3000\,\text{kg/h}}{(1100\,\text{kg/m}^3)(0.038\,\text{m}^2)}\left(\frac{1\,\text{h}}{3600\,\text{s}}\right) \approx 0.020\,\text{m/s}$$

When substituting these data in the former equation, the coefficient h_i can be obtained:

$$h_i = \frac{0.52\,\text{W/m}\,°\text{C}}{0.22\,\text{m}}\,4.9(3)^{0.57}(10{,}308)^{0.47}(1.104)^{0.17}\left(\frac{0.22}{L}\right)^{0.37}$$

Therefore,

$$h_i = 968 L^{-0.37}\,\text{W/m}^2\,°\text{C}$$

The calculation of L should be made by iteration:

$$\frac{1}{U_i} = \frac{1}{h_i} + 2.904\times10^{-3}$$

TABLE P.17.8
Variables of the Iterative Calculation Process of the Length of the Scraped-Surface Heat Exchanger

L_s (m)	h_i (W/m² °C)	U_i (W/m² °C)	L_c (m)	
6	499	204	7.39	Countercurrent
7.39	462	197	7.64	
7.64	456	196	7.68	
7.68	455	196	7.68	
7	471	199	7.66	Parallel
7.66	456	196	7.77	
7.77	453	196	7.78	

Since

$$L = \frac{w_c\, C_P)_c\, (t_{out} - t_{in})}{\pi\, d_i\, \Delta T_{ml}\, U_i}$$

In countercurrent

$$L = \frac{1507}{U_i}\,\text{m}$$

In parallel

$$L = \frac{1523}{U_i}\,\text{m}$$

Table P.17.8 presents the values of the variables of the different stages of the iterative process.

Hence, the length of the heat exchanger obtained is

Operation in parallel: $L = 7.68$ m
Operation in countercurrent: $L = 7.78$ m

17.9 A hot 40°Brix sugar solution is employed in one of the elaboration stages of preserved fruits. Heating is carried out by introducing 1000 kg of this solution in a cylindrical agitated vessel of 1 m of diameter, which is perfectly insulated. This vessel has a baffle of 30 cm diameter that rotates at 120 rpm. A helicoidal coil is submerged in the vessel; this coil consists of stainless steel tubes of 12 mm of internal diameter, 1 mm of wall thickness, and 15 m of total length. Saturated steam at 3 atm circulates inside the coil and condenses, with its convective heat-transfer coefficient being 9300 W/(m² °C). If the solution is initially at 16°C, calculate (a) the global heat-transfer coefficient, (b) the time required by the solution to reach 60°C, (c) the flow rate and quantity of steam required to carry out heating, and (d) the rate of temperature increase of the solution when it is at 50°C, and (e) if it is supposed that the global heat-transfer coefficient is constant, what is the temperature of the solution after 50 min?

Data and notes: The resistance of the coil surface to heat transfer can be considered as negligible.

Properties of the sugar solution: Thermal conductivity 0.814 W/(m °C); specific heat 2.85 kJ/(kg °C).

Viscosity	$\eta = 3.7 \times 10^{-7} \exp(2850/T)$ Pa s	T in Kelvin
Density	$\rho = 1.191 - 4.8 \times 10^{-4} \times T$ g/cm^3	T in°C

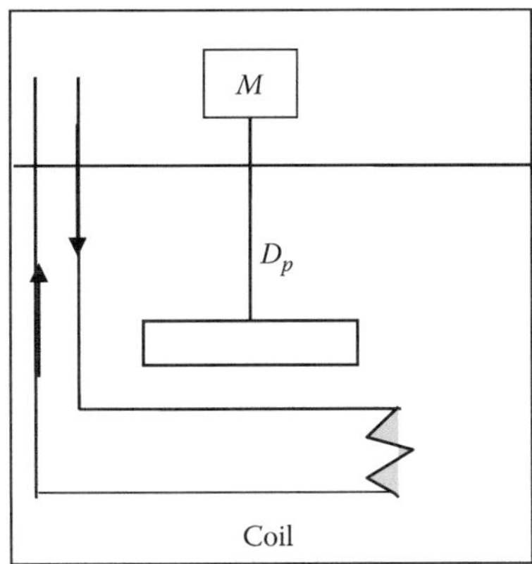

For condensed saturated steam at 3 atm, the following is obtained from tables:

$$T = 132.9°\text{C}$$

$$\left.\begin{aligned}\hat{H}_w &= 2721\,\text{kJ/kg}\\ \hat{h}_w &= 558\,\text{kJ/kg}\end{aligned}\right\}\ \lambda_w = 2163\,\text{kJ/kg}$$

$$D_T = 1\text{m} \qquad L = 15\text{ m}$$

$$D_P = 0.3\text{m} \qquad d_e = 14 \times 10^{-3}\text{ m}$$

Heating is done from 16°C to 60°C; therefore, the properties of the solution will be taken at a mean temperature t_m = 38°C.

Properties at 38°C

$$\eta \approx 3.53 \times 10^{-3}\text{ Pa s}$$

$$\rho \approx 1.173\,\text{g/cm}^3 = 1173\,\text{kg/m}^3$$

Calculation of the coefficient h_e

$$\frac{h_e D_T}{k} = 0.87\left(\frac{D_P^2 N\rho}{\eta}\right)^{0.62}\left(\frac{C_p\eta}{k}\right)^{1/3}\left(\frac{\eta}{\eta_W}\right)^{0.14}$$

$$(\text{Re}) = \frac{D_P^2 N\rho}{\eta} = \frac{(0.3^2\,\text{m}^2)(2\,\text{s}^{-1})(1173\,\text{kg/m}^3)}{5.53 \times 10^{-3}\text{ Pa s}} = 5.98 \times 10^4$$

$$(\text{Pr}) = \frac{\hat{C}_p\eta}{k} = \frac{(2.85\,\text{kJ/kg °C})(3.53 \times 10^{-3}\text{ Pa s})}{(0.814 \times 10^{-3}\text{ kJ/s m °C})} = 12.4$$

Substitution yields

$$\frac{h_e(1\,\text{m})}{0.814\,\text{W/m °C}} = 0.87\,(5.98\times 10^4)^{0.62}(12.4)^{1/3}(3.53\times 10^{-3})^{0.14}(\eta_W)^{-0.14}$$

obtaining

$$h_e = 680.6\,(\eta_W)^{-0.14}\ \text{W/(m}^2\,\text{°C)} \tag{P.17.1}$$

To obtain h_e, it is necessary to know T_W, the temperature of the coil surface, to calculate η_W, that is, the viscosity of the solution at the wall temperature.

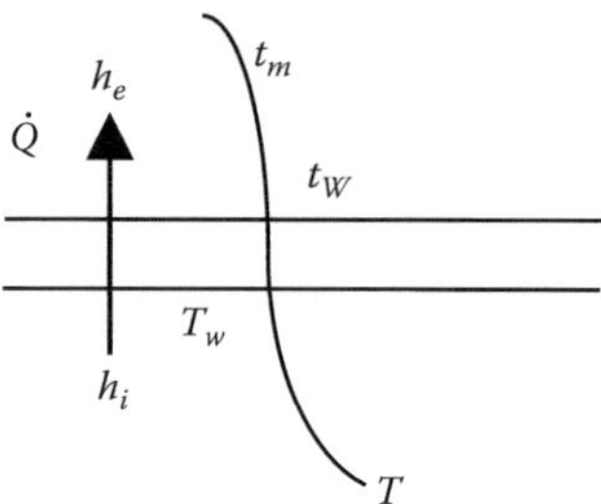

Heat transfer is

$$\dot{Q} = h_i A_i (T - T_W) = h_e \cdot A_e (t_W - t_m)$$

Since it is supposed that the surface does not offer resistance to heat transfer, $T_W \approx t_W$. Hence,

$$t_W = \frac{h_i A_i T + h_e A_e t_W}{h_i A_i + h_e A_e} = \frac{h_i d_i T + h_e d_e t_W}{h_i d_i + h_e d_e}$$

$$t_W = \frac{(9300)(12\times 10^{-3})(132.9) + h_e(14\times 10^{-3})(38)}{(9300)(12\times 10^{-3}) + h_e(14\times 10^{-3})}$$

obtaining

$$t_W = \frac{14{,}832 + 0.532 h_e}{112 + 14\times 10^{-3}\, h_e} \tag{P.17.2}$$

One iteration is needed to calculate h_e. A value is supposed for t_W, and the viscosity is calculated at this temperature, which allows us to calculate h_e using Equation P.17.1. t_W is calculated by substituting this value in Equation P.17.2. If the calculated value coincides with the supposed value, then the problem is solved; if not, the process is repeated until it coincides. The following table presents the results of the iteration:

Iteration	t_W (°C)	$\eta_W \times 10^4$ (Pa s)	h_e (W/(m² °C))	t_W (°C)
1	100	7.70	1857	114.6
2	114.6	5.78	1933	114.1
3	114.1	5.83	1931	114.1

Then,

$$h_e = 1931\,\text{W/m}^2\,°\text{C}$$

a. Calculation of the global coefficient U_e

$$\frac{1}{U_e} = \frac{1}{h_e} + \frac{1}{h_i(d_i/d_e)}$$

$$\frac{1}{U_e} = \frac{1}{1931} + \frac{1}{(9300)(12/14)} \quad \Rightarrow \quad U_e \approx 1554\,\text{W/m}^2\,°\text{C}$$

b. Energy balance
Accumulation term

$$A = m\hat{C}_p \frac{dt}{d\theta}$$

Input term

$$E = U_e A_e (T - t)$$

where
T is the temperature of condensing steam
t is the temperature of the solution in the tank
θ is the time

Equaling these two terms,

$$m\hat{C}_p \frac{dt}{d\theta} = U_e A_e (T - t)$$

This is a differential equation in separated variables, which when integrated on the boundary condition ($\theta = 0$, $t = t_0$) yields the following equation:

$$\ln\left(\frac{T - t_0}{T - t}\right) = \frac{U_e A_e}{m\hat{C}_p}\theta$$

This expression allows the calculation of the heating time for a determined temperature, or vice versa:

Time

$$\theta = \frac{m\hat{C}_p}{U_e A_e} \ln\left(\frac{T - t_0}{T - t}\right)$$

Temperature

$$t = T - (T - t_0)\exp\left(-\frac{U_e A_e \theta}{m C_p}\right)$$

where
$m = 1000$ kg
$T = 132.9°\text{C}$ $\quad$ $t_0 = 16°\text{C}$

$$A_e = \pi d_e L = 0.6527\,\text{m}^2$$

$$\widehat{C}_P = 2.85\,\text{kJ/kg}\,°\text{C}$$

For $t = 60°\text{C}$,

$$\theta = \frac{(10^3\,\text{kg})(2.85\ \text{kJ/kg}\,°\text{C})}{(1554\times 10^{-3}\,\text{kJ/s}\,\text{m}^2\,°\text{C})(0.6597\,\text{m}^2)}\ln\left(\frac{132.9-16}{132.9-60}\right) = 1313\,\text{s}$$

obtaining a time equal to $\theta = 1313$ s ≈ 22 min

c. Steam flow rate and quantity condensed

$$w_v\lambda_w = \frac{m}{\theta}\hat{C}_p(t-t_0)$$

$$w_v\left(2163\,\text{kJ/kg}\right) = \left(\frac{1000\,\text{kg}}{1313\,\text{s}}\right)\left(\frac{3600\,\text{s}}{1\,\text{h}}\right)(2.85\,\text{kJ/kg}\,°\text{C})(60°\text{C}-16°\text{C})$$

Steam flux

$$w_v \approx 159\,\text{kg/h}$$

Mass of steam

$$M = w_v\theta = 58\,\text{kg}$$

d. Velocity at which temperature increases

$$\frac{dt}{d\theta} = \frac{U_e A_e}{m\hat{C}_p}(T-t)$$

$$\frac{dt}{d\theta} = \frac{(1931\times 10^{-3}\,\text{kJ/s}\,\text{m}^2\,°\text{C})(3600\,\text{s/1}\,\text{h})(0.6597\,\text{m}^2)}{(1000\,\text{kg})(2.85\,\text{kJ/kg}\,°\text{C})}(132.9°\text{C}-50°\text{C}) \approx 133.4°\text{C/h}$$

e. Temperature at 50 min

$$t = 132.9-(132.9-16)\exp\left(-\frac{(1931\times 10^{-3}\,\text{kJ/s}\,\text{m}^2\,°\text{C})(0.6597\,\text{m}^2)(3000\,\text{s})}{(1000\,\text{kg})(2.85\,\text{kJ/kg}\,°\text{C})}\right)$$

Thus, temperature is $t = 102.3°\text{C}$.

Since it is assumed that processing is conducted at atmospheric pressure, water becomes steam, and there is no liquid phase. However, since a sugar solution is being used in this case, it is possible that it could boil at 100°C, due to the increase of the boiling point produced by the soluble solids.

17.10 12,000 kg/h of a fluid at 15°C, whose specific heat is 3.817 kJ/(kg °C), is being processed in a food industry. In order to improve energy performance, it is desired to install a heat exchanger to increase the temperature of the fluid up to 72°C, using a water stream of 20,000 kg/h that comes from the condensation of steam at 1 atm. Calculate the surface of a heat exchanger for circulation in parallel and in countercurrent, if the global heat-transfer coefficient is 1750 W/(m² °C).

Operation in parallel
Energy balance

$$\dot{Q} = w_c\,\hat{C}_p\Big)_c (t_{out} - t_{in}) = w_h\,\hat{C}_p\Big)_h (T_{in} - T_{out})$$

Since steam condenses at 1 atm, ⇒ T_{in} = 100°C.
Substitution yields

$$(12{,}000\text{ kg/h})(3.817\text{ kJ/kg °C})(72\text{°C} - 15\text{°C}) = (20{,}000\text{ kg/h})(4.185\text{ kJ/kg °C})(100 - T_{out})$$

The following is obtained: T_{out} = 68.8°C. This is impossible because $T_{out} < t_{out}$, and this is contrary to the laws of thermodynamics.

Operation in countercurrent

$$C_h = (20{,}000\text{ kg/h})(4.185\text{ kJ/kg °C}) = 83{,}700\text{ kJ/h °C}$$

$$C_c = (12{,}000\text{ kg/h})(3.817\text{ kJ/kg °C}) = 45{,}804\text{ kJ/h °C}$$

Then,

$$C_{min} = C_c \qquad C_{max} = C_h$$

Hence,

$$\beta = \frac{C_{min}}{C_{max}} = 0.547$$

The performance or efficiency of the heat transfer is

$$\varepsilon = \frac{C_c(t_{in} - t_{out})}{C_{min}(t_{in} - T_{in})} = \frac{15 - 72}{15 - 100} = \frac{57}{85} = 0.671$$

The value of NTU can be found from efficiency graphs:

$$\left.\begin{array}{l}\varepsilon = 0.671\\ \beta = 0.547\end{array}\right\} \xrightarrow{\text{Graph}} \text{NTU} \approx 1.5$$

Since

$$\text{NTU} = \frac{UA}{C_{min}}$$

the heat exchange area is

$$A = \frac{C_{min}\text{NTU}}{U} = \frac{(45{,}804\,\text{kJ/h}\,°\text{C})(1\,\text{h}/3600\,\text{s})(1.5)}{1.75\,\text{kJ/s}\,\text{m}^2\,°\text{C}} \approx 10.906\,\text{m}^2$$

The area can also be calculated from the heat-transfer velocity through the exchange area equation:

$$\dot{Q} = UA\Delta T_{ml}$$

$$\dot{Q} = (12{,}000\,\text{kg/h})(3.817\,\text{kJ/kg}\,°\text{C})(72°\text{C} - 15°\text{C}) = 2{,}610{,}828\,\text{kJ/h}$$

$$(\Delta T)_{ml} = \frac{(T_{in} - t_{out}) - (T_{out} - t_{in})}{\ln((T_{in} - t_{out})/(T_{out} - t_{in}))} = \frac{(100 - 72) - (68.8 - 15)}{\ln((100 - 72)/(68.8 - 15))}$$

obtaining

$$(\Delta T)_{ml} = 39.5°\text{C}$$

Since

U = 1750 W/(m^2 °C), then the area is

$$A = \frac{(2{,}619{,}828\,\text{kJ/h})(1\,\text{h}/3600\,\text{s})}{(1.75\,\text{kJ/s}\,\text{m}^2\,°\text{C})(39.5°\text{C})} = 10.492\,\text{m}^2$$

The small difference found can be due to the error made during the graphical determination of NTU.

PROPOSED PROBLEMS

17.1 A juice concentration plant pilot possesses a water steam generator to supply the energy necessities of this plant. For reasons of security, the generator has been installed far away from the plant, so it is necessary to install 100 m of 1 in. steel pipe to transport the vapor from the generation site to that application site. The generator produces saturated steam at 120°C, and the pipe is covered by 1 cm of insulating material, but the saturated steam arrives at its application site at 105°C, with a volumetric flow of 200 m^3/h. Because it is necessary for the steam to arrive at 110°C, the insulating thickness must be increased. If the room temperature is 20°C and the heat convection coefficient insulating material—air is 12 W/(m^2 °C), calculate (a) the insulating thermal conductivity, (b) the overall heat coefficient, and (c) the thickness of insulation necessary so that the water steam arrives at 110°C.

Data: Steel pipe properties: Thermal conductivity 40 W/(m °C). Internal diameter 26.7 mm. External diameter 33.4 mm.

Assume that the steam properties are constant, with values of thermal conductivity 0.025 W/(m °C), specific heat 2.15 kJ/(kg °C), and viscosity 0.0135 mPa s.

17.2 A canning industry requires a liquid with 17% soluble solid content at 98°C. The preparation of this solution is carried out at 22°C, and to raise its temperature, a concentric cylinder heat exchanger is used. The internal pipe possesses a nominal diameter of 2 in. type 40. The liquid flows through this pipe at 1.000 kg/h, while for the annular section, a steam is condensing at 105°C. It can be considered that the convection coefficient value is 5000 W/(m^2 °C). The pipe has a thermal conductivity of 40 W/(m °C). Calculate (a) the overall transmission of heat coefficient and (b) the heat exchanger length.

Data: Liquid properties: Thermal conductivity 0.20 W/(m °C). Specific heat 3.9 kJ/(kg °C). Viscosity 1.5 mPa s. Density 1020 kg/m^3.

17.3 A steam of 120°C is diverted from an evaporator to a condenser that consists of 20 lines of parallel pipe, and each line contains 50 tubes of commercial steel of 2 m length. An aqueous solution at 18°C is fed to the pipe at 0.8 m/s. There is a fouling layer of 1 mm thickness. If in the condenser there is a pressure of 1 kgf/cm^2, calculate (a) the overall transmission of heat coefficient, (b) the temperature of the condensed steam that exits the device, (c) the exit temperature of the aqueous solution, and (d) the condensing steam flow rate.

Data: Pipe characteristics: External diameter 203 mm. Wall thickness 1.5 mm. Thermal conductivity 45 W/(m °C). Thermal conductivity of fouling layer 2.2 W/(m °C). Heat convection coefficient for water condensing steam 8000 W/(m^2 °C).

Aqueous solution properties: Thermal conductivity 0.60 W/(m °C). Specific heat 3.95 kJ/(kg °C). Viscosity 1 mPa s. Density 1010 kg/m^3.

17.4 It is desired to heat an air stream from 25°C to 80°C in a shell and tube-heat exchanger 2/8. The air is flowing outside the pipes, while inside the pipes, 1000 kg/h of water at 95°C is being fed, and this water exits at 56°C. If the overall transmission of heat coefficient is 40 W/(m^2 °C), determine the total area of the heat exchanger.

17.5 In an integrated sterilization process, 1200 kg/h of milk is processed, and the holding steps are conducted at 120°C. This warm milk stream is used to preheat the milk from 4°C to 72°C, and this milk is later sterilized. This operation is carried out in a plate heat exchanger of stainless steel AISI 316 that operates with a disposition 1/1. Determine (a) the individual convection coefficient for cool and warm fluids and (b) the total thermal plates.

Data: Plate characteristics: 50 cm × 25 cm × 3 mm. Surface 0.135 m^2. Gasket thickness 4 mm. Thermal conductivity 16.3 W/(m °C).

Assume that the thermal properties of milk are independent temperatures. Milk properties: Thermal conductivity 0.58 W/(m °C). Specific heat 3.852 kJ/(kg °C). Viscosity 1.2 mPa s. Density 1030 kg/m^3.

17.6 In one of the beer elaboration steps, the must is pasteurized, using a plate-type heat exchanger 1/1 working in parallel operation. 10,000 kg/h of the must at 22°C is fed to the exchanger, using water at 95°C as heater fluid. If the must that exits from the exchanger is at 77°C and the water is at 82°C, calculate (a) the warm water flow rate, (b) the overall heat transmission coefficient, and (c) the number of thermal plates.

Data and notes: Assume that most properties are the same as those of the water and that the temperatures do not vary, with the mean values being thermal conductivity 0.58 W/(m °C), specific heat 4.185 kJ/(kg °C), viscosity 0.55 mPa s, and density 990 kg/m^3.

Plate characteristics: 1 m × 25 cm × 2 mm. Surface 0.27 m^2. Gasket thickness 5 mm. Thermal conductivity 40 W/(m °C).

17.7 10,000 kg/h of an aqueous solution cools from 60°C to 28°C in a plate-type heat exchanger 1/1, using a cold fluid 4,150 kg/h of water at 15°C. Determine (a) the total thermal plates for a series fluid circulation and (b) repeat for parallel circulation.

Data: Assume negligible fouling effects. Assume that properties are constant with temperature.

Water properties: Thermal conductivity 0.54 W/(m °C). Specific heat 4.185 kJ/(kg °C). Viscosity 1 mPa s. Density 1000 kg/m^3.

Aqueous solution properties: Thermal conductivity 0.40 W/(m °C). Specific heat 2 kJ/(kg °C). Viscosity 100 mPa s. Density 1350 kg/m^3.

Plate characteristics: Stainless steel AISI 316. Dimensions = 1.25 m × 50 cm × 1.5 mm. Surface 0.625 m^2. Gasket thickness 5 mm. Thermal conductivity 16 W/(m °C).

17.8 The final product obtained in the manufacture of a pear concentrated juice ends with the evaporation step at 71°Brix and 60°C. Before its storage, the concentrated juice should be cooled to 5°C, using two plate-type heat exchangers working in series. In the first example, water is used like a cold fluid, and in the second step, a 30% glycol solution is used. The necessary plate number for the first exchanger should be calculated, when the 2200 kg/h flow rate of concentrated juice, at the time the juice exits from the heat exchanger at 20°C. Calculate, also, the necessary water flow rate if the water temperature changes from 15°C to 45°C.

Data: Consider the fouling effects to be negligible. Plate characteristics: Stainless steel. Dimensions = 1.25 m × 22 cm × 12 mm. Gasket thickness 5 mm. Thermal conductivity 45 W/(m °C).

Cold water properties: Thermal conductivity 0.60 W/(m °C). Specific heat 4.185 kJ/(kg °C). Viscosity 1 mPa s. Density 1000 kg/m^3.

Juice properties: Thermal conductivity 0.25 W/(m °C). Specific heat 3.95 kJ/(kg °C). Viscosity 1 mPa s. Density 1358 kg/m^3. The juice viscosity varies with temperature in accordance with an exponential-type equation: $\eta = 8.87 \times 10^{-11} \exp(13.1/RT)$, where R is in kcal/molK, T in Kelvin, and η in Pa s.

17.9 In order to obtain a good-quality wine, in the fermentation process, it is necessary, but not sufficient, to use a controlled temperature. With the purpose of achieving an energy saving and better use of the refrigerating device, the must is usually refrigerated up to 17°C, previous to its introduction in the fermentation tank.

In a cellar that produces white wine, it is desirable to refrigerate 2700 kg/h of grape must from 22°C to 16°C using a plate-type heat exchanger 1/1 in parallel, with the cold fluid water at 7°C. If the water exits from heat exchanger at 11°C, calculate (a) the cold water flow rate, (b) the overall heat transmission coefficient, and (c) the number of thermal plates.

Data: Assume that the grape must properties are the same as that of water and that they do not vary with the temperature, with the mean values being thermal conductivity 0.55 W/(m °C), specific heat 4.185 kJ/(kg °C), viscosity 1.25 mPa s, and density 1020 kg/m^3.

Plate characteristics: 1.1 m × 54 cm × 1 mm. Surface 0.38 m^2. Gasket thickness 8 mm. Thermal conductivity 40 W/(m °C).

17.10 A grape must from the Macabeo variety is at 24°C. The goal is to cool the grape must to 17°C and then to transport it to a jacketed tank where a controlled fermentation at this temperature will take place. For the refrigeration, a plate-type heat exchanger 2/2 will be employed, using water at 5°C as a cooling fluid. If it is desired to process 9000 kg/h of must and the temperature at which the water exits the exchanger is 14°C, calculate (a) the cooling water flow rate, (b) the number of thermal plates for operation in series, and (c) the number of thermal plates for operation in parallel.

Data: Plate characteristics: Stainless steel AISI 316. Dimensions = 75 cm × 25 cm × 1.5 mm. Surface 0.18 m^2. Gasket thickness 5 mm. Thermal conductivity 20 W/(m °C).

Fluid properties

Fluid	k	η	C_p	ρ
	W/(m °C)	**(mPa s)**	**(kJ/kg °C)**	**(kg/m^3)**
Water	0.58	1	4.185	1000
Grape must	0.52	1.2	3.95	1030

17.11 In a process step for an egg product, it is desired to obtain egg yolk with 12% salt. Due to the great viscosity that salt confers to the yolk, it is difficult for the egg yolk to flow through the plate heat exchanger if its temperature is lower than 35°C. An industry is obtaining 2500 kg/h of pasteurized salted egg yolk; previously to that, its introduction in the pasteurization, its temperature raises from 15°C to 35°C, using a scraped concentric cylinder-type heat exchanger. Through the internal tube, the egg yolk circulates, and to avoid deterioration problems, the wall is scraped with eight blades inserted in a central axis that rotates at 3 rpm. In the annular section is the countercurrent flowing 7500 kg/h of hot water that enters at 95°C. Calculate the heat exchanger length long, if the internal pipe possesses 20 cm diameter and a wall thickness 8 mm, while the external tube has 30 cm internal diameter.

Data: Steel thermal conductivity 25 W/(m °C).

Salted egg yolk properties: Thermal conductivity 0.45 W/(m °C). Specific heat 3.2 kJ/(kg °C). Density 1100 kg/m^3.

17.12 A viscous fluid food is at 15°C, and it is desired to increase the temperature to 40°C in order to introduce 3000 kg/h in a plate heat exchanger. To carry out this heating, a scraped concentric cylinder-type heat exchanger will be used, with four blades inserted in a central axis that rotates at 6 rpm. For the annular section, 10,000 kg/h of hot water, at 98°C, is introduced. If the heat exchanger is perfectly isolated to avoid heat loss toward the exterior, calculate its length.

Data: Assume that the thickness of axis and blades is negligible.

Fluid food properties

Fluid	K	η	C_P	ρ
	(W/(m °C)	(mPa s)	(kJ/kg °C)	(kg/m^3)
Water	0.58	1	4.185	1000
Food	0.52	1.6	3.350	1100

The pipes are stainless steel constructed, with thermal conductivity of 23 W/(m °C). The internal pipe has 22 cm internal diameter and a wall thickness of 8 mm. The external pipe has 30 cm internal diameter.

17.13 A cylindrical tank of 1 m of diameter is built with stainless steel AISI 316 of 5 mm of thickness, and it is perfectly agitated. A 20 cm blade is rotating inside the tank at 180 rpm. The tank contains 1000 kg of a food solution that initially is at 15°C, and it is desired to raise its temperature to 50°C. The tank possesses a lateral jacket flowing condensing water steam at 125°C, which provides the necessary heat to carry out the heating of the food solution. Under working conditions, the condensing steam heat convective coefficient is 8500 W/(m^2 °C). The device is perfectly isolated, and it can also be assumed that the resistance that the lateral tank wall offers to the heat transmission is negligible. Calculate (a) the overall heat transmission coefficient and (b) the necessary time to carry out the desired heating. (c) If the food solution does not experience an increasing appreciable boiling point, determine the time at which the solution begins to boil.

Food solution properties: Thermal conductivity 0.90 W/(m °C). Specific heat 3.1 kJ/(kg °C). Density 1250 kg/m^3. Viscosity 2.5 mPa s.

18 Heat Transfer by Radiation

18.1 INTRODUCTION

Energy transfer by radiation is basically different from other energy transfer phenomena because it is not proportional to a temperature gradient, nor does it need a material medium to propagate. In addition, heat transfer by radiation is simultaneous with convective transfer.

Any molecule possesses translational, vibrational, rotational, and electronic energy, and all of those are done on quantum states, that is, discrete values of energy. Passing from one energy level to another implies an energy absorption or emission. Passing to a higher energy state implies energy absorption by a molecule; on the contrary, a molecule emits energy as radiation when passing to a lower energy level. Since the energy levels are quantized, the absorption or emission of energy is in the form of photons where the duality wave particle becomes relevant.

Any body at a temperature higher than absolute zero can emit radiant energy, and the amount of energy emitted depends on the temperature of the body. As the temperature of a body increases, energy levels are excited first, followed by the electronic level changes. A temperature increase implies that the radiation spectrum moves to shorter wavelengths or are more energetic.

The corpuscle theory states that radiant energy is transported by photons and also that it is a function of its frequency ν, according to the expression

$$E = h\nu \tag{18.1}$$

in which the proportionality constant h is the so-called Planck's constant, whose value is $h = 6.6262 \times 10^{-34}$ J s.

The wave theory considers radiation as an electromagnetic wave, relating frequency to wavelength according to the following equation:

$$\nu = \frac{c}{\lambda} \tag{18.2}$$

where

λ is the wavelength of the radiation

c is the value of light speed under vacuum (2.9979×10^8 m/s)

So-called thermal radiation, which includes the ultraviolet, visible, and infrared spectra, corresponds to wavelengths of 10^{-7}–10^{-4} m.

18.2 FUNDAMENTAL LAWS

Before describing the different laws that rule the radiation phenomenon, it is convenient to define what a *black body* is. The black body is a body that absorbs and emits the maximum quantity of energy at a determined temperature but reflects none.

18.2.1 Planck's Law

The energy emitted by a black body at a given temperature T is a function of the wavelength. In this way, if q_λ is considered as the spectral emissive power, expressed in J/(m^3 s), then

$$q_{n,\lambda}^{e} = \frac{C_1}{\lambda^5 \left[\exp\left(C_2 / \lambda T\right) - 1\right]} \tag{18.3}$$

in which $q_{n,\lambda\lambda}^{e}$ is the radiation energy emitted per unit time and volume at a wavelength λ by a black body that is at a temperature T. The superscript e refers to emission, while the subscripts n and λ refer to a black body and wavelength, respectively. The constants C_1 and C_2 are defined by

$$C_1 = 2\pi c^2 h = 3.742 \times 10^{-16} \text{ J/(m}^2\text{ s)}$$

$$C_2 = \frac{hc}{k} = 1.438 \times 10^{-2} \text{ m K}$$

in which k is the Boltzmann constant, which has a value of 1.3806×10^{-23} J/K.

18.2.2 Wien's Law

The emissive power of a black body at a determined temperature shows variation with the wavelength, in such a way that it passes through a maximum. Also, if the temperature of emission increases, it is observed that the maximum corresponds to a lower wavelength, while the value of the emissive power is higher.

In order to find the wavelength corresponding to the maximum emission, the derivative of Planck's law with respect to the wavelength (Equation 18.3) should be equal to zero:

$$\frac{d\left(q_{n,\lambda\lambda}^{e}\right)}{d\lambda} = 0$$

Hence,

$$\lambda_{max} T = 2.987 \times 10^{-3} \text{ m K} \tag{18.4}$$

This expression confirms what was indicated before.

18.2.3 Stefan–Boltzmann Law

The flux of the radiation energy or the total radiation energy per unit area and per unit of time, emitted by a black body, is obtained by integrating Planck's equation on all the wavelengths. This integration shows that such energy depends only on temperature, and it is proportional to its fourth power:

$$q_n^e = \sigma T^4 \tag{18.5}$$

being σ the Stefan–Boltzmann constant, having a value of

$$\sigma = \frac{2\pi^5 k^4}{15 c^2 h^3} = 5.67 \times 10^{-8} \text{ J/(s m}^2\text{ K}^4\text{)} \tag{18.6}$$

18.3 PHYSICAL PROPERTIES ASSOCIATED WITH RADIATION

As can be observed from the discussion in the last section, radiation energy depends on wavelength, thus influencing the behavior of bodies to emission and absorption of radiating energy dependent on this energy. The properties that show this dependence are called *monochromatic* properties. Also, the properties can depend on the direction in which radiation is transmitted; these are called *directional* properties. However, some mean *total* properties independent of wavelength and direction can be considered, simplifying the calculations and difficulties of some problems in which radiation energy intervenes.

18.3.1 Global Properties

It is considered that all properties have an equal mean value for all wavelengths and directions. Radiation received by a body can experience some of the following phenomena: part of it can be absorbed, part of it can be transmitted, and part of it can be reflected.

If q is the total flux of radiation received by a body, it can be written that

$$q = q^r + q^t + q^a \tag{18.7}$$

in which superscripts r, t, and a stand for reflected, transmitted, and absorbed, respectively.

The fraction of energy reflected, transmitted, or absorbed has different names, thus

- Reflectivity or reflection factor (r): Fraction of radiation received by a body that is reflected

$$r = \frac{q^r}{q} \tag{18.8}$$

- Transmissivity or transmission factor (t): Fraction of radiation received by a body that is transmitted

$$t = \frac{q^t}{q} \tag{18.9}$$

- Absorptivity or absorption factor (a): Fraction of radiation received by a body that is absorbed

$$a = \frac{q^a}{q} \tag{18.10}$$

The following expression can be easily obtained from the last equations: $r + t + a = 1$.

The emissivity of a body is defined as the ratio between the quantity of radiation emitted by such a body and a quantity that a black body could emit at the same temperature:

$$e = \frac{q^e}{q_n^e} = \frac{q^e}{\sigma T^4} \tag{18.11}$$

It should be pointed out that all of these properties have a mean value that is the same for all directions and wavelengths.

The following is assumed in those bodies in which all the energy that they receive is reflected: $r = 1$, $t = 0$, and $a = 0$; hence, they are called *perfect mirrors*. Those in which all the energy is transmitted ($r = 0$, $t = 1$, and $a = 0$) are known as *transparent bodies*, while those that absorb all the energy received ($r = 0$, $t = 0$, and $a = 1$) are denominated *black bodies*.

It is assumed that, in a black body, $q^e = q_n^e$. So, if this value is substituted in Equation 18.11, it is obtained that emissivity is equal to one. This means that in black bodies, emissivity and absorptivity are equal to one ($a = e = 1$).

18.3.2 Monochromatic Properties: Kirchhoff's Law

If the properties for each wavelength (λ) are considered instead of the mean value for all wavelengths (and independent of them), then the reflectivity, transmissivity, absorptivity, and emissivity are defined according to the following expressions:

$$r_\lambda = \frac{q_\lambda^r}{q_\lambda} \tag{18.12a}$$

$$t_\lambda = \frac{q_\lambda^t}{q_\lambda} \tag{18.12b}$$

$$a_\lambda = \frac{q_\lambda^a}{q_\lambda} \tag{18.12c}$$

The following is also observed with $r_\lambda + t_\lambda + a_\lambda = 1$.

Monochromatic emissivity can be defined in the same way as total emissivity:

$$e_\lambda = \frac{q_\lambda^e}{q_{b,\lambda}^e} \tag{18.13}$$

The mean value of each of these properties, as a function of the monochromatic properties, is obtained according to the following equations:

$$r = \frac{\int_0^\infty r_\lambda q_\lambda d\lambda}{\int_0^\infty q_\lambda d\lambda} \tag{18.14a}$$

$$t = \frac{\int_0^\infty t_\lambda q_\lambda d\lambda}{\int_0^\infty q_\lambda d\lambda} \tag{18.14b}$$

$$a = \frac{\int_0^\infty a_\lambda q_\lambda d\lambda}{\int_0^\infty q_\lambda d\lambda} \tag{18.14c}$$

$$e = \frac{\int_0^\infty e_\lambda q_{b,\lambda}^e \, d\lambda}{\int_0^\infty q_{b,\lambda}^e \, d\lambda} = \frac{1}{\sigma T^4} \int_0^\infty e_\lambda q_{b,\lambda}^e \, d\lambda \tag{18.14d}$$

Now, suppose a cavity at a temperature T that emits a flow density q, and inside the cavity there is a body with absorptivity a and emissivity e. When thermal equilibrium is reached, it is assumed that the energy absorbed and emitted by the body is equal. That is,

$$aq = q^e$$

If it is a black body, then $a = 1$, so $1\ q = q_b^e$. Thus, if these two equations are divided, then

$$a = \frac{q^e}{q_b^e}$$

When comparing this expression with Equation 18.11, one can see that it is the same, coinciding with the total emissivity ($a = e$). If monochromatic properties are used instead of total properties, then an analogous expression can be obtained ($a_\lambda = e_\lambda$).

As has already been introduced, it was stated that under thermal equilibrium in a black body, the emissivity is equal to the absorptivity, and this is known as Kirchhoff's law.

It should be pointed out that a body in which the monochromatic properties are constant for the whole wavelength range is called a *gray body*. Also, in gray bodies, it is always assumed that the absorptivity and emissivity are equal, even if they are not under thermal equilibrium.

18.3.3 Directional Properties

It is evident that a body that emits radiation does not necessarily have to do so in the same way in all directions. For this reason, the radiation properties of a body will be different depending on the direction considered.

If a point on a differential area dA that emits energy in a determined direction (Figure 18.1) is considered, then the solid angle ($d\omega$) displayed at such area is

$$d\omega = \frac{dA \cos\theta}{r^2} \tag{18.15}$$

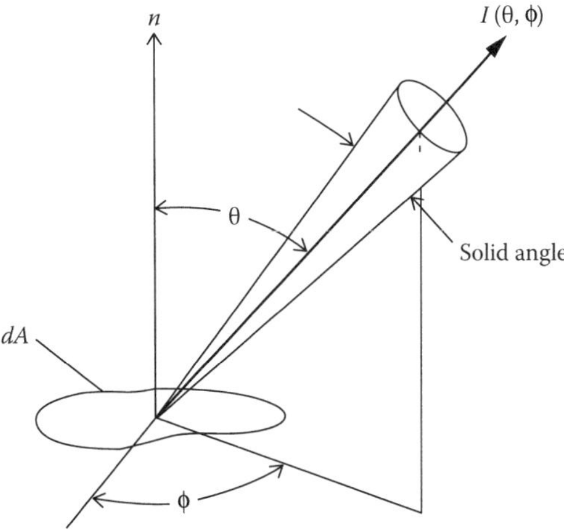

FIGURE 18.1 Solid angle displayed by dA.

in which θ is the angle formed by the director vector of surface dA and the vector of the direction from the emission point in such differential area, while r is the distance that separates a point from the area. If it is desired to obtain the solid angle displayed from the point to the whole space, then Equation 18.15 should be integrated for a whole sphere, obtaining that $\omega_{SPHERE} = 4\pi$.

It is convenient to define the intensity of radiation of an emitter, which depends on the direction considered. In this way, the intensity of radiation emitted by a point is defined as the radiant energy per unit time, unit solid angle, and unit area projected in a direction normal to the surface:

$$I(\theta,\phi) = \frac{dq}{\cos\theta\, d\omega} \quad (18.16)$$

where

dq is the flux emitted by the emitter
θ and ϕ are the angular coordinates that define the direction of emission
$d\omega$ is the solid angle

If it is desired to obtain the total flux emitted in all directions, the following integration should be conducted:

$$q = \int I(\theta,\phi)\cos\theta\, d\omega \quad (18.17)$$

In order to perform such integration, it is necessary to know the intensity of emission. Thus, if it is considered that the emitter behaves in such a way that the intensity of emission is the same in all directions, then the emission is *diffuse*. On the contrary, it is said that the emission is *specular* if its behavior is similar to a mirror, that is, it emits in specular directions with respect to the radiation received.

In the case of a diffuse emitter, the intensity of radiation is independent of the angular coordinates of emission. If the intensity of radiation is defined as $I = I(\theta, \phi)$, then the total flux emitted in all directions is

$$q = \int I(\theta,\phi)\cos\theta\, d\omega = \pi I \quad (18.18)$$

18.4 VIEW FACTORS

When the energy exchanged by any two bodies is studied, it is evident that the energy emitted by one is not completely absorbed by the other, since it can only absorb the portion it intercepts. For this reason, the view factors, also called direct view factors, shape factors, geometric factors, or angular factors, should be defined.

18.4.1 Definitions and Calculations

Consider two bodies of differential areas, dA_1 and dA_2 (Figure 18.2). The view factor of the first body with respect to the second is defined as the fraction of the total energy emitted by dA_1 that is intercepted by dA_2.

The following expression defines the view factor of dA_1 with respect to dA_2:

$$F_{dA_1,dA_2} = \frac{q_{1,2}^e}{q_1^e} = \frac{\cos\theta_1 \cos\theta_2 dA_2}{\pi r^2} \quad (18.19)$$

where θ_1 and θ_2 are the angles that each of the director vectors (n_1 and n_2) make with r, which is the straight line that connects the centers of the areas (dA_1 and dA_2).

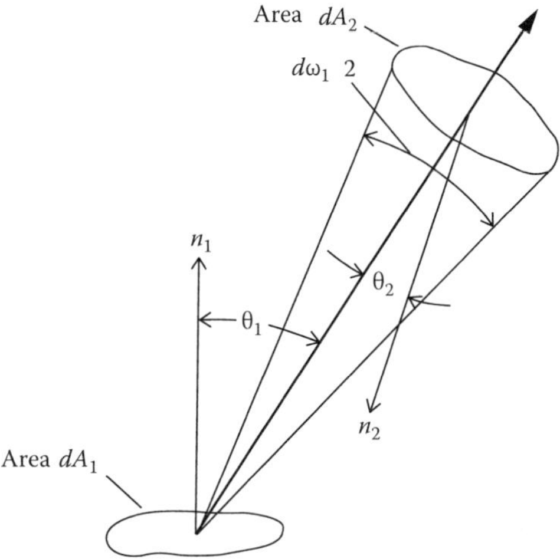

FIGURE 18.2 View factor.

If the second body is not a differential, but has an area A_2, the view factor between dA_1 and this area is calculated by the expression

$$F_{dA_1,A_2} = \int_{A_2} \frac{\cos\theta_1 \cos\theta_2 \, dA_2}{\pi r^2} \tag{18.20}$$

The view factor between two surfaces that have infinite areas is obtained from the expression

$$F_{A_1,A_2} = \frac{1}{A_1} \int_{A_1}\int_{A_2} \frac{\cos\theta_1 \cos\theta_2 \, dA_2 \, dA_1}{\pi r^2} \tag{18.21}$$

The mathematical calculation of view factors is difficult. However, graphs that facilitate this task can be found in the literature for specific cases (Figures 18.3 through 18.7).

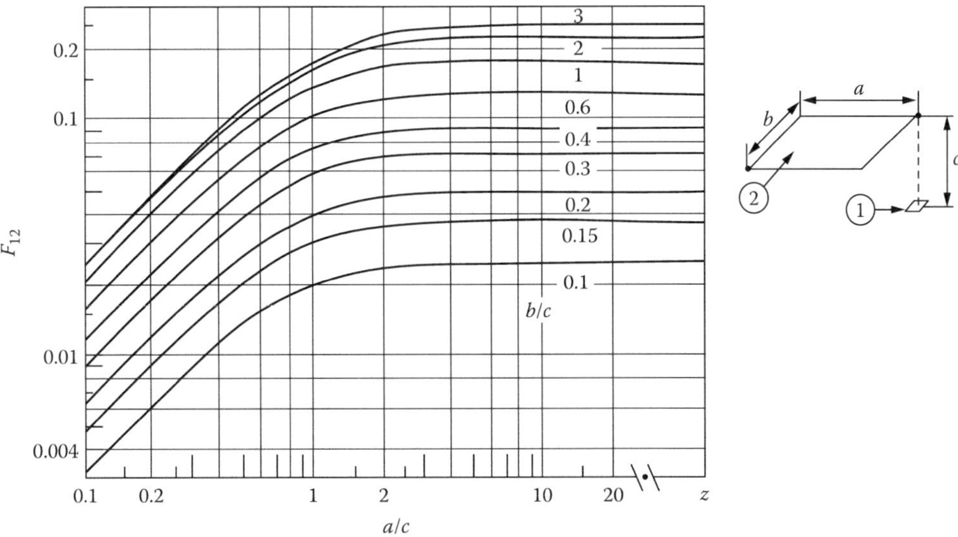

FIGURE 18.3 View factor from a differential area to a parallel rectangle. (Adapted from Costa, E. et al., *Ingeniería Química*. 4. *Transmisión de Calor*, Alhambra, Madrid, Spain, 1986b.)

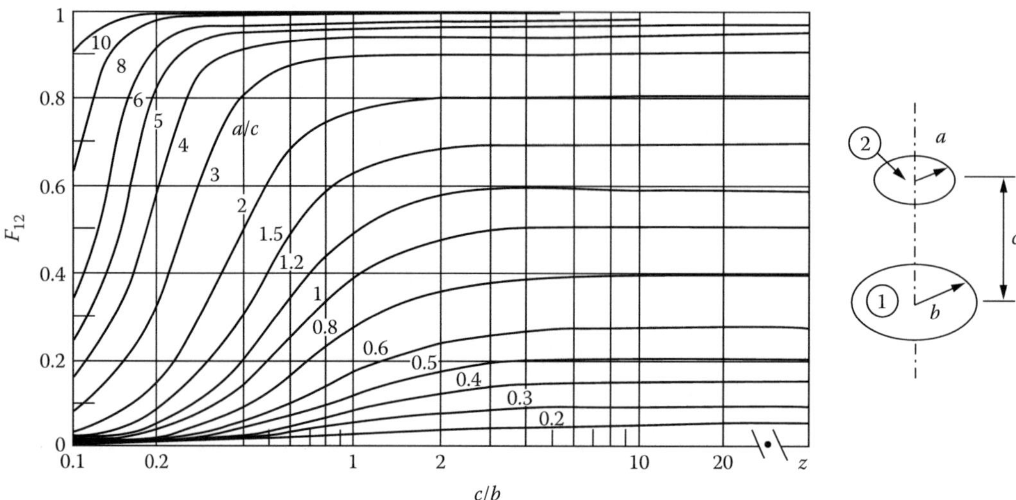

FIGURE 18.4 View factor between parallel circular surfaces. (Adapted from Costa, E. et al., *Ingeniería Química*. 4. *Transmisión de Calor*, Alhambra, Madrid, Spain, 1986b.)

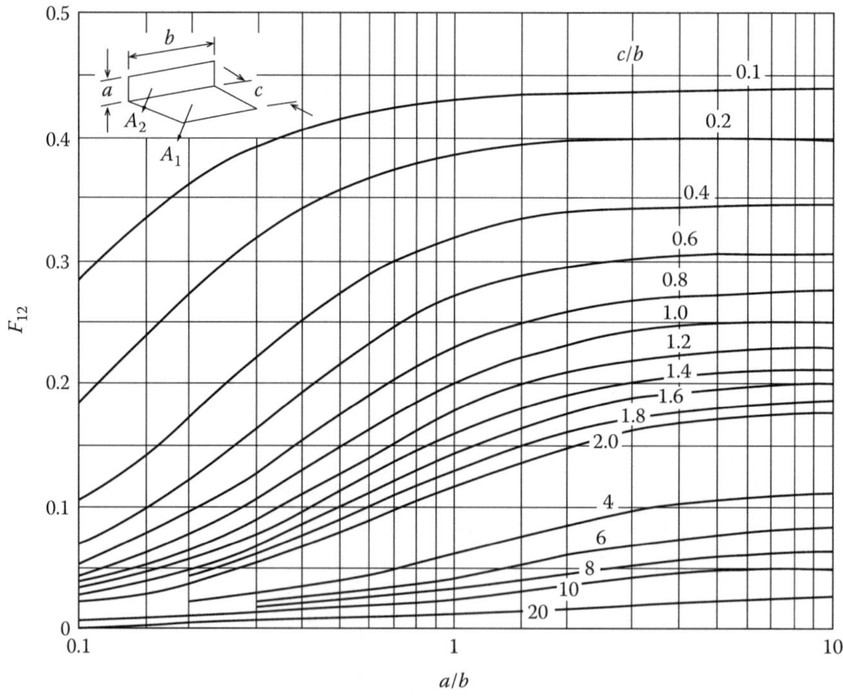

FIGURE 18.5 View factors for perpendicular rectangles with a common edge. (Adapted from Costa, E. et al., *Ingeniería Química*. 4. *Transmisión de Calor*, Alhambra, Madrid, Spain, 1986b.)

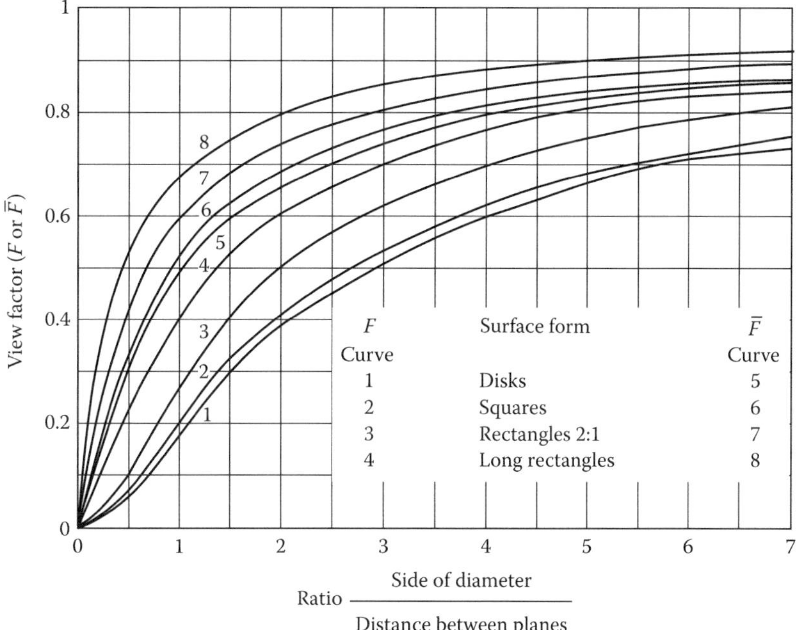

FIGURE 18.6 View and refractory factor between parallel planes. (Adapted from Costa, E. et al., *Ingeniería Química*. 4. *Transmisión de Calor*, Alhambra, Madrid, Spain, 1986b.)

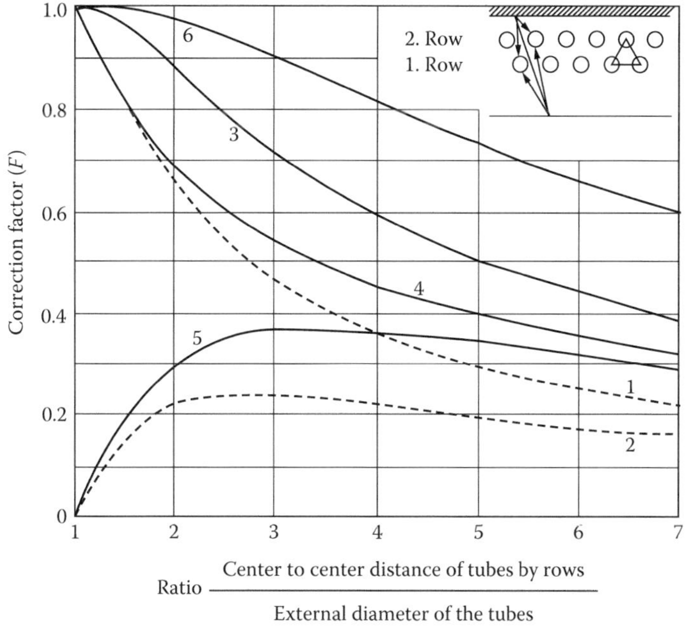

FIGURE 18.7 Correction factor F for radiation to a tube bundle: (1) Direct to the first row. (2) Direct to the second row. (3) Total to one row (one row). (4) Total to the first row (two rows). (5) Total to the second row (two rows). (6) Total to two rows (two rows). (Costa, E. et al., *Ingeniería Química*. 4. *Transmisión de Calor*, Alhambra, Madrid, Spain, 1986b.)

18.4.2 View Factor Properties

View factors have different properties that are listed next. These factors are useful for different applications to solve problems in which the exchange of radiant energy takes place:

a. *Reciprocity relationship.* For any two areas, it is assumed that the product of the area of the first body and the factor with which it views the second body are equal to the product of the area of the second body and the factor with which it views the first body:

$$A_i F_{i,j} = A_j F_{j,i} \tag{18.22}$$

b. *Conservation principle.* For a closed system formed by N surfaces, it is assumed that the addition of the geometric factors of a surface, with respect to all the surfaces that surround it, is equal to one, since the addition of all the fractions of energy emitted by such surface that are intercepted by the other surfaces is equal to one:

$$F_{i,1} + F_{i,2} + F_{i,3} + \cdots + F_{i,N} = 1 \tag{18.23}$$

For those bodies that cannot see themselves (plane or convex surfaces), the view factors with respect to themselves are null, since the radiation emitted cannot be intercepted by the body itself:

$$F_{i,i} = 0$$

The shape factor of a black surface completely surrounded by another black surface is equal to one, since all of the energy emitted by the first body is captured by the second body.

c. *Additive relationship.* The view factor of a single surface A_i with respect to a composite surface $A_{(jkl)}$ complies with

$$A_i F_{i,(jkl)} = A_i F_{i,j} + A_i F_{i,k} + A_i F_{i,l} \tag{18.24}$$

For example, in this case the second surface is composed by three other surfaces:

$$A_{(jkl)} = A_j + A_k + A_l$$

18.5 EXCHANGE OF RADIANT ENERGY BETWEEN SURFACES SEPARATED BY NONABSORBING MEDIA

Some cases of exchange of radiant energy between bodies where there is not a material medium separating them are presented next. That is, there is no possibility for the medium that separates two bodies to intercept the radiation emitted by one of the bodies.

18.5.1 Radiation between Black Surfaces

Suppose two black bodies with surface area A_1 and A_2 and at temperatures T_1 and T_2. The energy flow rate that is emitted from each of them and that is intercepted by the other one can be obtained from the expressions

$$\dot{Q}_{1,2} = F_{12} A_1 \sigma T_1^4$$

$$\dot{Q}_{2,1} = F_{21} A_2 \sigma T_2^4$$

The net flow rate received by each body is

$$\dot{Q}_{net} = F_{12}A_1\sigma\left(T_1^4 - T_2^4\right) \tag{18.25a}$$

$$\dot{Q}_{net} = F_{21}A_2\sigma\left(T_2^4 - T_1^4\right) \tag{18.25b}$$

Taking into account the reciprocity property, it can be seen that these expressions coincide, except that one will be positive and the other negative, depending on whether it is receiving or emitting net heat by radiation. Thus, if $T_1 > T_2$, then $\dot{Q}_1$ will be positive, meaning that the surface A_1 emits net radiation, while A_2 will have the same value but with a negative sign.

18.5.2 Radiation between a Surface and a Black Surface That Completely Surrounds the First One

Consider a body whose surface area A_1 is completely surrounded by a black surface A_2, T_1, and T_2 being the temperature of each of these bodies. The radiant energy emitted by each body and intercepted by the other body is

$$\dot{Q}_{1,2} = F_{12}A_1e_1\sigma T_1^4$$

$$\dot{Q}_{2,1} = F_{21}A_2a_1\sigma T_2^4$$

The net heat is

$$\dot{Q}_{net} = F_{12}A_1e_1\sigma T_1^4 - F_{21}A_2a_1\sigma T_2^4 \tag{18.26}$$

Taking into account the reciprocity property, and if surface area A_1 does not view itself ($F_{12} = 1$), the net heat emitted or captured by A_1 is

$$\dot{Q}_{net} = A_1\sigma\left(e_1T_1^4 - a_1T_2^4\right) \tag{18.27}$$

The following is assumed according to whether it was a gray body where $a = e$:

$$\dot{Q}_{net} = A_1\sigma e_1\left(T_1^4 - T_2^4\right)$$

In the case of a black body, the expression is simplified, since its emissivity is equal to one.

18.5.3 Radiation between Black Surfaces in the Presence of Refractory Surfaces: Refractory Factor

In general, in the design of industrial equipment (ovens, heaters, etc.), the radiation emitter and receptor are not alone inside the machinery, but often there is another type of surface called refractory constituting of the piece of equipment. Under steady state, refractory surfaces emit of all the radiation that they have absorbed, as long as there is no heat loss through the refractory surface. Therefore, there is no net heat flow associated with the radiation exchange in refractory surfaces.

A surface that emits all of the radiation it receives is called adiabatic. A refractory wall is adiabatic if all of the heat absorbed by radiation is emitted.

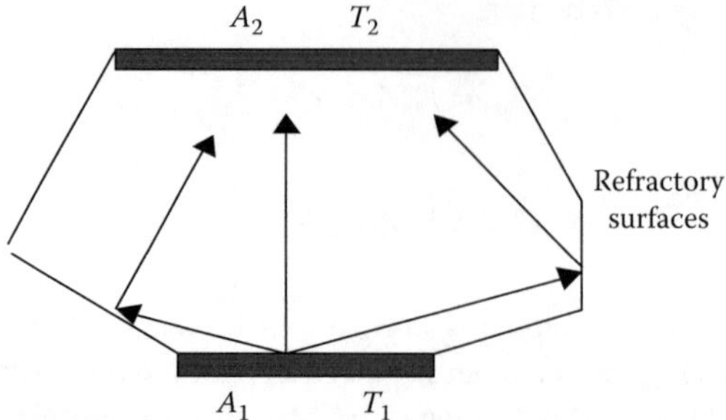

FIGURE 18.8 Radiation between two black surfaces separated by refractory surfaces.

Consider two surfaces A_1 and A_2 joined by refractory walls to study this case (Figure 18.8). Consider also that all of these surfaces maintain the same temperature. The transfer of radiant energy from A_1 to A_2, or vice versa, includes not only the direct transfer but also the heat transfer that comes from the refractory walls.

The heat emitted by A_1 that reaches A_2 is

$$\dot{Q}_{1,2} = F_{12}A_1\sigma T_1^4 + F_{1R}A_1\sigma T_1^4 \frac{F_{R2}}{1-F_{RR}}$$

where $F_{R2}/(1 - F_{RR})$ is a factor that represents the fraction of energy emitted by the refractory walls that reaches surface A_2.

A *refractory factor* $(\bar{F})$ between surfaces A_1 and A_2 is defined according to the expression

$$\bar{F}_{12} = F_{12} + \frac{F_{1R}F_{R2}}{1-F_{RR}} \tag{18.28}$$

Thus, the heat that comes out of one surface and that reaches the other surface will be expressed according to the equations

$$\dot{Q}_{1,2} = \bar{F}_{12}A_1\sigma T_1^4$$

$$\dot{Q}_{2,1} = \bar{F}_{21}A_2\sigma T_2^4$$

Therefore, the net heat flow rate exchanged by both surfaces

$$\dot{Q}_{net} = \bar{F}_{12}A_1\sigma T_1^4 - \bar{F}_{21}A_2\sigma T_2^4 \tag{18.29}$$

The refractory factors comply with the same properties as normal view factors, so

$$\dot{Q}_{net} = \bar{F}_{12}A_1\sigma\left(T_1^4 - T_2^4\right) \tag{18.30}$$

18.5.4 Radiation between Nonblack Surfaces: Gray Factor

In the case that the radiation energy is exchanged between bodies that are not black, it should be taken into account that the energy emitted by a body is calculated by the Stefan–Boltzmann equation multiplied by the emissivity. Also, the energy received by one of these bodies is affected by the absorption coefficient. All of these factors complicate the mathematical treatment, although in engineering it is assumed that the bodies are gray to simplify the problem. This is equivalent to considering that the absorption coefficient is independent of the wavelength of the incident radiation and therefore of the temperature and other characteristics of the emitter. The emissivity and absorption coefficient in gray bodies are equal.

For the case in which two surfaces A_1 and A_2 are joined by any number of refractory zones, the net radiant energy flow rate exchanged can be expressed according to Equation 18.31:

$$\dot{Q}_{net} = \Im_{12} A_1 \sigma \left(T_1^4 - T_2^4\right) \tag{18.31}$$

where $\Im_{12}$ is a shape factor, called *gray factor*, which depends on the view factors of the surfaces, on the refractory factor as well as on the emissivity and surface area of the bodies considered.

The gray factor is defined as

$$\Im_{12} = \frac{1}{(1/\bar{F}_{12}) + \left((1/e_1) - 1\right) + (A_1/A_2)\left((1/e_2) - 1\right)} \tag{18.32}$$

This equation allows us to obtain the value of the gray factor and is a general expression that in some cases may be simplified. Thus, for the case of two big parallel planes that exchange radiant energy, the refractory factor is equal to one, and the surface areas are equal, obtaining that

$$\Im_{12} = \frac{1}{(1/e_1) + (1/e_2) - 1}$$

18.6 RADIATION HEAT TRANSFER COEFFICIENT

In cases of heat transfer by conduction and convection, the heat flow rate is proportional to the temperature increase and to the transfer area, so the proportionality constant is the thermal conductivity in the case of heat conduction and the individual film coefficient in the case of heat convection. Thus,

- Heat conduction: $\dot{Q} = \dfrac{k}{e} A(T_0 - T_1)$
- Heat convection: $\dot{Q} = hA(T_P - T_f)$

In the case of radiation, the heat flow is calculated by the Stefan–Boltzmann equation. However, sometimes it is convenient to express it in a similar way to heat conduction and convection. In this way, the heat flow by radiation is

$$\dot{Q} = h_R A(T_1 - T_2) \tag{18.33}$$

in which h_R is called the coefficient of heat transfer by radiation and has the same units as film coefficients. Tables 18.1 and 18.2 show the values of the coefficient of heat transfer by radiation, along with convective coefficients, for the case of pipes whose surface can lose heat by both mechanisms.

In the specific case of black surfaces with a view factor equal to one, the net heat flow is given by

$$\dot{Q} = \sigma A(T_1^4 - T_2^4)$$

TABLE 18.1
Values of $(h_C + h_R)$ for Steel Pipes toward Their Surroundings[a]

	$(T_S - T_G)$ (°C)													
d_0 (cm)	10	25	50	75	100	125	150	175	200	225	250	275	300	325
2.5	11.03	11.66	12.68	13.76	14.96	16.29	17.08	19.18	20.92	22.61	24.44	26.34	28.29	30.28
7.5	10.00	10.54	11.01	12.49	13.66	14.87	16.13	17.87	19.93	21.12	22.92	24.77	26.67	28.67
12.5	9.51	10.04	10.93	11.91	13.12	14.44	15.57	17.18	18.79	20.49				
25.5	9.13	9.65	10.58	11.60	12.75	14.04	15.17	16.80	18.50					

Source: Costa, E. et al., *Ingeniería Química. 4. Transmisión de Calor*, Alhambra, Madrid, Spain, 1986b.

[a] Units of $(h_C + h_R)$ kcal/(h m^2 °C).

TABLE 18.2
Values of (h_C + h_R) for Steel Pipes toward Their Surroundings[a]

	(T_S − T_G) (°F)														
d_0 (in.)	**30**	**50**	**100**	**150**	**200**	**250**	**300**	**350**	**400**	**450**	**500**	**550**	**600**	**650**	**700**
1	2.16	2.26	2.50	2.73	3.00	3.29	3.60	3.95	4.34	4.73	5.16	5.60	6.05	6.51	6.99
3	1.97	2.05	2.25	2.47	2.73	3.00	3.31	3.69	1.03	4.73	4.85	5.26	5.71	6.19	6.08
5	—	1.95	2.15	2.36	2.61	2.90	3.20	3.54	3.90						
10	1.80	1.87	2.07	2.29	2.54	2.80	3.12	3.47	3.84						

Source: Perry, R.H. and Chilton, C.H., *Chemical Engineer's Handbook*, New York, McGraw-Hill, 1973.
[a] Units of (h_C + h_R) Btu/(h ft² °F).

When comparing this expression with Equation 18.33, an expression to calculate the transfer coefficient by radiation can be obtained:

$$h_R = \sigma \frac{T_1^4 - T_2^4}{T_1 - T_2} = \sigma\left(T_1^3 + T_1^2 T_2 + T_1 T_2^2 + T_2^3\right) \tag{18.34}$$

In the case of radiation between gray surfaces or surfaces whose view factor is different from one, the appropriate equations should be taken into account.

18.7 SIMULTANEOUS HEAT TRANSFER BY CONVECTION AND RADIATION

Heat transfer, in practice, occurs by more than one mechanism at the same time. Thus, in the case of heat transfer from a hot surface to the exterior, convection and radiation perform such transmission simultaneously. Consider a hot surface at a temperature T_S, which is surrounded by a fluid at a temperature T_G, T_W being the temperature of the walls. The heat transfer mechanisms are radiation and convection, so the heat flow transferred from the hot surface will be the sum of heat transferred by radiation plus the heat transferred by convection:

$$\dot{Q}_{TOTAL} = \dot{Q}_R + \dot{Q}_C$$

$$\dot{Q}_R = h_R A(T_S - T_W)$$

$$\dot{Q}_C = h_C A(T_S - T_G)$$

Hence,

$$\dot{Q}_{TOTAL} = h_R A\left(T_S - T_W\right) + h_C A\left(T_S - T_G\right)$$

In the case where the temperature of the fluid T_G is the same as the temperature of the wall,

$$\dot{Q}_{TOTAL} = \left(h_R + h_C\right) A\left(T_S - T_W\right) \tag{18.35}$$

The values of the coefficients h_R and h_C should have been calculated previously. The individual coefficient of heat transfer by convection h_C is calculated from graphs or equations obtained in an empirical way and based on a dimensional analysis. The coefficient h_R can be obtained from graphs or equations, as indicated in the previous section.

PROBLEMS

18.1 The use of solar energy to heat an airstream that will be used to dry barley is studied in a seed experimental center. Hence, a black solar collector (5 m × 10 m) was placed on the roof of the building. The surroundings can be considered as a black body that has an effective radioactive temperature of 32°C. Heat losses by conduction to the outside can be neglected, while losses due to convection can be evaluated from the individual coefficient of heat transfer, which in turn is calculated according to the expression $h = 2.4\ (\Delta T)^{0.25}$ W/(m^2 K). In this equation ΔT is the difference of temperatures between the surface of the collector and its surroundings.

If the incident solar energy produces in the collector a radiant flux of 800 W/m^2, calculate the temperature at which the thermal equilibrium in the collector is reached.

The temperature at which thermal equilibrium is reached is obtained when all of the heat that reaches the collector is equal to all of the heat that leaves the collector. Heat inlet will be by direct radiation of the sun and by the radiant energy coming from the surroundings, whereas heat by radiation and by convection will come out of the collector.

Suppose that the temperature of the collector at which equilibrium is reached is T_1 and the temperature of the surroundings is $T_2 = 32°\text{C}$ and that the area of the collector is A_1 and A_2 is the area of the surroundings. The energy balance in the collector leads to the expression

$$q_{sun}A_1a_1 + a_1F_{21}A_2e_2\sigma T_2^4 = a_2F_{12}A_1e_1\sigma T_1^4 + A_1h\left(T_1 - T_2\right)$$

The solar collector and the surroundings are black bodies, so their emissivities (e) and the absorption coefficient (a) have a value equal to one. The surroundings completely encircle the collector, so if the reciprocity property is applied, $F_{12}A_1 = F_{21}A_2$, and if it is supposed that the solar collector has a plane area value, $F_{12} = 1$ is obtained. Substitution in the last equation yields

$$q_{sun} + \sigma T_2^4 = \sigma T_1^4 + h\left(T_1 - T_2\right)$$

Rearrangement of terms and substitution of data give

$$800\ \text{W/m}^2 = \left(5.67\times10^{-8}\ \ \text{W/(m}^2\ \text{K}^4)\right)\left(T_1^4 - 305^4\right)\text{K}^4 + 2.1\left(T_1 - 305\right)^{1.25}\ \text{W/m}^2$$

This equation should be solved by iteration, obtaining that $T_1 = 359$ K (= 86°C).

18.2 It is desired to grill some meat fillets that have an emissivity of 0.45, using a grill made of a 50 cm × 90 cm metallic base, a grate placed 25 cm above the base and the opening between the grate and the base is completely closed by refractory sheets. Fillets are placed in a way such that they cover the whole grate, while vegetal charcoal is placed on the base, estimating that their absorptivity and emissivity are 1 and 0.85, respectively.

The temperature of charcoal during the grilling process is 800°C and the temperature of the environment is 25°C. Heat exchanged by convection between charcoal and fillets is 600 W. If heat transfer from the bottom part of the charcoal and the upper side of the fillets toward the surroundings is neglected, determine the temperature acquired by the fillets under thermal equilibrium.

When carrying out a heat balance in the meat, it is obtained that the heat that enters by radiation and convection from the charcoal is equal to the heat that exits the meat by radiation:

$$a_2\bar{F}_{12}A_1e_1\sigma T_1^4 + \dot{Q}_C = a_1\bar{F}_{21}A_2e_2\sigma T_2^4$$

Subscripts 1 and 2 refer to the charcoal and fillets, respectively. Due to the geometry of the system, $A_1 = A_2$ and $\overline{F}_{12} = \overline{F}_{21}$. Also, $a_1 = 1$, since charcoal is considered as a black body, and the fillets as gray ($a_2 = e_2$). Taking this into account, it is obtained that the temperature of the fillets is expressed according to the equation

$$T_2 = \sqrt[4]{e_1 T_1^4 + \frac{\dot{Q}_C}{\overline{F}_{12} A_1 e_2 \sigma}}$$

The refractory view factor $\overline{F}_{12}$ is calculated from Figure 18.6, where curve 7 is taken for a rectangle 2:1:

$$\text{Ratio} = \frac{\text{Side length}}{\text{Distance between planes}} = \frac{90 \text{ cm}}{25 \text{ cm}} = 3.6$$

Hence, $\overline{F}_{12} = 0.82$

Substitution of data yields

$$T_2 = \sqrt[4]{(0.85)(1073)^4 \text{ K}^4 + \frac{600 \text{ W}}{(0.82)(0.45 \text{ m}^2)(0.45)\left(5.67\times 10^{-8} \text{ W(m}^2 \text{ K}^4)\right)}}$$

$T_2 = 1044.6$ K (= 771.6°C).

18.3 Saturated steam at 2.1 kgf/cm^2 circulates inside a tube of 3 cm of external diameter and 2.5 mm of wall thickness, so the temperature outside the wall of the tube is 120°C. To avoid heat losses to the exterior that could cause steam to condense, the pipes are insulated with a 4 cm thick insulator, whose thermal conductivity is 0.1 kcal/(h m °C). If the pipe is inside a room at 25°C, calculate (a) the amount of heat dissipated per meter of pipe, when it is not insulated, (b) the temperature of the external surface of insulator that covers the pipe, and (c) the percentage of heat loss is saved by the insulation.

a. For the non-insulated pipe, the heat dissipated to the exterior is due to simultaneous convection and radiation, the exchange area being $A_e = \pi d_e L$. The heat flow rate exchanged per meter of pipe is

$$\frac{\dot{Q}}{L} = \pi d_e (h_C + h_R)(T_W - T_G)$$

where $T_W - T_G = 120°\text{C} - 25°\text{C} = 95°\text{C}$.

For the calculation of ($h_c + h_R$), it should be interpolated in Table 18.1, obtaining a value of

$$\left(h_C + h_R\right) = 14.59 \text{ kcal/(h m}^2 \text{ °C)}$$

Therefore, the heat flow rate per meter of pipe exchanged with the surroundings is

$$\frac{\dot{Q}}{L} = \pi(0.03 \text{ m})(14.59 \text{ kcal/(h m}^2 \text{ °C))}(95°\text{C}) = 130.6 \text{ kcal/(h m)}$$

b. When the pipe has an insulating cover, the heat per meter of pipe that flows through the insulation is the same amount that is dissipated to the outside by radiation and convection:

$$\frac{\dot{Q}}{L} = \frac{\pi\left(d_0 - d_e\right)}{\ln\left(d_0 / d_e\right)} \frac{k}{e_A}\left(T_W - T_0\right)$$

$$= \pi d_0\left(h_C + h_R\right)\left(T_W - T_G\right)$$

where d_0 and d_e are the external diameters of the pipe and the insulator ($d_0 = d_e + 2e_A$), e_A being the thickness of the insulator. T_P and T_0 are the temperatures of the external wall of the pipe and of the insulator, respectively.
Substitution yields

$$\frac{\pi(11-3)\times 10^{-2}\text{ m}}{\ln\left(11/3\right)} \frac{0.1\text{ kcal/(h m °C)}}{0.04\text{ m}}\left(120 - T_0\right)\text{°C} = \pi(0.11\text{ m})\left(h_C + h_R\right)\left(T_W - 25\right)\text{°C}$$

Rearranging,

$$T_0 = \frac{168 + 25\left(h_C + h_R\right)}{1.4 + \left(h_C + h_R\right)}$$

This equation is solved by iteration, so T_0 is supposed and $(T_0 - T_G)$ is calculated. This value is used to determine $(h_C + h_R)$ from Table 18.1, which in turn allows calculation of the temperature T_0 using the last equation. Thus, the following is obtained:

$$T_0 = 37\text{°C}$$

$$(h_C + h_R) = 9.73\text{ kcal/(h m}^2\text{ °C)}$$

Using these data, the heat flow lost per meter of pipe is

$$\frac{\dot{Q}}{L} = \pi\left(0.11\text{ m}\right)\left(9.73\text{ kcal/(h m}^2\text{ °C)}\right)\left(37 - 25\right)\text{°C} = 40.4\text{ kcal/(h m)}$$

c. When observing the heat losses per unit time and meter of pipe in the last two sections, it is easy to obtain that the percentage of loss saved with insulation is 69%.

18.4 It is desired to heat 10,000 kg of a product, using an electric oven whose base has the following dimension 3 m × 4 m. The roof of the oven contains a row of electric resistors of cylindrical shape (4 m in length and 2 cm in diameter) so that the distance between their centers is 6 cm. These resistors are contained in a plane at 2.5 m above the product to be heated, having a temperature of 1500°C. If the oven has refractory walls that are perfectly insulated from the exterior, determine the time needed to heat the product from 20°C to 500°C.

Data: emissivity of the resistance, 0.70; emissivity of the product, 0.90; specific heat of the product, 1.046 kJ/(kg°C).

First, the heat flow from the electric resistance to the product is calculated. Since the walls of the oven are refractory, T_1 is the temperature of the resistance and T the temperature of the product, then according to Equation 18.31,

$$\dot{Q}_{1P} = \Im_{1P} A_1 \sigma \left(T_1^4 - T^4\right)$$

in which $\Im_{1P}$ is the gray factor defined by Equation 18.32. The refractory factor that appears in this equation is calculated as if they were two parallel surfaces separated by refractors, although it should be corrected with a factor F, since actually the emission is by the resistance.

It is supposed that there exists an imaginary plane parallel to the product at 2.5 m. This implies calculation of the geometric factor between 3×4 m^2 parallel planes separated 2.5 m.

Using the ratio, smaller side/distance between planes = 3/2.5 = 1.2 and using Figure 18.6 curve 7, it is obtained that $\overline{F} \cong 0.64$.

The correction factor F is obtained from Figure 18.7 (radiation to a tube bundle). The radiation plane with respect to the tubes is supposed as

$$\frac{\text{Distance between the center of the tubes}}{\text{External diameter of the tubes}} = \frac{6\text{ cm}}{2\text{ cm}} = 3$$

Using this value in the abscissas and the total curve at one row (curve 3), it is obtained that $F = 0.72$

The value of the refractory factor is $\overline{F}_{1P} = (0.64)(0.72) = 0.46$

The areas are $A_1 = A_P = 12$ m^2. The gray factor is obtained by Equation 18.32:

$$\Im_{12} = \frac{1}{(1/0.46) + \left((1/0.7) - 1\right) + (12/12)\left((1/0.9) - 1\right)} = 0.369$$

The value of the heat flow rate exchanged by the resistance and the product is calculated using Equation 18.31:

$$\dot{Q}_{1P} = (0.369)(12\text{ m}^2)\left(5.67 \times 10^{-8}\text{ W/(m}^2\text{ K}^4)\right)(1773^4 - T^4)\text{K}^4$$

$$\dot{Q}_{1P} = 2.51 \times 10^{-7}\left(1773^4 - T^4\right)\text{W}$$

The heat irradiated by the resistance goes to the product that accumulates this heat by increasing its temperature. When performing an energy balance in the product, it is obtained that

$$\dot{Q} = m\hat{C}_P \frac{dT}{dt} = 2.51 \times 10^{-7}\left(1773^4 - T^4\right)\text{W}$$

where m is the mass of the product. Substitution of data given in the statement of the problem yields

$$(10{,}000\text{ kg})\left(1.046 \times 10^3\text{ J/(kg °C)}\right)\frac{dT}{dt} = 2.51 \times 10^{-7}\left(1773^4 - T^4\right)\text{W}$$

This expression allows us to obtain an equation in separated variables to calculate time:

$$t = 4.17 \times 10^{13} \int_{293}^{773} \frac{dT}{\left(1773^4 - T^4\right)}$$

Integration yields

$$t = \frac{4.17 \times 10^{13}}{2 \times 1{,}773^3} \left[\frac{1}{2} \ln\left(\frac{1{,}773 + T}{1{,}773 - T} \right) + arctg\left(\frac{T}{1{,}773} \right) \right]_{293}^{773} = 53{,}700 \text{ s}$$

Therefore, 14 h and 55 min are required to carry out the heating process indicated in the problem.

PROPOSED PROBLEMS

18.1 Obtain an equation that allows calculating the geometric factor between a differential area dA_1 and a circle (A_2) of radius R, in a parallel plane perfectly centered regarding the area dA_1. Between the differential area differential dA_1 and the geometric center of the area A_2, there is a distance d.

18.2 At the top of a building there is a terrace whose surface possesses an emissivity of 0.95 and an absorption coefficient of 0.9 for the sun radiation. The heat flux sun rays on the terrace is 800 W/(m^2), while the ambient air is at 30°C. The heat convection coefficient between terrace–air can be expressed as $h = 2.2\ (\Delta T)^{1/4}$ W/(m^2 °C), being ΔT the temperature difference between the terrace and the air expressed in °C. Assuming that there is no heat conduction toward the terrace, calculate the equilibrium temperature in the terrace surface.

18.3 There is a cylinder where the bases can be considered as black bodies and the lateral area is a refractory surface. Determine the expressions that allow calculating all geometric factors F_{ij}. Obtain an equation that allows the calculation of net energy exchanged between the two bases of the cylinder.

18.4 The walls of a bakery oven are built of refractory brick and they possess 30 cm of thickness. In order to observe the interior of the oven, a circular peephole of 15 cm diameter has been built. When the oven works in stationary state, the peephole surface temperature is 1100°C, while the external temperature is 20°C. If it is assumed that the peepholes act as black surfaces, calculate the heat losses that take place through the peephole.

18.5 Saturated water steam, at 4 kgf/cm^2, is flowing through a 10 cm of external diameter metallic pipe. The air room temperature is 25°C, and the resistances to heat transmission that the metallic wall and the condensed layer offer can be assumed negligible. To avoid great heat losses, the pipe is covered:

A. With 2 cm thickness of an insulating material with thermal conductivity is 0.40 W/(m °C) and its emissivity is 0.95.
B. With 1 cm thickness of an insulating material, thermal conductivity of 0.30 W/(m °C), and a low emissivity

Determine the following: (a) external temperature of the insulating material for both cases. (b) To avoid the heat losses, which of the two options A or B is better? (c) Water steam mass flow rate at which it can condense, if the total pipe length is 100 m.

19 Refrigeration
Chilling and Freezing

19.1 INTRODUCTION

Refrigeration can be defined as the process in which heat is eliminated from a material that is at a higher temperature than its surroundings. In general, refrigeration of a food product is a term used to denominate storage at temperatures lower than 15°C and above the freezing point. Refrigeration has been applied to food preservation for centuries. The so-called natural refrigeration has been used since ancient times taking advantage of the fact that snow, ice, and brines allow temperatures lower than where the food is at. Mechanical refrigeration, where both mechanical and electric devices are used to achieve temperatures lower than the surroundings, is discussed in this chapter.

The base of mechanical refrigeration is that heat is transferred from a region of low temperature to another region of higher temperature. To achieve such heat being transferred from a cold region toward a hotter one, it is necessary to add external work, which is generally accomplished through a compressor. The devices that carry out the refrigeration are denominated refrigerators, and they are based on the thermodynamic cycle that a refrigerant fluid follows. This thermodynamic cycle has two important applications, as part of refrigeration cycles and as heat pumps. In both cases, they operate in the same way, transferring heat from a cold region to a hotter one; however, the purpose is different. As refrigerator, the objective is to take heat out from the cold region, while in the heat pump, the objective is to introduce heat in the hot region (Figure 19.1). The refrigeration cycles are those where the refrigerant vaporizes and condenses alternately, and the external work is provided at the time of compression the refrigerant in its vapor state.

To evaluate the effectiveness of those refrigerators and heat pumps, the yield coefficient is used. It is defined in a different way depending on the objective that it is pursued in each case. For refrigeration, the refrigeration coefficient (ϕ_R) is defined as the quotient between the eliminated heat of the cold region and the network applied on the system, while for the heat pump, the heating coefficient (ϕ_H) is defined as the relationship between the heat given to the hot region and the network on the system

$$\text{Refrigeration coefficient } \phi_R = \frac{Q_F}{W_{net}} \tag{19.1}$$

$$\text{Heating coefficient } \phi_H = \frac{Q_H}{W_{net}} \tag{19.2}$$

Both coefficients can take higher values to the unit, and for processes without irreversibility, it is accomplished that

$$\phi_H = \phi_R + 1 \tag{19.3}$$

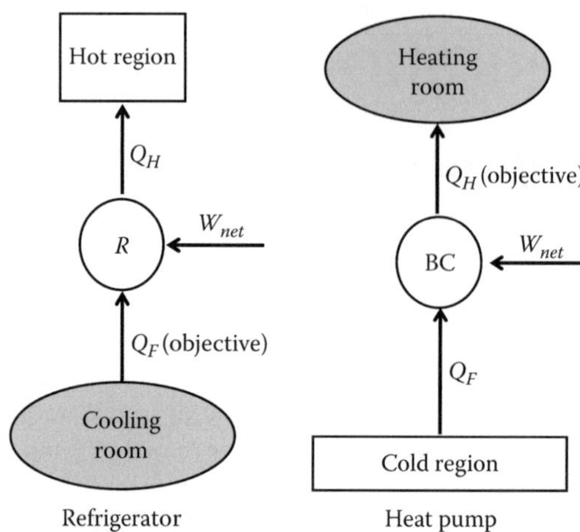

FIGURE 19.1 Scheme for refrigerator and heat pump.

19.2 REFRIGERANTS

Various fluids are used as refrigerant fluids. The fluids used in vapor compression systems are called primary refrigerants, while secondary refrigerants are those used for transportation at low temperatures from one place to another.

Generally, solutions with freezing temperatures below 0°C are used as secondary refrigerants. Aqueous solutions of ethylene glycol, propylene glycol, and calcium chloride are the most commonly used ones. The properties of these solutions are similar, although propylene glycol has the advantage of being innocuous if it comes into contact with food.

There are many primary refrigerants that have been used and are currently used in industry. Table 19.1 shows a list of some of these refrigerants. For the hydrocarbons and halocarbons, the numerical designation is determined as follows: The first digit from right to left is given by the number of fluorine atoms present in the molecule, the second digit (the middle) corresponds to the number of hydrogen atoms in the molecule plus one, and the third digit corresponds to the number of carbon atoms minus one. For inorganic compounds, the last two digits represent their molecular weight.

Selection of the type of refrigerant depends on the process in which the refrigerant will be used. Sometimes leaks can occur in a refrigeration system, and if the refrigerant comes into contact with the food, it can be contaminated. Thus, if a food is allowed to be in contact with ammonia for a long time, its taste and aroma may be affected. Although halocarbons do not cause serious problems when in contact with foods (Stoecker and Jones, 1982), these compounds produce deleterious effects on the atmospheric ozone layer; therefore, their use is limited, and some of them have been banned.

For instance, to choose a refrigerant, its certain properties should be kept in mind, such as chemical stability and lack of toxicity; it has to be neither corrosive nor inflammable and reasonably economic and also has a high vaporization enthalpy, which will allow working with a low flow of refrigerant.

Besides all the mentioned characteristics, the two fundamental parameters to choose a refrigerant are the two mean temperatures that interact with the refrigerant, that is to say, the temperatures of the hot and cold regions. In the condenser, the temperature of the refrigerant should be between 5°C and 10°C above the temperature of the hot region (T_H); meanwhile, in the evaporator, the temperature of the refrigerant should be between 5°C and 10°C lower than the temperature of the cold region (T_C). Setting these temperatures, it implies that the condenser and evaporation operation pressures are fixed.

In relation to the pressure that operates on the evaporator, it should be clear that it is convenient to work at the same pressure or greater than the atmospheric one, since if it is working at lower

TABLE 19.1
Some Types of Refrigerants

Type	Number	Chemical Name	Formula
Halocarbons	11	Trichlorofluoromethane	CCl_3F
	12	Dichlorodifluoromethane	CCl_2F_2
	13	Chlorotrifluoromethane	$CCLF_3$
	22	Chlorodifluoromethane	$CHClF_2$
	40	Chloromethyl	CH_3Cl
	113	Trichlorotrifluoroethane	CCl_2FCClF_2
	114	Dichlorotetrafluoroethane	$CClF_2CClF_2$
	134	Tetrafluoroethane	CH_2FCF
Hydrocarbons	50	Methane	CH_4
	170	Ethane	C_2H_6
	290	Propane	C_3H_8
Inorganic	717	Ammonia	NH_3
	718	Water	H_2O
	729	Air	—
	744	Carbon dioxide	CO_2
	764	Sulfur dioxide	SO_2

Source: Stoecker, W.F. and Jones, J.W., *Refrigeration and Air Conditioning*, McGraw-Hill Book Company, New York, 1982.

pressures, a danger of air entrance in the refrigerant circuit could exist. In this way, for example, if it is desired to keep the room at −5°C, the temperature of the refrigerant in the evaporator should be at least −15°C, for which a refrigerant should be chosen with a minimum saturation pressure at −15°C of 1 atm. In relation to the condenser, it is convenient that at pressures below the critical pressure of the refrigerant, the saturation temperature should be greater than the surrounding temperatures.

19.3 REFRIGERATION MECHANICAL SYSTEMS

The second law of thermodynamics indicates that heat will flow only in the direction of decreasing temperature. However, industrial refrigeration processes have the objective to eliminate transfer heat from low-temperature points to higher-temperature ones. In order to achieve this objective, the denominated refrigeration cycles with a circulating fluid in different stages are used. The most important refrigeration system is vapor compression. A simple scheme of this cycle is given in Figure 19.2. Vapor compression is a closed cycle in which the flowing liquid is called refrigerant. Supposing that the cycle begins at point 1, the compressor suction point, the fluid that is in a vapor state receives energy from the compressor, and it goes to point 2 through a polytropic compression in which the fluid increases its pressure and temperature and at the same time its enthalpy. This fluid decreases its energy content in a condenser, by an isobaric process, passing to a liquid state (point 3). Then it flows through an expansion valve, in an isentropic process, with a decrease in pressure (point 4), obtaining a liquid–gas mixture. This mixture changes to a saturated vapor state in an evaporator where the fluid receives heat, changing to an isobaric process and then changing to the conditions of point 1, where the cycle begins again. It should be pointed out that the global system takes heat from the environment in the evaporator and releases heat in the condenser.

The global process can be represented in a pressure–enthalpy diagram (Figure 19.3) or in a temperature–entropy diagram (Figure 19.4). The fluid receives the compression work during compression stage 1–2, and perhaps it is the most expensive step of all the system. This compression

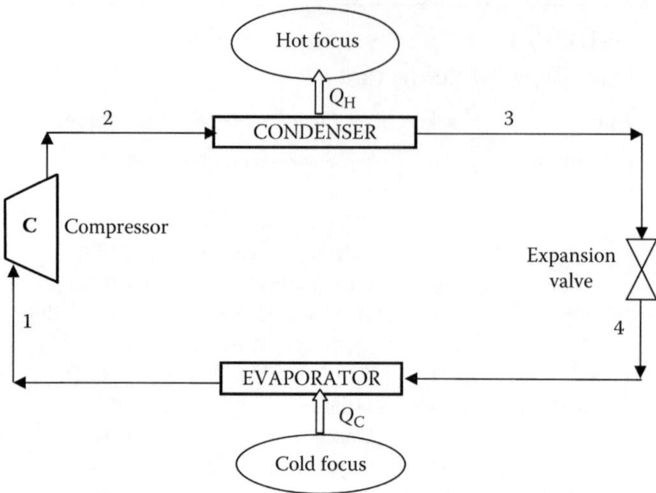

FIGURE 19.2 Refrigeration cycle system.

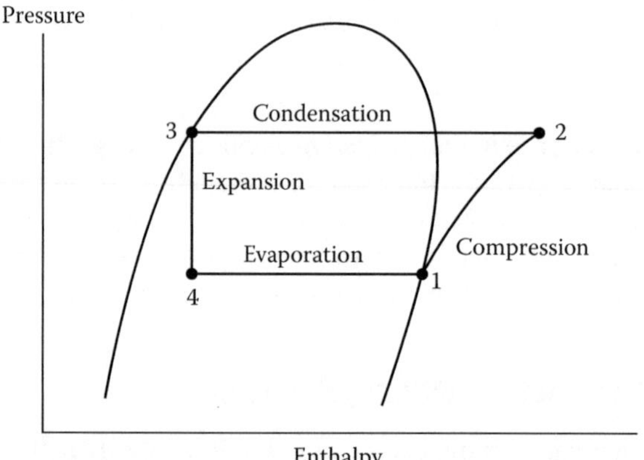

FIGURE 19.3 Vapor compression cycle, *P–H* diagram.

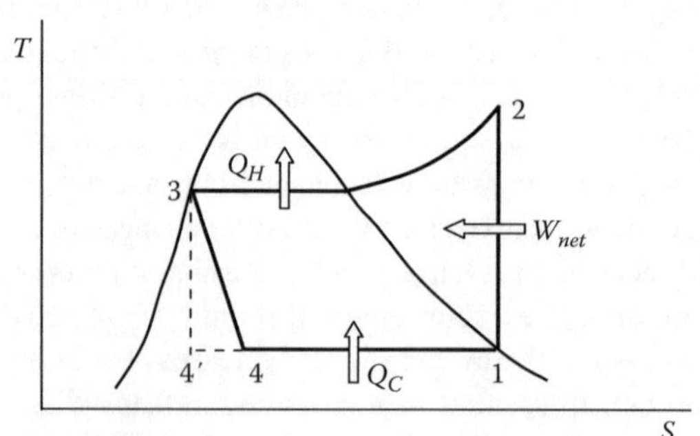

FIGURE 19.4 Vapor compression cycle, *T–S* diagram.

work can be evaluated by the enthalpy difference between the discharge and suction points of the compressor. Thus, the compression work per unit mass of fluid is

$$\hat{W}_C = \hat{H}_2 - \hat{H}_1 \qquad (19.4)$$

If it is desired to obtain the theoretical power of the compressor, work should be multiplied by the mass flow of the fluid that circulates through the system:

$$Pot = w\hat{W}_C = w\left(\hat{H}_2 - \hat{H}_1\right) \qquad (19.5)$$

As it can be observed, the circulation flow of the fluid determines to some extent the size of the compressor to be used, since the compression power required will be higher as the flow becomes greater.

The amount of energy released by the fluid in the condenser is determined from the enthalpies of points 2 and 3. In this way, the heat flow released by the fluid is

$$\dot{Q}_H = w\left(\hat{H}_3 - \hat{H}_2\right) \qquad (19.6)$$

where

w is the mass flow rate of the fluid
$\hat{H}$ the enthalpy per unit mass of fluid

At point 2, where the fluid leaves the compressor, the enthalpy is greater than in point 3, and therefore heat with negative sign is obtained. Hence, this heat is released by the fluid, and its value is used for the sizing of the condenser and the calculation of the amount of cooling fluid to be used.

Evaporation stage 4–1 represents the refrigerant effect of the system. At this stage, the heat transferred from the environment to the fluid is

$$\dot{Q}_C = w\left(\hat{H}_1 - \hat{H}_4\right) \qquad (19.7)$$

This heat is absorbed by the fluid and represents the refrigerant capacity, which is the essential aim of the process.

There is an important parameter in these processes, the *coefficient of performance* (ϕ), defined as the refrigerant effect divided by the work externally provided. Since this work is only that of compression, the performance coefficient of the refrigeration system is

$$\phi_R = \frac{\hat{H}_1 - \hat{H}_4}{\hat{H}_2 - \hat{H}_1} \qquad (19.8)$$

The power per kilowatt of refrigeration is the inverse of the coefficient of performance: an efficient refrigeration system should have a low-power value but a high-performance coefficient value (Stoecker and Jones, 1982).

In some refrigeration systems, the fluid that feeds the compressor is overheated steam, ensuring in this way that no refrigerant in a liquid state enters into the compressor. In order to accomplish this, the saturated vapor that leaves the evaporator passes through a heat exchanger, in which the fluid that gives heat is the flow that leaves the condenser. Besides the cited effect, the liquid of the condenser is subcooled, avoiding the formation of bubbles before it passes through the expansion valve. Figure 19.5 shows a scheme of the process, while in Figure 19.6, a cycle in T–S diagram is shown.

It seems that in the real cycles, the appearance of irreversibilities is generally due to the refrigerant friction, causing a decrease of pressure. Because of this, the irreversibility effects take place in the

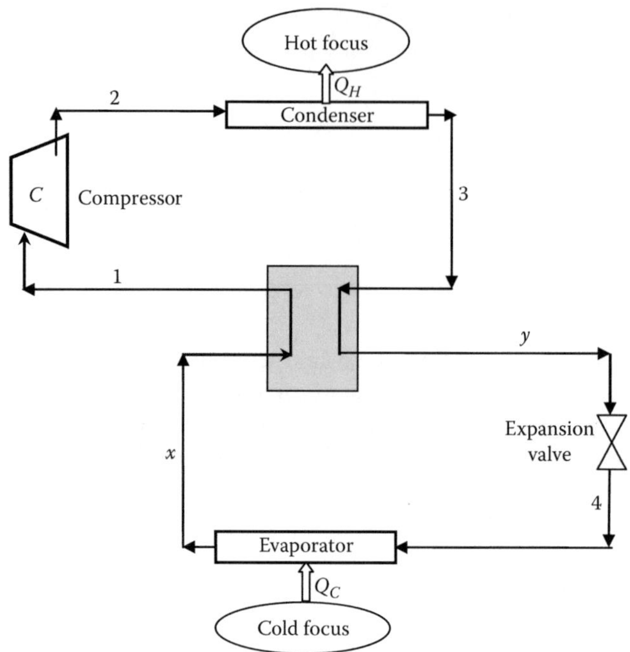

FIGURE 19.5 Refrigeration installation scheme with heat exchanger.

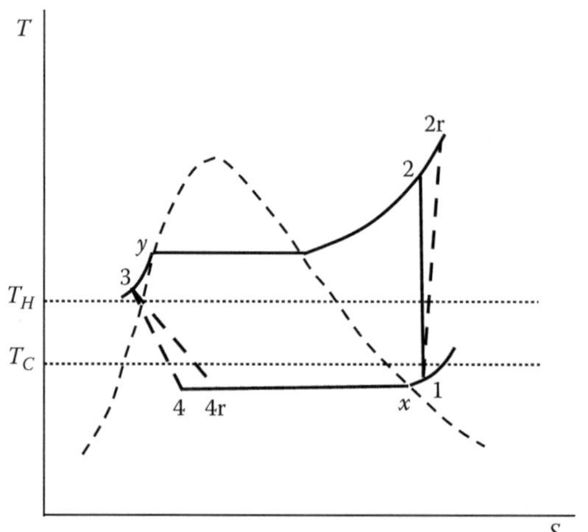

FIGURE 19.6 Real refrigeration cycle, *T–S* diagram.

whole cycle; however, in a general way, it is assumed that the most important irreversible processes are given in the compression step and in the expansion valve, so the real diagram for the refrigeration with heat exchanger is similar to the one presented in Figure 19.6. The more accused irreversibilities are given in the compressor. In an ideal cycle, the compression process is reversible and adiabatic, that is to say, isentropic. However, in the real processes, due to the friction, entropy increases and goes to point 2r instead of going to point 2, causing an increase in the compression work. Thus, the isentropic yield can be defined as the relationship between the ideal work of compression and the real one:

$$\eta_{Isentropic} = \frac{W_C^{ideal}}{W_C^{real}} \tag{19.9}$$

19.4 MULTIPRESSURE SYSTEMS

A multipressure system has two or more low pressure points. A low pressure point is the one between the expansion valve and the inlet to the compressor. Multipressure systems can be found in the dairy industry (Stoecker and Jones, 1982), in which an evaporator operates at −35°C to strengthen or harden ice cream, whereas another evaporator operates at 2°C to cool milk.

In refrigeration systems, a stream that is a mixture of liquid and vapor can be obtained between the condenser and the evaporator due to its passage through an expansion valve. The fraction of gas can be separated from the liquid fraction by a separation tank (Figure 19.7). Thus, point 1 indicates the conditions of a saturated liquid that, when expanded, produces a mixture 3, which can be separated into a liquid 4 and a vapor 6. For each kilogram of the mixture in 3, $(1-x)$ kg of liquid will be separated in the current 4 and x kg of vapor in the current 6. This vapor is fed into the compressor, while the liquid is driven through an expansion valve to the evaporator. Therefore, the system needs two compressors. Systems with only one compressor are not commonly used since they are not efficient (Stoecker and Jones, 1982).

When it is desired to compress a gas at the outlet of the compressor, the gas, besides having a greater pressure, increases its temperature as well. In many cases, with the objective of improving the process performance, two compressors with intermediate refrigeration of the vapor are used, reducing in this way the overheating produced and obtaining a saturated vapor. This method is usually replaced in refrigeration systems by an alternative one shown in Figure 19.8, in which liquid refrigerant from the condenser is used as refrigeration fluid for the intermediate refrigerator. The compressor of the first step discharges vapor that bubbles in the liquid of the intermediate refrigerator, leaving a saturated vapor (point 4), which has a lower temperature than the point 2 of discharge. In this type of system, if p_S is the suction pressure of the first compressor and p_D is the discharge of the second compressor, the intermediate pressure (p_i) is calculated as the geometric mean of these two pressures as

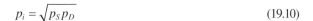

$$p_i = \sqrt{p_S p_D} \tag{19.10}$$

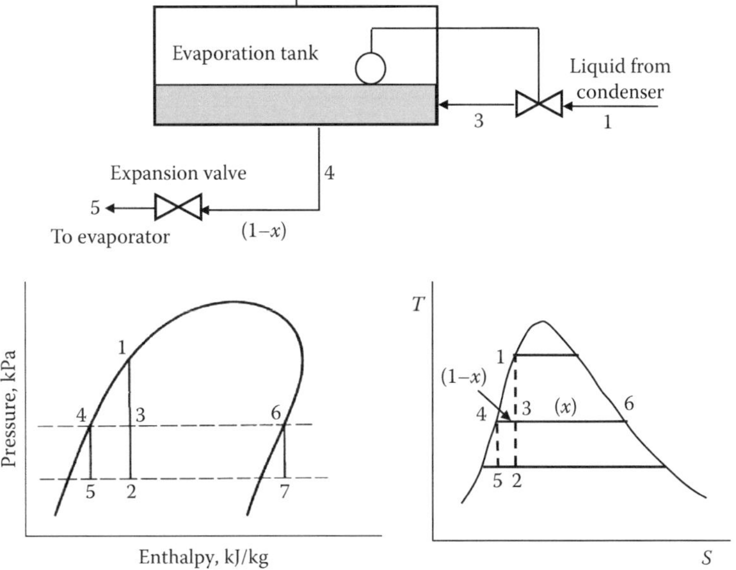

FIGURE 19.7 Expansion process with separation.

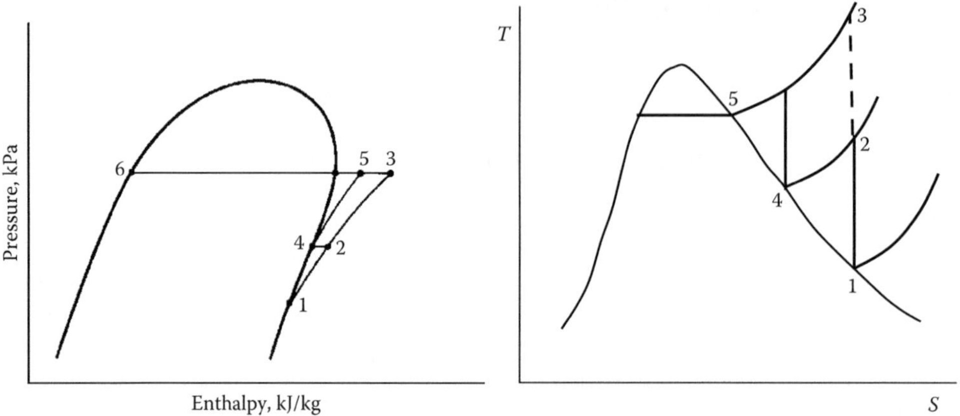

FIGURE 19.8 Intermediate refrigeration with liquid refrigerant.

This pressure is the discharge pressure of the first compressor and the suction pressure of the second compressor and also the pressure at which the intermediate refrigerator should operate in order to maintain the optimum global economy of the system.

19.4.1 Systems with Two Compressors and Evaporator

A common way to obtain a low-temperature evaporator is by using a two-stage compression with an intermediate refrigerator and separation of gas (Figures 19.8 and 19.9). This system requires less power than one with simple compression, and usually the power savings justify the extra cost of the equipment.

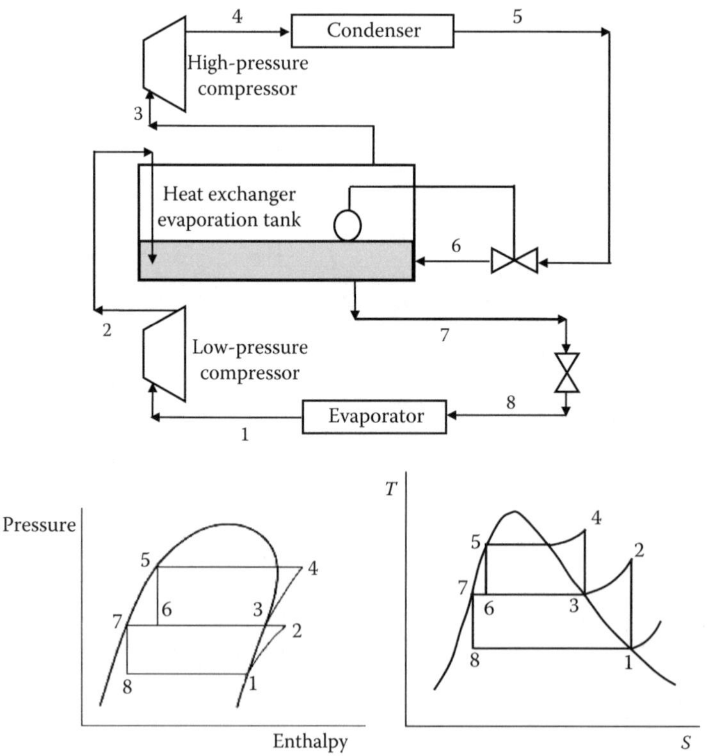

FIGURE 19.9 Operation with two compressors and one evaporator.

It can be observed that the fluid that circulates in different sections of the installation is the same; hence,

$$w_1 = w_2 = w_7 = w_8 \tag{19.11a}$$

$$w_3 = w_6 \tag{19.11b}$$

For the solution of problems with this type of system, different balances should be performed:

- Energy balance in the evaporator

$$\dot{Q}_E = w_1\left(\hat{H}_1 - \hat{H}_8\right) \tag{19.12}$$

- Balance in the intermediate refrigerator

$$w_2\hat{H}_2 + w_6\hat{H}_6 = w_3\hat{H}_3 + w_7\hat{H}_7 \tag{19.13}$$

Taking into account Equation 19.11, it is obtained that

$$w_3\left(\hat{H}_3 - \hat{H}_6\right) = w_2\left(\hat{H}_2 - \hat{H}_7\right) \tag{19.14}$$

Power of the compressors:

First compressor

$$Pow)_1 = w_1\hat{W}_1 = w_1\left(\hat{H}_2 - \hat{H}_1\right) \tag{19.15}$$

Second compressor

$$Pow)_2 = w_3\hat{W}_2 = w_3\left(\hat{H}_4 - \hat{H}_3\right) \tag{19.16}$$

19.4.2 Refrigeration Systems in Cascade

Systems that have two compressors and two evaporators operating at different temperatures are common in the refrigeration industry. Different refrigeration temperatures are often required in different processes of the same plant. Evaporators at two different temperatures can operate in an efficient way in a two-stage system that uses an intermediate refrigerator and vapor separator (Figure 19.10). It should be pointed out that the intermediate pressure of the system corresponds to the saturation temperature T_3 of the second evaporator and is fixed by this temperature.

The solution of problems should be carried out through mass and energy balances in the different parts of the system.

The following is complied within this system:

$$w_1 = w_2 = w_7 = w_8 \tag{19.17}$$

- Balances in the evaporators

$$\dot{Q}_{E1} = w_1\left(\hat{H}_1 - \hat{H}_7\right) \tag{19.18}$$

$$\dot{Q}_{E2} = w_6\left(\hat{H}_3 - \hat{H}_6\right) \tag{19.19}$$

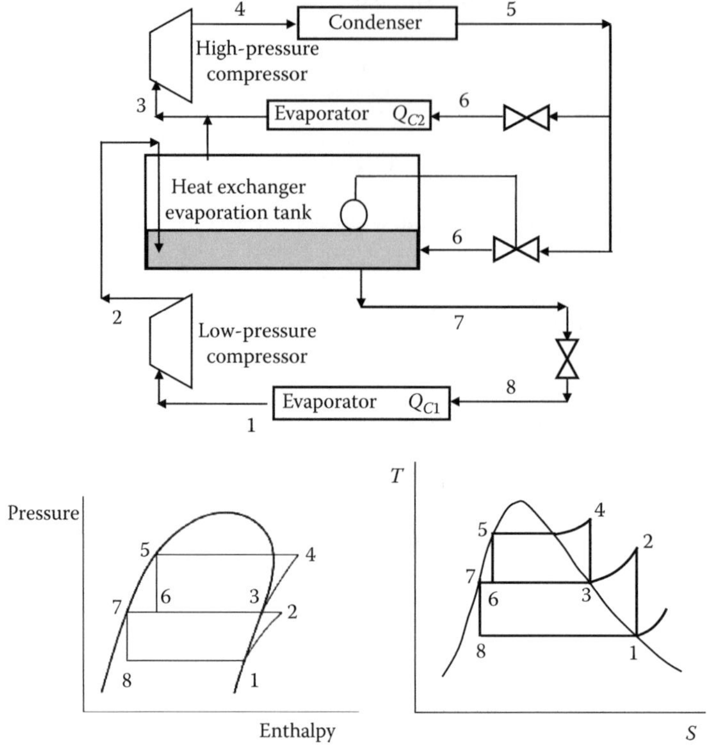

FIGURE 19.10 Operation with two compressors and two evaporators and flash chamber.

- Balances in the second evaporator and intermediate refrigerator:

$$w_2\hat{H}_2 + w_5\hat{H}_5 + \dot{Q}_{E2} = w_3\hat{H}_3 + w_7\hat{H}_7 \tag{19.20}$$

$$w_2 + w_5 = w_3 + w_7 \tag{19.21}$$

Since $w_2 = w_7$, it is also true that $w_3 = w_5$; hence,

$$w_3\left(\hat{H}_3 - \hat{H}_5\right) = w_2\left(\hat{H}_2 - \hat{H}_7\right) + \dot{Q}_{E2} \tag{19.22}$$

- Power of the compressors
  First compressor

$$Pow)_1 = w_1\hat{W}_1 = w_1\left(\hat{H}_2 - \hat{H}_1\right) \tag{19.23}$$

Second compressor

$$Pow)_2 = w_3\hat{W}_2 = w_3\left(\hat{H}_4 - \hat{H}_3\right) \tag{19.24}$$

There are systems in cascade in which a vapor separator does not exist, just as it is shown in the installation scheme (Figure 19.11) and in the T–S diagram (Figure 19.12). In this case, with two systems, the refrigerant circulates in a closed way in each one of them. There is an independent circuit for each one; this allows being able to work with two different types of refrigerants in each circuit. Both circuits share a closed heat exchanger, so the heat given by the refrigerant in the condensation stage of the inferior cycle is used to evaporate the refrigerant of the superior cycle. The refrigerant

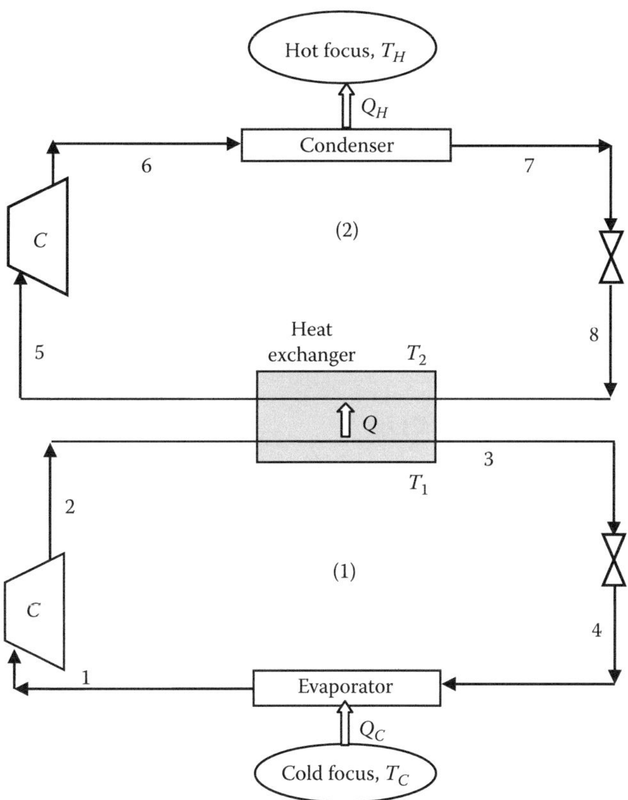

FIGURE 19.11 Scheme of cascade refrigeration cycle.

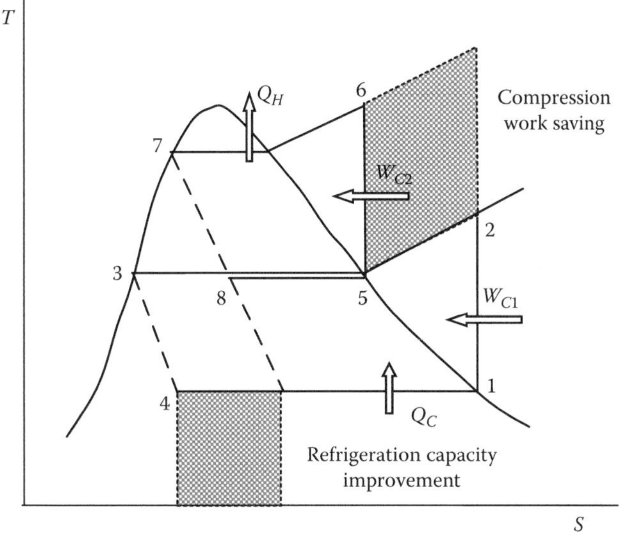

FIGURE 19.12 Cascade refrigeration system, T–S diagram.

effect takes place in the evaporator at low temperature, while in the condenser at high temperature, the heat is eliminated to the surroundings. In Figure 19.12, the compression energy saving can be observed, as well as the increment of refrigeration capacity that takes place in this type of facilities.

When carrying out a heat balance in the heat exchanger, it is obtained:

$$w_1\left(\hat{H}_2 - \hat{H}_3\right) = w_1\left(\hat{H}_5 - \hat{H}_8\right) \tag{19.25}$$

For the refrigeration coefficient, when applying Equation 19.1, it is obtained:

$$\varphi_R = \frac{\dot{Q}_F}{\dot{W}_{Total}} = \frac{w_1\left(\hat{H}_1 - \hat{H}_4\right)}{w_1\left(\hat{H}_2 - \hat{H}_1\right) + w_2\left(\hat{H}_6 - \hat{H}_5\right)} \tag{19.26}$$

19.5 GAS REFRIGERATION CYCLES

In the refrigeration cycles studied in the previous sections, the refrigerant experienced phase changes along the cycle. Besides these refrigerant cycles, there are some of those refrigerants that are always in form of gas. The gas cycles present an advantage over the vapor ones, since with a gas cycle lower refrigeration temperatures can be achieved. In the vapor cycles, the refrigerant extracts the heat from the cold focus to vaporize the entire refrigerant (change of state) plus the heat to increase the temperature of the vapor; however, in the gas cycles, the extracted heat only increases the temperature of the refrigerant that is in gas phase. Therefore, for the same refrigeration capacity, it is necessary that the gas cycle operates with a higher refrigerant flow, indicating that the installations are more voluminous or that it works at pressure in a tight system.

The experimental device of this type of facilities is similar to the one shown in Figure 19.2; however, due to the fact that in this case there is not a partial condensation of the refrigerant fluid (but rather the gas expansion work takes place), it is better to use a turbine than an expansion valve. In Figure 19.13, a scheme of a typical installation with refrigeration of gas is shown, while in Figure 19.14, the so-called Brayton's refrigeration cycle is shown in the *T–S* diagram. In this case, the network consumed in the cycle is equal to the work consumed by the compressor less the work that takes place in the turbine:

$$W_C = \hat{H}_2 - \hat{H}_1 \tag{19.27}$$

$$W_T = \hat{H}_3 - \hat{H}_4 \tag{19.28}$$

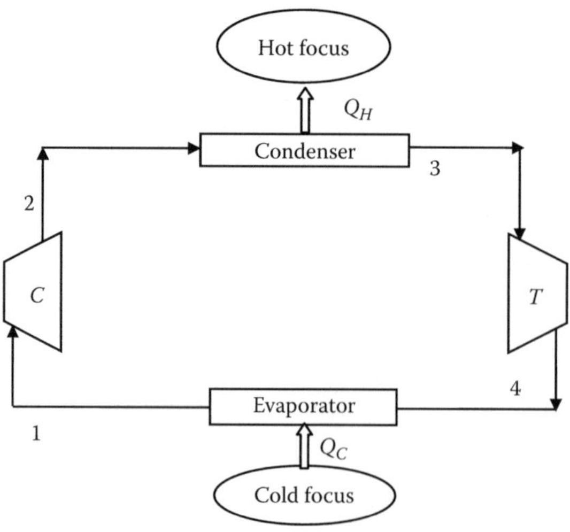

FIGURE 19.13 Scheme for a gas refrigeration cycle installation.

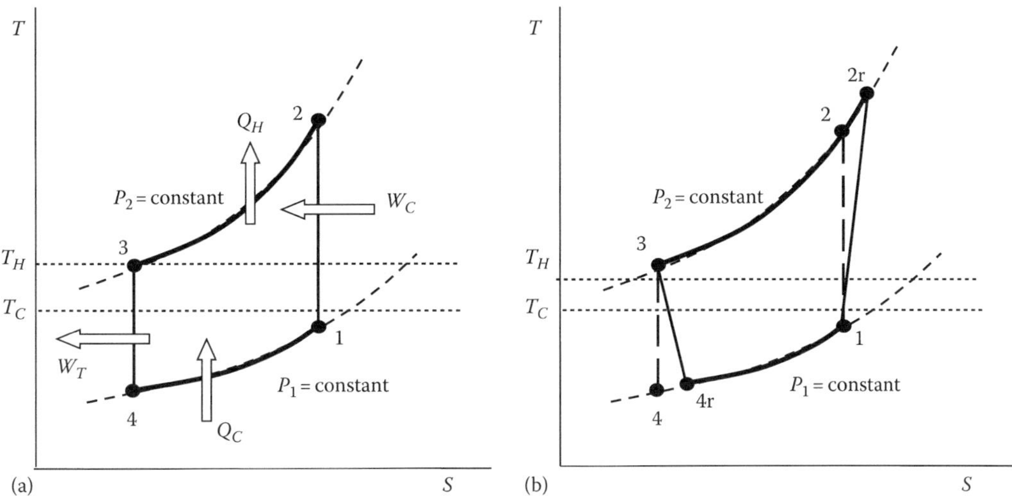

FIGURE 19.14 Gas refrigeration cycle *T–S* diagram: (a) ideal and (b) real.

The heat absorbed in the cold focus and given in the hot focus is

$$Q_C = \hat{H}_1 - \hat{H}_4 \tag{19.29a}$$

$$Q_H = \hat{H}_2 - \hat{H}_3 \tag{19.29b}$$

Thus, the refrigeration coefficient can be obtained from the following equation:

$$\varphi_R = \frac{\hat{H}_1 - \hat{H}_4}{\left(\hat{H}_2 - \hat{H}_1\right) - \left(\hat{H}_3 - \hat{H}_4\right)} \tag{19.29c}$$

19.6 FOOD CHILLING

Once foods are obtained from their natural source, they present problems such as being perishable and having a limited commercial life, since these foods can be altered mainly due to three mechanisms:

- Living organisms contaminate and deteriorate foods, not only microorganisms (parasites, bacteria, and molds) but also insects in different stages of their vital cycle.
- Biochemical activities performed within foods such as respiration, browning, and over-ripeness, which in most cases are due to the enzymes present in food. These activities can reduce the food's quality. They are present in vegetables after harvest and in animal foods after slaughter.
- Physical processes, such as loss of moisture that yields dehydration.

A decrease of storage temperature implies a decrease in the rate of deterioration reactions, so the food increases its shelf life. The rate of deterioration will be lower as the temperature is lower, which points out that it is desirable to decrease the temperature to the minimum whenever it is higher than freezing. However, this is not always possible. It has been observed in some foods that at low temperatures, even above freezing temperature, undesirable reactions can take place.

TABLE 19.2
Deterioration of Fruits and Vegetables under Storage Conditions

Product	Critical Temperature of Storage (°C)	Type of Deterioration between the Critical and Freezing Temperature
Olive	7	Internal browning
Avocado	4–13	Brown discoloration in pulp
Cranberry	2	Rubbery texture, red pulp
Eggplant	7	Scald on surface, rotting
Sweet potato	13	Internal discoloration, spot, rotting
Squash	10	Rotting
Green beans	7	Spot and redness
Lime	7–9	Spot
Lemon	14	Spot, red mark, membranous staining
Mango	10–13	Gray discoloration of peel, unequal ripeness
Melon	7–10	Spot, peel putrefaction, no ripeness
Apple	2–3	Internal and core browning, scald, moist breaking
Orange	3	Spot, brown marks
Papaya	7	Spot, rotting, no ripeness
Potato	3	Mahogany browning, sweetening
Cucumber	7	Spot, rotting
Pineapple	7–10	Green color upon ripening
Banana	12–13	Opaque color upon ripening
Grapefruit	10	Scald, spot
"Quingombó"	7	Discoloration, spot, embedded zone, rotting
Watermelon	4	Spot, unpleasant odor
Tomato		
Ripe	7–10	Rotting, softening, soaked water
Green	13	Light color upon ripening, rotting

Source: Lutz, J.M. and Hardenburg, R.E., *Agriculture Handbook*, vol. no. 66, U.S. Department of Agriculture, U.S. Government Printing Office, Washington, DC, 1968.

Thus, some fruits can present internal and core browning during storage, as in the case of pears. It is recommended to store them at 3°C–4°C, since if they are stored at 0°C browning can occur. It is not advisable to store potatoes below 3°C because imbalances in the starch–sugar system that produce the accumulation of sugars can occur, and consequently the potatoes will be deteriorated. Table 19.2 presents storage temperatures below which deterioration problems can occur in some types of foods.

Another important parameter to consider during food storage under refrigeration is the relative humidity. If humidity is lower than the so-called relative humidity of equilibrium, there is a water loss from the food to the exterior, leading to dehydration of the product. On the contrary, if the relative humidity of the environment is higher, then water condensation on the food surface can occur, facilitating in many cases microbial growth and deterioration of the product.

TABLE 19.3
Storage Temperature under Normal Atmosphere

T (°C)	Product
Just above the freezing point	*Animal tissues*—mammal meat, fish, chicken
	Fruits—apricots, lemons (yellow), pears, nectarines, oranges (Florida), plums, berries, apples (some varieties), peaches
	Vegetables—asparagus, beets, peas, radishes, broccoli, Brussels sprouts, carrots, celery, cauliflower, corn, spinach
	Milk
	Egg (with shell)
2°C–7°C	*Fruits*—melons, oranges (except Florida), apples (some varieties), ripe pineapples
	Vegetables—early potatoes
>7°C	*Fruits*—avocados, bananas, grapefruits, mangos, lemons (green), limes, green pineapples, tomatoes
	Vegetables—green beans, cucumbers, sweet potatoes, late potatoes

Source: Karel, M. et al., Preservation of food by storage at chilling temperatures, in *Principles of Food Science. Part II. Physical Principles of Food Preservation*, O.R. Fennema, ed., Marcel Dekker, New York, 1975a.

TABLE 19.4
Recommended Relative Humidity

Relative Humidity	Product
Less than 85%	Butter, cheese, coconut, dried fruits, garlic, nuts, dry onions, egg (shell), dates
85%–90%	*Animal tissues*—mammal meat (except veal), chicken
	Fruits—banana (yellow), citric, pineapples, nectarines, plums, peaches, tomatoes
	Vegetables—sweet potatoes, early potatoes
90%–95%	*Animal tissues*—veal, fish
	Fruits—apples (90%), berries, pears, bananas (green)
	Vegetables—green beans, cucumbers, sweet corn, late potatoes, vegetable leaves, green peas, edible roots

Source: Karel, M. et al., Preservation of food by storage at chilling temperatures, in *Principles of Food Science. Part II. Physical Principles of Food Preservation*, O.R. Fennema, ed., Marcel Dekker, New York, 1975a.

Tables 19.3 and 19.4 show temperatures and relative humidities for the optimum storage of different foods.

19.7 FREEZING

Freezing is one of the most widely used methods of food preservation, due to two main factors. The first is that many microorganisms cannot grow at low temperatures used in freezing. Also, when a food is frozen, part of the water is transformed into ice, thus decreasing the food's water activity (a_w). This decrease of a_w influences the growth of many microorganisms, since they cannot develop in low water activity conditions.

Food freezing can be carried out in different ways, and depending on the method, the quality of the frozen food will vary. Thus, if freezing is instantaneous, there will be many points in the food

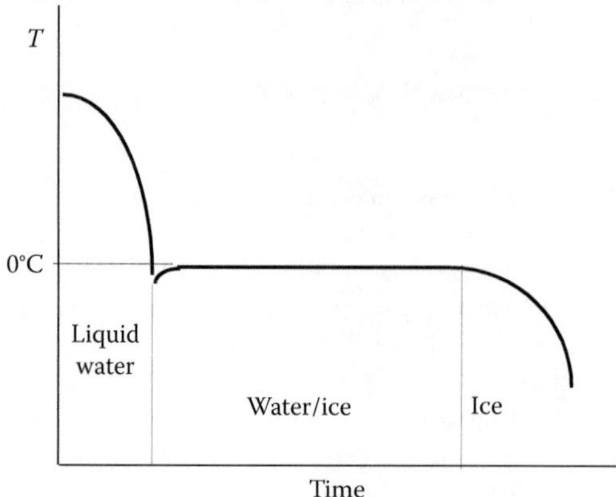

FIGURE 19.15 Pure water freezing.

where formation of ice begins. This means that there are many nucleation sites and the ice crystals formed are small; therefore, the food tissues are slightly affected. On the contrary, if freezing is slow, there are few nucleation sites, and the small number of ice crystals formed will grow with time. This will cause larger crystals to form, which can affect the final quality of the frozen product, mainly through a greater water loss during thawing.

When freezing a food, it is important to know how the temperature of the food varies along the freezing process. It is well known that the temperature of pure water varies with time as it freezes (Figure 19.15). The freezing temperature of pure water is 0°C, so if beginning with water at a higher temperature, initially, there exists a temperature decrease below 0°C. This means that there will be a subcooling, and later, due to the beginning ice formation, fusion heat is released and the temperature increases again to 0°C. At this point, temperature remains constant until all of the water becomes ice. Then, the temperature decreases again with a higher slope, since the thermal conductivity of ice is greater than that of liquid water. In foods, this process is different from the freezing of pure water (Figure 19.16). If T_F is the temperature at which the freezing of the product begins, initially, the temperature decreases below that temperature. Once the first ice crystals are formed, the temperature increases to a value equal to T_F. However, the temperature does not remain at that point, but rather there is a slight continuous decrease due to the transformation of water into ice and the increasing concentration of soluble solids coming from food in the unfrozen water fraction. There is a moment when the crystallization of any of the solids may begin and crystallization heat

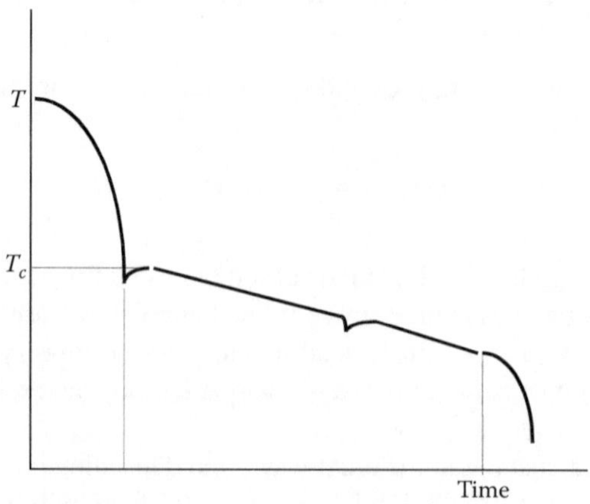

FIGURE 19.16 Freezing of a food product.

is released, in this way increasing the temperature. Finally, a temperature at which it is not possible to freeze more water will be reached, since the soluble solid content such that unattainably low temperatures are required. This is the final freezing point of the product, after which the product decreases in temperature until reaching the temperature of the freezing medium.

It is important to point out that in frozen products, not all the water can be frozen. There is a portion of unfrozen water that remains, and this portion is known as bound water. Also, the time that passes from when the food reaches the initial freezing temperature T_F until reaching the final freezing point is known as freezing time. This parameter is required for calculation of freezing processes, and it is needed for the design of freezers.

Different types of freezers in the food industry are used depending on the type of freezing desired. One type of freezer is the plate or contact freezer, in which the product to be frozen is shaped as a parallelepiped and placed between plates that are at lower temperatures than the freezing temperature of the product. Heat transfer from the food to the plates achieves the freezing of the product. In this type of freezer, the process is usually slow, so ice crystal growth will prevail over nucleation. Other freezers use refrigerant fluids in which the food to be frozen is placed. Freezing can be more or less rapid depending on the type of fluid and its temperature. Thus, if the fluid is liquid nitrogen, there will be a great number of nucleation sites, and small ice crystals will be formed; since the temperature is very low, nucleation will prevail over growth, and freezing times will be short. On the contrary, if cold air is used as fluid, freezing is slow, and crystal growth will prevail over nucleation. Therefore, the type of freezer to be used depends on need and on the desired quality of the product.

19.7.1 Freezing Temperature

Freezing temperature is the temperature at which the first ice crystals begin to form, that is, the temperature at which ice crystals and liquid water coexist in equilibrium. For pure water, this temperature corresponds to 0°C (273 K). However, the water in foods contains soluble solids, and it is known that these solids cause a decrease in the freezing point of water. For this reason, the temperature at which the freezing of a food begins is lower than 0°C. It is clear that the concentration of soluble solids in the unfrozen water portion of a food increases as freezing develops, causing the freezing temperature to vary with time. Therefore, for calculation purposes, it is common practice to take the temperature at which the first ice crystals appear as the initial freezing temperature.

At the beginning of freezing, the aqueous solution is diluted, so as a first approximation, the initial freezing temperature (T_F) can be calculated by applying Raoult's law, and the freezing point depression (ΔT_F) can be calculated by the following equation:

$$\Delta T_F = T_{0W} - T_F = K_A \frac{m_S}{M_S} \tag{19.30}$$

where

m_S is the g solute/100 g water
M_S is the molecular weight of the solute
K_A is 18.6 (cryogenic constant of water)

It should be pointed out that for foods M_S is an equivalent molecular weight of the solutes contained in the food.

Empirical equations exist in the literature that allows us to determine the initial freezing temperature of certain foods only as a function of their moisture content (Levy, 1979).

When the moisture content of a food is known, the molar fraction of water can be calculated. An equation that allows calculation of the initial freezing temperature is

$$T_F = \frac{T_{0W}\lambda}{\lambda - RT_{0W}\ln X_W} \tag{19.31a}$$

where

T_{0W} is the freezing temperature of pure water (273 K)
λ is the latent heat of freezing of water (6003 kJ/kmol)
R is the gas constant (8.314 kJ/kmol K)
X_W is the molar fraction of unfrozen water

Table 19.5 presents the values of the initial freezing temperature for some foods.

TABLE 19.5
Initial Freezing Temperature of Some Foods

Product	Moisture Content (%)	Freezing Temperature (°C)
Meat	55–70	−1.0 to −2.2
Fruits	87–95	−0.9 to −2.7
Cranberry	85.1	−1.11
Plum	80.3	−2.28
Raspberry	82.7	−1.22
Peach	85.1	−1.56
Pear	83.8	−1.61
Strawberry	89.3	−0.89
Egg	74	−0.5
Milk	87	−0.5
Fish	65–81	−0.6 to −2.0
Isotonic	−1.8 to −2.0	—
Hypotonic	−0.6 to −1.0	—
Vegetables	78–92	−0.8 to −2.8
Onion	85.5	−1.44
Asparagus	92.6	−0.67
Spinach	90.2	−0.56
Carrot	87.5	−1.11
Juices	—	—
Cranberry	89.5	−1.11
Cherry	86.7	−1.44
Raspberry	88.5	−1.22
Strawberry	91.7	−0.89
Apple	87.2	−1.44
Apple purée	82.8	−1.67
Apple concentrate	49.8	−11.33
Grape must	84.7	−1.78
Orange	89.0	−1.17
Tomato pulp	92.9	−0.72

Sources: Heldman, D.R., Food freezing, in *Handbook of Food Engineering*, D.R. Heldman and D.B. Lund, eds., Marcel Dekker, New York, 1992; Mafart, P., *Ingeniería Industrial Alimentaria*, ed., Acribia, Zaragoza, Spain, 1994; Plank, R., *El Empleo del Frío en la Industria de la Alimentación*., Ed., Reverté, Barcelona, Spain, 1980.

19.7.2 Unfrozen Water

Along the freezing process, there always exists a fraction of unfrozen water in a food. Also, as previously mentioned, there is bound water that is not in the form of ice at the final freezing point. It is important to determine the amount of unfrozen water, since this affects not only the properties of the product but also the enthalpy needed for freezing to occur.

If at a given instant the mass fractions of unfrozen water and solids in a food are x_W and x_S, respectively, then the molar fraction of nonfrozen water can be calculated from the following expression:

$$X_W = \frac{(x_W/18)}{(x_W/18)+(x_S/M_S)} \tag{19.32a}$$

In this equation, it is necessary to know the equivalent molecular mass of the solids. Therefore, the unfrozen water is

$$x_W = \frac{18 x_S X_W}{M_S(1-X_W)} \tag{19.32b}$$

Hence, the mass fraction of the ice formed (x_I) is the initial mass fraction of the moisture content of the food (x_{0W}) less the mass fraction of unfrozen water:

$$x_I = x_{0W} - x_W$$

The molar fraction of unfrozen water can be calculated using Equation 19.31a, which when appropriately expressed yields the molar fraction of unfrozen water as a function of the freezing temperature:

$$\ln X_W = \frac{\lambda}{R}\left[\frac{1}{T_{0W}} - \frac{1}{T_f}\right] \tag{19.31b}$$

19.7.3 Equivalent Molecular Mass of Solutes

The so-called equivalent molecular mass of solids appears in different equations presented in the previous section. It is necessary to determine this variable for freezing calculations. Equation 19.30 can be used in the case of considering an ideal diluted solution whenever the initial freezing temperature is known. In the same way, another equation that allows us to determine M_S is Equation 19.32b, if the mass and molar fractions of unfrozen water are known, as well as the mass fraction of solids.

Also, there are empirical equations (Chen, 1985) that allow the calculation of the equivalent molecular mass for specific foods:

$$\text{Orange and apple juices}: M_S = \frac{200}{1+0.25x_S}$$

$$\text{Meat}: M_S = \frac{535.4}{x_{0W}}$$

$$\text{Cod}: M_S = \frac{404.9}{x_{0W}}$$

19.8 THERMAL PROPERTIES OF FROZEN FOODS

The thermal properties of a food appear as variables in the different equations used to calculate freezing times and to design processing systems. Therefore, it is essential to know the value of these properties in order to solve the different problems that could be set up. In spite of the fact that there are numerous references on properties of frozen foods reported in literature, it is necessary to have equations that allow their calculation.

19.8.1 Density

When a food is frozen, its density decreases due to the fraction of ice it contains. An equation that allows calculation of the density of a frozen product is

$$\frac{1}{\rho} = \frac{x_W}{\rho_W} + \frac{x_I}{\rho_I} + \frac{x_S}{\rho_S} \tag{19.33}$$

in which x_W, x_I, and x_S are the mass fractions of unfrozen water, ice, and total solids, respectively. The densities of water, ice, and total solids can be calculated from the equations given by Choi and Okos (1986b). It should be pointed out that when calculating the density of the total solids, it is necessary to know the composition of this fraction.

19.8.2 Specific Heat

The specific heat of the frozen product is a function of its content of unfrozen water, ice, and total solids, according to the following expression:

$$\hat{C}_P)_F = \hat{C}_P)_W x_W + \hat{C}_P)_I x_I + \hat{C}_P)_s x_S \tag{19.34}$$

However, Choi and Okos (1986b) presented a general equation in which the specific heat of the product is expressed as the sum of the products of specific heats of each component and its mass fraction.

These equations should be used for calculation of the enthalpies of frozen foods. But in freezing, in general, the initial temperature of the food does not correspond to the freezing temperature but rather to a higher one. Therefore, it is also important to determine the specific heat of the unfrozen food. An equation to calculate the specific heat of the unfrozen food as a function of its moisture content is

$$\hat{C}_P)_{NF} = \hat{C}_P)_W (0.3 + 0.7 x_{0W}) \tag{19.35}$$

There are other equations in the literature, some of which are specific for some products (Levy, 1979; Mafart, 1994). However, it is appropriate to apply the equations of Choi and Okos whenever the composition of the food is known.

19.8.3 Thermal Conductivity

Calculation of the thermal conductivity of frozen foods may be more complicated than the calculation of density or specific heat, since thermal conductivity not only depends on the water content and conductivity of the aqueous and solid phases but also on the structure of the product. Thus, Kopelman (1966) considered that there could exist three different structure models in foods. In one model, it is considered that the food is a homogeneous system made of two components in a dispersed form. A second model considers a homogeneous system made of two components, in which the solid fraction is arranged in the form of fibers in two directions. The third model assumes a homogeneous system in which the solid fraction is arranged as parallel slabs in one direction. This author introduced different equations that allow calculation of the thermal conductivity of the product depending on the type of structure.

The conductivity of the liquid phase is usually higher than the conductivity of the fraction of solids. In this case, for a homogeneous system in which both phases are dispersed, the thermal conductivity of a food is calculated by the equation

$$k = k_L \frac{1-\left(X_S^V\right)^2}{1-\left(X_S^V\right)^2\left(1-X_S^V\right)} \tag{19.36}$$

where
k_L is the thermal conductivity of the liquid
X_S^V and X_W^V are the volumetric fractions of solids and water, respectively

If the conductivity of the liquid and solid phases is similar, the following expression should be used:

$$k = k_L \frac{1-Q}{1-Q\left(1-X_S^V\right)} \tag{19.37}$$

where

$$Q = \left(X_S^V\right)^2\left[1-\frac{k_S}{k_L}\right]$$

k_S being the thermal conductivity of the solid fraction.

For systems in which it is considered that the solids are in the form of striated fibers or slabs, conductivity depends on whether it is considered parallel or perpendicular to such fibers or slabs.

19.9 FREEZING TIME

Calculation of freezing time is one of the most important parameters within the design of freezing stages, since it represents the time the food will be inside the freezing equipment. In principle, it represents the time needed by the geometric center of the food to change its initial temperature down to a given final temperature lower than freezing temperature; that is also called *effective freezing time*. Sometimes, the so-called *nominal freezing time* is employed. Nominal freezing time is the time passed from the moment when the surface of the food reaches 0°C until the geometric center reaches a temperature 10°C lower than the initial freezing temperature.

Calculation of freezing time can be complex, since the freezing temperature of the food changes continuously along the process, as mentioned before. However, as a first approximation, the time elapsed from the moment when the food is at its freezing temperature until the whole food is frozen can be calculated.

In order to do this calculation, a slab with infinite dimensions, but of finite thickness, is supposed. This implies that the heat transfer will be only in one direction. This body is initially at a temperature T_F and is introduced into a freezer in which the external temperature is T_e. A freezing front is formed along the freezing process that advances from the surface at a temperature T_S toward the center of the slab (distance x) (Figure 19.17). When performing an energy balance, it is obtained that the outlet-heat term should be equal to energy dissipated because of freezing.

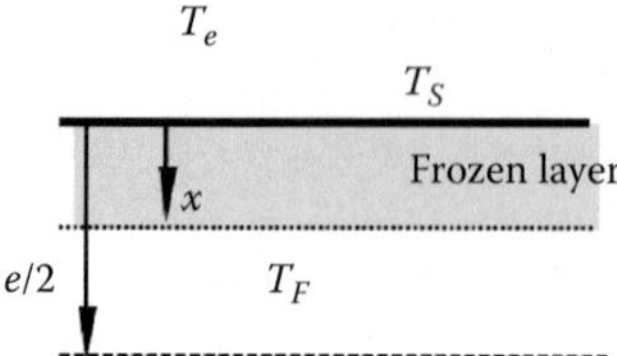

FIGURE 19.17 Freezing front of a slab.

Heat is transmitted through the frozen layer by conduction and from the surface to the exterior by convection, so the outlet-heat term per unit time can be expressed as

$$\dot{Q}_S = A\frac{k}{x}(T_F - T_S) = Ah(T_S - T_e) = A\frac{T_F - T_e}{(x/k) + (1/h)}$$

The term of energy dissipated by freezing is expressed as

$$\dot{Q}_D = A\rho\lambda_I \frac{dx}{dt}$$

In these equations
- A is the area of the slab
- k and ρ are the thermal conductivity and density of the frozen layer
- h is the coefficient of heat transfer by convection to the exterior

When equaling these equations, it is obtained that

$$\rho\lambda_I \frac{dx}{dt} = \frac{T_F - T_e}{(x/k) + (1/h)}$$

This equation, with separable variables, can be integrated on the following boundary conditions:

$$\text{For} \quad t = 0 \qquad x = 0$$
$$\text{For} \quad t = t_F \qquad x = e/2$$

where
- t_F is the freezing time
- e is the thickness of the slab

The integrated equation allows calculation of the freezing time:

$$t_F = \frac{\rho\lambda_I}{T_F - T_e}\left(\frac{e^2}{8k} + \frac{e}{2h}\right) \tag{19.38}$$

This expression is known as Planck's equation, and it is important to point out that it is valid only for the freezing period. This means that it is assumed that the food was initially at its freezing temperature. Also, λ_I is the latent heat of the frozen fraction, and it is calculated by multiplying the latent heat of pure water λ and the mass fraction of frozen water: $\lambda_I = x_I\lambda$.

For cylinders of infinite length and spheres, freezing time is obtained in an analogous way, although the resulting expression differs on the values of coefficients 8 and 2 that affect the conductivity and the coefficient of convection. Thus, for cylinders, these values are 16 and 4, while for spheres, such values are 24 and 6. Also, the characteristic dimension will not be the thickness but the radius.

If it is taken into account that the numbers of Fourier, Biot, and Stefan are defined by the equations,

$$\text{Fourier's number (Fo)} = \frac{k}{\rho\hat{C}_P(e)^2}$$

$$\text{Biot's number (Bi)} = \frac{he}{k}$$

$$\text{Stefan's number (Ste)} = \frac{\hat{C}_P(T_F - T_e)}{\lambda}$$

then the equation is transformed into an expression that correlates these three modules:

$$(\text{Fo}) = \frac{1}{8}\frac{1}{(\text{Ste})} + \frac{1}{2}\frac{1}{(\text{Bi})(\text{Ste})}$$

This equation is valid for infinite slabs with finite thickness but can be generalized for spheres and cylinders of infinite height:

$$(\text{Fo}) = \frac{R}{(\text{Ste})} + \frac{P}{(\text{Bi})(\text{Ste})} \qquad (19.39)$$

The values for the parameters *P* and *R*, in this equation, for slabs, infinite cylinders, and spheres are presented in Table 19.6.

The equation of Planck is an approximation for the calculation of freezing processes. Although it cannot be used for the exact calculation of freezing times, it is useful to obtain an approximation of these times.

In practice, there are cases in which the product to be frozen has a finite geometry. Equation 19.39 can be used in such cases, although the parameters *P* and *R* will differ from the values given in Table 19.6. The calculation of these parameters is very complex. Thus, for the case of a parallelepiped of thickness e, width a, and length l, the dimensionless length parameters $\beta_1 = a/e$ and $\beta_2 = l/e$ are defined, in such a way that the values of *P* and *R* are calculated from the following equations (Plank, 1980):

$$P = \frac{\beta_1\beta_2}{2\left(\beta_1\beta_2 + \beta_1 + \beta_2\right)} \qquad (19.40a)$$

$$R = \frac{(M-1)(\beta_1 - M)(\beta_2 - M)\ln(M/M-1) - (N-1)(\beta_1 - N)(\beta_2 - N)\ln(N/N-1)}{8L} + \frac{2\beta_1 + 2\beta_2 - 1}{72} \qquad (19.40b)$$

TABLE 19.6
Parameters *P* and *R* of Equation 19.39

Geometry	P	R	Dimension
Infinite slab	1/2	1/8	Thickness, e
Infinite cylinder	1/4	1/16	Radius, r
Sphere	1/6	1/24	Radius, r

$$L = \sqrt{(\beta_1 - \beta_2)(\beta_1 - 1) + (\beta_2 - 1)^2}$$

$$M = \frac{\beta_1 + \beta_2 + 1 + L}{3}$$

$$N = \frac{\beta_1 + \beta_2 + 1 - L}{3}$$

The equation of Plank is useful to determine the freezing times but only in an approximate way. The development of Plank's equation was made assuming that at the beginning of the freezing process, the food was at the freezing temperature. However, in general, it is not that way, since the food is usually at a temperature higher than the freezing temperature. The real time should be the sum of the time calculated by Plank's equation and the time needed by the surface of the product to decrease from an initial temperature down to the freezing temperature. The method described for heat transfer under unsteady state should be used to calculate the additional time, using the properties of the unfrozen food.

Some examples for calculating such time can be found in the literature. One of these examples is presented by Nagaoka et al. (1955), in which the calculation of freezing time is performed using the following expression:

$$t_F = \frac{\rho \Delta \hat{H}}{T_F - T_e}\left(\frac{\text{Re}^2}{k} + \frac{\text{Pe}}{h}\right)[1 + 0.008(T_i - T_F)] \tag{19.41}$$

where

T_i is the food's temperature at the start of freezing

$\Delta \hat{H}$ is the difference between the food's enthalpy at the initial temperature and its enthalpy at end of the freezing process

The enthalpy increase experienced by food during the freezing process can be evaluated by means of the equation:

$$\Delta \hat{H} = \hat{C}_P)_{NF}(T_i - T_F) + x_I \lambda_W + \hat{C}_P)_F(T_F - T_f) \tag{19.42}$$

In this expression, the first addend of the right-hand side of the equation represents the heat that should be eliminated from the product from the initial temperature to the freezing temperature. The second addend represents the heat released during the phase change of the frozen water fraction, while the third addend is the heat eliminated so the food can pass from the freezing temperature to the final T_f. This temperature does not necessarily have to coincide with the temperature of the freezing medium but can be slightly higher, although in case they coincide, it is accomplished that $T_f = T_e$. Also, λ_W is the latent heat of fusion of pure water, while the specific heat of the food before and after freezing can be calculated from Equations 19.35 and 19.34, respectively.

This is a simple method to calculate the increase of enthalpy. Empirical equations can be found in the literature that also allows calculation of such enthalpy variation (Levy, 1979; Succar and Hayakawa, 1983; Chen, 1985). Also, Riedel (1956, 1957a,b) developed diagrams that allow calculation of the enthalpies for different products (meat, eggs, fruits, and juices) as a function of the water content and the fraction of frozen water.

Another modification to Plank's equation is that given by Cleland and Earle (1976, 1982), in which they define a new dimensionless number:

$$\text{Plank's number (Pk)} = \frac{\hat{C}_P)_W\,(T_i - T_F)}{\Delta H}$$

where $\hat{C}_P)_W$ stands for the specific heat of the unfrozen water.

Plank's equation (Equation 19.39) is used to calculate the freezing time, in which the values of the parameters P and R depend on the type of geometry:

- The following equations are used for slabs:

$$P = 0.5072 + 0.2018(\text{Pk}) + (\text{Ste})\left[0.3224(\text{Pk}) + \frac{0.0105}{(\text{Bi})} + 0.0681\right] \tag{19.43a}$$

$$R = 0.1684 + (\text{Ste})[0.0135 + 0.274(\text{Pk})] \tag{19.43b}$$

- The following expressions are used for cylinders:

$$P = 0.3751 + 0.0999(\text{Pk}) + (\text{Ste})\left[0.4008(\text{Pk}) + \frac{0.071}{(\text{Bi})} - 0.5865\right] \tag{19.44a}$$

$$R = 0.0133 + (\text{Ste})[0.3957 + 0.0415(\text{Pk})] \tag{19.44b}$$

- For spherical geometry, the equations to be used are as follows:

$$P = 0.1084 + 0.0924(\text{Pk}) + (\text{Ste})\left[0.231(\text{Pk}) - \frac{0.3114}{(\text{Bi})} + 0.6739\right] \tag{19.45a}$$

$$R = 0.0784 + (\text{Ste})[0.0386(\text{Pk}) - 0.1694] \tag{19.45b}$$

In addition to these modifications, these authors introduced a parameter called equivalent dimension of heat transfer, which takes into account the shape of the product to be frozen. However, the correction due to this factor yields lower freezing times, so it might be better not to do such correction and to remain in a conservative position.

Freezing time is calculated for regular-shaped foods in all of the methods previously described. However, there can be cases of freezing of nonregular-shaped products. In order to solve this problem, a dimensionless factor is defined, which is a function of Biot's number and form factors β_i. The equations that allow the calculation of this factor can be found in the literature (Cleland et al., 1987a,b; Cleland, 1992). Thus, for example, in the case of an ellipsoid-shaped body, in which the axis has the dimensions r, $\beta_1 r$, and $\beta_2 r$, the parameter E is calculated according to the expression (Cleland, 1992):

$$E = 1 + \frac{1 + (2/\text{Bi})}{\beta_1^2 + (2\beta_1/\text{Bi})} + \frac{1 + (2/\text{Bi})}{\beta_2^2 + (2\beta_2/\text{Bi})} \tag{19.46}$$

The value of the characteristic dimension r of the irregular object is found by taking the shorter distance from the surface to the slowest cooling point. The parameters β_1 and β_2 are obtained from the following equations:

$$A_x = \pi\beta_1 r^2$$

$$V = \frac{4}{3}\pi\beta_1\beta_2 r^3$$

where

A_x is the area of the smaller transversal section that contains the thermal center

V is the volume of the considered body

The calculation of the freezing time of a food, whatever its geometry, is obtained by dividing the calculated freezing time by the value of this factor E.

19.10 DESIGN OF FREEZING SYSTEMS

In the design of freezing systems, it is necessary to know the amount of energy that should be eliminated from the food to change from the initial to the final temperature of the frozen product. To do this, it is necessary to know the enthalpy of the food at the beginning and at the end of the freezing process. Since enthalpy is a state function, it should be given in relation with a reference temperature that in the case of freezing processes is −40°C. This means that at this temperature, the enthalpy of any product is considered null.

In order to calculate the power needed to carry out the freezing process, it is necessary to determine the variation of enthalpy experienced by the product, since it is introduced in the freezer until the product reaches the final temperature. This can be achieved by using Equation 19.42.

Another factor that should be calculated is the power that the freezing equipment should have to perform a given process. Such power is the total energy to be eliminated from the food per unit time, and it is a measure of the capacity of the freezing system. Freezing power is calculated by the following equation:

$$Pow = \frac{m\Delta\hat{H}}{t} \tag{19.47}$$

where

m is the total quantity of food to be frozen

$\Delta\hat{H}$ is the increase of enthalpy experienced by the food from the initial to the final temperature

t is the time the food stays in the freezing equipment

This time usually coincides with freezing time calculated in the previous section.

If the equation given by Nagaoka et al. (Equation 19.41) is used to calculate the freezing time, power can be calculated by the following expression:

$$Pow = \frac{m(T_F - T_e)}{\rho((\mathrm{Re}^2/k) + (Pe/h))[1 + 0.008\,(T_i - T_F)]} \tag{19.48}$$

It can be observed that it is not necessary to know the enthalpy increase experienced by the food, and therefore neither the values of specific heat nor the unfrozen water fraction are needed.

PROBLEMS

19.1 A refrigeration system uses Freon R-12 as refrigerant fluid. It is assumed that it operates following an ideal vapor compression cycle between 0.1 and 0.9 MPa. The refrigerant mass flow is 100 kg/h. Calculate (a) the heat flow that can be removed from the cold focus and the heat flow given in the hot focus, (b) the power that should be applied in the compressor, and (c) the refrigeration coefficient.

As it is an ideal refrigeration cycle, the diagram of Figures 19.3 and 19.4 can be applied to this case. Also, Figure 19.2 shows a scheme of the system of this refrigeration cycle.

To calculate the different variables, it is necessary to determine the enthalpies of the states of the different points marked in Figures 19.2 through 19.4.

Point 1: The conditions of point 1 belong together to refrigerant in liquid state and to vapor with a pressure of 0.1 MPa. From tables of the refrigerant R-12, its enthalpy is calculated:

Pressure:	$P_1 = 0.1$ MPa	Entropy:	$\hat{S}_1 = 0.7171$ kJ/(kg K)
Temperature:	$T_1 = -30.1°C$	Enthalpy:	$\hat{H}_1 = 174.15$ kJ/kg

Point 2: Point 2 is reached after an isentropic compression process such that the refrigerant is under conditions of reheated vapor.

The conditions at point 2 are refrigerant in reheated vapor state and pressure of 0.9 MPa. As the compression process is isentropic, it is accomplished that $\hat{S}_2 = \hat{S}_1 = 0.7171$ kJ/(kg K). Enthalpy of refrigerant is calculated from tables of the refrigerant R-12. It is necessary to get the entropy values and to interpolate in tables of reheated vapor:

Pressure:	$P_2 = 0.9$ MPa	Entropy:	$\hat{S}_1 = 0.7171$ kJ/(kg K)
Temperature:	$T_1 = 51.5°C$	Enthalpy:	$\hat{H}_1 = 213.07$ kJ/kg

Point 3: The conditions of point 3 correspond to a refrigerant in liquid state and pressure of 0.9 MPa. From tables of the refrigerant R-12, its enthalpy is calculated:

Pressure:	$P_1 = 0.9$ MPa	Entropy:	$\hat{S}_1 = 0.2634$ kJ/(kg K)
Temperature:	$T_1 = 37.4°C$	Enthalpy:	$\hat{H}_1 = 71.93$ kJ/kg

Point 4: Point 4 can be reached through an isoenthalpic expansion, and the conditions correspond to refrigerant mixture of liquid–vapor and pressure of 0.1 MPa. From tables of the refrigerant R-12, its enthalpy is calculated:

Pressure:	$P_1 = 0.1$ MPa	Enthalpy:	$\hat{H}_1 = 71.93$ kJ/kg

a. The heat flow rate that is removed from the cold focus can be calculated using Equation 19.7. Initially, it is necessary to calculate the enthalpies at points 4 and 1, corresponding to the entrance and exit of the evaporator.

From the data gathered from points 1 and 4, it is possible to calculate the heat removed in the cold focus:

$$\dot{Q}_C = w\left(\hat{H}_1 - \hat{H}_4\right) = \left((100\ \text{kg/h})\frac{1\ \text{h}}{3600\ \text{s}}\right)(174.15 - 71.93)\ \text{kJ/kg} = 2.84\ \text{kW}$$

The heat flow rate given in the warm focus is obtained from Equation 19.6. Applying the values of the enthalpies of the points 2 and 3, it is obtained:

$$\dot{Q}_H = w\left(\hat{H}_2 - \hat{H}_3\right) = \left((100\ \text{kg/h})\frac{1\ \text{h}}{3600\ \text{s}}\right)(213.07 - 71.93)\ \text{kJ/kg} = 3.92\ \text{kW}$$

b. The power that should be applied in the compressor is obtained from the enthalpy increment between points 1 and 2:

$$\dot{W}_C = w\left(\hat{H}_2 - \hat{H}_1\right) = \left((100\ \text{kg/h})\frac{1\ \text{h}}{3600\ \text{s}}\right)(213.07 - 174.15)\ \text{kJ/kg} = 1.1\ \text{kW}$$

c. The refrigeration coefficient is obtained from Equation 19.8:

$$\varphi_R = \frac{\dot{Q}_C}{\dot{W}_C} = \frac{\hat{H}_1 - \hat{H}_4}{\hat{H}_2 - \hat{H}_1} = \frac{2.84\ \text{kW}}{1.1\ \text{kW}} = 2.58$$

This result indicates that for each energy unit consumed in the compression stage, approximately 2.6 energy units can be eliminated in the focus that is wanted to refrigerate desired for refrigeration.

19.2 An industrial facility requires 100 kW of freezing power. In order to meet such requirements, a standard system of the vapor compression cycle was installed, using ammonia as refrigerant fluid. This cycle operates at a condensation temperature of 30°C and an evaporation temperature of –10°C. Calculate (a) the circulation mass flow of refrigerant fluid, (b) the power needed by the compressor, (c) the coefficient of performance, and (d) the power per kilowatt of refrigeration.

This is a vapor compression cycle, so the global process is like the one shown in the pressure–enthalpy diagram of Figure 19.3.

The enthalpy at point 1 is calculated in the diagram or in the ammonia tables (see Appendix) and corresponds to a saturated vapor at –10°C:

$$\hat{H}_1 = 1450.2 \text{ kJ/kg} \quad p_1 = 291.6 \text{ kPa}$$

The enthalpy of point 2 is calculated by the isentropic line that passes over point 1 and cuts the isobaric line that corresponds to a temperature of 30°C:

$$\hat{H}_2 = 1650 \text{ kJ/kg} \quad p_2 = 1173 \text{ kPa}$$

The enthalpy of point 3 corresponds to saturated liquid at 30°C and can be obtained from tables: $\hat{H}_3 = 341.8$ kJ/kg. Also, $\hat{H}_4 = \hat{H}_3 = 341.8$ kJ/kg.

a. The refrigeration power represents the heat flow rate absorbed in the evaporator. From Equation 19.7,

$$w = \frac{\dot{Q}_E}{\hat{H}_1 - \hat{H}_4} = \frac{100 \text{ kW}}{(1450.2 - 341.8) \text{ kJ/kg}} = 0.090 \text{ kg/s} \approx 324.8 \text{ kg/h}$$

that is, the refrigerant mass flow rate.

b. The power needed by the compressor is calculated from Equation 19.5:

$$Pow = w\left(\hat{H}_2 - \hat{H}_1\right) = (0.09 \text{ kg/s})((1650 - 1450.2) \text{ kJ/kg}) \approx 18 \text{ kW}$$

c. The coefficient of performance is obtained from Equation 19.8:

$$\phi = \frac{\hat{H}_1 - \hat{H}_4}{\hat{H}_2 - \hat{H}_1} = \frac{w\left(\hat{H}_1 - \hat{H}_4\right)}{w\left(\hat{H}_2 - \hat{H}_1\right)} = \frac{100 \text{ kW}}{18 \text{ kW}} = 5.56$$

d. The refrigeration power or compressor power per kilowatt of refrigeration is the inverse of the coefficient of performance:

$$\text{Refrigeration power} = \frac{18 \text{ kW}}{100 \text{ kW}} = 0.18$$

19.3 The refrigeration requirements in an industry are of 400 kW, so a system of two compressors with intermediate refrigerator and vapor elimination is installed. The system uses ammonia as refrigerant fluid. If the evaporation temperature is –20°C and the condensation temperature is 30°C, calculate the power of the compressors.

The system installed is the one described in Figure 19.9.

According to the statement of the problem, the suction pressure of the first compressor, p_S, and the discharge pressure of the second compressor, p_D, correspond to the saturation pressure at the temperatures of the evaporator and condenser, respectively. From ammonia tables (see Appendix),

$$T_S = -20°\text{C} \quad \text{corresponds to:} \quad p_S = 190.7 \text{ kPa}$$
$$T_D = 30°\text{C} \quad \text{corresponds to:} \quad p_D = 1168.6 \text{ kPa}$$

The intermediate pressure for an economic optimum is obtained from Equation 19.25:

$$p_i = \sqrt{p_S p_D} = \sqrt{(190.7)(1168.6)} = 472.1 \text{ kPa}$$

The enthalpies for the different points representative of the system can be obtained from the tables and graphs for ammonia:

Point 1: Saturated vapor at –20°C has an enthalpy: $\hat{H}_1 = 1437.2$ kJ/kg
Point 2: is placed on the isentropic line that passes over point 1 and cuts the isobaric line $p_i = 472.1$ kPa. Hence, $\hat{H}_2 = 1557.5$ kJ/kg
Point 3: to saturated vapor at the pressure $p_i = 472.1$ kPa corresponds to a temperature $T_3 \approx 2.5$°C and an enthalpy: $\hat{H}_3 = 1464.3$ kJ/kg
Point 4: is placed on the isentropic line that passes over point 3 and cuts the isobaric line at the discharge pressure of the second compressor $p_D = 472.1$ kPa. The conditions of point 4 are obtained from the point where both lines cross, yielding an enthalpy: $\hat{H}_4 = 1590$ kJ/kg
Point 5: corresponds to saturated liquid at 30°C, and from tables, it is possible to obtain the value of its enthalpy: $\hat{H}_5 = 341.8$ kJ/kg
Point 6: is a liquid–vapor mixture at a pressure of 472.1 kPa but with the same enthalpy as point 5: $\hat{H}_6 = \hat{H}_5 = 341.8$ kJ/kg
Point 7: is a saturated liquid at 472.1 kPa. The value of its enthalpy can be obtained by interpolation in tables: $\hat{H}_7 = 211.6$ kJ/kg
Point 8: is a liquid–vapor mixture at –20°C but with the same enthalpy as that of point 7: $\hat{H}_8 = \hat{H}_7 = 211.6$ kJ/kg

Once the values of the different points representative of the system have been obtained, it is possible to determine the different variables that intervene.

The circulation mass flow rate of the refrigerant fluid in the first compressor is obtained from the energy balance in the evaporator (Equation 19.12):

$$w_1 = \frac{\dot{Q}_E}{\hat{H}_1 - \hat{H}_8} = \frac{400 \text{ kW}}{(1437.2 - 211.6) \text{ kJ/kg}} = 0.326 \text{ kg/s} \approx 1175 \text{ kg/h}$$

Also, it is known (Equation 19.11a and b) that there are different streams in the system that have the same circulation flow. Thus,

$$w_1 = w_2 = w_7 = w_8 \quad \text{and} \quad w_3 = w_6$$

When carrying out the energy balance in the intermediate refrigerator, Equation 19.14 is obtained, which allows calculations of the circulation flow of the refrigerant fluid in the second compressor:

$$w_3 = \frac{w_2\left(\hat{H}_2 - \hat{H}_7\right)}{\left(\hat{H}_3 - \hat{H}_6\right)} = \frac{0.326(1557.5 - 211.6)}{(1464.3 - 341.8)} = 0.391 \text{ kg/s} \approx 1409 \text{ kg/h}$$

The power of the compressors is obtained from Equations 19.15 and 19.16:

$$Pow)_1 = w_1\left(\hat{H}_2 - \hat{H}_1\right) = (0.326)(1557.5 - 1437.2) = 39.2 \text{ kW}$$

$$Pow)_2 = w_3\left(\hat{H}_4 - \hat{H}_3\right) = (0.391)(1590 - 1464.3) = 49.2 \text{ kW}$$

The total compression power for the system is $Pow = 88.4$ kW.

19.4 A 3 cm thick slab of lean meat is placed inside a freezer in which the temperature is −25°C. The coefficient of heat transfer by convection from the surface of the meat is 15 J/(s m^2 °C). Determine the time needed to freeze the meat slab if 70% of its weight is water.

Data: Properties of lean meat: Thermal conductivity 1.7 J/(s m °C).
Specific heat 2.1 kJ/(kg °C). Density 995 kg/m^3.

According to Table 19.5, the initial freezing temperature for lean meat is $T_C = -2.2$°C.

- The molar fraction of water at the freezing point is obtained from Equation 19.31b:

$$\ln X_{OW} = \frac{6003}{8.314}\left[\frac{1}{273} - \frac{1}{270.8}\right] \quad X_{OW} = 0.9787$$

The equivalent molecular mass of the solutes is calculated by Equation 19.32b:

$$M_S = \frac{(18)(0.3)(0.9787)}{(0.7)(1-0.9787)} = 355.18 \text{ kg/kmol}$$

- The molar fraction of unfrozen water at −25°C is obtained from Equation 19.31b:

$$\ln X_W = \frac{6003}{8.314}\left[\frac{1}{273} - \frac{1}{248}\right] \quad X_W = 0.7660$$

The mass fraction of unfrozen water is obtained from Equation 19.32b:

$$x_S = \frac{(18)(0.3)(0.766)}{(355.18)(1-0.766)} = 0.04976$$

The mass fraction of frozen water is

$$x_I = x_{Wi} - x_W = 0.65024$$

The effective latent heat is calculated from the ice fraction in the meat:

$$\lambda_I = x_I\lambda = (0.65024)(6003) = 3903.4 \text{ kJ/kmol} = 216.86 \text{ kJ/kg}$$

Then, the equation of Plank is applied to determine the freezing time. For this reason, it is necessary to determine previously the numbers of Biot and Stefan:

$$\text{Biot's number (Bi)} = \frac{he}{k} = \frac{(15)(0.03)}{(1.7)} = 0.2647$$

$$\text{Stefan's number (Ste)} = \frac{C_P(T_F - T_e)}{\lambda_I} = \frac{(2.1)(-2.2-(-25))}{(216.86)} = 0.2208$$

The Fourier number is obtained from Plank's equation for a slab:

$$(\text{Fo}) = \frac{1}{8}\frac{1}{(0.2208)} + \frac{1}{2}\frac{1}{(0.2647)(0.2208)} = 9.120$$

The freezing time is obtained by solving the Fourier number for t:

$$t = (\text{Fo})\frac{\rho C_P (e)^2}{k} = (9.12)\frac{(995)(2.1)(0.03)^2}{(0.0017)} = 10{,}089 \text{ s}$$

$$t = 2 \text{ h} \, 48 \text{ min}$$

19.5 4 kg blocks of beaten egg yolk contained in a rectangular geometry of 5 cm thickness, 20 cm width, and 40 cm length are introduced into a continuous belt freezer in which air circulates at −25°C. The yolks are introduced into the freezer at 20°C and leave at −10°C. Determine the time needed to freeze the yolks.

If each yolk block moves 1 m each minute inside the freezer, what length should the freezer's band have? Calculate the power that the freezer should have if 100 yolk blocks are treated simultaneously.

Data. Composition in weight of egg yolk: 48.4% water, 16% proteins, 34% lipids, 0.5% hydrocarbons, and 1.1% ash.

Individual coefficient of heat transfer by convection 20 J/(s m² °C).

The thermal properties of yolk at 20°C are calculated by the equations of Choi and Okos, so the properties of its components are determined first:

Component	k (J/s m °C)	$\hat{C}_P$ (kJ/kg °C)	ρ (kg/m³)	X_i^m	X_i^V
Water	0.6037	4.177	995.7	0.484	0.4931
Protein	0.2016	2.032	1319.5	0.160	0.1230
Fat	0.1254	2.012	917.2	0.340	0.3761
Hydrocarbons	0.2274	1.586	1592.9	0.005	0.0021
Ash	0.3565	1.129	2418.2	0.011	0.0070

$$k = (0.6037)(0.4931) + (0.2016)(0.123) + (0.1254)(0.3761) + (0.2274)(0.0021) + \cdots (0.3565)(0.007)$$

$$\hat{C}_P = (4.177)(0.484) + (2.032)(0.16) + (2.012)(0.34) + (1.586)(0.005) + \cdots (1.129)(0.011)$$

$$\rho = \left(\frac{1}{995.7}\right)(0.484) + \left(\frac{1}{1319.5}\right)(0.16) + \left(\frac{1}{917.2}\right)(0.34) + \left(\frac{1}{1592.9}\right)(0.005) + \cdots \left(\frac{1}{2418.2}\right)(0.011)$$

$$k = 0.3726 \text{ J/(s m °C)} \quad \hat{C}_P = 3.051 \text{ kJ/(kg °C)}$$

$$\rho = 1014.5 \text{ kg/m}^3$$

The properties of frozen yolk will be determined at −10°C. The fraction of unfrozen water should be calculated previously at this temperature. The temperature at which yolk begins to freeze is −0.5°C.

The molar fraction of water in yolk when freezing begins is calculated by Equation 19.31b:

$$\ln X_{OW} = \frac{6003}{8.314}\left[\frac{1}{273} - \frac{1}{272.5}\right] \quad X_{OW} = 0.9952$$

This value allows calculation of the equivalent molecular mass of the solutes by means of Equation 19.32b:

$$M_S = \frac{(18)(0.516)(0.9952)}{(0.484)(1-0.9952)} = 3944.8\,\text{kg/kmol}$$

- The molar fraction of unfrozen water at −10°C is obtained from Equation 19.31b:

$$\ln X_W = \frac{6003}{8.314}\left[\frac{1}{273} - \frac{1}{263}\right] \quad X_W = 0.9043$$

The mass fraction of unfrozen water is obtained from Equation 19.32b:

$$x_W = \frac{(18)(0.516)(0.9043)}{(3944.8)(1-0.9043)} = 0.0223$$

The mass fraction of frozen water is

$$x_I = x_{Wi} - x_W = 0.4617$$

The thermal properties of the components at −10°C are shown in the following table:

Component	k (J/s m °C)	$\hat{C}_P$ (kJ/kg °C)	ρ (kg/m³)	X_i^m	X_i^V
Water	0.5541	4.234	996.8	0.0223	0.0220
Protein	0.1666	1.996	1335.1	0.160	0.1177
Fat	0.2083	1.969	929.8	0.340	0.3591
Hydrocarbons	0.1871	1.529	1602.2	0.005	0.0031
Ash	0.3153	1.073	2426.6	0.011	0.0045
Ice	22.922	2.002	918.2	0.4617	0.4938

The properties of yolk at −10°C are obtained by applying the equations of Choi and Okos:

$$k = 1.240\,\text{J/(s m °C)} \qquad \hat{C}_P = 2.027\,\text{kJ/(kg °C)} \qquad \rho = 982\,\text{kg/m}^3$$

The generalized equation of Plank (Equation 19.39) will be used to calculate the freezing time, but the numbers of Biot and Stefan should be determined first using the properties of yolk at −10°C:

Biot's number

$$(\text{Bi}) = \frac{h\,e}{k} = \frac{(20)(0.05)}{(1.240)} = 0.8065$$

Stefan's number

The effective latent heat is calculated from the fraction of ice that the yolk contains:

$$\lambda_I = x_I \lambda_W = (0.4617)(6003)\left(\frac{1}{18}\right) = 154 \text{ kJ/kg}$$

so the module of Stefan is

$$(\text{Ste}) = \frac{\hat{C}_P\ (T_F - T_e)}{\lambda_I} = \frac{(2.027)(-0.5-(-25))}{(154)} = 0.3225$$

The parameters P and R are obtained from Equation 19.40, in which $\beta_1 = (20)/(5) = 4$ and $\beta_2 = (40)/(5) = 8$.

Hence,

$P = 0.3636$ and $R = 0.0993$

- Fourier's number is obtained from the equation of Plank for a slab:

$$(\text{Fo}) = \frac{0.0993}{(0.3225)} + \frac{0.3636}{(0.8065)(0.3225)} = 1.706$$

The freezing time is obtained when solving the equation of Fourier's number for t:

$$t_F = (\text{Fo})\frac{\rho\,\hat{C}_P\ (e)^2}{k} = (1.706)\frac{(982)(2.027)(0.05)^2}{(0.00117)} = 7256\,\text{s}$$

This time corresponds to the time needed to freeze the product if it is initially at the freezing temperature. For this reason, the time required by the surface of the product to change from 20°C to −0.5°C should be calculated. The rule of Newman should be applied to parallelepiped geometry. However, the width and length of the yolk blocks are really greater than their thickness. For this reason, it will be supposed that the heat transfer predominates on this direction. Using dimensionless modules,

$$Y = \frac{T_e - T_F}{T_e - T_i} = \frac{-25-(-0.5)}{-25-20} = 0.544$$

$$m = \frac{k}{h(e/2)} = \frac{(0.372)}{(20)(0.025)} = 0.744$$

$$n = 1$$

The dimensionless time module is obtained graphically: $\tau = 0.5$. This module allows the calculation of the time needed to decrease the temperature down to the initial point of freezing:

$$t = \frac{\tau(e/2)^2 \rho\,\hat{C}_P}{k} = \frac{(0.5)(0.025)^2(1014.5)(3.051\times 1000)}{0.372} = 2600\,\text{s}$$

The total time is

$$t_{TOTAL} = t_F + t = (7256) + (2600) = 9856\,\text{s} \approx 164\,\text{min}$$

The residence time of the food in the freezer should be at least the freezing time (164 min). If v is the velocity with which the yolk blocks move inside the freezer, then the length the egg yolk should cross in the freezer is

$$L = t_{TOTAL} v = (164)(1) = 164\ \text{m}$$

The total freezing time calculated from Plank's equation is an approximation, so it will be calculated from the equation of Nagaoka et al. (Equation 19.41). For this reason, the enthalpy variation experienced by the product should be previously calculated from Equation 19.42:

$$\Delta\hat{H} = (3.051)(20 - (-0.5)) + (154) + (2.027)(-0.5 - (-10)) = 235.8\ \text{kJ/kg}$$

Substitution of data in Equation 19.41 yields the total freezing time:

$$t_T = \frac{(982)(235.8 \times 10^3)}{-0.5 - (-25)} \left(\frac{(0.0993)(0.05)^2}{1.240} + \frac{(0.3636)(0.05)}{20} \right) [1 + 0.008(10 - (-0.5)]$$

Hence,

$$t_T = 12{,}202 \quad s \equiv 203.4\,\text{min}$$

Therefore, the total length is L = (203.4 min) (1 m/min) = 203.4 m.

It can be observed that the freezer's band length calculated by this method is greater and the one obtained by the Plank's equation is just estimated.

The total mass to freeze is m = (100)(4) = 400 kg.

The power of the freezer is calculated applying Equation 19.47:

$$Pow = \frac{m\Delta\hat{H}}{t_{TOTAL}} = \frac{(400)(235.8)}{(12{,}202)} = 7.73\ \text{kW}$$

PROPOSED PROBLEMS

19.1 In a certain food process, it is necessary to provide 200 kW of refrigerating power. In order to supply this power, a refrigeration system is installed. The refrigeration system is a standard vapor compression cycle, using ammonia as refrigerant fluid. The condensation temperature to which this cycle works is 35°C, while the evaporation temperature is −12°C. Calculate the required compression power and the refrigeration unit power.

19.2 A fruit packinghouse has installed a refrigeration system with two compressors with intermediate refrigerator and vapor elimination. The system uses ammonia as refrigerant fluid, and it produces 500 kW power. The evaporation and condensation temperatures are −20°C and 30°C, respectively. Determine the compression power for both compressors.

19.3 Determine the freezing time for a 10 cm thickness meat sheet wrapped with two 1 mm cardboard leaves, one to each side. The meat is placed in a plate freezer whose temperature is –34°C. What would be the freezing time if there was not cardboard?

Data. Assume that the heat convection coefficient in the freezer is 60 W/(m^2 °C). Freezing meat properties: Initial freezing temperature –2°C. Thermal conductivity 1.8 W/(m °C). Density 1090 kg/m^3. Latent heat 250 kJ/kg. Cardboard thermal conductivity 0.058 W/(m °C).

19.4 The soluble coffee freeze drying process begins with an extraction stage with hot water that contained coffee in crushed grains. The obtained solution cools down, and then it is frozen in a tray placed in a freeze dryer, in which the ice is eliminated by sublimation. The freeze dryer sublimation camera operates under vacuum in stationary state, it contains a 0.25 m^2 surface radiant plate, and parallel to it and at a distance of 17 cm, there is a tray of same surface that contains the sample to freeze drying. In an experiment, the tray has been loaded with a coffee solution that later has been freeze, being obtained a 15 mm frozen layer. The radiant plate is at 150°C, and it emits the necessary heat to be able to sublime the ice of the frozen solution. Calculate (a) the time to eliminate all the water in the frozen layer and (b) the temperature in the surface of the dried coffee.

Data. Under the freeze dryer condition work, the coffee sublimation temperature is –23°C. Thermal conductivity of the dried coffee 0.035 W/(m K). Density of frozen coffee solution 990 kg/m^3. Sublimation heat 2850 kJ/kg. Emissivity of freeze dried coffee 0.90.

19.5 A freeze dryer that operates under vacuum is used to elaborate soluble coffee. The freeze dryer possesses a 0.25 m^2 tray that contains a frozen coffee solution that reaches 15 mm height. Parallel to the tray, and at a distance of 17 cm, there is a radiant plate of same surface. This source emits the necessary heat to be able to sublimate the ice of the frozen solution. If after 24 h all the water has been eliminated, calculate (a) the surface temperature in the dried coffee layer, (b) the temperature to which the radiant plate should operate, and (c) the dried coffee layer thickness after 10 h.

Data. Under the freeze dryer condition work, the coffee sublimation temperature is –23°C. Thermal conductivity of the dried coffee 0.035 W/(m K). Density of frozen coffee solution 990 kg/m^3. Sublimation heat 2850 kJ/kg. Freeze dried coffee emissivity 0.90.

20 Thermal Processing of Foods

20.1 INTRODUCTION

One of the main problems in food engineering deals with the inactivation of microorganisms present in foods, not only to avoid their harmfulness but also with the objective of preserving foods for as long as possible. In order to achieve the inactivation of microorganisms, either in spore or vegetative form, foods are thermally treated inside containers or in a continuous fashion and packed later on in aseptic conditions. It is important to obtain a high-quality final product whether performing the treatment in one way or the other by minimizing losses in nutrients or sensorial properties.

The thermal processing of packed products is carried out in equipment that uses steam or hot water as the heating fluid. In aseptic processing, the products are first thermally treated, then carried to a container that was previously sterilized, and finally sealed under sterile environment conditions. This is a technique traditionally used in fluids such as milk and fruit juices, although recently it has also been applied to other specific food products. Aseptic processing presents different advantages when compared to traditional thermal treatment in containers, since the food undergoes less deterioration, processing times are shorter, energy consumption is reduced, and the quality of the treated product is improved and more uniform.

Pasteurization and sterilization can be distinguished within thermal treatments. The former refers to the thermal destruction of specific pathogen microorganisms, although the resulting product is unstable if not stored under refrigeration. The sterilization is a process by which products are made to be stable without storage refrigeration.

20.2 THERMAL DEATH RATE

The death rate for any microorganism in a specific medium and thermally treated at certain fixed temperature follows a first-order kinetics. Thus, if N is the number of microorganisms, its variation with time is expressed as

$$\frac{dN}{dt} = -kN \tag{20.1}$$

This equation can be integrated considering the boundary condition for the initial time, there are N_0 microorganisms, yielding

$$N = N_0 \exp(-kt) \tag{20.2}$$

where

N is the number of microorganisms present at a time t
k is the death rate constant

The value of the rate constant depends on the type of microorganism, on the medium, and on the temperature. Also, for a specific microorganism, the value depends on the state of the microorganism, vegetative or spores. These constants are much greater for vegetative form than for spore forms, indicating that spores are much more difficult to destroy. The values of the thermal death constant of vegetative forms are around 10^{10} min^{-1}, while spore form values are approximately 1 min^{-1}.

20.2.1 Decimal Reduction Time *D*

The *decimal reduction time* is usually employed in calculations of thermal treatment problems and can be defined as the treatment time required to reduce the number of microorganisms to the tenth part, represented as D_T. In thermal treatment calculations, it is assumed that this time is independent of the initial concentration of microorganisms but that it depends on temperature, type of microorganism, and culture or food media in which microorganisms grow.

The following expression is obtained from Equation 20.2:

$$D_T = \frac{2.303}{k}\log_{10}\left(\frac{N}{N_0}\right)$$

and since $N = 0.1N_0$, the decimal reduction time is expressed as a function of the rate constant of thermal death as

$$D_T = \frac{2.303}{k} \tag{20.3}$$

and if treatment time is expressed according to Equation 20.4:

$$t = D_T \log_{10}\left(\frac{N_0}{N}\right) \tag{20.4}$$

20.2.2 Thermal Death Curves

Equation 20.2 can become a linear equation if it is expressed in logarithmic form:

$$\ln\left(\frac{N}{N_0}\right) = -kt \tag{20.5}$$

If N/N_0 is plotted in semilogarithmic coordinates against time, a straight line with slope $-k$ and ordinate to the origin 1 is obtained. This straight line is called a *thermal death curve*, and for each microorganism it is determined by the treatment temperature, in such a way that if the temperature is different, the slope of the straight line will also be different, since the death rate constant varies. Thus, if temperature increases, the slope is greater (Figure 20.1).

This is common for thermal treatment of microorganisms, but it may not be true in the case of spore forms and when N/N_0 is plotted against time in semilogarithmic coordinates, curves, and not straight lines are obtained, as it can be seen in Figure 20.1.

Figure 20.2 shows thermal death curves for some microorganisms present in milk, while Figure 20.3 shows curves for thermal inactivation of some enzymes.

Figure 20.4 shows a thermal death curve that represents how to obtain the value of the decimal reduction time for a specific treatment temperature.

20.2.3 Thermal Death Time Constant *z*

Thermal treatments are carried out at different temperatures, depending in each case on the needs or facilities of each industry. For this reason, thermal treatments are not necessarily performed at the temperatures at which thermal death data were obtained. Therefore, a relationship between

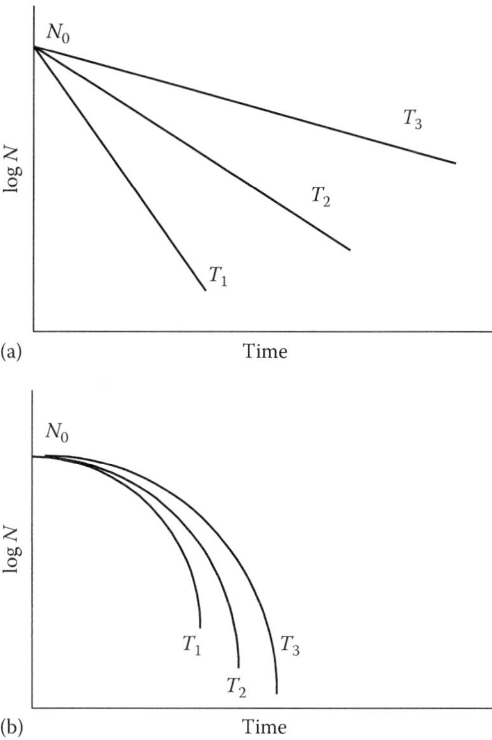

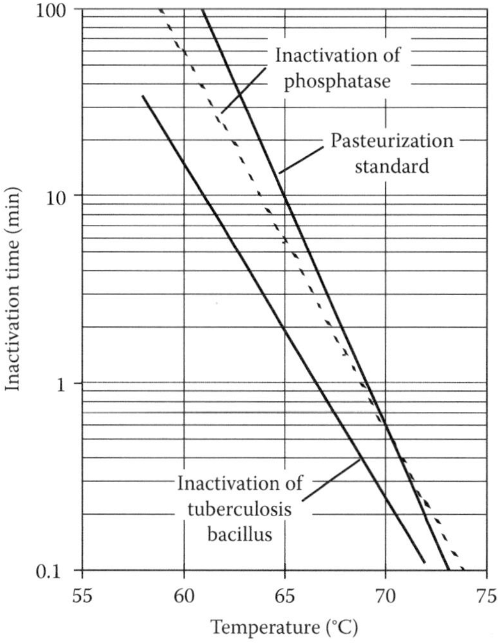

FIGURE 20.1 Thermal death curves for microorganisms. $T_1 > T_2 > T_3$: (a) vegetative form and (b) spore form.

FIGURE 20.2 Thermal death curves for milk microorganisms. (Adapted from Earle, R.L., *Ingeniería de los Alimentos*, 2nd edn., Acribia, Zaragoza, España, 1983.)

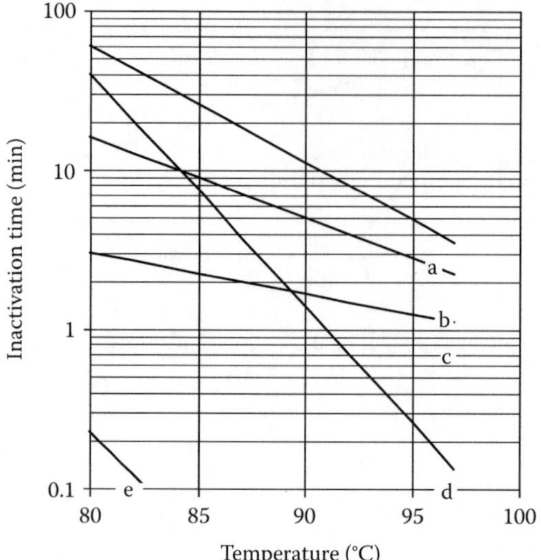

FIGURE 20.3 Thermal inactivation curves for biological factors: (a) pectinesterase (citric products), (b) polygalacturonase (citric products), (c) ascorbic oxidase (peach), (d) *Clostridium pasteurianum* (6D), and (e) molds and yeast (12D). (Adapted from Toledo, R.T. and Chang, S.-Y., *Food Technol.*, 44(2), 75, 1990.)

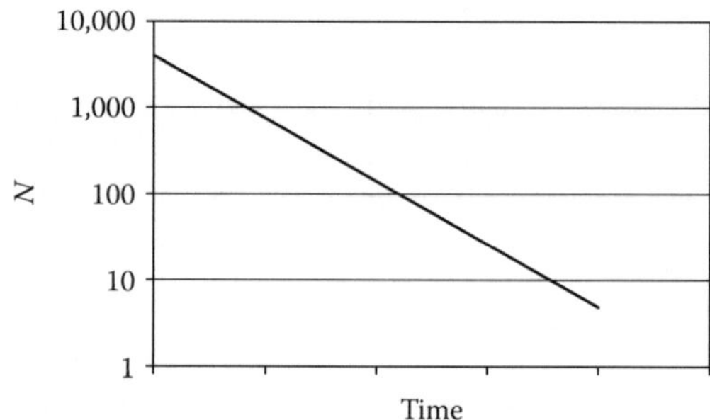

FIGURE 20.4 TDT curve to obtain decimal reduction time.

thermal death time (TDT) and temperature should be found. This relationship is given as a graph in Figure 20.5, where TDTs or decimal reduction (D_T) are plotted in semilogarithmic coordinates against temperature. It can be observed that a straight line with negative slope ($-m$) is obtained. The value of the slope is

$$m = \frac{\log D_{T_1} - \log D_{T_2}}{T_1 - T_2} \quad (20.6)$$

where

D_{T_1} is the decimal reduction time at temperature T_1
D_{T_2} is the time corresponding to temperature T_2

The parameter z is defined as the inverse of slope m and measures the variation of the thermal death rate with temperature, representing the temperature increase required to reduce the treatment time to the tenth part, or in turn D_T.

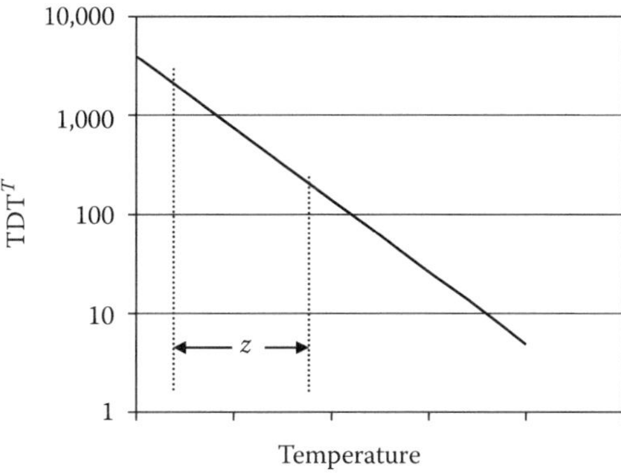

FIGURE 20.5 Thermal death curve TDT or D_T.

The relationship between two treatment times and their corresponding temperatures can be easily obtained from the figure

$$\frac{\log t_1 - \log t_2}{\log 10} = \frac{T_2 - T_1}{z} \tag{20.7a}$$

or as a function of the decimal reduction times

$$\frac{\log D_{T_1} - \log D_{T_2}}{\log 10} = \frac{T_2 - T_1}{z} \tag{20.7b}$$

If one of these temperatures is the one of reference, then

$$t_1 = t_R 10^{(T_R - T_1)/z} \tag{20.8a}$$

$$D_1 = D_R 10^{(T_R - T_1)/z} \tag{20.8b}$$

The value of z is used to calculate the lethal efficiency rate L that measures lethality at a temperature T with respect to the reference temperature T_R:

$$L = 10^{(T - T_R)/z} \tag{20.9}$$

Therefore, treatment time t_T, as well as decimal reduction time D_T at any temperature T, can be expressed as a function of the lethal rate:

$$t_T = \frac{t_R}{L} \tag{20.10a}$$

$$D_T = \frac{D_R}{L} \tag{20.10b}$$

20.2.4 Reduction Degree *N*

It can be easily observed in Equation 20.2 that an infinite time of treatment is required in order to obtain a null concentration of microorganisms at the end of a thermal treatment. This points out that it is impossible to achieve a complete sterilization of the product. For this reason, a final concentration N_F is defined to ensure that the treated product is *commercially* sterile. Therefore, complete sterilization of a product cannot be discussed; rather, we will focus on commercial sterility.

The decimal logarithm of the ratio between the initial amount of microorganisms N_0 and the amount of microorganisms at a given time, at which N_F is obtained, is called *reduction degree* of number of log cycles:

$$n = \log_{10}\left(\frac{N_0}{N_F}\right) \tag{20.11}$$

This n value is arbitrary and depends on the type of microorganism, although the safety of thermal treatment is greater as the value of the reduction degree is greater. Table 20.1 presents the values of n for different microorganisms that cause deterioration in foods. It can be observed that this value is different depending on the type of microorganism. Those microorganisms with higher values indicate that a stronger treatment is necessary to reduce the microorganisms' content to lower levels than those that present a lower reduction degree to ensure commercial sterility. Values of the rate parameters for degradation of food components can be found in the literature (Lund, 1975).

TABLE 20.1
Destruction or Thermal Degradation Time of Thermosensitive Microorganisms, Enzymes, Vitamin B$_1$, and Chlorophylls

Food	Microorganism or Thermosensitive Factor	T_R (°C)	D_T (min)	F_R (min)	z (°C)	n
Low acid (pH > 4.6)	*Bacillus stearothermophilus*	121	4–5	15	9.5–10	5
	Clostridium thermosaccharoliticum	121	2–5	15	10	5
	Clostridium nigrificans	121	2–5	15	10	5
	C. botulinum	121	0.1–0.3	2.5–3	10	12
	C. sporogenes	121	0.1–1.5	5	10	5
Acid	*B. coagulans*	121	0.01–0.07	—	10	5
(4 < pH < 4.6)	*Bacillus polymyxa* and *macerans*	100	0.1–0.5	1.3–10 (F_{93})	14–17.5	—
	C. pasteurianum	100	0.1–0.5	1.3–10 (F_{93})	8	5
Very acid (pH < 4)	*Lactobacillus, Leuconostoc*	65	0.5–6.0	0.1 (F_{93})	5–11	—
	Yeast and molds	121	5×10^{-8}	10^{-6}	5–8	20
Sweet corn and green beans						
	Peroxidase	121	1.22	5.1	11–52	4
Spinach	Catalase	121	2.3×10^{-7}	9.3×10^{-7}	8.3	4
Pea	Lipoxygenase	121	1.7×10^{-7}	7×10^{-6}	8.7	4
Pear	Polyphenol oxidase	121	3.2×10^{-8}	1.3×10^{-7}	5.6	4
Papaya	Polygalacturonase	121	4.4×10^{-6}	1.7×10^{-8}	6.1	4
Spinach and green beans	Vitamin B$_1$	121	140	5.6	25	0.04
	Chlorophyll a	121	12.8	0.50	51–87	0.04
	Chlorophyll b	121	14.3	0.57	98–111	0.04

20.2.5 Thermal Death Time *F*

The treatment time needed to reach a reduction degree n at a given temperature T is called TDT F_T. The value of this treatment time can be obtained from Equation 20.5:

$$F_T = nD_T \tag{20.12}$$

This means that the treatment time is obtained by multiplying the decimal reduction time by the reduction degree. This equation is known as the survival law, or the first law of thermal death of microorganisms, or the first law of degradation of one quality factor of a food that can be thermally destroyed. The value of this parameter F is called TDT.

The treatment time to reach a given reduction degree n depends on temperature in such a way that when it increases, the time required to achieve such reduction decreases. If treatment time and temperature are plotted in semilogarithmic coordinates, a straight line is obtained. If the reduction degree is higher, for example, $n + 1$, a straight line with the same slope as for the degradation degree n is obtained, although parallel and with a higher ordinate to the origin. On the contrary, if the reduction degree is $n - 1$, the straight line will have the same slope, although its ordinate to the origin will be lower.

Generally, the TDT F of a certain microorganism is given as a function of a reference temperature T_R and a reduction degree n.

The reference temperature usually taken is 121.1°C (250°F), labeling F_0 to the value of F corresponding to this temperature. This means that F_0 is the time required to reduce the given population of a microbial spore with a z of 10°C at 121.1°C. F_C is defined as the value of F in the center of the container, and F_S is the integrated lethality of the heat received by all the points of the container:

$$F_S = nD_R \tag{20.13}$$

in which D_R is the decimal reduction time at 121.1°C. In systems where the food to be treated is rapidly heated, the value of F_S can be considered as equal to F_0 or F_C. In practice, the value of F_0 is obtained by summing the lethal rates at 1 min intervals from the heating and cooling curves of a product during thermal processing. If the heating process is slow, the time required for the thermal processing is corrected by the following equation:

$$B = f_h \log\left(\frac{j_h I_h}{g}\right) \tag{20.14}$$

where

B is the time of thermal processing corrected for the time needed by the treatment device to reach the processing temperature

f_h are the minutes required by a semilogarithmic heating curve to cross one logarithmic cycle

j_h is the heating lag factor

I_h is the difference in temperature between the treatment device and the food at the beginning of the process

g is the difference between the temperature of the treatment device and the maximum temperature attained at a given point of the food, which is usually the slowest heating point (Karel et al., 1975; Stumbo et al., 1983)

Each of these parameters is defined next:

$$I_h = T_E - T_i \tag{20.15}$$

$$j_h = \frac{T_E - T_{ip}}{T_E - T_i} \tag{20.16}$$

$$g = T_E - T \tag{20.17}$$

where

T_E is the temperature of treatment or sterilizing device
T is the temperature at a given point of the food
T_i is the initial temperature of the food
T_{ip} is the temperature obtained at the intersection of the prolongation of the straight part of the semilogarithmic heating curve and the vertical line corresponding to the start of the process, called apparent initial temperature

Besides these parameters, analogous functions can be defined for the cooling stage. In this way, I_c is the difference between the temperature at the end of the heating process (T_f) and the temperature of the cooling water (T_w). The cooling lag factor j_c, corresponding to the cooling curve, is similar to the factor j_h and it is defined as

$$j_c = \frac{T_w - T_{ip}}{T_w - T_i} \tag{20.18}$$

where

T_i is the initial temperature of the product at the start of the cooling process
T_{ip} is the apparent initial temperature

20.2.6 Cooking Value C

Thermal processing not only affects the microorganisms present in food but also affects food's general quality. For this reason, a cook value C, which is a concept similar to that of lethality, is applied to degradation of sensory attributes. The reference temperature for this value is 100°C, with typical values of z within the range of 20°C–40°C:

$$C = C_{100} 10^{(100-T)/z} \tag{20.19}$$

20.2.7 Effect of Temperature on Rate and Thermal Treatment Parameters

The thermal death of microorganisms follows a first-order kinetics in such a way that when treatment temperature increases, the rate constant increases, and consequently the thermal death rate increases, too. The effect of temperature on the rate constant can be described by the Arrhenius equation:

$$k = K_0 \exp\left(\frac{-Ea}{RT}\right) \tag{20.20}$$

where

k is the rate constant
K_0 is the frequency factor
Ea is the activation energy
R is the gas constant
T is the absolute temperature

The relationship between two rate constants for two temperatures T_1 and T_2 can be expressed as

$$\frac{k_2}{k_1} = \exp\left[\frac{Ea}{R}\left(\frac{1}{T_1} - \frac{1}{T_2}\right)\right] \tag{20.21a}$$

or

$$\log\left(\frac{k_2}{k_1}\right) = \frac{Ea}{2.303R}\left(\frac{1}{T_1} - \frac{1}{T_2}\right) \tag{20.21b}$$

Since $k = 2.303/D$

$$\log\left(\frac{k_2}{k_1}\right) = \log\left(\frac{D_1}{D_2}\right)$$

From Equation 20.8b

$$\frac{D_1}{D_2} = 10^{(T_2 - T_1)/z}$$

Hence

$$\log\left(\frac{k_2}{k_1}\right) = 10^{(T_2 - T_1)/z}$$

The combination of these equations yields

$$\frac{T_2 - T_1}{z} = \frac{Ea}{2.303R}\left(\frac{1}{T_1} - \frac{1}{T_2}\right)$$

The value of z can be obtained as a function of the activation energy and the temperatures:

$$z = \frac{2.303R}{EaT_1T_2} \tag{20.22}$$

20.3 TREATMENT OF CANNED PRODUCTS

The importance of treatment temperature on TDT was explained in the previous section. In canning, treatments are performed in devices where the container goes from room temperature to a treatment temperature, so that after a given time adequate sterilization levels are achieved. In these devices, the product to be treated increases its temperature until reaching the process or holding temperature and then goes through a cooling stage.

20.3.1 Heat Penetration Curve

In thermal treatment it is important to know the point within the canned product with the lowest temperature, that is, the slowest heating point. This point receives the lower degree of thermal treatment and might not undergo an adequate thermal treatment. For this reason, it is necessary to know the heat penetration curve at this point, which shows its change of temperature with heating time.

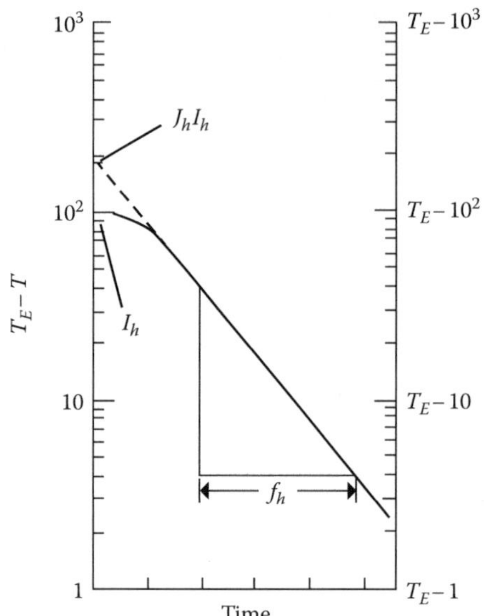

FIGURE 20.6 Heat penetration curve.

The construction of the heat penetration curve for a given product and container is usually obtained experimentally, although in some cases it can be generated by analytical methods. The latter assumes that heat penetrates the food by conduction, which is only true for solid foods, and calculates the temperature of the geometric center as a function of time. However, this is not complied for liquid foods, because heat transmission is not only carried out by conduction and its convective component should be taken into account. Also, the slowest heating point does not coincide with the geometric center.

When a container with food is placed in a thermal treatment device, which is at a temperature T_e, it can be observed that the temperature of the food increases gradually. It is important to know the evolution of the temperature at the slowest heating point (T_C), since this is the point that receives the lowest thermal treatment, and it should be ensured that the microbial load is adequately eliminated. When heat transfer is performed by conduction, the slowest heating point coincides with the geometric center. However, if during heating there are convective streams inside the container, the slowest heating point does not correspond to the geometric center; instead, it is located on the vertical axis but closer to the bottom of the container.

The plot of the variation in temperature of a food as a function of heating time is used to characterize heat penetration in foods. The temperature of the geometric center, T_C or T, is taken as the food temperature, and a linear function is obtained when plotting the logarithm of g $(T_e - T)$ versus heating time. In a semilogarithmic graph, $(T_e - T)$ is represented on the left ordinates axis and the values of $T = T_e - g$ on the right ordinates axis (Figure 20.6). Log cycles of $T_e - 1$ to $T_e - 10$ and to $T_e - 100$ appear on this last axis.

The temperature of the product against time can be plotted in semilogarithmic form based on this graph. This can be done by turning the semilogarithmic paper 180° and labeling the top line with a number equivalent $T_e - 1$. The next log cycle is labeled with a number equivalent to T_e – 10 and the third cycle with a number equivalent to $T_e - 100$. This type of graph is represented in Figure 20.7.

When canned foods are placed inside the thermal treatment equipment, there exists an induction period before the temperature of the food begins to rise. This makes the heat penetration curve in Figures 20.6 and 20.7 nonlinear at the beginning of the operation.

It is possible to determine the apparent initial temperature T_{ip} from Figure 20.7 by extending the straight line of the curve until it intersects with the ordinate axis. This temperature, along with the

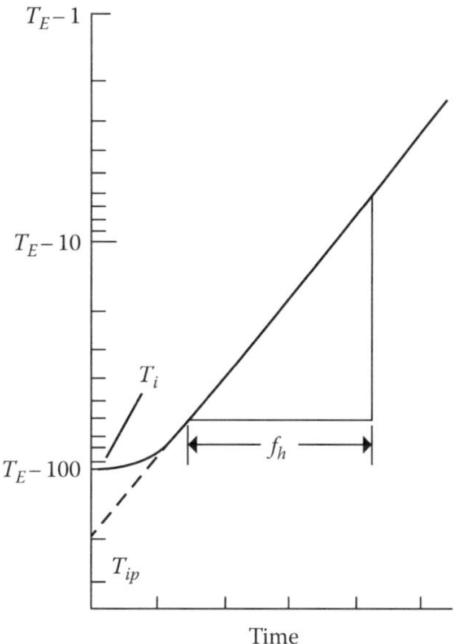

FIGURE 20.7 Graph of the heat penetration curve data.

initial temperature T_i and that of the treatment equipment T_e, is used to calculate j_h. The value of f_h is obtained from the straight part of the heat penetration curve taking the inverse of the slope of this straight line for one log cycle on the heating curve.

20.3.2 Methods to Determine Lethality

20.3.2.1 Graphical Method

The level of sterilization is expressed as a treatment time and temperature for each type of product, shape, and size of container. If the product is treated at a fixed temperature, the treatment time can be directly obtained from Equation 20.10. However, the temperature of the product varies, not only with position, but also with time. For this reason, it is commonly established that time should be measured from the instant when the work temperature is reached until the end of the heating process.

If temperature varies with treatment time, integration is needed to obtain the reduction degree required:

$$\log\left(\frac{C_0}{C}\right) = \int_0^t \frac{dt}{D_T} \tag{20.23}$$

It is necessary to know how the treatment or decimal reduction time varies, at each temperature, with heating time, in order to solve the integral term of Equation 20.23. Thus, it is necessary to previously know the variation of temperature with time (heat penetration curve). There exist different solving methods; however, only the TDT curve method will be used in this book.

In order to ensure an adequate thermal treatment, the integral term of Equation 20.23 should be greater than the reduction degree n previously established for each type of microorganism and product:

$$\int_0^t \frac{dt}{D_T} = \int_0^t \frac{L}{D_R}\,dt \geq n \tag{20.24a}$$

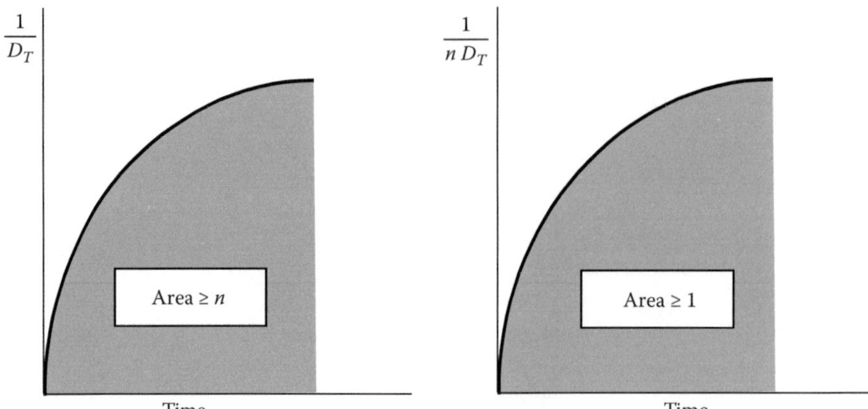

FIGURE 20.8 Graphical method to evaluate lethality.

Since the treatment or TDT at a determined temperature is a function of the decimal reduction time ($F_T = nD_T$), it is obtained that

$$\int_0^t \frac{dt}{F_T} = \int_0^t \frac{L}{F_R} dt \geq 1 \tag{20.24b}$$

The value of the integral term can be graphically solved by plotting $1/t_T$ versus time and obtaining the area below the curve between two times, in such a way that its value is at least 1. In the case of using the decimal reduction time, $1/D_T$ should be plotted against time, and the value of the area should be greater than the reduction degree n (Figure 20.8).

Equation 20.8a and b are used to obtain the values of treatment or decimal reduction times for each temperature.

20.3.2.2 Mathematical Method

The mathematical method uses Equation 20.14 to calculate the processing time. This time can be obtained if the values of f_h, of the temperature difference $T_e - T_i$, and of the thermal induction factor j_h, as well as the g value of the temperature difference between the treatment equipment T_e and the slowest heating point at the end of thermal processing, are known. The three first parameters can be obtained from the heat penetration curve (Figure 20.7), while factor g is not so easily obtained, since it cannot be known in advance which is the final temperature to ensure that an adequate food sterility level is achieved.

This method was originally developed by Ball (1923) and is based on the integration of the lethal effects produced by the time–temperature relationship. The method can be applied when the heat penetration curve, in semilogarithmic coordinates, is a straight line after an initial induction period. The equation developed by Ball (Equation 20.14) takes into account the lethal effects of the cooling stage. A parameter U is defined as the time required to accomplish a certain degree of microbial inactivation at the temperature of the treatment equipment, equivalent to the value F of the process:

$$U = FF_R = F10^{(T_R - T_e)/z} \tag{20.25}$$

where F_R is the time at temperature T_R, equivalent to 1 min at 121°C (Stumbo et al., 1983).

Stumbo and Longley (1966) suggested the incorporation of another parameter f_h/U, and tables showing this parameter as a function of g for different values of z have been published (Stumbo et al., 1983). Table 20.2 shows one example for the case of $z = 10$°C. To obtain these tables, Stumbo used data of different points of the container in order to have different values of j_c. It is assumed in

TABLE 20.2
Relationships f_h/U: g for Values of z = 10°C

f_h/U	Values of g (°C) When j Is								
	0.4	**0.6**	**0.8**	**1.0**	**1.2**	**1.4**	**1.6**	**1.8**	**2.0**
0.2	2.27×10^{-5}	2.46×10^{-5}	2.64×10^{-5}	2.83×10^{-5}	3.02×10^{-5}	3.20×10^{-5}	3.39×10^{-5}	3.58×10^{-5}	3.76×10^{-5}
0.4	7.39×10^{-3}	7.94×10^{-3}	8.44×10^{-3}	9.00×10^{-3}	9.50×10^{-3}	1.00×10^{-2}	1.06×10^{-2}	1.11×10^{-2}	1.16×10^{-2}
0.6	4.83×10^{-2}	5.24×10^{-2}	5.66×10^{-2}	6.06×10^{-2}	6.44×10^{-2}	6.83×10^{-2}	7.28×10^{-2}	7.67×10^{-2}	8.06×10^{-2}
0.8	0.126	0.136	0.148	0.159	0.171	0.182	0.194	0.205	0.217
1	0.227	0.248	0.269	0.291	0.312	0.333	0.354	0.376	0.397
2	0.85	0.92	1.00	1.07	1.15	1.23	1.30	1.38	1.45
3	1.46	1.58	1.69	1.81	1.93	2.04	2.16	2.28	2.39
4	2.01	2.15	2.30	2.45	2.60	2.74	2.89	3.04	3.19
5	2.47	2.64	2.82	3.00	3.17	3.35	2.53	3.71	3.88
6	2.86	3.07	3.27	3.47	3.67	3.88	4.08	4.28	4.48
7	3.21	3.43	3.66	3.89	4.12	4.34	4.57	4.80	5.03
8	3.49	3.75	4.00	4.26	4.51	4.76	5.01	5.26	5.52
9	3.76	4.03	4.31	4.58	4.86	5.13	5.41	5.68	5.96
10	3.98	4.28	4.58	4.88	5.18	5.48	5.77	6.07	6.37
20	5.46	5.94	6.42	6.89	7.37	7.84	8.32	8.79	9.27
30	6.39	6.94	7.56	8.11	8.72	9.33	9.89	10.5	11.1
40	7.11	7.72	8.39	9.06	9.72	10.4	11.1	11.7	12.4
50	7.67	8.39	9.11	9.83	10.6	11.3	12.0	12.7	13.4
60	8.22	8.94	9.72	10.5	11.2	12.0	12.7	13.5	14.3
70	8.67	7.78	10.2	11.1	11.8	12.6	13.4	14.2	15.0
80	9.01	9.89	10.7	11.6	12.3	13.2	14.0	14.8	15.6
90	9.44	10.28	11.2	12.0	12.8	13.7	14.5	15.3	16.2
100	9.78	10.7	11.6	12.4	13.3	14.1	15.0	15.8	16.7

these tables that heat transfer occurs only by conduction, and nonarbitrary assumptions regarding the shape of the temperature profiles at the cooling stage are made. The range of j_c and rounding errors in numeric calculations limits the use of this method. There are graphs in the literature that allow us to obtain the value of g at the end of the treatment. In these graphs, f_h/U is plotted versus log g in semilogarithmic coordinates for different values of z, and different curves can be obtained depending on the value of j (Toledo, 1980; Teixeira, 1992).

The processing time calculated by this method assumes that when the containers are placed in the treatment equipment, the temperature of the latter is T_e. This only occurs in equipment that operates in a continuous form. However, in batch processing there is a come-up time until the equipment reaches the treatment temperature. In order to take into account the contribution of this come-up time to global lethality, Ball (1923) assumed that 40% of this come-up period contributes to thermal processing. In this way the real processing time is

$$t_R = B - 0.4t_l \tag{20.26}$$

in which t_l is the come-up time required by the equipment to reach the treatment temperature T_e.

20.4 THERMAL TREATMENT IN ASEPTIC PROCESSING

Different types of treatment equipment are used for the aseptic packaging of food products, where food is thermally treated to reduce its microbial load in an adequate form and then packaged under aseptic conditions. This equipment consists of three stages. The first one is a heating stage, where the food goes from its initial temperature to the treatment temperature. In the second stage, the food receives a thermal treatment at a constant temperature, and this is called the holding stage. The third part is a cooling stage. Once the food is treated, it is placed in a sterile container that is sealed in an aseptic environment. Heat exchangers are used in the heating and cooling stages, and they can be one of the different types described in Chapter 13. During the holding stage, food usually circulates through a cylindrical pipe receiving heat from the wall of a tube of a heat exchanger, generally of concentric tubes.

In the treatment equipment, the residence time of a fluid in the treatment equipment should be at least the minimum time required, so as to reduce the microbial load in the reduction degree desired.

The residence time, in any treatment equipment, is the relationship between the volume of the device V and the volumetric flow rate q:

$$t = \frac{V}{q} \tag{20.27}$$

In the case of fluid foods circulating in tubular equipment, the residence time is obtained by the expression

$$t = \frac{L}{v} \tag{20.28}$$

where

L is the tube length

v is the linear circulation velocity of the fluid through the tube

Generally, in aseptic processing, it is necessary to calculate the length of the holding tube by using Equation 20.28. In this equation, the velocity depends on the circulation regime and type of fluid.

For Newtonian fluids that circulate under turbulent regime, a mean velocity (v_m) is used. However, if the fluid circulates under laminar regime (Re < 2100), the maximum velocity, which is a function of the mean velocity, should be used:

$$v_{max} = 2v_m \tag{20.29}$$

The v_m/v_{max} relationships for Newtonian and power law fluid can be obtained from Figures 20.9 and 20.10.

Calculation problems can arise in aseptic processing when the product to be treated consists of two phases, with particles suspended within a fluid, as in the case of soups. In these cases, the problem lies with the fact that when the fluid phase has already reached the processing temperature, the solid particles have a lower temperature. In order to ensure adequate thermal processing, calculations should be made with respect to the slowest heating points, which coincide with the

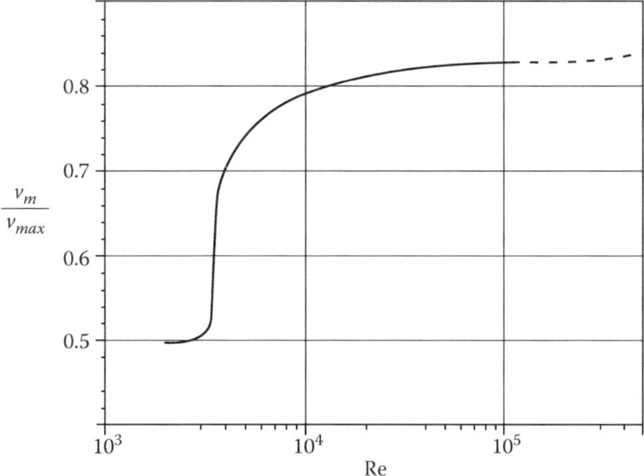

FIGURE 20.9 Variation of v_m/v_{max} with the Reynolds number for Newtonian fluids.

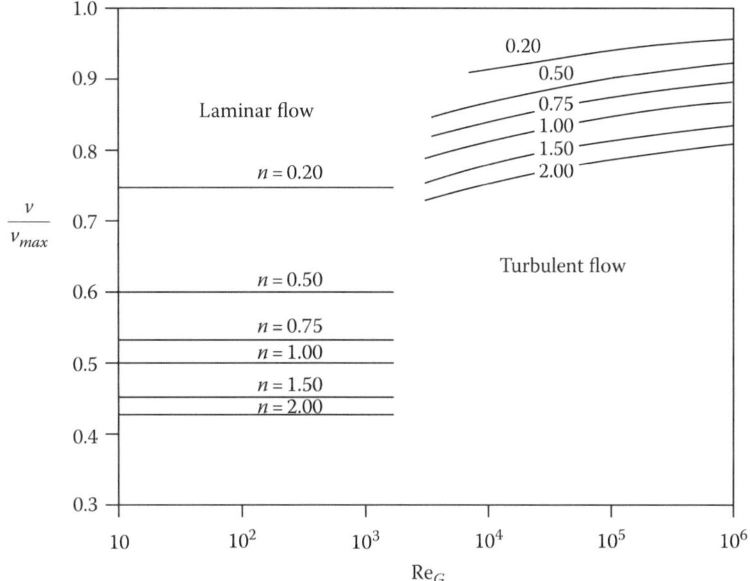

FIGURE 20.10 Variation of v_m/v_{max} with the generalized Reynolds number for power law fluids.

geometric center of the solid particles. This causes the residence time in the holding tube to be greater and particles to need a greater thermal treatment. Also, solid particles can become damaged (Ohlsson, 1994). It is important to know the heat transfer to the interior of the particles and to obtain the evolution of the temperature at their geometric center with processing time. This is a heating process under unsteady state, and one of the methods described in Section 12.3 should be applied. Therefore, the heat transmission coefficient from the fluid to the particles should be previously evaluated. In the literature there are equations that allow us to calculate heat transfer coefficients of fluids circulating by the exterior of solids. Thus, for fluids that circulate outside spheres, an expression that relates the Nusselt, Reynolds, and Prandtl numbers is (Ranz and Marshall, 1952a)

$$(\mathrm{Nu}) = 2 + 0.6(\mathrm{Re})^{0.5}(\mathrm{Pr})^{1/3} \tag{20.30}$$

If the fluid and the particles circulate at the same velocity, then the Reynolds number is cancelled and (Nu) = 2.

According to Chandarana et al. (1990), if the fluid is water, then the following expression can be employed:

$$(\mathrm{Nu}) = 2 + 1.33\times 10^{-3}(\mathrm{Re})^{1.08} \tag{20.31}$$

This equation is valid for $287 < (\mathrm{Re}) < 880$.

If the fluid that circulates outside the particles is a starch solution, then Equation 20.32 should be used:

$$(\mathrm{Nu}) = 2 + 2.82\times 10^{-3}(\mathrm{Re})^{1.16}(\mathrm{Pr})^{0.89} \tag{20.32}$$

This equation is valid for $1.2 < (\mathrm{Re}) < 27$ and $9.5 < (\mathrm{Pr}) < 376$.

In all these equations the velocity that should be used in the calculation of the Reynolds number is the relative velocity with which particles circulate inside the tube with respect to the fluid.

In the case of fluids with suspended particles, equations that allow the direct calculation of processing times can also be applied. Hence, the distribution of temperatures in the particles should be obtained first. The F value is related to the diffusivity, size of particle (R), and position in the tubing (r) by means of the equation

$$F(t) = j\exp(-Bt)$$

$$F_0 = \frac{E\left[A\exp(-Bt_p)\right]}{B}10^{(T_A - T_R)/z}$$

$$A = j2.3^{(T_A - T_R)/z} \tag{20.33}$$

where

T_A is the external temperature of the particle
T_R is the reference temperature
t_p is the processing time
$E(x)$ is the exponential integral, while j and B can be obtained by the relationships given in Table 20.3 for different types of particles, in which α is the thermal diffusivity of the particle and η_l is the viscosity of the carrying fluid

TABLE 20.3
Parameters for the Calculation of *F* in Equation 20.33

Geometry of the Particle	Position in the Particle	j	B
Sphere of radius R	Center	2	$\frac{\pi^2\alpha}{R^2}$
	Any	$\frac{2}{\pi}\frac{R}{r}\sin\frac{\pi r}{R}$	
Parallelepiped $(2X)(2Y)(2Z)$	Center	2.0641	
	Any	$2.0641\cos\frac{\pi}{2}\frac{x}{X}\cos\frac{\pi}{2}\frac{y}{Y}\cos\frac{\pi}{2}\frac{z}{Z}$	$\frac{\pi^2\alpha}{4}\left(\frac{1}{X^2}+\frac{1}{Y^2}+\frac{1}{Z^2}\right)$
Cylinder $(2R)(2L)$	Center	2.0397	$\frac{\alpha}{4}\left(\frac{\pi^2}{L^2}+\frac{\eta_i^2}{R^2}\right)$

20.4.1 Residence Time

In tubular devices in which the fluid circulates under piston flow, the residence time of the microorganisms on the fluid coincides with the mean residence time. For this reason, the calculation of thermal treatments will not present problems. However, it can occur that the fluid does not circulate under piston flow. In this case there will be microorganisms entering the equipment at the same time but with different residence time.

Thus, for example, if the fluid circulates under laminar regime through a cylindrical pipe, there exists a parabolic velocity profile in such way that those microorganisms that enter through the central stream will have a higher velocity than microorganisms traveling close to the walls, and the residence time of the former will be shorter. This can cause the treatment to be inadequate, since the residence time of these microorganisms is shorter than the mean residence time. For this reason, it is essential to know the maximum velocity and to use it for the calculations, since it corresponds to the minimum residence time in the treatment equipment.

Figure 20.11 shows different velocity profiles in tubular devices for different circulation regimes in pipes, and it presents a scheme of an agitated vessel.

20.4.2 Dispersion of Residence Time

In real thermal treatment equipment, the microorganisms that enter at the same instant do not stay in the equipment during the same time. It is convenient to know what is the dispersion of the residence time to perform an adequate calculation of treatment time.

Some parameters that will be used in other sections are defined next.

The *residence time* of a microorganism in the equipment is the time interval between the inlet and the outlet.

Age is the time that passes from the inlet to a given instant.

Function of distribution of internal ages (I) is the curve that represents the distribution of ages inside the treatment equipment.

Function of distribution of external ages (E) is the curve that represents the distribution of ages of the microorganisms that leave the treatment equipment, that is, the residence time.

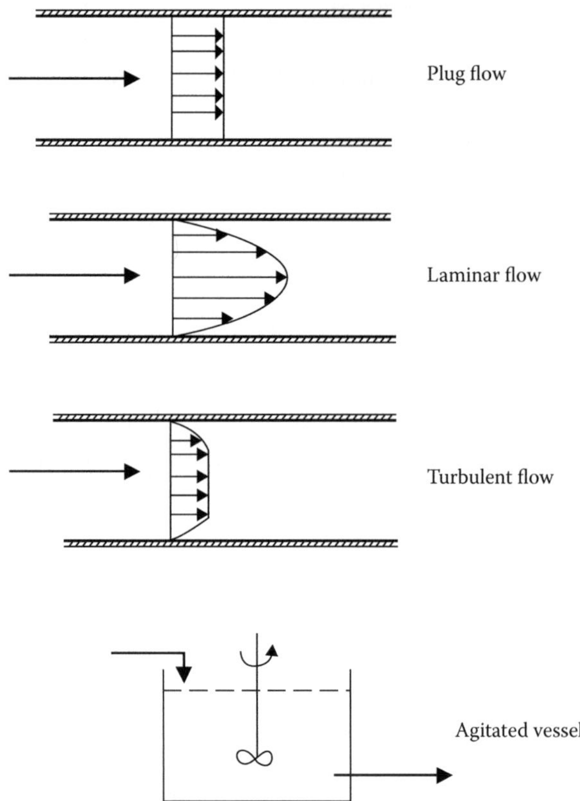

FIGURE 20.11 Flow of fluids in tubing and in an agitated vessel.

In order to determine these functions, it is necessary to use experimental techniques. One of these techniques, the *tracer response*, consists of introducing a tracer into the entering stream and measuring its concentration at the exit.

The more common entrance functions are those of step and of impulse or delta. In the first function, a tracer is introduced at a certain concentration at a given instant, and it is maintained along all the experiment. On the other hand, in the delta or impulse entrance, a certain level of tracer is introduced at once (Figure 20.12). In this figure, C_t is the concentration of the tracer at a given instant t, while C_0 is the total concentration of the tracer.

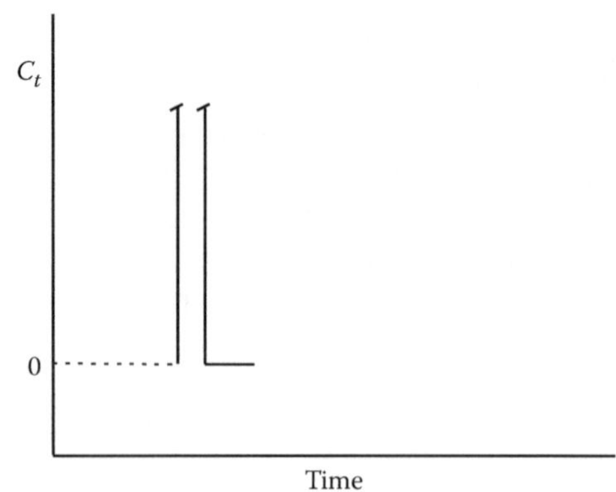

FIGURE 20.12 Entrance of tracer as impulse.

The *response functions* of each entrance function are represented by F and C, for step and impulse, respectively.

It should be pointed out that all the functions defined in this section should be normalized.

The relationships between the different functions of entrance and response are the following:

$$I = 1 - F \tag{20.34a}$$

$$E = C \tag{20.34b}$$

$$E = \frac{dF}{d\theta} = -\frac{dI}{d\theta} \tag{20.34c}$$

in which θ is a dimensionless time variable defined as the quotient between time t and the mean residence time t_M.

The function that will be used in the following sections for continuous thermal treatment calculations is the function of distribution of external ages E.

20.4.3 Distribution Function *E* under Ideal Behavior

The circulation of a food in a thermal treatment equipment is considered as ideal when such equipment is a perfect mixing agitated vessel or a tubular equipment with either laminar or piston flow through it.

The expressions for the function of distribution of external ages for each one of the equipment mentioned are given next as a function of the real and dimensionless time variable:

Perfect mixing agitated vessel:

$$E = \frac{1}{t_M}\exp\left(-\frac{t}{t_M}\right) \tag{20.35a}$$

$$E_\theta = \exp(-\theta) \tag{20.35b}$$

Tubular equipment with plug flow:

$$E = \delta(t - t_M)$$

which is known as delta distribution function or Dirac distribution function.

Tubular equipment with laminar flow:

$$E = \frac{t_M^2}{2t^3} \tag{20.36a}$$

$$E_\theta = \frac{1}{2\theta^3} \tag{20.36b}$$

Figure 20.13 presents all these functions E as a response of an impulse entrance.

It is possible to obtain distribution functions of external ages for *non-Newtonian* fluids. Thus, for fluids that follow the *power law*, the function is expressed as

$$E = \frac{2n}{3n+1}\frac{t_M^3}{t^3}\left(1 - \frac{n+1}{3n+1}\frac{t_M}{t}\right)^{(n-1)/(n+1)} \tag{20.37}$$

in which n is the flow behavior index.

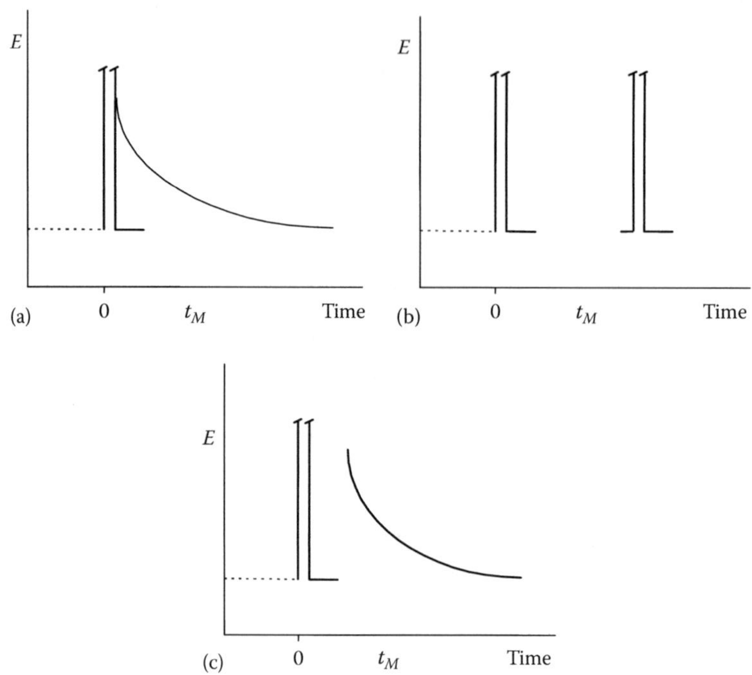

FIGURE 20.13 Distribution function of external ages for different types of flow: (a) agitated, (b) plug, and (c) laminar flow.

For Bingham plastics that circulate in pipes, a velocity profile has been created in which the central stream circulates under piston flow. The distribution function of external ages is

$$E = \frac{(1-m)t_0 t_M^3}{t^3}\left[1-m+\frac{m}{\sqrt{1-\sigma_0 (t_M/t)}}\right] \tag{20.38}$$

in which

$$m = \frac{r}{R} = \frac{\sigma_0}{\sigma_{WALL}} \quad \text{and} \quad t_0 = \frac{m^2+2m+3}{6}$$

where

σ_0 is the yield stress
σ_{WALL} is the value of the shear stress on the wall
t_0 is a parameter that represents the residence time of the fluid in the central stream under piston flow

20.4.4 Distribution Function *E* under Nonideal Behavior

The distribution function of external ages is the residence time of each element of the fluid in the treatment equipment. This function depends on the path followed by the fluid inside the container.

If a container for thermal treatment is considered where the food enters with a volumetric flow q, the distribution function of external ages can be represented according to what is indicated in Figure 20.14. In order to normalize this function, the area below the curve should have a value equal to one:

$$\int_0^\infty E\,dt = 1 \tag{20.39}$$

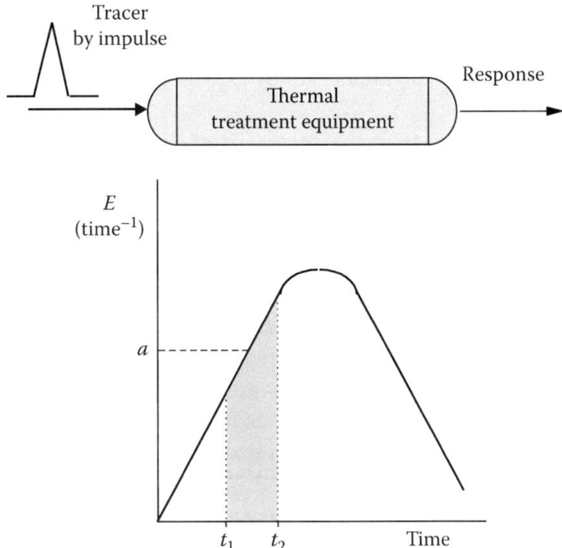

FIGURE 20.14 Obtaining of the curve of external ages.

The fraction of fluid that leaves the container between any two times t_1 and t_2 will be

$$E\Delta t = a\left(t_1 - t_2\right) \tag{20.40}$$

where a is the mean value of the distribution function of external ages in this time interval.

The fraction of fluid that leaves at a time smaller than t_1 is

$$\int_0^{t_1} E\,dt$$

while the fraction that exists times greater than t_1 is

$$\int_{t_1}^{\infty} E\,dt = 1 - \int_0^{t_1} E\,dt$$

As indicated before, the function E should be experimentally obtained by injecting a tracer and observing the response. For the specific case in which the tracer is introduced as an impulse function in the food stream, it should be operated as indicated next. If V is the volume of the thermal treatment device, through which a stream with volumetric flow q circulates, an amount m of tracer is introduced at a certain time. The amount of tracer is analyzed in the outlet stream. If C_i is this concentration, then it is tabulated or plotted against time, to obtain the distribution function of the tracer's concentration.

The area below this curve will be

$$\int_0^{\infty} C_i\,dt = \sum_i C_i \Delta t = \frac{M}{q} \tag{20.41}$$

in which M is the total quantity of tracer injected.

The curve E is obtained from this curve, but since it should be normalized, the concentration obtained at each time is divided by the area given by Equation 20.41:

$$E = \frac{C_i}{\int_0^{\infty} C_i \, dt} = \frac{qC_i}{M} \tag{20.42}$$

It is interesting to point out that many food fluids that are thermally treated contain suspended solid particles. In these cases, the distribution curves of external ages show a double node, in such a way that the response curve begins to appear at a time greater than half of the mean residence time.

Since in later calculations mean residence times are required, they have to be calculated. Also, the variance gives an idea of the dispersion of the residence times, since such dispersion is greater as the variance is greater. For this reason, equations that allow calculation of the mean and the variance will be stated.

The *mean* allows us to calculate the mean residence time t_M. In general, this value is calculated from the equation

$$t_M = \frac{\int_0^{\infty} t_i C_i \, dt}{\int_0^{\infty} C_i \, dt} \tag{20.43}$$

In case the number of measurements is discrete, the integrals can be substituted by summations

$$t_M = \frac{\sum_0^{\infty} t_i C_i \Delta t}{\sum_0^{\infty} C_i \Delta t} \tag{20.44}$$

For continuous functions with discrete measurements with equal time intervals, the distribution function of external ages can be used in the calculation of the mean

$$t_M = \int_0^{\infty} t_i E \, dt = \sum_0^{\infty} t_i E_i \Delta t \tag{20.45}$$

The *variance* can be obtained from the expression

$$\sigma^2 = \frac{\int_0^{\infty} t_i^2 C_i \, dt}{\int_0^{\infty} C_i \, dt} - t_M^2 \tag{20.46}$$

When the number of values is discrete, the integral can be substituted by a summation

$$\sigma^2 = \frac{\sum_0^{\infty} t_i^2 C_i \Delta t}{\sum_0^{\infty} C_i \Delta t} - t_M^2 \tag{20.47}$$

For continuous curves or discrete measurements at equal time intervals, the variance can be calculated by the following equation:

$$\sigma^2 = \int_0^\infty t_i^2 E\,dt - t_M^2 = \sum_0^\infty t_i^2 E_i \Delta t_i - t_M^2 \tag{20.48}$$

20.4.5 Application of the Distribution Model to Continuous Thermal Treatment

The destruction of microorganisms or of some thermosensitive factors of food occurs in the thermal treatment equipment. Such destruction follows a first-order kinetics; therefore the equipment can be considered as a reactor in which a first-order reaction is performed.

The following expression can be applied for the calculation of the mean concentration C_M in the stream that leaves the thermal treatment equipment as a function of the inlet concentration C_0:

$$C_M = C_0 \int_0^\infty \left(\frac{C}{C_0}\right)_t E\,dt \tag{20.49}$$

Since the destruction kinetics is of first order

$$C = C_0 \exp(-kt)$$

then, substitution in Equation 20.49 yields

$$C_M = C_0 \int_0^\infty \exp(-kt)_t\, E\,dt \tag{20.50}$$

When the number of values is discrete, the latter equation can be expressed using a summation

$$C_M = C_0 \sum_0^\infty \left[\exp(-kt)\right]_t E\Delta t \tag{20.51}$$

Equations 20.50 and 20.51 allow calculation of thermal treatments of products that circulate in a continuous form through the thermal treatment equipment.

For the case of ideal behavior (perfect mixing agitated vessel, piston flow, and laminar flow through pipes), there are analytical solutions to Equation 20.50. Such solutions are given next for these three cases.

The distribution function of external ages for a fluid that is treated in a *perfect mixing agitated vessel* is given by Equation 20.35a, which should be substituted in Equation 20.50, and the integration is performed:

$$C_M = C_0 \int_0^\infty \exp(-kt)_t \left[\frac{1}{t_M}\exp\left(-\frac{t}{t_M}\right)\right] dt \tag{20.52}$$

Integration yields

$$C_M = C_0 \frac{1}{1+kt} \tag{20.53}$$

For circulation under *plug flow*, the distribution function of external ages is the Dirac delta function that when substituted in Equation 20.50 yields

$$C_M = C_0 \int_0^\infty \exp(-kt)_t \left[\delta(t - t_M)\right] dt = C_0 \exp(-kt_M) \tag{20.54}$$

For circulation under *laminar flow*, the distribution function of external ages is given by Equation 20.36a. The following expression is obtained by substituting it in Equation 20.50:

$$C_M = C_0 \int_0^\infty \exp(-kt)_t \left(\frac{1}{t_M}\frac{t_M^2}{2t^3}\right) dt \tag{20.55}$$

PROBLEMS

20.1 A thermocouple located in the slowest heating point of a crushed tomato can give the following temperature–time variation:

Time (min):	0	10	30	40	50	60
Temperature (°C):	60	71	100	107	110	113

The can is placed in the center of a pile subjected to a sterilization process in an autoclave, in which the processing temperature is kept at 113°C. Determine the processing time, assuming that the lethal effect of the cooling period is negligible.

Data and notes. The temperature increment needed to decrease the treatment time to the tenth part (z) for a reduction degree of 12 for *Bacillus coagulans* is 10°C. The time required to obtain this reduction degree at 121°C is 3 min.

The treatment time for a given temperature is calculated from Equation 20.8, in which the following is complied, according to the data given in the statement:

$$T_R = 121°\text{C} \qquad F_R = 3\,\text{min} \quad z = 10°\text{C} \qquad n = 12$$

In order to solve the problem, the conditions given by Equation 20.24 should be complied with. First, the effect produced during the heating time has to be evaluated by calculating the values of process time required and decimal reduction at each temperature. It is known that for a given temperature Equation 20.8 is useful to calculate D_T from F_T.

Beginning with the data in the table of the problem statement and Equations 20.8 and 20.12, it is possible to obtain Table P.20.1:

TABLE P.20.1
Values of TDT and Decimal Reduction as a Function of Heating Time

$t_{heating}$ (min)	T (°C)	F_T (min)	D_T (min)	$1/F_T$ (min^{-1})	$1/D_T$ (min^{-1})
0	60	3.8×10^6	3.2×10^5	2.6×10^{-7}	3.2×10^{-6}
10	71	3×10^5	2.5×10^4	3.3×10^{-6}	4.0×10^{-5}
20	85	1.2×10^4	992	8.4×10^{-5}	10^{-3}
30	100	378	31.5	2.7×10^{-3}	0.032
40	107	75.4	6.3	0.013	0.159
50	110	37.8	3.2	0.027	0.312
60	113	18.9	1.6	0.053	0.625

The lethality for sterilization is obtained by plotting $1/F_T$ against time, or $1/D_T$ (Figure 20.8) against time, and graphically integrating the functions obtained. When performing this integration between the initial and final heating times (60 min), it is obtained that

$$\int_0^{60} \frac{dt}{F_T} = 0.65 < 1$$

$$\int_0^{60} \frac{dt}{D_T} = 7.8 < 12$$

This indicates that it is not possible to achieve the sterilization of the product with this heating time and it is required to continue heating at 113°C during a period of time such that the value of the lethality sterilization level complies with the conditions in Equation 20.24.

For F_T, 1 − 0.65 = 0.35 is needed to comply with Equation 20.24b, so the additional time is

$$t_{113} = (0.35)(18.9\text{ min}) = 6.62\text{ min}$$

If the decimal reduction time D_T were used instead, what is needed for sterilization is 12 − 7.8 = 4.2; therefore, the additional time is

$$t_{113} = (4.2)(1.58\text{ min}) = 6.64\text{ min}$$

20.2 A cylindrical can of 6 cm diameter and 15 cm height contains 250 g of pea cream and has a microbial load of 10^5 spores of *Clostridium sporogenes* per each kg of cream. In order to decrease the spore content, the can is subjected to a sterilization process in an autoclave that condenses steam at 121°C. The slowest heating point initially is at 71°C, and the temperature along the thermal treatment evolves in the following way:

t (s)	600	800	1000	1400	1600	2000	2200	2400	2600
T (°C)	94	103	109	113.5	116	118	110	82	65

At 121°C, the *C. sporogenes* spores present the following values: z = 10°C and F = 1.67 min. Calculate the following:

a. The viable spores per kg of cream that exist in the slowest heating point when it reaches the maximum temperature
b. Which is the microbial load at such point after 45 min of treatment?

The treatment time at the different temperatures that is being acquired by the product is calculated from Equation 20.8. According to the data given in the statement of the problem, it is found that

$$T_R = 121°\text{C} \qquad F_R = 3\text{ min} \qquad z = 10°\text{C}$$

Also, *C. sporogenes* $n = 5$

Since $D_T = F_T \cdot n$, it is possible to obtain Table P.20.2:

TABLE P.20.2
Values of TDT and Decimal Reduction as a Function of Heating Time

$t_{heating}$ (s)	T (°C)	F_T (min)	D_T (min)	$1/F_T$ (min^{-1})	$1/D_T$ (min^{-1})
600	94	837	167	1.2×10^{-3}	5.9×10^{-3}
800	103	105	21	9.5×10^{-3}	0.047
1000	109	26.5	5.3	0.038	0.189
1400	113.5	9.4	1.88	0.107	0.532
1600	116	5.3	1.06	0.189	0.943
2000	118	3.3	0.67	0.303	1.500
2400	82	13,265	2,653	7.5×10^{-5}	0.0004
2600	65				

The level of lethality for sterilization is obtained by integration from the beginning of heating up to 2000 s, which corresponds to the time of maximum temperature:

$$\int_0^{2000} \frac{dt}{D_T} = \log\left(\frac{C_0}{C}\right) = 13.29$$

Therefore, the concentration of *C. sporogenes* in the product after 2000 s is

$$C = C_0 \times 10^{-13.29} = 5.1 \times 10^{-9} \text{ spores/kg}$$

For a time equal to 45 min, integration should be performed between the initial time and 2700 s:

$$\int_0^{2700} \frac{dt}{D_T} = \log\left(\frac{C_0}{C}\right) = 16.34$$

Hence, $C = 4.6 \times 10^{-12}$ spores/kg

20.3 A packed food is thermally treated in equipment whose processing temperature is 130°C. It has been obtained that the evolution of temperature at the slowest heating point with processing time is

t (min)	0	6.5	10	12	16.5	23	28	36	44
T (°C)	30	40	50	60	80	100	110	120	125

If the initial microbial load is 10 CFU/container and it is desired to reduce it to 10^{-5} CFU/container, calculate the processing time. The target microorganism for this thermal process has a decimal reduction time of 2 min for a reference temperature of 121°C, and to reduce the treatment time to the tenth part it is required to increase the temperature to 10°C.

This problem will be solved by the method of Ball, using Equation 20.14. The heat penetration curve, similar to that in Figure 20.7, will be plotted in order to calculate the apparent initial temperature. The data of the table included in the statement of the problem are graphically represented in semilogarithmic coordinates as shown in Figure P.20.3.

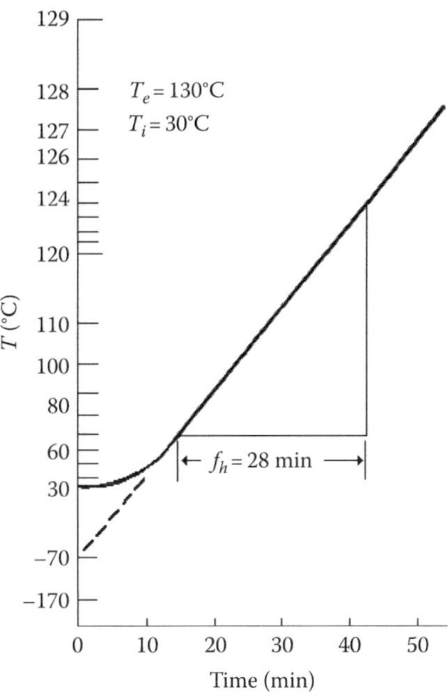

FIGURE P.20.3 Heat penetration curve (Problem 20.3).

Hence, T_{ip} = −70°C. I_h and j_h are calculated from this temperature and Equations 20.15 and 20.16:

$$I_h = T_e - T_i = 130 - 30 = 100°\text{C}$$

$$j_h = \frac{T_e - T_{ip}}{T_e - T_i} = \frac{130-(-70)}{130-30} = 2$$

The value of f_h can also be obtained from Figure P.20.3, since it is the inverse of the slope of the straight part of the heat penetration curve. Therefore, it is obtained that f_h = 28 min.

The value of F for this process must be calculated, then

$$F = D_T \log\left(\frac{N_0}{N}\right) = (2\,\text{min})\log\left(\frac{10}{10^{-5}}\right) = 12\,\text{min}$$

The value of the parameter U is calculated from Equation 20.25:

$$U = F10^{(T_R - T_E)/z} = (12\,\text{min})10^{(121-130)/10} = (12\,\text{min})(0.1259) = 1.51\,\text{min}$$

Therefore

$$\frac{f_h}{U} = 18.53$$

Interpolation in Table 20.2, using the latter quotient (18.53) and j = 2, yields a value of g = 8.84°C.

Now, Equation 20.14 can be applied, since all the variables are known:

$$B = (28\,\text{min})\log\left(\frac{(2)(100)}{8.84}\right) = (28\,\text{min})(1.355) \approx 38\text{ min}$$

This means that the processing time is 38 min.

20.4 A baby food based on apple purée is thermally treated at a rate of 1500 kg/h employing an aseptic packaging process. The product is heated from 22°C to 90°C in a plate heat exchanger. Then it is introduced into a concentric tube heat exchanger that has an internal diameter of 5 cm. Condensing steam circulates by the annular space that allows the maintenance of the product temperature at 90°C while it remains in the heat exchanger. Commercial sterilization of the product, at 90°C, is achieved if it remains for 90 s at such temperature. Determine the length that the residence tube should have to ensure that the food receives an adequate thermal treatment.

Data. Food properties: density 1200 kg/m³. It behaves as a power law fluid with a consistency index of 2.4 Pa s^n and presents a flow behavior index of 0.5.

The length of the residence tube is obtained from Equation 20.28, with a time of 90 s, which is the time needed to ensure an adequate thermal treatment for the product. The velocity to be used in this equation is the maximum velocity, which in turn is a function of the mean velocity at which the product circulates inside the tubing.

The mean circulation velocity of the product is obtained from the continuity equation

$$v_m = \frac{4w}{\rho\pi d^2} = \frac{4\left(1500\,\text{kg/h}\left(1\,\text{h}/3600\,\text{s}\right)\right)}{\left(1200\,\text{kg/m}^3\right)\pi\left(0.05\,\text{m}\right)^2} = 0.177\,\text{m/s}$$

Since it is a fluid that rheologically behaves according to the power law, the value of the generalized Reynolds number should be previously calculated using the following equation:

$$\text{Re}_G = \frac{d^n v^{2-n}\rho}{8^{n-1}k}\left(\frac{4n}{1+3n}\right)^n$$

$$\text{Re}_G = \frac{\left(0.05\,\text{m}\right)^{0.5}\left(0.177\,\text{m/s}\right)^{2-0.5}\left(1200\ \text{kg/m}^3\right)}{8^{0.5-1}\left(2.4\,\text{Pa s}^{0.5}\right)}\left(\frac{4(0.5)}{1+3(0.5)}\right)^{0.5} \approx 21$$

This value means that the product circulates inside the pipe under laminar regime. Since the product exhibits a power law rheological behavior, the maximum speed is obtained from Figure 20.10. For a flow behavior index $n = 0.5$, the mean velocity/maximum velocity relationship is

$$\frac{v_m}{v_{max}} = 0.6$$

Hence, the value of the maximum velocity is

$$v_{max} = 0.295\,\text{m/s}$$

Therefore, the length of the residence tube should be

$$L = \left(0.295\,\text{m/s}\right)\left(90\,\text{s}\right) = 26.55\,\text{m}$$

20.5 An aseptic processing system is used to treat a vegetable soup that contains small pieces of meat in suspension. It can be assumed that the carrier fluid is an aqueous solution and the meat particles are spheres with a diameter equal to 12 mm that circulate at a velocity of 0.002 m/s with respect to the carrier fluid. The fluid has a temperature of 150°C at the entrance of the residence tube, while the meat particles have a uniform temperature of 90°C. In order to adequately process the soup, a microbial reduction degree of 12 in the center of the particles should be attained. If the volumetric flow with which the soup circulates is 30 L/min, calculate the length that the residence tube should have in order to achieve the desired microbial reduction if the internal diameter of the tube is 4.5 cm.

Data. Properties of the aqueous solution: density 1000 kg/m^3. Thermal conductivity 0.58 W/(m °C). Viscosity 1.5 mPa s. Specific heat 4.1 kJ/(kg °C).

Thermal diffusivity of meat 1.3 × 10^{-7} m^2/s. Thermal conductivity for meat, assume the same water value.

The microorganism contained in the soup has a decimal reduction time of 1.5 s at 121°C, and in order to reduce the treatment time to the tenth part, it is necessary to increase the temperature to 10°C.

When the soup enters the residence tube, the carrier fluid has a temperature higher than the temperature of the meat particles, so inside the tube there will be heat transfer from the fluid to the particles. Thus, the temperature of such particles will progressively increase. It is required to calculate the evolution of temperature at the particles' geometric center in order to determine the lethality required for the product to be adequately processed.

The heating process occurs under unsteady state, so the temperature of the particles' center should be calculated according to the description in Chapter 16. Initially, the individual heat transfer coefficient will be calculated using Equation 20.32 since the fluid is an aqueous solution. The Reynolds, Prandtl, and Nusselt numbers are

$$(\text{Re}) = \frac{\left(1000\,\text{kg/m}^3\right)\left(0.002\,\text{m/s}\right)\left(0.012\,\text{m}\right)}{1.5\times10^{-3}\,\text{Pa s}} = 16$$

$$(\text{Pr}) = \frac{\left(4100\,\text{J/(kg°C)}\right)\left(1.5\times10^{-3}\,\text{Pa s}\right)}{0.58\,\text{J/(s m°C)}} = 10.8$$

$$(\text{Nu}) = 2 + 2.82\times10^{-3}\left(16\right)^{1.16}\left(10.6\right)^{0.89} = 2.575$$

The individual heat transfer coefficient by convection is

$$h = 249\,\text{W/(m}^2\text{°C)}$$

In order to apply the method described in Section 16.3, the following dimensionless numbers should be calculated:

$$m = \frac{k}{hr_0} = \frac{\left(0.58\,\text{W/(m °C)}\right)}{\left(249\,\text{W/(m}^2\,\text{°C)}\right)\left(0.006\,\text{m}\right)} \approx 0.4$$

$$n = \frac{r}{r_0} = 0$$

$$(\text{Fo}) = \frac{\alpha t}{\left(r_0\right)^2} = 3.417\times10^{-3}t, \text{ expressing the time } t \text{ in seconds}$$

Figure 16.14 is used to calculate the temperature of the slowest heating point. Table P.20.5 shows the different data obtained for the times assumed.

The graphical method described in Section 20.3.2 is used to calculate lethality. In order to calculate the TDTs, it is supposed that the desired reduction degree is $n = 12$; hence

$$F_R = nD_R = (12)(1.5\,\text{s}) = 18\ \text{s}$$

Calculation of lethal velocity is made by Equation 20.9, while the global lethality is calculated using Equation 20.24. The last columns of Table P.20.5 present the values of these parameters for the different processing times.

In this table it is observed that a time between 80 and 100 s yields an integral value equal to 1, and this time is 85 s:

$$\int_0^{85} \frac{L\,dt}{F_R} = 1$$

TABLE P.20.5
Calculation of the Global Lethality

Time (s)	(Fo)	T (°C)	$L = 10^{(T-121)/10}$	$\frac{L}{F_R}$ (s^{-1})	$\int_0^t \frac{L}{F_R}$
0	0	90			
30	0.103	102	0.0126	0.0007	
60	0.205	114	0.1995	0.0111	0.175
80	0.273	120	0.7943	0.0441	0.743
100	0.342	136	1.8197	0.1011	2.118

The linear circulation velocity in the residence tube is calculated from the continuity equation

$$v = \frac{4q}{\pi d^2} = \frac{4\left(30\,\text{L/min}\left(1\,\text{m}^3/10^3\,\text{L}\right)\left(1\,\text{min}/60\,\text{s}\right)\right)}{\pi\left(0.045\,\text{m}\right)^2} \approx 0.314\ \text{m/s}$$

Since it is a laminar regime, the relationship between the maximum and the mean velocities is (Equation 20.29) $v_{max} = 2v = 0.628\text{m/s}$

Thus, the length of the residence tube is

$$L = \left(0.628\,\text{m/s}\right)\left(85\,\text{s}\right) = 53.38\,\text{m}$$

20.6 Milk flowing at 1000 kg/h that contains the tuberculosis bacillus with a concentration equal to 10^8 CFU/cm^3 is fed into a tubular thermal treatment equipment having the objective of reducing the microorganism content down to 0.01 CFU/cm^3. The tubular section has an internal diameter of 1 in., and the density of milk is 1030 kg/m^3. The product is kept at 71°C during the thermal treatment. Determine the length of the tubular section of treatment if the distribution of external ages follows the δ function of Dirac.

What would be the final microbial load if the treatment equipment were a 500 L perfect mixing agitated vessel?

The treatment time, for a temperature of 71°C = 159.8°F and according to Figure 20.2, should be $F_{71°C}$ = 0.17 min = 10.2 s.

It is assumed that the reduction degree is n = *12*.

So the decimal reduction time is D_T = 10.2/12 = 0.85 s.

The destruction constant of the tuberculosis bacillus is

$$k = \frac{2.303}{D_T} = 2.71\ \text{s}^{-1}$$

The distribution function of external ages, for the tubular equipment, is the delta of Dirac, $\delta(t - t_M)$, while the mean concentration at the outlet is given by Equation 20.54:

$$10^{-2} = 10^{8}\exp\left(-2.71t_M\right)$$

Hence, t_M = 8.5 s, which is the mean residence time in the equipment.

The circulation velocity of milk is calculated from the continuity equation

$$v = \frac{4w}{\rho\pi d^2} = 0.532\ \text{m/s}$$

The length of the tubular equipment is

$$L = (0.532\ \text{m/s})(8.5\ \text{s}) = 4.52\ \text{m}$$

For the agitated vessel, the distribution function of external ages is given by Equation 20.35a, while the mean concentration at the exit of the vessel is calculated by Equation 20.53. In this case, the residence time is obtained by dividing the volume of the vessel by the fluid's circulation volumetric flow:

$$t = \frac{V}{q} = \frac{\rho V}{w} = 1854\ \text{s}$$

The concentration of the tuberculosis bacillus in the milk that leaves the agitated vessel is (Equation 20.53)

$$C = \frac{10^8\ \text{CFU/cm}^3}{1 + (2.71\ \text{s}^{-1})(1854\ \text{s})} = 2\times10^4\ \text{CFU/cm}^3$$

20.7 A food fluid that contains 10^8 CFU/cm^3 of a pathogenic microorganism is treated in a tubular equipment at 120°C. The fluid circulates under piston flow regime with a mean residence time equal to 15 s, achieving a reduction degree of 12. Determine the microbial load of the fluid that leaves the treatment equipment.

Calculate the microbial load of this fluid if it were treated in a sterilizing piece of equipment in which the distribution of external ages is given by the following table:

t (s)	0	5	10	15	20	25	30	35
E (s^{-1})	0	0.03	0.05	0.05	0.04	0.02	0.01	0

It can be obtained, from the definition of reduction degree $n = \log(C_0/C)$, for n = 12 and $C_0 = 10^8$ CFU/cm^3, that C = 10^{-4} CFU/cm^3.

Initially, the microorganism inactivation kinetics constant should be calculated. Since the equipment is tubular, Equation 20.54 is applied:

$$10^{-4} = 10^{8} \exp(-k \cdot 15\,\mathrm{s})$$

Hence, $k = 1.84\ \mathrm{s^{-1}}$

The mean residence time for the distribution of external ages given in the statement can be calculated from Equation 20.45:

$$t_M = 5 \times (5 \times 0.03 + 10 \times 0.05 + 15 \times 0.05 + 20 \times 0.04 + 25 \times 0.02 + 30 \times 0.01)$$
$$= 5 \times 3 = 15\ \mathrm{s}$$

Although a mean time $t_M = 15$ s is obtained, which is equal to that of the tubular equipment, the mean concentration at the outlet will be different, as can be observed next.

The mean concentration at the outlet is calculated using Equation 20.51. Results are shown in Table P.20.7:

TABLE P.20.7
Data for Evaluate the Mean Concentration C_M

t (s)	E_t (s^{-1})	$\exp(-kt)$	$\exp(-kt)E_t\Delta t$
5	0.03	10^{-4}	1.5×10^{-5}
10	0.05	10^{-8}	2.5×10^{-9}
15	0.05	10^{-12}	2.5×10^{-13}
20	0.04	10^{-16}	2.0×10^{-17}
25	0.02	10^{-20}	1.0×10^{-21}
30	0.01	10^{-24}	5.0×10^{-26}
			Total 1.50×10^{-5}

Therefore

$$C_M = \left(10^{8}\ \mathrm{CFU/cm^3}\right)\left(1.5 \times 10^{-5}\right) = 1500\ \mathrm{CFU/cm^3}$$

This concentration is practically due to the contribution of the fraction that leaves at 5 s. Although the mean residence time is 15 s, which is equal to the time of the piston flow, the number of CFU is markedly higher, which demonstrates the great influence of the distribution of the residence times.

20.8 A food fluid that contains 10^7 CFU of a pathogenic microorganism per liter is thermally treated in pasteurizing equipment that uses water as heating fluid. The hot water keeps the food's temperature at 75°C. It is known, from previous experiments, that at this temperature the time needed to reduce the initial microorganism population in the food fluid to the tenth part is 6 s. Calculate the concentration of microorganisms in the fluid that leaves the pasteurizing equipment, if it behaves in such a way that the distribution of external ages (E_t) is as follows:

a.

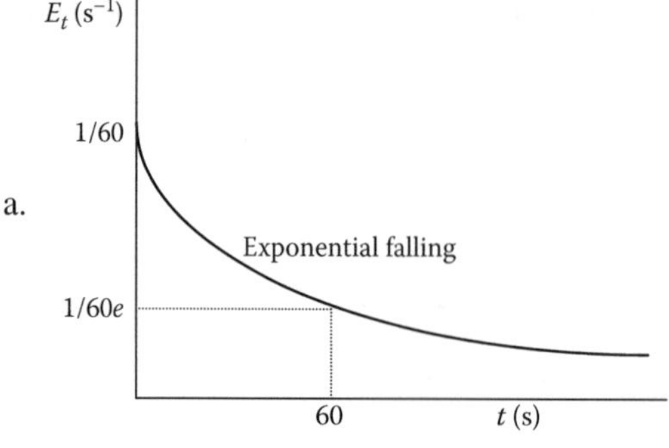

b. Dirac's function, in which the response time to impulse is 45 s

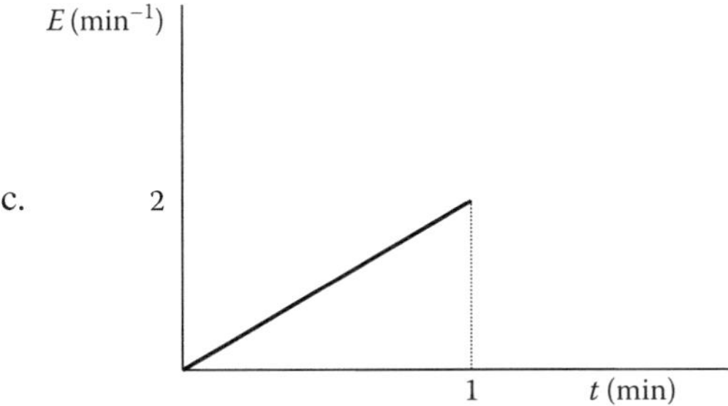

The kinetics of the thermal destruction of microorganisms is of first order, and it is known that the kinetics constant is related to the decimal reduction time by Equation 20.3, obtaining that

$$k = \frac{2.303}{D_T} = \frac{2.303}{6\,\text{s}} = 0.384\ \text{s}^{-1} \cong 23\ \text{min}^{-1}$$

a. In this case, the falling of E_t is exponential, so the type of equation is

$$E_t = a\exp(-bt)$$

in which a and b are constants that have to be determined.

$$\text{For } t = 0 \qquad E = a = \left(\frac{1}{60}\right)\text{s}^{-1}$$

$$\text{For } t = 60\,\text{s} \qquad \frac{1}{60e} = \left(\frac{1}{60}\right)\exp(-b60)$$

obtaining that $b = (1/60)\ \text{s}^{-1}$
The equation of distribution of external ages is

$$E = \left(\frac{1}{60}\right)\exp\left(-\frac{t}{60}\right) \qquad \text{if it is in expressed in s}^{-1}$$

$$E = \exp(-t) \qquad \text{if it is in expressed in min}^{-1}$$

When compared to Equation 20.35a, it is observed that the residence time is

$$t_M = 60\ \text{s} = 1\ \text{min},$$

and also it is a perfect mixing agitated vessel.
The mean concentration of microorganisms at the outlet is obtained from Equation 20.53:

$$C_M = C_0\frac{1}{kt_M + 1} = 10^7\frac{1}{(23\,\text{min}^{-1})(1\,\text{min}) + 1} = 4.17\times10^5\ \text{CFU/L}$$

b. For a distribution of external ages following the delta function of Dirac with $t_M = 45$ s, $\delta = t - 45$ if it is expressed in s^{-1} or $\delta = t - 0.75$ if it is expressed in min^{-1}. The mean concentration of microorganisms at the outlet is obtained by means of Equation 20.54:

$$C_M = \exp(-23\times0.75) = 0.32\ \text{CFU/L}$$

c. For the distribution given in the graph, it is obtained that it can be expressed according to the equation $E = 2t$ min^{-1}, for times between 0 and 1 min.

In this case, the mean concentration at the outlet is calculated by Equation 20.50:

$$\int_0^{\infty} \exp(-kt) 2t dt$$

Since after 1 min the microorganisms that have entered at the same time have already exited, this integration should be performed between 0 and 1 min. Thus, the following is obtained from this integration:

$$C_M = 2C_0 \left[\frac{1}{k^2} - \frac{k+1}{k^2} \exp(-k) \right]$$

$$C_M = 2\left(10^7 \text{ CFU/L}\right)\left(1.89 \times 10^{-3}\right) = 3.78 \times 10^4 \text{ CFU/L}$$

PROPOSED PROBLEMS

20.1 A canned food is thermally treated in a device whose processed temperature is 130°C, and it has been obtained that the temperature evolution for the smaller heating point with the processing time is

t (min)	0	6.5	10	12	16.5	23	28	36	44
T (°C)	30	40	50	60	80	100	110	120	125

If the initial microbial load is 10 CFU/can and it is wished to reduce it up to 10^{-5} CFU/can, calculate the processing time. The microorganism that impacts this thermal processing, for a reference temperature of 121°C, possesses 2 min decimal reduction time, and to reduce the treatment time to the tenth part, it is necessary to increase 10°C the temperature.

20.2 A milk pasteurization process consists essentially on three heating steps, the first one at 64°C during 2 min, the second one at 65°C during 3 min, and the third one at 66°C during 2 min. From the microbiological point of view, what should be said about the milk quality obtained in this pasteurization process? In the case that the milk had not received an appropriate treatment, what additional adjustments can you propose?

20.3 A fluid with similar properties to water is thermally treated in a concentric cylinder heat exchanger, in a 100 kg/h mass flow rate, flowing through a 1/4 in. pipe. Through the annular space, it condenses saturated water steam at 120°C. The fluid possesses certain content of *Clostridium botulinum* ($z = 10$°C and $F_0 = 2.45$ min). If the fluid is fed into the heat exchanger at 15°C, determine the length that should be used so there is an appropriate fluid for sterilization.

20.4 A tubular-type heat exchanger that works at 60°C is used to sterilize egg yolk, which is flowing through the heat exchanger in plug flow, with a residence time of 27 s. In this process it has been observed that the content in pantothenic acid (vitamin B_5) decreases by 13%. Calculate what content of this vitamin would be used if the velocity profile in the sterilizer was such that the residence time of the egg yolk is 30% of 50 s, 30% of 20 s, and 40% of 15 s.

20.5 In a milk sterilization process, carried out at 72°C, the milk is flowing through a tubular-type heat exchanger. Initially the milk contains 10^{11} spores/kg of a *Clostridium* thermal resistant; calculate the final population of this microorganism in the following cases: (a) The milk is flowing through the sterilizer in a plug flow regime, with a residence time of 30 s. (b) The flow presents a velocity distribution such a that 40% of the *fluid* remains 20 s in the sterilizer, 30% during 30 s, 20% during 40 s, and 10 remaining % during 50 s.

Data. Assume that for the milk at 72°C the necessary time to reduce to the tenth part of the initial population of the *Clostridium* is 2.5 s.

20.6 It is desirable to treat thermally, at 112°C, a fluid food, in order to reduce its microbial load. In previous experiments, it has been obtained that in order to reduce treatment time to the tenth part, it is necessary to increase the treatment temperature 12°C, while at 121°C 5 s, it is necessary to obtain a decimal reduction grade that allows assuring the product sterilization. Determine (a) the decrease of microbial load at the final treatment time, if it is carried out in a tubular device of 45 m length and the fluid is flowing at 0.6 m/s in plug flow. This treatment produces a 5% decrease in B_6 vitamin content. (b) Calculate the decrease of microbial load and percentage of vitamin B_6 loss if the treatment is carried out in a 100 L perfectly agitated tank, with a residence time of 15 min.

20.7 A fluid food contains 10^8 CFU/L of *B. coagulans*, and it is desired to process it thermally at 115°C in order to reduce the thermal load. This thermal treatment negatively impacts thiamine content. Calculate the microorganism's content and the thiamine percentage degradation if the device used in the thermal treatment is (a) a tubular device whose distribution function is a δ Dirac function, in which the time of answer to impulse is 210 s, or (b) a perfectly mixed tank with 60 L of capacity, with 28 min residence time.

Thermal Effect	T_R (°C)	z (°C)	D (min)
B. coagulans inactivation	121.1	10	0.1
Thiamine degradation	121.1	25	140

20.8 It is desirable to sterilize 2.000 kg/h of a fluid food that at 4°C possesses a density of 1050 kg/m³, using a plate-type heat exchanger to increase its temperature to 120°C. Then, in the holding step, a 2 cm of internal diameter tubular-type heat exchanger is used, working at 120°C by water condensing steam at 2.5 kgf/cm².

Before the treatment the fluid contains 10^8 CFU/L of pathogen microorganisms, and it is considered that the product has been well sterilized when it possesses 1 microorganism /L. In previous laboratory experiments, at 120°C, it has been obtained that the initial microbial load is reduced to the tenth part in 2 s.

a. Calculate the length that will have the tubular section (holding) to assure the sterilization, if the fluid behaves in such a way that the distribution function E is a δ Dirac function.
b. For the same tubular heat exchanger length calculated in the previous section, calculate the fluid microbial concentration that is leaving from the device, if the distribution function E is shown in Figure P.20.8:

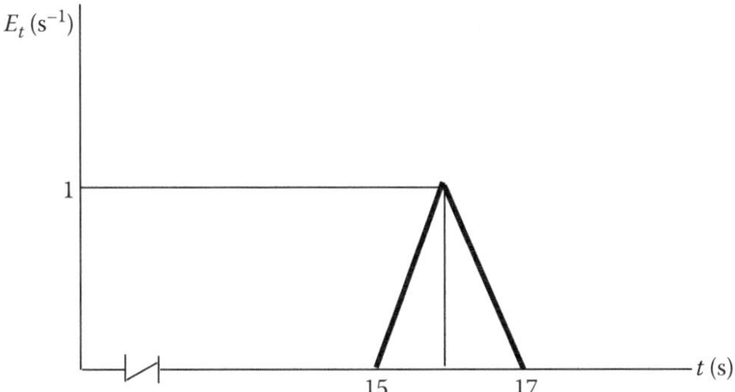

FIGURE P.20.8 Distribution function E (Problem 20.8b).

21 Emerging Technologies in Food Processing

21.1 INTRODUCTION

Foods in general are perishable products; to extend their life, it is necessary to carry out treatments to inactivate the deteriorative microorganisms. In addition, foods do not only carry deteriorative microorganisms but also pathogenic ones. In addition the presence of microorganisms, enzymatic reactions can also deteriorate the food. To avoid these problems, foods receive different types of treatment. The most commonly used treatments are thermal ones, but they have a negative effect on the overall quality of the processed food. It is worth mentioning that consumers demand foods that are healthy, easy to store and to prepare, with excellent texture and appearance, fresh-like, and with minimum or no preservative content.

In order to fulfill these requirements, research has been done in processing technologies besides conventional thermal treatments so as to develop the so-called emerging technologies, some of which are already in the adoption phase. Although most of them have already been applied for many years, it has not been until the last two decades that they have taken a growing interest in the food processing industry.

In this chapter, a brief description of different emerging technologies is presented. The purpose of these new technologies is to retain or improve the effectiveness in the preservation and development of safer high-quality foods. Most of these technologies are based on physical treatments, such as the ionizing irradiation with electron beam, gamma radiation, electric pulses, high pressure, radio-frequency (RF) treatment, or ohmic heating, among others.

21.2 IONIZING IRRADIATION

The irradiation of foods is not a new treatment technology, as its roots can be traced back to the late nineteenth century, although it is not until the 1940s that the term irradiation appears. Three clearly differentiated stages or periods in the history of food irradiation can be distinguished: the period from 1890 to 1940 represents the beginnings of the physics of irradiation and of the different sources used all of them tied to the first treatments of food with radiation. The period from 1940 to 1970 corresponds to a stage of intensive research and development in the application of radiation in food treatment and to the study of the healthiness of irradiated foods. Since 1970 a series of regulations for the safe control and application of irradiation have appeared.

The irradiation of foods is a technique that has been given an unfortunate name, as the word irradiation has been associated to nuclear energy. This has meant that many times the irradiated food has been mistaken for being radioactive. However, these are completely different terms, as the former is treatment with radiation, while the latter refers to foods with radioactive potential. For this reason this kind of processing must demonstrate that food treated with irradiation is safe, more than any other processing technique, even though there are numerous scientific tests that corroborate its healthiness. This kind of treatment has been attacked because it produces physical and chemical changes in foods, but if one thinks of thermal treatments, these produce very important alterations and are still accepted by the consumer.

Processed foods are those that reach the consumers, and it is the consumers who are the most concerned with their healthiness and safety. With regard to irradiated foods, as early as 1925, studies were initiated into their safety, and currently there are large numbers of publications that have been carried out on this subject. Thayer (1994) and Diehl and Josephson (1994) have performed reviews on the subject of healthiness and safety from radiological, microbiological, and toxicological viewpoints and the nutritional suitability of irradiated foods.

The Joint FAO/WHO/IAEA Expert Committee has examined 100 compounds of irradiated meat from cow, pig, and chicken, declaring that these foods are treated as healthy and safe. Their declaration of 1980 states: "the irradiation of any food product at an average general dose of 10 kGy presents no toxicological risk; therefore it is not necessary to carry out more toxicological trials on the foods treated in this way" (WHO, 1981).

21.2.1 Ionizing Radiation Fundamentals

The term ionizing radiation is given to the series of emissions of subatomic particles and electromagnetic radiation of nuclear or atomic origin, which when interacting with matter are capable of ionizing it. In other words, it is radiation that acts on matter, making it lose electrons, which leads to the production of ions.

Radiation is a form of energy, and every person receives natural radiation from the sun and other natural components of the environment. In the same way as other forms of radiant energy, the radiation waves used for food treatment are found inside the electromagnetic spectrum (Figure 21.1).

The emission of ionizing radiation is a common characteristic of many unstable atoms. These atoms, described as *radioactive*, transform themselves to become stable atoms, what is achieved by freeing energy in the form of radiation. The kind of radiation freed by the radioactive atoms can be of four different types:

1. Alpha particles (α): helium nucleuses, which contain two protons and two neutrons
2. Beta particles (β): electrons or positrons deriving from transformation in the nucleus
3. Gamma radiation (γ): electromagnetic radiation from the most energetic extreme of the radiation spectrum
4. Neutrons: chargeless particles

Radioactive activity is the speed at which the transformations are produced in a radioactive substance and measures the number of atoms that disintegrate in the unit of time. The unit of radioactive activity in the International System is the *becquerel* (Bq), defined as a disintegration by second, although sometimes the *curie* (Ci) is used, which is the activity existing in a gram of the ^{226}Ra atom, corresponding to 3.7×10^{10} disintegrations by second.

The Ci represents a considerable activity, while the Bq is a small unit. Radioactive substances with an activity below 100 Bq/g, or natural solid substances with an activity below 500 Bq/g, are considered to be harmless. Any human being possesses an activity of approximately 4000 Bq due to ^{40}K.

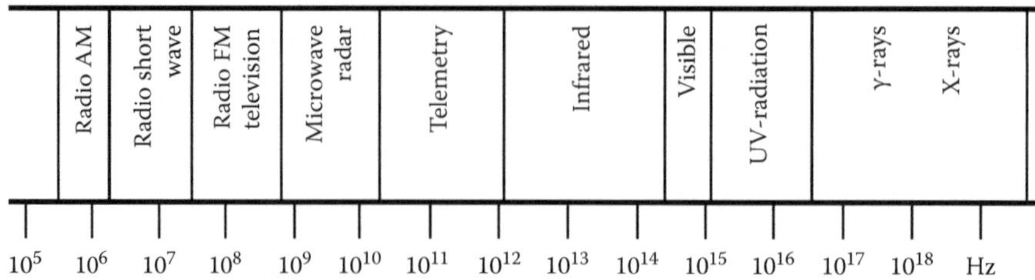

FIGURE 21.1 Radiation spectrum.

21.2.1.1 Interaction of Radiation with Matter

Radiation that affects matter experiments different kinds of interactions, depending on the nature of the radiation and the type of matter. Radiation that affects matter can cause two kinds of phenomena: atomic excitation or ionization. The former produces a thermal effect, while the latter causes the formation of ions. This is the reason why radiation is sometimes classified as thermal or ionizing, depending on the kind of interaction that it causes in the matter on which it acts.

α radiation consists of particles with two protons and two neutrons; hence, they are charged and considered heavy particles. Therefore, this is radiation with a limited power of penetration, of a few centimeters through the air or of a few microns in any tissue, and thus it is not able to penetrate the skin. However, it can produce a high concentration of ions, which makes these particles very dangerous, as they can cause severe cellular damage. β radiation consists of electrons, which are charged particles, and the interaction it experiments with matter responds to Coulomb's law of electrical charges, in the same way as for the α particles. The power of penetration of electrons is greater than that of α particles, being a few meters through air, capable of traversing human skin, although not the subcutaneous tissue. γ radiation consists of high-energy photons, which can be absorbed by matter according to three types of processes: the photoelectric effect, the Compton effect, and the production of electron–positron pairs ($e^- e^+$). Depending on the energy of the incident radiation, one type of interaction will occur, although the final effect of all of them is the production of charged particles. γ radiation possesses a power of penetration estimated at various hundreds of meters in the air and is capable of traversing the human body, metal sheets, and up to several centimeters of lead.

21.2.1.2 Absorbed Radiation Dose

When matter receives radiation, the incident energy of the radiation can cause ionization and/or excitation of the matter's atoms, although other effects may also appear such as different photoelectric effects, the Compton effects, and the formation of electron–positron pairs. Furthermore, part of the incident radiation may not interact with the matter, traversing it without producing any effect. Thus, it is necessary to measure the energy absorbed by the matter, as it is this that is able to cause ionization.

The *absorbed dose* (*D*) is the amount of energy absorbed per unit mass of the matter during the time that it is exposed to the radiation. In the International System, the absorbed dose is measured in *gray* (Gy), which is 1 J of energy absorbed per kilogram of mass of the matter. Sometimes a historical unit known as the *rad* (*radiation absorbed dose*) is still used, which corresponds to the energy of 10^{-2} J absorbed by each kg of irradiated matter. These measurement units are very small and often multiples are used. Thus, for example, the kGy is normally used, and to give an idea of the energy level it has, 1 kGy is equal to the amount of energy needed for a kg of water to increase its temperature by 0.25°C. Therefore, this type of process is a "cold" or "nonthermal" method of food treatment.

Besides the radiation absorbed by the matter (absorbed dose), the radiation type and the potential biological damage that it can cause should also be present. To accomplish this purpose, the *equivalent dose* is defined. Its measurement unit in the International System is the *sievert* (Sv), which is the relationship between the energy absorbed from 1 J for each kilogram of mass but taking into account the kind of radiation. In the same way as with the absorbed dose, for the equivalent dose, a historical unit known as the *rem* (*Roentgen equivalent man*) has been used, which is equivalent to 10^{-2} J for each kg of matter. Another variable used in the measurement of radiations is the effective dose (E), which takes into account the risk of developing cancers or hereditary effects and is measured in Sv. This effective dose is a weighted sum of the average doses received by the different tissues and organs of the human body.

21.2.2 Biological Effects of Ionizing Radiation

The biological effects produced by ionizing radiation depend on the type of matter interaction. The absorption of radiation by living organisms depends on the kind and quantity of the radiation, as well as the structure and the kind of absorbing matter. Thus, different kinds of effects may be shown, although in any case the incident radiation is an energy bearer, energy that is transferred to the absorbing medium either directly or indirectly, according to the mechanisms of excitation or ionization. When the absorbed radiation produces the effect of excitation of the matter's atoms and molecules, it can cause molecular changes if enough energy is absorbed, even if this is greater than that of the atomic bonds. If the ionization process is involved, the effect is more important, as changes are always produced in the atoms and it is capable of causing alterations in the structure of the molecules on which the radiation has fallen.

The biological effects of radiation can act at different levels, on cells, tissues, or whole organisms. The biological damage as well as the acceptable doses can vary greatly with each case. According to the molecular complexity of the living organisms, the biological effect is produced for different doses of radiation. Thus, for mammals the lethal dose is ranged from 0.005 to 0.01 kGy, for humans this dose is in the order of 4 Gy, for insects it is from 10 to 1000 Gy, and for plants it is 1 kGy. For bacteria, the lethal dose depends on if they are in vegetative or sporulated form; thus, in their vegetative form the lethal dose is 0.0–10.0 kGy, while for the sporulated forms it is ranged from 10 to 50 kGy. The most resistant organisms are the virus, whose lethal dose is ranged from 10 to 200 kGy. This indicates that the greater the molecular complexity, the smaller the dose required for producing biological effects. Low overall doses are capable of killing a person, or else causing significant damage, while in order to completely destroy insects, larvae, and eggs, the required doses are higher, and the doses needed to destroy bacteria, fungi, and yeasts are much higher still.

There are considered to be three different stages in the overall process, a physical stage, a chemical stage, and a biochemical stage. In the first physical stage, the radiation interacts with the matter, which can excite or ionize its atoms, with characteristic times of 10^{-15} and 10^{-17} s, respectively. In the chemical stage, free radicals are formed, with characteristic times in the order of 10^{-12} s. The last is a molecular or biochemical stage, in which the free radicals recombine and can form toxic molecules. The molecules formed by direct irradiation, or radicals, and those obtained indirectly are known as radio-induced substances.

21.2.3 Ionizing Radiation in the Food Industry

Food irradiation is considered as the process of applying high radiation energy to a food with the aim of pasteurizing, sterilizing, or prolonging its commercial life, eliminating microorganisms and insects.

The sources of ionizing radiation that are applied in the food industry are X-rays, electron beams, and γ radiation. X-rays constitute a much more energetic electromagnetic radiation, for which their power of penetration is higher, showing a continuous spectrum of radiation with a maximum value of 5 MeV. X-rays are usually obtained by bombarding a metal plate with a high potential electronic beam. γ radiation is produced with radioactive isotopes, which in the food industry are normally the radioisotopes of ^{60}Co and ^{137}Cs. At present ^{60}Co is the most commonly used one for irradiating foods with γ radiation, as it is relatively straightforward to obtain and it produces radiation with a greater power of penetration than that of ^{137}Cs. The energy spectrum of γ radiation is not continuous, but rather discreet, and it depends on the radioisotope used. The electron beam is a series of electrically charged particles of high energy, of up to 10 MeV. To ensure that electrons have a high energy level, they are led to a linear accelerator that confers them high voltages, thereby obtaining electrons with high speed, approaching that of light. The advantage of this radiation compared to γ rays is that the electronic beam is produced in an electric machine and can be turned on and off like a light bulb. Nevertheless, its power of penetration is low, from 5 to 10 cm.

For every food product the permitted doses of radiation depend on its characteristics and the aim of the treatment. This means that the dose for the elimination of insects, for pasteurization, and for

TABLE 21.1
Food Irradiation, Dose, and Applications

Dose (kGy)		Absorbed Dose (kGy)	Application
Low	<1	0.04–0.10	Sprout inhibition of tubers and bulbs
		0.03–0.20	Insects, grubs and eggs sterilization
		0.50–1.00	Fruit and vegetables ripening process control
Medium	1–10	1–3	Insect death
		1–7	Radicidation (pathogen elimination)
		2–10	Radurization (pasteurization)
High	10–50	15–50	Radapertization (sterilization)
		10–50	Spices and seasonings decontamination

sterilization will be different. Hence, three irradiation categories are considered according to the dose employed: low, medium, or high doses.

Low doses are those that do not exceed 1 kGy and are used in the control of insects in grain, in the control of trichina in pork, and in the inhibition of decomposition of fruits and vegetables. Medium doses are those in the range from 1 to 10 kGy and are applied in the control of pathogens in meat, poultry, and fish and also retard the growth of moulds on strawberries and other fruits. High doses exceed 10 kGy and are used to kill microorganisms and insects in spices and also when aiming to obtain commercially sterile foods. According to the dose of radiation, the treatment usually receives different names. Thus, the elimination of non-spore-producing pathogenic microorganisms and parasites to an imperceptible level is called *radicidation*. The treatment of foods with ionizing radiation aims at increasing their average life by reducing the number of modifying microorganisms (pasteurization) receives the name of *radurization*, while the elimination of microorganisms by irradiation to levels of sterilization is called *radapertization*. Table 21.1 shows the doses used to irradiate foods and the applications of each case.

Irradiated foods are treated at low levels of radiation, which means that only chemical changes are possible and that changes that would make them radioactive do not occur.

Irradiation is a *cold process*, which means that there is only a slight temperature rise of the food during processing. There is hardly any change in the physical appearance of the irradiated foods, which do not undergo the changes in texture and color shown by foods treated by heat pasteurization or by tinned and frozen foods. In irradiated foods some changes do occur, although they are not as important as those that occur with conventional cooking methods.

Retailed irradiated foods must bear the symbol *radura* (Figure 21.2), which identifies them as such. Furthermore, the sentence *treated with irradiation* must also appear. Manufacturers are permitted to add the objective of the treatment; thus, for example, it may be labeled as "treated with radiation to control deterioration."

21.2.4 Inactivation of Microorganisms by Radiation

Irradiation treatment is an efficient method for the destruction of parasites and both pathogenic and nonpathogenic bacteria and, to a lesser degree, viruses. The mechanisms of inactivation, as indicated previously, are due to the damage that irradiation produces in the genetic material, both directly and indirectly. Thus, ionizing radiation can collide directly with the cell's genetic material and damage its DNA, although it can also act on an adjacent molecule that subsequently reacts with the genetic material.

FIGURE 21.2 Radura symbol.

In thermal treatments the inactivation of microorganisms at a certain temperature follows first-order kinetics. In a similar way, it has been observed that the number of microorganisms present in an irradiated food decreases with the applied dose according to first-order kinetics:

$$N = N_0 \exp(-kD) \tag{21.1}$$

where

N_0 is the number of microorganisms initially present in the food
N is the number of microorganisms that survive
D is the dose applied
k is the kinetic constant of microorganism destruction by irradiation

For pasteurization treatment, doses of between 1 and 10 kGy are required, while for sterilization, the product doses of between 15 and 50 kGy are needed.

In food irradiation treatments, the variable D_{10} is usually applied. It is known as the decimal reduction dose, which represents the dose applied in order to reduce the number of microorganisms to one-tenth of the original. In this case $N = 0.1N_0$, so that

$$\ln\left(\frac{N_0}{0.1N_0}\right) = \ln(10) = kD_{10} \tag{21.2}$$

which is the same as

$$D_{10} = \frac{2.303}{k} \tag{21.3}$$

Table 21.2 shows the D_{10} values for some of the commonest bacteria in food. It can be seen that for the spore-producing forms the D_{10} values are higher, which indicates that they are more difficult to destroy. This difference compared to the vegetative forms can be explained by the fact that in the spore-producing forms the water content is much greater, which would minimize the secondary effects of the radiation, leading to an increase in the resistance to radiation. D_{10} data

TABLE 21.2
D_{10} Values for Selected Gram-Positive Bacteria

Bacteria	Medium	Conditions	D_{10} (kGy)
	Gram-Positive		
Spore formers			
Bacillus cereus	Distilled water	20°C–25°C; aerobic	1.6
	Mozzarella cheese	−78°C, aerobic; spores	3.6
	Yogurt	−78°C	4.0
C. botulinum	Beef stew	20°C–25°C; type E	1.4
Clostridium perfringens	Water	20°C–25°C	1.2–1.3
Non-spore formers			
Listeria monocytogenes	Chicken	2°C–4°C	0.77
	Chicken	12°C	0.49
	Ground beef	12°C	0.5–0.9
	Tryptic soy broth	0°C	0.21
	Phosphate buffer	0°C	0.18
	Ice cream	−78°C	2.0
Staphylococcus aureus	Poultry	10°C	0.42
	Meat	—	0.86
	Gram-Negative		
Aeromonas hydrophila	Ground fish	2°C	0.16
	Ground fish	−15°C	0.274
C. jejuni	Ground turkey	0°C–5°C, vacuum	0.19
E. coli O157:H7	Ground beef	−17°C	0.307
	Ground beef	2°C–5°C	0.241
Salmonella	Salsa	3°C; *S. typhimurium*	0.416
	Roast beef	3°C; *S. typhimurium*	0.567
	Ground beef	20°C; *S. typhimurium*	0.55
	Deboned chicken	−40°C; air; *S. typhimurium*	0.533
	Deboned chicken	−40°C; air; *S. enteritidis*	0.534
	Deboned chicken	−40°C; air; *S. newport*	0.436
	Deboned chicken	−40°C; air; *S. anatum*	0.542
	Liquid whole egg	Frozen; *S. seftenberg*	0.47
	Liquid whole egg	Frozen; *S. gallinarum*	0.57
Shigella	Oysters	*S. dysenteriae*	0.40
	Crabmeat	*S. dysenteriae*	0.35
	Oysters	*S. flexneri*	0.26
	Crabmeat	*S. flexneri*	0.22
	Oysters	*S. sonnei*	0.25
	Crabmeat	*S. sonnei*	0.27
Vibrio	Prawns	Frozen; *V. cholerae*	0.11
	Shrimps	Frozen; *V. parahaemolyticus*	0.1
Yersinia enterocolitica	Ground beef	25°C	0.2
	Ground beef	−30°C	0.39
	Minced meat	—	0.10–0.21

Source: Dickson, J.S. Radiation inactivation of microorganisms. In *Food Irradiation, Principles and Applications*, R. Molins (ed.) Wiley-Interscience, New York, 2001.

TABLE 21.3
D_{10} Values for Selected Virus

Virus	Medium	Conditions	D_{10} (kGy)
Coxsackie	Raw and cooked beef	−90°C to 16°C	6.8–8.1
Polio	Fish	0°C	3
Echovirus	MEM medium	—	4.3–5.5
Hepatitis A	Oysters	—	2
Rotavirus SA11	Oysters	—	2.4

Source: Dickson, J.S. Radiation inactivation of microorganisms. In *Food Irradiation, Principles and Applications*, R. Molins (ed.) Wiley-Interscience, New York, 2001.

are also available for some parasites and certain pathogenic viruses (Dickson, 2001) (Table 21.3). In the case of viruses, the contamination of the food generally originates from infected food handlers that can transmit the disease if they have contaminated the food they have prepared. The D_{10} values for viruses are greater than those for vegetative bacteria and are more similar to the spore-producing forms, which indicate greater resistance to radiation. Thus, the Coxsackie virus shows values of 7–8 kGy, the polio virus a value of 3, while the hepatitis A virus shows a value of 2 (Heildelbaugh and Giron, 1969; Sullivan et al., 1973; Mallet et al., 1991; Dickson, 2001).

21.2.5 Effect of Irradiation on Foods

The effects produced by irradiation depend on the type of food that is being treated, as well as the characteristics of the medium. These effects depend on radio-induced processes, as they are favored by the presence of oxygen, by an increase in pH, by an increase in temperature, and by the water content.

In dry and dehydrated food products, direct irradiation is the most effective, as there is less water available and therefore the formation of free radicals is reduced. It is also advisable to irradiate at freezing temperatures, as the water is not present in liquid form, which eliminates the process of radiolysis. Furthermore, it is better to irradiate vacuum-packed foods or those in a modified atmosphere, in the absence of oxygen, or with very low levels of this gas. With regard to pH, the irradiation of foods with a value of less than 4.5 is safer, thus requiring lower radiation doses than foods with higher pH.

It is important to underline that irradiation, like other treatments, is only effective with foods in a healthy and hygienic state. Irradiation will never improve already degraded foods.

The food's components will vary depending on the food under consideration. Thus, carbohydrates will be the majority components of foods of plant origin, while proteins and fats are predominant in foods of animal origin.

21.2.6 Food Irradiation Plants

Food irradiation plants, whichever the process they apply, consist of different elements that are common to all of them. Hence, the differential stage is usually that of the treatment applied. It is important to bear in mind that the zone where the products awaiting treatment are stored is separated from the zone where the already-treated products are stored; otherwise there may be cross-contamination. Figure 21.3 shows a typical treatment electron beam installation, while in Figure 21.4 an installation of gamma irradiation is shown.

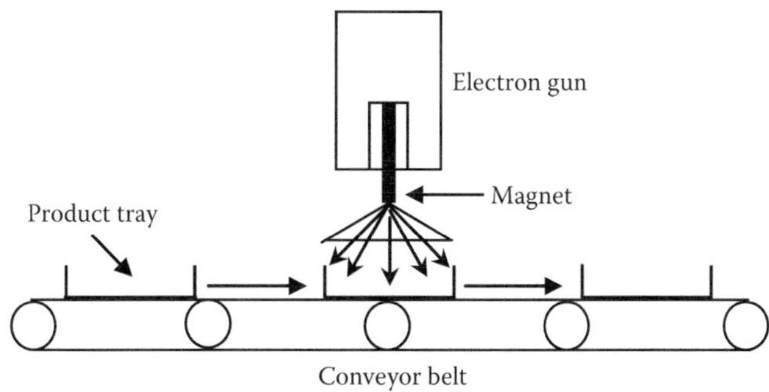

FIGURE 21.3 Electron beam device for food irradiation. (Adapted from Satin, M. *Food Irradiation. A Guide book*, 2nd edn. Technomic Publishing, Lancaster, PA, 1996.)

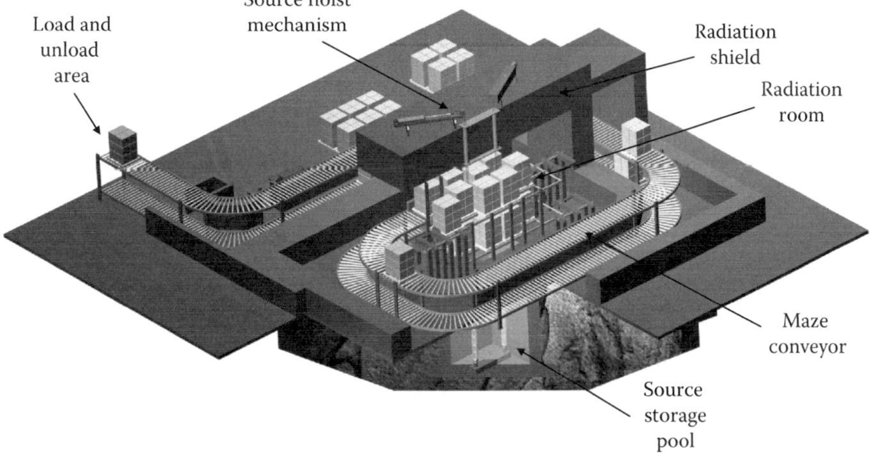

FIGURE 21.4 Pallet conveyor gamma plant irradiator. (Courtesy of MDS Nordion, Ottawa, Ontario, Canada.)

In any treatment plant the following elements can be distinguished:

1. Storage zone of the products awaiting treatment. Generally it is located near the loading area of the treatment devices.
2. Loading zone. In this zone the products awaiting treatment are loaded in crates or on suitable supports, which are then placed on a conveyor belt.
3. Conveyor belt. This is used to transport the products awaiting treatment from the storage zone to the treatment point. This belt also conveys the irradiated products to the treated products storage zone. The belt's movement should allow the products to receive the appropriate dose.
4. Irradiation zone. This is the main element of any plant, and as it is potentially the most dangerous area, it must be isolated from the plant operation personnel. For this reason the treatment chamber is covered with 2 m thick concrete walls that protect the operators. Inside the chamber, there is the source of radiation. Because water is one of the best protections against radiation deriving from the decay of ^{60}Co or ^{137}Cs, the radioactive source is submerged in a pool and is raised by remote control once the food to be treated is inside the treatment chamber. In the case of the electron accelerator, the machine comes on when the product enters the chamber. The trajectory followed by the product should ensure that it receives the dose calculated for the treatment's objectives; hence, the traveling speed of the conveyor belt should be such that the time spent by the product inside the treatment chamber is that required. When the product is large or very dense, it is turned over and irradiated again in order to ensure that it receives adequate treatment.

5. For plants that use an electron accelerator, it is necessary for them to be equipped with a chamber for the refrigeration circuit. These accelerators are usually compact, and there are different models on the market, with a differentiated range of potentials. One such model has a potential of 35 kW, which emits electrons with energies of between 3 and 10 MeV, although there are also models with potentials of 80 and 150 kW. As for the sources of ^{60}Co o ^{137}Cs, there should be a pool to store these sources, which consist of bars containing the capsules of radioactive material. These bars are kept in deionized water in a covered pool buried at a depth of approximately 4 m underground, in order to guarantee operator safety. When the product is treated, these bars rise vertically out of the pool and once the treatment is over they are resubmerged.
6. Treated product loading zone. It is located in a part of the plant away from the loading zone, where the treated product is received.
7. Storage zone. Before definitive storage, the product must be suitably labeled and the received dose should be measured. Then, it is taken to the warehouse where it is stored under suitable conditions, so that there is no recontamination of the product. For this reason this zone should be at some distance from the zone where the non-treated product is stored.
8. Control laboratory. This is where the dosimetry in all parts of the plant is controlled. It is necessary to measure the dose received by the products, although it is also essential to measure the radiation received by the personnel manning the plant.

In addition to all of these elements, the plant should have a control room, from where all the operations of the plant are controlled, including the speed of the conveyor belt, the systems for raising and lowering the bars of the radioactive source, and all the safety systems of the plant. It should be highlighted that this type of plant is completely automated and that all of the mechanisms can be controlled from the control room. Other elements that are found in irradiation plants are offices and auxiliary service rooms.

21.3 MICROWAVE AND RF HEATING

The fundamental base of this type of treatment is the dielectric heating, which is achieved by means of microwave or RF.

For some time, microwave ovens have been part of the appliances used in the kitchens of many houses. Mainly they are used to heat foods. But, at the moment, this technology is applied in an industrial way not only as a heating source, but rather they are used in pasteurization and sterilization processes, as well as in combined processes like vacuum drying with microwaves.

In the case of the RF waves, they have not had a clear industrial application in pasteurization and sterilization processes; however, they have been used in certain bakery processes. The treatment with this type of waves has the advantage that the heat is generated in the own product, because the effect of the conductivity and heat transfer coefficients does not play an important role. Also, the electromagnetic energy becomes heat where it is needed.

21.3.1 Dielectric Heating Fundamentals

The microwave/RF heating refers to the use of electromagnetic waves of certain wavelength in order to generate heat inside the food. The microwaves are electromagnetic waves with a band of frequency from 300 MHz to 300 GHz. The wavelength in the vacuum is in the interval from 1 m to 1 mm; however, inside the food its length enters in the range of micrometers. For industrial applications, the whole spectrum corresponding to the microwaves is not used, but rather certain bands, usually called industrial, scientific, and medical bands (ISM). The most used bands are 915 and 2450 MHz, among which the first one is used in the household ovens, while the industry uses both frequencies. The wavelength of the frequencies allowed by the ISM in the microwaves field is $\lambda_{915} = 0.3277$ m and $\lambda_{2450} = 0.1224$ m. The wave length used in the RF treatments is substantially bigger than the microwave ones, since their interval is between 11 m and 27.12 MHz. In this RF case, the waves allowed by ISM are 13.6, 27.1, and 40.7 MHz that belong together to the wavelength $\lambda_{13.6} = 22.06$ m, $\lambda_{27.1} = 11.07$ m, and $\lambda_{40.7} = 7.37$ m, respectively.

The heating with waves includes two main mechanisms: the dielectric mechanism and the ionic one. Foods possess high water content that is responsible for the dielectric heating. The water can be considered as a dipole, and because of that, under the electric field associated with the electromagnetic radiation, the water molecules try to align with the great frequency field that is oscillating. These oscillations cause the sample heating. The polar molecules rotate to maintain their alignment with the field polarity, existing friction among the molecules that causes heat generation. Besides this mechanism, the oscillating electric field causes the oscillatory migration of the food ions, whose consequence is the heat generation. In solutions with ions, there are collisions among them, transforming the kinetic energy in thermal heating.

The material dielectric properties come given by the real (ε') and imaginary (ε'') components of the complex permittivity ε^*. If ε_0 is the permittivity in empty space, the relative permittivity is defined as

$$k = \frac{\varepsilon}{\varepsilon_0} = \frac{\varepsilon'}{\varepsilon_0} - j\frac{\varepsilon''}{\varepsilon_0} = k' - jk'' \tag{21.4}$$

The k' constant is the dielectric constant relative to the food that represents the material capacity to absorb the wave and to store it in energy form. The k'' constant is called the loss factor or dielectric relative loss, which indicates the material capacity to dissipate energy. The tangent of the δ angle that is formed by both components is called the dissipation factor, and it is defined by

$$\tan\delta = \frac{k''}{k'} \tag{21.5}$$

The heat flow rate generated by volume unit ($\tilde{q}_G$) in a certain point of the food during the heating by waves can be expressed by (Decareau, 1992)

$$\tilde{q}_G = 2\pi\varepsilon_0\varepsilon'' f E^2 \tag{21.6}$$

If the dielectric conductivity (σ) is defined as

$$\sigma = 2\pi\varepsilon_0\varepsilon'' f \tag{21.7}$$

the generated heat can be expressed by

$$\tilde{q}_G = \sigma E^2 \tag{21.8}$$

If one keeps in mind that the permittivity of empty space $\varepsilon_0 = 8.85 \times 10^{-12}$ F/m and that the dielectric loss factor ε'' can be put on a k' function, when combining Equations 21.4 and 21.5, it is obtained that the generated heat in W/m³ can be expressed as (Decareau and Peterson, 1986; Singh and Heldman, 1993)

$$\tilde{q}_G = 55.61 \times 10^{-12} f\left(k' \tan\delta\right) E^2 \,(\text{W/m}^3) \tag{21.9}$$

where

- E (V/m) is the wave electric field force in the considered point
- f (Hz) is the frequency
- δ is the loss angle
- k' is the food dielectric relative constant

It can be observed that the heat generation is proportional to the material dielectric constant that in turn depends on the food composition, mainly on its water and salts content. The loss factor has tendency to increase with the moisture content, and it decreases when the temperature increases. Also, this loss factor increases with the salt content; however, fats and oils do not absorb well the microwaves, for what the foods rich in fats present low values of the loss factor. Globally, the greater the loss factor, the greater heat generation in the food, what will be translated in a temperature increase. This food temperature increase depends on the heating operation time.

The electric field, when penetrating in the food, could suffer attenuation as it is deepened in the food. According to Lambert's energy absorption law, the power P for a deep penetration d can be expressed according to the equation

$$P = P_0 \exp\left(-2\alpha d\right) \tag{21.10}$$

where

- P_0 is the incident power
- α is the so-called attenuation factor

The attenuation factor can be calculated by the equation

$$\alpha = \frac{2\pi}{\lambda}\left[\frac{k'}{2}\left(\sqrt{1+\tan^2\delta} - 1\right)\right]^{1/2} \tag{21.11}$$

where

- λ is the wavelength
- k' is the dielectric relative constant
- δ is the loss angle

The inverse of the attenuation is the depth inside the material in which the electric field intensity is $1/e$ of the corresponding to the free space. Therefore, the penetration depth in the point that the power decreases to $P = P_0/e$ is the corresponding distance that is obtained when substituting this value in Equation 21.7, giving a value of $d = 1/2\alpha$.

There is another definition of the penetration depth, and it is the depth to which the power decreases halfway the incident ($P = P_0/2$). With this new definition, it is obtained that the penetration depth presents a value of $d = 0.347/\alpha$.

As it has already been mentioned, the heat generation is inside the food, and it depends on the electric field in each considered point, since the penetration in the food makes all the points not to have the same generation. However, if it is considered that the heat generation due to the losses caused inside the food is uniform, the variation of temperature in different points of the sample, the heat fundamental conduction transfer equation in transient regime with heat generation could be obtained (Equation 16.1):

$$\rho \hat{C}_p \frac{\partial T}{\partial t} = k\nabla^2 T + \tilde{q}_G \quad (16.1)$$

This equation does not have analytic solution; however, for cases of heat transmission in a single direction, it can be solved by means of numerical methods (Mudgett, 1986).

If it is considered that the generated energy is uniform in the whole bulk food, the generated heat will make the product temperature increase in a uniform way; for this reason when carrying out a heat balance for all the sample volume (V), it is obtained that

$$\tilde{q}_G V = \frac{d}{dt}\left(m\hat{C}_p T\right) \quad (21.12)$$

where
T is the temperature
t is the time
m and V are the sample mass and volume, respectively
$\hat{C}_p$ is the specific heat

If there is no mass and volume variation, and the specific heat is considered to be constant and keeping in mind the definition of density (ρ), it is obtained that the product temperature variation velocity will be

$$\frac{dT}{dt} = \frac{\tilde{q}_G}{\rho \hat{C}_p} \quad (21.13)$$

This equation can be integrated with the boundary condition: for the initial time ($t = 0$), the temperature is the initial temperature of the sample ($T = T_0$). Knowing that the heat generation in the whole sample is given by Equation 21.6, the temperature variation with the treatment, it can be determined from the following equation:

$$T = T_0 + \frac{55.61 \times 10^{-12} f\left(k' \tan\delta\right) E^2}{\rho \hat{C}_p} t \quad (21.14)$$

21.3.2 Treatment Installations

The elements that configure a microwave device are the energy source, the magnetron, the waveguide, the agitator, and the oven (Figure 21.5). The energy source transforms the electric power into high voltage energy for the magnetron. This last one consists of an oscillator that transforms the energy into microwaves, emitting high-frequency radiant energy, generally 2450 or 915 MHz. The guide of waves transfers the energy generated by the magnetron to the oven cavity. The agitator distributes the energy transmitted by the whole cavity. The oven is the cavity where the food to be treated is placed, and it is where the waves are reflected in the metallic walls and the food absorbs the energy, transforming it in heat.

The RF devices consist essentially of a high-power electric valve that produces the RF power of the transmission lines that transport the RF energy and the RF energy applicators that are

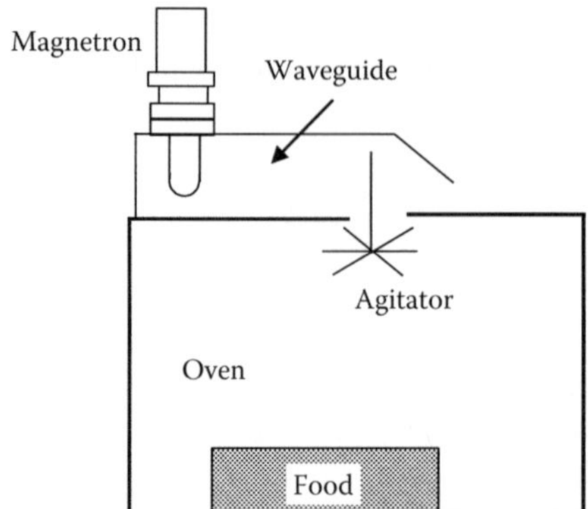

FIGURE 21.5 Microwave oven.

condensers. At the present time, these conventional devices are substituting the RF 50 devices, and they differ on the conventional ones in the fact that the RF generated is physically separated from the RF applicator by a high-power coaxial wire.

21.3.3 Microwave and RF Treatment of Foods

RF and microwaves are used in the heating of foods, which goes from the sterilization and pasteurization processes to some specific final product finishing.

In the pasteurization and sterilization process, the microwave heating is preferred to the conventional heating treatments because a quicker heating is obtained, and therefore a smaller time is needed to obtain the wished treatment temperature. It is possible to obtain a quicker microorganism inactivation, and it also decreases the thermal degradation of the desirable components in the food.

The microwaves absorbed by the food cause its heating, and this effect increases the temperature at a level so that the microorganism inactivation is effective enough to achieve its pasteurization or sterilization. Besides this heating effect, some authors (Burton, 1949; Fung and Cunningham, 1980; Cross and Fung, 1982) attribute the inactivation to other mechanisms, such as electroporation, rupture of the cellular membrane, and the lysis of the cell due to the coupling of the electromagnetic energy.

The RF treatments are used in the final stage of cereals baking, pastry, and bakery, in order to obtain more uniform final products. It is also used in the drying of ingredients, as herbs and spices, as well as in the drying of derived potato and pasta products, with the purpose of obtaining an auto-leveled drying, since in the most humid zones they vanish greater energy than in the driest points. In thawing processes, RF is used to defrost meat and fish, since it accelerates the global process, obtaining the product defrosted in a smaller time, with process times from 1 to 2 h. Also, this technology is used in the pasteurization and sterilization treatments of ready-to-eat foods, as well as in the pasteurization of containers. The RF has the advantage of carrying out a uniform heating of the product moisture, and for this reason it is often used in products finishing.

Microwaves processing is used industrially in baking and cooking of bread, pastries, and pasta, as finalization stage, and it has the advantage of quickly inactive α-amylase. It is also used as tempering step in the thermal treatment of frozen foods, since it substantially reduces the treatment time, and the heating is more uniform. In the drying processes, it is used at atmospheric pressure or under vacuum, mainly in the finishing stage, since it avoids the superficial hardening that can take place with a conventional hot-air drying. In the pasteurization and sterilization treatments, it is used to treat prepacked foods, as well as in the continuous milk pasteurization.

21.4 INFRARED HEATING

Infrared waves are electromagnetic ones whose wavelength is ranged from 0.76 m to 1 mm. Inside this variation range, infrared waves are divided in three types: short waves (from 0.76 to 2 m), medium waves (from 2 to 4 m), and long waves (from 4 m to 1 mm). Infrared waves are thermal waves and they are used in heating processes. This type of waves possesses lower wavelength than microwaves and RF ones and therefore greater energy.

21.4.1 Infrared Heating Fundamentals

As infrared waves are electromagnetic waves, the heat transmission toward the food is due to the exchange of radiant energy between the wave radiant surface and the food surface. The emission of waves depends on the radiating surface temperature, in such a way that the spectrum emissive power (q_λ) is expressed by Planck's equation (Equation 18.3):

$$q^{e}_{n,\lambda} = \frac{C_1}{\lambda^5\left[\exp(C_2/\lambda T)-1\right]} \tag{18.3}$$

This equation gives the emission spectrum of a black body in function of the temperature and wavelength.

According to this equation, the maximum of energy is obtained when deriving the expression regarding the wavelength and equaling to zero, obtaining Wien's law (Equation 18.4):

$$\lambda_{max} \cdot T = 2.987 \times 10^{-3} \text{ m K} \tag{18.4}$$

In this equation, it is observed that the greater the radiant temperature, the smallest the surface wavelength. Therefore, the great infrared waves (long waves) belong to lowest radiant temperatures, below 400°C, while the short waves of smaller wavelength belong to higher radiant temperatures, above 1000°C.

The total energy emitted by a black body is obtained when integrating Planck's equation in the whole spectrum of wavelength, obtaining the Stefan–Boltzmann equation (Equation 18.5). In the case that the radiant surface is not a black body, the radiant energy flow rate value should multiply for the emissivity (e) of the radiant body and for its area (A):

$$\dot{Q} = eA\sigma T^4 \tag{21.15}$$

where

σ (= 5.67×10^{-8} W/m² K) is the Stefan–Boltzmann constant

T is the temperature in Kelvin

The interaction of the incident radiant energy causes the wave energy to become mechanical energy, which gives place to an increment in the molecular vibrations. These vibrations cover stretching and shortening of the bonds between the atoms and molecules, as well as torsions of the same ones. For the infrared waves, there are intramolecular vibrations that affect to the chemical unions besides intermolecular vibrations that affect fundamentally the hydrogen bonds.

The infrared wave effect on the food heating can be considered from two points of view. One of them considers that the waves can penetrate in the food until a depth, and the global penetration process is determined by Lambert's equation (Equation 21.10). If this equation is expressed in the function of energy flux for a wavelength λ (I_λ),

$$I_\lambda = I_{\lambda 0}\exp\left(-2\alpha_\lambda x\right) \tag{21.16}$$

where

I_λ and $I_{\lambda 0}$ are the energy flux at the wavelength λ, at position x inside the product and the incident one, respectively

α_λ is the spectral attenuation factor

The total flux is obtained integrating this equation over the whole wavelength spectrum:

$$I = \int_0^\infty I_{\lambda 0} \exp(-2\alpha_\lambda x)\, d\lambda \tag{21.17}$$

The generated heating flow rate by volume unit inside the product can be obtained from the following expression:

$$\tilde{q}_G = -\frac{dI}{dx} = \int_0^\infty \alpha_\lambda I_{\lambda 0} \exp(-2\alpha_\lambda x)\, d\lambda \tag{21.18}$$

When problems about infrared treatments are set out, this generation term is the one that should be included in Equation 16.1 and to solve the equation as it has been described previously for the microwave heating.

Another approach to the problem is to assume that there is no infrared wave penetration inside the product, and then there is no heat generation, and the expression that governs the process can be determined by the equation

$$\frac{\partial T}{\partial t} = \frac{k}{\rho \hat{C}_p} \nabla^2 T \tag{21.19}$$

where $\nabla^2 T$ will have different expressions depending on the selected coordinate system. In this case, when integrating the previous equation, the exchanged radiant energy between the infrared radiant surface and the food surface will appear as boundary condition. The exchanged radiant energy net flow rate ($\dot{Q}_R$) will be governed by the Stefan–Boltzmann equation, and it will be given by

$$\dot{Q}_R = F_{12} A \sigma \left(e_1 T_h^4 - e_2 T_s^4 \right) \tag{21.20}$$

where it has been assumed that as much the radiant surface as the food surface are gray bodies being e_1 and e_2 the radiant and product emissivity, respectively, F_{12} the vision factor, A the exchange surface, T_h the temperature of the infrared heater, and T_s the surface product temperature. Once this heat arrives to the food surface, it penetrates by conduction inside the food, and depending on the geometry the solutions that have been given in Chapter 16 will be applied for heat conduction in transient regime. In the heat balance applied to the food, when it is the case, the convection heat losses toward the surrounding air must be considered, since their superficial temperature will be greater than external air.

The application of any of the two models leads to solutions that hardly differ one from another (Sakai and Mao, 2006), in such a way that the temperature profiles inside the infrared-treated product are similar in both models.

21.4.2 Infrared Treatment Devices

In infrared wave heating facilities, the wave radiant body depends on the radiant temperature, since it will give different wavelengths. As radiant, it can use radiators that are warmed with gases, and in this case the emitted infrared waves are of long type. Also, this radiator types are heated electrically. If the heater is metallic or ceramic, the produced waves are long; in the case of heaters of quartz tube, the emitted waves are in the interval of medium–short; finally, with halogen tubes heaters, the emitted infrared waves are ultrashort. For a more effective energy use, many facilities have reflectors that can help to concentrate the radiant energy on the product.

The infrared heating devices can work very well continuously and discontinuously. For treatments of big productions, those that work continuously are preferred, and they are generally chambers in which the infrared radiators are located. The radiation impacts on the product, which is placed on a conveyor belt that is moving through the treatment chamber.

21.4.3 Infrared Food Treatment

The treatment with infrared waves has been applied in different food industry sectors. In this way, in the bakery sector, it has been used with short-wave radiators, and combined with convection it gives good results in the surface drying of products like cookies, bread, and pizza.

Another one of the applications of infrared treatment is in toasting, applied in such products as coffee, green tea, sweet potatoes, and chestnuts. The temperature increase that the product experiences when it is treated by infrared waves is quicker than in the conventional toasting treatment. It seems that the obtained products have better quality, because the product warms in a more uniform way.

This technology can also be applied in drying processes of vegetables, fish, rice, and pasta. In the case of vegetable drying, it is observed that there is a greater retention of pigments and nutrients, compared with the conventional hot-air drying treatments. Also, the rehydration of the infrared dried products leads to rehydrated products with smaller damage in its structure and with a quicker rehydration time.

Infrared waves do not possess enough energy to directly damage or destroy the microorganisms contained in a food. Therefore, this type of waves applied to pasteurization processes can only destroy the microorganisms by heating effects. However, it seems that the treatments don't affect the enzymes in a direct way.

Another field in which the infrared heating can be used is in thawing processes. Generally, the thawing time of a frozen food is greater than in the freezing process, because in thawing there appears a water layer that surrounds the food, whose thermal conductivity is smaller than that of ice. The application of microwaves in thawing processes can accelerate the process time, although certain upper heating can exist in the food surface because the defrosted water layer absorbs more energy, due to the difference among the values of the dielectric properties of the frozen and thawed products. However, with infrared waves, this phenomenon is not presented due to the work wavelength, and the absorption coefficients for the ice and the water are approximately the same.

21.5 OHMIC HEATING

The ohmic heating is based on the fact that the food is part of a circuit, which behaves as heat generator due to an electric current passing through the circuit. The internal heat generation can be uniform in the whole sample volume, for what it should not have temperature gradients inside. This technology was already applied in the nineteenth century, where diverse processes were patented that used the electric current for the heating of different flowing materials. Already in the twentieth century, it was applied in the "electric" milk pasteurization, when making a current of milk pass between parallel plates in those that a voltage difference existed. It was also applied in the cooking of Frankfurt's sausages and in potato blanching. However, this treatment type presents an important concern associated to the electrodes material, since it can contaminate the treated product. This technology can be applied to particulate fluids, and it may happen that the particles reach greater temperatures than the fluid in which they are contained. Because the heating is not carried out from the surface, depositions and fouling on electrodes surfaces are avoided, also decreasing the possibility of burnt. The food deterioration is the lowest, with better nutrients and vitamins retention. The startup and the stops are instantaneous, only pressing the switch, and it is easy to be controlled.

21.5.1 Ohmic Heating Fundamentals

To generate heat inside the food, it is necessary to apply a potential difference (voltage) between the electrodes of the device. The current pass through the food causes the heat generation, due to the electric resistance of the own food. The heat flow rate ($\dot{Q}$) generated by the pass of a current of intensity (I) through the electric resistance (R) of the own food leads to

$$\dot{Q} = I^2 R \tag{21.21}$$

In an alternative way, if one knows the applied voltage gradient and the electric material conductivity (σ), the generated heat flow rate can be obtained according to the equation

$$\dot{Q} = \left|\nabla V\right|^2 \cdot \sigma \tag{21.22}$$

where the electric conductivity is a function of the position and temperature. There is a dependence of electric conductivity with regard to the position and it is because the foods are not totally homogeneous, as it is the case of the particulate fluids; on the contrary, the liquids are reasonably homogeneous.

The generated heat flow rate per volume unit can be obtained by dividing the previous expressions by the sample volume (V). And then, when applying energy balance to the sample,

$$\tilde{q}_G = \frac{\dot{Q}}{V} = \frac{I^2 R}{V} \tag{21.23}$$

In the case that the sample is a homogeneous system, it can be considered that its properties don't vary in all the volume. For that, when applying an energy balance to the sample,

$$\rho \hat{C}_p \frac{\partial T}{\partial t} = k\nabla^2 T + \frac{I^2 R}{V} \tag{21.24}$$

If the food sample to be treated is a particulate fluid, it will be necessary to carry out a balance in the fluid fraction and another one in the group of particles contained in the sample. In this case, the accumulation and generation terms will be similar to the ones showed in Equation 21.24, but for the entrance and exit energy terms, it can be expressed as Fourier's law, that is, a dissipation term due to heat convection from or toward the solid particles contained in the added sample (Fryer, 1995; Vicente et al., 2006).

This convection term appears due to the fact that the fluid and particle temperatures are different because the fluid has a greater temperature than particle surface. In this way, the heat flow rate that is transmitted from the fluid toward the solid particle surface can be expressed by the equation

$$\dot{Q}_{convection} = n_p A_p h_{fp} \left(T_f - T_p\right) \tag{21.25}$$

where

- n_p is the particle number container in the treated food volume
- A_p is the area of one particle
- h_{fp} is the fluid–particle heat convection coefficient
- T_f is the fluid temperature
- T_p is the surface particle temperature

In the case of a homogeneous food with a uniform heat volumetric generation in all the points of the sample, a balance similar to the one presented in Equation 21.12 can be carried out. This allows to obtain the temperature presented by the treated sample for a certain treatment time:

$$T = T_0 + \frac{I^2 R}{m\hat{C}_p} t \tag{21.26}$$

where
T_0 is the initial sample temperature
m is the total mass
R and $\hat{C}_p$ are the electric resistance and specific heat of the sample, respectively
I is the current intensity
t is the treatment time

21.5.2 Ohmic Treatment Devices

A conventional cell of ohmic treatment consists of a recipient in which the food is placed, and in the ends or the walls the electrodes are placed and connected to an alternating electric current supply that makes current pass through the food, generating heat by the resistance that the same sample offers. This simple device can work discontinuously, for what thc sample is loaded in the recipient and it is connected to the current, and heat is directly generated (Figure 21.6). In the case of fluid foods, this device can also work in a continuous way, so the fluid starts to circulate at a certain velocity through the treatment chamber (Figure 21.7). There are certain designs that involve

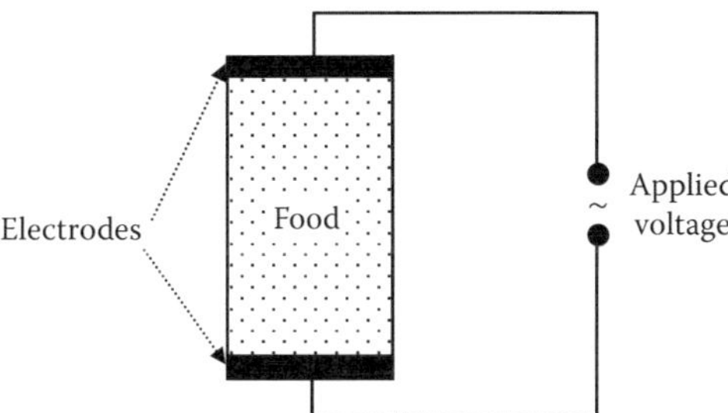

FIGURE 21.6 Ohmic heating principle.

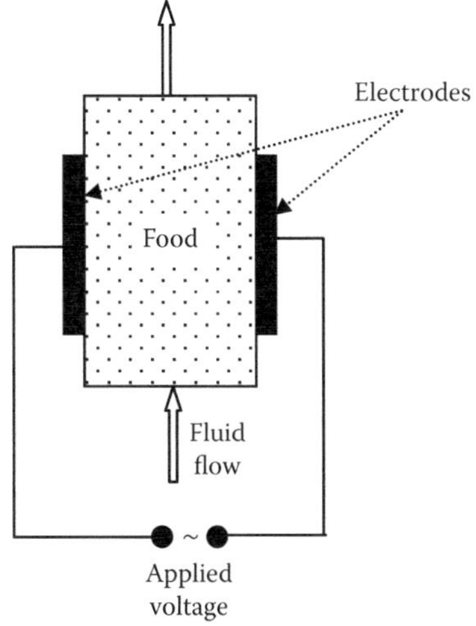

FIGURE 21.7 Continuous ohmic heating device.

several treatment cells in series. The electrodes can be placed in different positions. If they are placed in passing along the longitude of the fluid flow, it is said that the field is online, but if they are placed perpendicular to the flow the field is crossed, that makes them to differ in the distribution of electric field.

In this treatment type, it is essential not to have an electrolytic effect that can cause the breakup of the metallic electrodes while the contamination of the product takes place. This phenomenon was the greatest problem in the installation of this food treatment method. Also, a form of avoiding the breakup of the metal is to work with alternating frequencies above the 100 kHz (Ruan et al., 2001).

21.5.3 Ohmic Treatment of Foods

The ohmic heating has been used in the treatment of diverse types of foods. This way, it has been used in the cooking of meat pastes obtaining times of much smaller treatment than in conventional cooking, although with these treatment times it is not possible to reduce the microbial load at levels that guarantee their safety. On the other hand, this technology has also been used in thawing processes giving a quicker and uniform process as a result. In reference to the vegetable products, the ohmic treatment can be adapted as alternative to the conventional blanching technologies, since the treated products have a better textural quality. With regard to the derived fish products, the ohmic treatments have been mainly applied to the surimi production, obtaining a final product with a higher quality of the formed gel.

The ohmic heating can also apply in microorganisms and enzymes inactivation. With this treatment type, it seems that the inactivation effects on the microorganisms and the enzymes are not only due to the thermal level, but they also take place for electric effects. This way, it has been seen (Cho et al., 1999; Castro et al., 2003; Vicente et al., 2006) that the decimal reduction times to a certain treatment temperature decreased when the treatment was carried out with ohmic heating. This indicates that besides the inactivation effect for the thermal treatment, there is a coupled electric effect that reduces the treatment times.

21.6 HIGH-PRESSURE TREATMENT

High hydrostatic pressure is one of the new emerging technologies that are taking importance in the food sector due to the advantages in microorganisms and enzymes inactivation and in the high-quality food production, with novel structural properties. The technology of high pressure was used initially in ceramic industry, of steels and super alloys. The effect of this technology on the microorganism inactivation was already known at the beginning of the twentieth century; however, it was not until the final of last century that an intense investigation work began for its application in the food industry. A high-pressure treatment is uniform through the whole food, for what the treatment gives as a result a product with more uniform characteristics. Contrary to the thermal treatments, the high pressure does not depend on the relationship time/mass, what causes the processing times to decrease.

This technology is based on the fact that an isostatic pressure is applied in a uniform way to the whole product. The high-pressure technology applies pressures that ranged from 100 to 1000 MPa. The medium used to transmit the pressure is usually water, and it is necessary to highlight that the pressure is transmitted in isostatic way (uniform), and in an almost instantaneous way to all the food sites, independently of its composition, volume, and shape.

This avoids the product deformation, is a very homogeneous process, and does not present over-treated zones. Once pressurized, it is not necessary to add more energy to maintain the system at this pressure. As it has been commented, the used pressures are usually ranged in the interval from 100 to 1000 MPa, while the operation times can oscillate between few minutes and some hours, and the treatment temperature can have values between −20°C and 90°C.

21.6.1 High Hydrostatic Pressure Fundamentals and Equipment

A high-pressure device basically consists of a pressure chamber and its closing system, a pressure generation system, a temperature control system, and a handling product system. The pressure chamber is the most important component in a high-pressure installation. It is a cylinder built in steel alloys, and in those cases in which pressures greater than 600 MPa are required, the cylinder wall is built in multilayer type.

One of the key points of this technology is the pressure generation. It can be carried out in different ways, for example, heating the medium its temperature increases, and in a closed enclosure it causes the pressure to increase. However, the two more usual forms are for direct or indirect compression (Figures 21.8 and 21.9). In the first case, the pressure is generated directly by pressurization of a medium by a piston. For the indirect compression, an intensifier of high pressure is used to pump the pressure medium from a tank toward a closed pressure chamber until the wished pressure is reached.

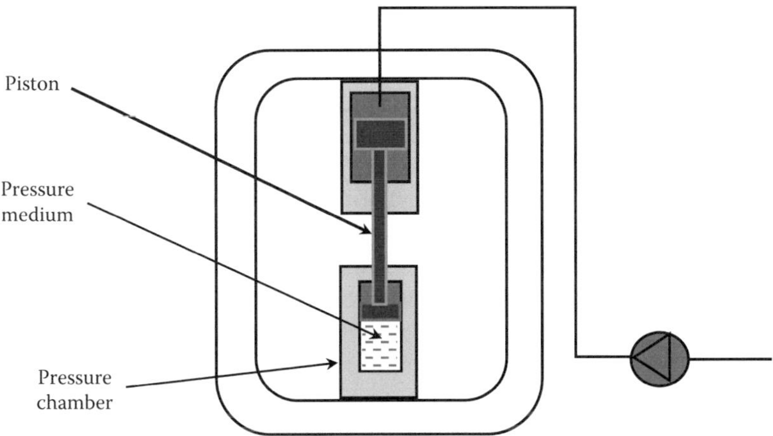

FIGURE 21.8 High-pressure generation by direct compression. (Adapted from Barbosa-Cánovas, G.V. et al., *Nonthermal Preservation of Foods*, Marcel Dekker, Inc., New York, 1998.)

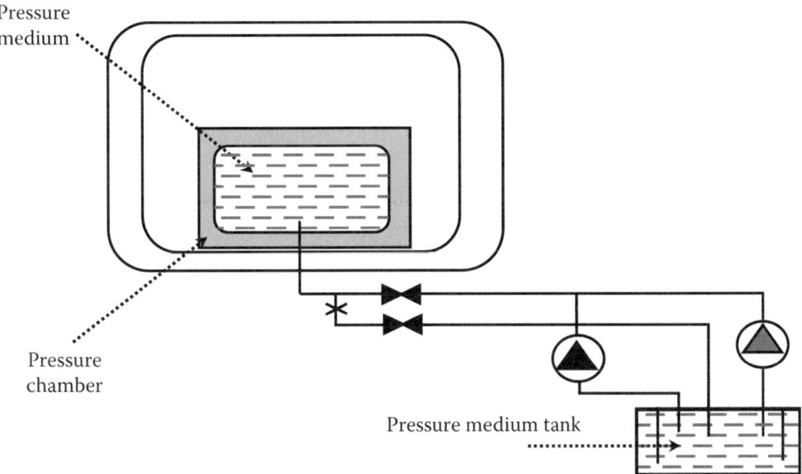

FIGURE 21.9 High-pressure generation by indirect compression. (Adapted from Barbosa-Cánovas, G.V. et al., *Nonthermal Preservation of Foods*, Marcel Dekker, Inc., New York, 1998.)

The high-pressure equipment can carry out the treatment in a continuous or discontinuous way. In discontinuous process systems, the food is pressurized by loads. The food to treat is packed and placed inside the pressurization chamber for a fixed time and pressure. This process type reduces the risk that great food amounts be contaminated by lubricant. In a discontinuous system, different food types can be processed without recontamination danger. In the case of fluid foods, the velocity treatment can be increased by carrying out semicontinuous treatments, in those that several pressure chambers are used that work in a sequential way.

Generally, the high-pressure chamber is in a discontinuous way. The food is placed in a container and it is sealed, being placed next in the pressurization chamber. Once the chamber is loaded with the packed food, it is filled by the pressure medium. In most of the equipments, the used pressure medium is blended water with a small quantity of soluble oil in order to lubricate and for anticorrosion purposes.

The bases for the application of the high pressure in foods are the water compression that surrounds the food. Because the liquid compression causes a small volume change, the high-pressure chambers use water, since they don't present the operation dangers of the chambers that use compressed gases. The food becomes low pressure during a certain period of time that depends on the food type and the process temperature. At the end of the process time, the chamber is decompressed to take out the treated load. Now the equipment is ready to begin a new treatment cycle.

21.6.2 Food Treatment by High Hydrostatic Pressure

The high pressure has been used to reduce the microbial load of the foods, whether the microorganism is in its vegetative phase or sporulated. Also, the high pressure can affect other food components, like enzymes, mainly proteins, and it makes the sensorial characteristics of the treated product different to the untreated food.

In the vegetative microorganisms, the high pressures induce morphological-, biochemical-, and genetic-type changes, and they take place in the membrane and in the cellular wall of the microorganisms. Most of bacteria are able to grow at pressures of 200–300 atm. The microorganisms that are able to grow at pressures higher than 400–500 atm are called barophiles. The barophobic organisms have difficultly growing at pressures higher than 300–400 atm. The microorganisms that can grow in the range of 1–500 atm are called barotolerants. The baroduric microorganisms survive pressures of 500–2000 atm, but they cannot grow.

One of the most difficult operations in the food preservation is the inactivation of microorganism spores. The sporulated forms are more resistant to pressure than the vegetative cells. However, it seems that moderate pressures facilitate spores germination; for it, it seems that the microorganism spores inactivation could be achieved through two phases. In a first stage at low pressure (between 20 and 200 MPa), the spore germination is obtained. Later on, if the pressure and/or temperature is sufficiently high, the germinated spore can be inactivated, because now it is pressure sensitive. The microorganisms from the germinated spores are much more sensitive to pressure and/or heat.

With regard to virus and parasites, they are few studies that have been carried out, although certain encapsulated viruses have been able to be inactivated at 300 MPa, and at 400 MPa, the inactivation power decreases in seven or four logarithmic cycles, respectively. It seems that the pressure breaks the viral capsule and it prevents the adhesion from the viral particles to the cells. This suggests that processes with high pressure in polluted biological samples with virus could be feasible. The possible use of a viral vaccine has also been studied by means of a partial inactivation by subjecting virus to high pressures. As for present parasites in diverse meats and fish, as the trichina, the tapeworm, and the nematodes, it is known that they are inactivated by means of high-pressure processes.

The behavior of the biochemical systems under pressure is governed by Le Chatelier's principle, which postulates that any phenomenon (phase change, changes in the molecular configuration, chemical reactions) accompanied by a volume reduction is increased by the pressure, and vice versa.

Therefore, the pressure favors those changes that go accompanied by volume decrease. It is necessary to highlight that the high pressure doesn't affect the covalent bonds, due to the compressibility of these bonds. High pressure up to about 100 MPa breaks up the hydrophobic interactions, and above this value the high pressure spreads to stabilize these interactions. When a charged group is solvated, the electrostriction phenomenon takes place, in which the water is organized in a more compact way around the ionic charges and the polarized centers of the molecule functional groups, and this phenomenon is favored by the high pressure; however, the formation of electrostatic interactions after the dehydration of charged groups represents a positive volume increment and it is under unfavorable conditions in front of the pressure.

Pressure induces protein denaturalization that could be explained for the hydrophobic bonds rupture and for the unfolding of the ionic couple of the peptidic chains. The application of high pressure to a system that contains a protein implies the displacement of the equilibrium of the native form of the protein and the so-called unfolded forms, that is to say, denaturalized forms. Once the system is depressurized, the proteins tend to be reorganized in a structure that does not depend on the effect of the pressure. The high pressures can modify the enzyme structure, since they are proteins, and therefore modify their activity, although it can also affect the substrate.

One of the main advantages of the high-pressure application in foods is that flavors, odors, and colors are retained. The sensorial quality of the products is quite greater than those obtained by conventional methods where the heat is used. The high pressure has been used in different food industries, like dairy- and milk-derived products, in meat and derived products, in fish and derived products, in ovoproducts, in vegetable products, and in alcoholic drinks, and it is also used such as one of the barriers in the combined methods.

21.7 HIGH-INTENSITY PULSED ELECTRIC FIELDS TREATMENT

This technology takes advantage of the property that the fluid foods consist mainly of water and nutrients that are very good electric drivers due to the high ion concentrations that they contain and to their capacity to transport electric charges. The food preservation requires pathogen and deteriorative microorganism inactivation, as well as the inactivation of enzymes that are responsible for nondesirable reactions. Pulsed electric fields can inactivate microorganisms and enzymes, whenever certain intensity threshold electric field is exceeded. The electric field is applied to fluid foods in form of short pulses with pulse duration between some few microseconds and milliseconds. The foods can be processed at room temperature or under refrigeration. In the treatment with pulsed electric fields, the food is processed in a short period of time and the energy lost by the food heating is minimal.

21.7.1 High Pulsed Electric Fields Fundamentals and Equipment

The application of high pulsed electric fields in microorganism destruction is based on destruction or deformation of the cellular wall due to the applied intensity of electric field. The applied electric field gives place to a potential difference at both membrane sides that causes damages in the cellular membrane. When the transmembrane difference of potential takes a certain value that depends on the microorganism type, it originates a pore in the wall that facilitates the permeabilization of the membrane. Depending on whether a certain electric field intensity threshold is exceeded, the process will be reversible or irreversible.

The application of a high voltage (20–80 kV/cm) during short times (microseconds order) can give place to the mechanical destruction of the cellular membrane. It can also cause electrolysis of substances, depending on the composition of the treated food and of the electrode material used. It can also generate heat due to the Joule effect. The microorganism destruction by electric fields depends on the intensity, treatment time, treatment temperature, conductivity, pH, ionic force and microorganism type, its concentration, and the growth step.

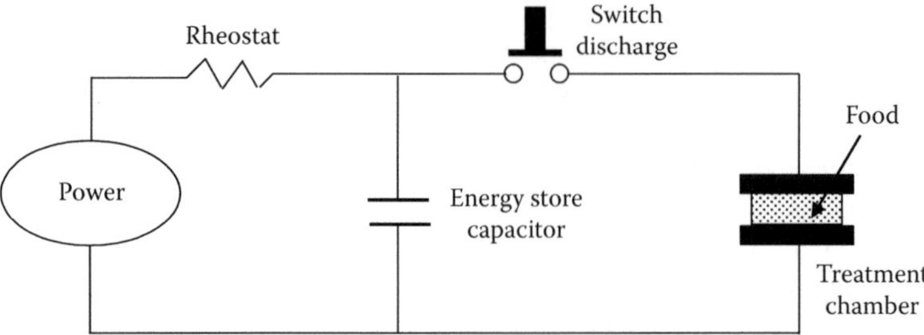

FIGURE 21.10 Pulsed electric fields basic circuit.

The basic components of a high pulsed electric fields process equipment are a high-voltage generator, a treatment chamber, and a control system for monitoring data, temperature, voltage, and intensity current probes, besides an aseptic packed equipment and a refrigeration chamber system. Figure 21.10 shows a typical scheme of a high pulsed electric fields treatment installation.

The high-voltage generator is composed by a current generator, a capacitor, and a switch. The tension pulse that is originated on the treatment chamber is due to the discharge process in the capacitor. This pulse can have different forms, such as exponential fall, square, oscillatory, and bipolar pulses. Among them, the most used ones and the ones that give better results are those of exponential fall and square wave. The energy that is discharged in each pulse depends on the capacitance (C) of capacitor and the voltage (V):

$$\text{Energy} = 0.5CV^2 \tag{21.27}$$

In general, in the treatments discharges are applied to the food at several pulses; the total energy that the food receives will be obtained when multiplying the energy discharged in a pulse (Equation 21.27) by the pulse number.

21.7.2 High Pulsed Electric Fields Food Processing

This technology has been focused mainly on the destruction of microorganisms and enzymes contained in fluid foods. Many of the initial studies were carried out with model fluids, although at the moment there are a great number of works that use different real fluids, such the ones that are derived of fruit, milk, and eggs, polluted with microorganisms. The microorganism inactivation using pulsed electric fields is more difficult in fluid foods than in model buffer solutions. In general, the bactericide effect of these treatments is inversely proportional to the ionic force and it grows with the electric resistivity. As it has been commented previously, the microorganism inactivation is based on the formation of pores in the microorganism cellular membrane.

The *Escherichia coli* inactivation has been studied in an ultrafiltrate milk model system, observing that the inactivation increases with the number of pulses and with the applied electric field. The ionic solution force also plays an important role in the inactivation, since when increasing the ionic force the electronic mobility through the solution also increases, giving a decrease of the inactivation as a result.

The inactivation of *E. coli* and *Bacillus subtilis* contained in pea soup depends on the field intensity, the pulse number, the pulsation velocity, and the flow rate. The inactivation increases with the number of pulses and the field intensity. The treatment temperature also affects the inactivation process, in such a way that the inactivation decreases when temperature decreases below 53°C. This can be attributed to the fact that the microorganism sensibility increases when temperature increases.

E. coli inoculated in liquid egg is inactivated in a 6D value by a treatment with pulsed electric fields (26 kV/cm). It was also observed that the pulses of 4 s were more effective than 2 s.

The electric resistance of the egg is low (1.9) compared with other foods, and the exposure of the liquid egg to a great number of pulses becomes necessary.

The treatment of commercial apple juice to different electric pulses does not show changes in the pH, in acidity, and in sugars and vitamin C content. The treatment field intensity, the time, and the number of pulses affect the inactivation of *Saccharomyces cerevisiae* contained in apple juice. The inactivation velocity increases with the increase of the field intensity and with the number of applied pulses.

The effect of the pulsed electric fields on enzymes also depends on the field intensity and on the number of pulses. When increasing the number of pulses, a greater inactivation is obtained; the same trend is observed when increasing the applied electric field value.

PROBLEMS

21.1 A microbiological analysis has been carried out on three food products: a sauce, a beef hamburger, and boneless chicken. In all the cases, a microbial load of *Salmonella typhimurium* has been found. It is wished to reduce the microbial load in 12 logarithmic cycles, for what there is a ^{60}Co irradiation source to treat these products. Determine the dose that these foods should receive to carry out the required microbial decrease. If the radiation rate received by the samples is of 1.5 kGy/s, how long should it be subjected to radiation (each one of the foods)?

For *S. typhimurium*, the decimal reduction dose D_{10} depends on medium growth (Table 21.2):

Product	D_{10} (kGy)
Sauce	0.416
Hamburger	0.550
Chicken	0.533

As a 12 logarithmic cycle reduction is wished,

$$\log\left(\frac{N_0}{N}\right) = 12$$

Therefore,

$$\ln\left(\frac{N_0}{N}\right) = \ln(10)\cdot\log\left(\frac{N_0}{N}\right) = (2.303)(12) = 27.63$$

Also, it is necessary to determine the kinetic constant of destruction that is obtained from D_{10} value:

$$k = \frac{2.303}{D_{10}}$$

For the three studied cases, the following is obtained:

Product	k (kGy^{-1})
Sauce	5.535
Hamburger	4.187
Chicken	4.320

The relationship between the initial and final microorganism content in the food is given by the equation

$$N = N_0 \exp\left(-kD\right)$$

and the absorbed dose can be expressed as

$$D = \frac{1}{k}\ln\left(\frac{N_0}{N}\right) = \frac{27.63}{k}$$

When substituting the microbial destruction kinetic constant values, the absorbed doses are obtained in each case:

Sauce:	$D_S = 4.992$ kGy
Hamburger:	$D_H = 6.600$ kGy
Chicken:	$D_P = 6.396$ kGy

The treatment device is able to give such radiation that the absorption rate is $\dot{D}$= 1.5 kGy/s, and keeping in mind that this rate is defined as the relationship between the absorbed dose and irradiation time, the following is obtained:

Sauce:	$t_s = 3.3$ s
Hamburger:	$t_h = 4.4$ s
Chicken:	$t_p = 4.3$ s

21.2 After carrying out an analysis in different products of a roasted bovine meat plant, it has been found that a batch of these products contains 10^8 spores of *Clostridium botulinum* for kg, while another one has a similar load of *Salmonella*. To reduce the microbial content of these products, they are treated with a beam electron device of 35 kW of power, the dose absorbed by each kg of product being 8% of the incident radiation. If the products are irradiated for 2 s, determine the microbial content that each product possesses leaving the irradiation step.

From Table 21.2, D_{10} values can be obtained:

C. botulinum	$D_{10} = 1.4$ kGy
Salmonella	$D_{10} = 0.567$ kGy

The inactivation kinetic constant is obtained from the expression

$$D_{10} = \frac{2.303}{k}$$

Therefore,

C. botulinum	$k_{Clostridium} = 1.645\ \text{kGy}^{-1}$
Salmonella	$k_{Salmonella} = 4.062\ \text{kGy}^{-1}$

It is observed that for *Salmonella* its destruction constant is higher than *C. botulinum*, what indicates that it will be easier to destroy.

The absorbed dose is

$$D=(0.08)(35\,\text{kW/kg})(2\,\text{s})=5.6\,\text{kJ/kg}=5.6\,\text{kGy}$$

The final microorganism concentration after having carried out the irradiation of the samples is obtained from the equation

$$N=N_0\exp(-kD)$$

In the *C. botulinum* case

$$N=10^8\exp\left[-(1.645\,\text{kGy})(5.6\,\text{kGy})\right]\approx 10^{-4}\,\text{CFU/kg}$$

Then, the reduction logarithmic cycles number obtained is

$$\log\left(\frac{N_0}{N}\right)=\log\left(\frac{10^8}{10^{-4}}\right)\approx 12$$

In the *Salmonella* case

$$N=10^8\exp\left[-(4.062\,\text{kGy})(5.6\,\text{kGy})\right]\approx 1.3\times 10^{-4}\,\text{CFU/kg}$$

Then, the reduction of logarithmic cycle number obtained is

$$\log\left(\frac{N_0}{N}\right)=\log\left(\frac{10^8}{1.3\times 10^{-10}}\right)\approx 17.9$$

21.3 An industry crushes turkey meat with the purpose of marketing it in form of hamburgers. Due to a hygienic problem in this industry, the crushed meat has been contaminated with *Campylobacter jejuni* bacteria. To assure the safety of this product, it is irradiated in an electron beam radiation plant of 35 kW of power. The absorbed dose by each kg of crushed meat is 5% of the incident. If it is wished to reduce the microbial load in 12 logarithmic cycles, calculate the time that the products should be exposed to the radiation. If the electron beam that leaves the gun embraces a lineal distance of 1 m, determine the velocity of the product transport conveyor.

For *C. jejuni*, the decimal reduction dose is $D_{10}=1.4$ kGy (Table 21.2).

The inactivation constant is

$$k=\frac{2.303}{D_{10}}=\frac{2.303}{0.19\,\text{kGy}}=12.12\,\text{kGy}^{-1}$$

The absorbed dose for each kg can be calculated from

$$D=\frac{1}{k}\ln\left(\frac{N_0}{N}\right)$$

As it is desired, a 12 logarithmic cycle reduction

$$\log\left(\frac{N_0}{N}\right)=12$$

Therefore,

$$\ln\left(\frac{N_0}{N}\right)=\ln(10)\log\left(\frac{N_0}{N}\right)=(2.303)(12)=27.63$$

The absorbed dose is

$$D = \left(\frac{1}{12.12\,\text{kGy}^{-1}}\right)(27.63) = 2.28\,\text{kGy}$$

The absorbed dose rate is

$$\dot{D} = (0.05)(35\,\text{kW/kg}) = 1.75\,\text{kJ/s kg} = 1.75\,\text{kGy/s}$$

The treatment time is

$$t = \frac{D}{\dot{D}} = \frac{2.28\,\text{kGy}}{1.75\,\text{kGy/s}} = 1.3\,\text{s}$$

Then, the conveyor belt velocity will be

$$v = \frac{L}{t} = \frac{1\,\text{m}}{1.3\,\text{s}} \approx 0.768\,\text{m/s}$$

21.4 In a microwave device that is working at a frequency of 2450 MHz, a beef meat piece is treated. At 25°C, the beef meat presents a dielectric constant value of 52.4, while the loss tangent value is 0.33. Calculate the attenuation factor, the penetration depth, and the depth for which the power is 25% of the incident power.

The attenuation factor can be calculated from Equation 21.11, where the wavelength $\lambda = c/f$ and c is the light velocity in free space (3×10^8 m/s):

$$\lambda = \frac{3 \times 10^8\,\text{m/s}}{2450 \times 10^6\,\text{s}^{-1}} = 0.122\,\text{m}$$

Substituting the known values in Equation 21.11

$$\alpha = \frac{2\pi}{(0.122)}\left[\frac{(52.4)}{2}\left(\sqrt{1+(0.33)^2} - 1\right)\right]^{1/2} = 71.3\,\text{m}^{-1}$$

The field penetration depth is defined as the inverse of the attenuation factor, then $Z = 0.014$ m = 1.4 cm.

To obtain the depth for which the power is 25% of the incident, Equation 21.10 is used, where $P = 0.25P_0$:

$$0.25P_0 = P_0 \exp(-2\alpha d)$$

Therefore,

$$d = -\frac{\ln(0.4)}{2\alpha} = -\frac{\ln(0.4)}{2(71.3)} = 6.4 \times 10^{-3}\,\text{m} = 0.64\,\text{cm}$$

21.5 Meat steak is placed in a microwave oven that is working with a 915 MHz frequency, in order to carry out the steak cooking. The steak has dimensions of 10 cm × 15 cm and 1 cm of thickness. It is considered that the heat generation inside the meat is uniform, and under the work conditions there can be considered average values of 50 and 0.60 for the dielectric constant and the tangent of losses, respectively. The steak is considered to have arrived to its cooking point when it reaches 70°C. If initially the meat is at 20°C, it is wished to know the required time to cook the meat.
Data: Microwave oven is working under 3 V/cm for electric field intensity. Meat properties: Density 1200 kg/m^3. Specific heat 3.35 kJ/(kg °C).

The heat generation can be obtained from Equation 21.9:

$$\tilde{q}_G = 55.61\times10^{-12}\left(915\times10^{6}\right)\left(50\times0.60\right)\left(300\right)^{2} = 1.37\times10^{6}\ \text{W/m}^3$$

Therefore, the heat generation flow rate is

$$\tilde{q}_G V = \left(1.37\times10^{6}\ \text{W/m}^3\right)\left(0.10\times0.15\times0.01\,\text{m}^3\right) = 205.5\,\text{W}$$

The steak mass is $m = \rho V = (1200)(0.1 \times 0.15 \times 0.01) = 0.18$ kg.

When considering that the generation energy is uniform in the whole steak, the generated heat will be producing a temperature increase, for what Equation 21.12 is applicable, and considering constant the volume, the mass, and the specific heat of the steak, the following is obtained:

$$\frac{dT}{dt} = \frac{\tilde{q}_G V}{m\hat{C}_p}$$

This equation that can be integrated with the boundary condition $t = 0$ implies $T = T_0$:

$$T - T_0 = \frac{\tilde{q}_G V}{m\hat{C}_p}t$$

Then, the processing time is

$$t = \left(T - T_0\right)\frac{m\hat{C}_p}{\tilde{q}_G V} = \left(70 - 20\right)\frac{\left(0.18\right)\left(3350\right)}{\left(205.5\right)} = 146.7\,\text{s} \approx 2.5\,\text{min}$$

21.6 It is wished to carry out the cooking of meat paste using an ohmic treatment chamber of 20 L. Initially the meat paste is at 22°C, and the treatment finishes when it has reached 65°C. Calculate the treatment time if through the circuit an intensity of 12 A is passing. Assume that the paste possesses dielectric properties that do not change with the temperature and they are uniform in all the chamber points.

Meat paste properties: Density 1250 kg/m^3. Specific heat 3.4 kJ/(kg °C). Electric resistance 20 Ω.

Mass paste contained in the treatment chamber: $m = \rho V = (1250)(0.020) = 25$ kg

As the food is homogeneous, it will be assumed that the heat volumetric generation is uniform in all the sample points. When carrying out a similar balance to the one presented in Equation 21.12, the following is obtained:

$$I^2R = \frac{d}{dt}\left(m\hat{C}_pT\right)$$

As the mass is constant and it is assumed that the specific heat does not vary, the following equation is obtained:

$$dT = \frac{I^2R}{m\hat{C}_p}dt$$

This equation can be integrated with the boundary condition: for $t = 0$, $T = T_0$. An equation similar to (21.26) is obtained, which allows obtaining the treatment time:

$$t = \left(T - T_0\right)\frac{m\hat{C}_p}{I^2R}$$

Substituting the problem data, the following is obtained:

$$t = \left(65 - 22\right)\frac{\left(25\right)\left(3400\right)}{\left(12\right)^2\left(20\right)} = 1269\,\text{s} = 21.1\,\text{min}$$

22 Concentration

22.1 EVAPORATION

Evaporation is a unit operation that consists of the elimination of water in a fluid food by means of vaporization or boiling. Several foods are obtained as aqueous solutions, and in order to facilitate and to make cheaper their preservation and transport, they are concentrated during a water elimination stage. This elimination can be performed in different ways, although evaporation is one of the most used methods. The equipment used to remove this water from the food product is called an evaporator.

An evaporator consists mainly of two chambers, one for condensation and another for evaporation. Steam condenses in the condensation chamber, giving off the latent heat of condensation, which is contained in the evaporation chamber. The evaporated water leaves the evaporation chamber at boiling temperature, obtaining at the same time a stream of concentrated solution.

Figure 22.1 shows a scheme of an evaporator. The mass flow rate of steam is w_V, while that of food is w_A, obtaining a stream of vapor V and another of concentrated solution (or liquor) w_C. The removed vapor V is driven to the condenser, where it condenses. It is important to point out that many food solutions are heat sensitive and can be adversely affected if they are exposed to high temperatures. For this reason, it is convenient to operate under vacuum in the evaporation chamber, which causes the boiling temperature of the aqueous solution to decrease and the fluid to be affected by heat to a lesser extent. If it is desired to operate under vacuum, a vacuum pump is needed. Also, a barometric column to compensate for the pressure difference with the exterior is needed in the condenser to condense the vapor released in the evaporation chamber.

The *capacity* of the evaporator (V) is defined as the amount of water evaporated from the food per unit time. The *consumption* (w_V) is the amount of heating steam consumed per unit time. The *economy* (E) is the amount of solvent evaporated per unit of heating steam:

$$E = \frac{\text{Capacity}}{\text{Consumption}} = \frac{V}{w_V} \tag{22.1}$$

22.2 HEAT TRANSFER IN EVAPORATORS

Figure 22.2 presents a scheme of a single effect evaporator, including the different variables of each stream. The condensation chamber is fed with a saturated vapor stream w_V that has a temperature T and an enthalpy $\hat{H}_W$. The vapor condenses, and the only heat given off is that of condensation, so a stream w_V of liquid water leaves this chamber at the condensation temperature T and with enthalpy $\hat{h}_W$, which corresponds to the enthalpy of water at the boiling point. The condensation heat $\dot{Q}$ is transferred through the exchange area of the evaporator to the food stream in the evaporation chamber.

A stream w_A is fed into the evaporation chamber at a temperature t_A, with enthalpy $\hat{h}_A$. Due to the heat released by the condensed vapor ($\dot{Q}$), a concentrated stream w_C is obtained, with temperature t_C and enthalpy $\hat{h}_C$. Also, a vapor stream V is obtained, at a temperature T_V and with enthalpy $\hat{H}_V$. It is important to point out that the temperatures of the concentrated and vapor streams are equal, and they correspond to the boiling temperature of the concentrated solution that leaves this chamber.

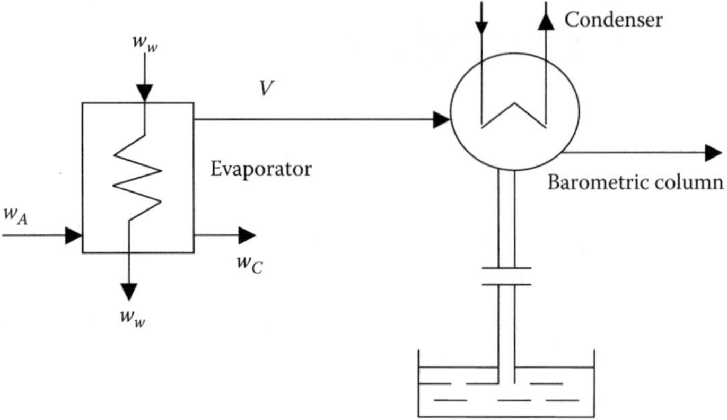

FIGURE 22.1 Scheme of the installation of an evaporator.

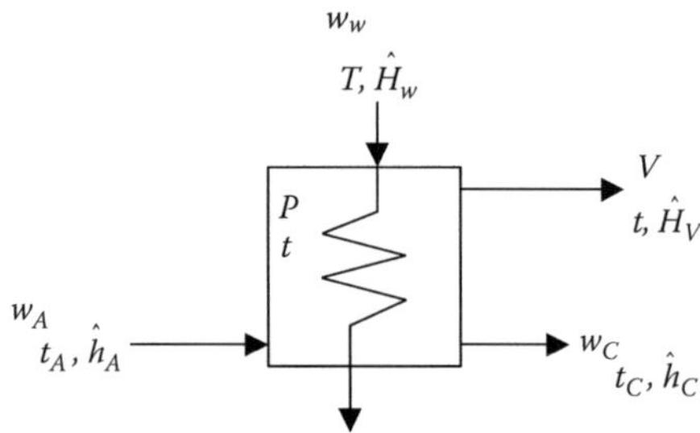

FIGURE 22.2 Simple evaporator.

The energy balances that should be performed are the following:

- Condensation chamber

$$w_V \hat{H}_w = w_V \hat{h}_w + \dot{Q} \tag{22.2}$$

- Evaporation chamber

$$w_A \hat{h}_A + \dot{Q} = w_C \hat{h}_C + V\hat{H}_V \tag{22.3}$$

- Exchange area

$$\dot{Q} = UA\Delta T = UA(T - t) \tag{22.4}$$

where
U is the global heat transfer coefficient
A is the area of the evaporator

22.2.1 Enthalpies of Vapors and Liquids

In the notation used here, the enthalpies per unit mass of vapor streams will be designated by $\hat{H}$ and those of liquid by $\hat{h}$.

The enthalpy per unit mass of vapor at a temperature T can be expressed as the summation of the enthalpy at saturation plus the integral between the boiling temperature T_b and the enthalpy at T, of the specific heat times dT:

$$\hat{H} = \hat{H}_{SATURATED} + \int_{T_b}^{T} \hat{C}_P)_V \, dT \qquad (22.5)$$

The term $\hat{H}_{SATURATED}$ is the enthalpy of the vapor at its condensation temperature. The specific heat of the water vapor $\hat{C}_P)_V$ depends on the pressure, although its value is close to 2.1 kJ/(kg °C).

Since enthalpy is a function, the state of the enthalpy of a liquid should be expressed as a function of a reference temperature. If this temperature is t^* and the liquid is at a temperature t, it is obtained that

$$\hat{h} = \int_{t^*}^{t} \hat{C}_P \, dT = \hat{C}_P\left(t - t^*\right) \qquad (22.6)$$

Tables used for the calculation of these enthalpies can be found in the literature. Generally, the reference temperature is the freezing temperature of water (0°C).

The enthalpy of the liquid at its boiling temperature is called $\hat{h}_{SATURATED}$. The latent heat of condensation or evaporation (λ) will be the difference between the saturation enthalpies of the vapor and the liquid, since the evaporation and condensation temperatures are the same:

$$\lambda = \hat{H}_{SATURATED} - \hat{h}_{SATURATED} \qquad (22.7)$$

The numerical values of the enthalpies of saturated vapor and of the liquid can be obtained from saturated water vapor tables, and the latent heat of condensation can be calculated. However, this value can be obtained in an approximate way from the equation of Regnault as follows:

$$\lambda = 2538 - 2.91T \text{ kJ/kg} \qquad (22.8)$$

where T is in °C.

The enthalpies of the liquid streams, food ($\hat{h}_A$), and concentrated ($\hat{h}_C$) that appear in Equation 22.3 are expressed as

$$\hat{h}_A = \int_{t^*}^{t_A} \hat{C}_P)_A dT = \hat{C}_P)_A\left(t_A - t^*\right) \qquad (22.9)$$

$$\hat{h}_C = \int_{t^*}^{t_C} \hat{C}_P)_C dT = \hat{C}_P)_C\left(t_C - t^*\right) \qquad (22.10)$$

The enthalpy of the vapor in Equation 22.3 will be different if the solution being concentrated presents or does not present a boiling point rise. If there is no increase in the boiling point of the concentrated solution, the enthalpy of the vapor will be the sum of the saturated liquid plus the latent heat:

$$\hat{H}_V)_{SAT} = \hat{C}_P\left(t_b - t^*\right) + \lambda \qquad (22.11)$$

where t_b is the boiling temperature of the solution.

If there is an increase in the boiling point, the boiling temperature of the solution (t) will be greater than that of pure water (t_b), so the vapor enthalpy will be

$$\hat{H}_V = \hat{C}_P(t_b - t^*) + \lambda + \hat{C}_P)_V(t - t_b) \tag{22.12}$$

In order to simplify the calculations, the reference temperature usually selected is the boiling point of pure water, $t^* = t_b$, which makes the enthalpy of the vapor leaving the evaporation chamber coincide with the latent heat of condensation if there is no increase in the boiling point. Also, the enthalpy of the concentrated stream will be annulled, since $t_C = t_b = t$.

22.2.2 Boiling Point Rise

Water boils at a fixed temperature whenever the pressure remains constant. If the pressure varies, the boiling point varies too. For aqueous solutions, the boiling temperature not only depends on pressure but also on the amount of solute they contain, in such a way that the presence of the solute causes the boiling temperature to increase. The determination of the boiling point rise presented by food solutions is very important for the calculation of evaporators. For this reason, expressions and means to calculate the increase in boiling temperature will be given next.

For diluted solutions that comply with Raoult's law, the boiling point rise can be calculated by the expression

$$\Delta T_e = \frac{1000 K_b X}{M_S} \tag{22.13}$$

where

- M_S is the molecular weight of the solute
- X is the ratio kg solute/kg solvent
- K_b is the so-called boiling constant of the solvent

For aqueous solutions, the following equation can be used:

$$\Delta T_e = 0.512X°\text{C} \tag{22.14}$$

where C is the molal concentration of the solute.

A general expression that allows the calculation of the boiling point rise, considering an ideal solution, is the equation

$$\Delta T_b = \frac{-t_b}{1 + (\lambda / (R t_b \ln X_w))} \tag{22.15}$$

If the solutions are diluted, the following equation can be used:

$$\Delta T_b = \frac{R t_b^2}{\lambda}(1 - X_w) \tag{22.16}$$

where

- X_w is the mass fraction of water
- λ is the latent heat of evaporation
- R is the gas constant
- t_b is the boiling temperature of pure water

For real solutions, the boiling point rise can be calculated by the empirical rule of Dühring, which states that the boiling point of the solution is a linear function of the boiling point of the pure solvent at the same pressure. For a given concentration of solute, the plots of the boiling temperatures of the solution against those corresponding to the pure solvent yield straight lines. Figures 22.3 and 22.4 present the diagram of Dühring for two aqueous systems.

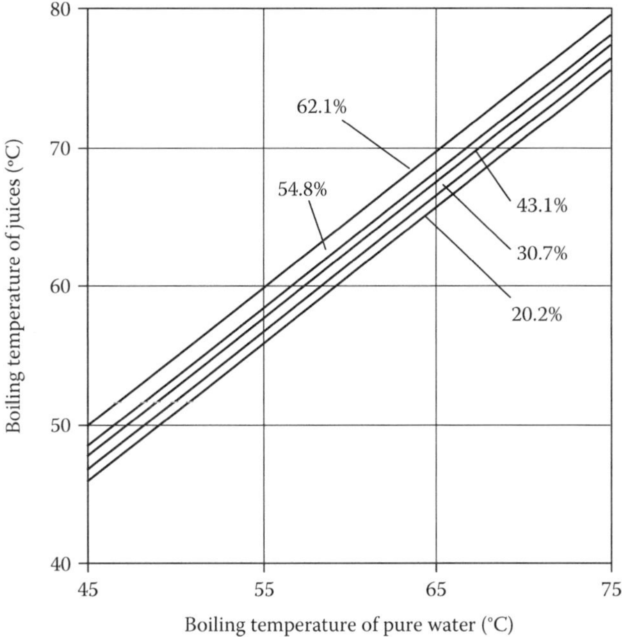

FIGURE 22.3 Diagram of Dühring for tamarind juices. (Adapted from Manohar, B. et al., *J. Food Eng.*, 13, 241, 1991.)

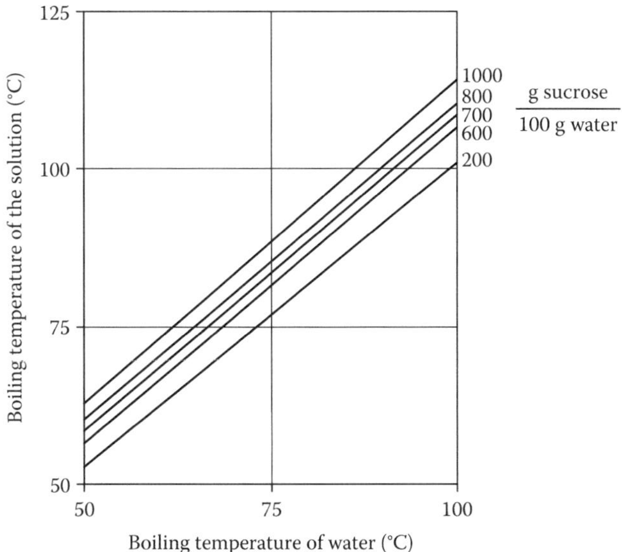

FIGURE 22.4 Diagram of Dühring for aqueous solutions of sucrose. (Adapted from Brennan, J.G., Butters, J.R., Cowell, N.D., and Lilly, A.E.V. *Las Operaciones de la Ingeniería de los Alimentos*. Acribia, Zaragoza, Spain, 1980.)

TABLE 22.1
Parameters α, β, δ, and γ

Sample	α × 10²	β	δ	γ × 10²
Sucrose	3.061	0.094	0.136	5.328
Reducing sugars	2.227	0.588	0.119	3.593
Juices	1.360	0.749	0.106	3.390

Source: Crapiste, G.H. and Lozano, J.E., *J. Food Sci.*, 53(3), 865, 1988.

In the case of sugar solutions, there exist empirical correlations that allow obtaining the boiling point increase of the solutions to be obtained. One of these expressions is (Crapiste and Lozano, 1988)

$$\Delta T_b = \alpha C^{\beta} P^{\delta} \exp(\gamma C) \tag{22.17}$$

where
C is the concentration of the solution in °Brix
P is the pressure in mbar
α, β, δ, and γ are empirical constants, whose values depend on the solute

Table 22.1 shows the values of these parameters for sucrose, reducing sugars and fruit juice solutions.

This equation has been modified for juices by adding a new term to the exponential term in such a way that the final expression is (Ilangantileke et al., 1991)

$$\Delta T_b = 0.04904 C^{0.029} P^{0.113} \exp\left(-0.03889 C + 6.52 \times 10^{-4} C^2\right) \tag{22.18}$$

22.2.3 Heat Transfer Coefficients

The calculation of the global heat transfer coefficient can be obtained from the expression:

$$\frac{1}{UA} = \frac{1}{h_C A'} + \frac{e_P}{k_P A_m} + \frac{1}{h_e A''} \tag{22.19}$$

where
h_C is the individual convective heat transfer coefficient for the condensing vapor
h_e is the coefficient corresponding to the boiling solution

The parameters e_P and k_P are the thickness of the solid through which heat transfer occurs and its thermal conductivity, respectively. In this type of operation, it is assumed that the areas are the same, so the expression is simplified to

$$\frac{1}{U} = \frac{1}{h_C} + \frac{e_P}{k_P} + \frac{1}{h_e} \tag{22.20}$$

In case there is deposition on the heat transfer surface, the resistance offered by such deposition (R_D) should be taken into account. Hence, the real global coefficient U_D is

$$\frac{1}{U_D} = \frac{1}{U} + R_D \tag{22.21}$$

TABLE 22.2
Global Heat Transfer Coefficients for Different Types of Evaporators

Evaporator	U (W/m² °C)
Long tube vertical	
Natural circulation	1,000–3,500
Forced circulation	2,300–12,000
Short tube	
Horizontal tube	1,000–2,300
Calandria type (propeller calandria)	800–3,000
Coiled tubes	1,000–2,300
Agitated film (Newtonian liquids)	
Viscosity	
1 mPa s	2,300
100 mPa s	1,800
10^4 mPa s	700

Source: McCabe, W.L. and Smith, J.C., *Operaciones Básicas de Ingeniería Química*, Reverté, Barcelona, Spain, 1968.

Despite the fact that the calculation of the theoretical global coefficient should be made by Equation 22.20, the values for this coefficient can be found in the literature, depending on the type of evaporator. Table 22.2 shows the values for this coefficient.

22.3 SINGLE EFFECT EVAPORATORS

Figure 22.5 shows a single effect evaporator with all the streams and variables. In order to perform the calculation of this type of evaporator, mass and energy balances should be conducted:

- *Mass balances*: A global balance and a component balance are carried out:

$$w_A = w_C + V \tag{22.22}$$

$$w_A X_A = w_C X_C \tag{22.23}$$

where X_A and X_C are the mass fractions of solute in the food and in the concentrated solution streams, respectively.

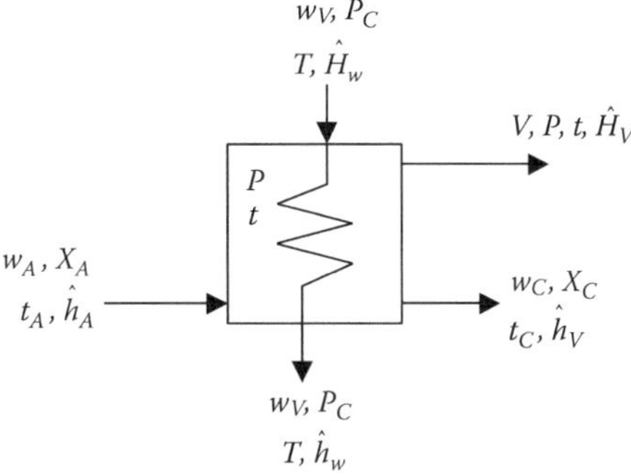

FIGURE 22.5 Single effect evaporator.

- *Energy balances*: Balances around the condensation and evaporation chambers are performed, in addition to the equation of heat transfer rate through the exchange area. These balances are similar to those performed previously and are given in Equations 22.2 through 22.4. If the expressions for the enthalpies of the liquid and the vapor given in Section 22.2.1 are taken into account, then
- Condensation chamber

$$w_V\left(\hat{H}_w - \hat{h}_w\right) = w_V\lambda_w = \dot{Q} \tag{22.24}$$

- Evaporation chamber

$$w_A\hat{C}_P)_A\left(t_A - t_b\right) + \dot{Q} = w_C\hat{C}_P)_C\left(t_C - t_b\right) + V\left[\lambda_V + \hat{C}_P)_V\left(t - t_b\right)\right] \tag{22.25}$$

- Exchange area

$$\dot{Q} = UA\Delta T = UA(T - t) \tag{22.4}$$

Note that $t_C = t$; that is, the temperatures of the streams leaving the evaporation chamber are equal, and the boiling point rise of the solution $\Delta T_b = t - t_b$. When combining Equations 22.24 and 22.25, it is obtained that

$$w_V\lambda_V = w_C\hat{C}_P)_C\Delta T_b + V\left[\lambda_V + \hat{C}_P)_V\Delta T_b\right] - w_A\hat{C}_P)_A\left(t_A - t_b\right) \tag{22.26}$$

If there is no boiling point rise ($\Delta T_b = 0$), the previous equation becomes

$$w_V\lambda_V = V\lambda_V - w_A\hat{C}_P)_A\left(t_A - t_b\right) \tag{22.27}$$

22.4 USE OF RELEASED VAPOR

The vapor released in the evaporation chamber contains energy that may be used for other industrial purposes. Such vapor has a temperature lower than that of steam, so its latent heat of condensation is greater. For this reason, it is very important to take advantage of this latent heat. There are different methods for using this energy, among which are vapor recompression, thermal pump, and multiple effects. The first and second methods mentioned will be briefly studied here, while in the following section, the multiple effect method will be studied in detail.

22.4.1 Recompression of Released Vapor

One way to use the energy contained in released vapor is to compress it and use it as heating steam. There are two methods, in practice, to compress this vapor: mechanical compression and thermocompression.

22.4.1.1 Mechanical Compression

This type of operation consists of compressing the vapor released in the evaporation chamber using a mechanical compressor. The vapor leaving the evaporation chamber at a temperature t_1 and a pressure P_1 is compressed to a pressure P_2, corresponding to the pressure of the steam used in the condensation chamber (Figure 22.6). In the enthalpy–entropy Mollier diagram for steam

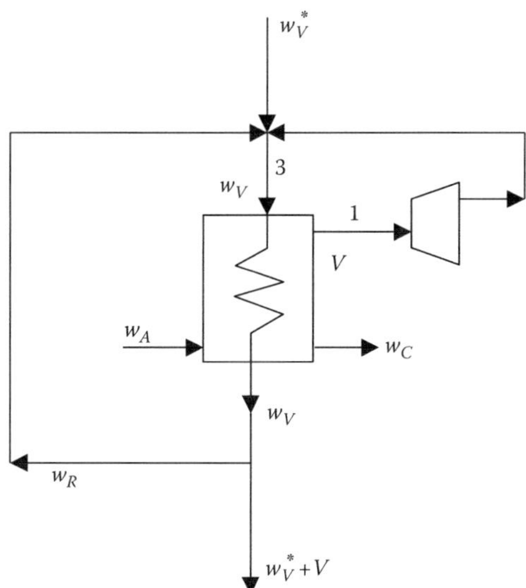

FIGURE 22.6 Simple evaporator with mechanical compression of released vapor.

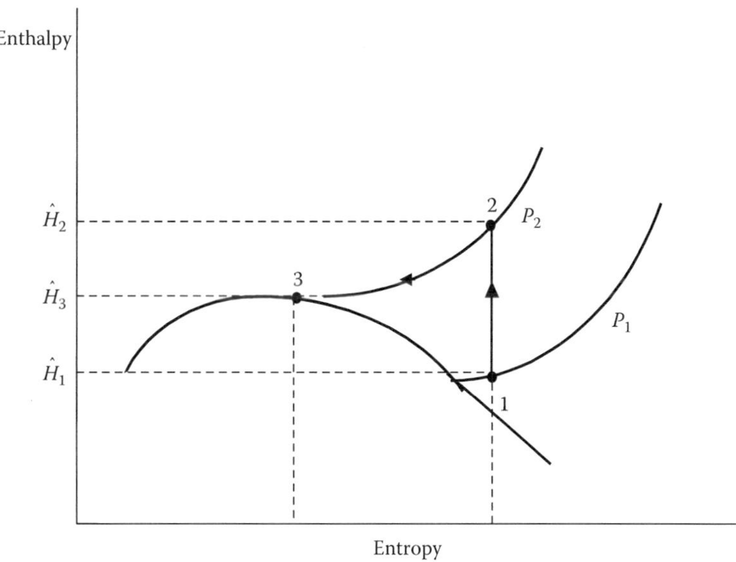

FIGURE 22.7 Evolution of released vapor during mechanical compression.

(Figure 22.7), the condition of the released vapor can be represented by point 1. Mechanical compression, generally, is an isentropic process, so a perpendicular straight line is followed until reaching the isobar corresponding to pressure P_2. The condition of this steam can be found in Mollier's diagram, where it is obtained that the outlet temperature of the compressor is t_2, its pressure is P_2, and its enthalpy is $\hat{H}_2$. It can be observed that the vapor obtained after compression is a reheated steam, so before mixing it with the saturated steam coming from the boiler, its temperature is lowered by recirculating a stream w_R. In this way, the condensation chamber can be fed with saturated steam.

As can be observed in Figure 22.6, in this type of operation, the balances around the evaporation chamber are not affected. However, additional balances should be performed in the condensation chamber.

- Energy and mass balances

$$w_V = V + w_R + \overset{*}{w_V} \tag{22.28}$$

$$w_V \hat{H}_w = V\hat{H}_1 + w_R \hat{h}_w + \overset{*}{w_V} \hat{H}_w \tag{22.29}$$

$$\dot{Q} = w_V \lambda_W \tag{22.30}$$

The calculation of the evaporator is similar to that described in the single effect, although in this case, these additional balances should be taken into account.

22.4.1.2 Thermocompression

Another way to use the energy of the released vapor is to use a jet that carries away part of the vapor and joins the steam coming from the boiler. Figure 22.8 represents a scheme of the evaporator jet system. The jet is a device that functions due to the Venturi effect, in such a way that a steam jet carries away part of the vapor released in the evaporation chamber. The vapor entering the condensation chamber is saturated, although its pressure is intermediate between steam and the released vapor. If the jet is fed with a steam flow w_V^* that has a pressure P_W and an enthalpy $\hat{H}^*$, then this steam carries away a fraction of vapor V, which is at pressure P_1 and at temperature t_1, and has an enthalpy $\hat{H}_1$. The vapor that leaves the jet will have a pressure P_C and an enthalpy $\hat{H}_w$, with a flow w_V. Similarly to mechanical compression, the balances around the evaporation and condensation chambers are unaltered. However, new balances should be carried out in the jet.

- Balances around the jet

$$\overset{*}{w_V} + aV = w_V \tag{22.31}$$

$$\overset{*}{w_V} \hat{H}^* + aV\hat{H}_1 = w_V \hat{H}_w \tag{22.32}$$

There is also an empirical equation that correlates the different variables and allows calculations of this type of compression. Thus, the expression to be used is (Vian and Ocón, 1967)

$$\frac{aV}{\overset{*}{w_V}} + 1 = R\,\frac{\log\left(P_w / P_1\right)}{\log\left(P_C / P_1\right)} \tag{22.33}$$

in which R is the thermal performance of the ejector.

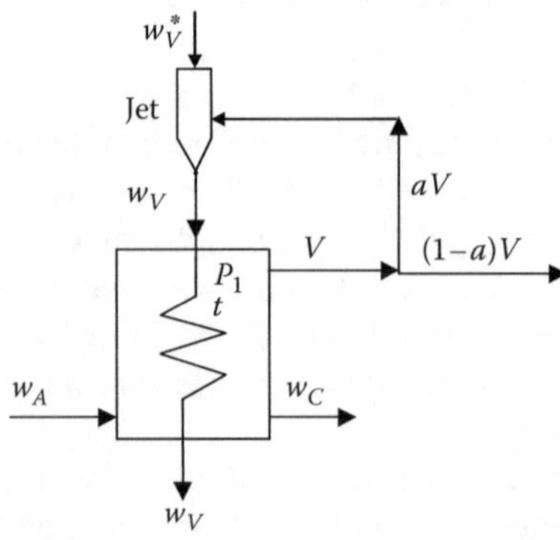

FIGURE 22.8 Simple evaporator with thermocompression of released vapor.

22.4.2 Thermal Pump

The so-called thermal pump is usually employed in heat-sensitive products in which high temperature can adversely affect the product. Low boiling temperatures can be achieved with this device. Figure 22.9 presents a scheme of this installation, while Figure 22.10 shows a temperature–entropy diagram for the heating fluid.

This installation has two evaporators. The evaporation chamber of the first evaporator is fed with the fluid to be concentrated, obtaining a vapor stream V that is used as heating fluid for the second evaporator. The condensation chamber of the first evaporator is fed with a vapor, which could be NH_3 that condenses. This liquid exits the condensation chamber (point 1) and is expanded in a valve (point 2). This liquid is employed to feed the evaporation chamber of the other evaporator, yielding a vapor stream (point 3) that is fed to a mechanical compressor, with the objective of raising its pressure and obtaining a more energetic vapor (point 4). This vapor is used as heating vapor for the first evaporator. It should be pointed out that the circuit followed by the heating vapor is a closed circuit.

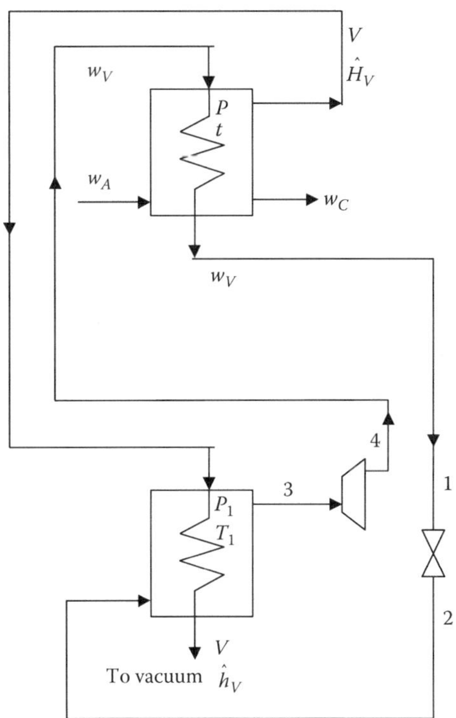

FIGURE 22.9 Installation with thermal pump.

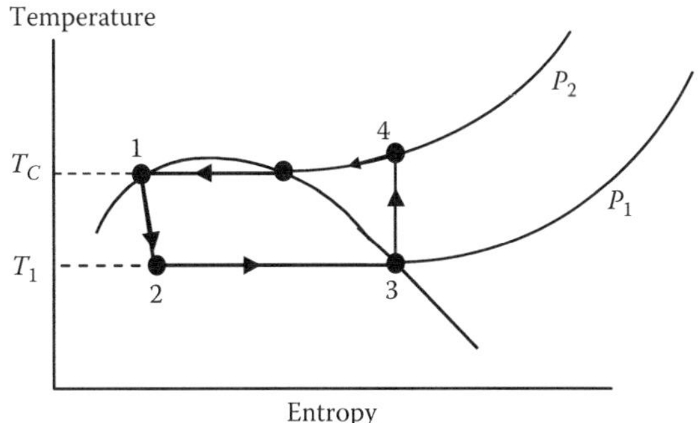

FIGURE 22.10 Evolution of the heating fluid in the temperature–entropy diagram for the thermal pump.

This type of installation is usually employed to concentrate juices, for example, orange juice, that are affected by high temperatures.

22.4.3 Multiple Effect

One of the most usual ways of employing the vapor released in the evaporation chamber is by using it as heating fluid in another evaporator. Figure 22.11 shows a scheme of a triple effect evaporation system. It can be observed that the vapor released in the first evaporator is used as heating fluid in the second one, while the vapor liberated in this effect is used to heat the third effect. Finally, the vapor released in the last effect is driven to the condenser.

For notation, the different streams have subscripts that correspond to the effect they are leaving. It is convenient to point out that the vapor released during the different effects is each time at a lower temperature and a lower pressure:

$$T > t_1 > t_{b1} > t_2 > t_{b2} > t_3 > t_{b3}$$

$$P_C > P_1 > P_2 > P_3$$

where

t_1, t_2, and t_3 are the boiling temperatures of the solutions leaving the evaporation chambers of the first, second, and third effects, respectively. The temperatures t_{b1}, t_{b2}, and t_{b3} are the boiling temperatures of pure water at pressures P_1, P_2, and P_3

The evaporation chamber of the first effect is at a pressure P_1 and a temperature t_1. The vapor that leaves this effect V_1 does so at these conditions and is used as heating fluid in the second effect, where it is supposed to enter in saturated conditions, that is, at its boiling temperature t_{b1}. The pressure in the condensation chamber of the second effect is still P_1, while the temperature is t_{b1}. The evaporation chamber of the second effect is at a pressure P_2 and at temperature t_2, the same as those of the vapor V_2 leaving this chamber. This vapor condenses in the condensation chamber of the second effect at a temperature t_{b2} and a pressure P_2. The evaporation chamber of the third effect is at a pressure P_3 and a temperature t_3; thus, the vapor V_3, leaving this effect, has the same characteristics. This vapor is driven to a condenser where it condenses at the temperature t_{b3}, which corresponds to that of pressure P_3.

If there is no boiling point rise of the solutions that pass through the evaporator, the temperatures of the evaporation chamber of one effect and that of condensation of the following effect will be equal ($t_i = t_{bi}$).

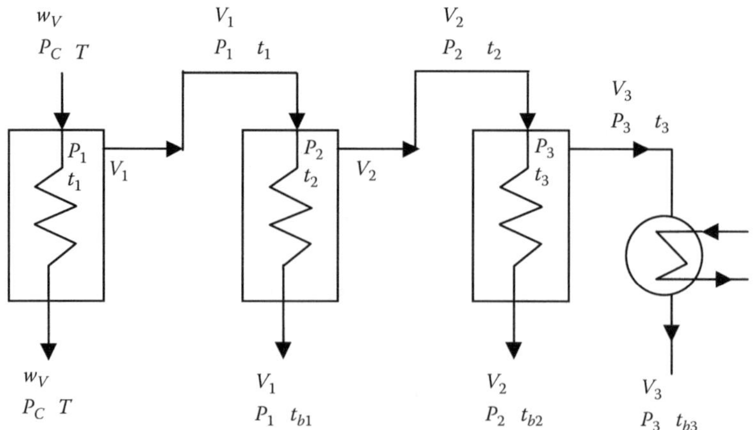

FIGURE 22.11 Scheme of released vapor use in a triple effect evaporator.

It is convenient to point out that in this type of installation, vacuum pumps are required to reach adequate temperatures in the chambers of each effect. A more detailed study, which allows the calculation of multiple effect evaporators, is performed in the following section.

22.5 MULTIPLE EFFECT EVAPORATORS

Only the case of a triple effect evaporator will be studied; however, the mathematical treatment in other cases for multiple effects is similar.

22.5.1 Circulation Systems of Streams

As previously mentioned, the vapor released in the evaporation chamber of one effect is employed as heating fluid for the next effect. However, depending on the circulation system of the solutions that will be concentrated, different passing systems, as explained next, are obtained:

a. *Parallel feed*
   The food is distributed in different streams that are fed into each of the effects (Figure 22.12a), while the concentrated solution stream of each effect is gathered in only one stream, which will be the final concentrate.
b. *Forward feed*
   The diluted stream is fed into the first effect, while the concentrate stream leaving each effect is fed into the following effect (Figure 22.12b). It can be observed that the vapor and

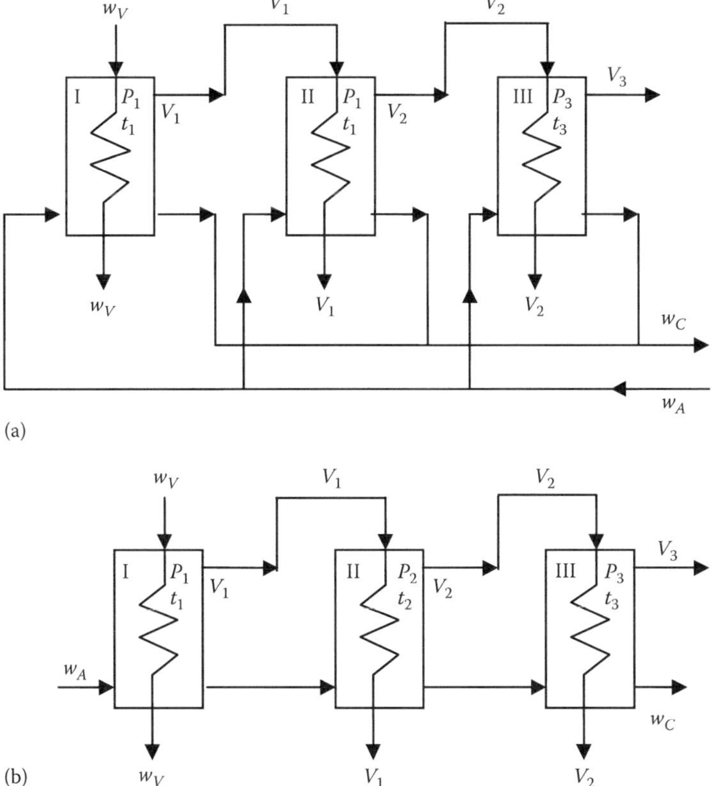

FIGURE 22.12 Circulation systems of fluid streams for triple effect evaporators. (a) Parallel feed, (b) forward feed.

(continued)

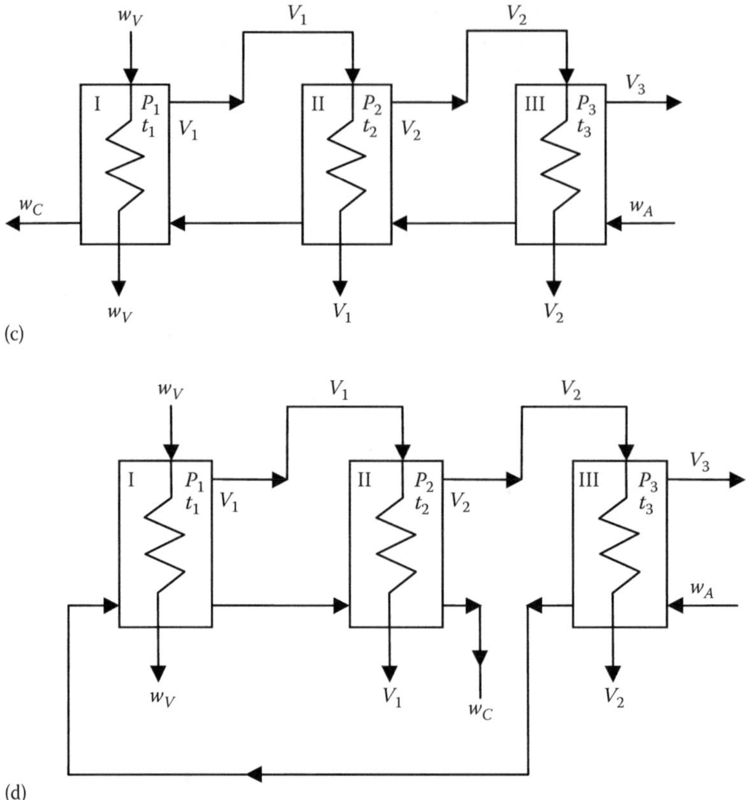

FIGURE 22.12 (continued) Circulation systems of fluid streams for triple effect evaporators. (c) backward feed, and (d) mixed feed.

concentrate streams of each effect are parallel flows. This pass system is frequently used for solutions that can be thermally affected, since the most concentrated solution is in contact with the vapor at the lowest temperature.

c. *Backward feed*
The flows of the solutions to concentrate and of vapor are countercurrent (Figure 22.12c). The diluted solution is fed into the last effect, where the vapor has less energy, and the concentrated solution leaving this effect is used to feed the previous effect and so on. This type of arrangement should be carefully used in the case of food solutions, since the solution with the highest concentration gains heat from the vapor at the highest temperature, and this may affect the food.

d. *Mixed feed*
In this type of arrangement, the diluted solution can be fed into any of the effects, while the concentrated solutions can be used to feed a previous or following effect. Figure 22.12d shows a mixed feed arrangement, in which the diluted solution is fed into the third effect, while the solution leaving this effect is used to feed the first one. The food stream feeding the second effect is the concentrated solution that leaves the first effect, obtaining the final concentrated solution at the second effect.

22.5.2 Mathematical Model

Among the different cases that can be studied, this section will examine only the triple effect evaporator, in which the pass system is a backward feed (Figure 22.13). The mathematical model here and its solution are similar for any type of circulation and number of effects.

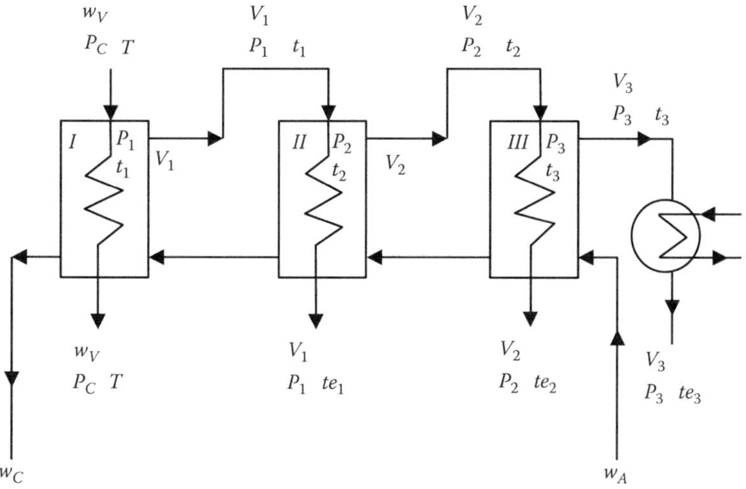

FIGURE 22.13 Backward feed triple effect evaporator.

In order to set up the mathematical model, the global and component mass balances should be performed, as well as the enthalpy balances and equations for rate of heat transfer through the exchange area in each effect.

The reference temperatures for calculating the enthalpies of the different streams are the boiling temperatures of pure water at the pressure in the evaporation chamber of each effect t_{bi}. The boiling point rise of each effect is the difference between the boiling temperature of the solution leaving the chamber of each effect and the boiling temperature of pure water at the pressure in this chamber $\Delta T_{bi} = t_i - t_{bi}$.

The notation used here implies that the streams leaving an effect will have the subscript corresponding to each effect:

- Mass balances

$$w_A = w_C + V_1 + V_2 + V_3 \tag{22.34}$$

$$w_A X_A = w_C X_C \tag{22.35}$$

$$w_2 = w_A - V_2 - V_3 \tag{22.36}$$

$$w_3 = w_A - V_3 \tag{22.37}$$

- Enthalpy Balances

  The enthalpy balances performed around each effect lead to the following equations:

$$w_V \hat{H}_W + w_2 \hat{h}_2 = w_V \hat{h}_W + w_C \hat{h}_C + V_1 \hat{H}_{V1}$$

$$V_1 \hat{H}_{V1} + w_3 \hat{h}_3 = V_1 \hat{h}_{V1} + w_2 \hat{h}_2 + V_2 \hat{H}_{V2}$$

$$V_2 \hat{H}_{V2} + w_A \hat{h}_A = V_2 \hat{h}_{V2} + w_3 \hat{h}_3 + V_3 \hat{H}_{V3}$$

Substitution of the expressions of the enthalpies of each stream and rearrangement yields

$$w_V\left(\hat{H}_w-\hat{h}_w\right)=w_C\hat{C}_P)_C\left(t_C-t_{b1}\right)+V_1\left[\lambda_{V1}+\hat{C}_P)_V\left(t_1-t_{b1}\right)\right]-w_2\hat{C}_P)_2\left(t_2-t_{b1}\right) \quad (22.38)$$

$$V_1\left[\lambda_{V1}+\hat{C}_P)_V\left(t_1-t_{b1}\right)\right]=w_2\hat{C}_P)_2\left(t_2-t_{b2}\right)+V_2\left[\lambda_{V2}+\hat{C}_P)_V\left(t_2-t_{b2}\right)\right]-w_3\hat{C}_P)_3\left(t_3-t_{b2}\right) \quad (22.39)$$

$$V_2\left[\lambda_{V12}+\hat{C}_P)_V\left(t_2-t_{b2}\right)\right]=w_3\hat{C}_P)_3\left(t_3-t_{b3}\right)+V_3\left[\lambda_{V3}+\hat{C}_P)_V\left(t_3-t_{b3}\right)\right]-w_A\hat{C}_P)_A\left(t_A-t_{b3}\right) \quad (22.40)$$

These are general equations; this means that it is assumed there is a boiling point rise. However, if there is no boiling point rise, the equations are simplified.

Even when the vapor leaving the evaporation chambers is reheated, it is assumed that when it enters the condensation chamber of the next effect, it enters as saturated vapor. This simplifies the latter equations to

$$w_V\lambda_w=w_C\hat{C}_P)_C\left(t_C-t_{b1}\right)+V_1\left[\lambda_{V1}+\hat{C}_P)_V\Delta T_{b1}\right]-w_2\hat{C}_P)_2\left(t_2-t_{b1}\right) \quad (22.41)$$

$$V_1\lambda_{V1}=w_2\hat{C}_P)_2\Delta T_{b2}+V_2\left[\lambda_{V2}+\hat{C}_P)_V\Delta T_{b2}\right]-w_3\hat{C}_P)_3\left(t_3-t_{b2}\right) \quad (22.42)$$

$$V_2\lambda_{V2}=w_3\hat{C}_P)_3\Delta T_{b3}+V_3\left[\lambda_{V3}+\hat{C}_P)_V\Delta T_{b3}\right]-w_A\hat{C}_P)_A\left(t_A-t_{b3}\right) \quad (22.43)$$

- Heat Rate Transfer Equations

  The heat transferred through the exchange area of each effect is obtained from the following equations:

$$\dot{Q}_1=w_V\lambda_w=U_1A_1\left(T-t_1\right) \quad (22.44)$$

$$\dot{Q}_2=V_1\lambda_{V1}=U_2A_2\left(t_{b1}-t_2\right) \quad (22.45)$$

$$\dot{Q}_3=V_2\lambda_{V2}=U_3A_3\left(t_{b2}-t_3\right) \quad (22.46)$$

It is assumed that the vapor entering the condensation chambers is saturated and the only heat that dissipates is the condensation heat.

22.5.3 Resolution of the Mathematical Model

Generally, the data available in evaporator problems are the flow of food to concentrate, as well as its composition and temperature. Also, the composition of the final concentrated solution is known. The characteristics of the vapor of the steam boiler are known; generally, the pressure and, since it is saturated vapor, its temperature and latent heat can be obtained from thermodynamic tables. The pressure of the evaporation chamber of the third effect is usually known; therefore, its characteristics are also known. It is possible to obtain the boiling point rise by Dühring diagrams or adequate equations once the composition of the streams leaving the evaporation chambers is known.

A 10-equation system is obtained from the mass and enthalpy balances as well as from the rate equations:

$$w_A=w_C+V_1+V_2+V_3 \quad (22.34)$$

$$w_AX_A=w_CX_C \quad (22.35)$$

$$w_2 = w_A - V_2 - V_3 \tag{22.36}$$

$$w_3 = w_A - V_3 \tag{22.37}$$

$$w_V \lambda_w = w_C \hat{C}_P)_C \left(t_C - t_{b1}\right) + V_1\left[\lambda_{V1} + \hat{C}_P)_V \Delta T_{b1}\right] - w_2 \hat{C}_P)_2\left(t_2 - t_{b1}\right) \tag{22.41}$$

$$V_1 \lambda_{V1} = w_2 \hat{C}_P)_2 \Delta T_{b2} + V_2\left[\lambda_{V2} + \hat{C}_P)_V \Delta T_{b2}\right] - w_3 \hat{C}_P)_3\left(t_3 - t_{b2}\right) \tag{22.42}$$

$$V_2 \lambda_{V2} = w_3 \hat{C}_P)_3 \Delta T_{b3} + V_3\left[\lambda_{V3} + \hat{C}_P)_V \Delta T_{b3}\right] - w_A \hat{C}_P)_A\left(t_A - t_{b3}\right) \tag{22.43}$$

$$\dot{Q}_1 = w_V \lambda_w = U_1 A_1 \left(T - t_1\right) \tag{22.44}$$

$$\dot{Q}_2 = V_1 \lambda_{V1} = U_2 A_2 \left(t_{b1} - t_2\right) \tag{22.45}$$

$$\dot{Q}_3 = V_2 \lambda_{V2} = U_3 A_3 \left(t_{b2} - t_3\right) \tag{22.46}$$

Since the number of unknowns is larger than the number of equations, there are infinite solutions. It is assumed that the area of each effect is equal in order to solve the problem. Also, the heat transferred through each effect is similar and can be assumed to be equal ($\dot{Q}_1 = \dot{Q}_2 = \dot{Q}_3$).
Since $\dot{Q}_i/A_i$ = constant, then

$$\frac{T - t_1}{1/U_1} = \frac{t_{b1} - t_2}{1/U_2} = \frac{t_{b2} - t_3}{1/U_3} \tag{22.47}$$

Due to the property of the ratios, they will be equal to the sum of the numerators divided by the sum of the denominators:

$$\frac{T - t_1}{1/U_1} = \frac{t_{b1} - t_2}{1/U_2} = \frac{t_{b2} - t_3}{1/U_3} = \frac{T - t_{b3} - \sum \Delta T_{bi}}{\Sigma(1/U_i)} \tag{22.48}$$

where ΔT_{bi} is the boiling point rise experienced by the solution in the ith effect.

These assumptions allow the solution of a system of ten equations, although an iterative process is needed to solve the mathematical model.

22.5.4 Calculation Procedure

The calculation procedure requires an iterative method, which is simpler when there is no boiling point rise.

22.5.4.1 Iterative Method When There Is Boiling Point Rise

The calculation steps are listed next:

1. Assume that the flow of heat transfer at each stage is the same. Also, assume that the exchange areas of the different stages are equal.
2. Determine w_C by Equations 22.34 and 22.35 as the total flow of released vapor ($V_1 + V_2 + V_3$).

3. Assume that the flows of eliminated vapor in each effect are equal: $V_1 = V_2 = V_3$.
4. Calculate the concentrations X_2 and X_3.
5. Use the concentration of each solution to determine its correspondent specific heat: $\hat{C}_P)_i$.
6. Calculate the boiling point rise in each effect. Such rises are calculated with the concentration of the solution leaving the effect.
7. Calculate the unknown temperatures of the evaporation and the condensation chambers using Equation 22.48.
8. Calculate the latent heat of condensations λ_w, λ_{V1}, λ_{V2}, and λ_{V3} of the saturated steams at temperatures T, t_{b1}, t_{b2}, and t_{b3}.
9. Solve the equation systems obtained from enthalpy and mass balances (Equations 22.34 through 22.43). This operation allows the determination of w_w, w_2, w_3, V_1, V_2, and V_3.
10. Obtain the areas of each effect A_1, A_2, and A_3 from the rate equations (Equations 22.44 through 22.46).
11. Check whether the areas obtained are different by less than 2% with respect to the mean value A_m, in which case the iterative process is finished.
12. If the areas are different, recalculate X_2 and X_3 using the values of V_2 and V_3 obtained in step 9.
13. Recalculate the boiling point rise with the new concentrations.
14. Determine the new temperatures of the different evaporation and condensation chambers. Use the following expressions to do this:

$$\left(T-t_1\right)_j = \left(T-t_1\right)_{j-1}\left(\frac{A_1}{A_m}\right) \tag{22.49}$$

$$\left(t_{b1}-t_2\right)_j = \left(t_{b1}-t_2\right)_{j-1}\left(\frac{A_2}{A_m}\right) \tag{22.50}$$

$$\left(t_{b2}-t_3\right)_j = \left(t_{b2}-t_3\right)_{j-1}\left(\frac{A_3}{A_m}\right) \tag{22.51}$$

These equations point out that the temperature rise between the condensation and evaporation chambers of each effect in an iterative stage j is equal to the rise existing between such chambers in the previous calculation stage $j - 1$, multiplied by the ratio between the area of each effect and the mean area. Two pairs of values are obtained for each unknown temperature, so the value taken is the arithmetic mean.

15. Continue the calculations, beginning with step 8, until the values of the areas of each effect coincide.

22.5.4.2 Iterative Method When There Is No Boiling Point Rise

The calculation steps are listed in the following:

1. Assume that the heat flows transferred in each stage are the same. Also, assume that the exchange areas of the different stages are equal.
2. Determine w_C and the total flow of released vapor ($V_1 + V_2 + V_3$) by Equations 22.34 and 22.35.
3. Assume that the flows of vapor eliminated in each effect are equal: $V_1 = V_2 = V_3$.
4. Calculate the concentrations X_2 and X_3.
5. Use the concentration of each solution to determine their correspondent specific heat: $\hat{C}_P)_i$.
6. Calculate the unknown temperatures of all evaporation and condensation chambers using Equation 22.48.

7. Use temperatures T, t_{b1}, t_{b2}, and t_{b3} to calculate the latent heat of condensation of saturated steams λ_w, λ_{V1}, λ_{V2}, and λ_{V3} at these temperatures.
8. Solve the equation systems obtained from the enthalpy and mass balances (Equations 22.34 through 22.43). This operation allows the calculation of w_w, w_2, w_3, V_1, V_2, and V_3.
9. Obtain the areas of each effect A_1, A_2, and A_3 from the rate equations (Equations 22.44 through 22.46).
10. Check whether the areas obtained are different by less than 2% with respect to the mean value A_m, in which case the iterative process is finished.
11. If the areas are different, recalculate X_2 and X_3 using the values of V_2 and V_3 obtained in step 9.
12. Calculate the new temperatures of the different evaporation and condensation chambers. Use Equations 22.49 through 22.51.
    Two pairs of values are obtained for each unknown temperature, so the value taken is the arithmetic mean.
13. Continue from step 7 until the values of the areas of each effect coincide.

22.6 EVAPORATION EQUIPMENT

Different types of equipment are used for evaporation processes: those in which the fluid circulates by pumps and those that do not require these devices. The former are called forced circulation evaporators, and the latter are called natural circulation evaporators. Also, there are long-tube evaporators of ascending and descending film, as well as plate and expanded flow evaporators. A brief description of each type is given next.

22.6.1 Natural Circulation Evaporators

There are different types of evaporators based on the natural circulation of fluids, with the simplest one being the open evaporator. There are also tube evaporators, usually short, based on this principle.

22.6.1.1 Open Evaporator

These evaporators are the simplest ones, consisting mainly of a container open to the atmosphere in which fluid is heated directly or by a heating coil or external jacket. Evaporators often have a low evaporation rate, hence showing poor thermal economy. Sometimes, and to allow operation under vacuum, the container may have a hermetic seal.

The main advantage of these evaporators is that they are useful when low capacity units are required. However, heating is not effective in large capacity units, since the ratio of heat transfer surface to volume of liquid is low. Also, heat transfer is reduced in units that have internal coils, since they make the circulation of liquid difficult.

This type of evaporator is used in the food industry to concentrate tomato pulp, to prepare soups and sauces, and to boil marmalades and confectionery products.

22.6.1.2 Short-Tube Horizontal Evaporator

These evaporators are formed by a chamber, the bottom of which is crossed by a bundle of interior horizontal tubes (Figure 22.14), which circulate steam that acts as heating fluid. Above the tubes, there is a space that allows the separation by gravity of drops carried away by the vapor released in the base.

Impact slabs are arranged in order to facilitate the separation and carrying away of drops. Since the tube bundle makes the circulation of liquid difficult, such evaporators present poor global heat transfer coefficients. They are usually employed for the concentration of low-viscosity liquids.

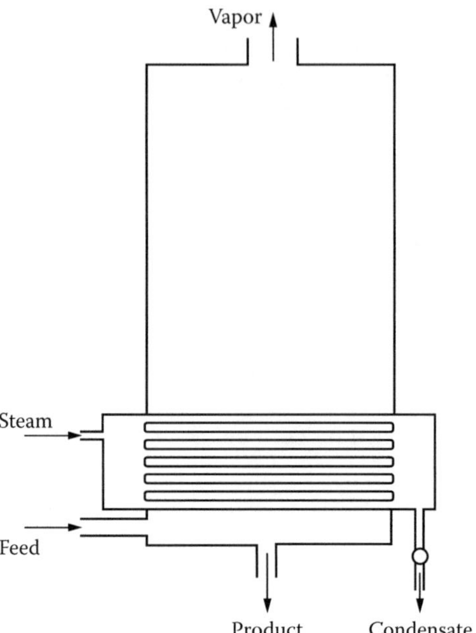

FIGURE 22.14 Short-tube horizontal evaporator. (Adapted from Brennan, J.G. et al., *Las Operaciones de la Ingeniería de los Alimentos*, Acribia, Zaragoza, Spain, 1980.)

22.6.1.3 Short-Tube Vertical Evaporator

Figure 22.15 shows a scheme of this type of evaporator in which the heating steam condenses outside tubes that are vertically arranged inside the evaporation chamber. The tube sheet, that is, the calandria, has a large central return tube through which a liquid that is colder than the liquid that circulates in the heating ascending tubes, thus forming natural circulation streams. The length of

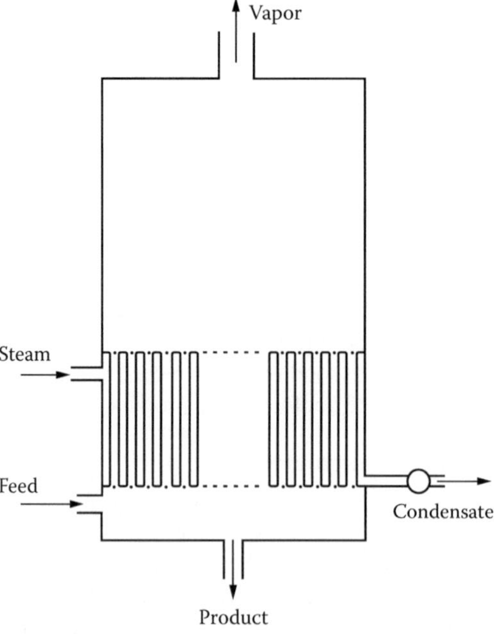

FIGURE 22.15 Short-tube vertical evaporator. (Adapted from Brennan, J.G. et al., *Las Operaciones de la Ingeniería de los Alimentos*, Acribia, Zaragoza, Spain, 1980.)

the tubes usually ranges between 0.5 and 2 m, with a diameter of 2.5–7.5 cm, while the central tube presents a transversal section between 25% and 40% of the total section occupied by the tubes.

These evaporators show adequate evaporation velocities for noncorrosive liquids with moderate viscosity. The units can be equipped with basket calandria that facilitates cleaning, since they can easily be dismounted.

Short-tube vertical evaporators are usually employed for the concentration of sugar cane and beet juices, as well as in the concentration of fruit juices, malt extracts, glucose, and salt.

22.6.1.4 Evaporator with External Calandria

In this type of evaporator (Figure 22.16), the tube bundle is located outside the vapor separator. They usually operate at reduced pressures and have easy access to the tube bundle. Also, the calandria can be substituted by a plate heat exchanger, which is useful in case crusts are formed, since the plates are easy to dismount and clean.

Since these evaporators can operate under vacuum, they are used to concentrate heat-sensitive foods such as milk, meat extracts, and fruit juices.

22.6.2 Forced Circulation Evaporators

Circulation in these evaporators is achieved by means of a pump that impels the food through the calandria into a separation chamber, where vapor and concentrate are separated (Figure 22.17). The pump causes the fluid to circulate at a velocity between 2 and 6 m/s; when it passes through the tube bundle, the fluid gains enough heat to be reheated, but the liquid is subjected to a static charge that prevents boiling inside the tubes. However, when the fluid reaches the chamber, there is a sudden evaporation, and the impact slab facilitates the separation of the liquid phase from vapor.

These evaporators are capable of concentrating viscous liquids when the pump impels the liquid at an adequate velocity. For this reason, centrifugal pumps are used if the liquids present low viscosity. If the liquids have a higher viscosity, then positive displacement pumps should be used.

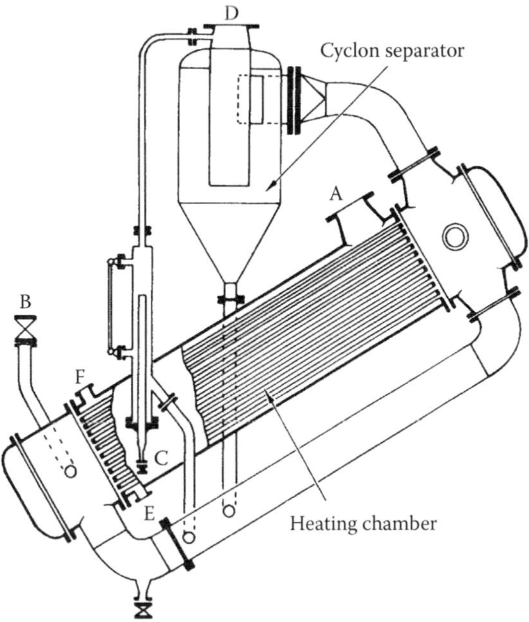

FIGURE 22.16 Evaporator with exterior calandria: A, vapor inlet; B, liquid feed inlet; C, concentrated liquid outlet; D, vapor outlet; E, condensate outlet; and F, noncondensable gases outlet. (Adapted from Brennan, J.G. et al., *Las Operaciones de la Ingeniería de los Alimentos*, Acribia, Zaragoza, Spain, 1980.)

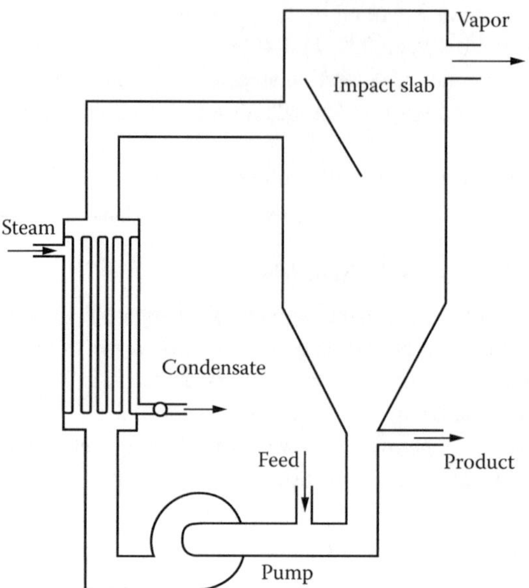

FIGURE 22.17 Forced circulation evaporator. (Adapted from Brennan, J.G. et al., *Las Operaciones de la Ingeniería de los Alimentos*, Acribia, Zaragoza, Spain, 1980.)

22.6.3 Long-Tube Evaporators

Some evaporators consist of a vertical chamber consisting of a tubular exchanger and a separation chamber. The diluted liquid is preheated to almost boiling temperature before entering the tubes. Once inside the tubes, the liquid begins to boil, and the expansion due to vaporization yields the formation of vapor bubbles that circulate at high velocity and carry away the liquid, which continues to concentrate as it moves forward. The liquid–vapor mixture then enters into the separation chamber, where baffle plates facilitate vapor separation. The concentrated liquid obtained can be directly extracted or mixed with nonconcentrated liquid and recirculated, or it can go into another evaporator where concentration can be increased.

Long-tube evaporators can be of ascending film, of falling film, or of ascending–falling film. In ascending film evaporators, the liquid enters the bottom of the tubes, and vapor bubbles that ascend through the center of the tube begin to form, creating a thin film on the tube wall that ascends at great velocity. In falling film evaporators, the feed is performed at the top of the tubes, so the vapor formed descends through the center of the tubes as a jet at great velocity. When high evaporation velocities are desired, ascending–falling film evaporators are used, where ascending film evaporation is used to obtain an intermediate concentration liquid with a high viscosity. This liquid is then further evaporated in tubes, where it circulates as falling film.

Generally, global heat transfer coefficients are high. In film evaporators, the residence time of the liquid being treated in the heating zone is short, since it circulates at great velocity. The product is thus not greatly affected by heat, and therefore these evaporators are useful for evaporation of heat-sensitive liquids. Descendent film evaporators are widely used to concentrate milk products.

22.6.4 Plate Evaporators

Plate evaporators consist of a set of plates distributed in units in which vapor condenses in the channels formed between plates. The heated liquid boils on the surface of the plates, ascending and descending as a film. The liquid and vapor mixture formed goes to a centrifugal evaporator.

These evaporators are useful to concentrate heat-sensitive products, since high treatment velocities are achieved, allowing good heat transfer and short residence times of the product in the evaporator.

Also, plate evaporators occupy little space on the floor and are easily manipulated for cleaning, since setup and dismount are easy and quick. Plate evaporators are usually employed to concentrate coffee, soup broth, light marmalades, and citrus juices.

Besides the evaporators described here, there are other types, such as expanded flow, scrape surface, and those based on the functioning of the thermal pump used for the evaporation of products that are very sensitive to heat.

22.7 FREEZE CONCENTRATION

Freeze concentration is a processing unit operation that consists of decreasing the temperature of the product to concentrate it by means of partial freezing. Since in the process there is not an interface of liquid–vapor, the loss of aromas and volatile flavors that occurs in the evaporation is avoided. These processes (freeze-drying and other similar processes) require refrigeration systems, since the product is at low temperatures when it exits the cryoconcentrator. In freeze concentration, the freezing point of water that contains soluble solids decreases, obtaining a more and more concentrated product in which water is eliminated in the form of ice. Thus, its freezing point diminishes as the process advances. Freeze concentration allows the concentration of heat-sensitive nutritional products, maintaining the quality of the final product.

The development of freeze concentration technology began in 1970 at the Grasso–Grenco group in the Netherlands. The focus was initially on coffee extract only, but soon other food liquids such as citrus juices, beer, and wine were applied worldwide. Freeze concentration operates at the freezing point of the product in a closed system, and it produces high-quality products since at low-temperature conditions there is no thermal damage to the product, there is no biological damage, and efficient water separation is achieved, and because there is no vapor phase, there is no loss of aroma or solids. In addition, the closed system prevents contact with oxygen.

Freeze concentration can be carried out in two fundamental ways (Figure 22.18): by means of freezing the bulk of the flowing food or with a freeze concentration layer. In the first method, ice crystals appear in the bulk of the fluid. These crystals will grow, and after reaching a certain size, they are removed in a continuous way; crystallization of the ice is carried out through high pressure, and the separation of the ice crystals is achieved by means of centrifuge, filters, or washing columns. In the progressive freeze concentration, ice is formed on a freezing surface, and the frozen layer advances toward the bulk solution, which becomes progressively more concentrated. This process is based on atmospheric pressure, and although the separation is not always necessary, there is a moment when the thickness of the layer is large enough that the process should be stopped to revert the cycle and to remove the ice layer that has been formed. The larger the ice formation speed, the more the retention of solids increases, while the speed of diffusion of the solutes decreases with the concentration. In Table 22.3, the main characteristics that distinguish the two suitable types of technologies are shown.

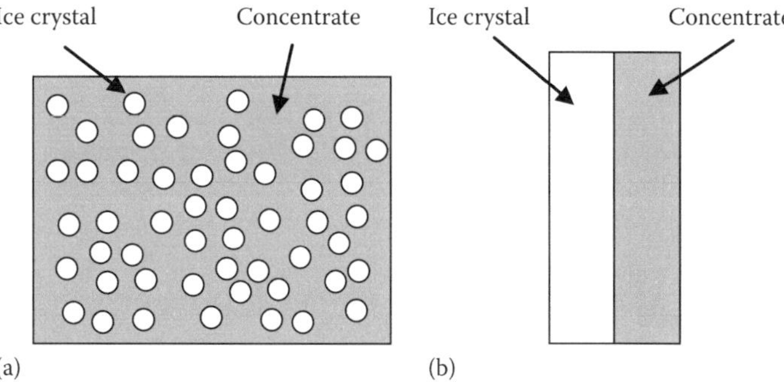

FIGURE 22.18 Two methods for concentration by freezing: (a) suspension crystallization and (b) progressive freeze concentration.

TABLE 22.3
Comparison of Two Methods for Concentration by Freezing

	Layer Crystallization	Suspension Crystallization
Heat removal	Through the crystallized surface	Through the solution
Ice growth rate	10^{-6}–10^{-7} m/s	10^{-7}–10^{-8} m/s
Equipment	No moving parts except in pumping equipment	Different moving parts
Ice/solution contact surface area	Low	High
Solid–liquid separation	Simple	Complex

Source: Hernández, E., Raventós, M., Auleda, J.M., and Ibarz, A. Concentration of apple and pear juices in a multi-plate freeze concentrator. *Innovative Food Science and Emerging Technologies,* 10, 348–355, 2009.

This technique has been used in different food processes. It is in the industry of fruit juices where the concentration technique by means of freezing is more established (Addison, 1986). There are numerous studies on the application of this technique in the concentration of kiwi juice (Valente and Duverneuil, 1986; Maltini and Mastrocola, 1999), forest fruits (Ghizzoni et al., 1995; Di Cesare et al., 2000), and apple and pear juices (Hernández et al., 2009), but the most important application is in citric fruits (Braddock, 1986). This concentration technique has been used in other food industries, as in the case of dairy (Buss, 1993), brewery (Putman et al., 1997), and wines and liquors (Patino et al., 1991; Di Cesare et al., 1993; Niro, 2004), as well as in the concentration of diluted solutions of tea and coffee with the purpose of freeze-drying it later on, thus avoiding loss of aroma and volatile compounds.

22.8 FREEZING TEMPERATURE

The freezing temperature begins with the first ice crystals, that is to say, the temperature in which ice crystals and liquid water coexist in balance. For pure water, this temperature corresponds to 0°C (273 K). However, the water in food contains soluble solids, and it is well known that the effect of these solids causes the freezing point to decrease. That is why the temperature at which foods begin to freeze is lower than 0°C. It is evident that for a food, the soluble solid concentration of unfrozen water increases as it advances in the freezing process, and what determines the freezing temperature is its variation with time. For this reason, from the point of view of later calculations, the initial temperature of freezing is usually used, corresponding to the appearance of the first ice crystals.

In Chapter 19, it was described that at the beginning of freezing, the aqueous solution is diluted, so as a first approximation, the initial freezing temperature (T_F) can be calculated by applying Raoult's law, and the freezing point depression (ΔT_F) can be calculated by Equation 19.30:

$$\Delta T_C = T_{0A} - T_C = K_A \frac{m_S}{M_S} \tag{22.52}$$

in which

m_S is the g solute/100 g water
M_S is the molecular weight of the solute
K_A is 18.6 (cryogenic constant of water)

It should be pointed out that for foods, M_S is an equivalent molecular weight of the solutes contained in the food.

For ideal solutions, the freezing temperature and the fraction molar of solutes contained in the solution are correlated by means of the equation:

$$\ln\left(1 - X_S\right) = \frac{\lambda}{R}\left[\frac{1}{T_{0W}} - \frac{1}{T_F}\right] \tag{22.53}$$

where

T_{0W} is the freezing temperature of pure water (273 K)
λ is the freezing latent heat of water (6003 kJ/kmol)
R is the gas constant (8.314 kJ/kmol K)
X_S is the solute molar fraction

In foods, the freezing process is different from that of pure water, in which the freezing temperature stays constant, while the change of phase of liquid water continues until it becomes ice. If T_F is the temperature at which the freezing of the product begins, initially, the temperature decreases below that temperature. Once the first ice crystals are formed, the temperature T_F increases until the value is reached. However, the temperature does not remain constant, but rather there is a small continuous decrease, because the water becomes ice, and the unfrozen water concentrates on the soluble solids that contain the food. A moment arrives at which the crystallization of some of the solutes can begin, and the crystallization heat is liberated, in this way increasing the temperature. Finally, there will be a temperature at which is not possible to freeze more water; the content in soluble solids is so high that very low temperatures would be needed. This point coincides with the so-called eutectic temperature, below which the solid phase possesses the same concentration as that of the liquid phase, since there cannot be more separation of water in the form of ice.

Aqueous solutions of glucose, fructose, and sucrose present a depression in their freezing point that depends on their sugar content. In Figure 22.19, the increment that experiences a depression in the freezing point of solutions of glucose, fructose, and sucrose is shown (Raventós et al., 2007). It can be observed that in these three cases, it has been obtained that the increment of depression of the freezing point varies with its concentration according to a power evolution. In this figure, it can be seen that for the case of sucrose, this depression is much smaller than in the other two sugars, which scarcely present a significant variation between them. This result occurs because the freezing temperature of a solution is mainly affected by the low molecular weight solutes' presence; for a certain weight, the molar fraction of these compounds is higher than the molar fraction of a solute with a high molecular weight. This way, in glucose and fructose solutions, the freezing point is much lower than the freezing point of a sucrose solution with the same percentage of solids in weight.

In Figure 22.20, the freezing temperatures for apple and pear juices are shown (Hernández et al., 2009) in those in which the proportions of glucose, fructose, and sucrose were 22, 62, and

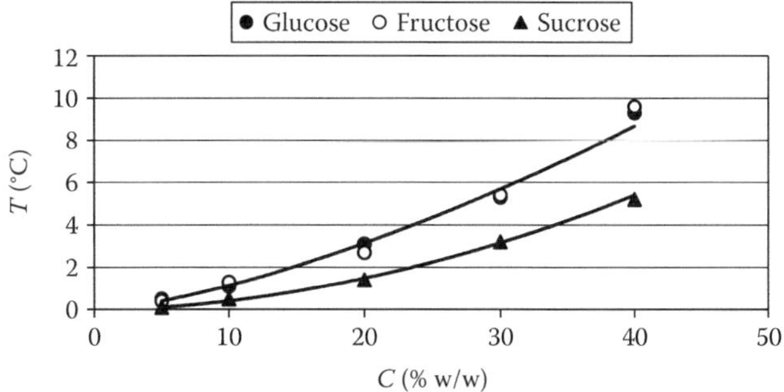

FIGURE 22.19 Freezing depression for sugar solutions. (Adapted from Hernández, E., Raventós, M., Auleda, J.M., and Ibarz, A. Concentration of aqueous sugar solutions in a multi-plate cryoconcentrator. *Journal of Food Engineering*, 79, 577–585, 2007.)

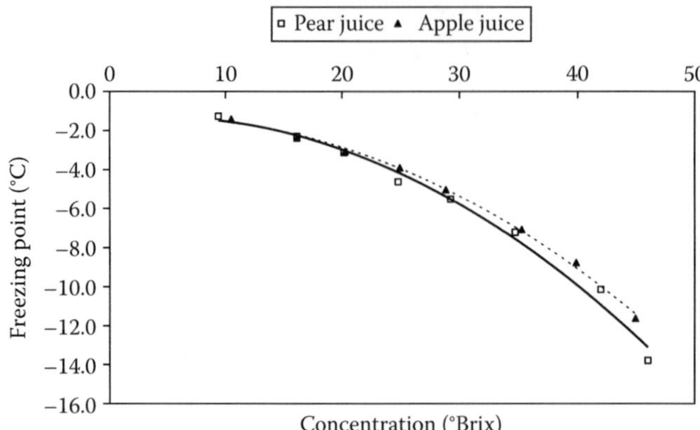

FIGURE 22.20 Freezing temperature for apple and pear juices. (Adapted from Hernández, E., Raventós, M., Auleda, J.M., and Ibarz, A. Concentration of apple and pear juices in a multi-plate freeze concentrator. *Innovative Food Science and Emerging Technologies*, 10, 348–355, 2009.)

16 (%, w/w), respectively, for apple juices and 39, 52.5, and 8.5 (%, w/w) for pear juices. The values of the freezing points for juices depend on the soluble solid content, so they decrease as their content increases. The differences between the two juices are due to the level in sugar content. The lowest freezing point belongs to the pear juice, due to the presence of a greater proportion of glucose and fructose in the apple juice. The freezing temperature of a solution is affected by the presence of low molecular weight solutes, since for a given weight the molar fraction of these compounds is greater than the molar fraction of a solute with a high molecular weight (Raventós et al., 2007).

22.9 ICE CRYSTAL FORMATION MECHANISMS

The freeze concentration process of food solutions is based on the ice glass formation that can be removed from the solution later on, giving a concentrated solution as a result.

The process begins with a decrease of the temperature below the freezing point of the solution. Once it has arrived at this point, the *nucleation* or formation of the ice glass nuclei begins. There are several types of nucleation, depending on temperature and of the levels of the solution concentration (Mullin, 1972). In case the solution is at a temperature quite below the freezing point, the nuclei are formed for a mechanism of homogeneous nucleation or molecular addition. For this to occur, it is necessary to achieve very low temperatures that require a high energy expense; thus, this mechanism is not usually the one that causes nucleation in food solutions. Most food solutions can contain particles that can catalyze the formation of ice crystal nuclei. This nucleation type is known as heterogeneous nucleation that takes place on a strange surface; the particles can exist in suspension on the wall of the recipient that contains the solution. Both homogeneous nucleation and heterogeneous nucleation constitute primary nucleation. However, the final mechanism of nucleation requires the presence of a crystal surface. Contact of the supercooling solution with this surface causes the formation of new nuclei by contact or by a shear mechanism of the fluid. In suspension crystallizers, the ice crystals are formed by both mechanisms. Another important mechanism for nucleation in the cryoconcentration systems for foods implies the formation of nuclei in heat scraped surface exchangers. A mechanism accepted for the formation of ice in scraped surface freezers is the following: the water freezes on the heat transfer surface for some heterogeneous mechanism, and later on the ice is removed by scraping with revolving knives. The size and distribution of the nuclei can be altered by the adaptation of the rotation index of the knife as well as of the yield index of the product. If the rotation speeds rise, they will cause smaller nuclei at the same time that the yield index will increase.

Once ice crystal nuclei have been formed, the following step is *crystal growth*. In the stage of crystal growth, different processes can be present that include the diffusion of water molecules from the bulk solution toward the ice crystal surface, with the subsequent incorporation of these molecules into the crystalline reticule that also includes the transfer of the formation of latent heat from the crystal surface toward the bulk solution. Also, a counterdiffusion of the solutes in the solution takes place from the surface of the crystal toward the solution. In general, it is accepted that in fluid foods, the stage of crystal growth is limited by a combination of heat transfer and counterdiffusion of the solute molecules (Hartel, 1992). In ice crystal growth, it has been observed that the growth rate increases with the increase of supercooling and, also, the increase of agitation has a tendency to increase the growth rate, probably due to the increment of the mass and heat transfer coefficients. In general, while the solute content increases, the crystal growth rate decreases, partly due to the effects of viscosity and partly due to the decrease of the incorporation kinetics to the surface. Also, the growth rate only shows a light dependence on crystal size.

The stage of crystal growth gives rise to a variety of sizes that makes the system thermodynamically unstable. The smallest ice crystals have a melting temperature slightly lower than the largest crystals. That means that when the small and large crystals mix in a suspension, the medium-size crystals settle down with an associated balance temperature. The smallest crystals in the balance will be unstable, and they will melt, while the biggest crystals will be supercooled, and consequently they will grow. In this way, the crystal size distribution will be reorganized according to the balance process, involving a distribution that increases the average size of the crystals as this process advances. The global process of reorganization of the crystal size is called the *crystal ripening* step.

Once the ice crystals are formed, they should be separated from the solution, obtaining a final product of higher concentration. The separation can be carried out in a continuous or discontinuous way by mechanical separation, such as press filters (filtration and filtration at pressure) and centrifuges, or by washing the crystals in a column. A combination of two of those operations is also a possibility.

It is interesting to carry out this stage with the most efficient method possible in order to avoid solute losses retained in the ice phase. A better separation is obtained when the viscosity of the concentrate solution is lower and when the size distribution is the most homogeneous possible and with large crystals.

In crystal suspension freeze concentration processes, washing columns are generally used. This system facilitates lower solute content of the crystals. It is important that the speed with which the fluid circulates through the column is adapted to obtain good efficiency. This speed can be calculated by means of the equation (van Pelt and Swinkels, 1986):

$$v = \frac{\varepsilon^3}{180\left(1-\varepsilon^2\right)} \frac{\left(\Gamma d_p\right)^2}{\eta} \frac{dP}{dz} \tag{22.54}$$

where

ε is the liquid fraction in the column
d_p is the diameter of the crystal
η is the liquid viscosity
Γ is the form factor of the area of the crystals, while dP/dz is the pressure gradient along the channel

22.10 FREEZE CONCENTRATION EQUIPMENT

There is a wide variety of freeze concentration systems whose components depend on the type of system that is used. Next, the devices based on the different solution freezing types are briefly described. Two essential parts can be distinguished in the cryoconcentrators: the crystallization units and the separation units. Initially, the food solution is taken to a temperature below its freezing point, and this supercooled solution is taken to a crystallization unit in which ice crystals are formed and grow until a uniform size that is the largest possible.

There are different crystallizer types, classified according to the freezing process in freezers of direct or indirect contact and by internal or external cooling, such as described in Table 22.2 (Hartel, 1992). In direct contact freezers, the solution to be frozen is mixed with the refrigerant, so when the refrigerant evaporates, it is possible to reduce the temperature of the product below its freezing point, which causes crystals to appear in the solution bulk. The appearance of ice crystals can also be achieved by subjecting the solution to vacuum, during which part of the water evaporates and the vaporization heat allows the temperature of the solution to reduce. This operation type can cause evaporation in the refrigerant fluid or for the evaporated water to carry out volatile compounds that affect the flavor and aroma of the food solution.

In the indirect contact freezers, the solution cools down by eliminating the heat through the heat exchanger wall. This heat exchanger can act directly on the solution or may be to an external element in which the solution cools down before being introduced to a crystallization tank.

The indirect contact cryoconcentration devices, with internal cooling that works with ice crystals in suspension, consist of a jacketed tank where a refrigerant fluid circulates, eliminating the heat of the solution (Figure 22.21). The same jacketed tank serves as a crystallizer, so ice crystals in suspension are formed in the solution bulk. This suspension is later pumped into the separation unit, with the purpose of obtaining a current of concentrated solution and a current of ice crystals.

Another cryoconcentrator type with indirect contact freezing consists of a revolving drum submerged in the solution in the interior of which the refrigerant fluid circulates. A layer of ice is formed over the drum surface that grows with time, while the solution concentrates. The ice layer can be removed by means of a scraping knife (Figure 22.22).

There are indirect contact cryoconcentrators that consist of a hollow plate inside which a refrigerant fluid circulates. On the exterior side, the solution to be concentrated circulates by gravity. In Figure 22.23, a typical installation of this operation type is shown. On the plate, an ice layer being formed grows with the operation time, obtaining at the end of the plate a concentrated solution.

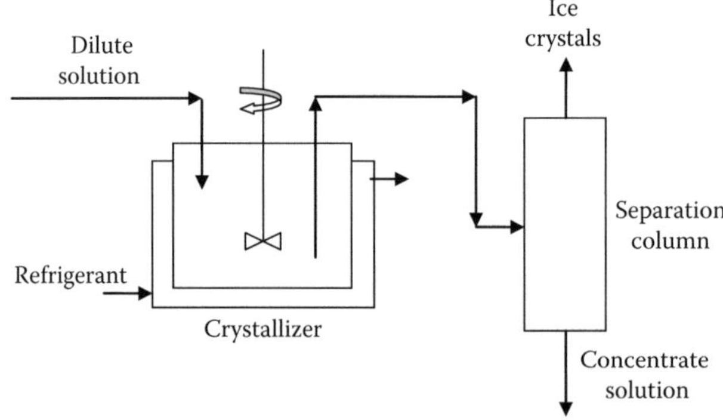

FIGURE 22.21 Indirect contact suspension cryoconcentrator.

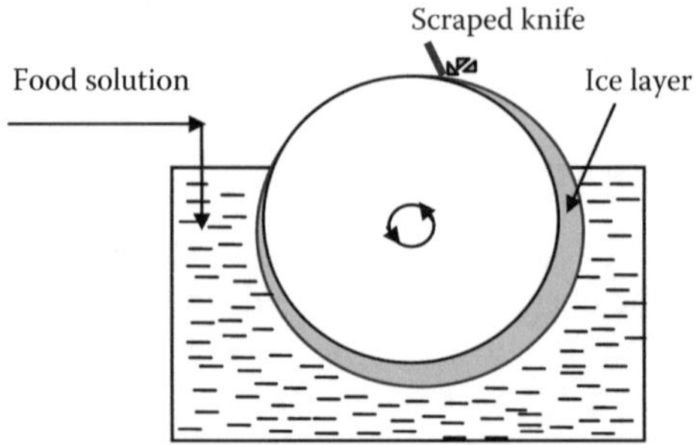

FIGURE 22.22 Indirect contact layer drum cryoconcentrator.

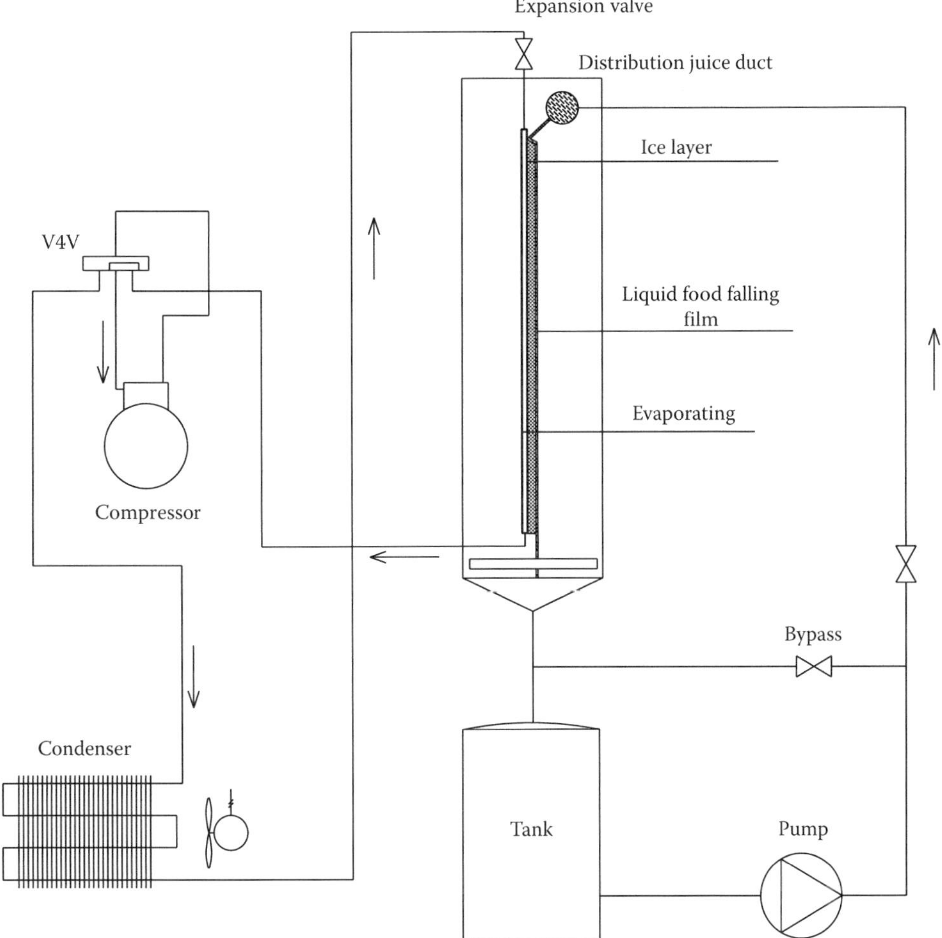

FIGURE 22.23 Basic scheme of layer cryoconcentrator pilot plant device. (From Hernández, E., Raventós, M., Auleda, J.M., and Ibarz, A. Concentration of apple and pear juices in a multi-plate freeze concentrator. *Innovative Food Science and Emerging Technologies*, 10, 348–355, 2009.)

To remove the ice layer formed on the plate, it is necessary to stop the operation and to invert the refrigeration cycle in order to allow the ice layer to be separated from the plate.

There are devices where the cooling of the solution is carried out using a heat exchanger that is usually of the scraped wall type (Figure 22.24). In these facilities, the diluted solution cools down below its freezing point in a scraped wall heat exchanger. The crystallization begins in this stage, and the supercooling solution is sent to a recrystallization tank, where the ripening of the ice crystals takes place. From this tank, it is sent to a separation column, where a current of concentrated solution and a current of ice crystals are obtained. This installation type is similar to the one shown in Figure 22.21 but with the difference that in this case the cooling and ripening are carried out in the same agitated tank, while in the other scenario, these processes are carried out in two successive stages.

The purpose of the separation units is the separation of the ice crystals from the solution that contains them. In food, cryoconcentration processes like separation units usually use centrifugal and washing columns. The type of centrifugation used is a filtrate operation, so by means of a centrifugal force, the solution passes through a filtrate mesh, while the ice crystals are retained in the mesh. The retained ice crystals usually retain solution in the hollow spaces among the crystals, and it is very usual to carry out a crystal washing operation, with the purpose of recovering the retained solution.

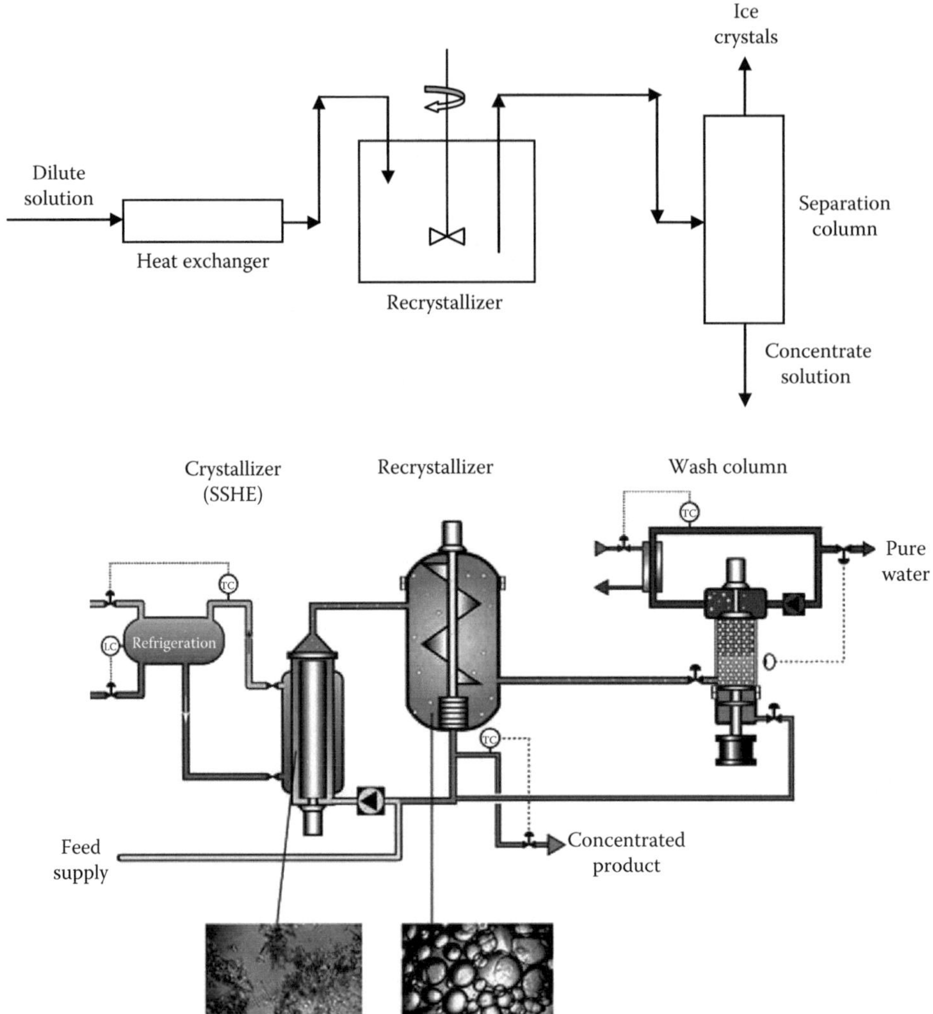

FIGURE 22.24 Freeze concentration process with external freezing. (Courtesy of GEA Niro.)

The washing columns are vertical columns in which the concentrated solution and ice currents are circulated in the countercurrent. In the superior part of the column, the ice current is obtained, while at the bottom of the column, the concentrated solution is extracted. At the bottom of the column, there is a mesh that prevents the exit of the ice crystals. Along the column, there is a certain melting of ice crystals, which causes a washing effect that gives a greater recovery of the solutes initially contained in the ice crystals.

22.11 EVALUATION OF FREEZE CONCENTRATION PROCESS EFFECTIVENESS

In order to measure the effectiveness of the freeze concentration processes, different variables or parameters can be used, such as relative impurity of the ice, concentration efficiency, and ice production.

Ice impurity is the quantity of soluble solids in the liquid phase that remains in the ice so that the greater the solute quantity retained, the lower the purity of the obtained ice. Ice purity must be as high as possible so that it retains the smallest quantity of solutes. To measure these impurities, it is defined that the relative cause of impurity of the ice is the relationship between the solute concentration in the ice (C_I) and the solute concentration in the concentrated solution (C_{FS}):

$$\text{Impurity ratio } (\%) = \frac{C_I}{C_{FS}} 100 \quad (22.55)$$

Another parameter to measure the effectiveness of the cryoconcentration process is the concentration efficiency that refers to the quantity of the increment of concentration of the solution in relation to the quantity of solutes that are retained in the ice crystals. The efficiency is calculated by means of the following equation:

$$\text{Efficiency } (\%) = \frac{C_{FS} - C_I}{C_{FS}} 100 \quad (22.56)$$

The quantity of produced ice is also a parameter that is useful to measure the effectiveness of the process. In the layer crystallization devices, a fundamental factor is the ice production by unit of heat exchange surface (Flesland, 1995). The ice productivity ($\overline{m}_u$) can be calculated by means of the following expression:

$$\overline{m}_u = \frac{M}{A \cdot t} \quad (22.57)$$

where

$\overline{m}_u$ is the ice productivity by surface unit and time (kg/m^2 s)
M is the net mass of ice (g)
A is the effective exchange area (m^2)
t is the time of each assay (s)

Ice production efficiency is another parameter that can measure the effectiveness of a cryoconcentration process (Raventós et al., 2007). Ice production efficiency is defined as the relationship between the quantity of ice formed in the solution and the quantity of ice that would be formed under identical conditions with pure water.

Separation units are an important consideration in the suspension of crystallization facilities. According to Fellows (2000), separation effectiveness is defined according to the expression:

$$\eta_{sep} = x_{mix} \frac{x_1 - x_i}{x_1 - x_j} \quad (22.58)$$

where

η_{sep} is the separation effectiveness
x_{mix} is the ice fraction weight in the frozen mixture before separation
x_1 is the solid fraction weight in the liquid after freezing
x_i is the solid fraction weight of solids in the ice after separation
x_j is the solution fraction weight of juice before freezing

PROBLEMS

22.1 A salt solution is concentrated from 5% to 40% in weight of salt. For this reason, 15,000 kg/h of the diluted solution is fed to a double effect evaporator that operates under backward feed. The steam used in the first effect is saturated at 2.5 atm, maintaining the evaporation chamber of the second effect at a pressure of 0.20 atm. If feed is at 22°C, calculate (a) the steam flow rate needed and economy of the system, (b) the heating area of each effect, and (c) the temperatures and pressures of the different evaporation and condensation chambers.

Data: Consider that only the 40% salt solution produces a boiling point rise of 7°C.

The specific heat of the salt solutions can be calculated by the expression $\hat{C}_p = 4.18 - 3.34X$ kJ/(kg°C), where X is the mass fraction of salt in the solution.

The global heat transfer coefficients of the first and second effect are, respectively, 1860 and 1280 W/(m^2 °C).

Specific heat of water is 2.1 kJ/(kg °C).

The scheme of the double effect evaporator is represented in Figure P.22.1.

- Properties of the saturated steam

$P_w = 2.5$ at $= 2452$ mbar $\qquad T = 126.8°C$

$$\hat{h}_w = 533 \text{ kJ/kg}$$

$$\hat{H}_w = 2716 \text{ kJ/kg}$$

$$\lambda_w = 2183 \text{ kJ/kg}$$

$P_2 = 0.2$ at $= 196$ mbar $\qquad t_{b2} = 59.7°C$

$$\hat{h}_{V2} = 250 \text{ kJ/kg}$$

$$\hat{H}_{V2} = 2609 \text{ kJ/kg}$$

$$\lambda_{V2} = 2359 \text{ kJ/kg}$$

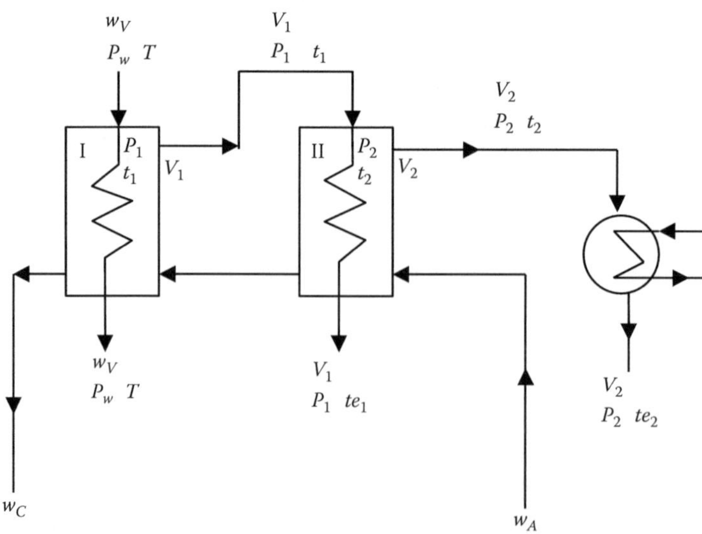

FIGURE P.22.1 Backward feed double effect evaporator.

- Global and component mass balances

$$15{,}000 = w_C + V_1 + V_2$$

$$(15{,}000)(0.05) = w_C\,(0.40)$$

obtaining

$$w_C = 1875 \text{ kg/h} \quad \text{and} \quad V_1 + V_2 = 13{,}125 \text{ kg/h}$$

Initially, it is assumed that $V_1 = V_2 = 6562.5$ kg/h, supposing that the composition of the stream w_2 is $X_2 = 0.09$.

- The specific heats of each stream are obtained from the equation given in the problem statement:
- For $X_A = 0.05$

$$\hat{C}_{PA} = 4.01 \text{ kJ/(kg °C)}$$

- For $X_C = 0.40$

$$\hat{C}_{PC} = 2.84 \text{ kJ/(kg °C)}$$

- For $X_1 = 0.09$

$$\hat{C}_{P1} = 3.88 \text{ kJ/(kg °C)}$$

According to the statement of the problem, there is only a boiling point rise in the first effect, while this factor can be neglected in the second effect, yielding that $t_2 = t_{b2} =$ 59.7°C.

In order to perform the calculation process, it is supposed that the areas and the heat flows transferred through these exchange areas are equal for the two effects, complying with (Equation 22.48)

$$\frac{\dot{Q}}{A} = \frac{T - t_{b2} - \Delta T_{b1}}{(1/U_1) + (1/U_2)} = \frac{\left(126.8 - 59.7 - 7\right)\text{°C}}{\left[(1/1860) + (1/1280)\right](\text{m}^2\text{ °C})/\text{W}}$$

Hence, $\dot{Q}/A = 45{,}569$ W/m^2.

The temperature t_{b1} is obtained from the heat transfer rate equation in the second effect:

$$\frac{\dot{Q}}{A} = U_2\left(t_{b1} - t_2\right) \qquad t_{b1} = 95.3\text{°C}$$

The boiling temperature in the first effect is

$$t_1 = t_{b1} + \Delta T_{b1} = 95.3 + 7 = 102.3\text{°C}$$

It is possible to find the properties of the saturated steam from the temperature $t_{b1} = 95.3°C$ and the saturated steam tables:

$$T_{b1} = 95.3°C \qquad P_1 = 855 \text{ mbar}$$

$$\hat{h}_{V1} = 399.3 \text{ kJ/kg}$$

$$\hat{H}_{V1} = 2668 \text{ kJ/kg} \qquad \lambda_{V1} = 2268.7 \text{ kJ/kg}$$

Enthalpy balances applied to both effects yield

1st effect

$$2183 w_V = (2268.7 + (2.1)\cdot(7))V_1 + (2.84)(1875)(102.3 - 95.3) - 3.88 w_2(59.7 - 95.3)$$

2nd effect

$$2{,}268.7\ V_1 = 2{,}359.4\ V_2 - (4.01)\ (15{,}000)\ (22 - 59.7)$$

Together with the following balance equations,

$$w_2 = w_A - V_2$$

$$V_1 + V_2 = 13{,}125$$

a four-equation system with four unknowns is obtained, which when solved yields

$$w_V = 8102 \text{ kg/h} \qquad w_2 = 9056.5 \text{ kg/h}$$

$$V_1 = 7181.5 \text{ kg/h} \qquad V_2 = 5943.5 \text{ kg/h}$$

The value of the areas through which heat is transferred can be obtained by means of the equations of heat transfer rate through such areas:

1st effect

$$8102\ (2183/3600) = 1.86\ A_1\ (126.8 - 102.3)$$

2nd effect

$$7181.5\ (2268.7/3600) = 1.28 \qquad A_2\ (95.3 - 59.7)$$

$$A_1 = 107.81 \text{ m}^2 \qquad A_2 = 99.32 \text{ m}^2$$

The mean area is $A_m = 105.56\ \text{m}^2$. Since these areas differ by more than 2%, the calculation procedure should begin again, rectifying the intermediate temperatures t_1 and t_{b1}, since the other temperatures remain the same:

$$T_{b1} - 59.7 = (95.3 - 59.7)(A_2/A_m) \qquad t_{b1} = 93.8°\text{C}$$

$$126.8 - t_1 = (126.8 - 102.3)(A_1/A_m) \qquad t_1 = 101.3°\text{C}$$

Hence, $t_{b1} = 101.3 - = 7 = 94.3°C$.
Since they are different, the mean value of each temperature is taken:

$$T_{b1} = 94°\text{C} \quad \text{and} \quad t_1 = 101°\text{C}$$

The new enthalpies for 94°C can be found in the saturated steam tables:

$$T_{b1} = 94°\text{C} \qquad P_1 = 815\ \text{mbar}$$

$$\hat{h}_{V1} = 393.8\ \text{kJ/kg}$$

$$\hat{H}_{V1} = 2666\ \text{kJ/kg} \qquad \lambda_{V1} = 2272.2\ \text{kJ/kg}$$

The four-equation system stated earlier is solved again using the new value of λ_{V1}, yielding

$$w_V = 8090\ \text{kg/h} \qquad w_2 = 9051\ \text{kg/h}$$

$$V_1 = 7176\ \text{kg/h} \qquad V_2 = 5949\ \text{kg/h}$$

The areas are recalculated from the velocity equations:

$$A_1 = 102.22\ \text{m}^2 \qquad A_2 = 103.16\ \text{m}^2$$

a. Economy of the system: $E = \dfrac{V_1 + V_2}{w_V} = \dfrac{13{,}125}{8{,}090} = 1.62$

b. Area per effect: $A_m = 102.7\ \text{m}^2$

c. 1st effect: $P_w = 2452$ mbar $\quad T = 126.8°\text{C}$

$P_1 = 815$ mbar $\quad t_1 = 101.0°\text{C}$

2nd effect: $P_1 = 815$ mbar $\quad t_{b1} = 94.0°\text{C}$

$P_2 = 196$ mbar $\quad t_2 = 59.7°\text{C}$

22.2 A double effect evaporator, operating under forward feed, is used to concentrate clarified fruit juice from 15°Brix to 72°Brix. The steam available from the boiler is saturated at 2.4, and the vacuum pressure in the evaporation chamber of the second effect is 460 mmHg. The diluted juice is fed into the evaporation chamber at a temperature of 50°C and a rate of 3480 kg/h. If the global heat transfer coefficients for the first and second effects are 1625 and 1280 W/(m² °C), respectively, determine (a) the steam flow rate from the boiler and the economy of the system, (b) the heating surface for each effect, and (c) the temperatures and pressures in the condensation and evaporation chambers for each effect.

Data: Properties of the fruit juices: The boiling point rise can be calculated according to the expression $\Delta T_b = 0.014\ C^{0.75}\ P^{0.1}\exp(0.034\ C)$ °C, where C is the soluble solid content in °Brix and P the pressure in mbar.

The specific heat is a function of the mass fraction of water according to the equation:

$$\hat{C}_P = 0.84 + 3.34X_{WATER} \text{ kJ/(kg °C)}$$

There is a vacuum pressure of 460 mmHg in the evaporation chamber of the second effect, so its pressure is $P_2 = 300$ mmHg.
Properties of the saturated steam

$P_w = 2.4$ atm $= 2353$ mbar $\quad T = 125.5$°C

$\hat{h}_w = 527$ kJ/kg

$\hat{H}_w = 2713$ kJ/kg $\quad \lambda_w = 2186$ kJ/kg

$P_2 = 300$ mm Hg $= 400$ mbar $\quad t_{b2} = 75.8$°C

$\hat{h}_{V2} = 317$ kJ/kg

$\hat{H}_{V2} = 2637$ kJ/kg $\quad \lambda_{V2} = 2320$ kJ/kg

Figure P.22.2 shows a scheme of the double effect evaporator that works under forward feed.

Global and component mass balances

$$3480 = w_C + V_1 + V_2$$

$$(3480)(0.15) = (0.72)w_C$$

yielding

$$w_C = 725 \text{ kg/h} \quad \text{and} \quad V_1 + V_2 = 2755 \text{ kg/h}$$

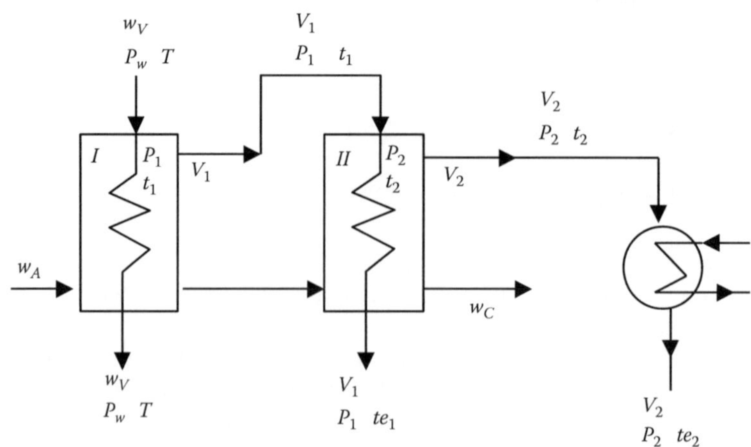

FIGURE P.22.2 Forward feed double effect evaporator.

Initially, it is supposed that $V_1 = V_2 = 1377.5$ kg/h, which makes us assume that the composition of the stream w_2 is $X_2 = 0.248$, corresponding to a content of 24.8°Brix.

The boiling point rises are calculated by the equation given in the data section. The concentration used is that of the stream leaving each effect:

- 1st effect: $C_1 = 24.8$°Brix $P_1 = ?$
- 2nd effect: $C_2 = 72.0$°Brix $P_2 = 400$ mbar

obtaining $\Delta T_{b1} = 0.36(P_1)^{0.1}$ $\Delta T_{b2} = 7.3$°C
The boiling temperature of the second effect is

$$t_2 = t_{b2} + \Delta T_{b2} = 75.8 + 7.3 = 83.1°\text{C}$$

P_1 should be known in order to obtain ΔT_{b1}, but it can be estimated. In case $P_1 = 1000$ mbar, the boiling point rise is $\Delta T_{b1} = 0.7$°C.
The specific heat of each stream is obtained from the equation given in the statement of the problem:

- For $C_A = 15$°Brix

$$\hat{C}_{PA} = 3.68 \text{ kJ/(kg °C)}$$

- For $C_C = 75$°Brix

$$\hat{C}_{PC} = 1.78 \text{ kJ/(kg °C)}$$

- For $C_1 = 24.8$°Brix

$$\hat{C}_{P1} = 3.35 \text{ kJ/(kg °C)}$$

In order to perform the calculation procedure, it is supposed that the areas and the heat flows transferred through these exchange areas are equal for both effects, complying with

$$\frac{\dot{Q}}{A} = \frac{T - t_{b2} - \Delta T_{b1} - \Delta T_{b2}}{(1/U_1) + (1/U_2)} = \frac{(125.5 - 75.8 - 0.7 - 7.3)°\text{C}}{((1/1625) + (1/1280))(\text{m}^2\ °\text{C})/\text{W}}$$

Hence, $\dot{Q}/A = 29{,}858$ W/m^2.

Temperature t_{b1} is obtained from the heat transfer rate equation for the second effect:

$$\frac{\dot{Q}}{A} = U_2(t_{b1} - t_2) \qquad t_{b1} = 106.4°\text{C}$$

and the boiling point in the first effect is

$$t_1 = t_{b1} + \Delta T_{b1} = 106.4 + 0.7 = 107.1°\text{C}$$

The properties of saturated steam can be found using the temperature $t_{b1} = 106.4°C$ from the saturated steam tables:

$$T_{b1} = 106.4°C \qquad P_1 = 1271 \text{ mbar}$$

$$\hat{h}_{V1} = 447 \text{ kJ/kg}$$

$$\hat{H}_{V1} = 2685 \text{ kJ/kg} \qquad \lambda_{V1} = 2238 \text{ kJ/kg}$$

If the boiling point rise in the first effect is recalculated using pressure $P_1 = 1271$ mbar, a slightly different value is obtained, so the same boiling point rise obtained before is taken.

The application of enthalpy balances to both effects yields

$$\text{1st effect: } (2186)\, w_V = (2238 + (2.1)\,(0.7)V_1 + w_1\,(3.35)\,(107.1 - 106.4) - (3480)\,(50 - 106.4)$$

$$\text{2nd effect: } (2238)\, V_1 = V_2\,(2319 + (2.1)\,(0.7) + (725)\,(1.78)\,(83.1 - 75.8) - w_1\,(3.35)\,(107.1 - 75.8)$$

With the equations of the balances,

$$w_1 = w_A - V_1$$

$$V_1 + V_2 = 2755$$

a four-equation system with four unknowns is obtained, which when solved yields

$$w_V = 1726 \text{ kg/h}$$

$$w_1 = 2120 \text{ kg/h}$$

$$V_1 = 1360 \text{ kg/h}$$

$$V_2 = 1395 \text{ kg/h}$$

These new values are used to calculate the soluble solid content in the stream w_1, obtaining 24.6°Brix, which allows the calculation of the specific heat and the boiling point rise: $\hat{C}_{P1}$ = 3.36 kJ/(kg °C) and $\Delta T_{b1} = 0.7°C$.

The value of the areas through which heat is transferred can be obtained by the equations of heat transfer rate through such areas:

$$\text{1st effect: } (1725) \times (2186)/(3600) = (1.625)\, A_1\,(125.5 - 107.1)$$

$$\text{2nd effect: } (1360) \times (2238)/(3600) = (1.28)\, A_2\,(106.4 - 83.1)$$

$$A_1 = 35.05 \text{ m}^2$$

$$A_2 = 28.35 \text{ m}^2$$

The mean area is $A_m = 31.70\ m^2$. Since these areas are different in more than 2%, the calculation procedure should begin again rectifying the intermediate temperatures t_1 and t_{b1} (Equations 22.49 through 22.51), as the other temperatures do not vary:

$$t_{b1} - 83.1 = (106.4 - 83.1)(A_2/A_m), \text{ then } t_{b1} = 103.9°C$$

$$125.5 - t_1 = (125.5 - 107.1)(A_1/A_m), \text{ then } t_1 = 105.2°C$$

With this temperature, it is obtained that $t_{b1} = 105.2 - 0.7 = 104.6°C$.

Since they are different, the mean value of each one is taken, so

$$t_{b1} = 104.2°C \quad \text{and} \quad t_1 = 104.9°C$$

The new enthalpies for 104.2°C can be found in the saturated steam tables:

$$t_{b1} = 104.2°C$$

$$P_1 = 1177 \text{ mbar}$$

$$\hat{h}_{V1} = 437 \text{ kJ/kg}$$

$$\hat{H}_{V1} = 2682 \text{ kJ/kg}$$

$$\lambda_{V1} = 2245 \text{ kJ/kg}$$

The four-equation system stated earlier is solved again, using the new value of λ_{V1}, yielding

$$w_V = 1720.8 \text{ kg/h}$$

$$w_2 = 2118.6 \text{ kg/h}$$

$$V_1 = 1361.4 \text{ kg/h}$$

$$V_2 = 1393.6 \text{ kg/h}$$

The areas are recalculated from the rate equations:

$$A_1 = 31.22 \text{ m}^2$$

$$A_2 = 31.43 \text{ m}^2$$

a. Mass flow rate of steam from the boiler

$$w_V = 1720.8 \text{ kg/h}$$

Economy of the system

$$E = \frac{V_1 + V_2}{w_V} = \frac{2755}{1720.8} = 1.6$$

b. Area per effect

$$A_m = 31.33 \text{ m}^2$$

c. 1st effect

$P_w = 2353$ mbar $T = 125.5°C$

$P_1 = 1177$ mbar $t_1 = 104.9°C$

2nd effect

$P_1 = 1177$ mbar $t_{b1} = 104.2°C$

$P_2 = 400$ mbar $t_2 = 83.1°C$

22.3 Tamarind is an important culinary condiment used as an acidifying ingredient. Due to the cost of transport, it is convenient to obtain tamarind as a concentrated juice through an evaporation stage. An Indian industry desires to obtain 1000 kg/h of a 62°Brix of concentrated juice beginning with 10°Brix. For this reason, the possibility of installing a single effect with mechanical compression of steam or a double effect operating under forward feed is studied. The global heat transfer coefficients of the first and second effects are 2100 and 1750 W/(m² °C), respectively. The food is at 22°C, while the 62°Brix juice cannot withstand temperatures higher than 70°C. The industry has a saturated steam stream at 1.8 kgf/cm² used to carry out the juice concentration. Calculate (a) the 10°Brix juice flow rate that can be concentrated and the consumption of steam at 1.8 kp/cm for both options; (b) the compression power, for the first option, if the isentropic performance of the compressor is 88%; and (c) the more profitable option, if the cost of each m² of evaporator is 22 USD, of each kW of compression power is 4 USD, of each kW h is 0.08 USD, and of each kg of steam at 1.8 kgf/cm² is 0.01 USD. Consider that the amortization of the equipment is estimated in 1 year.

Data and notes: Juices with soluble solid content lower than 18°Brix do not present an appreciable boiling point rise.

The plant functions 300 days a year, 16 h daily.

- The specific heat of the tamarind juices is

$$\hat{C}_P = 4.18 + (6.84 \times 10^{-5}\, T - 0.0503) X_S \text{ kJ/(kg K)}$$

where

X_S is the percentage of soluble solids
T is the temperature in Kelvin

Single effect evaporation with mechanical compression

The diagram of this type of installation corresponds to Figure 22.6.

The temperature in the evaporation chamber is $t_1 = 70°C$, while the boiling temperature of pure water is obtained from the graph of Dühring (Figure 22.3): $t_{b1} = 66°C (\Delta T_{b1} = 4°C)$.

The following conditions can be obtained from saturated steam tables:

$P_w = 1.8 \text{ at} = 1765 \text{ mbar}$ $\qquad T_C = 116.3°\text{C}$

$\hat{h}_w = 488 \text{ kJ/kg}$

$\hat{H}_w = 2700 \text{ kJ/kg}$

$\lambda_w = 2212 \text{ kJ/kg}$

$t_{b1} = 66°\text{C}$

$P_1 = 262 \text{ mbar} = 0.27 \text{ atm}$

$\hat{h}_{V1} = 276 \text{ kJ/kg}$

$\hat{H}_{V1} = 2619 \text{ kJ/kg}$

$\lambda_{V1} = 2343 \text{ kJ/kg}$

The vapor that leaves the evaporation chamber is reheated to a temperature of t_1 = 70°C, its enthalpy being

$$\hat{H}_1 = \hat{H}_{V1} + \hat{C}_{PV}\Delta T_{b1} = 2619 + (2.1)(4) = 2627.4 \text{ kJ/kg}$$

The compression of the vapor that leaves the evaporation chamber is an isentropic process (Figures 22.6 and 22.7), from point 1 at a 262 mbar isobar to point 2 at the 1765 mbar isobar. The conditions of point 2 corresponding to the vapor outlet at the compressor can be obtained by means of the graph:

$P_2 = P_w = 1765 \text{ mbar}$ $\qquad t_2 = 270°\text{C}$

$\hat{H}_2 = 3009 \text{ kJ/kg}$

Vapor (V) with these conditions is reheated and mixed with the vapor coming out of the boiler (w_V') that is saturated at the same pressure, yielding a reheated vapor mixture. In order to avoid this, part of the condensate (w_R) is recirculated to obtain a saturated vapor at point 3 (w_V), which is fed to the condensation chamber of the evaporator (Figures 22.6 and 22.7).
The specific heats of the food and concentrate streams are calculated at their respective concentrations:

- Food $\qquad C_A$ = 10°Brix

$\hat{C}_{PA} = 3.88 \text{ kJ/(kg °C)}$

- Concentrate $C_C = 62°\text{Brix}$

$\hat{C}_{PC} = 2.52 \text{ kJ/(kg °C)}$

Mass balances

$w_A = w_C + V$

$w_A = 1000 + V$

$w_A X_A = w_C X_C$

$w_A 0.1 = w_C 0.62$

Obtaining

$w_A = 6200 \text{ kg/h}$ $V = 5200 \text{ kg/h}$

If an enthalpy balance is performed in the evaporator, it is possible to obtain the amount of the vapor w_V that gets into the evaporation chamber:

$$2212 w_V = (5200)\left[(2343) + (2.1)(4)\right] + (1000)(2.52)(4) - (6200)(3.88)(22 - 66)$$

$$w_V = 6010.2 \text{ kg/h}$$

When performing the mass and enthalpy balances around point 3 where the streams of steam from the boiler (w'_V), compressed vapor ($V_2 = V$), and recirculated condensate (w_R) convene, it can be obtained that

$$6010.2 = w'_V + w_R + 5200$$

$$(6010.2)(2700) = w'_V(2700) + w_R(488) + (5200)(3009)$$

The solution of this system yields

$$w'_V = 83.8 \text{ kg/h}$$

$$w_R = 726.4 \text{ kg/h}$$

The area of the evaporator is obtained from the heat transfer rate equation:

$$w_V \lambda_w = U_1 A (T_C - t_1)$$

$$(6010.2)\ (2212)/3600 = (2.1)A\ (116.5 - 70)$$

Hence, the area is

$$A = 37.82 \text{ m}^2$$

The theoretical compression power is obtained from the expression:

$$Pow)_T = V\left(\hat{H}_2 - \hat{H}_1\right)$$

$$Pow)_T = (5200/3600)\,\text{kg/s}\,(3009 - 2627.4)\,\text{kJ/kg}$$

$$Pow)_T = 551.2\ \text{kJ/kg}$$

The real power is obtained by dividing the theoretical power by the isentropic performance:

$$Pow)_R = (551.2\,/\,0.88) = 626.4\ \text{kJ/kg}$$

The annual operation cost is calculated from the expression:

$$C = C_A A + C_P Pow)_R C_{POWxh} h_T + C_V w_V' h_T$$

where C_A, C_P, C_V, and C_{POWxh} are the cost per m^2 of evaporator area, cost per compression power installed, cost of the waste of steam coming from the boiler, and operation cost of the compressor, respectively, while h_T are the annual operation hours.

- Operation hours

$$h_T = (16)\,(300) = 4800\ \text{h}$$

The annual cost is

$$C = (22)\,(37.82) + (4)\,(626.4) + (0.08)\,(626.4)\,(4800) + (0.017)\,(83.8)\,(4800)$$

Hence,

$$C = \text{U.S. } \$250{,}713.$$

Forward feed double effect evaporator
This type of evaporator is similar to the one presented in Figure P.22.1. A vapor at a pressure P_w = 1.8 atm = 1765 mbar enters the condensation chamber of the first effect, while in the second effect the boiling temperature of the 62°Brix juice is t_2 = 70°C, so the boiling temperature of pure water is t_{b2} = 66°C (ΔT_{b2} = 4°C).

The following is obtained from the saturated steam tables:

$$P_w = 1.8\ \text{atm} = 1765\ \text{mbar} \qquad T_C = 116.3°\text{C}$$

$$\hat{h}_w = 488\ \text{kJ/kg}$$

$$\hat{H}_w = 2700\ \text{kJ/kg}$$

$$\lambda_w = 2212\ \text{kJ/kg}$$

$$t_{b2} = 66°\text{C}$$

$$P_2 = 262 \text{ mbar} = 0.27 \text{ atm}$$

$$\hat{h}_{V2} = 276 \text{ kJ/kg}$$

$$\hat{H}_{V2} = 2619 \text{ kJ/kg}$$

$$\lambda_{V2} = 2343 \text{ kJ/kg}$$

The global and component mass balances yield

$$w_A = 6200 \text{ kg/h}$$

$$V_1 + V_2 = 5200 \text{ kg/h}$$

Initially, it is supposed that $V_1 = V_2 = 2600$ kg/h, allowing the concentration of stream w_1 to be obtained, leaving the first effect: $C_1 = 17.2$°Brix, pointing out that there will not be an appreciable boiling point rise ($\Delta T_{b1} = 0$).

The specific heats of the different juice streams are

- Food $C_A = 10$°Brix

$$\hat{C}_{PA} = 3.88 \text{ kJ/(kg °C)}$$

- Stream $C_1 = 17.2$°Brix

$$\hat{C}_{P1} = 3.75 \text{ kJ/(kg °C)}$$

- Concentrate $C_C = 62$°Brix

$$\hat{C}_{PC} = 2.52 \text{ kJ/(kg °C)}$$

To perform the calculation process, it is supposed that the areas and heat flows transferred through such exchange areas are equal in both effects, complying with

$$\frac{\dot{Q}}{A} = \frac{T - t_{b2} - \Delta T_{b2}}{(1/U_1) + (1/U_2)} = \frac{(116.3 - 66 - 4)\text{°C}}{\left((1/2100) + (1/1750)\right) \text{W/(m}^2\text{ °C)}}$$

Thus,

$$\frac{\dot{Q}}{A} = 44{,}196 \text{ W/m}^2$$

Temperature t_{e1} is obtained from the heat transfer rate equation in the second effect:

$$\frac{\dot{Q}}{A} = U_2(t_{b1} - t_2) \qquad t_{b1} = 95.3°C$$

and the boiling temperature in the first effect is

$$t_1 = t_{b1} + \Delta T_{b1} = 95.3 + 0 = 95.3°C$$

It is possible to find the properties of saturated steam, using the temperature t_{b1} = 95.3°C, from saturated steam tables:

$$t_{b1} = 95.3°C \qquad P_1 = 855 \text{ mbar}$$

$$\hat{h}_{V1} = 399.3 \text{ kJ/kg}$$

$$\hat{H}_{V1} = 2668 \text{ kJ/kg}$$

$$\lambda_{V1} = 2268.7 \text{ kJ/kg}$$

Thus, applying enthalpy balances to both effects,

1st effect

$$(2212)w_V = (2268.7)V_1 + w_1(3.75)(0) - (6200)(3.88)(22 - 95.3)$$

2nd effect

$$V_1 2268.7 = V_2(2343 + (2.1)(4) + (1000)(2.52)(70 - 66) - w_1(3.75)(95.3 - 66)$$

Besides the equations of the balances,

$$w_1 = w_A - V_1 = 6200 - V_1$$

$$V_1 + V_2 = 2600$$

a four-equation system with four unknowns is obtained, which when solved yields

$$w_V = 3425 \text{ kg/h}$$

$$w_1 = 3637.8 \text{ kg/h}$$

$$V_1 = 2562.2 \text{ kg/h}$$

$$V_2 = 2637.8 \text{ kg/h}$$

The composition of stream w_1 is recalculated, obtaining $C_1 = 17°$Brix, so there is no boiling point rise in the first effect ($\Delta T_{b1} = 0$), and its specific heat is almost the same as the one calculated before.

The value of the areas through which heat is transferred can be obtained from the heat transfer rate equations through such areas:

1st effect

$$(3425)(2212)/(3600) = (2.16)A_1(116.3 - 95.3)$$

2nd effect

$$(2562.2)(2268.7)/(3600) = (1.28)A_2(95.3 - 70)$$

$$A_1 = 47.72 \text{ m}^2$$

$$A_2 = 36.47 \text{ m}^2$$

and a mean area $A_m = 42.10$ m^2. Since these areas differ by more than 2%, the calculation procedure should begin again, rectifying the intermediate temperatures t_1 and t_{b1}, since the other temperatures remain the same:

$$t_{b1} - 70 = (95.3 - 70)(A_2/A_m)\ t_{b1} = 92.0°\text{C}$$

$$116.3 - t_1 = (116.3 - 95.3)(A_1/A_m)\ t_1 = 92.5°\text{C}$$

Since $\Delta T_{b1} = 0$, it is complied that $t_{b1} = t_1$.

Since they are different, the mean value of both is taken; thus, $t_{b1} = t_1 = 92.3°$C. The new enthalpies, for the temperature 92.3°C, can be found in the saturated steam tables:

$$t_{b1} = 92.3°\text{C} \qquad P_1 = 770 \text{ mbar}$$

$$\hat{h}_{V1} = 387 \text{ kJ/kg}$$

$$\hat{H}_{V1} = 2662 \text{ kJ/kg}$$

$$\lambda_{V1} = 2275 \text{ kJ/kg}$$

The four-equation system stated earlier is solved using the new value of λ_{V1}, obtaining

$$w_V = 3405.4 \text{ kg/h}$$

$$w_1 = 3632.3 \text{ kg/h}$$

$$V_1 = 2567.7 \text{ kg/h}$$

$$V_2 = 2632.3 \text{ kg/h}$$

The areas are recalculated from the rate equations:

$$A_1 = 41.52\ \text{m}^2$$

$$A_2 = 41.58\ \text{m}^2$$

Thus, the mean area by effect is $A_m = 41.55\ \text{m}^2$.

The annual cost is obtained from the expression:

$$C = C_A 2A_m + w_V C_V h_T$$

Hence,

$$C = (22)\ 2\ (41.55) + (3405.4)(0.017)(4800)$$

$$C = \text{U.S. } \$279{,}709$$

According to the result obtained, it is better to install the effect with vapor recompression, since the annual cost is lower.

PROPOSED PROBLEMS

22.1 It is desired to concentrate 20,000 kg/h of a food solution, from 5% to 50% in soluble solids. A heating vapor at 121°C is used, while the condensing temperature for steam exiting from the last effect is 52°C. If the feeding solution temperature is 93°C, calculate for each one of the indicated cases: (a) heating area for each effect, (b) heating steam consumption by effect, and (c) evaporator economy.

Case 1: One effect. $U = 3800$ W/(m² °C).
Case 2: Two effects direct feed. $U_1 = 3600$ W/(m² °C). $U_2 = 3300$ W/(m² °C).
Case 3: Three effects direct feed. $U_1 = 3600$ W/(m² °C). $U_2 = 3300$ W/(m² °C). $U_3 = 3000$ W/(m² °C).

Data: Assume that specific heat is the same as for water for all solutions 4.185 kJ/(kg °C). Assume that boiling point rise is negligible. For multiple effects, assume that the area is the same for all effects.

22.2 An aqueous solution contains 5% of solids, and it is desired to concentrate it up to 40%. Of this solution, 18,000 kg/h is feeding to an evaporation system with mechanical recompression of the water steam removed in the evaporation chamber. The pressure in the evaporation chamber is 98 kPa, while that of condensation chamber operates at 147 kPa. If the economy of the vapor is 0.9 kg of vapor removed/kg feeding vapor and it is assumed that the boiling point rise is negligible, calculate the energy required to carry out the concentration of the solution.

22.3 Of a clarified apple juice, 10,000 kg/h is being concentrated from 15°Brix to 75°Brix using a simple evaporator with mechanical compression. To carry out this concentration, a 147 kPa water steam is used, and the evaporation chamber is working at 98 kPa. If the evaporator economy is of 0.906 kg vapor removed by kg that is feeding in the condensation chamber, determine (a) the temperatures in the evaporation and condensation chambers, (b) the steam flow rate from the boiler, (c) the area required if the overall coefficient heat transmission coefficients 2000 are W/m²/°C, and (d) the compressor power if their isentropic yield is 85%.

Data: Juice boiling point rise can be calculated from the following equation: $\Delta T_b = 0.05C^{0.83}P^{0.113}\exp(-0.04C + 0.00065C^2)$, where T is expressed in °C, C in °Brix, and P in mbar.

To calculate the specific heat for the juice, the following expression may be applied: $\hat{C}_P = 4.18 - 3.34X$ kJ/kg/°C, where X is the solid mass fraction. The water steam specific heat is 2.1 kJ/kg/°C.

22.4 A Catalonian company wishes to carry out an engineering project in Kuwait. This company needs 735 L/h of distilled water for different uses. To provide this volumetric flow rate, an evaporator has been installed that is fed by seawater at 22°C. In order to carry out better energy use, the steam removed in the evaporation chamber is compressed in a compressor whose isentropic yield is 80%. The evaporation chamber works at atmospheric pressure, while the condensing chamber works at 2.2 kgf/cm^2, conditions in which the overall coefficient heat transmission coefficients can be assumed as 18,000 W/(m^2 °C). The water steam exiting from the evaporation chamber is 60% of the feeding flow rate. If the boiler produces water steam at vapor 2.2 kgf/cm^2, determine (a) the consumption of boiler steam and seawater flow rate that can be treated, (b) the compressor power, and (c) the evaporator area, and (d) if the final distilled current is cooled down to 75°C in a heat exchanger, using the seawater feeding current as a coolant fluid, would it improve the system's economy?

Data: Assume that the specific heat of the water for all solutions is the same, that is, 4.185 kJ/(kg °C); the water steam specific heat is 2.1 kJ/(kg °C); and the boiling point rise is 1.2°C.

22.5 A triple effect evaporator in direct current is used to concentrate 15,000 kg/h of an aqueous solution, at 25°C, from 5% to 35%. The boiler water steam is 2 kgf/cm^2, and in the third effect, there is a vacuum in such a way that the evaporation chamber temperature is 30°C. The overall heat coefficients are 3300, 2900, and 2000 W/(m^2 °C), for the first, second, and third effect, respectively. Determine (a) the global evaporation economy, (b) the boiling temperature in each effect, and (c) the area for each effect.

Data: Assume that the boiling point rise is negligible. The specific heat is a function of the mass solute fraction (Xs), according to the expression:

$$Cp = 4.185 - 0.82\, Xs \text{ kJ/(kg °C)}$$

22.6 A simple effect evaporator is used in a concentrated pear juice from 12°Brix to 70°Brix. A water steam at 2 kgf/cm^2 is used as a heating fluid, and the evaporation chamber there is at 460 mm of Hg vacuum. Dilution and hydration effects are considered negligible. Calculate the water steam and the required area, if the juice is feeding to the evaporator at 20°C with a 1500 kg/h flow rate. Determine what would be more advantageous: (a) to recycle part of the vapor removed from the evaporation chamber by using an ejector or (b) to use a double effect evaporator.

Data: Overall heat transfer coefficient 2200 and 2000 W/(m^2 °C) for the first and second effects, respectively. Water steam specific heat 2.1 kJ/(kg °C). Assume that both specific heat and boiling point rise are linear, varying with soluble solid content. At 10°Brix and 70°Brix, the following values are as follows:

Concentration (°Brix)	Specific Heat kJ/(kg °C)	Boiling Point Rise (°C)
10	3.90	0.5
70	2.80	10.0

22.7 It is desired to concentrate a caustic solution, at 18°C, from 10% to 35%, using a simple effect evaporator that possesses 100 m^2 surface area and an overall heat transfer coefficient 2200 W/(m^2 °C). In order to carry out a better energy use, 80% of the vapor removed from the evaporation chamber is mechanically compressed, and it is mixed in a water steam boiler. 20% of the vapor from the evaporation chamber is taken to a condenser, in which the coolant fluid is the diluted caustic solution, before it is fed to the evaporator. If in the condensation evaporation chambers there are 3 and 1 kgf/cm^2, respectively, calculate (a) a process diagram; (b) the dilute solution mass flow rate and feeding temperature to the evaporation chamber; (c) the water steam boiler mass flow and steam leaving from evaporation chamber, along with the evaporator economy; and (d) the compressor power if its isentropic yield is 85%; (e) if it is desirable to duplicate feeding solution mass flow rate, is it possible to use this installation? Justify your answer.

Data: Water steam specific heat 2.1 kJ/(kg °C). The specific heat from caustic solution can be obtained from equation $Cp = 0.82 + 3.3X$ kJ/(kg °C), where X is the water mass fraction.

22.8 It is desirable to concentrate 10,000 kg/h of a clarified peach juice at 20°C, from 14.5°Brix to 71°Brix, using a vertical pipe evaporator. To improve the process yield, there are two possible alternatives: using an ejector to recycle water steam removed in the evaporation chamber or using a countercurrent double effect evaporator.

The steam leaving from the evaporation chamber is saturated at 0.8 kgf/cm², while in the condensation chamber, there is a 2.6 kgf/cm² pressure. Calculate (a) the water steam boiler mass flow rate and the system's economy if a simple effect evaporator is used; (b) when an ejector is used whose thermal yield is of 80%, what quantity of saturated vapor at 2.6 8 kgf/cm² is necessary to sweep out the 70% of water steam removed from evaporation chamber, if inside this chamber there is 1.16 8 kgf/cm² of pressure? What is the system's economy? (c) Repeat section (a) for the case of a double effect evaporator; (d) from the system economy point of view, which installation is more advisable?

Data: Overall heating coefficient values 1700 and 1300 W/(m² °C) for the first and second effects, respectively. The specific heat for juice can be calculated from equation $Cp = 0.82 + 3.3X$ kJ/(kg °C), where X is the water mass fraction.

22.9 Of a clarified peach juice, 16,000 kg/h, at 22°C, is fed into a countercurrent double effect evaporator in order to concentrate the juice from 12°Brix to 71°Brix. The water steam boiler used is saturated at 3 kgf/cm², while the pressure in the evaporation chamber in the second effect is 0.20 kgf/cm². The two evaporators possess the same heating area, their overall heat transfer coefficients being 1800 and 1400 W/(m² °C) for the first and second effects, respectively. Determine (a) the water steam boiler mass flow rate and the vapor mass flow rate removed in each effect, (b) the area required and system economy, and (c) the boiling temperature and pressure for the evaporation chamber in the first effect.

Data: Juice properties: Specific heat $Cp = 0.82 + 3.3X$ kJ/(kg °C). Boiling point rise $\Delta T_b = 17\,X$ °C, where X is the water mass fraction in the juice.

22.10 A solution current of 10,000 kg/h, at 25°C, that contains 7% salt is fed to a double effect evaporator in order to obtain a concentrated solution with 35% salt. The water steam boiler that is feeding to the first effect is saturated at 2.4 kgf/cm², while in the second effect evaporation chamber, there is a pressure of 0.15 kgf/cm². If the heating surface area of the second effect is double that of the first one, calculate (a) the water steam mass flow rate removed in each effect, (b) the water steam boiler mass flow rate required, (c) the heating area in each effect, (d) the boiling temperature in the evaporation chambers, and (e) the economy of the system.

Data: Overall heat transfer coefficients 1700 and 1200 W/(m² °C) for the first and second effects, respectively. Water steam specific heat 2.1 kJ/(kg °C). Salt solution properties: Specific heat $Cp = 4.18 - 3.2X$ kJ/(kg °C). Boiling point rise $\Delta T_b = 27\,X$°C, where X is the salt mass fraction in the solution.

22.11 In a two-effect direct current evaporation system, 20,000 kg/h of a food fluid, at 25°C, is concentrated from 5% to 60%. The water steam removed in the second effect condenses at 55°C. The specific heats and the boiling point rise of the different currents vary linearly with the soluble solid content. At 5% and 65%, the following values are obtained:

Concentration (%)	Specific Heat (kJ/(kg °C))	Boiling Point Rise (°C)
5	3.95	0.5
60	2.80	10.0

The overall heat transfer coefficients are 2800 and 2200 W/(m² °C) for the first and second effects, respectively. The water steam boiler is 2.4 kgf/cm². Calculate (a) the water steam boiler and steam mass flow rate removed in the evaporation chambers and (b) temperatures and pressures for the different condensation and evaporation chambers.

22.12 A current of 10,000 kg/h of a 15°Brix tamarind fruit juice at 22°C is fed into a double effect evaporator working in the direct current way. The juice is concentrated to 62°Brix using a water steam boiler at 1.4 kgf/cm^2; it is known that the vapor removed in the last effect condenses at 65°C. Calculate (a) the temperature and pressure in the evaporation and condensation chambers of each effect, (b) the water steam boiler mass flow rate and steam removed in each effect, and (c) the heating area for each effect and the system economy.

Data: Overall heat transfer coefficients 1800 and 1200 W/(m^2 °C) for the first and second effects, respectively. For juices whose soluble solid content is lower than 25°Brix, the boiling point rise is negligible. Water steam specific heat 2.1 kJ/(kg °C).

22.13 A 14°Brix clarified apple juice is concentrated up to 41°Brix on a double effect evaporator working in a direct current way. A 5000 kg/h of juice, at 50°C, is fed to the evaporator, using saturated water steam at 1.2 kgf/cm^2 for the heating process. The pressure in the evaporation chamber of the second effect is 0.20 kgf/cm^2. If the heating areas are the same in both effects and the heating losses toward the exterior are negligible, calculate (a) the heating area for the surface of each effect, (b) the boiling temperature in both effects, and (c) the water steam boiler mass flow rate and system economy.

Data: Overall heat transfer coefficients 1700 and 1200 W/(m^2 °C) for the first and second effects, respectively. The specific heat for juice can be calculated from equation $Cp = 0.82 + 3.3X$ kJ/(kg °C), where X is the water mass fraction. Assume that for 41°Brix juice the boiling point rise is 2°C.

23 Dehydration

23.1 INTRODUCTION

Dehydration, or drying, is one of the unit operations most commonly used for food preservation. Our ancestors, since prehistoric times, used it because it allowed them to obtain food products with a longer shelf life. As time went by, the demand for food increased because of larger populations. This higher food demand has impacted quite significantly the relevance of preservation by drying. The development of the drying industry has been linked to foods to feed soldiers stationed in war zones around the world. Dehydration is useful for military purposes, as well as for the general public since it reduces the weight and size of foods. Advances achieved within the military field have been transferred to the whole drying industry, yielding significant advances to this very relevant unit operation within the food sector.

Water in food is eliminated during dehydration processes to a greater or lesser extent, achieving better microbiological preservation, as well as retarding many undesirable reactions. Although food preservation has great importance, dehydration also can decrease packaging, handling, storage, and transport costs, since it decreases the food's weight and in some cases its volume.

Although the terms drying and dehydration may be used, technically speaking, they are not. A food is considered to be dehydrated if it does not contain more than 2.5% water, while a dried food may contain more than 2.5% water (Barbosa-Canovas and Vega-Mercado, 1996). Except for freeze-drying, osmotic drying, and vacuum drying, the removal of water from a food is achieved in most cases by application of a dry air flow, which eliminates water from the surface of the product by releasing it into the air stream. The food drying process not only decreases the water content of the food but can also affect other physical and chemical characteristics, such as destruction of nutrients and enzymatic and nonenzymatic reactions, among others.

In the drying process, it is important to know the mechanisms related to the movement of water both inside and outside the food. This movement can be due to capillary forces, water diffusion due to concentration gradients, diffusion on the surface, diffusion of water vapor in pores filled with air, flow due to pressure gradients or to vaporization, and condensation of water.

23.2 MIXING OF TWO AIR STREAMS

In air-drying processes, the gaseous stream that leaves the dryer usually has energy content that makes reuse possible, even when its moisture content is higher than the moisture content of the air entering the dryer. For this reason the recirculation of the air leaving the dryer is common practice and also allows the global drying process to be less expensive. In most cases, a hot and wet air stream is partially recirculated and mixed with a fresh air stream. The heat and material balances for this operation can be expressed as

$$w_C X_C + w_H X_H = (w_C + w_H) X_m \tag{23.1}$$

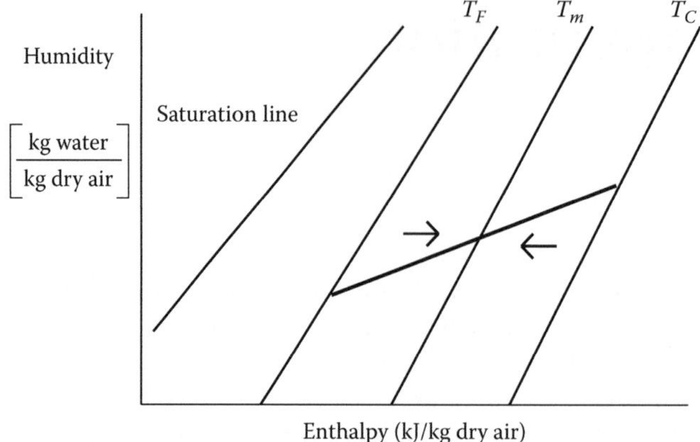

FIGURE 23.1 Representation of the mixing of two air streams.

$$w_C H_C + w_H H_H = (w_C + w_H) H_m \tag{23.2}$$

where

- w is the air flow rate
- X is the moisture content
- H is enthalpy
- subscripts C and H are the cold and hot, respectively
- subscript m is the conditions of the mixture

This type of process can be seen in the moisture-enthalpy diagram (Figure 23.1), pointing out that the differences in enthalpy between the mixed stream and the initial streams are proportional to the mass fluxes.

23.3 MASS AND HEAT BALANCES IN IDEAL DRYERS

23.3.1 Continuous Dryer without Recirculation

A dryer of this type consists mainly of a chamber in which the air and the solids to be dried flow in countercurrent (Figure 23.2). The solids are introduced at flow rate w_S (kg of dry solids/h), with a moisture content of Y_E at a temperature T_{SE}, leaving the dryer at a temperature T_{SS} with a moisture

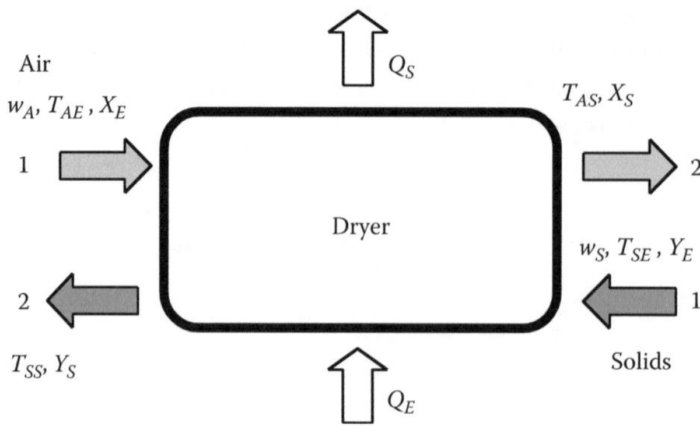

FIGURE 23.2 Continuous ideal dryer without recirculation.

content Y_S. The air stream enters the dryer with a mass flow rate w'_A (kg of dry air/h), at a temperature T_{AE} with a humidity X_E (kg water/kg dry air), and it leaves the dryer at a temperature T_{AS}, with a humidity X_S.

The following is obtained when performing a mass balance for water:

$$w_S(Y_E - Y_S) = w'(X_S - X_E) \tag{23.3}$$

while the energy balance yields the expression

$$\dot{Q}_E + w'\hat{i}_E + w_S\hat{h}_E = \dot{Q}_S + w'\hat{i}_S + w_S\hat{h}_S \tag{23.4}$$

In this equation $\dot{Q}_E$ and $\dot{Q}_S$ are the heat flow rates given and lost in the dryer, respectively, with $\hat{h}$ being the enthalpy of the solids, while $\hat{i}$ is the enthalpy of the air given by Equation 8.16:

$$\hat{i} = \hat{s}(T - T^*) + \lambda_0 X = (1 + 1.92X)(T - T^*) + \lambda_0 X \tag{23.5}$$

where $\hat{s}$ is the wet specific heat of air. The enthalpy of the solids is

$$\hat{h} = \hat{C}_P)_S\,(T - T^*) + Y\hat{C}_P)_w(T - T^*) \tag{23.6}$$

where

$\hat{C}_P)_S$ is the specific heat of the solids

$\hat{C}_P)_A$ is the specific heat of the water in the product

In these last equations, T^* is a reference temperature, which is usually taken as 0°C.

23.3.2 Continuous Dryer with Recirculation

The air stream leaving the dryer, as described in the last section, contains more water than at the inlet, but its temperature is still high. For this reason, the energy contained in this stream is usually used by recirculation. However, since its water content would not permit good application in the dryer, this stream is mixed with a stream of fresh dry air, and before introducing the mixture into the dryer, it passes through a heater to increase its enthalpy content (Figure 23.3).

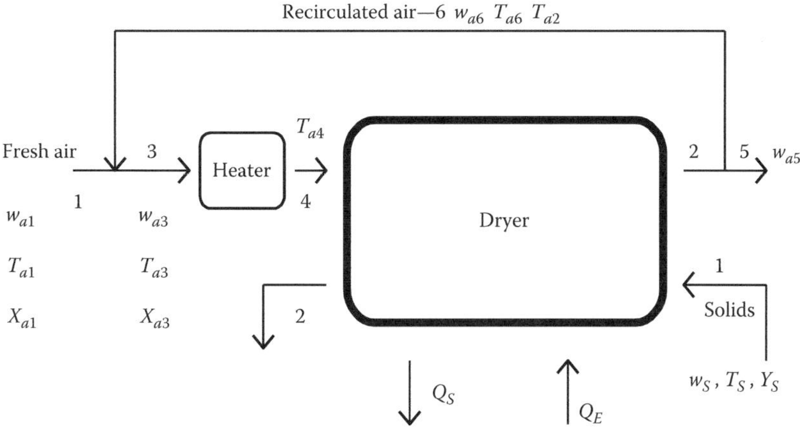

FIGURE 23.3 Ideal continuous dryer with recirculation.

A mass balance in the heater yields

$$w_1'X_1 + w_6'X_2 = (w_1' + w_6')X_4 \tag{23.7}$$

where

w_1' is the fresh dry air flow rate
w_6' is the recirculated air
X_1 is the moisture content of fresh air
X_2 is the moisture content of the recirculated air
X_4 is the moisture content of the mixture leaving the heater

The mass balance in the dryer leads to the expression

$$(w_1' + w_6')X_4 + w_S'Y_E = (w_1' + w_6')X_2 + w_S'Y_S \tag{23.8}$$

The enthalpy balances can be performed in the same way for the heater, the dryer, or the complete system.

23.4 DEHYDRATION MECHANISMS

Drying is defined as the removal of moisture from a product, and in most practical situations the main stage during drying is internal mass transfer. The mechanisms of water transfer in the product during the drying process can be summarized as follows (Van Arsdel and Copley, 1963): water movement due to capillary forces, diffusion of liquid due to concentration gradients, surface diffusion, water vapor diffusion in pores filled with air, flow due to pressure gradients, and flow due to water vaporization–condensation. In the pores of solids with rigid structure, capillary forces are responsible for the retention of water, whereas in solids formed by aggregates of fine powders, osmotic pressure is responsible for water retention within the solids as well as on the surface.

The type of material to be dried is an important factor to consider in all drying processes, since its physical and chemical properties play a significant role during drying due to possible changes that may occur and because of the effect that such changes may have in the removal of water from the product. A hygroscopic material is one that contains bound water that exerts a vapor pressure lower than the vapor pressure of liquid water at the same temperature. It is expected that products made mainly of carbohydrates will behave in a hygroscopic way, since the hydroxyl groups around the sugar molecules allow formation of hydrogen bonds with water molecules. The interaction between the water molecules and the hydroxyl groups causes solvation (dissolution) or solubilization of sugars. In water-soluble proteins, as in most globular proteins, the polar amino acids are uniformly distributed on the surface, while the hydrophobic groups are located toward the inside of the molecule. This arrangement allows the formation of hydrogen bonds with water.

23.4.1 Drying Process

In drying processes, data are usually obtained as the change in the weight of the product over time (Figure 23.4). However, sometimes drying data can be expressed in terms of drying rate, as follows.

The moisture content of the product is defined as the relationship between the amount of water in the food and the amount of dry solids, and is expressed as

$$Y_t = \frac{w_T - w_S'}{w_S'} \tag{23.9}$$

where

w_T is the total weight of the material at a given time
w_S' is the weight of dry solids
Y_t is the moisture expressed as water weight/dry solid weight

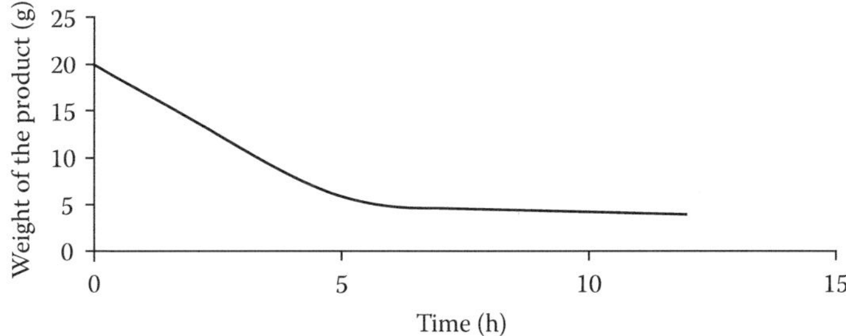

FIGURE 23.4 Variation of the product's weight in a drying process. (Adapted from Barbosa-Cánovas, G.V. and Vega-Mercado, H., *Dehydration of Food*, Chapman & Hall, New York, 1996.)

A very important variable in the drying process is the so-called free moisture content, *Y*, which is defined as

$$Y = Y_t - Y_{eq} \tag{23.10}$$

In this equation Y_{eq} is the moisture content when equilibrium is reached. A typical drying curve is obtained by plotting the free moisture content against drying time (Figure 23.5).

The drying rate, *R*, is proportional to the change in moisture content over time:

$$R \propto \frac{dY}{dt} \tag{23.11}$$

The value of *dY/dt* for each point in the curve can be obtained from Figure 23.5 by the value of the tangent to the curve at each of the points.

The drying rate can be expressed as (Geankoplis, 1993)

$$R = -\frac{w'_S}{A}\frac{dY}{dt} \tag{23.12}$$

where

w'_S is the flow rate of dry solids
A is the area of the drying surface

When plotting the drying rate versus time, a curve similar to the curve in Figure 23.6 is obtained.

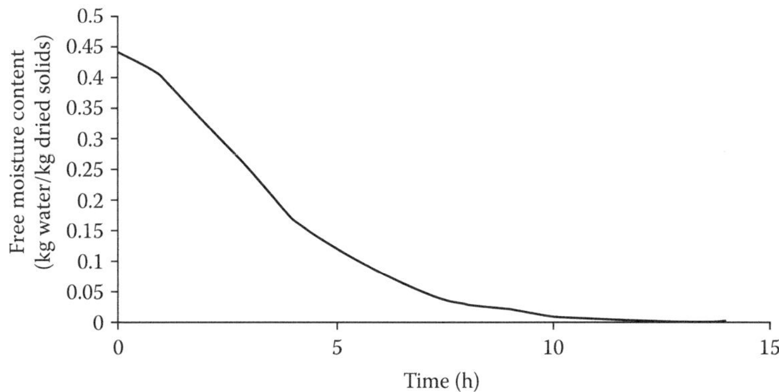

FIGURE 23.5 Free moisture content as a function of drying time. (Adapted from Barbosa-Cánovas, G.V. and Vega-Mercado, H., *Dehydration of Food*, Chapman & Hall, New York, 1996.)

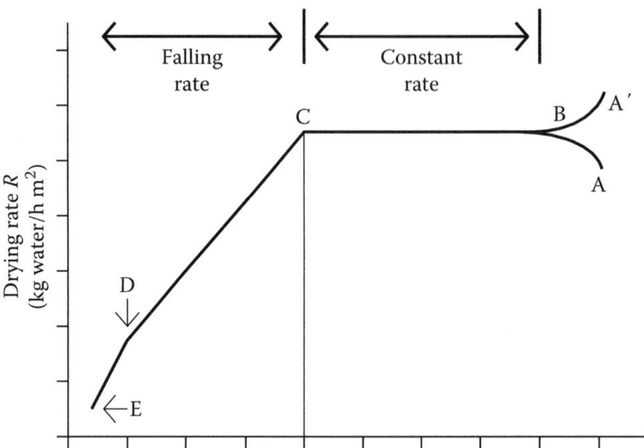

FIGURE 23.6 Drying rate curve. (Adapted from Barbosa-Cánovas, G.V. and Vega-Mercado, H., *Dehydration of Food*, Chapman & Hall, New York, 1996.)

The drying process of a material can be described by a series of stages in which the drying rate plays a key role. Figure 23.6 shows a typical drying rate curve in which points A and A′ represent the initial point for a cold and a hot material, respectively. Point B represents the condition of equilibrium temperature of the product surface. The elapsed time from point A or A′ to B is usually low, and it is commonly neglected in the calculation of drying time. The section B–C of the curve is known as the constant drying rate period and is associated with the removal of unbound water in the product. In this section water behaves as if the solid were not present. Initially, the surface of the product is very wet, having a water activity value close to one. In porous solids, the water removed from the surface is compensated by the flow of water from the interior of the solid. The constant rate period continues, while the evaporated water at the surface can be compensated by the internal water. The temperature at the surface of the product corresponds approximately to the wet bulb temperature (Geankoplis, 1993).

The falling rate period begins when the drying rate no longer be kept constant and begins to decrease, and the water activity on the surface becomes smaller than one. In this case, the drying rate is governed by the internal flow of water and water vapor. Point C represents the start of the falling rate period, which can be divided into two stages. The first stage occurs when the wet points on the surface decrease continuously until the surface is completely dry (point D), while the second stage of the falling rate period begins at point D, where the surface is completely dry and the evaporation plane moves to the interior of the solid. The heat required to remove moisture is transferred through the solid to the evaporation surface, and the water vapor produced moves through the solid in the air stream going toward the surface. Sometimes there are no marked differences between the first and second falling rate periods. The amount of water removed during this period may be small, while the time required could be long, since the drying rate is low.

23.4.2 Constant Rate Drying Period

The transport phenomena that take place during the constant drying rate period are mass transfer of steam to the environment (from the surface of the product through an air film that surrounds the material) and heat transfer through the solid. The surface of the material remains saturated with water during the drying process, since the rate at which water moves from the interior of the solid is fast enough to compensate the water evaporated at the surface. If it is assumed that there is only heat transfer by convection from the hot air to the surface of the solid and mass transfer from the surface to the hot air (Figure 23.7), it is obtained that

$$q = hA(T - T_W) \tag{23.13}$$

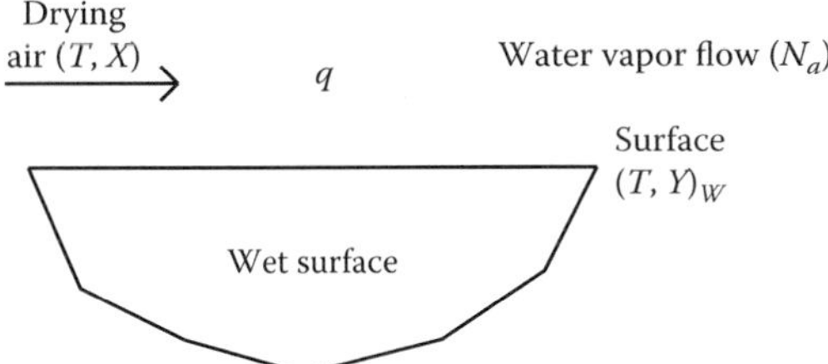

FIGURE 23.7 Heat and mass transfer during drying. (Adapted from Barbosa-Cánovas, G.V. and Vega-Mercado, H., *Dehydration of Food*, Chapman & Hall, New York, 1996.)

$$N_a = k_y(X_W - X) \tag{23.14}$$

where
- h is the heat transfer coefficient
- A is the area that is being dried
- T_W is the wet bulb temperature
- T is the drying temperature
- N_a is the flux of water vapor
- X_W is the moisture content of air at the solid surface
- X is the moisture content of the dry air stream
- k_y is the mass transfer coefficient

The heat needed to evaporate water at the surface of the product can be expressed as

$$q = N_a \lambda_W A \tag{23.15}$$

where λ_W is the latent heat of vaporization at the temperature T_W.

The drying rate during the constant rate period is expressed as (Okos et al., 1992)

$$R_C = k_y(X_W - X) \tag{23.16}$$

or

$$R_C = \frac{h(T - T_W)}{\lambda_W} = \frac{q}{\lambda_W A} \tag{23.17}$$

where
- X_W is the humidity corresponding to the wet bulb temperature
- X is the humidity of the air within the gaseous stream

If there is no heat transfer by conduction or radiation, the temperature of the solid at the wet bulb temperature is equal to the temperature of the air during the constant rate drying period.

In drying calculations, it is essential to know the mass transfer coefficient, which can be evaluated by the following expression (Okos et al., 1992):

$$\frac{k_y l}{D_{AB}} = 0.664\,(\text{Re})^{1/2}(\text{Sc})^{1/3} \tag{23.18}$$

This equation is valid for laminar flow parallel to a flat plate, l being the length of the plate in the flow direction; the Reynolds and Schmidt numbers are defined by the expressions

$$(\mathrm{Re})=\frac{\rho v d}{\eta} \quad (\mathrm{Sc})=\frac{\eta}{\rho D_{AB}} \tag{23.19}$$

where
- D_{AB} is the molecular diffusivity of the air–water mixture
- d is the characteristic length or diameter
- v is the velocity of the fluid
- ρ is the density
- η is the viscosity

The heat transfer coefficient can be obtained by the following equation (Geankoplis, 1993):

$$\text{Parallel flow: } h = 14.28(G)^{0.8} \tag{23.20a}$$

$$\text{Perpendicular flow: } h = 24.06(G)^{0.37} \tag{23.20b}$$

In this expression G is the mass flux of air expressed in kg/(m^2 s), yielding the heat transfer coefficient in W/(m^2 °C). The heat transfer coefficient in a slab can be expressed as a function of the Nusselt number, according to an expression of the type (Chirife, 1983)

$$(\mathrm{Nu})=\frac{hd}{k}=2+\alpha(\mathrm{Re})^{1/2}(\mathrm{Pr})^{1/3} \tag{23.21}$$

The Prandtl number is defined as

$$(\mathrm{Pr})=\frac{\hat{C}_P\eta}{k} \tag{23.22}$$

where
- k is the thermal conductivity
- α is a constant
- $\hat{C}_p$ is the specific heat

23.4.3 Falling Rate Drying Period

This period occurs after drying at a constant rate and, as its name indicates, the drying rate R decreases when the moisture content is lower than the critical moisture content Y_C. Equation 23.12 should be solved by integration in order to solve this type of problem, and this can be calculated by a graphic integration method when plotting $1/R$ versus Y.

The movement of water in the solid can be explained by different mechanisms (Barbosa-Canovas and Vega-Mercado, 1996) such as diffusion of liquid due to concentration gradients, diffusion of water vapor due to partial vapor pressure, movement of the liquid due to capillary forces, movement of the liquid due to gravity forces, or surface diffusion. The movement of water through the food depends on the food's porous structure as well as on the interactions of water within the food matrix. Some of the theories listed will be described next. Further information about all of these theories can be found in the literature (Barbosa-Canovas and Vega-Mercado, 1996; Chen and Johnson, 1969; Bruin and Luyben, 1980; Fortes and Okos, 1980; Geankoplis, 1993).

23.4.3.1 Diffusion Theory

The main mechanism in the drying of solids is water diffusion in solids of fine structure and in capillaries, pores, and small holes filled with water vapor. The water vapor diffuses until it reaches the surface, where it passes to the global air stream. Fick's law, when applied to a system like the one shown in Figure 23.8, can be expressed as

$$\frac{\partial Y}{\partial t} = D_{effect} \frac{\partial^2 Y}{\partial x^2} \tag{23.23}$$

where
Y is the moisture content of the product
t is the time
x is the dimension in the direction of transfer
D_{effect} is the coefficient of diffusion

Depending on the type of geometry considered, the solution to Fick's equation takes different forms. The solutions for simple geometry such as slab, cylinder, and sphere are given next.

Slab

$$\Gamma = \frac{Y - Y_S}{Y_0 - Y_S} = \frac{8}{\pi^2} \sum_{n=1}^{\infty} \frac{1}{h_n^2} e^{((-h_n^2 \pi^2 D_{effect} / 4L^2)t)} \tag{23.24a}$$

$$h_n = 2n - 1$$

where
Y is the moisture content at the time t
Y_0 is the initial moisture content
Y_S is the moisture at the surface
L is the thickness of the slab

Cylinder

$$\Gamma = \frac{Y - Y_S}{Y_0 - Y_S} = \frac{4}{r_a^2} \sum_{n=1}^{\infty} \frac{1}{\beta_n^2} e^{\left(-\beta_n^2 D_{effect} t\right)} \tag{23.24b}$$

where
r_a is the radius of the cylinder
β_n is the roots of the Bessel function of first type and order zero

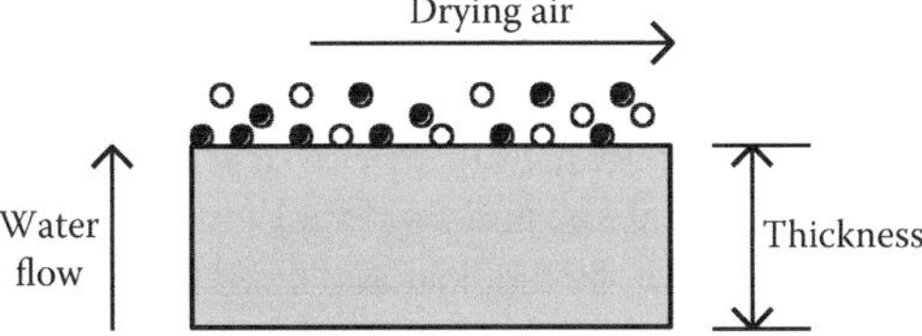

FIGURE 23.8 Surface diffusion and water vapor transport mechanism. (Adapted from Bruin, S. and Luyben, K.Ch.A.M., Drying of food materials, in *Advances in Drying*, Mujumdar, A.S., Ed., Vol. 1, Hemisphere Publishing, New York, 1980.)

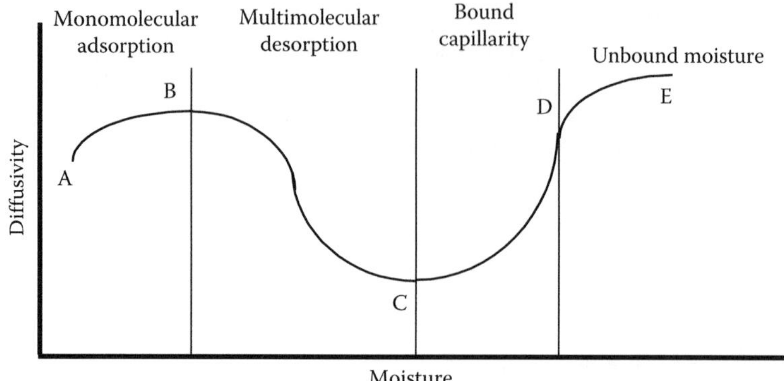

FIGURE 23.9 Relationship between moisture content and diffusivity. (Adapted from Barbosa-Cánovas, G.V. and Vega-Mercado, H., *Dehydration of Food*, Chapman & Hall, New York, 1996.)

Sphere

$$\Gamma = \frac{Y - Y_S}{Y_0 - Y_S} = \frac{6}{\pi^2}\sum_{n=1}^{\infty}\frac{1}{n^2}e^{\left(\left(-n^2 D_{effect}^2/r^2\right)t\right)} \quad (23.24c)$$

where r is the radius of the sphere.

The effective diffusion coefficient (D_{effect}) is determined experimentally from drying data; when the term $\ln\Gamma$ is plotted against time, the slope of the linear section gives the value of D_{effect} (Okos et al., 1992).

The relationship between diffusivity and moisture is presented in Figure 23.9. Region A–B represents the monomolecular adsorption at the surface of the solid and consists of the movement of water by diffusion of the vapor phase. Region B–C involves multimolecular desorption, where moisture begins to move in the liquid phase. Microcapillarity plays an important role in region C–D, where moisture easily emigrates from water filled pores. In region D–E, moisture exerts its maximum vapor pressure and the migration of moisture is essentially due to capillarity.

Effective diffusivity values for some food products are given in Table 23.1.

23.5 CHAMBER AND BED DRYERS

The main objective of food dehydration is to lengthen the commercial life of the final product. For this reason, moisture content is reduced to levels so as to limit microbial growth and to delay deteriorating chemical reactions. Hot air is used in most drying processes, and this type of operation has been used since historic times.

The basic configuration of an atmospheric dryer is a chamber in which the food is introduced and equipped with a fan and conduits that allow the circulation of hot air through and around the food. Water is eliminated from the surface of the food and is driven outside the dryer together with the air stream that exits in a simple operation. Air is heated at the inlet of the dryer by heat exchangers, or directly, with a mixture of combustion gases. This type of dryer is widely used in the production of cookies, dried fruits and chopped vegetables, and food for domestic animals.

In general, the drying process depends on the heat and mass transfer characteristics of the drying air and food. There are two types of phenomena involved in the drying process in an atmospheric dryer: the heating of the product and the reduction of its moisture content, with both as a function of time. Figure 23.10 presents moisture and temperature profiles as a function of drying time.

TABLE 23.1
Effective Diffusivity of Some Food Products

Food	T (°C)	D_{effect} (m²/s)
Whole milk, foam	50	2.0E–9
	40	1.4E–9
	35	8.5E–10
Apples	66	6.40E–9
Freeze-dried apples	25	2.43E–10
Raisins	25	4.17E–11
Potatoes	54	2.58E–11
	60	3.94E–11
	65.5	4.37E–11
	68.8	6.36E–11
Pears (slabs)	66	9.63E–10
Veal, freeze-dried powder	25	3.07E–11
Carrot cubes	40	6.75E–11
	60	12.1E–11
	80	17.9E–11
	100	24.1E–11

Source: Okos, M.R. et al., Food dehydration, in *Handbook of Food Engineering*, Heldman, D.R.Y. and Lund, D.B., Eds., Marcel Dekker, New York, 1992.

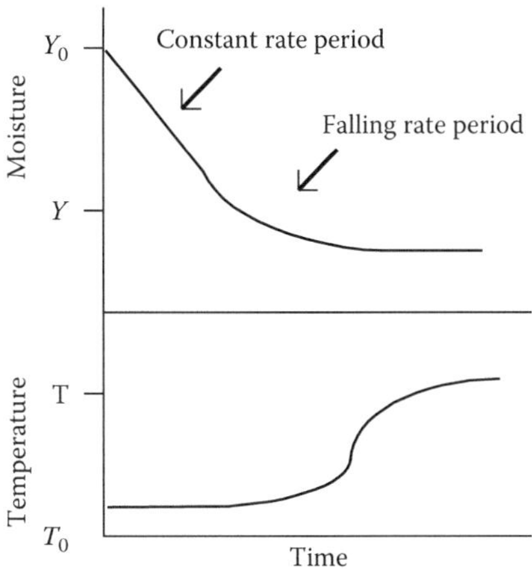

FIGURE 23.10 Moisture and temperature profiles in food dehydration.

Some types of dryers expose the food to a direct hot air stream that heats the product and eliminates water vapor. However, the nature of some foods does not allow direct exposure to hot air, and heating is carried out by heat exchangers that prevent direct contact between the product and the heating medium. The first type of dryer is called a direct dryer, while the second type is called an indirect dryer. In atmospheric drying operations, direct dryers are usually employed.

23.5.1 Components of a Dryer

The basic configuration of a dryer consists of a feeder, a heater, and a collector. The final arrangement of these components is characteristic of each type of dryer. Figure 23.11 shows a basic scheme of an atmospheric dryer.

Feeder: The most common feeders for foods are screw conveyors, rotary slabs, vibrating trays, and rotary air chambers. In some cases special feeders are needed, as in the case of bed dryers in which it is necessary to ensure a uniform distribution of material.

Heater: In some direct heaters, air is heated by mixing with combustion gases. In indirect heaters the air and the product are heated in a heat exchanger. The cost of direct heating is lower than indirect heating, but some products can be damaged by gases. The maximum possible temperature of air in a direct heater ranges from 648°C to 760°C, while for an indirect heater the maximum temperature is 425°C.

Collector: The separation of products as powder or particulate in the air stream can be achieved by cyclones, bag filters, or wet washers.

23.5.2 Material and Heat Balances

23.5.2.1 Discontinuous Dryers

The conditions of the air do not remain constant in a drying chamber or tray dryer during the drying process. Heat and mass balances are used to estimate the conditions of the air leaving the drier.

A heat balance for a tray drier such as the one shown in Figure 23.12, considering a differential length dL_t and a section of thickness z, can be expressed as

$$dq = -G\hat{C}_S(zb)dT \tag{23.25}$$

where

G is the mass flux of air
b is the distance between trays
z is the thickness of the trays
q is the flux of heat
T is the temperature
$\hat{C}_S$ is the specific wet heat of the air–water mixture

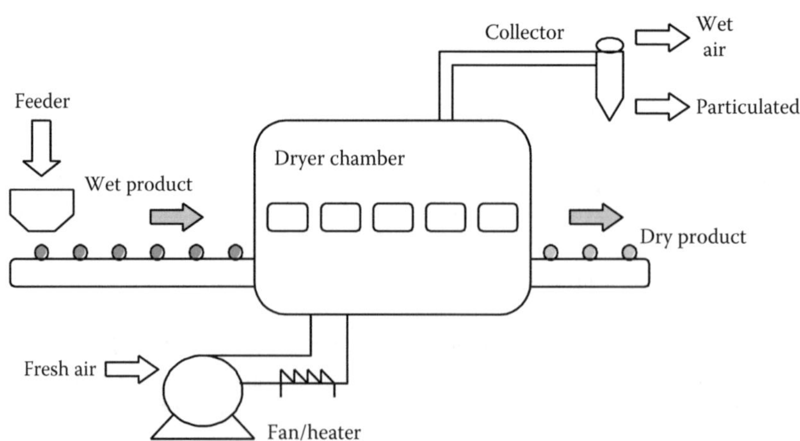

FIGURE 23.11 Basic configuration of an atmospheric dryer. (Adapted from Barbosa-Cánovas, G.V. and Vega-Mercado, H., *Dehydration of Food*, Chapman & Hall, New York, 1996.)

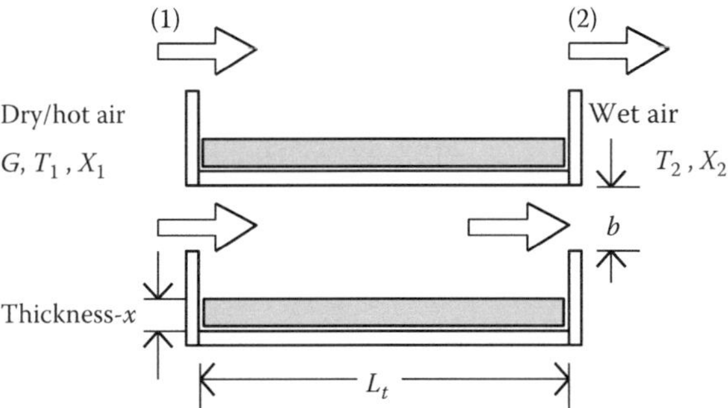

FIGURE 23.12 Tray dryer. (Adapted from Barbosa-Cánovas, G.V. and Vega-Mercado, H., *Dehydration of Food*, Chapman & Hall, New York, 1996.)

The flux of heat can also be expressed as

$$dq = h(zdL_t)(T - T_w) \tag{23.26}$$

where
- h is the heat transfer coefficient
- T_W is the wet bulb temperature
- L_t is the length of the tray

Assuming that h and $\hat{C}_S$ are constants, when combining these two equations, the following can be obtained by integration:

$$\frac{hL_t}{G\hat{C}_S b} = \ln\left(\frac{T_1 - T_W}{T_2 - T_W}\right) \tag{23.27}$$

in which T_1 and T_2 are the temperatures of air at the inlet and outlet of the tray, respectively. The mean logarithmic temperature is defined by

$$\Delta T_{ml} = (T - T_W)_{ml} = \frac{(T_1 - T_W) - (T_2 - T_W)}{\ln(T_1 - T_W / T_2 - T_W)} \tag{23.28}$$

Combining Equations 23.27 and 23.28,

$$(T - T_W)_{ml} = \frac{(T_1 - T_W)(1 - \exp(-hL_t / G\hat{C}_S b))}{(hL_t / G\hat{C}_S b)} \tag{23.29}$$

The heat flow rate reaching the product surface from the hot air can be expressed as

$$\dot{Q} = \frac{Q}{t} = hzL_t \Delta T_{ml} \tag{23.30}$$

This heat is used to evaporate water from the surface of the food. The total heat required to change from initial moisture content Y_1 to a final moisture content corresponding to the critical moisture content Y_C is

$$Q = (zL_t x\rho)\lambda_W (Y_1 - Y_C) \tag{23.31}$$

When equaling the last two equations and taking into account Equation 23.29, it is obtained that the drying time for the constant rate period is given by

$$t_c = \frac{x\rho_S L_t \lambda_W (Y_1 - Y_c)}{G\hat{C}_S b(T_1 - T_w)(1 - \exp(-hL_t / G\hat{C}_S b))} \tag{23.32}$$

where

Y_1 is the initial moisture content of the product
Y_c is the critical moisture content
x is the thickness of the bed
ρ_s is the density of the solid
λ_W is the latent heat of vaporization at the temperature T_W

Calculation of the drying time for the falling rate period is determined as explained next. The equation that describes the drying rate in this period is

$$R = \frac{-w_S}{A}\frac{dY}{dt} \tag{23.33}$$

in which w_S is the amount of solids. This rate can also be expressed according to the equation (Barbosa-Canovas and Vega-Mercado, 1996)

$$R = \frac{h}{\lambda_W}(T - T_w)_M \tag{23.34}$$

in which the drying rate is expressed as a function of the mean temperature increase. Combining these equations and assuming that the drying velocity is a linear function of Y, when integrating on the boundary condition $t = 0$, $Y = Y_C$, and $t = t_D$, $Y = Y_F$, the drying time for the falling rate is obtained (Geankoplis, 1993):

$$t_D = \frac{w_S \lambda_W Y_C \ln(Y_C / Y_F)}{Ah(T - T_W)_M} \tag{23.35}$$

If the mean difference of temperatures is logarithmic, it can be substituted in Equation 23.29 and it is obtained that

$$t_D = \frac{x\rho_S L_t \lambda_W Y_C \ln(Y_C / Y_F)}{G\hat{C}_S b(T_1 - T_W)(1 - \exp(-hL_t / G\hat{C}_S b))} \tag{23.36}$$

where Y_F represents the final moisture of the product.

The total drying time to change from a moisture content Y_1 to a final moisture content Y_F is obtained by adding the times calculated with Equations 23.32 and 23.36, the total drying time being equal to $t_S = t_C + t_D$.

23.5.2.2 Discontinuous Dryers with Air Circulation through the Bed

Another type of discontinuous dryer is one in which drying air is circulated through a bed of food. Figure 23.13 presents this type of dryer. It is supposed that the system is adiabatic, that there are no heat losses, and that the air circulates with a mass flow density G, entering at a temperature T_1, with a humidity X_1, while at the outlet air has a temperature T_2 and a humidity X_2.

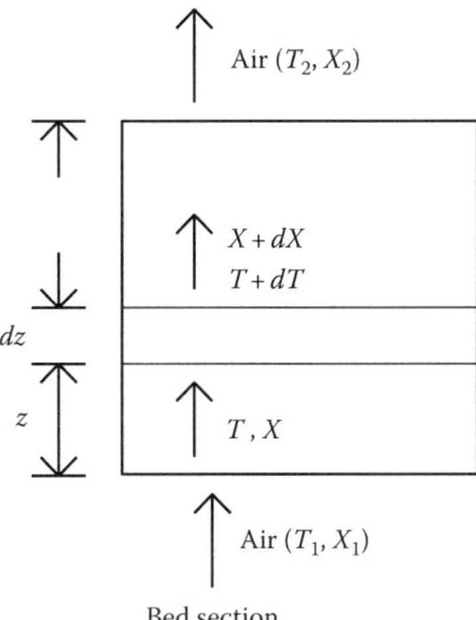

FIGURE 23.13 Drying due to air circulation through the bed. (Adapted from Barbosa-Cánovas, G.V. and Vega-Mercado, H., *Dehydration of Food*, Chapman & Hall, New York, 1996.)

The global drying velocity can be expressed as

$$R = G(X_2 - X_1) \tag{23.37}$$

while for a bed's height differential dz, the heat flow transmitted by air is

$$dq = G\hat{C}_S A\, dT \tag{23.38}$$

Here, A is the crossing section. The heat transferred to the solid can be expressed according to the equation

$$dq = haA(T - T_W)dz \tag{23.39}$$

where a is the specific surface of the bed particles. Such specific surface is determined depending on the type of particle; thus,

$$\text{Spherical particles: } a = \frac{6(1-\varepsilon)}{D_P} \tag{23.40}$$

$$\text{Cylindrical particles: } a = \frac{6(1-\varepsilon)(1+0.5D_C)}{D_C l} \tag{23.41}$$

where
- l is the length of the particle
- D_C is the diameter of the cylinder
- D_P is the diameter of a sphere
- ε is the fraction of holes in the solid

Supposing that h and $\hat{C}_S$ are constants, when equaling Equations 23.38 and 23.39 and integrating, it is obtained that

$$\frac{haz}{G\hat{C}_S b} = \ln\left(\frac{T_2 - T_w}{T_1 - T_w}\right) \tag{23.42}$$

Considering $w_s = A\rho_s/a$, the expressions for drying times are (Geankoplis, 1993; Barbosa-Canovas and Vega-Mercado, 1996)

Constant rate period

$$t_C = \frac{\rho_S \lambda_w (Y_l - Y_c)}{hA(T - T_w)_M} \tag{23.43}$$

or

$$t_C = \frac{\rho_S (Y_l - Y_c)}{aK_y M_B (X_w - X)} \tag{23.44}$$

where

K_y is the mass transfer coefficient
M_B is the molecular weight
X_W is the humidity of air at the temperature T_W

Falling rate period

$$t_D = \frac{\rho_S \lambda_W Y_C \ln(Y_c/Y_F)}{ha(T - T_w)_M} \tag{23.45}$$

or

$$t_D = \frac{\rho_S Y_C \ln(Y_c/Y_F)}{aK_y M_B (X_W - X)} \tag{23.46}$$

The difference of temperatures through the bed can be taken as the logarithmic mean (Equation 23.29), which when substituted in Equations 23.43 and 23.45 yields

Constant rate period

$$t_C = \frac{x\rho_S \lambda_W (Y_l - Y_c)}{G\hat{C}_S (T_1 - T_W)(1 - \exp(-hax/G\hat{C}_S))} \tag{23.47}$$

Falling rate period

$$t_D = \frac{x\rho_S \lambda_W Y_C \ln(Y_c/Y_F)}{G\hat{C}_S (T_1 - T_W)(1 - \exp(-hax/G\hat{C}_S))} \tag{23.48}$$

where x is the thickness of the bed.

The heat transfer coefficient for circulation of drying air can be evaluated by equations (Geankoplis, 1993)

$$h = 0.151 \frac{(G_t)^{0.59}}{(D_P)^{0.41}} \quad \frac{D_P G_t}{\eta} > 350 \quad \text{for SI} \tag{23.49}$$

$$h = 0.214 \frac{(G_t)^{0.49}}{(D_P)^{0.51}} \quad \frac{D_P G_t}{\eta} < 350 \quad \text{for SI} \tag{23.50}$$

The equivalent diameter (D_P) for a cylindrical particle is

$$D_P = \left(D_C l + 0.5 D_C^2\right)^{1/2} \tag{23.51}$$

23.5.2.3 Continuous Dryers

The equation that allows the calculation of drying time for the constant rate period in a dryer in which the solid food and air circulate in countercurrent is (Geankoplis, 1993)

$$t = \left(\frac{G}{w_S}\right)\left(\frac{w_S}{A}\right)\left(\frac{1}{K_y M_B}\right)\ln\left(\frac{X_W - X_c}{X_W - X_1}\right) \tag{23.52}$$

in which A/w_S is the surface exposed to drying. This equation can be expressed as

$$t = \left(\frac{G}{w_S}\right)\left(\frac{w_S}{A}\right)\left(\frac{1}{K_y M_B}\right)\left(\frac{X_1 - X_C}{\Delta X_{ml}}\right) \tag{23.53}$$

Here, ΔX_{ml} is the logarithmic mean difference of moisture content:

$$\Delta X_{ml} = \frac{(X_1 - X_w) - (X_c - X_w)}{\ln(X_w - X_c / X_w - X_1)} \tag{23.54}$$

in which the critical moisture content is (Geankoplis, 1993)

$$X_C = X_2 + \frac{w_S}{G}(Y_c - Y_2) \tag{23.55}$$

The drying time for the falling rate period is obtained from the equation

$$t_D = \left(\frac{G}{w_s}\right)\left(\frac{w_s}{A}\right)\left(\frac{Y_c}{Y_2 + (X_w - X_2)(G/w_s)K_y M_B}\right)\ln\left(\frac{Y_c(X_w - X_c)}{Y_2(X_w - X_1)}\right) \tag{23.56}$$

In the case of concurrent flow, as shown in Figure 23.14, hot air comes into contact with the wet food at the inlet, so the mass balance is expressed as

$$GX_1 + w_S Y_1 = GX_2 + w_S Y_2 \tag{23.57}$$

or

$$G(X_2 - X_1) = w_S(Y_1 - Y_2) \tag{23.58}$$

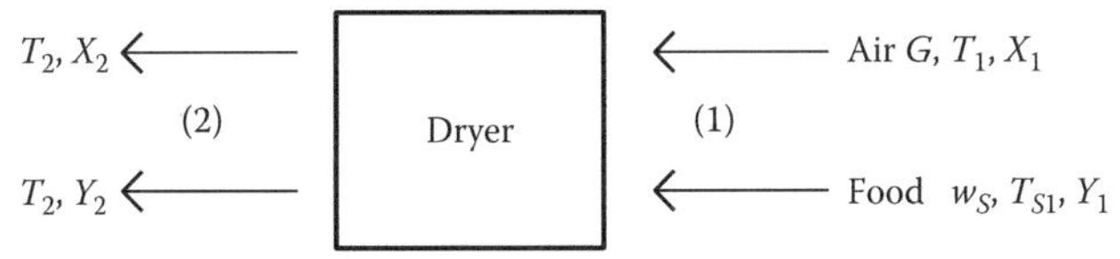

T_2, X_2 ← (2) Dryer (1) ← Air G, T_1, X_1

T_2, Y_2 ← ← Food w_S, T_{S1}, Y_1

FIGURE 23.14 Drying operation for concurrent flow.

These equations are expressed as follows in terms of critical values:

$$G(X_C - X_1) = w_S(Y_1 - Y_C) \tag{23.59}$$

$$X_C = X_1 + \frac{w_S}{G}(Y_1 - Y_C) \tag{23.60}$$

23.6 SPRAY DRYING

This type of drying is used for foods dissolved in water and includes the formation of droplets that, when later dried, yield dry food particles. Initially, the fluid food is transformed into droplets that are dried by spraying a mist into a continuous hot air medium. Open cycles are most commonly used in this type of drying, as shown in Figure 23.15. The drying air is heated using a dry medium and, after drying, it is cleaned using cyclones before releasing it into the atmosphere. In this type of operation, the air leaving the system can still contain heat. A second type of arrangement is the use of a closed circuit with a heating medium (air, CO_2, etc.). Air is used in the drying process, then is cleaned, dried, and used again in a continuous process. The efficiency of this type of drying is higher than for open systems. In a closed-circuit system, the dry product is the only one leaving the system, while in open circuits hot air is also released to the exterior, and it may contain microparticles.

This type of drying includes atomization of food into a drying media in which moisture is eliminated by evaporation. Drying is performed until the moisture content fixed for the product is reached. This type of drying is controlled by the flow and temperature conditions of the product as well as those of the air at the inlet. Spray drying was used for the first time around 1900 to dry milk, and it was later applied to eggs and coffee.

The most important characteristic of spray drying is the formation of droplets and their contact with air. Breaking the food stream into droplets produces the atomization of food. Different types of atomizers will be explained next (Barbosa-Canovas and Vega-Mercado, 1996).

23.6.1 Pressure Nozzles

Pressure nozzles are used to form droplets, since it is possible to control the food flow and the atomization characteristics by varying the pressure. The mean size of the droplets formed is proportional to the food flow and to its viscosity. Figure 23.16 shows one type of pressure nozzle.

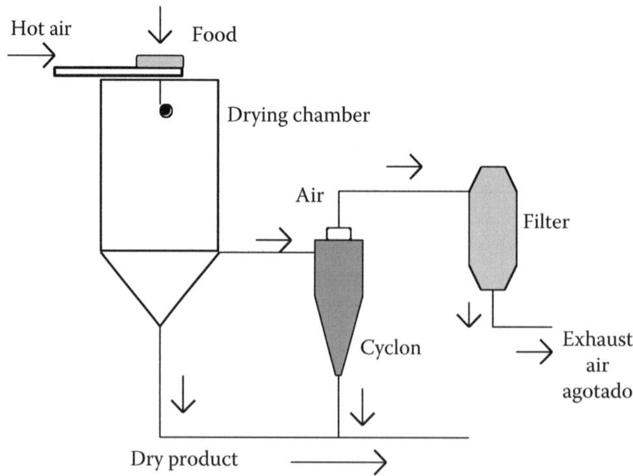

FIGURE 23.15 Open cycle; spray drying under concurrent flow. (Adapted from Barbosa-Cánovas, G.V. and Vega-Mercado, H., *Dehydration of Food*, Chapman & Hall, New York, 1996.)

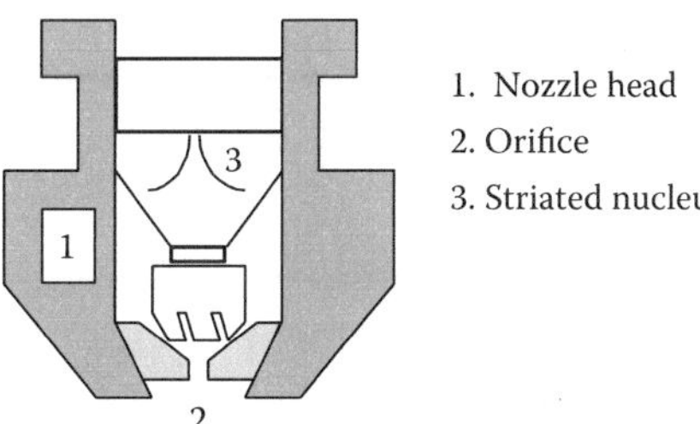

FIGURE 23.16 Striated pressure nozzle. (Adapted from Masters, K., *Spray Drying Handbook*, 5th edn., Longman Group Limited, London, U.K., 1991.)

The fundamental principle of this type of nozzle is the conversion of pressure energy into kinetics energy. The liquid layers are broken under the influence of the physical properties of the liquid and by the friction effects when they come into contact with air. The power required by a pressure nozzle is proportional to the feed rate and to the pressure of the nozzle (Barbosa-Canovas and Vega-Mercado, 1996):

$$P_k = 0.27\frac{\Delta P}{\rho} \tag{23.61}$$

where
ΔP is the total pressure drop
ρ is the density of food

The conversion of pressure into kinetic energy in a centrifugal pressure nozzle results in a rotational movement of the liquid that can be expressed as (Marshall, 1954)

$$E_h = 19.2w\Delta P \tag{23.62}$$

where
w is the mass flow rate
E_h is the energy or power

The flow of liquid in the orifice of a centrifugal pressure nozzle can be expressed as

$$2\left(\pi r_1^2 V_{inlet}\right) = 2\left(\pi b r_2 U_r\right) \tag{23.63}$$

$$V_{inlet} = \frac{w_1}{2\pi r_1^2 \rho} \tag{23.64}$$

or

$$\frac{U_r}{V_{inlet}} = \frac{r_1^2}{r_2 b} \tag{23.65}$$

where
b is the thickness of the liquid film in the orifice
r_1 is the radius of the inlet channel
r_2 is the radius of the orifice
V_{inlet} is the velocity of the liquid at the inlet
U_r is the vertical component of the atomization velocity
w_1 is the mass flow of the liquid

The velocity of the liquid leaving the nozzle is expressed as

$$V_{outlet} = \sqrt{U_h^2 + U_v^2} \tag{23.66}$$

When expressed in terms of pressure drop through the nozzle is

$$V_{outlet} = C_v(2gh)^n = C_v\left(2g\frac{\Delta P}{\rho}\right)^n \tag{23.67}$$

where

U_h and U_v are the horizontal and vertical components of velocity
$n = 0.5$ for the turbulent flow
C_v is the velocity coefficient
g is the gravitational constant
n is a constant
h is the head pressure

The performance of a pressure nozzle is affected by pressure and by the density and viscosity of the liquid. Masters (1991) proposed a correlation between flow changes through the injector and the pressure and density changes according to the expression

$$\frac{w_2}{w_1} = \left(\frac{P_2}{P_1}\right)^{0.5} = \left(\frac{\rho_1}{\rho_2}\right)^{0.5} \tag{23.68}$$

The effect of viscosity on the flow is not clearly defined, although it can be determined in an experimental form. The effect of process variables such as capacity of the nozzle, atomization angle, pressure, viscosity, surface tension, and diameter of the orifice on droplet size is given in Table 23.2.

Industrial dryers containing multi-nozzles are installed to permit high feeding velocities and to provide equal conditions in each nozzle for a better uniformity of atomization. The configurations of the nozzles should present the following conditions: easy access to remove the nozzles, uniform distribution, possibility of isolation, and visibility of each nozzle. Some of the possible configurations are shown in Figure 23.17.

TABLE 23.2
Effect of Some Process Variables on Droplet Size

Variable	Effect
Nozzle capacity	
Feed rate below the designed rate	Incomplete atomization
Feed rate below minimum	Size of droplets decreases
Specified feed rate	Size of droplets increases
Large spraying angle	Small droplets
Pressure increase	Size of droplets decreases
Viscosity	
Increase	Thick atomization
Very high	Impossible operation
High surface tension	Makes atomization difficult
Orifice size	Droplet size = kD^2, D = orifice diameter, k = constant

Source: Barbosa-Cánovas, G.V. and Vega-Mercado, H., *Dehydration of Food*, Chapman & Hall, New York, 1996.

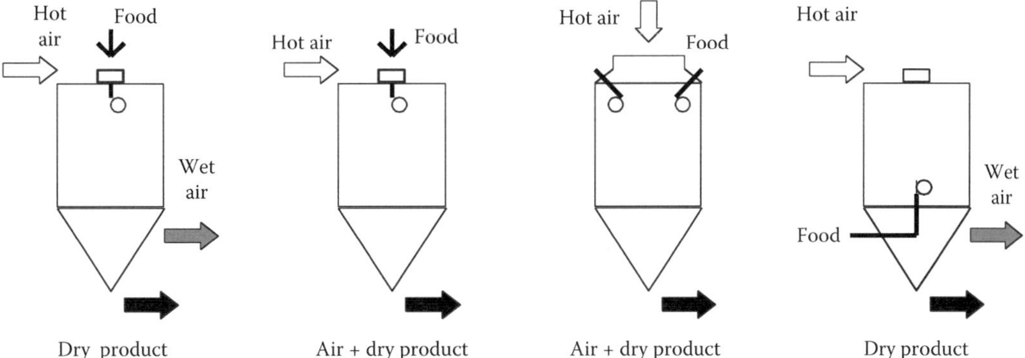

FIGURE 23.17 Configuration of pressure nozzles in industrial dryers. (Adapted from Barbosa-Cánovas, G.V. and Vega-Mercado, H., *Dehydration of Food*, Chapman & Hall, New York, 1996.)

23.6.2 Rotary Atomizers

Rotary atomizers differ from pressure nozzles because liquid achieves velocity without high pressure. Also, the feed rate can be controlled with disks, while in the case of nozzles, both pressure drop and the diameter of the orifice change simultaneously. Figure 23.18 shows the physical properties of the food for the atomization mechanism in the case of disks. The formation and releasing of droplets from the edge of the disk, considering a low velocity of feed and disk, are shown in Figure 23.18a. Atomization consists of one droplet and two satellites. An increase in the velocity of the disk and the food produces a change in the mechanism of formation of droplets (Figure 23.18b). The arrangement of the liquid in layers (Figure 23.18c) appears when the bindings of the liquid join each other and extend beyond the edge of the disk.

In disks with blades (Figure 23.19a), the disintegration of the liquid takes place on the edge of the disk due to the friction effect between the air and the surface of the liquid. The liquid emerges as a thin film from the blade. The optimum size of a droplet for a given feed depends on the following conditions: rotation without vibration, centrifugal force, soft and complete wetting of the blade's surface, and a uniform distribution and feed.

The acceleration along the blade stops when the liquid reaches the edge of the disk, so the velocity of the liquid can be expressed as (Barbosa-Canovas and Vega-Mercado, 1996)

$$U_r = 0.0024\left(\frac{\rho\pi^2 N^2 D w^2}{\eta h^2 n^2}\right)^{0.33} \tag{23.69}$$

$$U_t = \pi D N \tag{23.70}$$

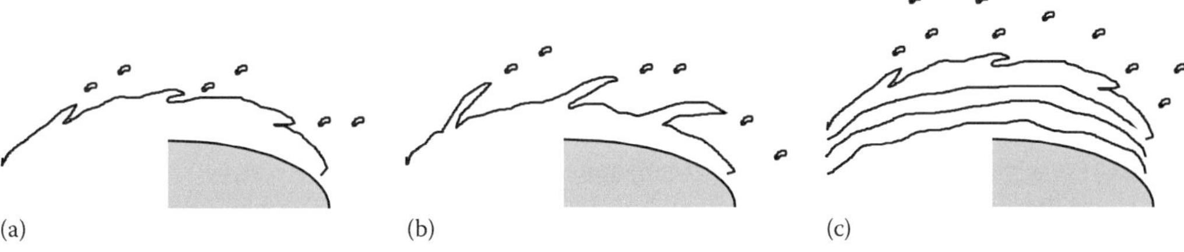

FIGURE 23.18 Atomization with disks without blades. (a) Droplets from the edge, (b) binding formation, and (c) liquid sheets. (Adapted from Barbosa-Cánovas, G.V. and Vega-Mercado, H., *Dehydration of Food*, Chapman & Hall, New York, 1996.)

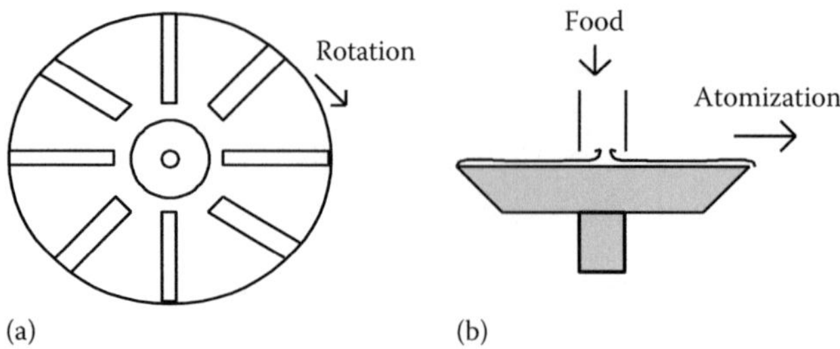

FIGURE 23.19 Rotary atomizers: (a) disk with blades and (b) plane disk with sharp edge. (Adapted from Barbosa-Cánovas, G.V. and Vega-Mercado, H., *Dehydration of Food*, Chapman & Hall, New York, 1996.)

$$U_{res} = \sqrt{U_r^2 + U_t^2} \tag{23.71}$$

$$\alpha = \tan^{-1}\left(\frac{Ur}{Ut}\right) \tag{23.72}$$

where
- U_r is the radial component of velocity
- U_t is the tangential component
- U_{res} is the result of stopping velocity
- α is the angle of the released liquid
- D is the diameter of the disk
- N is the rotation velocity of the atomizer
- n is the number of blades
- h is the height of blades
- η is the viscosity

The effect of process variables such as disk speed, feed rate, liquid viscosity, surface tension, and density of the liquid on the size of the droplet for a rotary atomizer is given in Table 23.3.

The distribution size of atomization in a rotary atomizer can be expressed as

$$D_{mean} = \frac{Kw^a}{N^b d^{0.6} (nh)^d} \tag{23.73}$$

in which the values of K, a, b, and d are a function of the disk speed and the load rate of the blade. Table 23.4 presents the values of these constants.

Rotary atomizers are generally installed in the center of the roof in spray dryers to allow enough contact time between the droplet and the hot air pass and to allow evaporation of the liquid (Shaw, 1994).

The relationship between the size of the wet droplet and the size of the dry particle is expressed according to the equation

$$D_{WET} = \beta D_{DRY} \tag{23.74}$$

where
- D_{WET} is the size of the droplet during atomization
- D_{DRY} is the size of the dry particle
- β is the change of shape factor

This last factor is a function of the type of product and temperature and it is useful to change the scale of a dryer (Masters, 1991).

TABLE 23.3
Effect of Process Variables on the Size of Droplets for a Rotary Atomizer

Variable		Effect
Disk speed Constant feed	$\frac{D_1}{D_2}=\left[\frac{N_2}{N_1}\right]^p$	D = Diameter of the disk N = Disk speed, rpm p = 0.55–0.80
Feed rate Constant disk speed	$\frac{D_1}{D_2}=\left[\frac{q_1}{q_2}\right]^s$	q = Feed rate s = 0.1–0.12
Viscosity of the liquid	$\frac{D_1}{D_2}=\left[\frac{\eta_1}{\eta_2}\right]^r$	μ = Viscosity r = 0.2
Surface tension	$\frac{D_1^2}{D_2}=\left[\frac{\sigma_1}{\sigma_2}\right]^s$	σ = Surface tension s = 0.1–0.5
Density of the liquid	$\frac{D_1}{D_2}=\left[\frac{\rho_2}{\rho_1}\right]^t$	ρ = Density t = 0.5

Source: Barbosa-Cãnovas, G.V. and Vega-Mercado, H., *Dehydration of Food*, Chapman & Hall, New York, 1996.

TABLE 23.4
Values of Constants of Equation 23.73

Disk Speed (m/s)	Load Rate of the Blade (kg/h m)	*a*	*b*	*d*	*K**10⁴
Normal	Low				
85–115	250	0.24	0.82	0.24	1.4
Normal-high	Normal				
85–180	250–1500	0.2	0.8	0.2	1.6
Very high	Normal-high				
180–300	1000–3000	0.12	0.77	0.12	1.25
Normal - high	Very high				
85–140	3000–60,000	0.12	0.8	0.12	1.2

Source: Masters, K., *Spray Drying Handbook*, 5th edn., Longman Group Limited, London, U.K., 1991.

23.6.3 Two-Fluid Pneumatic Atomizers

The atomization of a liquid using a gas at high velocity is known as pneumatic atomization. The mechanism involves high velocity of a gas that allows creation of high friction forces, causing the liquid to break into droplets. The formation of droplets takes place in two stages: first, the liquid is broken into filaments and long drops and then the filaments and large drops are broken into droplets. This process of formation of droplets is affected by the properties of the liquid (surface tension, density, and viscosity), as well as by those of the gas flow (velocity and density).

Air and vapor are the primary gas media used in pneumatic atomization. In the case of closed systems, inert gases are usually employed. In order to achieve optimum friction conditions, high velocities between air and liquid are required. Such conditions are obtained by expansion of the gas phase at sonic and ultrasonic velocities, before contact with the liquid or by direct gas flow on a thin film of liquid in the nozzle. Pneumatic nozzles include an internal mixing and an internal/external combined mixing. The power requirement for an isentropic expansion is expressed as

$$P = 0.402 w_A T \left\{ 0.5 M_a^2 + 2.5 \left[1 - \left(\frac{P_1}{P_2} \right)^{0.286} \right] \right\} \tag{23.75}$$

where
- w_A is the mass flow rate of air
- T is the absolute temperature
- M_a is the module of Mach
- P_1 and P_2 are the initial and final pressures, respectively

The mean atomization size obtained in a pneumatic atomizer can be expressed as

$$D = \frac{A}{\left(V^2 \rho_a\right)^{\alpha}} + B \left(\frac{w_{air}}{w_{liq}} \right)^{-\beta} \tag{23.76}$$

where
- V is the relative velocity between air and liquid
- α and β are the function of the nozzle
- A and B are the constants
- w_{air} is the mass flow of air
- w_{liq} is the mass flow of liquid

Table 23.5 shows the effects of the process variables on droplets for pneumatic atomizers.

23.6.4 Interaction between Droplets and Drying Air

The distance traveled by a drop until it is completely affected by air depends on its size, shape, and density. While the common atomizers are independent of airflow, in fine atomizers, such flow should

TABLE 23.5
Effect of Process Variables on Droplets for Pneumatic Atomizers

Variable	Effect
Mass rate air/liquid	
Rate increase	Decrease of droplet size
$w_{air}/w_{liq} < 0.1$	Deteriorating atomization
$w_{air}/w_{liq} \geq 10$	Upper limit for effective rate
	Increase to create particles of smaller size
Relative rate	
Increase of air rate	Decrease of droplet size
Viscosity	
Increase of fluid viscosity	Increase of droplet size
Increase of air viscosity	Decrease of droplet size

Source: Masters, K., *Spray Drying Handbook*, 5th edn., Longman Group Limited, London, U.K., 1991.

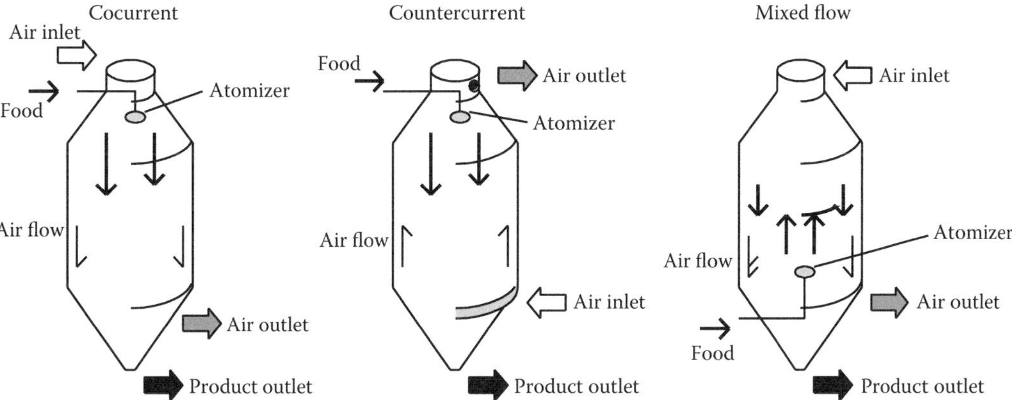

FIGURE 23.20 Classification of dryers according to atomization movement. (Adapted from Barbosa-Cánovas, G.V. and Vega-Mercado, H., *Dehydration of Food*, Chapman & Hall, New York, 1996.)

be taken into account (Barbosa-Canovas and Vega-Mercado, 1996). Atomization movement can be classified according to the drier design as concurrent, countercurrent, or mixed flow (Figure 23.20).

Atomization movement can be explained for a simple droplet. The forces that act on a droplet are

$$\frac{\pi}{6}D^3\rho_w\frac{dV}{dt}=\frac{\pi}{6}D^3(\rho_w-\rho_a)g-0.5C_d\rho_a V_r^2A \quad (23.77)$$

where

D is the diameter of the droplet
C_d is the dragging coefficient
V_r is the relative velocity of the droplet with respect to the air
A is the area of the droplet
ρ_w is the density of the droplet
ρ_a is the density of air

Masters (1991) discussed atomization movement under different flow conditions.

The temperature profile inside the dryer is an important aspect and is a function of the type of flow (Masters, 1991; Barbosa-Canovas and Vega-Mercado, 1996).

23.6.5 Heat and Mass Balances

In food drying by atomization, the liquid that should be removed in most cases is water, although the removal of organic solvents in closed cycle operations is also common. If a system like the one shown in Figure 23.21 is considered, the heat and mass balances yield the following equations.

A mass balance applied to the whole system yields

$$w_S(Y_{S1}-Y_{S2})=G_a(X_{a2}-X_{a1}) \quad (23.78)$$

where

w_S is the flow of dry solids
Y_{S1} is the moisture content of the solid that leaves the dryer
Y_{S2} is the moisture content of the solid leaving the drier
G_a is the flow of dry air
X_{a1} is the humidity of the air that entering the dryer
X_{a2} is the humidity of the air leaving the dryer

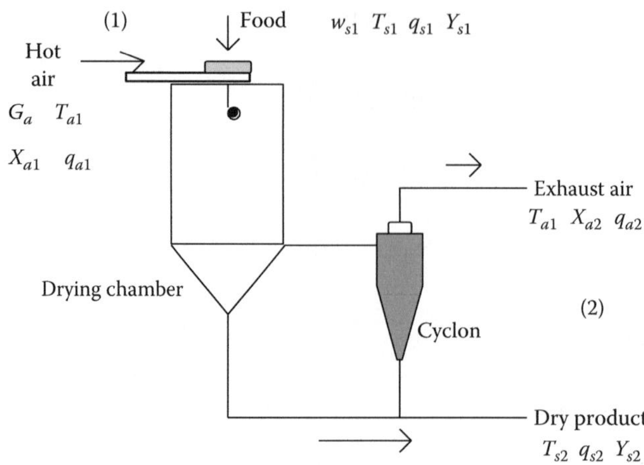

FIGURE 23.21 Data of a dryer for heat and mass balances (w_S, dry solids rate; T_s, solids temperature; q_s, solids enthalpy; Y_s, solids moisture content; G_a, dry air flow rate; T_a, air temperature; X_a, air moisture content; q_a, enthalpy).

A balance of enthalpies yields the equation

$$w_S q_{S1} + G_a q_{a1} = w_S q_{S2} + G_a q_{a2} + q_L \tag{23.79}$$

where
q_{S1} and q_{S2} are the enthalpies of the solid at the inlet and outlet, respectively
q_{a1} and q_{a2} are the enthalpies of air at the inlet and outlet of the dryer
q_L is the heat losses

The performance of a spray dryer is measured in terms of thermal efficiency, which is related to the heat input required to produce a unit weight of dry product with the desired specifications. Global thermal efficiency (ϕ_{global}) is defined as the fraction of total heat supplied to that used in the dryer during the evaporation process:

$$\phi_{global} = 100\frac{(T_1 - T_2)}{(T_1 - T_0)} \tag{23.80}$$

where
T_1 is the temperature of the hot air at the inlet
T_2 is the temperature corresponding to the outlet
T_0 is the temperature of the atmospheric air

The evaporation efficiency ($\phi_{evaporation}$) is defined as the rate of actual evaporation capacity to the capacity obtained in an ideal case of air exhausting at the saturation temperature:

$$\phi_{evaporation} = 100\frac{(T_1 - T_2)}{(T_1 - T_{sat})} \tag{23.81}$$

23.7 FREEZE-DRYING

Freeze-drying was developed to overcome the loss of compounds responsible for aroma in foods, which were lost in conventional drying operations (Barbosa-Canovas and Vega-Mercado, 1996). The freeze-drying process consists mainly of two stages: (1) the product is frozen and (2) the product

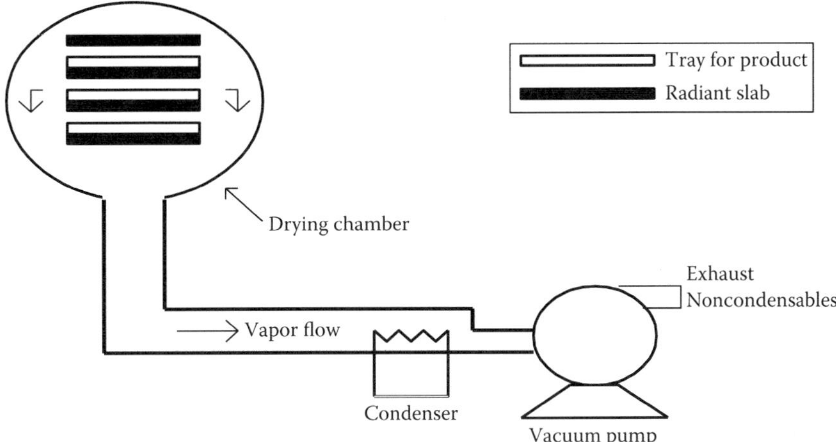

FIGURE 23.22 Freeze-drying basic system. (Adapted from Barbosa-Cánovas, G.V. and Vega-Mercado, H., *Dehydration of Food*, Chapman & Hall, New York, 1996.)

is dried by direct sublimation of ice under reduced pressure. This type of drying was initially introduced on a large scale in 1940 to produce dry plasma and blood products. After that, antibiotics and biological materials were prepared on an industrial scale by freeze-drying. Figure 23.22 shows the basic schema of a freeze-drying system.

Freeze-drying has been shown to be an effective method for extending the mean life of foods, and it consists of two important characteristics:

1. Absence of air during processing. The absence of air and low temperature prevents deterioration due to oxidation or modifications of the product.
2. Drying at a temperature lower than room temperature. Products that decompose or experience changes in their structure, texture, appearance, and/or aroma as a consequence of high temperatures can be dried under vacuum with minimum damage.

Freeze-dried products that have been adequately packaged can be stored for an almost unlimited time, maintaining most of the physical, chemical, biological, and sensorial properties of the fresh product. Quality losses due to enzymatic and nonenzymatic browning reactions are also reduced. However, the oxidation of lipids, caused by the low moisture levels achieved during drying, is higher in freeze-dried products. Packing the products in packages impermeable to oxygen can control this lipid oxidation. Nonenzymatic browning occurs slightly during drying, since the reduction of the moisture content of the product during the process is almost instantaneous. The use of low temperatures also reduces the denaturalization of proteins in this type of drying (Okos et al., 1992).

Freeze-dried products can recover their original shape and structure by the addition of water. The spongy structure of freeze-dried products allows their rapid rehydration. The characteristics of the rehydrated product are similar to those of a fresh product. The porosity of freeze-dried products allows a more complete and rapid rehydration than in air-dried foods. However, one of the greatest disadvantages of freeze-drying is the energy cost and long drying time period. Some commercial products obtained by freeze-drying are coffee and tea extracts, vegetables, fruits, meats, and fish. These products have 10%–15% of their original weight and do not require refrigeration. Products with moisture content lower than 2% can also be obtained.

As indicated previously, the freeze-drying process consists of two stages: freezing and drying. Freezing must be very quick, with the objective of obtaining a product with small ice crystals and in an amorphous state. The drying stage is performed at low pressures in order to sublimate the ice. Figure 23.23 presents a diagram of water phases, while Figure 23.24 presents the drying stages of freeze-drying.

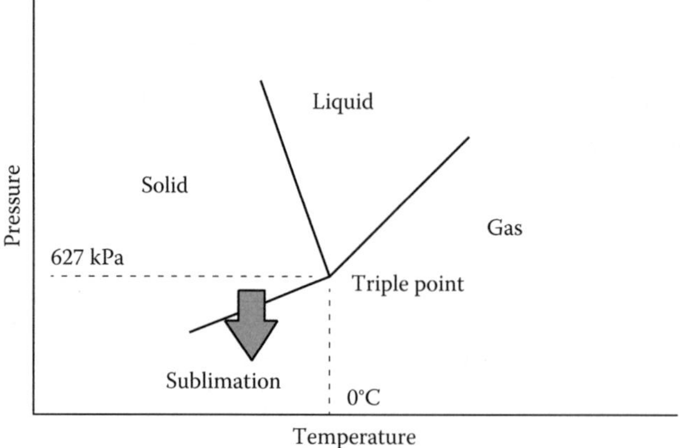

FIGURE 23.23 Diagram of water phases. (Adapted from Barbosa-Cánovas, G.V. and Vega-Mercado, H., *Dehydration of Food*, Chapman & Hall, New York, 1996.)

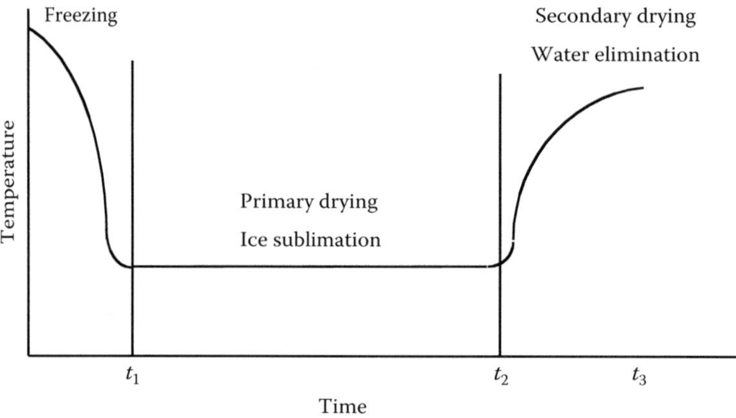

FIGURE 23.24 Freeze-drying stages. (Adapted from Barbosa-Cánovas, G.V. and Vega-Mercado, H., *Dehydration of Food*, Chapman & Hall, New York, 1996.)

Three important design variables should be considered for freeze-drying: (1) vacuum inside the drying chamber, (2) radiant energy flow applied to the food, and (3) the temperature of the condenser. The initial drying rate is high, since the resistance to mass and energy flow is low. However, a thin layer around the frozen product accumulates and causes the drying rate to decrease as it progresses. This layer acts as an insulator and affects the heat transfer toward the ice front. Also, the mass transfer from the ice front decreases as the thickness of the dry layer increases. This result is due to the reduction of diffusion from the sublimation interface toward the surface of the product.

23.7.1 Freezing Stage

The freezing temperature and time of food products is a function of the solutes in the solution that it contains. The freezing temperature for pure water remains constant at the freezing point until the water is frozen. The freezing temperature of foods is lower than for pure water, since the solutes concentrate in the nonfrozen water fraction and the freezing temperature continuously decreases until the solution is frozen. The whole mass of the product becomes rigid at the end of the freezing process, forming a eutectic that consists of ice crystals and food components. It is necessary to reach the eutectic state to assure the removal of water by sublimation only and not because of a combination of sublimation and evaporation.

The permeability of the frozen surface can be affected by the migration of soluble components during the freezing stage. However, removal of the thin layer on the surface of the frozen product, or freezing under conditions that inhibit the separation of the concentrated phase, gives place to better drying rates (Barbosa-Canovas and Vega-Mercado, 1996).

23.7.2 Primary and Secondary Drying Stages

Two stages can be distinguished during the freeze-drying process. The first stage involves ice sublimation under vacuum. Ice sublimates when the energy corresponding to the latent heat of sublimation is supplied. The vapor generated in the sublimation interface is eliminated through the pores of the product due to the low pressure in the drying chamber. The condenser prevents the vapor from returning to the product. The driving force of sublimation is the pressure difference between the water vapor pressure in the ice interface and the partial water vapor pressure in the drying chamber. The energy to sublimate ice is supplied by radiation or conduction through the frozen product or by irradiation of water molecules using microwaves.

The second drying stage begins when the ice in the product has been removed and moisture exudes from water partially bound to the material that is being dried. The heating rate should decrease at this moment in order to keep the temperature of the product under 30°C–50°C, which will prevent the material from collapsing. If the solid part of the material is too hot, the structure collapses, resulting in a decrease of the ice sublimation rate in the product (Barbosa-Canovas and Vega-Mercado, 1996).

23.7.3 Simultaneous Heat and Mass Transfer

The mass and heat transfer phenomena during freeze-drying can be summarized in terms of diffusion of vapor from the sublimation front and heat radiation and conduction from the radiation slab. A steady-state model will be supposed in the development of this section in order to facilitate calculations.

The energy required to maintain sublimation is assumed to be equal to the radiant or conductive flow due to the temperature gradient between the frozen product and the heat source in the drying chamber. Water sublimates below the triple point under pressures of 627 Pa or less. The sublimation interface is located above the ice front, and the elimination of water takes place close to or at the sublimation interface. Figure 23.25 shows the heat and mass flows during drying of frozen slabs.

The heat flux due to convection or conduction at the sublimation surface for Figure 23.25a can be expressed as (Barbosa-Canovas and Vega-Mercado, 1996)

$$q = h(T_e - T_s) = k\frac{(T_s - T_f)}{(L_2 - L_1)} \tag{23.82}$$

where

q is the heat flux
h is the external coefficient of heat transfer by convection
T_e is the external temperature of vaporization of the gas
T_s is the temperature of the surface of the dry solid
T_f is the sublimation temperature of the ice front
k is the thermal conductivity of the dry solid
$(L_2 - L_1)$ is the thickness of the dry layer

Thermal conductivity and sublimation temperature values for some freeze-dried products are given in Table 23.6.

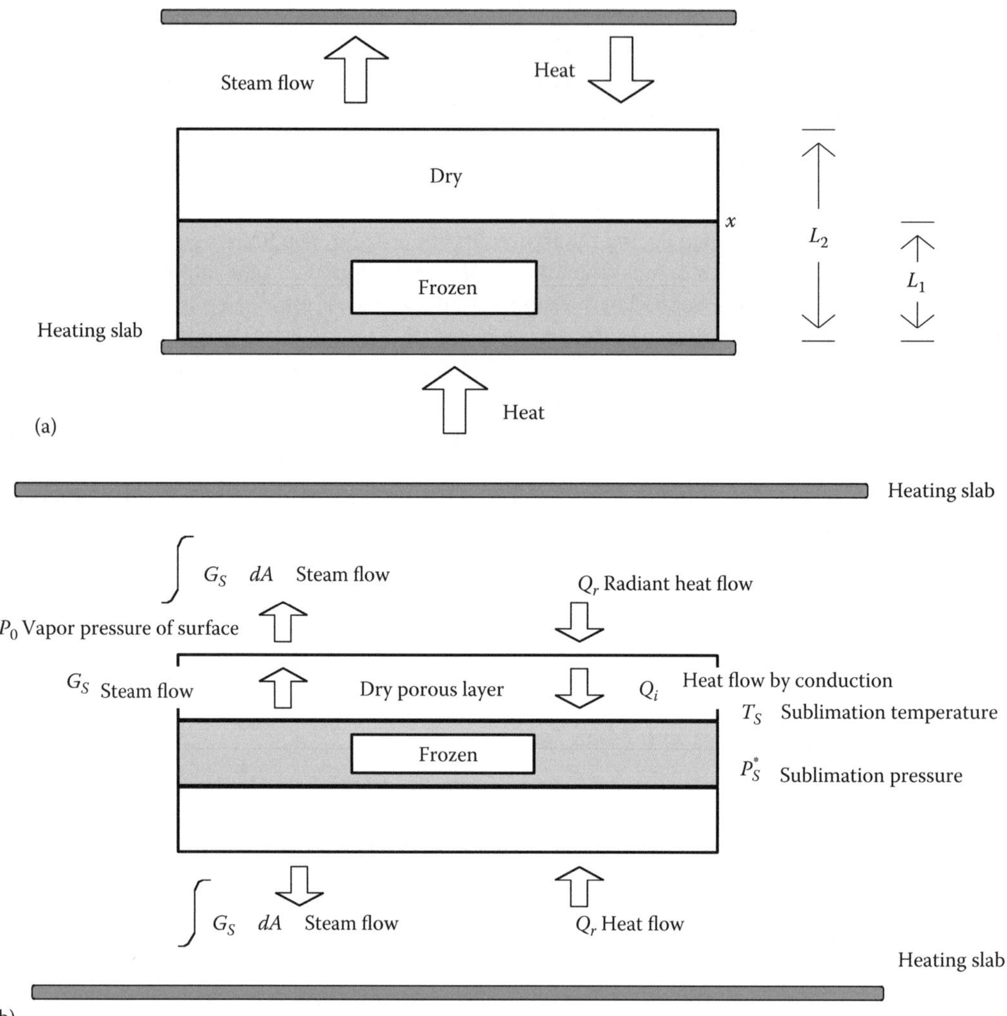

FIGURE 23.25 Heat and mass flows during freeze-drying: (a) drying on one side and (b) symmetric arrangement. Drying on both sides. (Adapted from Barbosa-Cánovas, G.V. and Vega-Mercado, H., *Dehydration of Food*, Chapman & Hall, New York, 1996.)

The flux of vapor from the sublimation front is given by Okos et al. (1992):

$$N_a = \frac{D'(P_{fw} - P_{sw})}{RT(L_2 - L_1)} = K_g(P_{sw} - P_{ew}) \tag{23.83}$$

where

N_a is the flux of vapor
D' is the mean effective diffusivity of vapor into the dry layer
R is the gas constant
T is the mean temperature of the dry layer
P_{fw} is the partial vapor pressure in equilibrium with the ice sublimation front
P_{sw} is the partial vapor pressure at the surface
P_{ew} is the partial vapor pressure within the external gas phase
K_g is the external mass transfer coefficient

TABLE 23.6
Thermal Conductivity and Sublimation Temperature of Freeze-Dried Products

Product	k (W/m K)	Product	T_{sub} (°C)
Coffee extract—25%	0.033	Coffee	−23
Gelatin	0.016	Shrimp	−18
Milk	0.022	Whole egg	−17
Apple	0.016–0.035	Apple	−7
Peach	0.016	Chicken	−21
Turkey	0.014	Salmon	−29
Mushrooms	0.010	Veal	−14
Veal	0.035–0.038	Carrot	−25

Source: Schwartzberg, H., *Freeze Drying—Lecture Notes*, Food Engineering Department, University of Massachusetts, Amherst, MA, 1982.

Equations 23.82 and 23.83 can be combined, expressing q and N_a in terms of the external operation conditions (Barbosa-Canovas and Vega-Mercado, 1996):

$$q = \frac{T_e - T_f}{((1/h) + ((L_2 - L_1)/k))} \tag{23.84}$$

and

$$N_a = \frac{(P_{fw} - P_{ew})}{(1/K_g) + (RT(L_2 - L_1)/D')} \tag{23.85}$$

where

h and K_g depend on the velocity of the gas and the dryer
k and D' depend on the nature of the dry material
T_e and P_{ew} are given by the operation conditions

These last two equations can be related through the latent heat of sublimation (ΔH_S) as

$$q = \Delta H_S N_a \tag{23.86}$$

Combination of Equations 23.82, 23.85, and 23.86 yields

$$\frac{k(T_s - T_f)}{(L_2 - L_1)} = \frac{\Delta H_s(P_{fw} - P_{ew})}{((1/K_g) + RT(L_2 - L_1)/D')} \tag{23.87}$$

or

$$h(T_e - T_s) = \frac{\Delta H_s(P_{fw} - P_{ew})}{((1/K_g) + RT(L_2 - L_1)/D')} \tag{23.88}$$

An increase of T_e or T_s causes an increase in the drying rate, as can be observed from Equations 23.87 and 23.88. The temperature T_S is limited by the sensitivity to heat of the material and T_f should be lower than the collapse temperature of the material. Sensibility is defined in terms of the degradation reactions, while the collapse temperature is defined in terms of deformation of the porous structure of the dry layer.

The drying rate can be expressed as

$$N_a = \left(\frac{L}{2M_aV_s}\right)\frac{-dx}{dt} \tag{23.89}$$

where

L is the total thickness of the product
x is the thickness of the dry layer
t is the time
M_a is the molecular weight of water
V_s is the volume of occupied solid per unit of mass of water (per kg of water), expressed as $V_S = 1/(Y_0\rho_s)$
Y_0 is the initial moisture content
ρ_s is the density of the dry solid

An analogous deduction was considered by Schwartzberg (1982) to describe the freeze-drying process, in which he considered the mass and heat transfer through both faces of the product (Figure 23.25b). The vapor flow in the system can be expressed as

$$G_s = \frac{K_P\left(P_s^*-P_0\right)}{x} \tag{23.90}$$

$$= \frac{K_p\left(P_s^*-P_c^*\right)}{x} \tag{23.91}$$

$$= \rho\frac{(Y_0-Y_f)}{1+Y_0}\frac{dx}{dt} \tag{23.92}$$

where

P_s^* is the vapor pressure in the sublimation interface
P_0 is the partial water pressure at the surface
P_c^* is the pressure in the condenser that should be equal to P_0, unless non-condensable components are introduced in the drier
K_P is the permeability of the dry layer
ρ is the density of the frozen layer of the slab
Y_0 is the initial moisture content of the food (mass of water/mass of dry solid)
Y_f is the final moisture content
x is the thickness of the dry layer
t is the drying time

The first part of Equation 23.92 represents the change in water content per unit of volume of frozen product. The change of thickness of the dry layer is a function of time, keeping constant the surface area. Table 23.7 presents values of permeability for some freeze-dried foods.

The drying time is obtained by integration of Equation 23.92, yielding

$$t_S = \rho\frac{(Y_0-Y_f)}{2K_p\left(1+Y_0\right)}\frac{a^2}{\left(P_s^*-P_0\right)} \tag{23.93}$$

where a is the thickness of half of the slab.

TABLE 23.7
Permeability of Freeze-Dried Foods

Product	Permeability (10^{-9} kg/m s μmHg)
Coffee 20% solids	4.0–8.6
Coffee 30% solids	3.0
Whole milk	2.7–5.3
Apple	3.3–6.0
Potato	1.3
Fish	8.7
Banana	1.1
Veal	0.7–4.4
Tomato 22°Brix	2.1
Carrot	2.0–5.6

Source: Schwartzberg, H., *Freeze Drying—Lecture Notes*, Food Engineering Department, University of Massachusetts, Amherst, MA, 1982.

The heat required for sublimation is supposed to be equal to the radiant energy and can be expressed as

$$Q_r = Q_i = G_s H_s \tag{23.94}$$

where

Q_i is the internal flow inside the slab
H_s is the mean latent heat of vapor (Barbosa-Canovas and Vega-Mercado, 1996)

The internal heat flow can be expressed as

$$Q_i = \frac{K_t(T_0 - T_s)}{x} \tag{23.95}$$

where

K_t is the thermal conductivity of the dry layer
T_0 is the temperature of the surface of the slab
T_s is the sublimation temperature

Substitution of Equation 23.92 by G_S in Equation 23.95 yields

$$\frac{K_t(T_0 - T_s)}{x} = \rho \frac{(Y_0 - Y_f)}{(1 + Y_0)} H_s \frac{dx}{dt} \tag{23.96}$$

Assuming that T_0, T_s, and K_t remain constant,

$$t_S = \rho \frac{(Y_0 - Y_f)}{(1 + Y_0)} H_s \frac{a^2}{2K_t(T_0 - T_s)} \tag{23.97}$$

The combination of mass and heat transfer relationships for drying time, t_S, yields a relation that is a function of the properties of the dry layer and the operation conditions during freeze-drying:

$$\left(P_s^*-P_0\right)=\left(\frac{-K_t}{K_pH_s}\right)(T_s-T_0) \tag{23.98}$$

where
P_s^* and T_S are considered as variables
T_0 and P_0 are fixed
K_t and K_p are independent of P_s^* and T_s (Schwartzberg, 1982)

The values of P^* can be calculated from the sublimation equation as follows:

$$\ln P^* = 30.9526 - \frac{6153.1}{T} \tag{23.99}$$

where T is the absolute temperature. Equations 23.98 and 23.99 can be used together to define the final operation conditions during the freeze-drying process.

Another very important variable is the surface temperature, T_0. The value of such variable is controlled by the heat transfer rate from the heating slab, whose temperature is T_p:

$$Q_r = F_{op}\sigma\left(T_p^4 - T_0^4\right) \tag{23.100}$$

In this equation σ is the Stefan–Boltzmann constant and F_{op} is the shape or vision factor, defined as

$$F_{op}=\frac{1}{((1/\varepsilon_o)+(1/\varepsilon_p)-1)} \tag{23.101}$$

where
ε_0 is the emissivity of the surface of the product
ε_p is the emissivity of the radiant slab

Assuming that $\varepsilon_0 \approx 1$, it is obtained that $F_{op} \approx \varepsilon_0$, and the combination of Equations 23.95, 23.97, and 23.100 yields

$$T_p^4 = T_0^4 + \frac{K_t(T_0 - T_s)}{((2K_t((T_0 - T_s)(1+Y_0)/H_s\rho(Y_0 - Y_f))t + x^2)^{0.5}} \tag{23.102}$$

23.8 OTHER TYPES OF DRYING

23.8.1 Osmotic Dehydration

The concentration of foods by immersion in a hypertonic solution is known as osmotic dehydration. Osmosis consists of molecular movement of certain components of a solution through a semipermeable membrane toward a less concentrated solution.

Water loss in food during osmotic dehydration can be divided into two periods (Barbosa-Canovas and Vega-Mercado, 1996): (1) a period that lasts about two hours with a high water removal rate and (2) a period, from two to six hours, with a falling rate of water removal rate. The temperature and the concentration of the osmotic solution affect the water loss rate of the product. Compared to air-drying or freeze-drying, osmotic dehydration is quicker since water elimination occurs without phase change.

The difference in chemical potential through the semipermeable membrane between the product and the osmotic solution is the driving force for mass transfer. The chemical potential μ_i is related to water activity according to the expression

$$\mu_i = \mu_i^o + RT \ln a_w \tag{23.103}$$

where

μ_i^o is the reference chemical potential
R is the gas constant
T is the absolute temperature

Mass transfer continues until the water activities of the osmotic solution and of the food are equal. The main mechanism through which mass transfer takes place is by diffusion due to the concentration gradient that exists between the food and the osmotic solution. The diffusion rate of water can be estimated by the modified Fick's law (Barbosa-Canovas and Vega-Mercado, 1996) and depends on the geometry of the product to be dried.

The temperature variable has a great effect on the osmotic dehydration process, since an increase of temperature intensifies the water removal and the penetration of the osmotic solution into the tissue. The water and solute content of the food is a function of time; Figure 23.26 shows this variation for fruit dehydration, in which a decrease in water content and an increase in sugar content can be observed over time.

The selection of the solute of the osmotic solution is also very important, and three main factors should be taken into account: (1) the sensory characteristics of the product, (2) the cost of the solute, and (3) the molecular weight of the solute. In general, the solutes mostly used in osmotic dehydration processes are sodium chloride, sucrose, lactose, fructose, and glycerol. Table 23.8 presents the uses and advantages of some osmotic solutes.

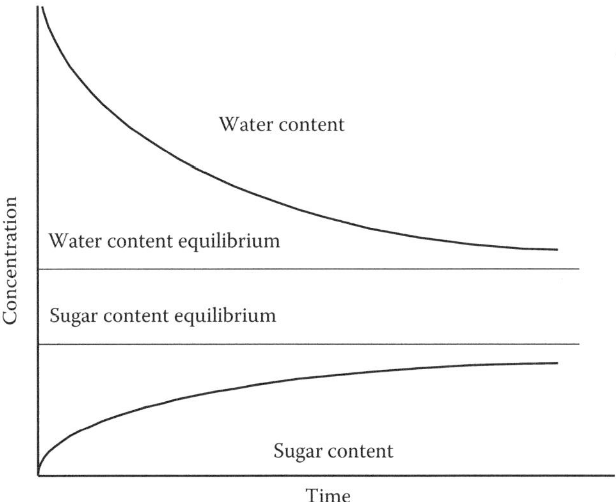

FIGURE 23.26 Water and sugar contents during osmotic dehydration. (Adapted from Barbosa-Cánovas, G.V. and Vega-Mercado, H., *Dehydration of Food*, Chapman & Hall, New York, 1996.)

TABLE 23.8
Uses and Advantages of Some Osmotic Solutes

Name	Uses	Advantages
Sodium chloride	Meats and vegetables Solutions higher than 10%	High a_w depression capacity
Sucrose	Fruits	Reduces browning and increases retention of volatiles
Lactose	Fruits	Partial substitution of sucrose
Glycerol	Fruits and vegetables	Improves texture
Combination	Fruits, vegetables, and meats	Adjusted sensorial characteristics; combines high a_w depression capacity of salts with high capacity of water elimination of sugar

Source: Barbosa-Cánovas, G.V. and Vega-Mercado, H., *Dehydration of Food*, Chapman & Hall, New York, 1996.

23.8.2 Solar Drying

The practice of drying harvested foods by scattering them in thin layers exposed to the sun is called open solar drying or natural solar drying. This technique is used to process grapes, figs, plums, coffee beans, cocoa beans, sweet peppers, pepper, and rice, among others. This type of drying has some limitations, such as

- Lack of control in the drying process, which can lead to excessive drying of the food, and loss of grains or beans due to germination and nutritional changes
- Lack of uniformity in drying
- Contamination due to molds, bacteria, rodents, birds, and insects

For this reason, solar dryers have been developed based on the use of energy coming from the sun and also the use of hot air to dry the food. Thus, natural convection solar dryers exist that do not require any type of mechanical or electric energy, and there also exist forced convection dryers that require the use of fans to blow hot air.

23.8.3 Drum Dryers

These dryers consist of hollow metal cylinders that rotate on their horizontal axis and are internally heated with steam, hot water, or another heating medium. Drum dryers are used to dry pastes and solutions. Potato flakes are obtained using these dryers. Figure 23.27 shows different types of drum dryers.

The global drying rate of the food film placed on the drum surface can be expressed as (Heldman and Singh, 1981)

$$\frac{dX}{dt} = \frac{UA\Delta T_m}{\lambda} \tag{23.104}$$

where

ΔT_m is the mean logarithmic difference between the drum surface and the product
U is the global heat transfer coefficient
A is the area of the drying surface
λ is the latent heat of vaporization at the temperature of the drying surface
X is the moisture content

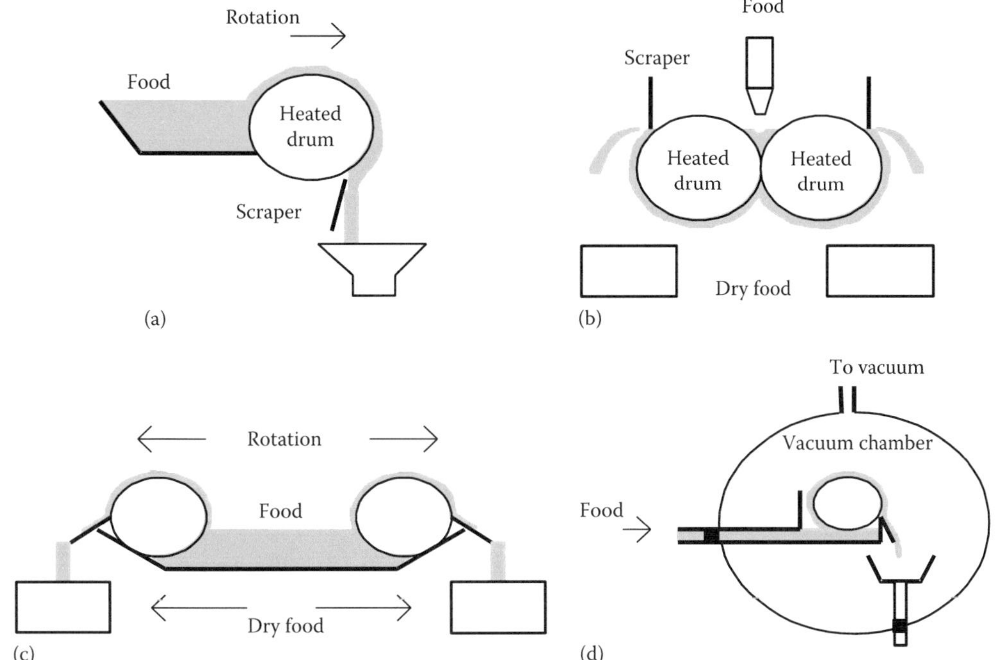

FIGURE 23.27 Drum dryers. (a) Simple dryer, (b) double dryer, (c) twin drums, and (d) vacuum drum. (Adapted from Barbosa-Cánovas, G.V. and Vega-Mercado, H., *Dehydration of Food*, Chapman & Hall, New York, 1996.)

23.8.4 Microwave Drying

Microwaves are high-frequency waves. The advantages of microwave heating over convection or conduction heating include

- Only the product to be heated absorbs energy
- No losses due to heating of the surrounding media (air and walls)
- Deep penetration of the heating source resulting in more uniform and effective heating

Microwaves are used in the food industry for the following products: drying of potato chips, blanching of vegetables, quick thawing of frozen fish, precooking of chicken and bacon, and elimination of molds in dry fruits and milk products.

23.8.5 Fluidized Bed Dryers

The particles that form a bed can be fluidized if the pressure that is dropped through the bed is equal to the weight of the bed, reaching the expansion and suspension of the particles into the air stream. The system behaves as a fluid when its Froude's number is smaller than one (Karel et al., 1975a,b), in which the air velocity is generally within the 0.005–0.075 m/s range.

In fluidized beds, the particles do not present contact points between them, which facilitates a more uniform drying.

PROBLEMS

23.1 It is desired to dry a solid that contains 0.075 kg of water/kg of dry solid down to a moisture content of 0.005 kg of water/kg of dry solid. Thus, 645 kg/h of a solid is fed to a dryer that is completely thermally insulated, in which air flows under countercurrent. Air is introduced at 100°C and at a humidity of 0.010 kg of water/kg of dry air and leaves the dryer at 45°C.

If the solids are introduced at 25°C and leave the dryer at 70°C, calculate the air flow needed to carry out this drying operation. The specific heat of the solids is 1.465 kJ/kg K.

The type of dryer used is continuous with recirculation, such as the one shown in Figure 23.2. If the dryer is thermally insulated, it can be assumed that there are no heat intakes or losses to the exterior. This means

$$\dot{Q}_E = \dot{Q}_S = 0$$

The dry solids flow rate entering into the dryer is

$$w_s = \frac{645}{(1+0.075)} = 600 \text{ kg dry solid/h}$$

The enthalpies of the air streams are obtained from Equation 23.5:

$$\hat{i}_E = (1+1.92\times 0.010)(100-0) + 2490\times 0.01 = 126.82 \text{ kJ/kg dry air}$$

$$i_S = (1+1.92X_S)(45-0) + 2490X_S = 4 + 2576.4X_S \text{ kJ/kg dry air}$$

The enthalpies of the solids are obtained from Equation 23.6:

$$\hat{h}_E = 1.465(25-0) + 0.06\times 4.185(25-0) = 44.47 \text{ kJ/kg dry solid}$$

$$\hat{h}_S = 1.465(70-0) + 0.005\times 4.185(70-0) = 104.02 \text{ kJ/kg dry solid}$$

Substitution in Equation 23.4 yields

$$w'126.82 + (600)(44.47) = w'(45 + 2576.4X_S) + (600)(104.2)$$

It is obtained from the mass balance (Equation 23.3) that

$$600(0.075-0.005) = w'(X_S - 0.01)$$

When solving the last two equations, it is obtained that the humidity of the air that leaves the dryer is $X_S = 0.0264$ kg of water/kg dry air, while the flow of dry air that should be introduced into the dryer is $w' = 2567$ kg dry air/h.

23.2 A 60 × 60 cm and 3 cm depth tray contains a granular wet product that is desired to be dried with an air stream. The air stream is hot and provides enough heat to dry the product by a convection mechanism. The air at 65°C flows at a rate of 5 m/s, its humidity being 0.02 kg of water/kg of dry air. If it is considered that the sides and bottom of the tray are completely insulated, determine the constant drying rate.

The following properties are obtained for air at 65°C and absolute humidity of 0.02 kg of water/kg of dry air from the psychrometric diagram:

Temperature $T_w = 32.5$°C $X_w = 0.034$ kg of water/kg of dry air

The wet volume is obtained from Equation 5.8:

$$V_H = \left(\frac{1}{28.9} + \frac{0.02}{18}\right)\frac{(0.082)(273+65)}{1} = 0.9898 \text{ m}^3\text{/kg dry air}$$

so the density of humid air is

$$\rho = \frac{1+0.02}{0.9898} = 1.031 \text{ kg/m}^3$$

The mass flux is

$$G = \rho v = (1.031)(6) = 6.186 \text{ (kg/m}^2\text{s)}$$

The convective heat transfer coefficient can be calculated from Equation 23.20a:

$$h = 14.28(6.186)^{0.8} = 61.35 \text{ W/(m}^2\text{°C)}$$

For $T_W = 32.5$ °C, the latent heat is $\lambda_W = 2423.4$ kJ/kg.

The drying rate can be calculated from Equation 23.17:

$$R = \frac{h(T - T_W)}{\lambda_W} = \frac{(61.35)(3600)(65-32.5)}{(2423.4)(1000)} = 2.96(\text{kg/h m}^2)$$

Since the surface being dried has an area of 0.36 m^2, the total evaporation rate is 1.066 kg water/h.

23.3 A porous solid with a critical moisture content of 0.22 kg of water/kg dry solid is subjected to a drying process to reduce its moisture from 0.22 to 0.15 kg of water/kg solid in 4 h. The thickness of the solid is 6 cm and the drying process only takes place on one of the faces of the solid. Calculate the drying time required by a solid with similar characteristics, but with a thickness of 8 cm and with the drying process taking place simultaneously on the two faces exposed to the air stream.

Initially, the value of the effective diffusivity through the solid should be calculated. The solution to Fick's equation given in Equation 23.24a can be used for a solid with such geometry. It is considered that $Y_S = 0$ and that all the water that reaches the surface evaporates immediately:

$$\Gamma = \frac{Y}{Y_o} = \frac{8}{\pi^2} e^{\left(-h_n^2 \pi^2 D_{effect} t / 4L^2\right)}$$

where $h_n = 1$, $Y = 0.15$; $Y_0 = Y_C = 0.22$; $L = 0.06$ m; and $t = 14{,}400$ s. The effective diffusivity can be obtained by substituting these data in the last equation:

$$D_{effect} = 1.75 \times 10^{-8} \text{ m}^2/\text{s}$$

For the second part of the problem, since the type of product is the same, the value obtained for diffusivity can be employed as well as the expression given before. However, it should be taken into account that drying takes place on both faces of the slab; therefore, the thickness should be half of its value, that is, $L = 0.04$ m. Calculations using the new thickness and the value of the effective diffusivity obtained previously yield a drying time of 6419 s or 1.78 h.

23.4 A food product is obtained in the form of spherical particles of 15 mm diameter and moisture content of 1.5 kg water/kg dry solid. The equilibrium moisture content is 0.01 kg water/kg dry solid. In order to reduce the moisture content of this product to 0.2 kg water/kg dry solid, it is placed in a dryer on a porous mesh forming a bed 5 cm thick. The apparent density of the bed is 560 kg/m^3, while the dry solids have a density of 1400 kg/m^3. Air at a velocity of 0.8 m/s is circulated through the bed. Air at the inlet is 120°C with a moisture content of 0.05 kg water/kg dry air. If the critical moisture content is 0.5 kg water/kg dry solids, calculate the total drying time.

Free moisture content

$$\text{Initial: } Y_1 = 1.5 - 0.01 = 1.59 \text{ kg water/kg dry solid}$$

$$\text{Final: } Y = 0.02 - 0.01 = 0.01 \text{ kg water/kg dry solid}$$

$$\text{Critical: } Y_C = 0.5 - 0.01 = 0.49 \text{ kg water/kg dry solid}$$

Conditions of air at the inlet (from the psychrometric chart)

$$T_1 = 120°\text{C} \quad X_1 = 0.05 \text{ kg water/kg dry air}$$

$$T_W = 49°\text{C} \quad X_W = 0.083 \text{ kg water/kg dry air} \quad \lambda_W = 2382 \text{ kJ/kg}$$

Humid air volume (Equation 5.8)

$$V_H = \left(\frac{1}{28.9} + \frac{X}{18}\right)\frac{RT}{P} = \left(\frac{1}{28.9} + \frac{0.05}{18}\right)\frac{(0.082)(273+120)}{1}$$

$$= 1.205 \text{ m}^3/\text{kg dry air}$$

Density of the humid air entering

$$\rho = \frac{(1+0.05)}{1.205}((\text{kg/kg dry air})/(\text{m}^3/\text{kg dry air})) = 0.872 \text{ (kg dry air + water)/m}^3$$

The air flux is

$$G = v\rho = 0.8 \text{ (m/s)} 0.872 \text{ (kg dry air + water/m}^3\text{)(1 kg dry air/1.05 kg dry air + water)}$$

$$G = 0.6644 \text{ (kg dry air/s} \cdot \text{m}^2) = 2392 \text{ (kg dry air/h} \cdot \text{m}^2)$$

The air that circulates through the bed gains water, and this content is greater at the outlet than at the inlet. A mean moisture content value is estimated, which can be 0.07 kg water/kg dry air, so the mean mass flow density for air is

$$G_t = G(1+0.07) = 2559 \text{ (kg dry air + water/h} \cdot \text{m}^2)$$

Also, the humid heat is calculated with this assumed mean moisture:

$$\hat{s} = 1 + (1.92)(0.07) = 1.1344 \text{ kJ/(kg dry air°C)}$$

The hollow fraction should be calculated from the density of the particles and the apparent density of the bed:

$$\varepsilon = 1 - \frac{\rho_a}{\rho_S} = 1 - \frac{560}{1400} = 1 - 0.4 = 0.6$$

The specific surface of the bed is calculated by Equation 23.40:

$$a = 6\frac{(1-0.6)}{0.015} = 160\ (\text{m}^2/\text{m}^3)$$

In order to estimate the viscosity of the air, a mean temperature of 93°C is considered, yielding a value of $\eta = 2.15 \times 10^{-5}$ Pa·s

The Reynolds number is

$$(\text{Re}) = \frac{G_t D_P}{\eta} = \frac{((2559/3600)\ (\text{kg/s}\cdot\text{m}^2))(0.015\ \text{m})}{2.15\times10^{-5}\ \text{Pa}\cdot\text{s}} = 496$$

The convective heat transfer coefficient is calculated from Equation 23.49:

$$h = 0.151\frac{(2559)^{0.59}}{(0.015)^{0.41}} = 86.6\ \text{W/(m}^2\cdot{}^\circ\text{C)}$$

The drying time is obtained from Equations 23.47 and 23.48:

- Constant rate period: $t_C = 1.44$ h
- Falling rate period: $t_D = 0.66$ h

So, the total drying time is $t = 2.1$ h.

23.5 A 2 cm thick veal slab is dried using a freeze-drying process. Initially, the product has a moisture content of 75%, and it is desired to dry it until it reaches 5% moisture content. The initial density of the veal is 1050 kg/m³. If the sublimation pressure is kept at 260 μmHg, and a pressure of 100 μmHg is maintained in the condenser, calculate the drying time. Suppose that $K_P = 0.75 \times 10^{-9}$ kg/(m/s μmHg).

According to the problem statement,

$$a = 0.01\ m \qquad P_0 = P_0^*$$

$$Y_0 = 0.75/0.25 = 3 \qquad Y_f = 0.05/0.95 = 0.0526$$

Using Equation 23.93

$$t_S = \rho\frac{(Y_0 - Y_f)}{2K_p(1+Y_0)}\frac{a^2}{(P_s^* - P_0)} = 1050\frac{(3-0.0526)}{2(0.75\times10^{-9})\ (1+3)}\frac{(0.01)^2}{(260-100)}$$

Hence, the drying time is

$$t_S = 322372\ s = 89.55\ \text{h}.$$

PROPOSED PROBLEMS

23.1 In a dryer, 300 kg/h of air, at 60°C and 70% relative humidity, is recycled, and it is mixed with 100 kg/h of air at 25°C 45% relative humidity, with the purpose of obtaining an air current for drying a food, previous to a heating stage. Determine the humidity and temperature of the air mixture.

23.2 In a dryer, 1000 kg/h is processed from a humid solid to ambient temperature in order to reduce its humidity from 9% to 1%. An air stream of 2320 kg/h at 80°C, obtained starting from ambient air, is used. The external air is at 20°C and contains 0005 kg water/kg dried air. In the dryer, 100 kW is supplied. Calculate the following: (a) the removed water flow rate;

(b) the temperature of air and solid at the dryer exit, assuming that the solid and the air leave from the dryer at the same temperature; and (c) the bulb temperatures for external air, for the air that is feeding to the dryer and for air that is leaving from the dryer.

Data. Water vaporization latent heat at 0°C = 2490 kJ/kg. Specific heat for liquid water = 4.18 kJ/(kg·°C). Specific heat for dried solid = 2.3 kJ/(kg·°C).

23.3 A granular spherical particle product of 7.5 mm contains 1.5 kg of water for each kg of dry solid. In laboratory experiments, it has been determined that the equilibrium moisture is of 0.01 kg water/kg dry solid. The particles are placed on a tray whose bottom is a porous mesh forming a layer of 5 cm of depth, and then they are introduced into a dryer in order to reduce its moisture until a final content 0.3 kg water/kg dry solid is obtained. Air is flowing through the porous mesh and the particle bed at 1 m/s and at 125°C whose moisture is 0.05 kg of water/kg dry air. The solid particles possess a density of 1500 kg/m^3, while the bed possesses an apparent density 560 kg/m^3. Calculate the drying time if the critical moisture is 0.4 kg water/kg dry solid.

23.4 It is wished to reduce the water content of a piece of meat that is 3 cm thick that contains 76% water. A freeze-dryer is used, where the sublimation pressure is 35 Pa, while in the condenser the pressure is 13 Pa. The density of the meat piece is 1040 kg/m^3, and it can be assumed that $K_P = 5.62 \times 10^{-9}$ kg/(m·s·Pa). Calculate the required time so that the meat moisture is 6%.

24 Hygienic Design of Food Processes

24.1 INTRODUCTION

The presence of impurities in foods can cause problems not only in the sensorial quality of the product but also in terms of food safety. In the elaboration process of a given food, it is essential that hygienic conditions are consistently maintained in order to obtain an appropriate final product and, above all, to ensure that it offers proven safety for consumers' health.

Impurities in foods can be various and of different origins. An impurity can even be a component that has been obtained in a certain step of the process, which should not be found in the final product. The residuals of these components that are deposited on the equipment that are part of the different stages of the process can contaminate a product that is being developed in a new operation. It is necessary to eliminate the soil deposited in the equipment before proceeding to a new operation, since not only is there a potential for polluting the new product, but it can also be a substrate for microbial growth, such as pathogen and spoilage microorganisms. The same food products, when they arrive at the factory for processing, may contain soil that must be eliminated before entering the processing line. Thus, fruits may contain leaves, dirt, or pesticides that are necessary to remove by means of a washing step.

The biggest hygienic problem while processing foods is due to microbial contamination. Deposits of food material that exist in the processing line serve as a culture broth for the growth of microorganisms, and it is necessary to remove these deposits. The deposited impurities constitute what could be denominated *dirty equipment* used in processing. The soil forms layers of dirt that contain different types of components. Depending on the type of component contained in the soil layers, the elimination will be more or less difficult. Generally, soil layers that contain sugar can cause caramelization problems in thermal treatment processes; because the sugars are soluble in water, they can be removed easily. In the case of fat, heat can cause polymerization, and as fat is not soluble in water, it is difficult to remove; however, alkalis and acids are somewhat soluble, and in the presence of surfactant agents, they can be eliminated. If the soil layer contains proteins, these can be denaturalized with heat; in addition, they are not very soluble in water, so they are difficult to eliminate, although they are soluble in alkalis and their elimination is easier. The mineral salts of monovalent cations are soluble as much in water as in an acid medium, so there is no problem in their elimination; but in the case of mineral salts of polyvalent cations, if the solubility product is exceeded, they can precipitate, since they are insoluble in water, which makes their elimination difficult, although they are soluble in acid, which facilitates their elimination.

Besides the elimination of soil, it is indispensable that the processing line is exempt of pathogens and damaging microorganisms. To do so, it will be necessary to use cleaning products in order to remove the soil and it will be necessary to use disinfectants that will be removed later on. When the equipment surface is free from soil, microorganisms, as well as cleaning and sanitizer agents, it is necessary to carry out an antidirt treatment on the clean surface with the purpose of inhibiting the absorption of components during the new processing cycle.

24.2 DISINFECTION AND CLEANING KINETICS

The soil deposited on the surface of different processing devices is due to adherence forces, depending on the individual nature of the materials as well as the surface ruggedness. Thus, on flat and polished surfaces, it is more difficult for soil to adhere, and it is also easier to remove it. Microorganisms can grow better on the deposited soil and in the pores of the material. The action on adherence by cleaning agents will depend on these factors.

The amount of soil that can be removed by means of solutions containing cleaning agents can be considered to be proportional to the grade of soil deposited on the surface. If a mass balance for the removed soil is carried out, there are only the accumulation and exit terms; thus, this can be mathematically expressed according to the following equation:

$$\frac{dm}{dt} = -N_S A \qquad (24.1)$$

where

m is the amount of removed soil
N_S is the mass flux of eliminated material
A is the contact area between the soil and the cleaning solution
t is the time

The mass flux of the material that has come off the soil layer is transferred from the surface to the cleaning solution bulk. Considering that the soil amount that exists in the cleaning solution bulk is negligible, the mass flux N_S can be expressed as

$$N_S = k(m - 0) \qquad (24.2)$$

where k is a mass transfer constant that depends on the soil type, the type and concentration of the cleaning solution, the temperature, and the agitation.

Combining Equations 24.1 and 24.2,

$$\frac{dm}{dt} = -Akm \qquad (24.3)$$

This is a differential equation in a detachable variable that can be integrated with the condition limit that for the initial time ($t = 0$), the soil amount is m_0. The integration leads to the expression

$$\ln\left(\frac{m}{m_0}\right) = -kAt \qquad (24.4)$$

If $\ln(m)$ is plotted against the time t (Figure 24.1), a straight line is obtained, whose slope is $-(kA)$ and the intercept $\ln(m_0)$. The value of the constant k gives an idea of the ease with which the soil is removed, so the greater this value, the greater the slope of the straight line of Equation 24.4, which indicates that the soil can be eliminated with a lower treatment time. This equation indicates that the speed of soil elimination in the cleaning process can be described by means of a logarithmic-type variation (Jennings, 1965). This model assumes that the cleaning kinetics follows a first-order model, which indicates that to achieve a total cleaning an infinite time would be needed, leading to the result that from the practical point of view a final quantity of permissible soil that remains on the surface of the processing devices must be defined.

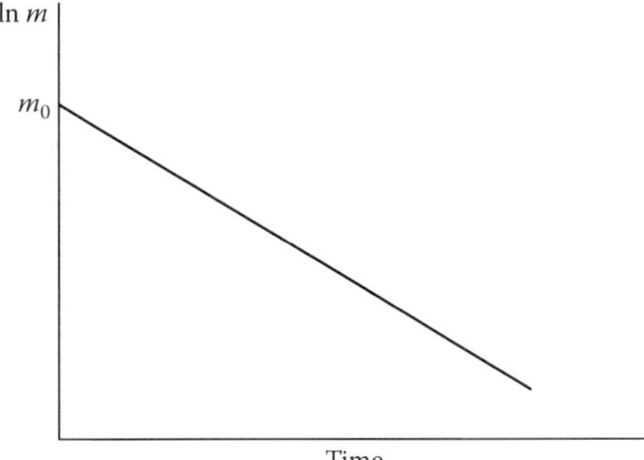

FIGURE 24.1 Ideal evolution of soil elimination.

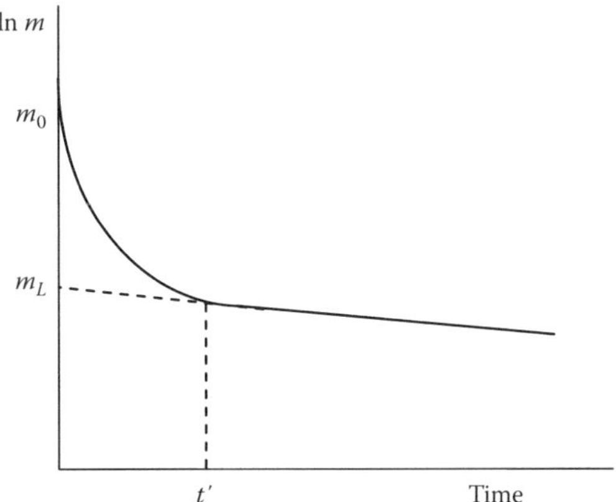

FIGURE 24.2 Non-ideal evolution of soil elimination.

There are some cases in which carrying out this representation a straight line is not obtained, but rather initially there is a pronounced fall in the value of the soil amount, as shown in Figure 24.2, where it is observed that from a given time (t'), the evolution of ln(m) with time is lineal, so the continuation of this line intercepts to the y axis for a value m_L, lower than the initial soil m_0. The curved part between the initial time and t' corresponds to a soil decrease from m_0 to m_L. This stretch would belong to a soil that is easily removable so that it seems that there is a part of soil that is not strongly bound to the equipment surface. The amount that can be more easily cleaned is given by the difference ($m_0 - m_L$), so as the time t' decreases, it is more easily removed. This amount that can be easily removed depends on the temperature at which the cleaning solution acts, so when increasing the temperature the soil amount removed in this initial stretch also increases, and the time t' is small.

Besides this first-order model, other models have also been used to describe soil elimination kinetics with cleaning agents, one of which assumes the following mechanism (Plett, 1992):

$$\text{Soil} + \text{Cleaning agent} \underset{k_d}{\overset{k_1}{\rightleftarrows}} \text{Complex} \xrightarrow{k_2} \text{Suspension}$$

In this kinetic model, it is assumed that there is a first reversible step where the soil and the cleaning agent form a complex that can revert to the initial components. The formed complex can become part of the cleaning solution in an irreversible step.

In the same way that it is important to remove the soil in the processing devices, the elimination of the microbial load deposited on the equipment surface is also of supreme importance. In the elimination of soil, cleaning solutions are used that contain agents that facilitate this cleaning.

Also, in the elimination of the microbial load, chemical agents are used to destroy microorganisms. The chemical elimination rate of microorganisms is considered to be proportional to the number of those that occur in a certain process. This can be mathematically expressed as

$$\frac{dC}{dt} = -kC \tag{24.5}$$

where
- C is the concentration of microorganisms by unit of soil mass
- k is the microbial destruction kinetic constant
- t is the time

The constant k depends on the type of microorganism, the type of chemical agent and its concentration, the temperature, and the composition of the means that supports the microbial load. An increase of the temperature treatment makes the value of the constant greater, and thus the destruction of the microorganisms will be easier.

This equation can be integrated with the boundary condition that for the initial time ($t = 0$), the microbial load is C_0. The integration leads to the expression

$$\ln\left(\frac{C}{C_0}\right) = -kt \tag{24.6}$$

The necessary time to destroy the microorganisms depends on the concentration of the chemical agent that is used, so when increasing this concentration the time decreases. The relationship between the concentration and the destruction time of the microbial load is of the following type:

$$t = \frac{k'}{\left(C_B\right)^n} \tag{24.7}$$

where
- k' is a proportionality constant
- C_B is the germicidal agent concentration, with n being a potential parameter

When comparing these last two equations, it is observed that the constant of microbial chemical destruction k is proportional to $(C_B)^n$. The value of the parameter n depends on the germicidal agent type used as the microorganism.

24.3 DISINFECTION AND CLEANING PRODUCTS

For cleaning and disinfection agents, a great diversity of products are used, including acids and alkalis, which are products that can react with the soil, although surfactants are probably the products that are used the most in the cleaning of processing equipment. Next, a description of some of the products used as agents of cleaning, such as disinfectants, will be discussed.

24.3.1 Cleaning Agents

The chemical agents used in the elimination of soil are known as detergents. The substances that possess this detergent property vary from alkalis and simple acids to molecules of more complicated structures such as surfactants.

Surfactants are substances that influence the surface between two phases in contact by means of the surface tension change. Such products are able to modify the properties of the surface or interface aqueous solutions. They are amphiphilic molecules that are made up of a hydrophobic nonpolar part and one or several polar hydrophilic groups. According to the load that possesses the part that modifies the superficial property, they are anionic, cationic, or amphoteric.

Surfactants can be classified as ionic or nonionic. The polar part of the surfactant makes it soluble in water, while the nonpolar part makes it soluble in fat. Anionic surfactants can be made into soaps that contain alkaline salts of fatty acids; they are classified as hard soaps if sodium salts of fatty acids are used and as soft soaps if potassium salts of fatty acids are used. Due to the limitations of soaps in hard water, the synthetic detergents that can be used are composed of sodium salts of sulfates or alkyl benzene sulfonic acids. The detergents act similarly to soaps, although they do not have problems with calcium and magnesium cations, since the alkyl sulfates and the alkyl sulfonates of these cations are soluble in water. Cationic surfactants are not usually used as detergents, although they can be used because of the germicidal power that some of them possess. Ethoxylated amines are the result of the fixation of one or two molecules of ethylene oxide on primary amines, and their main property is that they are metal corrosion inhibitors. Quaternary ammonium derivatives can present germicidal properties, as is the case of dodecyl-dimethyl-benzyl ammonium chloride and (p-(diisobutyl)phenoxy)ethoxy)ethyldimethylbenzylammonium chloride. These compounds can destroy microorganisms because of their more intense germicidal effect in an alkaline medium.

Detergents in aqueous solution act in such a way that the polar part bonds to the water, while the nonpolar part will bind to the soil, causing formation of micelles. It has been shown that detergent action is necessary for the critical concentration of micelle formation that favors the elimination of soil. The decrease of tension in the interface favors the formation of emulsions that facilitate the soil elimination.

In relation to the detergent power of alkalis, it is necessary to mention that caustic soda gives very good results, although it can produce corrosion in some types of surfaces. Caustic soda acts directly on the soil, reacting with impurities, and it also presents an additional action because some formed products possess surfactant properties. Also, caustic soda has germicidal properties that are increased if it acts when it is hot.

For stainless steel equipment, cleaning diluted solutions of nitric acid are used. Organic acids (citric, lactic, sulfanilic, gluconic, etc.) possess a lower cleaning capacity than inorganic acids, although some of them can be used together with quaternary ammonium products for the purpose of obtaining a germicidal action.

Other substances used as cleaning agents are products with chelating action that form complexes with soil components, avoiding their deposit on the equipment surfaces. Among the chelating agents, polyphosphates prevent the precipitation of calcium and magnesium carbonates in hard waters. Ethylenediaminetetraacetic acid (EDTA) and their salts are also used for their complex power on calcium and magnesium salts.

Other components used in the cleaning of processing equipment are corrosion inhibitors, such as sodium silicate or sodium sulfite, antifoam agents, and enzymes such as proteases.

24.3.2 Sanitizers

For the elimination of microbial load in food processing lines, heat treatments or disinfectant chemical agents can be used. In heat treatments, water is usually used in the form of steam or hot liquid that in many cases is combined with disinfectant compounds in different stages of washing. It has already been mentioned that some detergent products also possess germicidal capacity; however, in the following discussion, the disinfectants that are more used in the food industry are enumerated.

Strong alkalis and mineral acids are agents that are used for their germicidal effect, besides being able to inactivate spores. Quaternary ammonium derivatives are cationic surfactants that are used for germicidal purposes, as in the case of alkyl dimethyl benzyl ammonium chloride or cetyl trimethyl ammonium bromide. Other compounds used as germicides are those that produce active chlorine, such as sodium hypochlorite and sodium phosphate hypochlorite, as well as organic compounds that liberate chlorine such as sodium dichloroisocyanurate. The germicidal action of chlorine is influenced by the presence of organic matter. Thus, bacteria can be destroyed with concentrations of active chlorine of the order of 5 ppm; however, in the presence of milk residues, it is necessary to use concentrations of active chlorine at least 10 times greater. The action of chlorine in high concentrations can cause the corrosion of stainless steel and other metals. Iodine can also be used as a disinfectant, although its germicidal power is lower than that of chlorine; however, in the presence of organic residues, its disinfectant action can be greater than that of chlorine.

Ethylene oxide has been used as a sterilizer in spices, although it can also be used as a disinfectant in processing facilities. Due to its high inflammability, it is usually blended with carbon dioxide. Formaldehyde, blended with a surfactant compound, is used in the disinfection of brewer facilities.

Hydrogen peroxide has been used as a milk preservative, and it can be used in the disinfection of surfaces, although because its germicidal power is low, it should be used in solutions in a high concentration. Ozone is another compound used in disinfection by its oxidative power. Traditionally, in the oenology industry, sulfurous anhydride has been used. It provides germicidal action in an acid medium at SO_2 concentrations greater than 200 ppm, although it presents bacteriostatic action on aerobic microorganisms. The β-propiolactone in a gas phase acts as a germicidal, showing greater effects than formaldehyde or ethylene oxide, although in the aqueous phase it is not effective as a disinfectant.

24.4 DISINFECTION AND CLEANING PROCESSES

In the cleaning and disinfection operations of food processing plants, soil that acts as the microbial load of the different devices that compose the processing stages is eliminated. Generally, these processes are composed of three successive steps: soil cleaning, destruction of microorganisms, and washing of the different types of equipment and devices. On some occasions, the steps of soil elimination and disinfection are carried out in a single step, depending on the type of chemical agent used.

The *cleaning process* usually begins with a washing step with water. Later, detergent solutions adapted to eliminate the soil are used, following a washing stage that carries away the detergent solution containing the eliminated soil. Next, the destruction of microorganisms is carried out by means of germicidal agents or using a heat treatment, and finally the remains of the different chemical agents used in the previous stages are removed with sterile water.

These cleaning techniques can be used in an open or closed path. Thus, in the case of external equipment surfaces and big storage and transport tanks, external washing techniques are used. For pipes, tanks, and different processing devices, *cleaning-in-place* (CIP) techniques are usually used in those in which the cleaning and disinfection steps are carried out by means of pumping of the pertinent solutions through the processing equipment.

In the equipment cleaning step, detergent solutions are used, but at the same time a mechanical action that favors soil elimination is carried out. As previously mentioned, the chemical agents used in the cleaning are acids, alkalis, surfactants, phosphate alkaline salts, and chelating agents, among others. As for the mechanical action that facilitates the cleaning and better penetration of cleaning agents, brushes, jets of cleaning solution, or circulation of these solutions through the processing system can be used. The action of brushes is usually carried out manually when tanks are cleaned and pipes are disassembled. The effectiveness of the manual use of brushes depends on the individual using them, and for that reason jets are usually directed toward the points desired to be cleaned. The cleaning solution jets can be carried out in a manual or automatic way, and high pressures are used, greater than 1 MPa. In these jets, the cleaning solution is sometimes hot, which can improve the cleaning process, but it can also cause the appearance of vapors. Another form of operation in the cleaning step is to circulate the solutions through the pipes and equipment. For greater effectiveness, it is recommended that the solution circulate in a turbulent regime, since this facilitates soil extraction and the removal of the solution containing the soil. It is not enough that the solution contacts the soil, but rather it should possess a turbulent motion to improve the extraction, mainly in compromised installation including fissures, gaskets, faucets, or dead points.

The *disinfection step* requires a good cleaning of the installation. On some occasions, the use of certain cleaning agents also acts as a disinfectant, as in the case of caustic soda and mineral acids in hot solution, in which case a second disinfection step would not be necessary. However, in most cases, after the cleaning stage, a disinfection step is applied. Disinfection can be carried out by means of thermal or chemical destruction of the microorganism. Thermal destruction can be carried out by means of water steam jets, with hot water or pressurized vapor. For this method, hot water is circulated through the installation, assuring a temperature higher than 90°C and that the contact time is enough to assure microbial destruction. Vapor jets are usually not effective, since the temperature that reaches the surface that contains the microorganisms is low, and it is not enough to achieve microbial reduction to the desired levels. On the contrary, pressurized vapor treatment can achieve the total elimination of the microorganisms, enough treatment time is allowed, and condensed water and noncondensable gases are completely drained off.

When disinfection chemical agents are used, it is convenient to carry out a good choice of germicidal agent. If the food products that will be processed have a pH lower than 4.5, quaternary ammonium products can be used, since the sporulated microorganisms do not proliferate. For pH greater than 4.5, spores of pathogen microorganisms such as *Clostridium* and *Bacillus* can be developed, and in these cases active chlorine can act in an effective way in the disinfection of equipment. However, because active chlorine can react with organic matter, it is indispensable that for good disinfection the soil is previously eliminated. With iodine solutions something similar can happen, and the use of halogen compounds can also be associated with the danger of corrosion of the equipment. The use of quaternary ammonium compounds should be carried out at temperatures near 90°C and assuring enough contact time. This way the disinfectant effect is adequate and it is essential that the germicidal solutions are in contact with all areas of the processing equipment. After the residue elimination and disinfection steps, it is necessary to carry out a washing step, making sure that the processing equipment is not recontaminated. This washing is carried out with sterile water. To obtain this type of sterile water, the water undergoes a chlorination treatment, generally with the addition of hypochlorous acid. For pure water, the hypochlorite addition progressively increases the active chlorine content; however, if the initial water contains a certain quantity of organic matter, the addition of active chlorine causes chloramine formation that later on, with more addition of chlorine, changes to dichloramine, which causes the active chlorine to decrease until a minimum level, known as the "break point," starting from which new additions of chlorine cause a progressive increase of the active chlorine. Water is considered as sterile when the active chlorine content is around 5 ppm, a concentration that does not cause corrosion of the equipment. Besides the washing step, chlorinated water is also used in the cleaning of premises and facilities as well as foods, since this method avoids the development of microorganisms.

Many food products are packaged, and this is why it is necessary that the containers have minimal microbial contamination. In many cases, the containers are washed before being used, mainly when they are recycled. Since these containers have already been used, such as bottles for milk, wine, beer, or soda, generally contain dry residues, with high microbial loads, or even small rodents or insects. First, washing the bottles with the purpose of eliminating soil and sludge is carried out, for which the bottles are placed in an inverted position and injection with hot water is applied.

Later the bottles are passed through alkaline washing baths. Then cleaning is carried out in hot liquid while avoiding thermal crashes that could cause breakage of the bottles. Finally, bottles are rinsed with cold and temperate water until obtaining a clean bottle that, once drained, is ready for use. In the immersion baths of the bottles, a solution of around 2% in caustic soda is usually used, while the final washing solutions can be a nonionic surfactant agent so that the solution drainage does not break the aqueous film, avoiding the formation of drops and stains in the bottles when drying. In the case of bottles with stuck deposited residuals, brushes are used for elimination of the deposited soil. In metallic recipients, care should be taken with the use of caustic soda, since it can attack aluminum and tinned steel. In this case, alkaline salt solutions are usually used, to which surfactants can be added. In new containers, it is not necessary to carry out an in-depth cleaning, although they are usually cleaned with an injection of saline solution and surfactants. In aseptic filling, the internal surface of laminate cardboard containers can be treated with hydrogen peroxide and surfactants, and then hydrogen peroxide can be eliminated by blowing water steam.

25 Packaging of Foods

25.1 INTRODUCTION

Packaging is an important aspect of the food manufacturing process, since its main purpose is the preservation of food over a long period of time. Also, another purpose of food containers is to indicate information about different aspects and characteristics of the food they contain.

The main function of a container is that of containing the food, although another important function is to provide protection against possible physical, chemical, and microbial deterioration. Historically, foods were preserved in different types of containers made out of wood, metal, or glass. However, since the 1960s with the arrival of polymers, a great amount and diversity of compounds began to appear that are currently being applied in packaging technologies and which present a great variety of properties.

Generally, the functions of food packaging are intended to contain and protect the food, as well as to identify and provide information, not only as to the maker, but also in reference to content, composition, expiration date, nutritional content, etc. The container serves as a barrier to prevent the entrance of water and oxygen that can deteriorate the food. It also serves as a barrier to prevent the escape of some components of the food such as aromatic compounds.

25.2 PACKAGING MATERIALS

At the present time, a great variety of materials are used for packaging foods. According to the function they carry out, these materials can be classified as containers used in transportation and containers used for the sale to the consumer (Fellows, 1994). Containers for transportation protect the food during its transport and distribution, as in the case of wooden and metal boxes, barrels, drums, and bags, among others. Containers for retail sale to consumers are those that contain the product in small quantities, and in addition to protecting the food, they also provide information about its content, as it is the case of glass bottles, plastic containers, trays, bags, and wrappings.

The materials used in the different types of containers are very diverse and their use depends on the desired purpose. The main materials that are used are metal, glass, wood, paper, cardboard, and a great variety of polymers.

25.2.1 METALS

In the production of metallic containers, steel and aluminum are generally used. These types of materials are used in the canning of foods and drinks. When considering the main advantages of these types of containers, it is necessary to highlight that they offer good protection of the food they contain, since they are shock- and high-temperature resistant, besides being impermeable to vapors and gases. Due to their resistance to high temperatures, steel and aluminum are suitable materials for the pasteurization and sterilization processes of packed products. On the other hand, this type of material does not allow the passage of light and avoids the deterioration that takes place due to photochemical reactions, although, on the contrary, they do not allow visualization of the internal contents. Concerning other disadvantages, it can be mentioned that metallic containers are heavy, and this can increase the cost of transport; in addition, corrosion can also be present. Internal corrosion is a serious problem, since it can negatively affect the interior of the contained food and oxidation products can also contaminate the food. External corrosion can cause holes to occur in the container, resulting in the total loss of the container and the contained product.

A steel sheet with low carbon content constitutes the tin used in containers with a thickness from 0.15 to 0.5 mm that is covered with a tin layer on both sides, with the purpose of avoiding corrosion. In addition, to avoid the transfer of metals into the food, tin sheets are re-covered with a polymeric lacquer film.

Aluminum is usually used in the manufacture of cans for beer and other beverages. Aluminum presents the same advantages and disadvantages as steel, but it is lighter in weight, although it is more expensive. Another important use of aluminum is to obtain aluminum paper, with different thicknesses, which is used to form trays for frozen and precooking of foods and in caps for bottles and other recipients. Aluminum is also used to metalize different flexible materials and in compound laminate materials. If its thickness is greater than 0.015 mm, it is impermeable to humidity, gases, and microorganisms and also reflects light, in this way avoiding the reactions of deterioration caused by light.

Metallic receptacles can have different forms and sizes, with cylindrical shapes being the most common one. Cans are usually manufactured in two or three pieces. In the case of cans that consist of three pieces, this construction begins with coils in cylindrical form, then sticking or welding the borders. For the bottom and cover, circular foils are used that bend for the borders for stamping and riveting from the cover to the body of the can. In cans that are constructed from two pieces, a cylinder is formed directly with the bottom, and the cover is then inserted. In this case, most of the caps are "easy opening" and they are often provided with a ring to facilitate easy removal of the cover.

25.2.2 Glass

Glass is a material that is often used in the manufacture of food containers, since different properties present advantages compared to other materials. Thus, glass containers are impermeable to gases, water, and microorganisms. But it also presents other prominent advantages, such as the fact that they are already inert and they do not react with food components; they are transparent to light and microwaves, allowing the contents to be seen; they can undergo thermal treatments, being easy to wash and thus able to be reused; and they can be manufactured in a variety of ways, being rigid, and so they can be piled up. There are, however, disadvantages, such as the fact that they weigh more than other types of containers, and they are fragile and can be broken, so pieces of glass can appear in foods, which obviously creates a risk for consumers.

Because glass containers are transparent to light, photochemical reactions can cause deterioration of the food, for which reason colored glass is sometimes used, as in the case of amber bottles used for bottling beer or green bottles that are used for wines.

Due to their use and reuse, glass containers and bottles usually support wearing away of the surface by abrasion when the containers come into contact with other containers. Also, in bottles, the cylindrical area that is bound to the base often presents breakage problems, which is why at this position there is a slight inward bend to minimize contact among containers.

In those cases in which glass containers are used to contain foods that need to be sterilized or pasteurized in the same container, it is necessary to keep in mind that a sharp thermal crash can cause the container to break. Therefore, it is necessary to use containers with a greater wall thickness, since they are more resistant, and also, with the purpose of avoiding sudden thermal crashes, the heating should be carried out slowly.

In practice, many foods are packed in glass containers, such as milk, beer, wine, nonalcoholic beverages, sauces, vinegar, pickles, and fruit juices, among others.

25.2.3 Wood

Wood is a material that has been used in the production of boxes and pallets that are generally used when a high grade of mechanical protection in the transport and storage of food products is necessary. However, recently, boxes and crates used in the transport and storage of fruits and vegetables are being manufactured from plastic, which has caused a lower use of wood. Other containers that are traditionally made of wood include barrels to contain wine, liquors, and beer. For wines of high quality, oak wood is usually used, although chestnut is also sometimes used. The use of wooden barrels for the storage of wines and liquors has continued because the aromatic compounds of each wood impart a better quality to the wine or liquor that they contain.

25.2.4 Paper, Cardboard, and Plant Tissues

Paper is a very often used material in food packaging in the form of cardboard, laminate and corrugated cardboard. Its cost is low in comparison with other materials, it has a good rigidity, and it can be printed on. On the contrary, paper is very sensitive to humidity, and environmental humidity can affect it negatively. Generally, paper is made from wood, although in the past several years recycled paper has acquired importance in packaging; however, it is often not advisable to wrap food with recycled paper products. In order to improve certain properties of paper, different treatments are carried out to create waxed and glazed papers; other paper materials such as paper silk and parchment paper will improve resistance to water penetration.

Paper is used for individual packaging, although bags can be manufactured to contain powdered foods, such as flour, sugar, and custards. In some cases it is necessary to use bags with greater resistance, and such packaging should be manufactured with several sheets of paper; and if the bag needs protection from the transfer of water vapor, it can be coated with wax or other impermeable materials.

Cardboard is usually used in the production of boxes and can be of different quality, depending on the product that will be contained in the package. If fatty or wet foods will be contained in the package, the cardboard can be covered with wax or polymers. As with paper, cardboard is manufactured from wood pulp, which allows the pulp to be molded, with the purpose of adapting it to the form of the food to be packed, such as in the case of trays for eggs and fruits. Food can be contained within cardboard boxes without further packaging, although on occasion they are used for products that are already packed, canned, or bottled in other materials.

In the aseptic packaging of sterilized foods, a laminated cardboard is used, with sheets combined in an alternate way to form layers of paper cardboard, aluminum, Surlyn® resin, and/or polyethylene. In this way, greater impermeable resistance of the container is obtained.

Tissues manufactured with jute and cotton are also used in food packaging. Jute is usually used in the production of bags to store and transport bulk foods, as in the case of sugar, grains, flour, and salt, among others. Cotton bags are also used in the packaging of the same types of products mentioned earlier. Containers manufactured with plant tissues are not impermeable to vapor nor to gases, which is why they are often substituted by plastic materials and multilaminated paper.

25.2.5 Polymers

Polymers are materials that are more and more often being used in food packaging, substituting other types of materials. Polymers present diverse advantages, including their shape versatility and wide spectrum of properties. However, depending on the polymer type, the possibility of interactions with food may exist and contamination of the food may occur. Polymers are quite impermeable to gases, water vapor, and oxygen. Most polymers are thermoplastic, which indicates that they are able to be thermo sealed.

Polymers are used in the production of flexible films, that is, for those to which some substance can be added to confer specific properties. Many polymeric films are transparent, although pigments can be added to improve retention properties or reflection of light, as well as to obtain bright and colorful products. Other additives are plasticizer substances, antioxidative stabilizers, and slippery agents that confer their characteristic properties to polymeric films. Flexible films adapt to the form of the contents, allowing for more space during storage and transport.

Polymers are substances of very high molecular weight and are formed by the repetition of smaller units called monomers, linked by chemical reactions. There are natural polymers, such as starch, pectins, and rubbers, of organic origin, although those most often used in food packaging are synthetic.

Generally, these films are obtained from granular polymers that are melted in an extrusion process to obtain sheets or tubes in the final shape of the desired container. The most important films used in food packaging are obtained from cellulose, polypropylene, polyester, polyethylene, polyvinylidene chloride, polystyrene, polyvinyl chloride, and polyamides, among others. All of these polymers form simple films, although some films can be covered with other polymers or with aluminum that allows obtaining composite films that improve their impermeability, or become able to be thermo sealed. The process of covering a polymer film with a very thin layer of aluminum is called metallization. This process obtains a film that is impermeable to light, gases, oils, water vapor, and even to odors. Metalized films are more flexible than aluminum paper and provide the same impermeability, thereby providing a better application for packaging material.

Different types of containers are manufactured from flexible films composed of simple compounds. Laminate film consisting of two or more films that allow an increase in impermeability and mechanical resistance is often used in the production of containers. The manufacture of laminate films begins with simple films that are combined by pressure between two rollers by means of an adhesive compound.

From different coextruded polymers, films of two or more layers can be elaborated. In the manufacture of this type of films, olefins, styrene, and polymers of polyvinyl chloride are usually used. In order to assure that in the extrusion process the sheets are properly adhered, it is necessary that the polymers possess a similar chemical structure and viscosity. The obtained sheets are very impermeable, and they possess little thickness, similar to the simple sheets, and these sheets are used in the production of containers and trays.

Certain types of polymers are used to obtain a rigid or semirigid final product that can be used in the production of trays, bottles, and other containers. These polymers present diverse advantages, such as containers that are less heavy, which is positive from the point of view of transport and distribution; are resistant to corrosion, unbreakable, and can be easily sealed; and can be easily shaped. On the contrary, they are less resistant to heat than glass and metals and are nonreusable.

25.3 PACKAGING MATERIAL CHOICE

When considering the choice of packaging for a certain food, it should be kept in mind that different factors can affect food quality, including both microbial and sensorial characteristics.

In the manipulation and transport of some foods such as fruits, vegetables, and cookies, among others, mechanical damages can occur, and that is why it is essential to utilize adequate packaging techniques in order to minimize this type of deterioration. For this reason, it is necessary to choose a primary rigid and resistant packing material such as glass, metal, or rigid plastics. For external packaging, wood and cardboard can be used, and also containers of certain foods can be better protected from impact and crushing if they incorporate certain adsorption materials such as foams and expanded plastics that have a padding function. For some foods such as eggs and fruits, shaped trays and containers can be used to contain the food individually.

One of the main functions of a container is to avoid microbial contamination of the food it contains. Thus, it is crucial that the material of the container does not allow the passage of microorganisms. For pasteurized and sterilized foods, aseptic packing is essential, so plastic containers that allow the use of thermo seals improve the prevention of microbial contamination.

Light is a factor that negatively affects foods that are photosensitive and can cause damage reactions such as vitamin loss, rancid appearance of fat, and discoloration. To avoid these problems, opaque materials should be used. However, if it is desirable to see the turbidity or transparency of a liquid food, it is possible to use pigmented glass such as the bottles used for wines and beers.

Another important factor to keep in mind is temperature. The container should be able to withstand the same temperature changes as those that the food undergoes. For example, in the case of frozen foods, it should be assured that the material of the container does not break when it is subjected to low freezing temperatures. Also, there are many products that are packed previous to pasteurization or sterilization thermal treatment; in this case, the container should be able to resist the high temperatures to which it is subjected. If glass containers are used, care must be taken during the heating and cooling stages that are carried out to avoid thermal crashes that could cause breakage of the container. Also, precooked foods need to be warmed in a microwave oven prior to ingestion, and in these cases it is advisable to choose a container that supports the required heating temperatures.

An important characteristic of container materials is the permeability they present to water vapor, to volatile compounds that compose the aroma of foods, and to different gases, such as oxygen, nitrogen, and carbon dioxide. In the case of foods that have a high water content, it can be transferred to the atmosphere, with the final result being a food whose aspect and texture are damaged, besides exhibiting certain weight losses. On the contrary, foods with low water activity tend to absorb water, as in the case of food powders, where agglomerates could be formed. For example, ice tea powder could become an array of clumps. Also, water gain can provide some favorable conditions for the growth of microorganisms that can affect both the quality and the safety of the food. In these cases it will be necessary to use materials of low water permeability; however, in the case of certain fruits and vegetables, the respiration effect causes the release of water vapor that can condense on the container surface, causing the possibility of microbial growth; so, in this case it would be convenient to use a water vapor permeable material.

In fresh foods packed in an impermeable container, the respiration effect causes the atmosphere that surrounds the food inside the container to be rich in carbon dioxide, and such anaerobic conditions can damage the food. It would be convenient, therefore, to use packing materials that allow the exit of carbon dioxide and the entrance of oxygen.

Ethylene is a compound that appears in climacteric fruit respiration, which in turn causes an increase in the ripening speed of the fruits. Thus, it is convenient that the container material is permeable to this compound, allowing the exit of ethylene toward the outside. For products of high respiration rate, it is sometimes convenient to use perforated films that allow elimination of the compounds that can affect the food. If it is desirable to retain the aromatic compounds of certain packed foods, it will be convenient to use impermeable materials. Thus, in the case of coffee, sealed glass or metal containers can be used. Foods that contain appreciable fat content or oils are products that can cross the containers and negatively affect the external aspect of the container and label. In this way, in products with low water content, such as chocolate and powdered milk, fat impermeable papers are used. Also, in fish and meat that possess high water content, hydrophilic films can be used.

When environmental conditions are favorable, problems with insect and rodent infestation can be present. Such problems can damage all types of containers, with the exception of metallic and glass. For this reason, opportune hygienic measures should be taken to avoid the presence of rodents; traps and poisons can be used. However, care must be taken so that the poisons do not contaminate the food. Likewise, insects such as moths, mites, and beetles can invade the food if they can perforate the container material or take advantage of existing openings in the package to infest the food. Therefore, resistant materials should be used to avoid infestation. The use of substances like pesticides is limited, since they can contaminate the food; however, if legislation allows it, pesticides can be incorporated in multilaminated containers between the adhesives of the sheets that form the container material.

A problem that can be present during the storage of packed products is the migration of certain components of the material packaging toward the food. In the case of fruits and fruit juices, the pH of the food is very acid, and this can cause corrosion of metallic containers and contamination of the food; in this case, it is necessary that the internal surface of the recipient is covered with lacquer with a resistant compound to avoid acid attack.

Polyvinyl chloride containers can allow migration of the monomer toward the food; even the preservatives that are added to the polymers, such as coloring, plasticizers, and other elements, can contaminate the contained food due to migration, which can represent a danger to the consumer's health. Therefore, before using a specific material in food packaging, experiments concerning migration of the material components should be carried out, with the purpose of determining their harmlessness. Some packaging materials, such as papers and cardboards, can give off certain fragrant volatile compounds that, if they are adsorbed by the food, can confer an unpleasant odor. Likewise, in the sealing of metal and glass containers, lacquers are used that can remove odors, representing an additional contamination focus. With the purpose of minimizing this contamination of food as much as possible, it is important that the choice of packaging material should represent these problems of possible migration of components from the packaging material to the food.

25.4 FOOD PACKAGING AND MODIFIED ATMOSPHERE

The main purpose of food packaging is to avoid deterioration, which can have many different causes, including physical damage and chemical and microbial reactions. In the commercialization of fresh fruits and vegetables, it is important that the food preserve its fresh characteristics for the longest possible time, in many cases using what is called modified atmosphere packaging. For the purpose of extending the commercial life of many products, it is necessary to carry out thermal treatments that inactivate pathogens such as deteriorative microbial flora. In these cases the treatment can be carried out in the product once it has been packaged, or the thermal treatment of the product may be initially carried out and then it is packaged in an aseptic manner. In this section, some of these packaging types that have differentiated characteristics will be addressed.

Aseptic packaging presents certain advantages over the treatments of previously packaged foods, since final products are obtained with lower food deterioration and are therefore more uniform and quality; the processing times are shorter, which translates to lower energy consumption. The packaging type will depend on the pH of the food; for acidic foods with pH values lower than 4.5, the thermally treated food is introduced in a container that should be clean; it is then closed and left for some minutes at a temperature close to 100°C. This method is sufficient to inactivate any microorganisms that can be contained inside the container. On the contrary, for foods that are not very acidic, it is necessary that prior to the filling operation the containers be sterilized, and most of the filling operations such as sealing of the recipient should be carried out under aseptic conditions. The filling installation and the metallic or glass containers are sterilized with superheated vapor with the goal of assuring that there is no microbial contamination of the product that has been previously thermally treated. When using laminate materials for the containers, the usual method is to combine heat treatment with chemical treatments, in which hydrogen peroxide, ultraviolet radiation, and sterile air are used. Hydrogen peroxide solutions at temperatures close to 100°C are effective for the inactivation of thermo-resistant and sporulated microorganisms, and this is a compound that is very often used in aseptic treatment in laminate materials.

In the packaging of some foods, such as meat and cheese, the internal air of the container that wraps the food can cause deteriorative reactions, such as rancidity due to lipid oxidation, or there can be a problem like superficial drying. To avoid these problems, the food can be packaged under vacuum, where the air of the interior of the container is removed before proceeding to sealing.

In certain cases *modified atmosphere* packaging is used, in which, prior to sealing, the air of the interior of the container is substituted by a gas mixture. The modified atmosphere can be obtained by vacuuming out the container and then filling it with the gas mixture, or the air can be replaced by passing through a continuous gas mixture current. For fruit and vegetable packaging, food respiration causes a change in the air composition of the container, in which case a decrease of oxygen and an increase of carbon dioxide is observed. Generally, the gases used to create modified atmospheres that are used in commercial packaging are usually nitrogen, oxygen, and carbon dioxide. Nitrogen is usually used to displace oxygen to avoid problems of oxidation reactions, as in the case of fats. The nitrogen itself does not directly affect the microorganisms. Oxygen can be used in red meat packaging; since its presence avoids myoglobin deterioration, it maintains a better aspect of the meat color. It can also be used to avoid botulism problems in white fish packaging. Carbon dioxide can inhibit the growth of certain aerobic bacteria and molds, while many types of yeast are resistant to carbon dioxide action, with little affect on anaerobic bacteria. On the contrary, carbon dioxide is dissolved in water to produce carbonic acid that reduces a food's pH. However, it is vital to maintain good temperature control throughout the packaging period, keeping in mind that the inhibition power of carbon dioxide is greater as much as greater is its concentration and the lower is the storage temperature. The materials used in modified atmosphere packaging should be chosen depending on the permeability they present to these gases and taking care that the composition inside the container is appropriate throughout the entire storage time.

Active packaging consists of the incorporation of specific preservatives in the packaging material or inside the containers with the purpose of extending commercial life and thus preserving the quality of the food. In this packaging type, preservatives are used to eliminate gases like carbon dioxide, oxygen, ethylene, and water vapor, as well as odors and flavors that can affect the food. Additionally, the material of the container can also liberate such substances as sorbate, ethanol, antioxidants, and preservatives.

In active packaging, eliminators of carbon dioxide can be used, depending on the proposed purpose. To eliminate carbon dioxide, bags of calcium oxide and active coal can be used, although under conditions of high humidity calcium hydroxide can be used to capture carbon dioxide and transform it into calcium carbonate. Oxygen can cause many undesirable reactions in the contained food, decreasing its quality, so the elimination of oxygen can decrease the metabolic rate of food respiration, thus decreasing the lipid oxidation that causes rancidity, as well as inhibiting the oxidation of vitamins and pigments and allowing control of discoloration processes. For oxygen elimination, bags containing metallic compounds that adsorb the oxygen, fixing it in the form of stable metallic oxides, are generally used. Due to the possibility of contamination by metals, nonmetallic oxygen eliminators have been developed, such as catechol, ascorbic acid, and ascorbates; although eliminators based on enzymes such as ethanol oxidase and glucose oxidase can also be used. In some cases, when the oxygen of the container is removed, it can create a vacuum, so systems of double action have been developed that remove oxygen and emit carbon dioxide to compensate for the oxygen loss.

As has been already introduced, ethylene is a compound that acts to accelerate the respiration rate of fruits and vegetables, causing an acceleration of the senescence of the mentioned products. In some cases, as in the color development of tomatoes, bananas, and citric fruits, the action of ethylene is desirable; however, in most fruit and vegetable cool preservations, it is desirable to reduce or eliminate the ethylene generated in the respiration process of the stored products. In the elimination of ethylene, potassium permanganate is used, which is usually immobilized on an inert substrate as silica gel or alumina. Permanganate is placed in bags that allow oxidation of the ethylene to acetate and ethanol. In removing ethylene, bags with activated-carbon incorporating metallic catalysts can also be used.

Packaging films that contain antioxidant and antimicrobial compounds that are favorable to extend the commercial life of many foods are also very interesting. One compound that has been incorporated in a packaging film is silver zeolite, which can slowly liberate silver ions and which has antimicrobial properties. Other compounds with antimicrobial action that can be incorporated into films are organic acid, bacteriocins, spice and herb extracts, enzymes, chelating agents, and antifungicidals that can be used in the packaging of meats, fish, cheese, bread, fruits, and vegetables. Films impregnated with vitamin E that are used for their antioxidative properties have also been highlighted. Also, the advantages of other products used in the lining of films such as butylhydroxytoluene (BHT) and butylated hydroxyanisole (BHA) have been presented, due to the controversy over the harmlessness of these products.

Excessive humidity can cause alteration of packaged foods, and to maintain the quality of water, adsorbent materials can be used that allow extension of the commercial life of the food. Humidity-adsorbent bags composed of silica gel, calcium oxide, clays, and other minerals can be used. Active coal may be introduced to the bag's composition in order to take advantage of its ability to adsorb the substances that confer offensive odors and flavors to the food. These humidity adsorbents are used in many foods, such as snacks and cereals. In products such as meats, fish, poultry, fruits, and vegetables, because there can be an expression of water, anti-dripping towels are usually used to control the exudated water. An excess of water can exist on the surface of packaged foods, which can facilitate microbial growth, and it is important to be able to reduce the relative humidity of the contained air in the recipient. Thus, in packaging films water-adsorbent products can be placed so that they avoid depositing liquid on the food surface.

Ethanol is a made up of antimicrobial properties that are mainly designed to fight molds, although it can also inhibit the growth of bacteria and yeasts. The pulverization with ethanol of bakery products can extend the commercial life of these products. However, an alternative to this pulverization would be to use bags and films impregnated with ethanol that slowly release ethanol.

One type of active packaging to highlight is the use of auto-heating and auto-cooling cans. Also, materials can be used that can serve to insulate heat transmission during the storage and distribution of refrigerated foods. The auto-heating cans are steel and aluminum containers that warm when using exothermic reactions similar to the mixture of lime (CaO) with water in the bottom of the can. In the same way, if endothermic reactions are used, such as that which is in contact with ammonium nitrate and chloride with water, auto-cooling cans can be obtained. The idea of active packaging opens new perspectives for packaging; however, the main objective should be the maintenance of microbial, nutritional, and sensorial quality and extending the commercial life of food.

25.5 MASS TRANSFER THROUGH PACKAGING MATERIALS

Packaging materials serve the purpose of maintaining the conditions of the contained food inside the container. Humidity retention is indispensable so that fresh foods are not dried out, while for dry foods such as powders and cookies, it is necessary that the packaging material is water vapor impermeable so that the food is not crowded and does not lose its crispy property. However, the materials do not present a total impermeability to humidity or different gases, such as oxygen, nitrogen, carbon dioxide, and odor and flavor compounds.

Permeability is a characteristic of the packaging materials that measures the ability of gases to pass through them. The mechanism of transfer of these gases through the packaging material is carried out by a diffusion mechanism in that the transfer driving is the pressure difference between both faces of the material layer. The transfer process for a given component is carried out in successive steps, in which the component is first transferred from the gas or liquid mixture of gas where its concentration is higher until it reaches the wall of the film, after which it is dissolved in the film, and then it diffuses through the film until it reaches the other face. Finally, it passes through to be part of the mixture where there is a lower component concentration.

The overall transfer process can be similar to that of a plane layer, as described in Chapter 7. In this way, the mass flux with which a component is transferred through a film of the container can be expressed according to Equation 7.19:

$$N_A = -D_A \frac{dC_A}{dz} = D_A \frac{C_{A_1} - C_{A_2}}{e} \tag{7.19}$$

For the gas diffusion, it is convenient to define the permeability that the packaging material presents to a certain component. As described in Chapter 7, the permeability can be expressed according to Equation 7.23:

$$P_M = D_A s \tag{7.23}$$

where
P_M is the permeability
D_A is the diffusivity
s is the solubility, expressed as the inverse of Henry's constant

The values of these three coefficients for different packaging materials and different gas components that diffuse through them can be found in the literature (Hernandez, 1997; Singh and Heldman, 1998). The mass flux for a given component can be expressed as a function of the permeability starting from Equation 7.24:

$$N_A = P_M \frac{p_{A_1} - p_{A_2}}{e} \tag{7.24}$$

It is observed that in this case the diffusion driving force is the pressure difference between both faces of the container film, while the amount of component transmitted through the film is inversely proportional to the film thickness and directly proportional to the permeability. For a better understanding of the component diffusion through a packaging film, refer to Problem 7.2.

Appendix

TABLE A.1
Properties of Saturated Steam

	Vapor	Specific Volume (m³/kg)		Enthalpy (kJ/kg)		Entropy (kJ/kg K)	
T (°C)	Pressure (kPa)	Liquid	Vapor	Liquid	Vapor	Liquid	Vapor
0.01	0.611	0.0010002	206.14	0.00	2501.4	0.0000	9.1562
3	0.758	0.0010001	168.132	12.57	2506.9	0.0457	9.0773
6	0.935	0.0010001	137.734	25.20	2512.4	0.0912	9.0003
9	1.148	0.0010003	113.386	37.80	2517.9	0.1362	8.9253
12	1.402	0.0010005	93.784	50.41	2523.4	0.1806	8.8524
15	1.705	0.0010009	77.926	62.99	2528.9	0.2245	8.7814
18	2.064	0.0010014	65.038	75.58	2534.4	0.2679	8.7123
21	2.487	0.0010020	54.514	88.14	2539.9	0.3109	8.6450
24	2.985	0.0010027	45.883	100.70	2545.4	0.3534	8.5794
27	3.567	0.0010035	38.774	113.25	2550.8	0.3954	8.5156
30	4.246	0.0010043	32.894	125.79	2556.3	0.4369	8.4533
33	5.034	0.0010053	28.011	138.33	2561.7	0.4781	8.3927
36	5.947	0.0010063	23.940	150.86	2567.1	0.5188	8.3336
40	7.384	0.0010078	19.523	167.57	2574.3	0.5725	8.2570
45	9.593	0.0010099	15.258	188.45	2583.2	0.6387	8.1648
50	12.349	0.0010121	12.032	209.33	2592.1	0.7038	8.0763
55	15.758	0.0010146	9.568	230.23	2600.9	0.7679	7.9913
60	19.940	0.0010172	7.671	251.13	2609.6	0.8312	7.9096
65	25.03	0.0010199	6.197	272.06	2618.3	0.8935	7.8310
70	31.19	0.0010228	5.042	292.98	2626.8	0.9549	7.7553
75	38.58	0.0010259	4.131	313.93	2635.3	1.0155	7.6824
80	47.39	0.0010291	3.407	334.91	2643.7	1.0753	7.6122
85	57.83	0.0010325	2.828	355.90	2651.9	1.1343	7.5445
90	70.14	0.0010360	2.361	376.92	2660.1	1.1925	7.4791
95	84.55	0.0010397	1.982	397.96	2668.1	1.2500	7.4159
100	101.35	0.0010435	1.673	419.04	2676.1	1.3069	7.3549
105	120.82	0.0010475	1.419	440.15	2683.8	1.3630	7.2958
110	143.27	0.0010516	1.210	461.30	2691.5	1.4185	7.2387
115	169.06	0.0010559	1.037	482.48	2699.0	1.4734	7.1833
120	198.53	0.0010603	0.892	503.71	2706.3	1.5276	7.1296
125	232.1	0.0010649	0.771	524.99	2713.5	1.5813	7.0775
130	270.1	0.0010697	0.669	546.31	2720.5	1.6344	7.0269
135	313.0	0.0010746	0.582	567.69	2727.3	1.6870	6.9777
140	316.3	0.0010797	0.509	589.13	2733.9	1.7391	6.9299
145	415.4	0.0010850	0.446	610.63	2740.3	1.7907	6.8833
150	475.8	0.0010905	0.393	632.20	2746.5	1.8418	6.8379
155	543.1	0.0010961	0.347	653.84	2752.4	1.8925	6.7935
160	617.8	0.0011020	0.307	675.55	2758.1	1.9427	6.7502
165	700.5	0.0011080	0.273	697.34	2763.5	1.9925	6.7078

(continued)

TABLE A.1 (continued)
Properties of Saturated Steam

	Vapor	Specific Volume (m³/kg)		Enthalpy (kJ/kg)		Entropy (kJ/kg K)	
T (°C)	Pressure (kPa)	Liquid	Vapor	Liquid	Vapor	Liquid	Vapor
170	791.7	0.0011143	0.243	719.21	2768.7	2.0419	6.6663
175	892.0	0.0011207	0.217	741.17	2773.6	2.0909	6.6256
180	1002.1	0.0011274	0.194	763.22	2778.2	2.1396	6.5857
190	1254.4	0.0011414	0.157	807.62	2786.4	2.2359	6.5079
200	1553.8	0.0011565	0.127	852.45	2793.2	2.3309	6.4323
225	2548	0.0011992	0.078	966.78	2803.3	2.5639	6.2503
250	3973	0.0012512	0.050	1085.36	2801.5	2.7927	6.0730
275	5942	0.0013168	0.033	1210.07	2785.0	3.0208	5.8938
300	8581	0.0010436	0.022	1344.0	2749.0	3.2534	5.7045

TABLE A.2
Water Properties of Super Heated Vapor

T (°C)	Specific Volume (m³/kg)	Enthalpy (kJ/kg)	Entropy (kJ/kg K)	T (°C)	Specific Volume (m³/kg)	Enthalpy (kJ/kg)	Entropy (kJ/kg K)
	p = 0.006 MPa = 0.06 bar				**p = 0.035 MPa = 0.35 bar**		
	T_{sat} = 36.16°C				**T_{sat} = 72.69°C**		
Sat	23.739	2567.4	8.3304	Sat	4.562	2631.4	7.7158
80	27.132	2650.1	8.5804	80	4.625	2645.6	7.7564
120	30.219	2726.0	8.7840	120	5.163	2723.1	7.9644
160	33.302	2802.5	8.9693	160	5.696	2800.6	8.1519
200	36.383	2879.7	9.1398	200	6.228	2878.4	8.3237
240	39.462	2957.8	9.2982	240	6.758	2956.8	8.4828
280	42.540	3036.8	9.4464	280	7.287	3036.0	8.6314
320	45.618	3116.7	9.5859	320	7.815	3116.1	8.7712
360	48.696	3197.7	9.7180	360	8.344	3197.1	8.9034
400	51.774	3279.6	9.8435	400	8.872	3279.2	9.0291
440	54.851	3362.6	9.9633	440	9.400	3362.2	9.1490
500	59.467	3489.1	10.1336	500	10.192	3488.8	9.3194
	p = 0.007 MPa = 0.70 bar				**p = 0.10 MPa = 1.0 bar**		
	T_{sat} = 89.95°C				**T_{sat} = 99.63°C**		
Sat	2.365	2660.0	7.4797	Sat	1.694	2675.5	7.3594
100	2.434	2680.0	7.5341	100	1.696	2676.2	7.3614
120	2.571	2719.6	7.6375	120	1.793	2716.6	7.4668
160	2.841	2798.2	7.8279	160	1.984	2796.2	7.6596
200	3.108	2876.7	8.0012	200	2.172	2875.3	7.8343
240	3.374	2955.5	8.1611	240	2.359	2954.5	7.9949
280	3.640	3035.0	8.3162	280	2.546	3034.2	8.1445
320	3.905	3115.3	8.4504	320	2.732	3114.6	8.2849
360	4.170	3196.5	8.5828	360	2.917	3195.9	8.4175
400	4.434	3278.6	8.7086	400	3.103	3278.2	8.5435
440	4.698	3361.8	8.8286	440	3.288	3361.4	8.6636
500	5.095	3488.5	8.9991	500	3.565	3488.1	8.8342
	p = 0.15 MPa = 1.5 bar				**p = 0.30 MPa = 3.0 bar**		
	T_{sat} = 111.37°C				**T_{sat} = 133.55°C**		
Sat	1.159	2693.6	7.2233	Sat	0.606	2725.3	6.9919
120	1.188	2711.4	7.2693	120			
160	1.317	2792.8	7.4665	160	0.651	2782.3	7.1276
200	1.444	2872.9	7.6433	200	0.716	2865.5	7.3115
240	1.570	2952.7	7.8052	240	0.781	2947.3	7.4774
280	1.695	3032.8	7.9555	280	0.844	3028.6	7.6299
320	1.819	3113.5	8.0964	320	0.907	3110.1	7.7722
360	1.943	3195.0	8.2293	360	0.969	3192.2	7.9061
400	2.067	3277.4	8.3555	400	1.032	3275.0	8.0330
440	2.191	3360.7	8.4757	440	1.094	3358.7	8.1538
500	2.376	3487.6	8.6466	500	1.187	3486.0	8.3251
600	2.685	3704.3	8.9101	600	1.341	3703.2	8.5892

(continued)

TABLE A.2 (continued)
Water Properties of Super Heated Vapor

T (°C)	Specific Volume (m³/kg)	Enthalpy (kJ/kg)	Entropy (kJ/kg K)	T (°C)	Specific Volume (m³/kg)	Enthalpy (kJ/kg)	Entropy (kJ/kg K)
	p = 0.50 MPa = 5.0 bar				**p = 0.70 MPa = 7.0 bar**		
	T_{sat} = 151.86°C				**T_{sat} = 164.97°C**		
Sat	0.3749	2748.7	6.8213	Sat	0.2729	2763.5	6.7080
180	0.4045	2812.0	6.9656	180	0.2847	2799.1	6.7880
200	0.4249	2855.4	7.0592	200	0.2999	2844.8	6.8865
240	0.4646	2939.9	7.2307	240	0.3292	2932.2	7.0641
280	0.5034	3022.9	7.3865	280	0.3594	3017.1	7.2233
320	0.5416	3105.6	7.5308	320	0.3852	3100.9	7.3697
360	0.5796	3188.4	7.6660	360	0.4126	3184.7	7.5063
400	0.6173	3271.9	7.7938	400	0.4397	3268.7	7.6350
440	0.6548	3356.0	7.9152	440	0.4667	3353.3	7.7571
500	0.7109	3483.9	8.0873	500	0.5070	3481.7	7.9299
600	0.8041	3701.7	8.3522	600	0.5738	3700.2	8.1956
700	0.8969	3925.9	8.5952	700	0.6403	3924.8	8.4391
	p = 1.0 MPa = 10.0 bar				**p = 1.5 MPa = 15.0 bar**		
	T_{sat} = 179.91°C				**T_{sat} = 198.32°C**		
Sat	0.1944	2778.1	6.5865	Sat	0.1318	2792.2	6.4448
200	0.2060	2827.9	6.6940	200	0.1325	2796.8	6.4546
240	0.2275	2920.4	6.8817	240	0.1483	2899.3	6.6628
280	0.2480	3008.2	7.0465	280	0.1627	2992.7	6.8381
320	0.2678	3093.9	7.1962	320	0.1765	3081.9	6.9938
360	0.2873	3178.9	7.3349	360	0.1899	3169.2	7.1363
400	0.3066	3263.9	7.4651	400	0.2030	3255.8	7.2690
440	0.3257	3349.3	7.5883	440	0.2160	3342.5	7.3940
500	0.3541	3478.5	7.7622	500	0.2352	3473.1	7.5698
540	0.3729	3565.6	7.8720	540	0.2478	3560.9	7.6805
600	0.4011	3697.9	8.0290	600	0.2668	3694.0	7.8383
640	0.4198	3787.2	8.1290	640	0.2793	3783.8	7.9391
	p = 2.0 MPa = 20.0 bar				**p = 3.0 MPa = 30.0 bar**		
	T_{sat} = 212.42°C				**T_{sat} = 233.90°C**		
Sat	0.0996	2799.5	6.3409	Sat	0.0667	2804.2	6.1869
240	0.1085	2876.5	6.4952	240	0.0682	2824.3	6.2265
280	0.1200	2976.4	6.6828	280	0.0771	2941.3	6.4462
320	0.1308	3069.5	6.8452	320	0.0850	3043.4	6.6245
360	0.1411	3159.3	6.9917	360	0.0923	3138.7	6.7801
400	0.1512	3247.6	7.1271	400	0.0994	3230.9	6.9212
440	0.1611	3335.5	7.2540	440	0.1062	3321.5	7.0520
500	0.1757	3467.6	7.4317	500	0.1162	3456.5	7.2338
540	0.1853	3556.1	7.5434	540	0.1227	3546.6	7.3474
600	0.1996	3690.1	7.7024	600	0.1324	3682.3	7.5085
640	0.2091	3780.4	7.8035	640	0.1388	3773.5	7.6106
700	0.2232	3917.4	7.9487	700	0.1484	3911.7	7.7571

TABLE A.2 (continued)
Water Properties of Super Heated Vapor

T (°C)	Specific Volume (m³/kg)	Enthalpy (kJ/kg)	Entropy (kJ/kg K)	T (°C)	Specific Volume (m³/kg)	Enthalpy (kJ/kg)	Entropy (kJ/kg K)
	p = 4.0 MPa = 40.0 bar				p = 6.0 MPa = 60.0 bar		
	T_{sat} = 250.4°C				T_{sat} = 275.64°C		
Sat	0.04978	2801.4	6.0701	Sat	0.03244	2784.3	5.8892
280	0.05546	2901.8	6.2568	280	0.03317	2804.2	5.9252
320	0.06199	3015.4	6.4553	320	0.03876	2952.6	6.1846
360	0.06788	3117.2	6.6215	360	0.04331	3071.1	6.3782
400	0.07341	3213.6	6.7690	400	0.04739	3177.2	6.5408
440	0.07872	3307.1	6.9041	440	0.05122	3277.3	6.6853
500	0.08643	3445.3	7.0901	500	0.05665	3422.2	6.8803
540	0.09145	3536.9	7.2056	540	0.06015	3517.0	6.9999
600	0.09885	3674.4	7.3688	600	0.06525	3658.4	7.1677
640	0.1037	3766.6	7.4720	640	0.06859	3752.6	7.2731
700	0.1110	3905.9	7.6198	700	0.07352	3894.1	7.4234
740	0.1157	3999.6	7.7141	740	0.07677	3989.2	7.5190
	p = 8.0 MPa = 80.0 bar				p = 10.0 MPa = 100.0 bar		
	T_{sat} = 295.06°C				T_{sat} = 311.06°C		
Sat	0.02352	2758.0	5.7432	Sat	0.01803	2724.7	5.6141
320	0.02682	2877.2	5.9489	320	0.01925	2781.3	5.7103
360	0.03089	3019.8	6.1819	360	0.02331	2962.1	6.0060
400	0.03432	3138.3	6.3634	400	0.02641	3096.5	6.2120
440	0.03742	3246.1	6.5190	440	0.02911	3213.2	6.3805
480	0.04034	3348.4	6.6586	480	0.03160	3321.4	6.5282
520	0.04313	3447.7	6.7871	520	0.03394	3425.1	6.6622
560	0.04582	3543.3	6.9072	560	0.03619	3526.0	6.7862
600	0.04845	3642.0	7.0206	600	0.03837	3625.3	6.9029
640	0.05102	3738.3	7.1283	640	0.04048	3723.7	7.0131
700	0.05481	3882.4	7.2812	700	0.04358	3870.5	7.1687
740	0.05729	3978.7	7.3782	740	0.04560	3968.1	7.2670
	p = 12.0 MPa = 120.0 bar				p = 140.0 MPa = 14.0 bar		
	T_{sat} = 324.75°C				T_{sat} = 336.75°C		
Sat	0.01426	2684.9	5.4924	Sat	0.01149	2476.8	5.3717
360	0.01811	2895.7	5.8361	360	0.01422	2816.5	5.6602
400	0.02108	3051.3	6.0747	400	0.01722	3001.9	5.9448
440	0.02355	3178.1	6.2586	440	0.01954	3142.2	6.1474
480	0.02576	3293.5	6.4154	480	0.02157	3264.5	6.3143
520	0.02781	3401.8	6.5555	520	0.02343	3377.8	6.4610
560	0.02977	3506.2	6.6840	560	0.02517	3486.0	6.5941
600	0.03164	3608.3	6.8037	600	0.02683	3591.1	6.7172
640	0.03345	3709.0	6.9164	640	0.02843	3694.1	6.8326
700	0.03610	3858.4	7.0749	700	0.03075	3846.2	6.9939
740	0.03781	3957.4	7.1746	740	0.03225	3946.7	7.0952

(continued)

TABLE A.2 (continued)
Water Properties of Super Heated Vapor

T (°C)	Specific Volume (m³/kg)	Enthalpy (kJ/kg)	Entropy (kJ/kg K)	T (°C)	Specific Volume (m³/kg)	Enthalpy (kJ/kg)	Entropy (kJ/kg K)
	p = 16.0 MPa = 160.0 bar				p = 18.0 MPa = 180.0 bar		
	T_{sat} = 347.44°C				T_{sat} = 357.06°C		
Sat	0.00931	2580.6	5.2455	Sat	0.00749	2509.1	5.1044
360	0.01105	2715.8	5.4614	360	0.00809	2564.5	5.1922
400	0.01426	2947.6	5.8175	400	0.01190	2887.0	5.6887
440	0.01652	3103.7	6.0429	440	0.01414	3062.8	5.9428
480	0.01842	3234.4	6.2215	480	0.01596	3203.2	6.1345
520	0.02013	3353.3	6.3752	520	0.01757	3378.0	6.2960
560	0.02172	3465.4	6.5132	560	0.01904	3444.4	6.4392
600	0.02323	3573.5	6.6399	600	0.02042	3555.6	6.5696
640	0.02467	3678.9	6.7580	640	0.02174	3663.6	6.6905
700	0.02674	3833.9	6.9224	700	0.02362	3821.5	6.8580
740	0.02808	3935.9	7.0251	740	0.02483	3925.0	6.9623
	p = 20.0 MPa = 200.0 bar				p = 24.0 MPa = 240.0 bar		
	T_{sat} = 365.81°C						
Sat	0.00583	2409.7	4.9269				
400	0.00994	2818.1	5.5540	400	0.00673	2639.4	5.2393
440	0.01222	3019.4	5.8450	440	0.00929	2923.4	5.6506
480	0.01399	3170.8	6.0518	480	0.01100	3102.3	5.8950
520	0.01551	3302.2	6.2218	520	0.01241	3248.5	6.0842
560	0.01689	3423.0	6.3705	560	0.01366	3379.0	6.2448
600	0.01818	3537.6	6.5048	600	0.01481	3500.7	6.3875
640	0.01940	3648.1	6.6286	640	0.01588	3616.7	6.5174
700	0.02113	3809.0	6.7993	700	0.01739	3783.8	6.6847
740	0.02224	3914.1	6.9052	740	0.01835	3892.1	6.8038
800	0.02385	4069.7	7.0544	800	0.01974	4051.6	6.9567
	p = 28.0 MPa = 280.0 bar				p = 140.0 MPa = 14.0 bar		
400	0.00383	2330.7	4.7494	400	0.00236	2055.9	4.3239
440	0.00712	2812.6	5.4494	440	0.00544	2683.0	5.2327
480	0.00885	3028.5	5.7446	480	0.00722	2949.2	5.5968
520	0.01020	3192.3	5.9566	520	0.00853	3133.7	5.8357
560	0.01136	3333.7	6.1307	560	0.00963	3287.2	6.0246
600	0.01241	3463.0	6.2823	600	0.01061	3424.6	6.1858
640	0.01338	3584.8	6.4187	640	0.01150	3552.5	6.3290
700	0.01473	3758.4	6.6029	700	0.01273	3732.8	6.5203
740	0.01558	3870.0	6.7153	740	0.01350	3847.8	6.6361
800	0.01680	4033.4	6.8720	800	0.01460	4015.1	6.7966
900	0.01873	4298.8	7.1084	900	0.01633	4285.1	7.0372

TABLE A.3
Water Properties of Super Cooled Vapor

T (°C)	(Specific Volume) × 10^3 (m^3/kg)	Enthalpy (kJ/kg)	Entropy (kJ/kg K)	T (°C)	(Specific Volume) × 10^3 (m^3/kg)	Enthalpy (kJ/kg)	Entropy (kJ/kg K)
	p = 2.5 MPa = 25.0 bar				**p = 5.0 MPa = 50.0 bar**		
	T_{sat} = 223.99°C				**T_{sat} = 263.99°C**		
20	1.0006	86.30	0.2961	20	0.9995	88.65	0.2956
40	1.0067	169.77	0.5715	40	1.0056	171.97	0.5705
80	1.0280	336.86	1.0737	80	1.0268	338.85	1.0720
100	1.0423	420.85	1.3050	100	1.0410	422.72	1.3030
140	1.0784	590.52	1.7369	140	1.0768	592.15	1.7343
180	1.1261	763.97	2.1375	180	1.1240	765.25	2.1341
200	1.1555	852.8	2.3294	200	1.1530	835.9	2.3255
220	1.1898	943.7	2.5174	220	1.1866	944.4	2.5128
Sat	1.1973	962.1	2.5546	Sat	1.2859	1154.2	2.9202
	p = 7.5 MPa = 75.0 bar				**p = 10.0 MPa = 100.0 bar**		
	T_{sat} = 290.59°C				**T_{sat} = 311.06°C**		
20	0.9984	90.99	0.2950	20	0.9972	93.33	0.2945
40	1.0045	174.18	0.5696	40	1.0034	176.38	0.5686
80	1.0256	340.84	1.0704	80	1.0245	342.83	1.0688
100	1.0397	424.62	1.3011	100	1.0385	426.50	1.2992
140	1.0752	593.78	1.7317	140	1.0737	595.42	1.7292
180	1.1219	766.55	2.1308	180	1.1199	767.84	2.1275
220	1.1835	945.1	2.5083	200	1.1805	945.9	2.5039
260	1.2696	1134.0	2.8763	220	1.2645	1133.7	2.8699
Sat	1.3677	1292.2	3.1649	Sat	1.4524	1407.6	3.3596
	p = 15.0 MPa = 150.0 bar				**p = 20.0 MPa = 200.0 bar**		
	T_{sat} = 342.24°C				**T_{sat} = 365.81°C**		
20	0.9950	97.99	0.2934	20	0.9928	102.62	0.2923
40	1.0013	180.78	0.5666	40	0.9992	185.16	0.5646
80	1.0222	346.81	1.0656	80	1.0199	350.80	1.0624
100	1.0361	430.28	1.2955	100	1.0337	434.06	1.2917
140	1.0707	598.72	1.7242	140	1.0678	602.04	1.7193
180	1.1159	770.50	2.1210	180	1.1120	773.20	2.1147
220	1.1748	947.5	2.4953	220	1.1693	949.30	2.4870
260	1.2550	1133.4	2.8576	260	1.2462	1133.5	2.8459
300	1.3770	1337.3	3.2260	300	1.3596	1333.3	3.2071
Sat	1.6581	1610.5	3.6848	Sat	2.036	1826.3	4.0139
	p = 25.0 MPa = 250.0 bar				**p = 30.0 MPa = 300.0 bar**		
20	0.9907	107.24	0.2911	20	0.9886	111.84	0.2899
40	0.9971	189.52	0.5626	40	0.9951	193.89	0.5607
100	1.0313	437.85	1.2881	100	1.0290	441.66	1.2844
200	1.1344	862.8	2.2961	200	1.1302	865.3	2.2893
300	1.3442	1330.2	3.1900	300	1.3304	1327.8	3.1741

TABLE A.4
Physical Properties of Water at Saturation Pressure

T (°C)	Density (kg/m³)	$\beta \times 10^{-4}$ (K^{-1})	$\hat{C}_p$ (kJ/kg K)	k (W/m K)	$\alpha \times 10^{-6}$ (m^2/s)	$\eta \times 10^{-6}$ (Pa s)	$\upsilon \times 10^{-6}$ (m^2/s)
0	999.9	−0.7	4.226	0.558	0.131	1793.64	1.79
5	1000.0		4.206	0.568	0.135	1534.74	1.54
10	999.7	0.95	4.195	0.577	0.137	1296.44	1.30
15	999.1		4.187	0.587	0.141	1135.61	1.15
20	998.2	2.1	4.182	0.597	0.143	993.41	1.01
25	997.1		4.178	0.606	0.146	880.64	0.88
30	995.7	3.0	4.176	0.615	0.149	792.38	0.81
35	994.1		4.175	0.624	0.150	719.81	0.73
40	992.2	3.9	4.175	0.633	0.151	658.03	0.66
45	990.2		4.176	0.640	0.155	605.07	0.61
50	988.1	4.6	4.178	0.647	0.157	555.06	0.56
55	985.7		4.179	0.652	0.158	509.95	0.52
60	983.2	5.3	4.181	0.658	0.159	471.65	0.48
65	980.6		4.184	0.663	0.161	435.42	0.44
70	977.8	5.8	4.187	0.668	0.163	404.03	0.42
75	974.9		4.190	0.671	0.164	376.58	0.37
80	971.8	6.3	4.194	0.673	0.165	352.06	0.36
85	968.7		4.198	0.676	0.166	328.52	0.34
90	965.3	7.0	4.202	0.678	0.167	308.91	0.33
95	961.9		4.206	0.680	0.168	292.24	0.31
100	958.4	7.5	4.211	0.682	0.169	277.53	0.29
110	951.0	8.0	4.224	0.684	0.170	254.97	0.27
120	943.5	8.5	4.232	0.685	0.171	235.36	0.24
130	934.8	9.1	4.250	0.686	0.172	211.82	0.23
140	926.3	9.7	4.257	0.684	0.172	201.04	0.21
150	916.9	10.3	4.270	0.684	0.173	185.35	0.20
160	907.6	10.8	4.285	0.680	0.173	171.62	0.19
170	897.3	11.5	4.396	0.679	0.172	162.29	0.18
180	886.6	12.1	4.396	0.673	0.172	152.00	0.17
190	876.0	12.8	4.480	0.670	0.171	145.14	0.17
200	862.8	13.5	4.501	0.665	0.170	139.25	0.16
210	852.8	14.3	4.560	0.655	0.168	131.41	0.15
220	837.0	15.2	4.605	0.652	0.167	124.54	0.15
230	827.3	16.2	4.690	0.637	0.164	119.64	0.15
240	809.0	17.2	4.731	0.634	0.162	113.76	0.14
250	799.2	18.6	4.857	0.618	0.160	109.83	0.14

Note: β is the volumetric thermal expansion coefficient, $\hat{C}_p$ is the specific heat, k is the thermal conductivity, α is the thermal diffusivity, η is the absolute viscosity, and υ is the kinematic viscosity.

TABLE A.5
Properties of Liquid and Saturated Vapor of Ammonia

		Enthalpy (kJ/kg)		Entropy (kJ/kg K)		Specific Volume (L/kg)	
T (°C)	p (kPa)	Liquid $\hat{h}$	Vapor $\hat{H}$	Liquid $\hat{s}$	Vapor $\hat{S}$	Liquid $\hat{v}$	Vapor $\hat{V}$
−60	21.99	−69.5330	1373.19	−0.10909	6.6592	1.4010	4685.080
−55	30.29	−47.5062	1382.01	−0.00717	6.5454	1.4126	3474.220
−50	41.03	−25.4342	1390.64	0.09264	6.4382	1.4245	2616.510
−45	54.74	−3.3020	1399.07	0.19049	6.3369	1.4367	1998.910
−40	72.01	18.9024	1407.26	0.28651	6.2410	1.4493	1547.360
−35	93.49	41.1883	1415.20	0.38082	6.1501	1.4623	1212.490
−30	119.90	63.5629	1422.86	0.47351	6.0636	1.4757	960.867
−28	132.02	72.5387	1425.84	0.51015	6.0302	1.4811	878.100
−26	145.11	81.5300	1428.76	0.54655	5.9974	1.4867	803.761
−24	159.22	90.5370	1431.64	0.58272	5.9652	1.4923	736.868
−22	174.41	99.5600	1434.46	0.61865	5.9336	1.4980	676.570
−20	190.74	108.599	1437.23	0.65436	5.9025	1.5037	622.122
−18	208.26	117.656	1439.94	0.68984	5.8720	1.5096	572.875
−16	227.04	126.729	1442.60	0.72511	5.8420	1.5155	528.257
−14	247.14	135.820	1445.20	0.76016	5.8125	1.5215	487.769
−12	268.63	144.929	1447.74	0.79501	5.7835	1.5276	450.971
−10	291.57	154.056	1450.22	0.82965	5.7550	1.5338	417.477
−9	303.60	158.628	1451.44	0.84690	5.7409	1.5369	401.860
−8	316.02	163.204	1452.64	0.86410	5.7269	1.5400	386.944
−7	328.84	167.785	1453.83	0.88125	5.7131	1.5432	372.692
−6	342.07	172.371	1455.00	0.89835	5.6993	1.5464	359.071
−5	355.71	176.962	1456.15	0.91541	5.6856	1.5496	346.046
−4	369.77	181.559	1457.29	0.93242	5.6721	1.5528	333.589
−3	384.26	186.161	1458.42	0.94938	5.6586	1.5561	321.670
−2	399.20	190.768	1459.53	0.96630	5.6453	1.5594	310.263
−1	414.58	195.381	1460.62	0.98317	5.6320	1.5627	299.340
0	430.43	200.000	1461.70	1.00000	5.6189	1.5660	288.880
1	446.74	204.625	1462.76	1.01679	5.6058	1.5694	278.858
2	463.53	209.256	1463.80	1.03354	5.5929	1.5727	269.253
3	480.81	213.892	1464.83	1.05024	5.5800	1.5762	260.046
4	498.59	218.535	1465.84	1.06691	5.5672	1.5796	251.216
5	516.87	223.185	1466.84	1.08353	5.5545	1.5831	242.745
6	535.67	227.841	1467.82	1.10012	5.5419	1.5866	234.618
7	555.00	232.503	1468.78	1.11667	5.5294	1.5901	226.817
8	574.87	237.172	1469.72	1.13317	5.5170	1.5936	219.326
9	595.28	241.848	1470.64	1.14964	5.5046	1.5972	212.132
10	616.25	246.531	1471.57	1.16607	5.4924	1.6008	205.221
11	637.78	251.221	1472.46	1.18246	5.4802	1.6045	198.580
12	659.89	255.918	1473.34	1.19882	5.4681	1.6081	192.196
13	682.59	260.622	1474.20	1.21515	5.4561	1.6118	186.058
14	705.88	265.334	1475.05	1.23144	5.4441	1.6156	180.154
15	729.79	270.053	1475.88	1.24769	5.4322	1.6193	174.475
16	754.31	274.779	1476.69	1.26391	5.4204	1.6231	169.009
17	779.46	279.513	1477.48	1.28010	5.4087	1.6269	163.748
18	805.25	284.255	1478.25	1.29626	5.3971	1.6308	158.683

(continued)

TABLE A.5 (continued)
Properties of Liquid and Saturated Vapor of Ammonia

		Enthalpy (kJ/kg)		Entropy (kJ/kg K)		Specific Volume (L/kg)	
T (°C)	p (kPa)	Liquid $\hat{h}$	Vapor $\hat{H}$	Liquid $\hat{s}$	Vapor $\hat{S}$	Liquid $\hat{v}$	Vapor $\hat{V}$
19	831.69	289.005	1479.01	1.31238	5.3855	1.6347	153.804
20	858.79	293.762	1479.75	1.32847	5.3740	1.6386	149.106
21	886.57	298.527	1480.48	1.34452	5.3626	1.6426	144.578
22	915.03	303.300	1481.18	1.36055	5.3512	1.6466	140.214
23	944.18	308.081	1481.87	1.37654	5.3399	1.6507	136.006
24	974.03	312.870	1482.53	1.39250	5.3286	1.6547	131.950
25	1004.60	317.667	1483.18	1.40843	5.3175	1.6588	128.037
26	1035.90	322.471	1483.81	1.42433	5.3063	1.6630	124.261
27	1068.00	327.284	1484.42	1.44020	5.2953	1.6672	120.619
28	1100.70	332.104	1485.01	1.45604	5.2843	1.6714	117.103
29	1134.30	336.933	1485.59	1.47185	5.2733	1.6757	113.708
30	1168.60	341.769	1486.14	1.48762	5.2624	1.6800	110.430
31	1203.70	346.614	1486.67	1.50337	5.2516	1.6844	107.263
32	1239.60	351.466	1487.18	1.51908	5.2408	1.6888	104.205
33	1276.30	356.326	1487.66	1.53477	5.2300	1.6932	101.248
34	1313.90	361.195	1488.13	1.55042	5.2193	1.6977	98.391
35	1352.20	366.072	1488.57	1.56605	5.2086	1.7023	95.629
36	1391.50	370.957	1488.99	1.58165	5.1980	1.7069	92.958
37	1431.50	375.851	1489.39	1.59722	5.1874	1.7115	90.374
38	1472.40	380.754	1489.76	1.61276	5.1768	1.7162	87.875
39	1514.30	385.666	1489.10	1.62828	5.1663	1.7209	85.456
40	1557.00	390.587	1490.42	1.64377	5.1558	1.7257	83.115
41	1600.60	395.519	1490.71	1.65924	5.1453	1.7305	80.848
42	1645.10	400.462	1490.98	1.67470	5.1349	1.7354	78.654
43	1690.60	405.416	1491.21	1.69013	5.1244	1.7404	76.528
44	1737.00	410.382	1491.41	1.70554	5.1140	1.7454	74.468
45	1784.30	415.362	1491.58	1.72095	5.1036	1.7504	72.472
46	1832.60	420.358	1491.72	1.73635	5.0932	1.7555	70.536
47	1881.90	425.369	1491.83	1.75174	5.0827	1.7607	68.660
48	1932.20	430.399	1491.88	1.76714	5.0723	1.7659	66.840
49	1983.50	435.450	1491.91	1.78255	5.0618	1.7712	65.075
50	2035.90	440.523	1491.89	1.79798	5.0514	1.7766	63.361
51	2089.20	445.623	1491.83	1.81343	5.0409	1.7820	61.697
52	2143.60	450.751	1491.73	1.82891	5.0303	1.7875	60.081
53	2199.10	455.913	1491.58	1.84445	5.0198	1.7931	58.511
54	2255.60	461.112	1491.38	1.86004	5.0092	1.7987	56.985
55	2313.20	466.353	1491.12	1.87571	4.9985	1.8044	55.502

TABLE A.6
Properties of Refrigerant R-12 Saturated

		Specific Volume (m^3/kg)		Enthalpy (kJ/kg)		Entropy (kJ/kg K)	
T (°C)	p (bar)	Liquid × 10^3	Vapor	Liquid	Vapor	Liquid	Vapor
−40	0.6417	0.6595	0.24191	0.00	169.59	0.0000	0.7274
−35	0.8071	0.6656	0.19540	4.42	171.90	0.0187	0.7219
−30	1.0041	0.6720	0.15938	8.86	174.20	0.0371	0.7170
−28	1.0927	0.6746	0.14728	10.65	175.11	0.0444	0.7153
−26	1.1872	0.6773	0.13628	12.43	176.02	0.0517	0.7135
−25	1.2368	0.6786	0.13117	13.33	176.48	0.0552	0.7126
−24	1.2880	0.6800	0.12628	14.22	176.93	0.0589	0.7119
−22	1.3953	0.6827	0.11717	16.02	177.83	0.0660	0.7103
−20	1.5093	0.6855	0.10885	17.82	178.74	0.0731	0.7087
−18	1.6304	0.6883	0.10124	19.62	179.63	0.0802	0.7073
−15	1.8260	0.6926	0.09102	22.33	180.97	0.0906	0.7051
−10	2.1912	0.7000	0.07665	26.87	183.19	0.1080	0.7019
−5	2.6096	0.7078	0.06496	31.45	185.37	0.1251	0.6991
0	3.0861	0.7159	0.05539	36.05	187.53	0.1420	0.6965
4	3.5124	0.7227	0.04895	39.76	189.23	0.1553	0.6946
8	3.9815	0.7297	0.04340	43.50	190.91	0.1686	0.6929
12	4.4962	0.7370	0.03860	47.26	192.56	0.1817	0.6913
16	5.0591	0.7446	0.03442	51.05	194.19	0.1948	0.6898
20	5.6729	0.7525	0.03078	54.87	195.78	0.2078	0.6884
24	6.3405	0.7607	0.02759	58.73	197.34	0.2207	0.6871
26	6.6954	0.7650	0.02614	60.68	198.11	0.2271	0.6865
28	7.0648	0.7694	0.02478	62.63	198.87	0.2335	0.6859
30	7.4490	0.7739	0.02351	64.59	199.62	0.2400	0.6853
32	7.8485	0.7785	0.02231	66.57	200.36	0.2463	0.6847
34	8.2636	0.7832	0.02118	68.55	201.09	0.2527	0.6842
36	8.6948	0.7880	0.02012	70.55	201.80	0.2591	0.6836
38	9.1423	0.7929	0.01912	72.56	202.51	0.2655	0.6831
40	9.6065	0.7980	0.01817	74.59	203.20	0.2718	0.6825
42	10.088	0.8033	0.01728	76.63	203.88	0.2782	0.6820
44	10.587	0.8086	0.01644	78.68	204.54	0.2845	0.6814
48	11.639	0.8199	0.01488	82.23	205.83	0.2973	0.6802
52	12.766	0.8318	0.01349	87.06	207.05	0.3101	0.6791
56	13.972	0.8445	0.01224	91.36	208.20	0.3229	0.6779
60	15.259	0.8581	0.01111	95.74	209.26	0.3358	0.6765
112	41.255	1.792	0.00179	183.35	183.35	0.5687	0.56987

TABLE A.7
Properties of Refrigerant R-12 Saturated

		Specific Volume (m^3/kg)		Enthalpy (kJ/kg)		Entropy (kJ/kg K)	
p (bar)	T (°C)	Liquid × 10^3	Vapor	Liquid	Vapor	Liquid	Vapor
0.6	–41.42	0.6578	0.2575	–1.25	168.94	–0.0054	0.7290
1.0	–30.10	0.6719	0.1600	8.78	174.15	0.0368	0.7171
1.2	–25.74	0.6776	0.1349	12.66	176.14	0.0526	0.7133
1.4	–21.91	0.6828	0.1168	16.09	177.87	0.0663	0.7102
1.6	–18.49	0.6876	0.1031	19.18	179.41	0.0784	0.7076
1.8	–15.38	0.6921	0.09225	21.98	180.80	0.0893	0.7054
2.0	–12.53	0.6962	0.08354	24.57	182.07	0.0992	0.7035
2.4	–7.42	0.7040	0.07033	29.23	184.32	0.1168	0.7004
2.8	–2.93	0.7111	0.06076	33.35	186.27	0.1321	0.6980
3.2	1.11	0.7177	0.05351	37.08	188.00	0.1457	0.6960
4.0	8.15	0.7299	0.04321	43.64	190.97	0.1691	0.6928
5.0	15.60	0.7438	0.03482	50.67	194.02	0.1935	0.6899
6.0	22.00	0.7566	0.02913	56.80	196.57	0.2142	0.6878
7.0	27.65	0.7686	0.02501	62.29	198.74	0.2324	0.6860
8.0	32.74	0.7802	0.02188	67.30	200.63	0.2487	0.6845
9.0	37.37	0.7914	0.01942	71.93	202.29	0.2634	0.6832
10.0	41.64	0.8023	0.01744	76.26	203.76	0.2770	0.6820
12.0	49.31	0.8237	0.01441	84.21	206.24	0.3015	0.6799
14.0	56.09	0.8448	0.01222	91.46	208.22	0.3232	0.6778
16.0	62.19	0.8660	0.01054	98.19	209.81	0.3329	0.6758

TABLE A.8
Properties of Refrigerant R-12, Super Heated Vapor

T (°C)	Specific Volume (m³/kg)	Enthalpy (kJ/kg)	Entropy (kJ/kg K)	T (°C)	Specific Volume (m³/kg)	Enthalpy (kJ/kg)	Entropy (kJ/kg K)
	p = 0.06 MPa = 0.6 bar				p = 0.10 MPa = 1.0 bar		
	T_{sat} = −41.42°C				T_{sat} = −30.10°C		
Sat	0.2575	168.94	0.7290	Sat	0.1600	174.15	0.7171
−40	0.2593	169.72	0.7324	−40			
−20	0.2838	180.94	0.7785	−20	0.1677	179.99	0.7406
0	0.3079	192.52	0.8225	0	0.1827	191.77	0.7854
10	0.3198	198.45	0.8439	10	0.1900	197.77	0.8070
20	0.3317	204.47	0.8647	20	0.1973	203.85	0.8281
30	0.3425	210.57	0.8852	30	0.2045	210.02	0.8488
40	0.3552	216.77	0.9053	40	0.2117	216.26	0.8691
50	0.3670	223.04	0.9251	50	0.2188	222.58	0.8889
60	0.3787	229.41	0.9444	60	0.2260	228.98	0.9084
80	0.4020	242.37	0.9822	80	0.2401	242.01	0.9464
	p = 0.14 MPa = 1.4 bar				p = 0.18 MPa = 1.8 bar		
	T_{sat} = −21.91°C				T_{sat} = −15.38°C		
Sat	0.1168	177.87	0.7102	Sat	0.0922	180.80	0.7054
−20	0.1179	179.01	0.7147	−20			
−10	0.1235	184.97	0.7378	−10	0.0925	181.03	0.7181
0	0.1289	190.99	0.7602	0	0.0991	190.21	0.7408
10	0.1343	197.08	0.7821	10	0.1034	196.38	0.7630
20	0.1397	203.23	0.8035	20	0.1076	202.60	0.7846
30	0.1449	209.46	0.8243	30	0.1118	208.89	0.8057
40	0.1502	215.75	0.8447	40	0.1160	215.23	0.8263
50	0.1553	222.12	0.8648	50	0.1201	221.64	0.8464
60	0.1605	228.55	0.8844	60	0.1241	228.12	0.8662
80	0.1707	241.64	0.9225	80	0.1322	241.27	0.9045
100	0.1809	255.00	0.9593	100	0.1402	254.69	0.9414
	p = 0.20 MPa = 2.0 bar				p = 0.24 MPa = 2.4 bar		
	T_{sat} = −12.53°C				T_{sat} = −7.42°C		
Sat	0.0835	182.07	0.7035	Sat	0.0703	184.32	0.7004
0	0.0886	189.08	0.7325	0	0.0729	188.99	0.7177
10	0.0926	196.02	0.7548	10	0.0763	195.29	0.7404
20	0.0964	202.28	0.7766	20	0.0796	201.63	0.7624
30	0.1002	208.60	0.7978	30	0.0828	208.01	0.7838
40	0.1040	214.97	0.8184	40	0.0860	214.44	0.8047
50	0.1077	221.40	0.8387	50	0.0892	220.92	0.8251
60	0.1114	227.90	0.8585	60	0.0923	227.46	0.8450
80	0.1187	241.09	0.8969	80	0.0985	240.71	0.8836
100	0.1259	254.53	0.9339	100	0.1045	254.20	0.9208
120	0.1331	268.21	0.9696	120	0.1105	267.93	0.9566

(continued)

TABLE A.8 (continued)
Properties of Refrigerant R-12, Super Heated Vapor

T (°C)	Specific Volume (m³/kg)	Enthalpy (kJ/kg)	Entropy (kJ/kg K)	T (°C)	Specific Volume (m³/kg)	Enthalpy (kJ/kg)	Entropy (kJ/kg K)
	p = 0.28 MPa = 2.8 bar				p = 0.32 MPa = 3.2 bar		
	T_{sat} = −2.93°C				T_{sat} = 1.11°C		
Sat	0.06076	186.27	0.6980	Sat	0.05351	188.00	0.6960
0	0.06166	188.15	0.7049	0			
10	0.06464	194.55	0.7279	10	0.05590	193.79	0.7167
20	0.06755	200.97	0.7502	20	0.05852	200.30	0.7393
30	0.07040	207.42	0.7718	30	0.06106	206.82	0.7612
40	0.07319	213.91	0.7928	40	0.06355	213.36	0.7824
50	0.07594	220.44	0.8134	50	0.06600	219.94	0.8031
60	0.07865	227.02	0.8334	60	0.06841	226.57	0.8233
80	0.08399	240.34	0.8722	80	0.07314	239.96	0.8623
100	0.08924	253.88	0.9095	100	0.07778	253.55	0.8997
120	0.09443	267.65	0.9455	120	0.08236	267.36	0.9358
	p = 0.40 MPa = 4.0 bar				p = 0.50 MPa = 5.0 bar		
	T_{sat} = 8.15°C				T_{sat} = 15.60°C		
Sat	0.04321	190.97	0.6928	Sat	0.03482	194.02	0.6899
10	0.04363	192.21	0.6972	10			
20	0.04584	198.91	0.7204	20	0.03565	197.08	0.7004
30	0.04797	205.58	0.7428	30	0.03746	203.96	0.7235
40	0.05005	212.25	0.7645	40	0.03922	210.81	0.7457
50	0.05207	218.94	0.7855	50	0.04091	217.64	0.7672
60	0.05406	225.65	0.8060	60	0.04257	224.48	0.7881
80	0.05791	239.19	0.8454	80	0.04578	238.21	0.8281
100	0.06173	252.89	0.8831	100	0.04889	252.05	0.8662
120	0.06546	266.79	0.9194	120	0.05193	266.06	0.9028
140	0.06913	280.88	0.9544	140	0.05492	280.23	0.9379
	p = 0.60 MPa = 6.0 bar				p = 0.70 MPa = 7.0 bar		
	T_{sat} = 22.00°C				T_{sat} = 27.65°C		
Sat	0.02913	196.57	0.6878	Sat	0.02501	198.74	0.6860
30	0.03042	202.26	0.7968	0	0.02535	200.46	0.6917
40	0.03197	209.31	0.7297	10	0.02676	207.73	0.7153
50	0.03354	216.30	0.7516	20	0.02810	214.90	0.7378
60	0.03489	223.27	0.7729	30	0.02939	222.02	0.7595
80	0.03765	237.20	0.8135	40	0.03184	236.17	0.8008
100	0.04032	251.20	0.8520	50	0.03419	250.33	0.8398
120	0.04291	265.32	0.8889	60	0.03646	264.57	0.8769
140	0.04545	279.58	0.9243	80	0.03867	278.92	0.9125
160	0.04794	294.01	0.9584	120	0.04085	293.42	0.9468

TABLE A.8 (continued)
Properties of Refrigerant R-12, Super Heated Vapor

T (°C)	Specific Volume (m³/kg)	Enthalpy (kJ/kg)	Entropy (kJ/kg K)	T (°C)	Specific Volume (m³/kg)	Enthalpy (kJ/kg)	Entropy (kJ/kg K)
	p = 0.80 MPa = 8.0 bar				p = 0.90 MPa = 9.0 bar		
	T_{sat} = 32.74°C				T_{sat} = 37.37°C		
Sat	0.02188	200.63	0.6845	Sat	0.01942	202.29	0.6832
40	0.02283	206.07	0.7021	40	0.01974	204.32	0.6897
50	0.02407	213.45	0.7253	50	0.02091	211.92	0.7136
60	0.02525	220.72	0.7474	60	0.02201	219.37	0.7363
80	0.02748	235.11	0.7894	80	0.02407	234.03	0.7790
100	0.02959	249.44	0.8289	100	0.02601	248.54	0.8190
120	0.03162	263.81	0.8664	120	0.02785	263.03	0.8569
140	0.03359	278.26	0.9022	140	0.02964	277.58	0.8930
160	0.03552	292.83	0.9367	160	0.03138	292.23	0.9276
180	0.03742	307.54	0.9699	180	0.03309	307.01	0.9609
	p = 1.00 MPa = 10.0 bar				p = 1.20 MPa = 12.0 bar		
	T_{sat} = 41.64°C				T_{sat} = 49.31°C		
Sat	0.01744	203.76	0.6820	Sat	0.01441	206.24	0.6799
50	0.01837	210.32	0.7026	50	0.01448	206.81	0.6816
60	0.01941	217.97	0.7259	60	0.01546	214.96	0.7065
80	0.02134	232.91	0.7695	80	0.01722	230.57	0.7520
100	0.02313	247.61	0.8100	100	0.01881	245.70	0.7937
120	0.02484	262.25	0.8482	120	0.02030	260.63	0.8326
140	0.02647	276.90	0.8845	140	0.02172	275.51	0.8696
160	0.02807	291.63	0.9193	160	0.02309	290.41	0.9048
180	0.02963	306.47	0.9528	180	0.02443	305.37	0.9385
200	0.03116	321.42	0.9851	200	0.02574	320.44	0.9711
	p = 1.40 MPa = 14.0 bar				p = 1.60 MPa = 16.0 bar		
	T_{sat} = 56.09°C				T_{sat} = 62.19°C		
Sat	0.01222	208.22	0.6778	Sat	0.01054	209.81	0.6758
60	0.01258	211.61	0.6881	60			
80	0.01425	228.06	0.7360	80	0.01198	225.34	0.7209
100	0.01571	243.69	0.7791	100	0.01337	241.58	0.7656
120	0.01705	258.96	0.8189	120	0.01461	257.22	0.8065
140	0.01832	274.08	0.8564	140	0.01577	272.61	0.8447
160	0.01954	289.16	0.8921	160	0.01686	287.88	0.8808
180	0.02071	304.26	0.9262	180	0.01792	303.14	0.9152
200	0.02186	319.44	0.9589	200	0.01895	318.43	0.9482
220	0.02299	334.70	0.9905	220	0.01996	333.78	0.9800

TABLE A.9
Properties of Refrigerant R-134a, Saturated

		Specific Volume (m^3/kg)		Enthalpy (kJ/kg)		Entropy (kJ/kg K)	
T (°C)	*p* (bar)	Liquid × 10^3	Vapor	Liquid	Vapor	Liquid	Vapor
−40	0.5164	0.7055	0.3569	0.00	222.88	0.0000	0.9560
−36	0.6332	0.7113	0.2947	4.73	225.40	0.0201	0.9506
−32	0.7704	0.7172	0.2451	9.52	227.90	0.0401	0.9456
−28	0.9305	0.7233	0.2052	14.37	230.38	0.0600	0.9411
−26	1.0199	0.7265	0.1882	16.82	231.62	0.0699	0.9390
−24	1.1160	0.7296	0.1728	19.29	232.85	0.0798	0.9370
−22	1.2192	0.7328	0.1590	21.77	234.08	0.0897	0.9351
−20	1.3299	0.7361	0.1464	24.26	235.31	0.0996	0.9332
−18	1.4483	0.7395	0.1350	26.77	236.53	0.1094	0.9315
−16	1.5748	0.7428	0.1247	29.30	237.74	0.1192	0.9298
−12	1.8540	0.7498	0.1068	34.39	240.15	0.1388	0.9267
−8	2.1704	0.7569	0.0919	39.54	242.54	0.1583	0.9239
−4	2.5274	0.7644	0.0794	44.75	244.90	0.1777	0.9213
0	2.9282	0.7721	0.0689	50.02	247.23	0.1970	0.9190
4	3.3765	0.7801	0.0600	55.35	249.53	0.2162	0.9169
8	3.8756	0.7884	0.0525	60.73	251.80	0.2354	0.9150
12	4.4294	0.7971	0.0460	66.18	254.03	0.2545	0.9132
16	5.0416	0.8062	0.0405	71.69	256.22	0.2735	0.9116
20	5.7160	0.8157	0.0358	77.26	258.36	0.2924	0.9102
24	6.4566	0.8257	0.0317	82.90	260.45	0.3113	0.9089
26	6.8530	0.8309	0.0298	85.75	261.48	0.3208	0.9082
28	7.2675	0.8362	0.0281	88.61	262.50	0.3302	0.9076
30	7.7006	0.8417	0.0265	91.49	263.50	0.3396	0.9070
32	8.1528	0.8473	0.0250	94.39	264.48	0.3490	0.9064
34	8.6247	0.8530	0.0236	97.31	265.45	0.3584	0.9058
36	9.1168	0.8590	0.0223	100.25	266.40	0.3678	0.9053
38	9.6298	0.8651	0.0210	103.21	267.33	0.3772	0.9047
40	10.164	0.8714	0.0199	106.19	268.24	0.3866	0.9041
42	10.720	0.8780	0.0188	109.10	269.14	0.3960	0.9035
44	11.299	0.8780	0.0177	112.22	270.01	0.4054	0.9030
48	12.526	0.8989	0.0159	118.35	271.68	0.4243	0.9017
52	13.851	0.9142	0.0142	124.58	273.24	0.4432	0.9004
56	15.278	0.9308	0.0127	130.93	274.68	0.4622	0.8990
60	16.813	0.9488	0.0114	137.42	275.99	0.4814	0.8973
70	21.162	1.0027	0.0086	154.35	278.43	0.5302	0.8918
80	26.324	1.0766	0.0064	172.71	279.12	0.5814	0.8827
90	32.435	1.1949	0.0046	193.69	272.32	0.6380	0.8655
100	39.742	1.5443	0.0027	224.74	259.13	0.7196	0.8117

TABLE A.10
Properties of Refrigerant R-134a, Saturated

		Specific Volume (m^3/kg)		Enthalpy (kJ/kg)		Entropy (kJ/kg K)	
p (bar)	*T* (°C)	Liquid × 10^3	Vapor	Liquid	Vapor	Liquid	Vapor
0.6	−37.07	0.7097	0.3100	3.46	224.72	0.0147	0.9520
0.8	−31.21	0.7184	0.2366	10.47	228.39	0.0440	0.9447
1.0	−26.43	0.7258	0.1917	16.29	231.35	0.0678	0.9395
1.2	−22.36	0.7323	0.1614	21.32	233.86	0.0879	0.9354
1.4	−18.80	0.7381	0.1395	25.77	236.04	0.1055	0.9322
1.6	−15.62	0.7435	0.1229	29.78	237.97	0.1211	0.9295
1.8	−12.73	0.7485	0.1098	33.45	239.71	0.1352	0.9273
2.0	−10.09	0.7532	0.0993	36.84	241.30	0.1481	0.9253
2.4	−5.37	0.7618	0.0834	42.95	244.09	0.1710	0.9222
2.8	−1.23	0.7697	0.0719	48.39	246.52	0.1911	0.9197
3.2	2.48	0.7770	0.0632	53.31	248.66	0.2089	0.9177
3.6	5.84	0.7839	0.0564	57.82	250.58	0.2251	0.9160
4.0	8.93	0.7904	0.0509	62.00	252.32	0.2399	0.9145
5.0	15.74	0.8056	0.0409	71.33	256.07	0.2723	0.9117
6.0	21.58	0.8196	0.0341	79.48	259.19	0.2999	0.9097
7.0	26.72	0.8328	0.0292	86.78	261.85	0.3242	0.9080
8.0	31.33	0.8454	0.0255	93.42	264.15	0.3459	0.9066
9.0	35.53	0.8576	0.0226	99.56	266.18	0.3656	0.9054
10.0	39.39	0.8695	0.0202	105.29	267.97	0.3838	0.9043
12.0	46.32	0.8928	0.0166	115.76	270.99	0.4164	0.9023
14.0	52.43	0.9159	0.0140	125.26	273.40	0.4453	0.9003
16.0	57.92	0.9392	0.0121	134.02	275.33	0.4714	0.8982
18.0	62.91	0.9631	0.0105	142.22	276.83	0.4954	0.8959
20.0	67.49	0.9878	0.0093	149.99	277.94	0.5178	0.8934
25.0	77.59	1.0562	0.0069	168.12	279.17	0.5687	0.8854
30.0	86.22	1.1416	0.0053	185.30	278.01	0.6156	0.8735

TABLE A.11
Properties of Refrigerant R-134a, Super Heated Vapor

T (°C)	Specific Volume (m³/kg)	Enthalpy (kJ/kg)	Entropy (kJ/kg K)	T (°C)	Specific Volume (m³/kg)	Enthalpy (kJ/kg)	Entropy (kJ/kg K)
	p = 0.06 MPa = 0.6 bar				p = 0.10 MPa = 1.0 bar		
	T_{sat} = −37.07°C				T_{sat} = −26.43°C		
Sat	0.31003	224.72	0.9520	Sat	0.19170	231.35	0.9395
−20	0.33536	237.98	1.0062	−20	0.19770	236.54	0.9602
−10	0.34992	245.96	1.0371	−10	0.20686	244.70	0.9918
0	0.36443	254.10	1.0675	0	0.21587	252.99	1.0227
10	0.37861	262.41	1.0973	10	0.22473	261.43	1.0531
20	0.39279	270.89	1.1267	20	0.23349	270.02	1.0829
30	0.40688	279.53	1.1557	30	0.24216	278.76	1.1122
40	0.42091	288.35	1.1844	40	0.25076	287.66	1.1411
50	0.43487	297.34	1.2126	50	0.25930	296.72	1.1696
60	0.44879	306.51	1.2405	60	0.26779	305.94	1.1977
70	0.46266	315.84	1.2681	70	0.27623	315.32	1.2254
80	0.47650	325.34	1.2954	80	0.28464	324.87	1.2528
90	0.49031	335.00	1.3224	90	0.29302	334.57	1.2799
	p = 0.14 MPa = 1.4 bar				p = 0.18 MPa = 1.8 bar		
	T_{sat} = −18.80°C				T_{sat} = −12.73°C		
Sat	0.13945	236.04	0.9322	Sat	0.10983	239.71	0.9273
−10	0.14549	243.40	0.9606	−10	0.11135	242.06	0.9362
0	0.15219	251.86	0.9922	0	0.11678	250.69	0.9684
10	0.15875	260.43	1.0230	10	0.12207	259.41	0.9998
20	0.16520	269.13	1.0532	20	0.12723	268.23	1.0304
30	0.17155	277.97	1.0828	30	0.13230	277.17	1.0604
40	0.17783	286.96	1.1120	40	0.13730	286.24	1.0898
50	0.18404	296.09	1.1407	50	0.14222	295.45	1.1187
60	0.19020	305.37	1.1690	60	0.14710	304.79	1.1472
70	0.19633	314.80	1.1969	70	0.15193	314.28	1.1753
80	0.20241	324.39	1.2244	80	0.15672	323.92	1.2030
90	0.20846	334.14	1.2516	90	0.16148	333.70	1.2303
100	0.21449	344.04	1.2785	100	0.16622	343.63	1.2573
	p = 0.20 MPa = 2.0 bar				p = 0.24 MPa = 2.4 bar		
	T_{sat} = −10.09°C				T_{sat} = −5.37°C		
Sat	0.09933	241.30	0.9253	Sat	0.08343	244.09	0.9222
−10	0.09938	241.38	0.9256	−10			
0	0.10438	250.10	0.9582	0	0.08574	248.89	0.9399
10	0.10922	258.89	0.9898	10	0.08993	257.84	0.9721
20	0.11394	267.78	1.0206	20	0.09399	266.85	1.0034
30	0.11856	276.77	1.0508	30	0.09794	275.95	1.0339
40	0.12311	285.88	1.0804	40	0.10181	285.16	1.0637
50	0.12758	295.12	1.1094	50	0.10562	294.47	1.0930
60	0.13201	304.50	1.1380	60	0.10937	303.91	1.1218
70	0.13639	314.02	1.1661	70	0.11307	313.49	1.1501
80	0.14073	323.68	1.1939	80	0.11674	323.19	1.1780
90	0.14504	333.48	1.2212	90	0.12037	333.04	1.2055
100	0.14932	343.43	1.2483	100	0.12398	343.03	1.2326

TABLE A.11 (continued)
Properties of Refrigerant R-134a, Super Heated Vapor

T (°C)	Specific Volume (m³/kg)	Enthalpy (kJ/kg)	Entropy (kJ/kg K)	T (°C)	Specific Volume (m³/kg)	Enthalpy (kJ/kg)	Entropy (kJ/kg K)
	p = 0.28 MPa = 2.8 bar				**p = 0.32 MPa = 3.2 bar**		
	T_{sat} = −1.23°C				**T_{sat} = 2.48°C**		
Sat	0.07193	246.52	0.9197	Sat	0.06322	248.66	0.9177
0	0.07240	247.64	0.9238	0			
10	0.07613	256.76	0.9566	10	0.06576	255.65	0.9427
20	0.07972	265.91	0.9883	20	0.06901	264.95	0.9749
30	0.08320	275.12	1.0192	30	0.07214	274.28	1.0062
40	0.08660	284.42	1.0494	40	0.07518	283.67	1.0367
50	0.08992	293.81	1.0789	50	0.07815	293.15	1.0665
60	0.09319	303.32	1.1079	60	0.08106	302.72	1.0957
70	0.09641	312.95	1.1364	70	0.08392	312.41	1.1243
80	0.09960	322.71	1.1644	80	0.08674	322.22	1.1525
90	0.10275	332.60	1.1029	90	0.08953	332.15	1.1802
100	0.10587	342.62	1.2193	100	0.09229	342.21	1.2076
110	0.10897	352.78	1.2461	110	0.09503	352.40	1.2345
120	0.11205	363.08	1.2727	120	0.09774	362.73	1.2611
	p = 0.40 MPa = 4.0 bar				**p = 0.50 MPa = 5.0 bar**		
	T_{sat} = 8.93°C				**T_{sat} = 15.74°C**		
Sat	0.05089	252.32	0.9145	Sat	0.04086	256.07	0.9117
10	0.05119	253.35	0.9182	10			
20	0.05397	262.96	0.9515	20	0.04188	260.34	0.9264
30	0.05662	272.54	0.9837	30	0.04416	270.28	0.9597
40	0.05917	282.14	1.0148	40	0.04633	280.16	0.9918
50	0.06164	291.79	1.0452	50	0.04842	290.04	1.0229
60	0.06405	301.51	1.0748	60	0.05043	299.95	1.0531
70	0.06641	311.32	1.1038	70	0.05240	309.92	1.0825
80	0.06873	321.23	1.1322	80	0.05432	319.96	1.1114
90	0.07102	331.25	1.1602	90	0.05620	330.10	1.1397
100	0.07327	341.38	1.1878	100	0.05805	340.33	1.1675
110	0.07550	351.64	1.2149	110	0.05988	350.68	1.1949
120	0.07771	362.03	1.2417	120	0.06168	361.14	1.2218
130	0.07991	372.54	1.2681	130	0.06347	371.72	1.2484
140	0.08208	383.18	1.2941	140	0.06523	382.42	1.2746
	p = 0.60 MPa = 6.0 bar				**p = 0.70 MPa = 7.0 bar**		
	T_{sat} = 21.58°C				**T_{sat} = 26.72°C**		
Sat	0.03408	259.19	0.9097	Sat	0.02918	261.85	0.9080
30	0.03581	267.89	0.9388	30	0.02979	265.37	0.9197
40	0.03774	278.09	0.9719	40	0.03157	275.93	0.9539
50	0.03958	288.23	1.0037	50	0.03324	286.35	0.9867
60	0.04134	298.35	1.0346	60	0.03482	296.69	1.0182
70	0.04304	308.48	1.0645	70	0.03634	307.01	1.0487
80	0.04469	318.67	1.0938	80	0.03781	317.35	1.0784
90	0.04631	328.93	1.1225	90	0.03924	327.74	1.1074
100	0.04790	339.27	1.1505	100	0.04064	338.19	1.1358

(continued)

TABLE A.11 (continued)
Properties of Refrigerant R-134a, Super Heated Vapor

T (°C)	Specific Volume (m³/kg)	Enthalpy (kJ/kg)	Entropy (kJ/kg K)	T (°C)	Specific Volume (m³/kg)	Enthalpy (kJ/kg)	Entropy (kJ/kg K)
	p = 0.60 MPa = 6.0 bar				p = 0.70 MPa = 7.0 bar		
	T_{sat} = 21.58°C				T_{sat} = 26.72°C		
110	0.04946	349.70	1.1781	110	0.04201	348.71	1.1637
120	0.05099	360.24	1.2053	120	0.04335	359.33	1.1910
130	0.05251	370.88	1.2320	130	0.04468	370.04	1.2179
140	0.05402	381.64	1.2584	140	0.04599	380.86	1.2444
150	0.05550	392.52	1.2844	150	0.04729	391.79	1.2706
160	0.05698	403.51	1.3100	160	0.04857	402.82	1.2963
	p = 0.80 MPa = 8.0 bar				p = 0.90 MPa = 9.0 bar		
	T_{sat} = 31.33°C				T_{sat} = 35.53°C		
Sat	0.02547	264.15	0.9066	Sat	0.02255	266.18	0.9054
40	0.02691	273.66	0.9374	40	0.02325	271.25	0.9217
50	0.02846	284.39	0.9711	50	0.02472	282.34	0.9566
60	0.02992	294.98	1.0034	60	0.02609	293.21	0.9897
70	0.03131	305.50	1.0345	70	0.02732	303.94	1.0214
80	0.03264	316.00	1.0647	80	0.02861	314.62	1.0521
90	0.03393	326.52	1.0940	90	0.02980	325.28	1.0819
100	0.03519	337.08	1.1227	100	0.03095	335.96	1.1109
110	0.03642	347.71	1.1508	110	0.03207	346.68	1.1392
120	0.03762	358.40	1.1784	120	0.03316	357.47	1.1670
130	0.03881	369.19	1.2055	130	0.03423	368.33	1.1943
140	0.03997	380.07	1.2321	140	0.03529	379.27	1.2211
150	0.04113	391.05	1.2584	150	0.03633	390.31	1.2475
160	0.04227	402.14	1.2842	160	0.03736	401.44	1.2735
170	0.04340	413.33	1.3098	170	0.03838	412.68	1.2992
180	0.04452	424.63	1.3351	180	0.03939	424.02	1.3245
	p = 1.00 MPa = 10.0 bar				p = 1.20 MPa = 12.0 bar		
	T_{sat} = 39.39°C				T_{sat} = 46.32°C		
Sat	0.02020	267.97	0.9043	Sat	0.01663	270.99	0.9023
40	0.02029	268.68	0.9066	40			
50	0.02171	280.19	0.9428	50	0.01712	275.52	0.9164
60	0.02301	291.36	0.9768	60	0.01835	287.44	0.9527
70	0.02423	302.34	1.0093	70	0.01947	298.96	0.9868
80	0.02538	313.20	1.0405	80	0.02051	310.24	1.0192
90	0.02649	324.01	1.0407	90	0.02150	321.39	1.0503
100	0.02755	334.82	1.1000	100	0.02244	332.47	1.0804
110	0.02858	345.65	1.1286	110	0.02335	343.52	1.1096
120	0.02959	356.52	1.1567	120	0.02423	354.58	1.1381
130	0.03058	367.46	1.1841	130	0.02508	365.68	1.1660
140	0.03154	378.46	1.2111	140	0.02592	376.83	1.1933
150	0.03250	389.56	1.2376	150	0.02674	388.04	1.2201
160	0.03344	400.74	1.2638	160	0.02754	399.33	1.2465
170	0.03436	412.02	1.2895	170	0.02834	410.70	1.2724
180	0.03528	423.40	1.3149	180	0.02912	422.16	1.2980

TABLE A.11 (continued)
Properties of Refrigerant R-134a, Super Heated Vapor

T (°C)	Specific Volume (m³/kg)	Enthalpy (kJ/kg)	Entropy (kJ/kg K)	T (°C)	Specific Volume (m³/kg)	Enthalpy (kJ/kg)	Entropy (kJ/kg K)
	p = 1.40 MPa = 14.0 bar				p = 1.60 MPa = 16.0 bar		
	T_{sat} = 52.43°C				T_{sat} = 57.92°C		
Sat	0.01405	273.40	0.9003	Sat	0.01208	275.33	0.8982
60	0.01495	283.10	0.9270	60	0.01233	278.20	0.9069
70	0.01603	295.31	0.9658	70	0.01340	291.33	0.9457
80	0.01701	307.10	0.9997	80	0.01435	303.74	0.9813
90	0.01792	318.63	1.0319	90	0.01521	315.72	1.0148
100	0.01878	330.02	1.0628	100	0.01601	327.46	1.0467
110	0.01960	341.32	1.0927	110	0.01677	339.04	1.0773
120	0.02039	352.59	1.1218	120	0.01750	350.53	1.1069
130	0.02115	363.86	1.1501	130	0.01820	361.99	1.1357
140	0.02189	375.15	1.1777	140	0.01887	373.44	1.1638
150	0.02262	386.49	1.2048	150	0.01953	384.91	1.1912
160	0.02333	397.89	1.2315	160	0.02017	396.43	1.2181
170	0.02403	409.36	1.2576	170	0.02080	407.99	1.2445
180	0.02472	420.90	1.2834	180	0.02142	419.62	1.2704
190	0.02541	432.53	1.3088	190	0.02203	431.44	1.2960
200	0.02608	444.24	1.3338	200	0.02263	443.11	1.3212

TABLE A.12
Normalized Dimensions of Steel Pipe

Catalog Number	Nominal Diameter (in.)	External Diameter (cm)	Internal Diameter (cm)	Wall Thickness (cm)
40	1/8	1.029	0.683	0.173
80	1/8	1.029	0.546	0.241
40	¼	1.372	0.925	0.224
80	¼	1.372	0.767	0.302
40	3/8	1.715	1.252	0.231
80	3/8	1.715	1.074	0.320
40	½	2.134	1.580	0.277
80	½	2.134	1.387	0.373
40	¾	2.667	2.093	0.287
80	¾	2.667	1.885	0.391
40	1	3.340	2.664	0.338
80	1	3.340	2.431	0.455
40	1¼	4.216	3.505	0.356
80	1¼	4.216	3.246	0.485
40	1½	4.826	4.089	0.368
80	1½	4.826	3.810	0.508
40	2	6.033	5.250	0.391
80	2	6.033	4.925	0.554
40	2½	7.303	6.271	0.516
80	2½	7.303	5.900	0.701
40	3	8.890	7.793	0.549
80	3	8.890	7.366	0.762
40	3½	10.16	9.012	0.574
80	3½	10.16	8.545	0.808
40	4	11.43	10.226	0.602
80	4	11.43	9.718	0.856
40	5	14.23	12.819	0.655
80	5	14.13	12.225	0.953
40	6	16.83	15.405	0.711
80	6	16.83	14.633	1.097
40	8	21.91	20.272	0.818
80	8	21.91	19.368	1.270
40	10	27.31	25.451	0.927
80	10	27.31	24.287	1.509
40	12	32.39	30.323	1.031
80	12	32.39	28.890	1.748

TABLE A.13
Bessel Functions $J_0(x)$ y $J_1(x)$

x	$J_0(x)$	$J_1(x)$	x	$J_0(x)$	$J_1(x)$	x	$J_0(x)$	$J_1(x)$
0.0	1.0000	0.0000	4.5	−0.3205	−0.2311	9.0	−0.0903	0.2453
0.1	0.9975	0.0499	4.6	−0.2961	−0.2566	9.1	−0.1142	0.2324
0.2	0.9900	0.0995	4.7	−0.2693	−0.2791	9.2	−0.1367	0.2174
0.3	0.9776	0.1483	4.8	−0.2404	−0.2985	9.3	−0.1577	0.2004
0.4	0.9604	0.1960	4.9	−0.2097	−0.3147	9.4	−0.1768	0.1816
0.5	0.9385	0.2423	5.0	−0.1776	−0.3276	9.5	−0.1939	0.1613
0.6	0.9120	0.2867	5.1	−0.1443	−0.3371	9.6	−0.2090	0.1395
0.7	0.8812	0.3290	5.2	−0.1103	−0.3432	9.7	−0.2218	0.1166
0.8	0.8463	0.3668	5.3	−0.0758	−0.3460	9.8	−0.2323	0.0928
0.9	0.8075	0.4059	5.4	−0.0412	−0.3453	9.9	−0.2403	0.0684
1.0	0.7652	0.4401	5.5	−0.0068	−0.3414	10.0	−0.2459	0.0435
1.1	0.7196	0.4709	5.6	0.0270	−0.3343	10.1	−0.2490	0.0184
1.2	0.6711	0.4983	5.7	0.0599	−0.3241	10.2	−0.2496	−0.0066
1.3	0.6201	0.5220	5.8	0.0917	−0.3110	10.3	−0.2477	−0.0313
1.4	0.5669	0.5419	5.9	0.1220	−0.2951	10.4	−0.2434	−0.0555
1.5	0.5118	0.5579	6.0	0.1506	−0.2767	10.5	−0.2366	−0.0789
1.6	0.4554	0.5699	6.1	0.1773	−0.2559	10.6	−0.2276	−0.1012
1.7	0.3980	0.5778	6.2	0.2017	−0.2329	10.7	−0.2164	−0.1224
1.8	0.3400	0.5815	6.3	0.2238	−0.2081	10.8	−0.2032	−0.1422
1.9	0.2818	0.5812	6.4	0.2433	−0.1816	10.9	−0.1881	−0.1603
2.0	0.2239	0.5767	6.5	0.2601	−0.1538	11.0	−0.1712	−0.1768
2.1	0.1666	0.5683	6.6	0.2740	−0.1250	11.1	−0.1528	−0.1913
2.2	0.1104	0.5560	6.7	0.2851	−0.0953	11.2	−0.1330	−0.2039
2.3	0.0555	0.5399	6.8	0.2931	−0.0652	11.3	−0.1121	−0.2143
2.4	0.0025	0.5202	6.9	0.2981	−0.0349	11.4	−0.0902	−0.2225
2.5	−0.0484	0.4971	7.0	0.3001	−0.0047	11.5	−0.0677	−0.2284
2.6	−0.0968	0.4708	7.1	0.2991	0.0252	11.6	−0.0446	−0.2320
2.7	−0.1424	0.4416	7.2	0.2951	0.0543	11.7	−0.0213	−0.2333
2.8	−0.1850	0.4097	7.3	0.2882	0.0826	11.8	0.0020	−0.2323
2.9	−0.2243	0.3754	7.4	0.2786	0.1096	11.9	0.0250	−0.2290
3.0	−0.2601	0.3391	7.5	0.2663	0.1352	12.0	0.0477	−0.2234
3.1	−0.2921	0.3009	7.6	0.2516	0.1592	12.1	0.0697	−0.2157
3.2	−0.3202	0.2613	7.7	0.2346	0.1813	12.2	0.0908	−0.2060
3.3	−0.3443	0.2207	7.8	0.2154	0.2014	12.3	0.1108	−0.1943
3.4	−0.3643	0.1792	7.9	0.1944	0.2192	12.4	0.1296	−0.1807
3.5	−0.3801	0.1374	8.0	0.1717	0.2346	12.5	0.1496	−0.1655
3.6	−0.3918	0.0955	8.1	0.1475	0.2476	12.6	0.1626	−0.1487
3.7	−0.3992	0.0538	8.2	0.1222	0.2580	12.7	0.1766	−0.1307
3.8	−0.4026	0.0128	8.3	0.0960	0.2657	12.8	0.1887	−0.1114
3.9	−0.4018	−0.0272	8.4	0.0692	0.2708	12.9	0.1988	−0.0912
4.0	−0.3971	−0.0660	8.5	0.0419	0.2731	13.0	0.2069	−0.0703
4.1	−0.3887	−0.1033	8.6	0.0146	0.2728	13.1	0.2129	−0.0489
4.2	−0.3766	−0.1386	8.7	−0.0125	0.2697	13.2	0.2167	−0.0271
4.3	−0.3610	−0.1719	8.8	−0.0392	0.2641	13.3	0.2183	−0.0052
4.4	−0.3423	−0.2028	8.9	−0.0653	0.2559	13.4	0.2177	0.0166

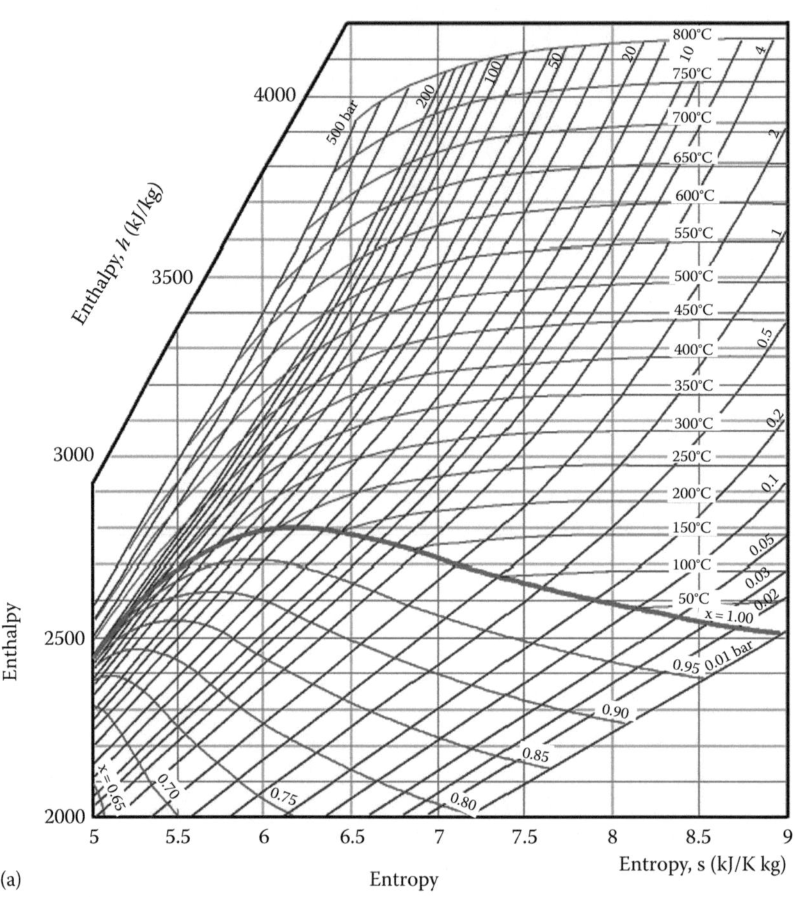

FIGURE A.1 Mollier diagram for water steam.

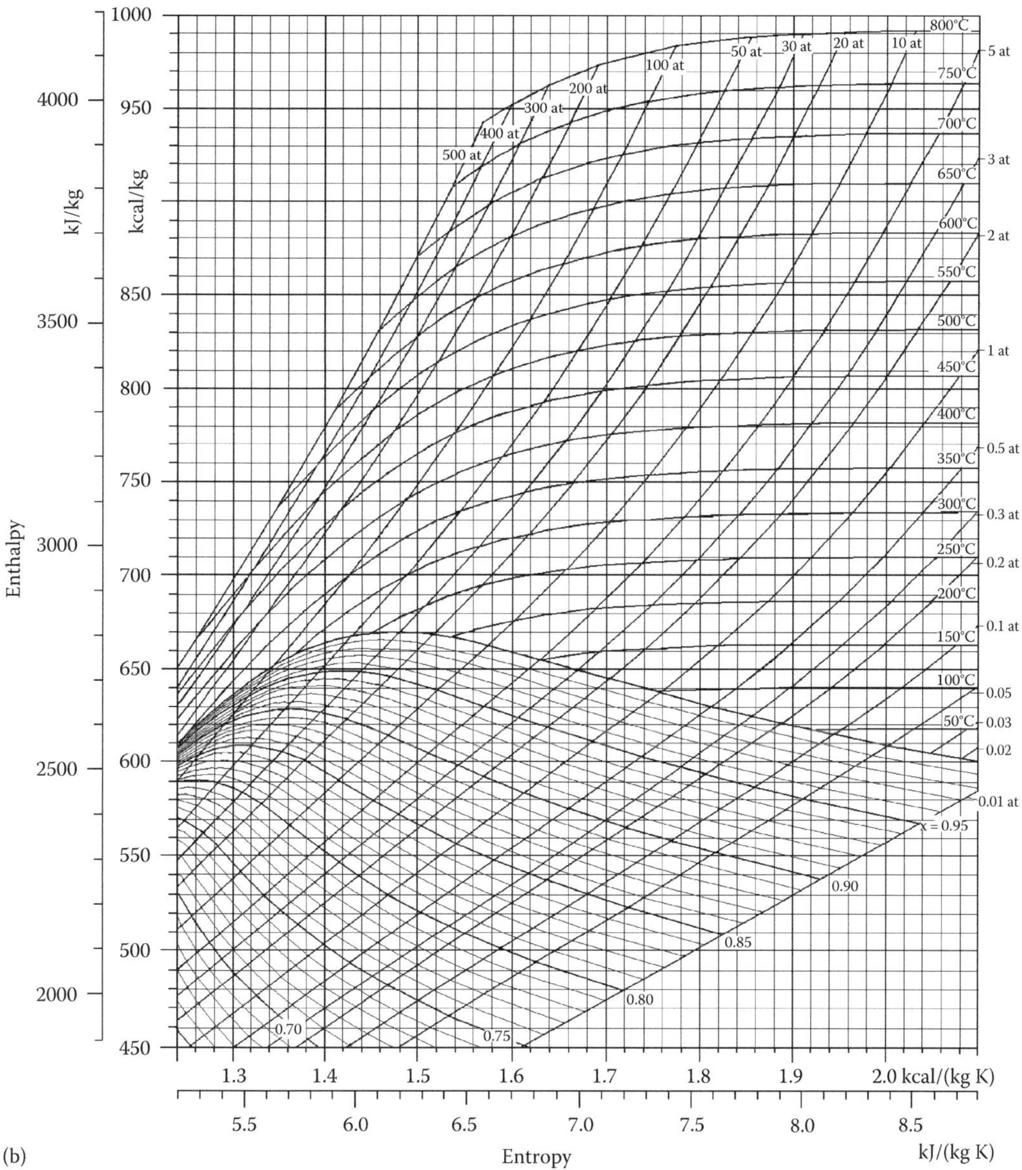

FIGURE A.1 (continued) Mollier diagram for water steam.

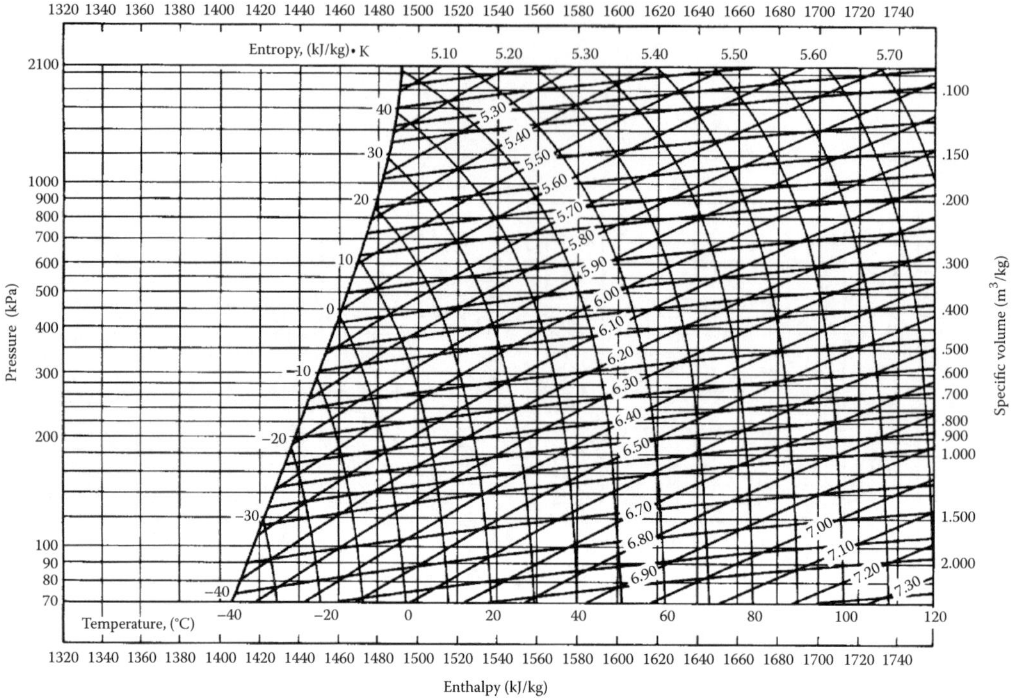

FIGURE A.2 Diagram for ammonia.

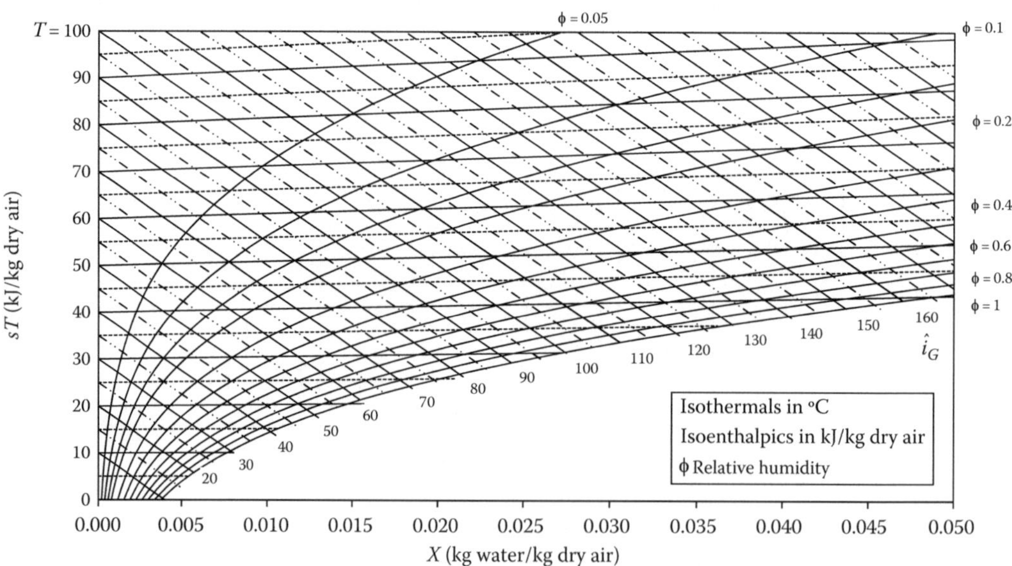

FIGURE A.3 Psycrometric chart $\hat{s}T$–X.

References

Abdullah, E.C. and Geldart, D. (1999). The use of bulk density measurements as flowability indicators. *Powder Technol.*, 102, 151–165.

Abrahamsen, A.R. and Geldart, D. (1980). Behavior of gas-fluidized beds of fine powders, Part I. Homogeneous expansion. *Powder Technol.*, 26, 35–46.

Adams, M.J., Mullier, M.A., and Seville, J.P.K. (1994). Agglomerate strength measurement using a uniaxial confined compression test. *Powder Technol.*, 78, 5–13.

Adams, N. and Lodge, A.S. (1964). Rheological properties of concentrated polymer solutions II. A cone-and-plate and parallel-plate pressure distribution apparatus for determining normal stress differences in steady shear flow. *Phil. Trans. R. Soc. Lond. A*, 256, 149–184.

Addison, W. (1986). Recent trends in fruit juice concentration. International symposium on fruit and vegetables for processing, (Cape Town: South Africa). *ISHS Acta Hort.*, 194, 241–248.

Adhikari, B., Howes, T., Bhandari, B.R., and Truong, V. (2001). Stickiness in foods: A review of mechanisms and test methods. *Int. J. Properties*, 4(1), 1–33.

Aguado, M.A. and Ibarz, A. (1988). Variación de la densidad de un zumo de manzana con la temperatura y concentración. *Aliment. Equipos Tecnol.*, 2(88), 209–216.

Aguilera, J.M., Valle, J.M., and Karel, M. (1995). Review: Caking phenomena in food powders. *Trends Food Sci. Technol.*, 6, 149–154.

Alfa-Laval, (1971). *Heat Exchanger Guide*. Alfa-Laval AB-2nd, Lund, Sweden.

Alvarado, J.D. and Romero, C.H. (1989). *Latin Am. Appl. Res.*, 19, 15–21.

American Society for Testing and Materials (ASTM). (2002). *Standard Test Method of Tumbler Test for Coal, D441-86*. ASTM, Baltimore, MD.

Andrianov, Y.P., Toerdokhleb, G.V., and Makarova, E.P. (1968). *Moloch. Prom.*, 29(8), 25–28.

APV. (1989). *Dryer Handbook (DRH-889)*, APV Crepaco, Inc., Rosemont, IL.

Aronson, M.H. and Nelson, R.C. (1964). *Viscosity Measurements and Control*. Instrument Publishing Company, Inc., Pittsburgh, PA.

ASTM. (1986). *Compilation of ASTM Standard Definitions*, 6th edn. ASTM, Baltimore, MD.

Atkins, A.G. (1987). The basic principles of mechanical failure in biological systems. In *Food Structure and Behavior* (J.M.V. Blanshard and P.J. Lillford, Eds.), pp. 149–176. Nottingham University Press, Nottingham, England.

Badger, W.L. and Banchero, J.T. (1970). *Introducción a la Ingeniería Química*. McGraw-Hill, D.F. México.

Ball, C.O. (1923). *Bull. Natl. Res. Council*, 7(1), 37.

Baquero, J. and Llorente, V. (1985). *Equipos para la Industria Química y Alimentaria*. Alhambra, Madrid, Spain.

Barbosa-Cánovas, G.V., Ibarz, A., and Peleg, M. (1993). Reología de alimentos fluidos. Revisión. *Alimentaria*, 241, 39–89.

Barbosa-Cánovas, G.V., Malavé-López, J., and Peleg, M. (1987). Density and compressibility of selected food mixtures. *J. Food Eng.*, 10, 1–19.

Barbosa-Cánovas, G.V., Pothakamury, U.R., Palou, E., and Swanson, B.G. (1998). *Nonthermal Preservation of Foods*. Marcel Dekker, New York.

Barbosa-Cánovas, G.V. and Juliano, P. (2005). Physical and chemical properties of food powders. In *Encapsulated and Powdered Foods* (C. Onwulata, Ed.), pp. 39–71. CRC Press, Boca Raton, FL.

Barbosa-Cánovas, G.V. and Peleg, M. (1982). Propiedades de flujo de alimentos líquidos y semi-líquidos. *Rev. Tecnol. Aliment. México*, 17(2), 4.

Barbosa-Cánovas, G.V. and Peleg, M. (1983). Flow parameters of selected commercial semi-liquid food products. *J. Texture Stud.*, 14, 213.

Barbosa-Cánovas, G.V. and Vega-Mercado, H. (1996). *Dehydration of Food*. Chapman & Hall, New York.

Barletta, B.J., Knight, K.M., and Barbosa-Cánovas, G.V. (1993a). Review: Attrition in agglomerated coffee. *Rev. Esp. Cienc. Tecnol. Aliment.*, 33(1), 43–58.

Barletta, B.J., Knight, K.M., and Barbosa-Cánovas, G.V. (1993b). Compaction characteristics of agglomerated coffee during tapping. *J. Text. Stud.*, 24, 253–268.

Barletta, B.J. and Barbosa-Cánovas, G.V. (1993a). Fractal analysis to characterize ruggedness changes in tapped agglomerated food powders. *J. Food Sci.*, 58(5), 1030–1046.

Barletta, B.J. and Barbosa-Cánovas, G.V. (1993b). An attrition index to assess fines formation and particle size reduction in tapped agglomerated food powders. *Powder Technol.*, 77, 89–93.

Beddow, J.K. (1997). *Image Analysis Sourcebook.* American Universities Science and Technology Press, Santa Barbara, CA.

Bemrose, C.R. and Bridgwater, J. (1987). A review of attrition and attrition test methods. *Powder Technol.*, 49, 97–126.

Bengtsson, N.E. and Risman, P.O. (1971). Dielectric properties of foods at 3 GHz, as determined by a cavity perturbation technique, II. Measurement of food materials. *J. Microw. Power*, 6(2), 107–123.

Bennet, C.O. and Meyers, J.E. (1979). *Transferencia de Cantidad de Movimiento, Calor y Materia.* Reverté, Barcelona, Spain.

Bernardini, E. (1981). *Tecnología de Aceites y Grasas.* Alhambra, Madrid, Spain.

van der Berg, C. (1985). Water activity. In *Concentration and Drying Foods* (D. MacCarthy, Ed.). Elsevier Applied Science Publishers, New York.

van der Berg, C. and Bruin, S. (1981). Water activity and its estimation in food systems: Theoretical aspects. In *Water Activity: Influences on Food Quality* (L.B. Rockland and G.F. Steward, Eds.). Academic Press, New York.

Bertsch, A.J., Bertsch, A.J., Bimbenet, J.J., Cerf, O., Lelubre, A., Vermeire, D., Degas, A., and Cavarroc, M. (1982). La masse volumique du lait et de crèmes de 65ºC à 140ºC. *Le Lait.* 62 (615–616), 250–264.

Bhattacharya, K.R. and Sowbhagya, C.M. (1978). On viscograms and viscography. *J. Texture Stud.*, 9, 341.

Bika, D.F., Gentzler, M., and Michaels, J.N. (2001). Mechanical properties of agglomerates. *Powder Technol.*, 117, 98–112.

Billmeyer, F.W. (1971). *Textbook of Polymer Science.* Wiley-Interscience, New York.

Bingham, E.C. (1922). *Fluidity and Plasticity.* McGraw-Hill Book Co., New York.

Bird, R.B., Armstrong, R.C., and Hassager, O. (1977). *Dynamics of Polymeric Liquids-Fluid Mechanics*, Vol. 1. John Wiley & Sons, New York.

Bird, R.B., Stewart, W.E., and Lightfoot, E.N. (1960). *Transport Phenomena.* John Wiley & Sons, New York.

Blanshard, J.M.V. (1993). The glass transition, its nature and significance in food processing. In *Glassy State in Foods* (J.M.V. Blanshard and P.J. Lillford, Eds.), pp. 18–48. Nottingham University Press, Nottingham, England.

Bondi, A. (1956). Theories of viscosity. In *Rheology*, Vol. 1 (F.R. Eirich, Ed.), p. 132. Academic Press, New York.

Bone, D.P., Shannon, E.L., and Ross, K.D. (1975). The lowering of water activity by order of mixing in concentrated solutions. In *Water Relations of Foods* (R.B. Duckworth, Ed.). Academic Press, London, U.K.

Boquet, R., Chirife, J., and Iglesias, H.A. (1978). Equations for fitting water sorption isotherms of foods II. Evaluation of various two-parameter models. *Int. J. Food Sci. Technol.*, 13, 319–327.

Braddock, R.J. (1986). Quality of freeze concentrated grapefruit juice. In: *Proceeding XIX International of the International Federation of Fruit Juice Producers*, The Hague, the Netherlands.

Brennan, J.G., Butters, J.R., Cowell, N.D., and Lilly, A.E.V. (1980). *Las Operaciones de la Ingeniería de los Alimentos.* Acribia, Zaragoza, Spain.

Bromley, L.A. (1973). Thermodynamic properties of strong electrolytes in aqueous solutions. *AIChE J.*, 19(2), 313–320.

Brown, A.I. and Marco, S.M. (1970). *Transmisión de Calor.* Cecsa, México.

Bruin, S. and Luyben, K.Ch.A.M. (1980). Drying of food materials. In *Advances in Drying,* Vol. 1 (A.S. Mujumdar, Ed.). Hemisphere Publishing, New York.

Brunauer, S., Emmet, P.H., and Teller, E. (1938). Adsorption of gases in multimolecular layers. *J. Am. Chem. Soc.*, 60, 309–319.

Brunner, E. (1949). The correlation between viscosity and energy of evaporation. *J. Chem. Phys.*, 17, 346.

Buma, T.J. (1971). Free fat in spray-dried whole milk 5. Cohesion: Determination, influence of particle size, moisture content and free-fat content. *Neth. Milk Dairy J.*, 25, 107–122.

Buonopane, R.A., Troupe, R.A., and Morgan, J.C. (1963). Heat transfer design method for plate heat exchangers. *Chem. Eng. Prog.*, 59(7), 57–61.

Burton, H. (1949). A survey of literature on bacterial effects of short electromagnetic waves. National Institute for Research in Dairying Shinfield, Shinfield, England. N.I.R.D. Paper No. 1041.

Buss (1993). Milk goes artic. *Food Process.*, 54(12), 62–64.

Campanella, O.H. and Peleg, M. (1987a). Squeezing flow viscosimetry of peanut butter. *J. Food Sci.*, 52(1), 180.

Campanella, O.H. and Peleg, M. (1987b). Determination of the yield stress of semi-liquid foods from squeezing flow data. *J. Food Sci.*, 52(1), 214–215, 217.

Carleton, A.J., Cheng, D.C., and Whittaker, W. (1974). Determination of the rheological properties and start-up pipeline flow characteristics of waxy crude and fuel oils. *Inst. Petrol. Technol.*, Paper No. 9.

Carman, P.C. (1956). *Flow of Gases through Porous Media*. Butterworths, London, U.K.

Carr, E. (1965). Evaluating flow properties of solids. *Chem. Eng.*, 163–168.

Carr, R.L. (1976). Powder and granule properties and mechanics. In *Gas-Solids Handling in the Processing Industries* (J.M. Marchello and A. Gomezplata, Eds.). Marcel Dekker, New York.

Carreau, P.J. (1972). Rheological equations from molecular network theories. *J. Rheol.*, 16(1), 99–127.

Carstensen, J.T. and Hou, X.P. (1985). The Athy-Heckel equation applied to granular agglomerates of basic tricalcium phosphate [$3Ca_3PO_4{\cdot}Ca(OH)_2$]. *Powder Technol.*, 42, 153–157.

Casson, N. (1959). A flow equation for pigment-oil suspensions of the printing ink type. In *Rheology of Disperse Systems* (C.C. Hill, Ed.), p. 82. Pergamon Press, New York.

Castaldo, D., Palmieri, L., Lo Voi, A., and Costabile, P. (1990). Flow properties of Babaco (*Carica pentagona*) purees and concentrates. *J. Texture Stud.*, 21, 253–264.

Castro, I., Teixeira, J.A., and Vicente, A.A. (2003). The influence of the presence of an electrical field on lipoxygenase and β-galactosidase inactivation kinetics. Paper presented at the *Proceedings of NFIF 2003: New Functional Ingredients and Foods: Safety, Health and Convenient*, Copenhagen, Denmark, 2003.

Cathala, J. (1951). Le genie chimique: Historique de son développement—Son objet et ses méthodes—Programme de notre revue. *Chem. Eng. Sci.*, 1, 1–7.

Caurie, M. (1970). A new model equation for predicting safe storage moisture levels for optimum stability of dehydrated foods. *Int. J. Food Sci. Technol.*, 5, 301–307.

CDOChE—Committee on Dynamic Objectives for Chemical Engineering. (1961). *Chem. Eng. Progr.*, 57, 69.

Chandarana, D.I., Gavin, A. III, and Wheaton, F.W. (1990). Particle/fluid interface heat transfer under UHT conditions at low particle/fluid relative velocities. *J. Food Process Eng.*, 13(3), 191.

Charm, S.E. (1960). Viscometry of non-Newtonian food materials. *Food Res.*, 25, 351.

Charm, S.E. (1963a). The direct determination of shear stress-shear rate behavior of foods in the presence of yield stress. *J. Food Sci.*, 28, 107–113.

Charm, S.E. (1963b). Effect of yield stress on the power law constants of fluid materials determined in low shear rate viscometer. *Ing. Eng. Chem.* (*Proc. Des. Dev.*), 2, 62.

Charm, S.E. (1971). *Fundamentals of Food Engineering*. AVI Publishing Co., Westport, CT.

Chen, C. (1985). Thermodynamic analysis of freezing and thawing of foods enthalpy and apparent specific heat. *J. Food Sci.*, 50, 1158–1162.

Chen, C.S. and Johnson, W.H. (1969). Kinetics of moisture movement in hygroscopic materials. I. Theoretical consideration of drying phenomena. *Trans. ASAE.*, 12, 109–113.

Cheng, D.C. (1986). Yield stress: A time dependent property and how to measure it. *Rheol. Acta*, 25, 542.

Cheng, D.C. and Evans, F. (1965). Phenomenological characterization of rheological behavior of inelastic reversible thixotropic and antithixotropic fluids. *Br. J. Appl. Phys.*, 16, 1599.

Cheryan, M. (1986). *Ultrafiltration Handbook*. Technomic Publishing Co., Lancaster, PA.

Cheryan, M. (1992). Concentration of liquid foods by reverse osmosis. In *Handbook of Food Engineering* (D.R. Heldman and D.B. Lund, Eds.). Marcel Dekker, New York.

Chikazawa, M. and Takei, T. (1997). III. Fundamental properties of powder beds adsorption characteristics. In *Powder Technology Handbook* (K. Gotoh, H. Masuda, and K. Higashitani, Eds.), pp. 245–264. Marcel Dekker, New York.

Chirife, J. (1983). Fundamentals of the drying mechanism during air dehydration of foods. In *Advances in Drying,* Vol. 2 (A.S. Mujumdar, Ed.). Hemisphere Publishing, New York.

Chirife, J., Ferro-Fontán, C., and Benmergui, E.A. (1980). The prediction of water activity in aqueous solutions in connection with intermediate moisture foods. IV. Aw prediction in aqueous non-electrolyte solutions. *J. Food Technol.*, 15, 59–70.

Chirife, J. and Favetto, G.J. (1992). Fundamental aspects of food preservation by combined methods. International Union of Food Science and Technology, CYTED D Univ. de las Américas, Puebla, México.

Chirife, J. and Iglesias, H.A. (1978). Equations for fitting water sorption isotherms of foods: Part 1—A review. *Int. J. Food Sci. Technol.*, 13(2), 159–174.

Cho, H.Y., Yousef, A.E., and Sastry, S.K. (1999). Kinetics of inactivation of *Bacillus subtilis* by continuous or intermittent ohmic and conventional heating. *Biotechnol. Bioeng.*, 62, 368–372.

Choi, Y. and Okos, M.R. (1986a). Thermal properties of liquid foods: Review. In *Physical and Chemical Properties of Foods* (M.R. Okos, Ed.). ASAE, New York.

Choi, Y. and Okos, M.R. (1986b). Effects of temperature and composition on the thermal properties of foods. In *Food Engineering and Process Applications*, Vol. 1, *Transport Phenomenon* (L. Maguer and P. Jelen, Eds.), pp. 93–101. Elsevier, New York.

Christiansen, E., Ryan, N., and Stevens, W. (1955). Pipe-line design for non-Newtonian fluids in streamline flow. *AIChE J.*, 1, 544–548.

Clark, D.F. (1974). Plate heat exchanger design and recent development. *Chem. Eng.*, 285, 275–279.

Cleland, A.C. (1992). Dynamic modelling of heat transfer for improvement in process design—Case studies involving refrigeration. In *Advances in Food Engineering* (R.P. Singh and M.A. Wirakartakusumah, Eds.). CRC Press, Boca Raton, FL.

Cleland, A.C. and Earle, R.L. (1976). A new method for prediction of surface heat transfer coefficients in freezing. *Bull. I.I.R.*, Annexe-1, 361.

Cleland, A.C. and Earle, R.L. (1979). A comparison of methods for predicting the freezing times of cylindrical and spherical foodstuffs. *J. Food Sci.*, 44, 964.

Cleland, A.C. and Earle, R.L. (1982). Freezing time prediction of foods: A simplified procedure. *Inst. J. Refrig.*, 5, 134–140.

Cleland, D.J., Cleland, A.C., and Earle, R.L. (1987a). Prediction of freezing and thawing times for multi-dimensional shapes by simple methods. Part I—Regular shapes. *Inst. J. Refrig.*, 10, 156–164.

Cleland, D.J., Cleland, A.C., and Earle, R.L. (1987b). Prediction of freezing and thawing times for multi-dimensional shapes by simple methods. Part II—Irregular shapes. *Inst. J. Refrig.*, 10, 234–240.

Constenla, D.T., Lozano, J.E., and Crapiste, G.H. (1989). Thermophysical properties of clarified apple juice as a function of concentration and temperature. *J. Food Sci.*, 54(3), 663–668.

Cooper, A. (1974). Recover more heat with plate heat exchangers. *Chem. Eng.*, 285, 280–285.

Cornford, S.J., Parkinson, T.L., and Robb, J. (1969). Rheological characteristics of processed whole egg. *J. Food Technol.*, 4, 353.

Costa, E., Calleja, G., Ovejero, G., de Lucas, A., Aguado, J., and Uguina, M.A. (1986a). *Ingeniería Química. 1. Conceptos Generales*. Alhambra, Madrid, Spain.

Costa, E., Calleja, G., Ovejero, G., de Lucas, A., Aguado, J., and Uguina, M.A. (1986b). *Ingeniería Química. 4. Transmisión de Calor*. Alhambra, Madrid, Spain.

Costa, J., Cervera, S., Cunill, F., Esplugas, S., Mans, C., and Mata, J. (1984). *Curso de Química Técnica*. Reverté, Barcelona, Spain.

Costell, E. and Duran, L. (1979). Esterilización de conservas. Fundamentos teóricos y cálculo del tiempo de esterilización. Instituto de Agroquímica y Tecnología de Alimentos (CSIC), Información Técnica General no. 66.

Coulson, J.M. and Richardson, J.F. (1979–1981). *Ingeniería Química*, Vols. I a VI. Reverté, Barcelona, Spain.

Couroyer, C., Ning, Z., and Ghadiri, M. (2000). Distinct element analysis of bulk crushing: Effect of properties and loading rate. *Powder Technol.*, 109, 241–254.

Cowan, C.T. (1975). Choosing materials of construction for plate heat exchangers. II. *Chem. Eng.*, 82, 102–104.

Crandall, P.G., Chen, C.S., and Carter, R.D. (1982). Models for predicting viscosity of orange juice concentrate. *Food Technol.*, May, 245–252.

Crapiste, G.H. and Lozano, J.E. (1988). Effect of concentration and pressure on the boiling point rise of apple juice and related sugar solutions. *J. Food Sci.*, 53(3), 865–868.

Cross, G.A. and Fung, D.Y.C. (1982). The effect of microwaves on nutrient value of foods. *CRC Crit. Rev. Food Sci. Nutr.*, 16, 355–381.

Cross, M.M.J. (1965). Rheology of non-Newtonian fluids: A new flow equation for pseudoplastic systems. *J. Colloid Sci.*, 20, 417–437.

Davis, R.B., De Weese, D., and Gould, W.A. (1954). Consistency measurement of tomato puree. *Food Technol.*, 8, 330.

Decareau, R.V. (1992). *Microwave Foods: New Product Development*. Food & Nutrition Press, Inc, Trumbull, CO.

Decareau, R.V. and Peterson, R.A. (1986). *Microwave Processing and Engineering*. Ellis Horwood Ltd., Chichester, England.

Di Cesare, L.F., Cortesi, P., and Maltini, E. (1993). Studies on the concentration of model solutions and fruit distillates by "Freezing out". *Fruit Process.*, 3(12), 442–445.

Di Cesare, L.F., Nani, R., Brambilla, A., Tessari, D., and Fussari, E.L. (2000). Volatile compounds and soluble substances in freeze concentrated feijoa juices [Feijoa sellowiana Berg]. *Industrie delle Bevande*, 29(166), 125–128.

Dick, R.I. (1970). Role of activated sludge final settling tanks. *J. San. Eng. Div. ASCE*, 96, 423–436.

Dickerson, R.W. (1969). Thermal properties of foods. In *The Freezing Preservation of Foods,* Vol. 2, 4th edn. (D.K. Tressler, W.B. Van Arsdel, and M.J. Copley, Eds.). AVI Publishing Co., Westport, CT.

Dickey, D.S. and Fenic, J.G. (1976). Dimensional analysis for fluid agitation systems. *Chem. Eng.*, 5, 139–145.

Dickie, A. and Kokini, J.L. (1981). Transient viscoelastic flow of fluid and semi-solid food materials. Presented at *Proceedings of the 52nd Annual Meeting of the Society of Rheology*, Williamsburg, VA.

Dickson, J.S. (2001). Radiation inactivation of microorganisms. In *Food Irradiation. Principles and Applications*, R. Molins (ed.). Wiley-Interscience, New York.

Diehl, J.F. and Josephson, E.S. (1994). Assessment of wholesomeness of irradiated foods (a review). *Acta Aliment.*, 23(2), 195–214.

Dodge, D.W. and Metzner, A.B. (1959). Turbulent flow of non-Newtonians systems. *AIChE J.*, 5(2), 189–204.

Duberg, M. and Nyström, C. (1986). Studies of direct compression of tablets. XVII. Porosity-pressure curves for characterization of volume reduction mechanisms in powder compression. *Powder Technol.*, 46, 67–75.

Earle, R.L. (1983). *Ingeniería de los Alimentos*, 2nd edn. Acribia, Zaragoza, España, Spain.

Edwards, M.F., Changal, A.A., and Parrot, D.L. (1974). Heat transfer and pressure drop characteristics of a plate heat exchanger using non-Newtonian liquids. *Chem. Eng.*, 285, 286–288 and 293.

Ehlerrmann, D.A.E. and Schubert, H. (1987). Compressibility characteristics of food powders: Characterizing the flowability of food powders by compression tests. In *Physical Properties of Foods-2 Cost 90bis Final Seminar Proceedings* (R. Jowitt, F. Escher, M. Kent, B. McKenna, and M. Roques, Eds.). ECSC, EEC, EAEC, Brussels, Belgium.

Einstein, A. (1911). Berichtigung zu meiner arbeit: Eine neue bestimmung der moleküldimensionen. *Ann. Phys.*, 339, 591–592.

Elliot, J.H. and Ganz, A.J. (1977). Salad dressings-preliminary rheological characterization. *J. Texture Stud.*, 8, 359–371.

Elliot, J.H. and Green, C.E. (1972). Modification of food characterization with cellulose hydrocolloids—II. The modified Bingham body—A useful rheological model. *J. Texture Stud.*, 3(2), 194–205.

Emmet, P.H. and de Witt, T. (1941). Determination of surface areas. *Ind. Eng. Chem. (Anal.)*, 13, 28.

Eolkin, D. (1957). The plastometer—A new development in continuous recording and controlling consistometers. *Food Technol.*, 11, 253.

Eyring, H. (1936). Viscosity, plasticity, and diffusion as examples of absolute reaction rates. *J. Chem. Phys.*, 4, 283–291.

Falguera, V. and Ibarz, A. (2010). A new model to describe flow behaviour of concentrated orange juice. *Food Biophys.*, 5, 114–119.

Fellows, P.J. (2000). *Food Processing Technology. Principles and Practice*. CRC Press LLC, Boca Raton, FL.

Fernández-Martín, F. (1972a). Influence of temperature and composition on some physical properties of milk concentrates. II. Viscosity. *J. Dairy Res.*, 39, 75–82.

Fernández-Martín, F. (1972b). Influence of temperature and composition on some physical properties of milk concentrates. I. Heat capacity. *J. Dairy Res.*, 39(1), 65–73.

Fernández-Martín, F. (1982). Las propiedades físicas de alimentos y la ingeniería de procesos. I. Productos lácteos. *Aliment. Equipos Tecnol.*, 2(82), 55–63.

Fernández-Martín, F. and Montes, F. (1972). Influence of temperature and composition on some physical properties of milk concentrates. III. Thermal conductivity. *Milchwiss*, 27(12), 772–776.

Fernández-Martín, F. and Montes, F. (1977). Thermal conductivity of creams. *J. Dairy Res.*, 44(1), 103–109.

Ferro-Fontán, C., Benmergui, E.A., and Chirife, J. (1980). The prediction of water activity in aqueous solutions in connection with intermediate moisture foods. III. Aw prediction in multicomponent strong electrolyte aqueous solutions. *J. Food Technol.*, 15, 47–58.

Ferro-Fontán, C., Chirife, J., and Boquet, R. (1981). Water activity in multicomponent non-electrolyte solutions. *J. Food Technol.*, 18, 553–559.

Ferro-Fontán, C. and Chirife, J. (1981). Technical note: A refinement of Ross's equation for predicting the water activity of non-electrolyte mixtures. *J. Food Technol.*, 16, 219–221.

Figoni, P.I. and Shoemaker, C.F. (1983). Characterization on time dependent properties of mayonnaise under steady shear. *J. Texture Stud.*, 14(4), 431–442.

Fiszman, S.M., Costell, E., Serra, P., and Durán, L. (1986). Relajación de sistemas viscoelásticos. Comparación de métodos de análisis de las curvas experimentales. *Rev. Agroquím. Tecnol. Aliment.*, 26(1), 63–71.

Fortes, M. and Okos, M.R. (1980). Drying theories. In *Advances in Drying*, Vol. 1 (A.S. Mujumdar, Ed.). Hemisphere Publishing, New York.

Foust, A.S., Wenzel, L.A., Clump, C.W., Maus, L., and Andersen, L.B. (1960). *Principles of Unit Operations*. John Wiley & Sons, New York.

Frederickson, A.G. (1970). A model for the thixotropy of suspensions. *AIChE J.*, 16(3), 436–441.

Fryer, P. (1995). Electrical resistance heating of foods. In *New Methods of Food Preservation* (G.W. Gould, Ed.). Blackie Academic & Professional, Glasgow, U.K.

Fung, D.Y.C. and Cunningham, F.E. (1980). Effect of microwaves on microorganisms in foods. *J. Food Protect.*, 43(8), 641–650.

Gal, S. (1983). The need for, and practical applications of, sorption data. In *Physical Properties of Foods* (R. Jowitt, F. Escher, B. Hallstrom, H.F. Meffert, W. Spiess, G. Vos, Eds.), pp. 13–25. Applied Science Publishers, Ltd., London, U.K.

Garcia, E.J. and Steffe, J.F. (1987). Comparison of friction factor equations for non-Newtonian fluids in pipe flow. *J. Food Process Eng.*, 9, 93–120.

García, E.J. and Steffe, J.F. (1986). Optimum economic pipe diameter for pumping Herschel-Bulkley fluids in pipe flow. *J. Food Process Eng.*, 8, 117–136.

Geankoplis, C.J. (1993). Drying of process materials. In *Transport Processes and Unit Operations*, 3rd edn. Prentice-Hall, Inc., Englewood Cliffs, NJ.

Geldart, D., Harnby, N., and Wong, A.C.Y. (1984). Fluidization of cohesive powders. *Powder Technol.*, 37, 25–37.

Gerritsen, A.H. and Stemerding, S. (1980). Crackling of powdered materials during moderate compression. *Powder Technol.*, 27, 183–188.

Ghadiri, M., Yuregir, K.R., Pollock, H.M., Ross, J.D.J., and Rolfe, N. (1991). Influence of processing conditions on attrition of NaCl crystals. *Powder Technol.*, 65, 311–320.

Ghadiri, M. and Ning, Z. (1997). Effect of shear strain rate on attrition of particulate solids in a shear cell. Paper presented at *Proceedings of the Third International Conference on Powders & Grains*, Durham, NC, 1997.

Ghizzoni, C., Del Popolo, F., and Porretta, S. (1995). Quality of cryoconcentrated red fruit juices as a function of volatile fraction. *Riv. Ital. Eppos*, 15, 5–21.

Greensmith, H.W. and Rivlin, R.S. (1953). *Phil. Trans. R. Soc. Lond. A*, 245, 399.

Gregorig, R. (1968). *Cambiadores de Calor*. Urmo, Bilbao, Spain.

Gromov, M.A. (1974). Formula for calculating the heat conductivity coefficient of cream. *Moloch. Prom.*, 35(2), 25–27.

Gromov, M.A. (1979). Thermophysical characteristics of milk plasma. *Moloch. Prom.*, 40(4), 37–39.

Hahn, S.J., Ree, T., and Eyring, H. (1959). Flow mechanism of thixotropic substances. *Ind. Eng. Chem.*, 51, 856.

Halsey, G. (1948). Physical adsorption on non-uniform surface. *J. Chem. Phys.*, 16(10), 931–937.

Hamano, M. and Aoyama, Y. (1974). Caking phenomena in amorphous food powders. *Trends Food Sci. Technol.*, 6, 149–155.

Hanks, R.W. and Ricks, B.L. (1974). Laminar-turbulent transition in flow of pseudoplastic fluids with yield stresses. *J. Hydronautics*, 8(4), 163.

Harper, J.C. (1976). *Elements of Food Engineering*. AVI Publishing Co., Westport, CT.

Harper, J.C. and El Sahrigi, A.F. (1965). Viscometric behavior of tomato concentrates. *J. Food Sci.*, 30, 470.

Harper, J.C. and Leberman, K.W. (1962). Rheological behavior of pear purees. In *Proceedings of the First International Congress of Food Science and Technology*, Vol. I, p. 719. Gordon & Breach Science Publishers, Newark, NJ.

Harris, J. (1967). A continuum theory of time-dependent inelastic flow. *Rheol. Acta*, 6, 6–12.

Hartel, R.W. (1992). Evaporation and freeze concentration. In *Handbook of Food Engineering* (D.R. Heldman and D.B. Lund, Eds.), Marcel Dekker, New York.

Hartley, P.A., Parfitt, G.D., and Pollack, L.B. (1985). The role of van der Waals force in agglomeration of food powders containing submicron particles. *Powder Technol.*, 42, 35–46.

Hayes, G.D. (1987). *Food Engineering Data Handbook*. John Wiley & Sons, New York, p. 83.

Head, K.H. (1982). *Manual of Soil Laboratory Testing*, Vol. 2. Pentech Press, London, U.K., pp. 581–585.

Heildelbaugh, N.D. and Giron, D.J. (1969). Effect of processing on recovery of poliovirus from inoculated food. *J. Food Sci.*, 34, 239–241.

Heldman, D.R. (1975). *Food Process Engineering*. AVI Publishing Co., Westport, CT.

Heldman, D.R. (1992). Food freezing. In *Handbook of Food Engineering* (D.R. Heldman and D.B. Lund, Eds.). Marcel Dekker, New York.

Heldman, D.R. and Lund, D.B. (1992). *Handbook of Food Engineering*. Marcel Dekker, New York.

Heldman, D.R. and Singh, R.P. (1981). *Food Process Engineering*. AVI Publishing Co., Westport, CT.

Henderson, S.M. (1952). A basic concept of equilibrium moisture. *Agric. Eng.*, 33, 29–32.

Hernández, E., Raventós, M., Auleda, J.M., and Ibarz, A. (2009). Concentration of apple and pear juices in a multi-plate freeze concentrator. *Innov. Food Sci. Emerg. Technol.*, 10, 348–355.

Herranz, J. (1979). *Procesos de Transmisión de Calor*. Castillo, Madrid, Spain.

Higgs, S.J. and Norrington, R.J. (1971). Rheological properties of selected foodstuffs. *Proc. Biochem.*, 6(5), 52.

Hildebrant, J.H. and Scott, R.L. (1962). *Regular Solutions*. Prentice-Hall, Inc., Englewood Cliffs, NJ.

Holdsworth, S.D. (1971). Applicability of rheological models to the interpretation of flow and processing behaviour of fluid food products. *J. Texture Stud.*, 2, 393.

Hollenbach, A., Peleg, M., and Rufner, R. (1982). Effect of four anticaking agents on the bulk characteristics of ground sugar. *J. Food Sci.*, 47, 538–544.

Hollman, J.P. (2001). *Experimental Methods for Engineers*. McGraw-Hill, New York, 340p.

Huggins, M.L. (1942). The viscosity of dilute solutions of long-chain molecules. IV. Dependence on concentration. *J. Am. Chem. Soc.*, 64, 2716.

Ibarz, A. (1986a). *Extracción Sólido-Líquido*. ETSEAL, Universitat Politècnica de Catalunya, Barcelona, Spain.

Ibarz, A. (1986b). Intercambiadores de calor de placas. *Aliment. Equipos Tecnol.*, 3(86), 119–129.

Ibarz, A. (1987). Un método de diseño de intercambiadores de calor de placas. *Aliment. Equipos Tecnol.*, 2(87), 187–193.

Ibarz, A. (2007). Ionizing irradiation of foods. In *New Food Engineering Research Trends* (A.P. Urwaye, Ed.). Nova Publishers, New York.

Ibarz, A., Giner, J., Pagán, J., and Gimeno, V. (1991). Influencia de la temperatura en la reología de zumos de kiwi. Presented at *III Congreso Mundial de Tecnología de Alimentos*, Barcelona, Spain.

Ibarz, A., González, C., Esplugas, S., and Vicente, M. (1992a). Rheology of clarified fruit juices. I: Peach juices. *J. Food Eng.*, 15, 49–61.

Ibarz, A., Pagán, J., and Miguelsanz, R. (1992b). Rheology of clarified fruit juices. II: Blackcurrant juices. *J. Food Eng.*, 15, 63–73.

Ibarz, A., Pagán, J., Gutiérrez, J., and Vicente, M. (1989). Rheological properties of clarified pear juice concentrates. *J. Food Eng.*, 10, 57–63.

Ibarz, A., Vicente, M., and Graell, J. (1987). Rheological behaviour of apple juice and pear juice and their concentrates. *J. Food Eng.*, 6, 257–267.

Ibarz, A. and Miguelsanz, R. (1989). Variation with temperature and soluble solids concentration of the density of a depectinised and clarified pear juice. *J. Food Eng.*, 10, 319–323.

Ibarz, A. and Pagán, J. (1987). Rheology of raspberry juices. *J. Food Eng.*, 6, 269–289.

Ibarz, A. and Sintes, J. (1989). Rheology of egg yolk. *J. Texture Stud.*, 20, 161–167.

IDF. (1979). *International IDF Standard 87:1979*. International Dairy Federation, Brussels, Belgium.

Ilangantileke, S.G., Ruba, Jr., A.B., and Joglekar, H.A. (1991). Boiling point rise of concentrated Thai tangerine juices. *J. Food Eng.*, 15, 235–243.

Iveson, S. (1997). Fundamentals of granule consolidation and deformation. PhD Thesis, Department of Chemical Engineering, The University of Queensland, St. Lucia, Queensland, Australia.

Jackson, B.W. and Troupe, R.A. (1964). Laminar flow in plate exchanger. A correlation predicting the average film coefficient for heating or cooling at Reynolds numbers below 400. *Chem. Eng. Prog.*, 60(7), 62–65.

Jacob, M. (1957). *Heat Transfer*. John Wiley & Sons, New York.

Jenson, V.G. and Jeffreys, G.V. (1969). *Métodos Matemáticos en Ingeniería Química*. Alhambra, Madrid, Spain.

Jones, S. (1983). The problem of closure in the Zahn-Roskies method of shape description. *Powder Technol.*, 34, 93–94.

Joye, D.D. and Poehlein, G.W. (1971). Characteristics of thixotropic behavior. *Trans. Soc. Rheol.* 15, 51–61.

Kamath, S. (1996). Constitutive parameter determination for food powders using triaxial and finite element analysis of incipient flow from hopper bins. Doctoral thesis, Pennsylvania State University, Pennsylvania, PA.

Kapsalis, J.G. (1987). Influences of hysteresis and temperature on moisture sorption isotherms. In *Water Activity: Theory and Applications to Food* (L.B. Rockland and L.R. Beuchat, Eds.). Marcel Dekker, New York.

Karel, M., Fennema, O.R., and Lund, D.B. (1975a). Preservation of food by storage at chilling temperatures. In *Principles of Food Science. Part II. Physical Principles of Food Preservation* (O.R. Fennema, Ed.). Marcel Dekker, New York.

Karel, M., Fennema, O.R., and Lund, D.B. (1975b). Protective packaging of foods. In *Principles of Food Science. Part II. Physical Principles of Food Preservation* (O.R. Fennema, Ed.). Marcel Dekker, New York.

Kemblowski, Z. and Petera, J. (1980). A generalized rheological model of thixotropic materials. *Rheol. Acta*, 19, 529.

Kern, D.Q. (1965). *Procesos de Transferencia de Calor*. Cecsa, México.

Kimball, D.A. (1986). Volumetric variations in sucrose solutions and equations that can be used to replace specific gravity tables. *J. Food Sci.*, 51(2), 529–530.

King, A.H. (1995). Encapsulation of food ingredients. In *Encapsulation and Controlled Release of Food Ingredients* (S.J. Risch and G.A. Reineccius, Eds.), Chapter 3, pp. 26–39. American Chemical Society, Washington, DC.

King, C.J. (1980). *Procesos de Separación*. Reverté, Barcelona, Spain.

Kokini, J.L. (1992). Rheological properties of foods. In *Handbook of Food Engineering* (D.R. Heldman and D.B. Lund, Eds.), p. 1. Marcel Dekker, New York.

Kokini, J.L. and Plutchok, G.J. (1987). Viscoelastic properties of semisolid foods and their biopolymeric components. *Food Technol.*, 41(3), 89.

Konstance, R.P., Onwulata, C.I., and Holsinger, V.H. (1995). Flow properties of spray-dried encapsulated butteroil. *J. Food Sci.*, 60(4), 841–844.

Kopelman, I.J. (1966). Transient heat transfer and thermal properties in food systems. PhD thesis, Michigan State University, Hickory Corners, MI.

Kostelnik, M.C. and Beddow, J.K. (1970). New techniques for tap density. In *Modern Developments in Powder Metallurgy* (H.H. Hausner, Ed.). Plenum Press, New York.

Kraemer, E.O. (1938). Molecular weights of cellulose and cellulose derivatives. *Ind. Eng. Chem.*, 30, 1200.

Kramer, A. and Twigg, B.A. (1970). *Quality Control for the Food Industry*, Vol. 1. AVI Publishing Co., Westport, CT.

Kreith, F. and Black, W.Z. (1983). *La Transmisión del Calor. Principios Fundamentales*. Alhambra, Madrid, Spain.

Kumar, M. (1973). Compaction behavior of ground corn. *J. Food Sci.*, 38(5), 877–878.

Kurup, T.R.R. and Pipel, N. (1978). Compression characteristics of pharmaceutical powder mixtures. *Powder Technol.*, 19, 147–155.

Lana, E.P. and Tischer, R.A. (1951). Evaluation of methods for determining quality of pumpkins for canning. *Proc. Am. Soc. Hort. Sci.*, 38, 274.

Lang, K.W. and Steinberg, M.P. (1981). Predicting water activity from 0.30 to 0.95 of a multicomponent food formulation. *J. Food Sci.*, 46, 670–672, 680.

Langmuir, I. (1918). The adsorption of gases on plane surfaces of glass, mica and platinum. *J. Am. Chem. Soc.*, 40(9), 1361–1403.

Lee, K.H. and Brodkey, R.S. (1971). Time-dependent polymer rheology under constant stress and under constant shear conditions. *Trans. Soc. Rheol.*, 15, 627–646.

Letort, M. (1961). La Génie Chimique. *Génie Chim.*, 15, 627.

Leung, H.K. (1986). Water activity and other colligative properties of foods. In *Physical and Chemical Properties of Foods* (M.R. Okos, Ed.). American Society of Agricultural Engineers, St. Joseph, MI.

Levenspiel, O. (1986). *El Omnilibro de los Reactores Químicos*. Reverté, Barcelona, Spain.

Levenspiel, O. (1993). *Flujo de Fluidos. Intercambio de Calor*. Reverté, Barcelona, Spain.

Levy, F. (1979). Enthalpy and specific heat of meat and fish in the freezing range. *J. Food Technol.*, 14, 549–560.

Lewis, M.J. (1990). *Physical Properties of Foods and Food Processing Systems*. CRC Press, Boca Raton, FL.

Li, F. and Puri, V.M. (1996). Measurement of anisotropic behavior of dry cohesive and cohesionless powders using a cubical triaxial tester. *Powder Technol.*, 89, 197–207.

Lin, O.C.C. (1975). Thixotropic behavior of gel-like systems. *J. Appl. Pol. Sci.*, 19, 199–214.

Lockemann, C.A. (1999). A new laboratory method to characterize the sticking property free flowing solids. *Chem. Eng. Process.*, 38, 301–306.

Lombardi, G. and Moresi, M. (1987). Sviluppo di un modello di simulazione per la concentracione di soluzione zuccherine mediante osmosi inversa. *Ind. Aliment.*, 3, 205.

Loncin, M. and Merson, R.L. (1979). *Food Engineering. Principles and Selected Applications*. Academic Press, New York, pp. 229–271.

Longree, K., Beaver, S., Buck, P., and Nowrey, J.E. (1966). Viscous behavior of custard systems. *J. Agric. Food Chem.*, 14, 653.

Lund, D. (1975). Heat transfer in foods. In *Principles of Food Science Part 2: Physical Principles of Food Preservation* (O. Fennema, Ed.), pp. 11–30. Marcel Dekker, New York.

Lutz, J.M. and Hardenburg, R.E. (1968). *Agricultural Handbook*, No. 66. U.S. Department of Agriculture, U.S. Government Printing Office, Washington, DC.

Ma, L., Davis, D.C., Obaldo, L.G., and Barbosa-Cánovas, G.V. (1997). Mass and spatial characterization of biological materials. In *Engineering Properties of Foods and Other Biological Materials*. Washington State University Publisher, Pullman, WA.

Mafart, P. (1994). *Ingeniería Industrial Alimentaria*. Acribia, Zaragoza, Spain.
Malavé-López, J., Barbosa-Cánovas, G.V., and Peleg, M. (1985). Comparison of the compaction characteristics of selected food powders by vibration, tapping and mechanical compression. *J. Food Sci.*, 50, 1473–1476.
Mallet, J.C., Beghian, L.E., Metcalf, T.G., and Kaylor, J.D. (1991). Potential of irradiation technology for improved shellfish sanitation. *J. Food Saf.*, 11, 231–245.
Maltini, E. and Mastrocola, D. (1999). Preparazione di succo integrale crioconcentrato di kiwifruit. *Industrie delle Bevande*, 28(159), 6–9.
Mandelbrot, B.P. (1977). *Fractals, Form, Chance and Dimension*. Freeman, San Francisco, CA.
Mandelbrot, B.P. (1982). *The Fractal Geometry of Nature*. Freeman, San Francisco, CA.
Manohar, B., Ramakrishna, P., and Udayasankar, K. (1991). Some physical properties of tamarind (*Tamarindus indica L.*) juice concentrates. *J. Food Eng.*, 13, 241–258.
Marriott, J. (1971). Where and how to use plate heat exchangers. *Chem. Eng.*, 8, 127–134.
Marshall, W.R. (1954). Atomization and spray drying. *Chem. Eng. Prog. Monogr. Ser.*, 50(2), 50–56.
Martens, T. (1980). Mathematical model of heat processing in flat containers. Doctoral thesis, Catholic University, Leuven, Belgium.
Mason, J.M. and Wiley, R.C. (1958). Quick quality test for lima beans. Maryland Processor's Department 4, 1. University of Maryland, College Park, MD.
Mason, P.L., Puoti, M.P., Bistany, K.L., and Kokini, J.L. (1982). A new empirical model to simulate transient shear stress growth in semi-solid foods. *J. Food Proc. Eng.*, 6(4), 219.
Masters, K. (1985). *Spray Drying Handbook*, 4th edn. George Godwin, London, U.K.
Masters, K. (1991). *Spray Drying Handbook*, 5th edn. Longman Group Limited, London, U.K.
McAdams, W.H. (1964). *Transmisión de Calor*. Castillo, Madrid, Spain.
McCabe, W.L., Smith, J.C., and Harriott, P. (1985). *Unit Operations of Chemical Engineering*. McGraw-Hill Book Company, Singapore.
McCabe, W.L., Smith, J.C., and Harriott, P. (1991). *Operaciones Unitarias en Ingeniería Química*. McGraw-Hill/Interamericana de España, S.A. Madrid, Spain.
McCabe, W.L. and Smith, J.C. (1968). *Operaciones Básicas de Ingeniería Química*. Reverté, Barcelona, Spain.
McKennell, R. (1960). The influence of viscometer design on non-Newtonian measurements. *Anal. Chem.*, 31(11), 1458.
Mezger, T. (2006). *The Rheology Handbook*. Vincentz Network, Hannover, Germany.
Michalski, M.C., Desobry, S., and Hardy, J. (1997). Food material adhesion: A review. *Crit. Rev. Food Sci. Nutr.*, 37(7), 591–619.
Miranda, L. (1975). *Ingeniería Química*, Agosto, pp. 81–90.
Mizrahi, S. and Berk, Z. (1972). Flow behavior of concentrated orange juice: Mathematical treatment. *J. Texture Stud.*, 3, 69.
Mohsenin, N.N. (1984). *Electromagnetic Radiation Properties of Food and Agricultural Products*. Gordon & Breach Science Publishers, Inc., New York.
Mohsenin, N.N. (1986). *Physical Properties of Plant and Animal Materials*, 2nd edn. Gordon & Breach Science Publishers, Inc., New York.
Molina, M., Nussinovitch, A., Normand, M.D., and Peleg, M. (1990). Selected physical characteristics of ground roasted coffees. *J. Food Process. Preserv.*, 14, 325–333.
Molins, R. (2001). Food irradiation. In *Principles and Applications* (R. Molins, Ed.). Wiley-Interscience, New York.
Money, R.W. and Born, R. (1951). Equilibrium humidity of sugar solutions. *J. Sci. Food Agric.*, 2, 180–185.
Moore, F. (1959). The rheology of ceramic slips and bodies. *Trans. Proc. Ceram. Soc.*, 58, 470.
Moresi, M. and Spinosi, M. (1984). Engineering factors in the production of concentrated fruit juices. II. Fluid physical properties of grape juices. *Int. J. Food Sci. Technol.*, 19, 519–533.
Moreyra, R. and Peleg, M. (1980). Compressive deformation patterns of selected food powders. *J. Food Sci.*, 45, 864–868.
Morris, E.R. and Jackson, J. (1953). *Absorption Towers*. Butterworths, London, U.K.
Morris, E.R. and Ross-Murphy, S.B. (1981). Chain flexibility of polysaccharides and glycoproteins from viscosity measurements. In *Techniques in Carbohydrate Metabolism*. North Holland Scientific Publishers, Ltd., Amsterdam, the Netherlands.
Mort, P.R., Sabia, R., Niesz, D.E., and Rimon, R.E. (1984). Automated generation and analysis of powder compaction diagram. *Powder Technol.*, 46, 67–75.
Mudgett, R.E. (1986). Microwave properties and heating characteristics of foods. *Food Technol.*, 40, 3–24.
Muller, H.G. (1973). *An Introduction to Food Rheology*. Crane, Russak & Company, Inc., New York.

Mullin, J.W. (1972). *Crystallization*. Butterworth, London, U.K.

Munro, J.A. (1943). The viscosity and thixotropy of honey. *J. Econ. Entomol.*, 36, 769.

Mylius, E. and Reher, E.O. (1972). Modelluntersuchungen zur Charakterisieriung thixotroper Medien und ihre Anwendung für verfahrenstechnische Prozessberechnungnen. *Plaste Kautschuk*, 19, 420.

Nagaoka, J., Takigi, S., and Hotani, S. (1955). Experiments on the freezing of fish in air-blast freezer. Presented at *Proceedings of the 9th International Congress of Refrigeration*, Paris, France, pp. 4, 105.

Ndife, M., Sumnu, G., and Bayindirli, L. Dielectric properties of six different species of starch at 2450 MHz. *Food Res. Int.*, 31, 43–52.

Niro Process Technology, B.V. (2004).

Norrish, R.S. (1966). An equation for the activity coefficients and equilibrium relative humidities of water in confectionery syrups. *J. Food Technol.*, 1, 25–39.

Nuebel, C. and Peleg, M. (1994). A research note: Compressive stress-strain relationships of agglomerated instant coffee. *J. Food Process Eng.*, 17, 383–400.

Nyström, C. and Karehill, P.-G. (1996). The importance of intermolecular bonding forces and the concept of bonding surface area. In *Pharmaceutical Powder Compaction Technology* (G. Alderborn and G. Nytröm, Eds.), p. 17. Marcel Dekker, New York.

Ocón, J. and Tojo, G. (1968). *Problemas de Ingeniería Química*. Aguilar, Madrid, Spain.

Ohlsson, T. (1994). Progress in pasteurization and sterilization. In *Developments in Food Engineering* (T. Yano, R. Matsuno, and K. Nakamura, Eds.). Chapman & Hall, London, U.K.

Oka, S. (1960). The principles of rheometry. In *Rheology* (F.R. Eirich, Ed.), Vol. 3, p. 18. Academic Press, New York.

Okos, M.R., Narsimhan, G., Singh, R.K., and Weitnaver, A.C. (1992). Food dehydration. In *Handbook of Food Engineering* (D.R. Heldman and D.B. Lund, Eds.). Marcel Dekker, New York.

Olivares-Francisco, C. and Barbosa-Cánovas, G.V. (1990). Characterization of the attrition process in agglomerated coffee by natural fractals. Presented at the *IFT Annual Meeting*, Anaheim, CA.

Onwulata, C.I., Konstance, R.P., and Holsinger, V.H. (1996). Flow properties of encapsulated milkfat powders as affected by flow agent. *J. Food Sci.*, 1(6), 1211–1215.

Onwulata, C.I., Smith, P.W., and Holsinger, V.H. (1998). Properties of single- and double-encapsulated butteroil powders. *J. Food Sci.*, 63(1), 100–103.

Osorio, F.A. (1985). Back extrusion of power law; Bingham plastic and Herschel-Bulkley fluids. MS thesis. Michigan State University, East Lansing, MI.

Osorio, F.A. and Steffe, J.F. (1984). Kinetic energy calculations for non-Newtonian fluids in circular tubes. *J. Food Sci.*, 49, 1295–1296, 1315.

Osorio, F.A. and Steffe, J.F. (1985). Back extrusion of Herschel-Bulkley fluids—Example problem. *Am. Soc. Agric. Eng.*, Paper No. 85-6004.

Osorio, F.A. and Steffe, J.F. (1987). Back extrusion of power law fluids. *J. Texture Stud.*, 18, 43–63.

Oswin, G.R. (1946). The kinetics of packaged life. *Int. Chem. Ind.*, 65, 419–421.

Papadakis, S.E. and Bahu, R.E. (1992). The sticky issue of drying. *Drying Technol.*, 10(4), 817–837.

Parfitt, G.D. and Sing, K.S.W. (1976). *Characterization of Powder Surfaces*. Academic Press, New York.

Patino, H., Knudsen, F.B., Gress, H.S., and Heard, G.E. (1991). Use of freeze concentration for preparing malt liqueurs. *Tech. Q. Master Brew. Assoc. Am.*, 28(3), 108–110.

Paulov, K.F., Ramakov, P.G., and Noskov, A.A. (1981). *Problems and Examples, for a Course in Basic Operations and Equipment in Chemical Technology*. Mir, Moscow, Russia. (Cited by O. Levenspiel (1993). Flujo de Fluidos e Intercambio de Calor. Reverté, Barcelona, Spain.)

Peleg, M. (1977). Flowability of food powders and methods for its evaluation and methods for its evaluation. A review. *J. Food Process. Eng.*, 1, 303–328.

Peleg, M. (1983). Physical characteristics of powders. In *Physical Properties of Foods* (M. Peleg and E.B. Bagley, Eds.), pp. 293–324. Van Nostrand Reinhold/AVI Publishing Co., New York.

Peleg, M. (1993). Glass transition and physical stability of food powders. In *Glassy State in Foods* (J.M.V. Blanshard and P.J. Lillford, Eds.), pp. 18–48. Nottingham University Press, Nottingham, England.

Peleg, M. and Hollenbach, A.M. 1984. Flow conditioners and anticaking agents. *Food Technol.*, 38, 93–102.

Peleg, M. and Mannheim, H. (1973). Effect of conditioners on the flow properties of powdered sucrose. *Powder Technol.*, 7, 45–50.

Peleg, M. and Normand, M.D. (1985). Mechanical stability as the limit to the fractal dimension of solid particle silhouettes. *Powder Technol.*, 43, 187–188.

Perry, R.H. and Chilton, C.H. (1973). *Chemical Engineer's Handbook*. McGraw-Hill, New York.

Peter, S. (1964). Zur Theorie der rheopexie. *Rheol. Acta*, 3, 178–180.

Petrellis, N.C. and Flumerfelt, R.W. (1973). Rheological behavior of shear degradable oils: Kinetic and equilibrium properties. *Can. J. Chem. Eng.*, 51, 291.

Pietsch, W. (1999). Readily engineer agglomerates with special properties from micro- and nanosized particles. *Chem. Eng. Prog.*, 8, 67–81.

Pitzer, K.S. and Kim, J.J. (1974). Thermodynamics of electrolytes. IV. Activity and osmotic coefficients for mixed electrolytes. *J. Am. Chem. Soc.*, 96, 5701–5707.

Pitzer, K.S. and Mayorga, G. (1973). Thermodynamics of electrolytes. II. Activity and osmotic coefficients for strong electrolytes with one or both ions univalent. *J. Phys. Chem.*, 77(19), 2300–2308.

Plank, R. (1980). *El Empleo del Frío en la Industria de la Alimentación.* Reverté, Barcelona, Spain.

Prentice, J.H. (1968). Measurements of some flow properties of market cream. In *Rheology and Texture Foodstuffs*. SCI Monograph, No. 27, p. 265. Society of Chemical Industry, London, U.K.

Pryce-Jones, J. (1953). The rheology of honey. In *Foodstuffs: Their Plasticity, Fluidity and Consistency* (G.W. Scott Blair, Ed.), p. 148. North Holland, Amsterdam, the Netherlands.

Putman, R., Vanderhasselt, B., and Vanhamel, S. (1997). Process for the preparation of beer of the "pilsener" or "lager" type. PCT—International-Patent-Application, Publication Number WO1997038081 A1. Belgium.

Rahman, M.S. and Labuza, T. (2002). Secado and conservación de alimentos. In *Manual de Conservación de los Alimentos* (M.S. Rahman, Ed.). Acribia, Zaragoza, Spain.

Rahman, S. (1995). *Food Properties Handbook*. CRC Press, Boca Raton, FL.

Raju, K.S.N. and Chand, J. (1980). Consider the plate heat exchanger. *Chem. Eng.*, 87 (August 11), 133–144.

Rambke, K. and Konrad, H. (1970), Physikalische Eigenschaften flüssiger Milchprodukte 1. Mitt. Dichte von Milch, Rahm und Milchkonzentraten. *Nahrung*, 14, 137–143.

Ramos, A.M. and Ibarz, A. (1998). Density of juice and puree of fruits as a function of soluble solids content and temperature. *J. Food Eng.*, 35, 57–63.

Ranz, W.E. and Marshall, Jr., W.R. (1952a). Evaporation from drops. Part I. *Chem. Eng. Prog.*, 48(3), 141–146.

Ranz, W.E. and Marshall, Jr., W.R. (1952b). Evaporation from drops. Part II. *Chem. Eng. Prog.*, 48(4), 173–180.

Rao, M.A. (1977). Rheology of liquid foods. A review. *J. Texture Stud.*, 8, 135.

Rao, M.A. (1980). Flow properties of fluid foods and their measurements. Paper presented at *Proceedings of the 89th National Meeting of AIChE*, August 17–20, Portland, OR.

Rao, M.A. (1986). Rheological properties of fluid foods. In *Engineering Properties of Foods* (M.A. Rao and S.S.H. Rizvi, Eds.), pp. 1–48. Marcel Dekker, New York.

Rao, M.A. (1987). Predicting the flow properties of food suspensions of plant origin. *Food Technol.*, 41(3), 85.

Rao, M.A. (1992). Transport and storage of food products. In *Handbook of Food Engineering* (D.R. Heldman and D.B. Lund, Eds.). Marcel Dekker, New York.

Rao, M.A., Cooley, H.J., and Vitali, A.A. (1984). Flow properties of concentrated juices at low temperatures. *Food Technol.*, 38(3), 113–119.

Rao, M.A., Otoya Palomino, L.N., and Bernhardt, L.W. (1974). Flow properties of tropical fruit purees. *J. Food Sci.*, 39, 160–161.

Rao, V.N.N., Hamann, D.D., and Humphries, E.G. (1975). Flow behavior of sweet potato puree and its relation to mouthfeel quality. *J. Texture Stud.*, 6, 197–209.

Raventós, M., Hernández, E., Auleda, J., and Ibarz, A. (2007). Concentration of aqueous sugar solutions in a multi-plate cryoconcentrator. *J. Food Eng.*, 79, 577–585.

Ree, F., Ree, T., and Eyring, H. (1958). Relaxation theory of transport problems in condensed systems. *Ind. Eng. Chem.*, 50(7), 1036–1040.

Ree, T. and Eyring, H. (1958).In *Rheology*. (F.R. Eirich, Ed.), Academic Press, New York.

Reidy, G.A. (1968). Thermal properties of foods and methods of their determination. M.S. thesis, Food Science Department, Michigan State University, Hickory Corners, MI.

Reiner, M. (1971). *Advanced Rheology*. H.K. Lewis, London, U.K.

Rennie, P.R., Chen, X.D., Hargreaves, C., and Mackereth, A.R. (1999). A study of the cohesion of diary powders. *J. Food Eng.*, 39, 277–284.

Rha, C.K. (1975). Theories and principles of viscosity. In *Theory Determination and Control of Physical Properties of Food Materials* (C.K. Rha, Ed.), Vol. 1, p. 7. D. -Reidel Publishing Company, Dordrecht, the Netherlands.

Rha, C.K. (1978). Rheology of fluid foods. *Food Technol.*, 32, 77.

Ricks, N.P., Barringer, S.A., and Fitzpatrick, J.J. (2002). Food powder characteristics important to nonelectrostatic and electrostatic coating and dustiness. *J. Food Sci.*, 67(6), 2256–2263.

Riedel, L. (1949). Wärmeleitfähigkeitsmessungen an zuckerlösungen fruchtsäften und milch. *Chem. Ing. Technol.*, 21, 340–341.

Riedel, V.L. (1956). Calorimetric studies of the freezing of fresh meat. *Kaltetechnik*, 8(12), 374–377.

Riedel, V.L. (1957a). Calorimetric studies of the meat freezing process. *Kaltetechnik*, 9, 38–40.

Riedel, V.L. (1957b). Calorimetric studies of the freezing of egg white and egg yolk. *Kaltetechnik*, 9(11), 342–345.

Risch, S.J. (1995). Review of patents for encapsulation and controlled release of food ingredients. In *Encapsulation and Controlled Release of Food Ingredients* (S.J. Risch and G.A. Reineccius, Eds.), pp. 197–203. American Chemical Society, Washington, DC.

Ritter, R.A. and Govier, G.W. (1970). The development and evaluation of a theory of thixotropic behavior. *Can. J. Chem. Eng.*, 48, 505.

Rockland, L.B. and Nishi, S.K. (1980). Influence of water activity on food product quality and stability. *Food Technol.*, 34(4), 42–51.

Rodrigo, M., Lorenzo, P., and Safon, J. (1980a). Optimización de las técnicas de esterilización por calor. I. Planteamientos generales. *Rev. Agroquím. Tecnol. Aliment.*, 20(2), 149–158.

Rodrigo, M., Lorenzo, P., and Safon, J. (1980b). Optimización de las técnicas de esterilización por calor. II. Concepto actualizado de la esterilización por calor y efectos de la misma sobre los alimentos. Cinética y parámetros. *Rev. Agroquím. Tecnol. Aliment.*, 20(4), 425–443.

Roebuck, B.D., Goldblith, S.A., and Westphal, W.B. (1972). Dielectric properties of carbohydrate water mixtures at microwave frequencies. *J. Food Sci.*, 37, 199–204.

Roos, Y. and Karel, M. (1991). Plasticizing effect of water on thermal behavior and crystallization of amorphous food model. *J. Food Sci.*, 56(1), 38–43.

Roos, Y. and Karel, M. (1993). Effects of glass transitions on dynamic phenomena in sugar containing food systems. In *Glassy State in Foods* (J.M.V. Blanshard and P.J. Lillford, Eds.), pp. 207–222. Nottingham University Press, Nottingham, England.

Rosen, J.B. (1952). Kinetics of a fixed bed system for solid diffusion into spherical particles. *J. Eng. Chem.*, 20, 387.

Rosen, J.B. (1954). General numerical solutions for solid diffusion in fixed beds. *Ind. Eng. Chem.*, 46, 1590.

Ross, K.D. (1975). Estimation of ether activity in intermediate moisture foods. *Food Technol.*, 29(3), 26–30.

Ruan, R., Ye, X., Chen, P., Doona, C.J., and Tabú, I. (2001). Calentamiento óhmico. In *Tecnologías Térmicas para el Procesado de Alimentos* (P. Richardson, Ed.). Acribia, Zaragoza, Spain.

Rumpf, H. (1961). Problemstellungen und neuere Ergebnisse der Bruchtheorie. *Materialprüfung*, 3, 253–265.

Rumpf, H. (1962). The strength of granules and agglomerates. In *Agglomeration* (W. Knepper, Ed.), pp. 379–418. Interscience Publisher, New York.

Rutgus, R. (1958). Consistency of starch milk. *J. Sci. Food Agric.*, 9, 61.

Sahin, S. and Sumnu, S.G. (2006). *Physical Properties of Foods*. Springer, New York.

Sakai, N. and Mao, W. (2006). Infrared heating. In *Thermal Food Processing*. CRC Press, Taylor & Francis Group, Boca Raton, FL.

Salwin, H. and Slawson, V. (1959). Moisture transfer in combination of dehydrated foods. *Food Technol.*, 13, 715–717.

Saravacos, G.D. (1968). Tube viscometry of fruit purees and juices. *Food Technol.*, 22, 585.

Saravacos, G.D. (1970). Effect of temperature on viscosity of fruit juices and purees. *J. Food Sci.*, 35, 122.

Saravacos, G.D. and Moyer, J.C. (1967). Heating rates of fruit products in an agitated kettle. *Food Technol.*, 21, 372.

Satin, M. (1996). *Food Irradiation. A Guidebook*, 2nd edn. Technomic Publishing, Lancaster, PA.

Saunders, S.R., Hamann, D.D., and Lineback, D.R. (1992). A systems approach to food material adhesion. *Lebensam. Wiss. Technol.*, 25, 309–315.

Sawistowski, H. and Smith, W. (1967). *Métodos de Cálculo en los Procesos de Transferencia de Materia*. Alhambra, Madrid, Spain.

Schlichting, H. (1955). *Boundary Layer Theory*, McGraw-Hill, New York.

Schowalter, W.R. (1978). *Mechanics of Non-Newtonian Fluids*. Pergamon Press, New York.

Schubert, H. (1980). Processing and properties of instant powdered food. In *Food Process Engineering*, Vol. 1 (P. Linko, Y. Mälkki, J. Olkku, and J. Larinkari, Eds.), pp. 675–684, Applied Science Publishers, London, U.K.

Schubert, H. (1981). Principles of agglomeration. *Int. Chem. Eng.*, 6, 1–32.

Schubert, H. (1987). Food particle technology. Part I: Properties of particles and particulate food systems. *J. Food Eng.*, 6, 22–26.

Schurer, K. (1985). Comparison of sensors for measurement o fair humidity. In *Properties of Water in Foods* (D. Simatos and J.L. Moulton, Eds.). Martinus Nijhoff Publishing, Leiden, the Netherlands.

Schwartzberg, H. (1982). *Freeze Drying—Lecture Notes*. Food Engineering Department, University of Massachusetts, Amherst, MA.

Scott-Blair, G.W. (1958). Rheology in food research. In *Advances in Food Research*, Vol. VIII. Academic Press, New York.

Shahidi, F. and Han, X.-Q. (1993). Encapsulation of food ingredients. *Crit. Rev. Food Sci. Hum. Nutr.*, 33(6), 501–547.

Shanahan, M.E.R., and Carre, A. (1995) Viscoelastic dissipation in wetting and adhesion phenomena. *Langmuir*, 11, 1396–1402.

Shaw, F.V. (1994). Fresh options in drying. *Chem. Eng.*, 101(7), 76–84.

Sherman, P. (1966). The texture of ice cream, III: Rheological properties of mix and melted ice cream. *J. Food Sci.*, 31, 707–716.

Sherman, P. (1970). *Industrial Rheology*. Academic Press, New York.

Shoemaker, C.F., Lewis, J.I., and Tamura, M.S. (1987). Instrumentation for rheological measurements of food. *Food Technol.*, 41(3), 80.

Siebel, J.E. (1982). Specific heat of various products. *Ice Refrig.*, 2, 256–257.

Simons, S.J.R. (1996). Modeling of agglomerating systems: From spheres to fractals. *Powder Technol.*, 87, 29–41.

Singh, R.K. and Nelson, P.E. (1992). *Advances in Aseptic Processing Technologies*. Elsevier, London, U.K.

Singh, R.P. (1982). Thermal diffusivity in food processing. *Food Technol.*, 36(2), 87–91.

Singh, R.P. (1992). Heating and cooling processes for foods. In *Handbook of Food Engineering* (D.R. Heldman and D.B. Lund, Eds.), pp. 247–276. Marcel Dekker, New York.

Singh, R.P. and Heldman, D.R. (1993). *Introduction to Food Engineering*. Academic Press, New York.

Singh, R.P. and Lund, D.B. (1984). *Introduction to Food Engineering*. Academic Press, New York.

Skelland, A.P.H. (1967). *Non-Newtonian Flow and Heat Transfer*. John Wiley & Sons, New York.

Slade, L., Levine, H., Ievolella, J., and Wang, M. (1993). The glassy state phenomenon in applications for food industry: Application of food polymer science approach to structure-function relationships of sucrose in cookie and cracker systems. *J. Sci. Food Agric.*, 63, 133–176.

Smith, J.M. and Van Ness, H.C. (1975). *Introduction to Chemical Engineering Thermodynamics*. McGraw-Hill, New York.

Smith, P.R. (1947). The sorption of water vapor by high polymers. *J. Am. Chem. Soc.*, 69(3), 646–651.

Steffe, J.F. (1992a). *Rheological Methods in Food Process Engineering*. Freeman Press, East Lansing, MI.

Steffe, J.F. (1992b). Yield stress: Phenomena and measurement. In *Advances in Food Engineering* (R.P. Singh and M.A. Wirakartakusumah, Eds.), p. 363. CRC Press, Boca Raton, FL.

Steffe, J.F., Mohamed, I.O., and Ford, E.W. (1984). Pressure drop across valves and fittings for pseudoplastic fluids in laminar flow. *Trans. ASAE*, 27, 616–619.

Steffe, J.F. and Morgan, R.G. (1986). Pipeline design and pump selection for non-Newtonian fluid foods. *Food Technol.*, 40(12), 78–85.

Steffe, J.F. and Osorio, F.A. (1987). Back extrusion of non-Newtonian fluids. *Food Technol.*, 41(3), 72.

Stoecker, W.F. and Jones, J.W. (1982). *Refrigeration and Air Conditioning*. McGraw-Hill Book Company, New York.

Strong, D.H., Foster, E., and Duncan, G. (1970). Influence of water activity on the growth of *Clostridium perfringens*. *J. Appl. Microbiol.*, 19, 980–987.

Stumbo, C.R. (1973). *Thermobacteriology in Food Processing*, 2nd edn. Academic Press, New York.

Stumbo, C.R., Purohit, K.S., Ramakrishna, T.V., Evans, D.A., and Francis, F.J. (1983). *Handbook of Lethality Guides for Low-Acid Canned Foods, Vol. I: Conduction-Heating*. CRC Press, Boca Raton, FL.

Stumbo, C.R. and Longley, R.E. (1966). *Food Technol.*, 20, 109.

Succar, J. and Hayakawa, K. (1983). Empirical formulae for predicting thermal physical properties of foods at freezing and defrosting temperatures. *Lebensm. Wiss. Technol.*, 16, 326–331.

Sullivan, R., Scarpino, P.V., Fassolitis, A.C., Larkin, E.P., and Peeler, J.T. (1973). Gamma radiation inactivation of coxsackievirus B-2. *Appl. Microbiol.*, 22, 61–65.

Sweat, V.E. (1974). *J. Food Sci.*, 39(6), 1080.

Sáenz, C. and Costell, E. (1986). Comportamiento reológico de productos de limón. Influencia de la temperatura y de la concentración. *Rev. Agroquím. Tecnol. Aliment.*, 26(4), 581–588.

Teixeira, A. (1992). Thermal process calculations. In *Handbook of Food Engineering* (D.R. Heldman and D.B. Lund, Eds.). Marcel Dekker, Inc., New York.

Teixeira, A.A. and Shoemaker, C.F. (1989). *Computerized Food Processing Operations*. Van Nostrand Reinhold, New York.

Teng, T.T. and Seow, C.C. (1981). A comparative study of methods for prediction of water activity of multicomponent aqueous solutions. *J. Food Technol.*, 16, 409–419.

Teunou, E., Fitzpatrick, J.J., and Synnott, E.C. (1999). Characterization of food powder flowability. *J. Food Eng.*, 39, 31–37.

Thayer, D.W. (1994). Wholesomeness of irradiated foods. *Food Technol.*, 48(5), 132–135.

Thomson, F.M. (1997). Storage and flow of particulate solids. In *Handbook of Powder Science & Technology* (M.E. Fayed and L. Otten, Eds.), pp. 389–436. Chapman & Hall, New York.

Tiu, C. and Boger, D.V. (1974). Complete rheological characterization of time-dependent food products. *J. Texture Stud.*, 5, 329.

Toledo, R.T. (1980). *Fundamentals of Food Process Engineering*. AVI Publishing Co., Westport, CT.

Toledo, R.T. (1993). *Fundamentals of Food Process Engineering*. Chapman & Hall, New York.

Toledo, R.T. and Chang, S.-Y. (1990). Advantages of aseptic processing of fruits and vegetables. *Food Technol.*, 44(2), 75.

Troller, J.A. (1983). Methods to measure water activity. *J. Food Protect.*, 46, 129–134.

Troller, J.A. and Christian, J.H.B. (1978). *Water Activity and Food*. Academic Press, New York.

Troupe, R.A., Morgan, J.C., and Prifiti, J. (1960) The plate heater versatile chemical engineering tool. *Chem. Eng. Prog.*, 56(1), 124–128.

Tscheuschner, H.D. (1994). Rheological and processing properties of fluid chocolate. *Appl. Rheol.*, 94, 83–88.

Tung, M.A., Richards, J.F., Morrison, B.C., and Watson, E.L. (1970). Rheology of fresh, aged and gamma-irradiated egg white. *J. Food Sci.*, 35, 872.

U.S. Department of Agriculture. (1953). *U.S. Standards for Grades of Tomato Catsup*. Agricultural Marketing Service, Washington, DC.

Usher, J.D. (1970). Evaluating plate heat exchangers. *Chem. Eng.*, 23, 90–94.

Valente, M., Duverneuil, G., and Nicolas, J. (1986). Studies on freeze concentration of whole kiwi fruit. In International symposium on fruit and vegetables for processing, (Cape Town, South Africa). *ISHS Acta Hort.*, 194, 249–260.

Van Arsdel, N.B. and Copley, M.J. (1963). *Food Dehydration*. AVI Publishing Co., Westport, CT.

Van Pelt, W.H.J.M. and Swinkels, W.J. (1986). Recent developments in freeze concentration. In *Food Engineering and Process Applications*, Vol. 2, *Unit Operations*, (M. LeMaguer and P. Jelen, Eds.), Elsevier, London, U.K.

Van Wazer, J.R., Lyons, J.W., Kim, K.Y., and Colwell, R.D. (1963). *Viscosity and Flow Measurements. A Laboratory Handbook of Rheology*. Interscience Publishers, New York.

Ventakesh, M.S. and Raghavan, G.S.V. (2004). An overview of microwave processing and dielectric properties of agri-food materials. *Biosyst. Eng.*, 88, 1–18.

Vian, A. and Ocón, J. (1967). *Elementos de Ingeniería Química*. Aguilar, Madrid, Spain.

Vicente, A.A., de Castro, I., and Teixeira, J.A. (2006). Ohmic heating for food processing. In *Thermal Food Processing*. CRC Press, Taylor & Francis Group, Boca Raton, FL.

Vitali, A., Roig, S.M., and Rao, M.A. (1974). Viscosity behavior of concentrated passion fruit juice. *Confructa*, 19, 201–206.

Vitali, A.A. and Rao, M.A. (1984a). Flow properties of low-pulp concentrated orange juice: Effect of temperature and concentration. *J. Food Sci.*, 49, 882–888.

Vitali, A.A. and Rao, M.A. (1984b). Flow properties of low-pulp concentrated orange juice: Serum viscosity and effect of pulp content. *J. Food Sci.*, 49(3), 876–881.

Wallack, D.A. and King, C.J. (1988). Sticking and agglomeration of hygroscopic, amorphous carbohydrate and food powders. *Biotechnol. Prog.*, 4(1), 31–35.

Walters, K. (1975). *Rheometry*. John Wiley & Sons, New York.

Watson, E.L. (1968). Rheological behavior of apricot purees and concentrates. *Can. Agric. Eng.*, 10, 1.

Webb, A.P. and Orr, C. (1997). *Analytical Methods in Fine Particle Technology*. Micrometrics Instrument Corp., Norcross, GA.

Weber, W.J. (1979). *Control de la Calidad del Agua. Procesos Fisicoquímicos*. Reverté, Barcelona, Spain.

Welti, J. and Vergara, F. (1997). Actividad de Agua. Concepto y aplicación en alimentos con alto contenido de humedad. In *Temas en Tecnología de Alimentos* (J.M. Aguilera, Ed.). CYTEDPrograma Iberoamericano de Ciencia y Tecnología para el Desarrollo, México.

Weltmann, R.N. (1943). Breakdown of thixotropic structure as function of time. *J. Appl. Phys.*, 14, 343.

Welty, J.R., Wicks, Ch.E., and Wilson, R.E. (1976). *Fundamentals of Momentum, Heat and Mass Transport*. John Wiley & Sons, New York.

White, G.W. (1970). Rheology in food research. *Chem. Eng.*, 285, 289.

White, H.J. and Eyring, H. (1947). The adsorption of water by swelling high polymeric materials. *Textile Res. J.*, 17(10), 523–553.

WHO. (1981). *WHO, Wholesomeness of Irradiated Foods*. Technical Report Series 659. WHO, Geneva, Switzerlan.

Wilkinson, W.L. (1974). Flow distribution in plate exchanger. *Chem. Eng.*, 285, 289–293.

Windsor, M. and Barlows, S. (1984). *Introducción a los Subproductos de Pesquería*. Acribia, Zaragoza, Spain.

Yan, H., Barbosa-Cánovas, G.V, and Swanson, B.G. (2001). Density changes in selected agglomerated food powders due to high hydrostatic pressure. *Lebensm. Wiss. Technol.*, 34(8), 495–501.

Yan, H. and Barbosa-Cánovas, G.V. (1997). Compression characteristics of agglomerated food powders: Effect of agglomerated size and water activity. *Food Sci. Technol. Int.*, 3, 351–359.

Yan, H. and Barbosa-Cánovas, G.V. (2000). Compression characteristics of selected food powders: The effect of particle size, mixture composition, and compression cell geometry. Doctoral thesis. Biological Systems Engineering Department, Washington State University, Washington, DC, Chapter 4, pp. 82–113.

Yan, H. and Barbosa-Cánovas, G.V. (2001a). Attrition evaluation for selected agglomerated food powders: The effect of agglomerate size and water activity. *J. Food Process Eng.*, 24, 37–49.

Yan, H. and Barbosa-Cánovas, G.V. (2001b). The effect of padding foam on the compression characteristics of some agglomerated food powders. *Food Sci. Technol. Int.*, 7(5), 417–423.

Yasuda, K. (1979). Investigation of the analogies between viscometric and linear viscoelastic properties of polystyrene fluids. PhD thesis, Department of Chemical Engineering, Massachusetts Institute of Technology, Cambridge, MA.

Yoshioka, N., Hotta, Y., Tanaka, S., Naito, S., and Tsugami, S. (1957). Continuous thickening of homogeneous flocculated slurries. *Chem. Eng. Jpn.*, 21, 66–75.

Zimm, B.H. and Grothers, D.M. (1962). Simplified rotating cylinder viscometer for DNA. *Proc. Natl. Acad. Sci.*, 48, 905.

Zitny, R., Rieger, F., Houska, M., and Sestak, J. (1978). Paper presented at *6th International. CHISA Congress*, Prague, Czech Republic.

Index

A

B

C

D

J

K

L

M

S

For Product Safety Concerns and Information please contact our
EU rcpresentative GPSR@taylorandfrancis.com Taylor & Francis
Verlag GmbH, Kaufingerstraße 24, 80331 München, Germany